T0280604

Landslides

Landslides cause tens of billions of dollars' worth of damage throughout the world every year, and losses are increasing due to a growing population and new development in potentially unstable areas. Fundamentally they have geological causes but can have natural triggers such as rainfall, snowmelt, erosion, and earthquakes, or can be triggered by human actions such as agriculture and construction. To reduce the threat that landslides pose to public safety and property, research aimed at providing a better understanding of slope stability and failure has accelerated in recent years. This acceleration has been accompanied by basic field research and numerical modeling of slope failure processes, mechanisms of debris movement, and landslide causes and triggers.

Written by 78 of the leading researchers and practitioners in the world, this book provides a state-of-the-art summary of landslide science. It features both field geology and engineering approaches, as well as modeling of slope failure and runout using a variety of numerical codes. The book is illustrated with international case studies that integrate geological, geotechnical, and remote sensing studies, and include recent slope investigations in North America, Europe, and Asia.

This comprehensive and complete one-stop synthesis of current landslide research forms an essential reference for researchers and graduate students in geomorphology, engineering geology, geotechnical engineering, and geophysics, as well as professionals in the field of natural hazard analysis.

JOHN J. CLAGUE is the Canada Research Chair in Natural Hazard Research at Simon Fraser University and also, at the same institution, Director of the Centre for Natural Hazard Research. He has published over 250 papers in 45 different journals on a range of earth science disciplines, including glacial geology, geomorphology, stratigraphy, sedimentology, and natural hazards. Professor Clague's other principal professional interest is improving public awareness of earth science by making relevant geoscience information available: he has written two popular books on the geology and geologic hazards of southwest British Columbia and a textbook on natural hazards. He is the recipient of the Geological Society of America Burwell Award, the Royal Society of Canada Bancroft Award, the Geological Association of Canada's (GAC) 2006 E.R.W. Neale Medal and GAC's 2007 Logan Medal. He was the 2007/8 Richard Jahns Distinguished Lecturer for the Geological Society of America and the Association of Environmental and Engineering Geology.

DOUGLAS STEAD has over 30 years' experience in rock and soil slope stability in industry, government and academia in the UK, Zambia, Hong Kong, Papua New Guinea, and Canada. He is now Professor and Chair in Resource Geoscience and Geotechnics at Simon Fraser University. He has published extensively in the areas of rock mechanics and engineering geology with application to landslides, and to surface and underground mining. Dr. Stead has a strong commitment to continuing development courses for professional engineers and geoscientists – delivering courses on methods of data collection and numerical modeling of rock slopes. He is a Professional Engineer in British Columbia and a Chartered Engineer in the UK and is currently a member of the *Engineering Geology* Editorial Board and an Associate Editor of the *Canadian Geotechnical Journal*. He is a recipient of the Canadian Geotechnical Society Thomas Roy Award for Engineering Geology (2008) and the John Franklin Award for Rock Mechanics (2009).

Landslides

Types, Mechanisms and Modeling

Edited by

JOHN J. CLAGUE
Simon Fraser University, British Columbia

DOUGLAS STEAD
Simon Fraser University, British Columbia

CAMBRIDGE
UNIVERSITY PRESS

CAMBRIDGE
UNIVERSITY PRESS

University Printing House, Cambridge CB2 8BS, United Kingdom

One Liberty Plaza, 20th Floor, New York, NY 10006, USA

477 Williamstown Road, Port Melbourne, VIC 3207, Australia

4843/24, 2nd Floor, Ansari Road, Daryaganj, Delhi - 110002, India

79 Anson Road, #06-04/06, Singapore 079906

Cambridge University Press is part of the University of Cambridge.

It furthers the University's mission by disseminating knowledge in the pursuit of education, learning and research at the highest international levels of excellence.

www.cambridge.org
Information on this title: www.cambridge.org/9781108446815

First published 2012
Reprinted 2013
First paperback edition 2017

A catalogue record for this publication is available from the British Library

Library of Congress Cataloging in Publication data
Landslides : types, mechanisms, and modeling / edited by John J. Clague, Douglas Stead.
p. cm.
Includes bibliographical references and index.
ISBN 978-1-107-00206-7 (hardback)
1. Landslide hazard analysis. 2. Slopes (Soil mechanics)–Stability. 3. Landslides–Mathematical models. 4. Slopes (Soil mechanics)--Mathematical models.
I. Clague, J. J. (John Joseph), 1946– II. Stead, D. (Douglas)
QE599.2.L365 2012
551.3′07–dc23
2012014480

ISBN 978-1-107-00206-7 Hardback
ISBN 978-1-108-44681-5 Paperback

Contents

Contributors

Federico Agliardi
Dipartemento Scienze Geologiche e Geotecnologie, University of Milano-Biococca, Piazza della Scienza 4, 20126 Milan, Italy

Andrea Alpiger
Engineering Geology, ETH Zurich, 8092 Zurich, Switzerland

Gianluca Bianchi Fasani
CERI Research Center on Prediction, Prevention and Mitigation of Geological Risks, Sapienza University of Rome, Piazza Umberto Pilozzi 9, 00038 Valmontone, Rome, Italy

Lars Harald Blikra
Åknes/Tafjord Early Warning Center, Ødegårdsvegen 176, 6200 Stranda, Norway

Brian D. Bornhold
Coastal and Ocean Resources Inc., A-759 Vanalman Avenue, Victoria, BC, Canada V8Z 3B8

Edward N. Bromhead
Centre for Earth and Environmental Science Research, Kingston University, Penrhyn Road, Kingston upon Thames, Surrey, KT1 2EE, UK

Marko H. K. Bulmer
Geophysical Flow Observatory, Joint Center for Earth Systems Technology, University of Maryland – Baltimore County, 1000 Hilltop Circle, MD 21250, USA

D. Calvin Campbell
Geological Survey of Canada (Atlantic), Bedford Institute of Oceanography, PO Box 1006, Dartmouth, NS, Canada B2Y 4A2

Marie Charrière
Water Resources Section, Civil Engineering and Geosciences Faculty, Delft University of Technology, 2600 AA Delft, The Netherlands, *Previously at:* Institute of Geomatics, University of Lausanne, Institut de Géomatique et d'Analyse du Risque – IGAR, Lausanne, Switzerland

Masahiro Chigira
Disaster Prevention Research Institute, Kyoto University, Gokasho, Uji 611–0011, Japan

John J. Clague
Department of Earth Sciences, Simon Fraser University, Burnaby, BC, Canada V5A 1S6

John Coggan
Camborne School of Mines, College of Engineering, Mathematics and Physical Sciences, University of Exeter, Cornwall Campus, Penryn, TR10 9EZ, UK

Giovanni B. Crosta
Dipartemento Scienze Geologiche e Geotecnologie, University of Milano-Biococca, Piazza della Scienza 4, 20126 Milan, Italy

Tim Davies
Department of Geological Sciences, University of Canterbury, Private Bag 4800, Canterbury, New Zealand

Marc-Henri Derron
Institut de Géomatique et d'Analyse du Risque – IGAR, Amphipôle 338, University of Lausanne, CH-1015 Lausanne, Switzerland

Mark Diederichs
Department of Geological Sciences and Geological Engineering, Queen's University, Kingston, ON, Canada K7L 3N6

Erik Eberhardt
Geological Engineering, Department of Earth and Ocean Sciences, University of British Columbia, Vancouver, BC, Canada V6T 1Z4

Carlo Esposito
Department of Earth Sciences, Sapienza University of Rome, Piazzale Aldo Moro 5, 00185 Rome, Italy

Robin Fell
School of Civil and Environmental Engineering, University of New South Wales, c/o 75D Roland Avenue, Wahroonga 2076, Australia

Paolo Frattini
Dipartemento Scienze Geologiche e Geotecnologie, University of Milano-Biococca, Piazza della Scienza 4, 20126 Milan, Italy

Corey R. Froese
Alberta Geological Survey/Energy Resources Conservation Board, 4th Floor, Twin Atria Building, 4999-98 Avenue, Edmonton, Canada T6B 2X3

Monica Ghirotti
Dipartimento di Scienze della Terra e Geologico-Ambientali, Alma Mater Studiorum – Università di Bologna, Via Zamboni 67, 40127 Bologna, Italy

Valentin Gischig
Matousek, Baumann & Niggli AG, Baden, Switzerland

James S. Griffiths
University of Plymouth, Drake Circus, Plymouth PL4 8AA, UK

Stephen R. Hencher
Halcrow China Ltd., Level 10, Millennium City 6, Kwun Tong Road, Kowloon, Hong Kong

Reginald L. Hermanns
International Centre for Geohazards, Norwegian Geotechnical Institute, Leiv Eirikssons vei 39, NO-7491 Trondheim, Norway

Kris Holm
BGC Engineering Inc., 1045 Howe Street, Vancouver, BC, Canada V6Z 2A9

Seyyedmahdi Hosseyni
Islamic Azad University, Azadshahr Branch, Iran

Niels Hovius
Department of Earth Sciences, University of Cambridge, Cambridge, UK

Christian Huggel
Glaciology, Geomorphodynamics & Geochronology, Department of Geography, University of Zurich, CH-8057 Zurich, Switzerland

Florian Humair
Institut de Géomatique et d'Analyse du Risque – IGAR, University of Lausanne, Lausanne, Switzerland

Oldrich Hungr
Geological Engineering, Department of Earth and Ocean Sciences, University of British Columbia, Vancouver, BC, Canada V6T 1Z4

D. Jean Hutchinson
Department of Geological Sciences and Geological Engineering, Queen's University, Kingston, ON, Canada K7L 3N6

Michel Jaboyedoff
Institut de Géomatique et d'Analyse du Risque – IGAR, University of Lausanne, Amphipôle 338, CH-1015 Lausanne, Switzerland

Matthias Jakob
BGC Engineering Inc., 1045 Howe Street, Vancouver, BC, Canada V6Z 2A9

Julien Jakubowski
Institut de Géomatique et d'Analyse du Risque – IGAR, University of Lausanne, Amphipôle 338, CH-1015 Lausanne, Switzerland

Randall W. Jibson
US Geological Survey, Box 25046, MS 966 Denver Federal Center, Denver, CO 80225, USA

Katherine S. Kalenchuk
Department of Geological Sciences and Geological Engineering, Queen's University, Kingston, ON, Canada K7L 3N6

Nikolay Khabarov
International Institute for Applied Systems Analysis, Laxenburg, Austria

Oliver Korup
Institute of Earth and Environmental Sciences, University of Potsdam, Karl-Liebknechtstrasse 24 (HS 27), D–14476, Potsdam, Germany

Luca Lenti
French Institute of Science and Technology for Transport, Development and Networks (IFSTTAR), Paris East University, 58 Boulevard Lefebvre, 75732 Paris Cedex 15, France

Serge Leroueil
Department de Génie Civil, Université Laval, Québec, PC, Canada G1K 7P4

Simon Loew
Engineering Geology, ETH Zurich, 8092 Zurich, Switzerland

Oddvar Longva
International Centre for Geohazards, Norwegian Geotechnical Institute, Leiv Eirikssons vei 39, NO-7491 Trondheim, Norway

Patrick MacGregor
Consulting Engineering Geologist, Glenside, South Australia, Australia

Andrew W. Malone
Room 410, James Lee Building, Department of Earth Sciences, University of Hong Kong, Pokfulam Road, Hong Kong

Salvatore Martino
Department of Earth Sciences, Sapienza University of Rome, Piazzale Aldo Moro 5, 00185 Rome, Italy

Scott McDougall
BGC Engineering Inc., Suite 500–1045 Howe St., Vancouver, BC, Canada V6Z 2A9

Mika McKinnon
Geological Engineering, Department of Earth and Ocean Sciences, University of British Columbia, Vancouver, BC, Canada V6T 1Z4

Mauri McSaveney
GNS Sciences, 1 Fairway Drive, Avalon 5010, PO Box 30-368, Lower Hutt 5040, New Zealand

Patrick Meunier
Laboratoire de Géologie, Ecole Normale Supérieure de Paris, Paris, France

Dennis Moore
Dennis Moore Engineering Inc., Burnaby, BC, Canada

Jeffrey R. Moore
Engineering Geology, ETH Zurich, 8092 Zurich, Switzerland

David C. Mosher
Geological Survey of Canada (Atlantic), Bedford Institute of Oceanography, PO Box 1006, Dartmouth, NS, Canada B2Y 4A2

Michael Obersteiner
International Institute for Applied Systems Analysis, Laxenburg, Austria

Lucio Olivares
Dipartimento di Ingegneria Civile, Seconda Università di Napoli, Aversa via Roma 29, 81031 Aversa, Italy

Thierry Oppikofer
Norwegian Geotechnical Institute, Leiv Eirikssons vei 39, NO-7491 Trondheim, Norway

Luca Pagano
Dipartimento di Ingegneria Idraulica, Geotecnica e Ambientale, Università di Napoli Federico II, Naples, Italy

Massimo Pecci
Italian Mountain Institute, Piazza dei Caprettari 70, 00186 Rome, Italy

Andrea Pedrazzini
Institut de Géomatique et d'Analyse du Risque – IGAR, Amphipôle 338, University of Lausanne, CH-1015 Lausanne, Switzerland

David Petley
Institute of Hazard, Risk and Resilience, Durham University, Durham DH1 3LE, UK

Luciano Picarelli
Dipartimento di Ingegneria Civile, Seconda Università di Napoli, Aversa via Roma 29, 81031 Aversa, Italy

David J. W. Piper
Geological Survey of Canada (Atlantic), Bedford Institute of Oceanography, PO Box 1006, Dartmouth, NS, Canada B2Y 4A2

John Psutka [deceased]
Engineering, Aboriginal Relations and Generation (EARG), Dam Safety, BC Hydro, Burnaby, BC, Canada

Nicholas J. Roberts
Department of Earth Sciences, Simon Fraser University, Burnaby, BC, Canada V5A 1S6

Gabriele Scarascia Mugnozza
Department of Earth Sciences, Sapienza University of Rome, Piazzale Aldo Moro 5, 00185, Rome, Italy

David Stapledon
Consulting Engineering Geologist, Mannering Park, New South Wales, Australia

Douglas Stead
Department of Earth Sciences, Simon Fraser University, Burnaby, BC, Canada V5A 1S6

Richard E. Thomson
Fisheries and Oceans Canada, 9860 West Saanich Road, PO Box 6000, Sidney, BC, Canada V8L 4B2

Paolo Tommasi
Istituto di Geologia Ambientale e Geo-Ingegneria, Consiglio Nazionale delle Ricerche, Rome, Italy

J. Kenneth Torrance
Geography and Environmental Studies, Carleton University, Ottawa, ON, Canada K1S 5B6

Nobuyuki Torii
Kobe City College of Technology, Kobe, Japan

Gianfranco Urciuoli
Dipartimento di Ingegneria Idraulica, Geotecnica e Ambientale, Università di Napoli Federico II, Naples, Italy

Gonghui Wang
Disaster Prevention Research Institute, Kyoto University, Gokasho, Uji 611-0011, Japan

Christopher F. Waythomas
Alaska Volcano Observatory, US Geological Survey, 4210 University Drive, Anchorage, AK 99508, USA

Malcolm Whitworth
School of Earth and Environmental Sciences, University of Portsmouth, Burnaby Building, Burnaby Road, Portsmouth PO1 3QL, UK

Heike Willenberg
Matousek, Baumann & Niggli AG, Baden, Switzerland

Xiyong Wu
Southwest Jiaotong University, Chengdu 610031, China

Preface

JOHN J. CLAGUE AND DOUGLAS STEAD

A series of well-publicized disasters and catastrophes during recent years, including a cyclone in Myanmar, earthquakes in China, Haiti, Chile, and New Zealand, and an earthquake and tsunami in Japan, underscore the importance of efforts to reduce risk from natural hazards around the world.

Landslides, floods, drought, wildfire, storms, tsunamis, and earthquakes continue to take a heavy toll in lives and infrastructure. In the past decade, disasters have killed over 750,000 people and caused damage costing hundreds of billions of dollars (Centre for Research on the Epidemiology of Disasters, 2011). Most of the loss of life was the result of earthquakes, tsunamis, and tropical storms; loss of life from landslides was only a small percentage of the total. Nevertheless, landslides are responsible for much primary and secondary economic damage – in the USA alone, landslides cause damage costing $1–2 billion and more than 25 fatalities each year (US Geological Survey, 2011).

Fig. 1. Rockslide on the "Sea-to-Sky Highway" (Highway 99) north of Vancouver, British Columbia. This landslide interrupted traffic along a major transportation corridor connecting Vancouver and Whistler, site of the 2010 Winter Olympics. (Photo courtesy of Erik Eberhardt.)

As the global population continues to increase, more people will be at risk from landslides. Even small slope failures are a threat to transportation infrastructure, as they disrupt the movement of goods and are costly to clear (Fig. 1). However, the economic costs of landslides are not limited to roads and railways. Underwater landslides have destroyed coastal infrastructure; catastrophic landslides may enter the sea and lakes, triggering destructive tsunamis (Fig. 2); landslides have blocked rivers, producing upstream flooding and reservoirs (Fig. 3), which are subject to sudden emptying and resulting downstream floods; and landslides may enter settled areas, causing death and injury.

Because of the threat that landslides pose to public safety and infrastructure, research aimed at better understanding slope stability and failure has accelerated in recent years. This acceleration is reflected in more basic field research, numerical modeling of slope failure processes, and improvements in understanding the mechanisms of debris movement, and landslide causes and triggers. This book summarizes recent advances in the study of landslides, written by 78 leading specialists from around the world.

The book is broadly divisible into three parts. The first part of the book comprises 12 chapters that deal with landslide types and mechanisms. John Clague and Nick Roberts provide an overview of landslide hazard and risk, set in the context of other hazardous natural processes. This chapter is followed by an overview of landslides in the Earth system by Oliver Korup, which sets the stage for a series of chapters that deal with different types of landslides and that collectively showcase recent developments and emerging technologies. Niels Hovius and Patrick Meunier examine patterns of landslides resulting from different earthquake ground motions. Chris Waythomas then discusses the factors responsible for large-scale instabilities on active stratovolcanoes. Tim Davies and Mauri McSaveney explore theories proposed to explain the long runout of rock avalanches, arguing that only dynamic rock fragmentation can account for the high mobility of this group of landslides.

Fig. 2. The Chehalis Lake rockslide (3 million m³) occurred in December 2007. It entered the lake and triggered a tsunami that removed forest up to 30 m above the lakeshore.

Fig. 3. Lake Sarez, impounded by the Usoi landslide dam in Tajikistan. The lake is 56 km long and holds about 16 km³ of water. The landslide dam (arrowed) is 567 m high and is formed of approximately 2 km³ of rock debris emplaced during an earthquake on February 18, 1911. Usoi Dam is the tallest dam in the world, either natural or engineered. Geologists are concerned that the dam might fail during future large earthquakes or might be overtopped by a displacement wave produced by a landslide into the lake. In either case, a catastrophic flood would devastate the heavily populated Murghab River valley below the dam. (Google Earth image.)

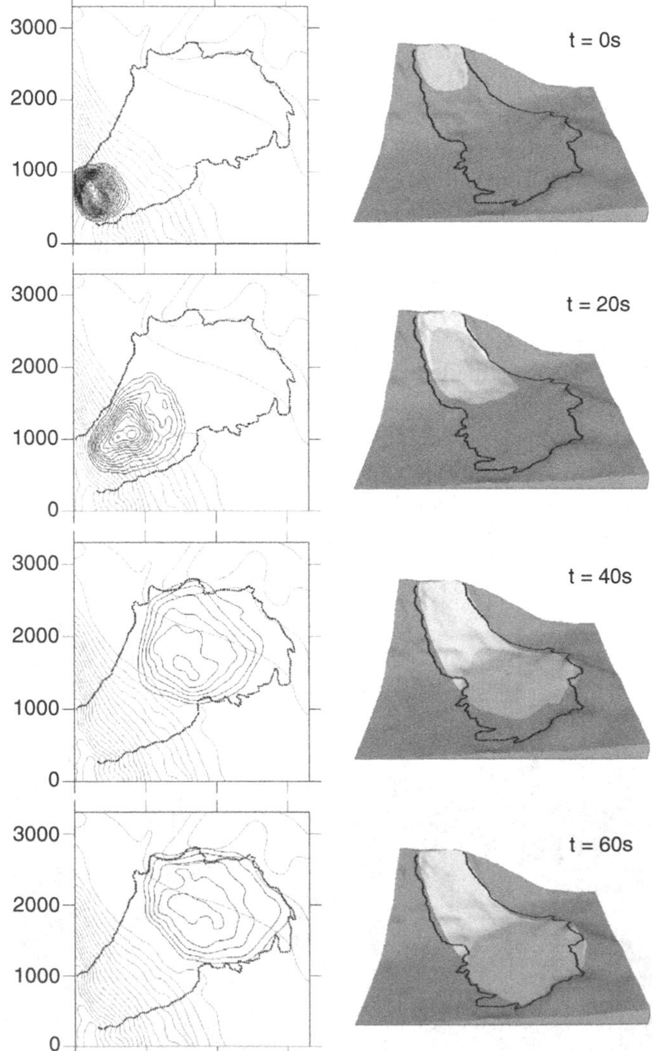

Fig. 4. Three-dimensional analysis of the Frank Slide. Plan (left) and oblique (right) views of the simulated moving mass at 20-s intervals. The flow-depth contours are 5 m, and the sliding surface contours are 50 m. The thick solid line demarcates the real extent of the landslide. (After McDougall and Hungr, 2004.)

Reginald Hermanns and Oddvar Longva continue the discussion of large rock-slope failures, drawing upon their considerable experience in the Andes. The next two chapters deal with more fluidized mass movements. Matthias Jakob and Kris Holm turn our attention to debris flows, focusing specifically on methods for assessing the risk from this type of mass movement. Kenneth Torrance describes different types of quick clay failures and the factors responsible for their sudden onset and retrogressive behavior. David Piper and colleagues discuss controls on different types of submarine landslides on the Canadian Atlantic continental margin. Brian Bornhold and Richard Thomson examine tsunami hazards related to landslides in coastal waters. Christian Huggel and colleagues review current understanding of the effects of climate change on the occurrence of landslides and debris flows in cold, temperate, and tropical mountains. The final chapter in the first part of the book, by Robin Fell and colleagues, deals with the geologic environments of landslides. They demonstrate that the geologic environment has a major influence on the likelihood and mechanisms of landsliding, the hydrogeology as it affects landsliding, and the strength of potential rupture surfaces in rock and soil.

The second part of the book consists of 10 chapters with a focus on numerical modeling of slope failure and new engineering measures aimed at reducing or eliminating landslide risk. Doug Stead and John Coggan introduce this section of the book with a summary of the state of the art in the numerical modeling of rock-slope instability. The next chapter, authored by David Petley, reviews remote sensing techniques applicable to landslides, including new technologies such as InSAR, LIDAR, and digital photogrammetry that are finding widespread use in characterizing slope instabilities. James Griffith and Malcolm Whitworth illustrate the importance of engineering geomorphology in the study of landslides, providing a review of mapping techniques and the engineering geomorphological aspects of landslide classification. Scott McDougall and colleagues discuss developments in runout prediction and the numerical methods used in current practice (Fig. 4). Randall Jibson then reviews methods of assessing the stability of slopes during earthquakes. Federico Agliardi and colleagues focus on slow, deep-seated rock-slope movements, commonly known by their German name, sackung. Although not as obvious as most other types of landslides, sackung are common around the world and can be very large. Numerical modeling is increasingly important in understanding this phenomenon. Erik Eberhardt then describes how landslide monitoring can be used, both as an early warning of failure and to improve our understanding of landslide failure mechanisms. Luciano Picarelli and colleagues use engineering case studies to illustrate the importance of groundwater in soil and rock slopes. The practical and theoretical concepts behind successful soil slope stabilization are then described by Edward Bromhead and colleagues. In the final chapter of this section of the book, Paolo Frattini and colleagues provide a state-of-the-art review of rockfall modeling.

The third part of the book comprises studies of specific landslides that integrate geologic, geotechnical, and remote sensing data. The case studies include: the 2006 Eiger rockslide, Switzerland (Michel Jaboyedoff and colleagues); the 2005 Randa landslides, Switzerland (Simon Loew and colleagues); instability on Turtle Mountain, Alberta (Corey Froese and colleagues); the Åknes rockslide, Norway (Lars Harald Blikra); a rockfall on Corno Grande in the Italian Apennines in 2006 (Gianluca Bianchi Fasani and colleagues); the Downie landslide, British Columbia (Katherine Kalenchuk and colleagues); the 1963 Vaiont Slide, Italy (Monica Ghirotti); landslides in Hong Kong (Steve Hencher and Andrew Malone); and landslides triggered by the 2008 Wenchuan earthquake, China (Masahiro Chigira and colleagues). The final chapter in the book, by Marko Bulmer, provides examples of landslides on other bodies in the solar system, including Mars, Venus, and Io, a moon of Jupiter. The presence of landslides on other planets, moons, and large asteroids illustrates the range of atmospheric and gravitational conditions in which mass movements can occur.

REFERENCES

Centre for Research on the Epidemiology of Disasters (2011). *EM-DAT: The OFDA/CRED International Disaster Database*. Université Catholique de Louvin, Centre for Research on the Epidemiology of Disasters, Brussels [available at www.emdat.be/database, accessed December 13, 2011].

McDougall, S. and Hungr, O. (2004). A model for the analysis of rapid landslide motion across three-dimensional terrain. *Canadian Geotechnical Journal*, 41, 1084–1097.

US Geological Survey (2011). *Landslides Hazards Program*. US Department of the Interior [available at http://landslides.usgs.gov, accessed November 16, 2011].

1 Landslide hazard and risk

JOHN J. CLAGUE AND NICHOLAS J. ROBERTS

ABSTRACT

Each year, landslides are responsible for hundreds of millions of dollars' worth of damage and, on average, claim more than 1000 lives around the world. Although most common in mountainous areas, landslides can occur anywhere with enough local relief to generate gravitational stresses capable of causing rock or soil to fail. In recent decades, research rooted in engineering and the physical sciences, new technologies, and improvements in computational power have greatly advanced our understanding of the causes, triggers, and mechanics of landslides. However, these improvements and advances bear on only part of the landslide risk equation – hazard and exposure; other factors that affect risk are much less understood. Notably, vulnerability and coping capacity, two concepts most developed in the social sciences, play an important – but poorly understood – role in landslide risk. We provide an example of an attempt to estimate landslide risk, which illustrates the difficulty of adequately quantifying vulnerability. We also argue that landslide risk will almost certainly increase over the rest of this century, due to a large increase in global population, settlement and development of previously sparsely populated landslide-prone regions, and climate change.

1.1 INTRODUCTION

Landslides are one of the most damaging and deadly of natural hazards. Data from the Centre for Research on the Epidemiology of Disasters (CRED) suggest that landslides were responsible for over 10,000 deaths and left 2.5 million people homeless over the past decade (2001–2010) (CRED, 2011). However, the true loss of life and incidence of injury may be much larger, due to the under-reporting of small events in many parts of the world, the exclusion of events in the database that are below predefined loss thresholds, and the misattribution of some landslide events to the seismic or hydrologic events that triggered them.

Although most common in mountainous areas, landslides are by no means restricted to them. They also occur in incised valleys in areas of otherwise low relief and are common in many lakes, in fjords, and on the seafloor at the edges of continental shelves. Irrespective of relief, water, and discontinuities in earth materials are critical determinants of slope stability.

Any discussion of landslide hazard and risk must recognize the variety of mass-movement processes and the range of geologic, topographic, and climatic environments in which they occur. Geoscientists distinguish landslides that occur in rock from those that occur in fine- and coarse-textured unconsolidated sediments (soils). They further categorize landslides according to failure mechanisms (falls, topples, slides, spreads, and flows), water content, and speed (Fig. 1.1; Varnes, 1978; Cruden and Varnes, 1996). A large percentage of landslides, however, do not lend themselves to being pigeonholed into these groups. Varnes (1978) terms these "complex landslides": mass movements that have a particular initial failure mechanism but one or more different styles of subsequent movement (Fig. 1.1). Examples include rockfalls that evolve into rock avalanches, and rockslides that transform into large debris flows (Hungr and Evans, 2004). The only commonality to landslides is captured in their generally accepted definition: the downslope movements of earth material under the influence of gravity. Some researchers exclude from the definition of "landslides" debris flows and creep; the latter occurs at very low velocities (millimeters per year). We will not dwell on the semantics of "landslides" here, but instead point out that they encompass a wide variety of phenomena and thus constitute a diverse group of hazards, with major implications for the risk they pose to people and property.

Landslides: Types, Mechanisms and Modeling, ed. John J. Clague and Douglas Stead. Published by Cambridge University Press.

MOVEMENT TYPE	MATERIAL TYPE: Rock (bedrock)	Debris (predominantly coarse soil)	Earth (predominantly fine soil)
Fall	Rockfall	Debris fall	Earth fall
Topple	Rock topple	Debris topple	Earth topple
Slide*	Rockslide	Debris slide	Earth slide
Spread	Rock spread	Debris spread	Earth spread
Flow	Solifluction flow	Debris flow	Earth flow
Complex	e.g. rock avalanche	e.g. debris slide-debris flow	e.g. earth slide -earth flow

* Slide includes translational and rotational slides. Slumps are rotational slides.

Fig. 1.1. Landslide classification scheme (adapted from Cruden and Varnes, 1996).

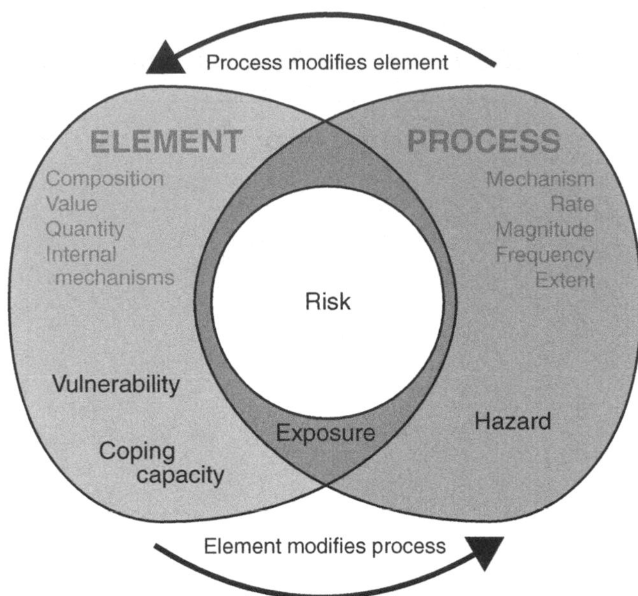

Fig. 1.2. Conceptual model of risk, showing its four components and their relationships. Risk occurs at the interface (exposure) between a process producing hazard and an element or elements characterized by vulnerability and coping capacity. The process and elements can influence each other.

In this chapter, we explore issues of landslide hazard and risk, the latter from both physical science and social science perspectives. We forecast trends in both hazard and risk over the remainder of this century and briefly consider strategies for reducing landslide risk.

1.2 HAZARD AND RISK

Before discussing the issues related to landslide hazard and risk, we define the key terms that we use.

- *Hazard* is the probability that a specific damaging event will happen within a specific area in a particular period of time (ISO/TMB/RMWG, 2007). This definition of hazard is common to both the natural and social sciences, but natural hazard analysis lies largely within the fields of engineering and the physical sciences, specifically geology and physical geography.
- *Risk*, on the other hand, is more commonly a subject of the social sciences, because it is rooted not only in hazard but also in vulnerability and coping capacity (Fig. 1.2; O'Keefe *et al.*, 1976; Chambers, 1989; United Nations Department of Humanitarian Affairs, 1992; Watts and Bohle, 1993; Bohle, 2001; International Strategy for Disaster Reduction, 2004; Birkmann, 2006; Villagrán de León, 2006). Definitions of risk are legion, but for our purposes, it can be expressed by the following function:

$$\text{Risk} = f(\text{hazard, exposure, vulnerability, coping capacity}). \qquad (1.1)$$

Although risk can be conceptualized as a function of these four components, their interrelationships cannot be described in mathematical terms or even fully understood. Vulnerability and coping capacity are latent states of an element at risk and are only manifested through the occurrence of a hazardous event (ISO/TMB/RMWG, 2007). Here we define "element" as a physical or social feature that can be affected by a process to which it is exposed.

- *Vulnerability* is the susceptibility of an element to a hazardous event and is commonly thought of as having technological and human dimensions. Technological aspects include damage and loss of life, which are subjects studied by engineers and geoscientists. Human aspects relate to a wide range of social issues, including, in addition to loss of life, loss of livelihood, physical displacement, and psychological and environmental impacts of hazardous events.
- *Coping capacity* is the ability of an element to respond to and reduce the negative effects of a hazardous event.

- *Exposure* is the overlap in space and time of a hazardous process and infrastructure or population.

The physical and social sciences each have strengths in conceptualizing and evaluating certain components of risk, but separately they are likely to oversimplify other components. The focus in the physical sciences is on hazard and exposure. In this paradigm, the human system is typically viewed as static or passive. In most engineering literature, the role of human behavior in mediating the consequences of hazards is not considered. In contrast, the focus in social sciences is on vulnerability and coping capacity, and hazard is viewed as a static process that reveals vulnerability and coping capacity. In reality, all four components of risk – hazard, exposure, vulnerability, and coping capacity – are dynamic and can vary greatly over a range of temporal and spatial scales.

In the social sciences, attention is directed to factors that limit the ability of individuals and society to contend with hazardous processes, rather than the negative impacts following a disaster. In studies carried out by the United Nations (Birkmann, 2007) and the World Bank (Arnold *et al.*, 2006), vulnerability is assessed based on socio-economic indicators, for example gross domestic product (GDP) per inhabitant, human poverty index (HPI), inflation rate, and population characteristics such as density, growth, age, life expectancy at birth, and literacy rate. Arguably, however, these indicators are poor measures of vulnerability and coping capacity. Perhaps it is for this reason that engineers and physical scientists focus on hazard and exposure, which are more easily quantified.

1.3 EVALUATING HAZARD

The first step in assessing landslide risk[1] is to understand and, if possible, quantify the hazard. Landslide hazard is analyzed first by understanding, as well as possible, the process that gives rise to the hazard, and second by deriving a frequency–magnitude model for the hazard (Moon *et al.*, 2005).

Different types of landslides pose different hazards. Most rockfalls affect relatively small areas directly below their source cliffs, which can be delineated using the "rockfall shadow" concept (Evans and Hungr, 1993) or numerical models of specific rockfall scenarios (Agliardi *et al.*, 2012, Chapter 18, this volume). Most rainfall-triggered debris flows also affect small areas, typically fans or cones onto which streams with steep, debris-laden channels flow. It is possible to identify areas that are likely to be affected based on the geology and topography of the watersheds that generate the debris flows and on sediment availability (Hungr *et al.*, 2005). Slow-moving earthflows may damage roads, buildings, and other engineered structures located on them, but they rarely injure or kill people because of their very low speeds; exceptions are slow-moving rock slopes that spawn rockslides or rock avalanches. Large rapid mass movements, including rockslides and rock avalanches, are much more difficult to forecast than smaller landslides, mainly because the state of stress deep within a slope prior to failure cannot yet be easily or reliably determined. Furthermore, the area impacted by a large rockslide or rock avalanche depends critically on the failure location, volume of the failed rock mass, and topography. Nevertheless, reasonable estimates of scenario landslide runouts can be made with state-of-the-art numerical codes, assuming that failure locations and volumes can be determined (McDougall *et al.*, 2012, Chapter 16, this volume).

The second step in landslide risk assessment is to establish a reliable frequency–magnitude model. Historic records on which such a model might be based generally do not extend far enough back in time to establish a robust and statistically reliable relationship, particularly for less frequent, larger magnitude events. The alternative is to supplement historic records with geologic data. The latter, however, are generally incomplete and temporally biased, limiting the frequency–magnitude analyses on which they are based. Nevertheless, records based on tree damage over several centuries may yield good estimates of magnitude and frequency for small debris flows (Stoffel *et al.*, 2005; Stoffel, 2006; Jakob and Friele, 2010). Similarly, the frequency of large landslides whose scars and deposits persist in the landscape (Guthrie and Evans, 2007) may also be reasonably estimated. In contrast, the deposits of medium-sized landslides are easily eroded or buried, and their frequency is commonly underestimated. As records of past events are nearly always incomplete, the formulation and use of frequency–magnitude plots must involve expert judgment.

1.4 FROM HAZARD TO RISK

An analysis of risk can proceed once a reliable frequency–magnitude model has been established. The frequency–magnitude model is only useful, however, if the hazardous process is well understood. It is not sufficient, for example, to know that, at a particular site, a 10^6 m^3 landslide has an average recurrence of 1000 years. The type of landslide (e.g., debris flow, rockslide, or rock avalanche) and the area of impact must be known. Only then can the next component of risk – exposure – be incorporated. Probabilities of injury and loss of life, and estimates of property damage are associated with a given hazardous event, which has a defined likelihood of occurrence, albeit with considerable and inevitable uncertainties. Potential impacts from all hazardous events in the frequency–magnitude model can then be examined to identify the events that carry the greatest risk. It is these events that are the basis for possible mitigation measures, within the context of both a cost–benefit analysis and a consideration of societally acceptable risk. It is common to

[1] *Risk analysis* is the process of formal risk characterization involving estimation and analysis of hazard, exposure, vulnerability, and coping capacity. *Risk assessment* is the process of comparing risk analysis results for risk mitigation (International Society for Soil Mechanics and Geotechnical Engineering, 2004).

find that the greatest risk reduction is achieved by planning for moderate-sized events with intermediate return periods. Very large events, although highly destructive, are rare; and very small events, although much more common, may cause little or no loss.

At this point in risk analysis, the physical scientist considers his job done. The analysis, however, is far from complete, because two other components that are critical to all considerations of risk and that are dynamic properties of a social system have not been considered – human vulnerability and coping capacity. As mentioned earlier, these two components have traditionally been examined within the social sciences.

Conceptual models of vulnerability may explain and communicate the process and components of risk. They can also facilitate the choice of appropriate indicators of vulnerability and, therefore, are integral to vulnerability analysis. Chambers (1989) introduced an early formal definition of vulnerability with a social science context: "exposure to contingencies and stresses and the difficulty which some communities experience while coping with such contingencies and stresses." He also identified two general types of vulnerability: external vulnerability, which relates to external shocks (that is, impacts due to sudden-onset events and stresses); and internal vulnerability, which relates to defenselessness or the inability to cope. Watts and Bohle (1993) examined vulnerability within the economic, political, and institutional capabilities of people, and concluded that it results from three factors: exposure, coping capacity, and recovery potential. Bohle (2001) later provided a clearer graphic representation of the model, in which coping capacity is explicitly included as a component of vulnerability. Exposure in the Bohle model is not equivalent to the concept of technological vulnerability inherent in the natural science paradigm. Rather, it represents people's ability to resist initial impact of stresses or shocks, and is determined by population dynamics and capacities, entitlement (ability to access and manage assets), and social and economic inequalities. The Bohle model thus places exposure under the umbrella of vulnerability. Within the so-called "disaster risk community," vulnerability is considered within the broader context of risk; vulnerability, coping capacity, and exposure are separate components that, together, produce risk and, potentially, disasters (Birkmann, 2006).

A difficulty in applying these concepts is that they are virtually impossible to quantify. How does one quantify human vulnerability or coping capacity? Clearly, individuals or societies with a limited ability to absorb external shocks are more vulnerable and less able to cope with hazardous processes, but most social measures of quality of life, such as per capita income, access to health assistance, equality, and access to social resources, are only rough indicators of vulnerability and coping capacity. These issues are intimately linked to the concept of individual and societal "acceptable risk" or risk tolerance. In developed countries, notably Japan, New Zealand, Australia, and those of North America and Europe, societal tolerance of injury from landslides and other hazardous phenomena is far lower than that in countries with a low standard of living. A consequence is that governments in developed countries invest heavily in mitigation to minimize hazard (e.g., slope stabilization) and vulnerability (e.g., public education). In addition, coping capacity after disasters in developed countries is generally high due to access to resources, although it also depends on social capital such as social networks. Accordingly, risk in developed countries is much lower than that in less developed countries. Nevertheless, coping capacity and vulnerability cannot yet be integrated in a quantitative way with the more easily measured factors – hazard and exposure. Thus, in the example that follows, we present a quantitative estimate of landslide risk based largely on hazard and exposure.

1.5 AN EXAMPLE OF LANDSLIDE RISK EVALUATION

Friele *et al.* (2008) evaluated the debris-flow risk to the communities of Pemberton and Mount Currie in Lillooet River valley, southwest British Columbia, Canada (Fig. 1.3). The hazard derives from large landslides at Mount Meager, a Quaternary volcano in the upper part of the watershed.

The settled area of Lillooet Valley can be divided into two zones with different population densities. Pemberton Meadows, 32–55 km downstream from Mount Meager, is primarily agricultural and has a population of about 200 people (average population density 5 persons/km^2). Pemberton and Mount Currie, 55–75 km downstream, have about 3800 and 1000 residents, respectively (average population density of 125 persons/km^2).

Drilling in Lillooet River valley has documented valley-wide sheets of debris-flow deposits derived from Mount Meager that are of Holocene age, 2–8 m thick, and 32–55 km downstream from the source (Friele *et al.*, 2005; Simpson *et al.*, 2006). The debris flows that left these deposits had velocities of 10–15 m s^{-1} in the upper Pemberton Meadows area and 3–6 m s^{-1} at Pemberton (Friele *et al.*, 2008). A hyperconcentrated flow or debris flow traveling at these velocities would destroy most residential buildings in the valley. Some people might survive a class 8 (10^7–10^8 m^3) debris flow by climbing into large standing trees or reaching higher ground, but death would be likely for class 9 (10^8–10^9 m^3) events.

Friele *et al.* (2008) established a frequency–magnitude model for debris flows from Mount Meager based on historic events and on prehistoric events inferred from a rich body of geologic evidence (Fig. 1.4). They used the method of Fell *et al.* (2005) to analyze the landslide risk of residents in the Lillooet River valley. They restricted their analysis to loss of life. The variables used in their analysis are:

P_H the probability of the hazard
P_{LOL} the annual probability of loss of life for an individual
$P_{S:H}$ the spatial probability that the event will reach the individual
$P_{T:S}$ the temporal probability of impact (the percentage of time the individual occupies the hazard area, in this case the affected part of the valley)

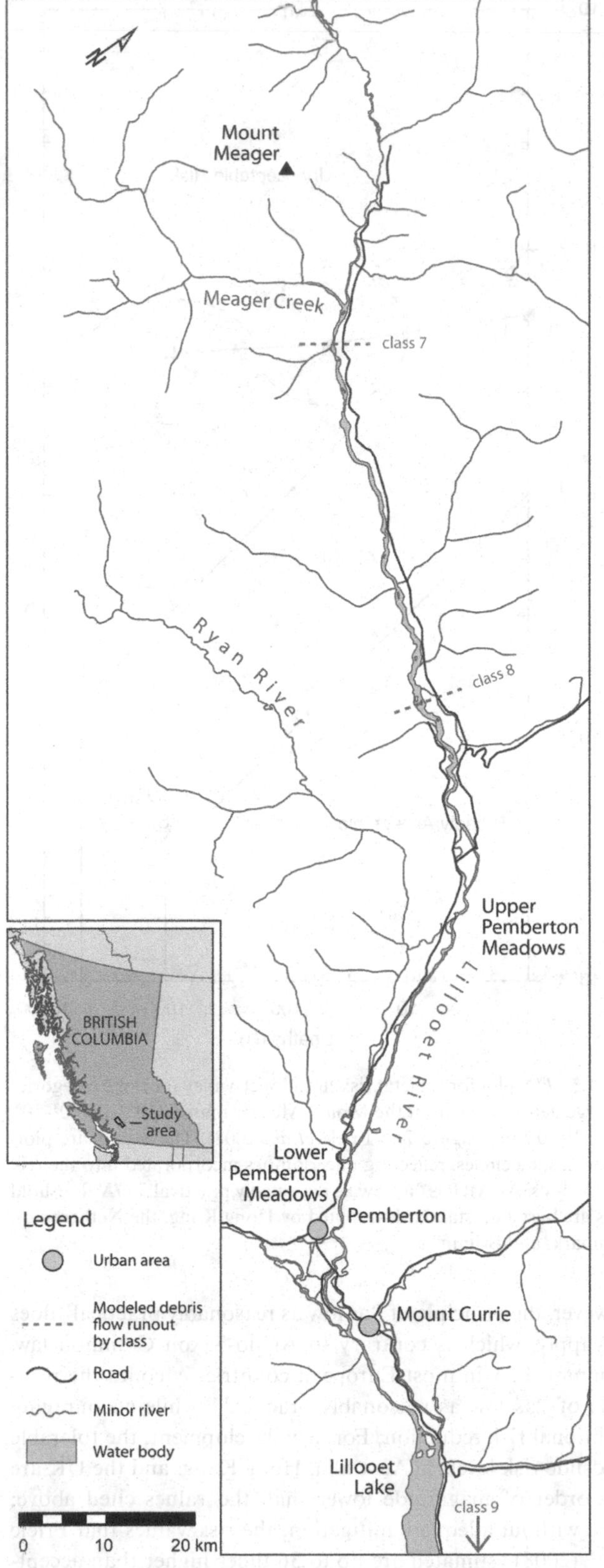

Fig. 1.3. Map of Lillooet River valley showing the location of Mount Meager, the communities of Pemberton and Mount Currie, and downstream limits of class 7 (10^6–10^7 m^3), class 8 (10^8–10^9 m^3), and class 9 (10^9–10^{10} m^3) debris flows from the Mount Meager massif.

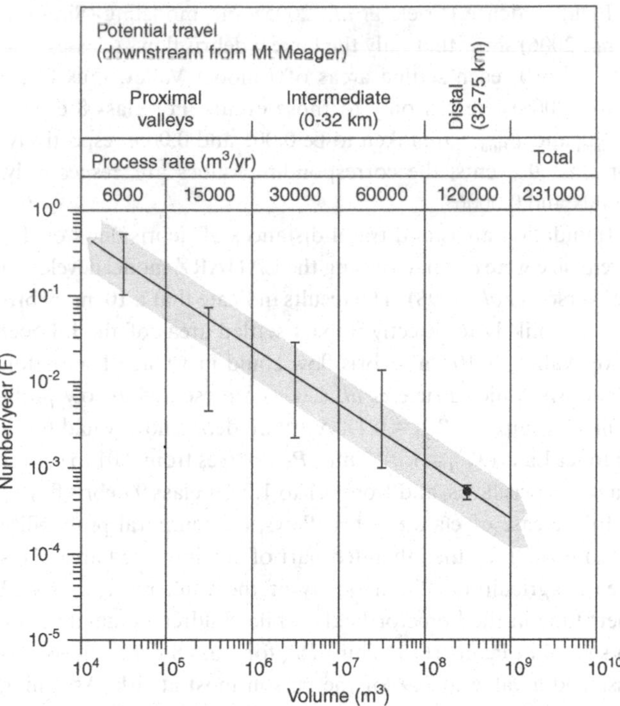

Fig. 1.4. Frequency–magnitude diagram for landslides at the Mount Meager massif, British Columbia (figure 4 in Friele *et al.*, 2008). Uncertainty limits (black bars) for landslides are derived from the historic record (upper bound) and geologic data (lower bound). The gray band represents the most likely uncertainty bounds for the frequency–magnitude model and takes into account data censoring.

V the likelihood of loss of life should the individual be affected by the hazardous phenomenon, which is a function of the intensity of the process at that location

E the element of concern, in this case the number of lives potentially at risk.

Risk can be quantified for individuals or for groups. Risk to individuals is commonly related to the person most at risk. Individual risk is generally compared to some socially accepted or tolerable risk threshold. Friele *et al.* (2008) estimated annual risk of loss of life to an individual (P_{LOL}) from the relation:

$$P_{LOL} = P_H \times P_{S:H} \times P_{T:S} \times V. \quad (1.2)$$

Societies are more tolerant of individual loss of life than to the simultaneous death of a large number of people (Ale, 2005). Group risk can be estimated by plotting the annual frequency (*F*) of one or more deaths from a particular hazard or suite of hazards against the expected number of fatalities (*N*), where *F* is defined, according to Fell *et al.* (2005), as:

$$F = P_H \times P_{S:H} \times P_{T:S} \quad (1.3)$$

and *N* is the product of the number of elements at risk (*E*) and their vulnerability (*V*) to the hazard under consideration. On the *F*/*N* plot of Friele *et al.* (2008), the total risk is the sum of partial risks from different magnitude classes.

Field evidence (Friele *et al.*, 2005) and modeling (Simpson *et al.*, 2006) show that only the largest debris flows (classes 8–9; 10^7–10^9 m^3) reach settled areas of Lillooet Valley, thus Friele *et al.* (2008) referred only to those events. For class 8 events, $P_{H(min)}$ and $P_{H(max)}$ are taken to be 0.001 and 0.005, respectively; for class 9 events, the corresponding values are, respectively, 0.0004 and 0.0006.

Inundation areas and travel distances of debris flows of different size were estimated using the LAHARZ model developed by Iverson *et al.* (1998). The results indicate that a 10^7 m^3 debris flow is unlikely to directly impact settled areas of the Lillooet River valley. A 10^8 m^3 debris flow could just reach Pemberton Meadows, which Friele *et al.* (2008) represented as low probability of impact ($P_{S:H} = 0.1$). A 10^9 m^3 debris flow would reach Lillooet Lake ($P_{S:H} = 1.0$). Thus, $P_{S:H}$ ranges from 0.01 to 0.1 for class 8 debris flows, and from 0.1 to 1.0 for class 9 debris flows.

In the case of class 8 debris flows, the temporal probability ($P_{T:S}$) is high for the inhabited part of the impacted area. This area is agricultural; the majority of the adult residents spend their time in the home or fields, while children commute daily to school in Pemberton. Thus, $P_{T:S}$ for class 8 debris flows was assigned a value of 0.9 for the person most at risk. Assuming a family of two adults and two children, with the children present at school 8 hours per day, $P_{T:S}$ for the average individual is 0.8. Class 9 debris flows travel farther, reaching areas occupied by farmers, First Nation residents, and service sector workers, some of whom commute daily to Whistler. Lacking detailed occupational statistics, Friele *et al.* (2008) assumed that 50 percent of those people live and work/school locally, and 50 percent commute out of the valley and are absent 12 hours per day. $P_{T:S}$ for those staying in the valley was assumed to be 0.9, and for commuters 0.5; the average value is 0.7.

Friele *et al.* (2008) defined vulnerability as the likelihood of death should a building or site be impacted directly by a debris flow or debris flood. They acknowledged that any estimate of vulnerability has a large degree of uncertainty, because it is affected by parameters that are poorly known or highly variable, for example the location of individuals within a building, the intensity of impact, and the ability of a building to withstand impact without incurring structural damage that could lead to death. Uncertainty is built into the vulnerability estimate by defining lower and upper bounds, V_{min} and V_{max}. Allowing for some possibility of survival, V_{min} was assumed to be 0.5 for a class 8 debris flow and 0.9 for a class 9 debris flow. V_{max} was assigned a value of 1.0.

The range of estimated annual debris-flow risk to an individual residing in Lillooet Valley is 5×10^{-6} to 5.0×10^{-4} deaths per year. Governments in Australia, Hong Kong, and England have defined the tolerable landslide risk level to be 10^{-4} annual probability for existing development and 10^{-5} annual probability of death for new development (Fell *et al.*, 2005; Leroi *et al.*, 2005). For Lillooet Valley, individual risk is up to 5.4 times higher than acceptable levels for Australia, Hong Kong, and the UK, and up to 54 times higher than acceptable risk for individuals in the Netherlands (Ale, 2005). In the Netherlands,

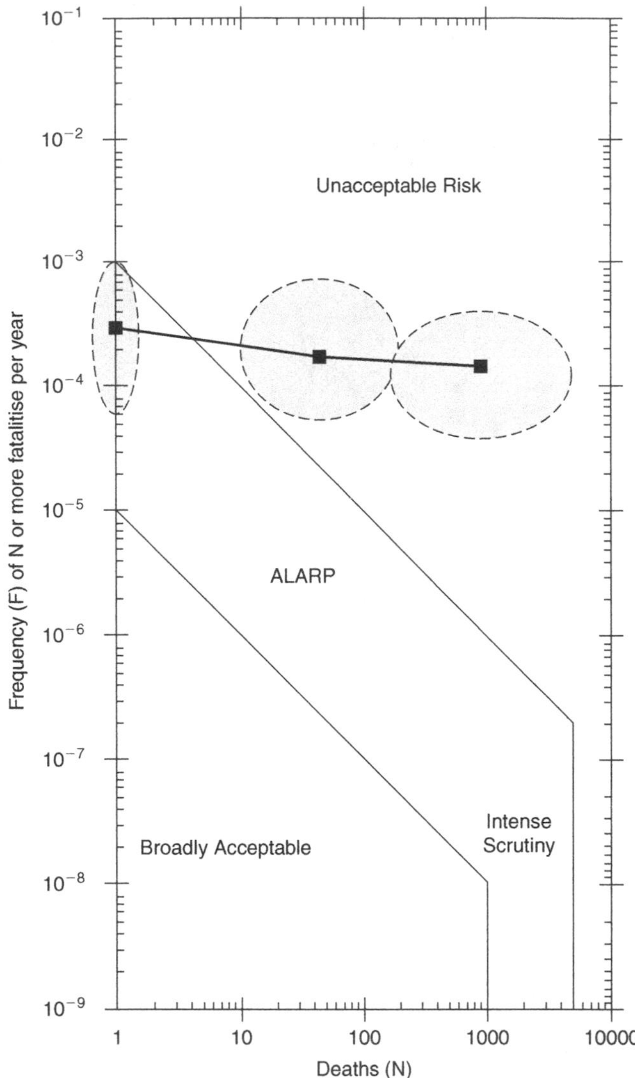

Fig. 1.5. *F*/*N* plot for societal risk in Lillooet Valley for three categories of large debris flows from the Mount Meager massif (10^6–10^7, 10^8–10^9, and 10^9–10^{10} m^3) (figure 7 in Friele *et al.*, 2008). The results are plotted as dashed circles, reflecting uncertainties incorporated into the risk calculations. ALARP is "as low as reasonably practical." *F*/*N* threshold lines are based on standards adopted by Hong Kong, the Netherlands, Denmark, and Britain.

however, the principle of "as low as reasonably practical" does not apply, which is contrary to Anglo-Saxon Common law. Common Law in most European countries encodes the principle of "as low as reasonably practical," while encouraging additional risk reduction. For new development, the tolerable landslide risk levels in Australia, Hong Kong, and the UK are one order of magnitude lower than the values cited above; thus, without adequate mitigation, the risk values that Friele *et al.* (2008) estimated are up to 36 times higher than acceptable levels.

Societal risk is quantified for each debris-flow class as *F*–*N* pairs in Figure 1.5. This figure shows evaluation criteria that are gaining acceptance in Australia, the UK, and recently in

Canada (Fell *et al.*, 2005; Porter *et al.*, 2007). The *F–N* plot is subdivided into four zones:

1. *unacceptable* – risk is generally considered unacceptable by society and requires mitigation;
2. *as low as reasonably practical* – risk from a hazard should, wherever possible, be reduced;
3. *broadly acceptable* – risk from a hazard is within the range that society can tolerate;
4. *intense scrutiny region* – the potential for large loss of life is low, but careful consideration is required.

In the Lillooet Valley study, uncertainties in hazard, exposure, and vulnerability required that the risk for each class be plotted as a zone rather than a point (Fig. 1.5). The plot shows that risk to groups in Lillooet Valley is unacceptable for both class 8 and 9 debris flows, based on international standards. Friele *et al.* (2008) therefore recommended that mitigation measures be taken to reduce risk to the "as low as reasonably practical" region of the *F–N* plot. They further recommended restricting development to areas where risk could be reduced to the "acceptable" level.

1.6 LANDSLIDE RISK IN THE FUTURE

We hypothesize that societal landslide risk will increase in the future and offer three reasons for this assertion. First, the human population will increase, perhaps by as much as 50 percent over the remainder of this century. Second, and more specifically, populations will increase in landslide-prone regions, notably in mountainous areas. Third, forecast climate change may increase the incidence of landslides in many areas. Collectively, these three factors will more than offset risk reductions achieved through improved scientific understanding of landslides and better-informed land-use decisions.

Global population reached 7 billion in 2011 and will exceed 9 billion by the middle of this century. Forecasts for the remainder of the twenty-first century are less certain, but a population of around 10 billion by the year 2100 is possible. Most of the additional 3 billion people will live in Asia and Africa, but almost all countries in North America and Europe will also experience population increases due to a combination of domestic growth and immigration. Development pressures related to larger populations and improved standards of living may result in the settlement of hazardous land. The percentage of the global population living in cities will rise, such that the total urban population will exceed the rural population by the middle of this century. Even with the concentration of people in large urban areas, more remote mountainous areas will also experience substantial absolute increases in population, due in part to the increase in the areas of cities and in part to recreational and resource opportunities that mountainous areas afford. Notable large metropolitan areas that will occupy more space and thus expand within or into mountainous areas include Vancouver, Calgary, Seattle, Portland, Denver, Mexico City, San Salvador, Bogotá, Quito, Santiago, La Paz, Chengdu, Kabul, Tehran, Rawalpindi, Dushanbe, Tashkent, Katmandu, Ankara, Milan, Turin, Addis Ababa, and Nairobi. However, as noted earlier, landslides also occur outside mountainous areas, thus increased damage and injury can be expected in cities built on lower-relief surfaces as their footprints increase.

It is widely recognized that water plays a definitive role in most landslides. All other things being equal, landslides are more frequent in humid environments than in dry ones. Until recently, however, the possibility that climate change might alter the frequency of landslides within a specific region had not been widely considered. It is now evident that climate has changed significantly over the past century, and the scientific community has achieved consensus that it will change even more over the remainder of this century (Solomon *et al.*, 2007). Since the late nineteenth century, the average surface temperature of Earth has increased about 0.8°C, and it is forecast to increase by 2–5°C over the next 90 years (Solomon *et al.*, 2007). Temperature increases at high latitudes and in many mountain ranges are likely to increase considerably more than the global average. Two consequences of such change are the melting of alpine glaciers and thawing of permafrost, both of which may destabilize slopes. Of greater significance for slope stability, however, are the attendant spatial and temporal redistribution of precipitation and a possible increase in extreme precipitation events. A warmer atmosphere will hold more moisture, and warmer oceans are likely to produce stronger cyclonal storms. Long-term or seasonal increases in rainfall, especially in coastal mountains, would lead to more frequent landslides, probably resulting in increased damage and loss of life.

An increase in landslide risk can be partially countered through land-use planning and hazard mitigation. Expansion into mountainous areas and onto slopes outside mountains can be controlled in order to reduce the exposure of people and infrastructure to landslides. Engineered protective works can be built to provide protection when such slopes are developed. Engineering, however, cannot eliminate all risk and is generally ineffectual in stabilizing large, unstable rock slopes and in protecting people from large landslides. Furthermore, engineered reductions in the risk of life loss are only possible in societies with low vulnerabilities due to their wealth and access to resources. Countries with limited resources are less able to implement policies and other measures required to significantly lower risk.

Although we are not confident that landslide risk, or for that matter risk from most other hazardous natural processes, can be significantly reduced in the short term, considerable progress could be made by increasing the coping capacity of the most vulnerable populations, specifically those of impoverished countries that have limited resources to support their citizens. Economic and social equity among nations would go a long way in reducing the loss of life from natural disasters, not to mention easing many seemingly intractable problems that we face today.

1.7 CONCLUSIONS

Landslides are natural processes that shape the Earth's surface and redistribute mass from high elevations to lower ones. They also pose threats to people and infrastructure. Physical scientists and engineers have spent considerable time, energy, and resources studying landslide processes, partly in order to provide better guidance for reducing landslide risk. Although the new scientific insights they have provided enable better estimates of the frequency, magnitude, and potential physical impacts of different types of landslides, this type of work is not adequate, on its own, to reduce risk. Rather, it must be integrated with research on the dynamic properties of social systems performed by social scientists. Specifically, the issues of vulnerability and coping capacity must be incorporated into hazard analysis.

Landslide risk is likely to increase through the remainder of the twenty-first century due to a 50 percent increase in global population, an increase in the number of people living in landslide-prone areas, and a warmer and locally wetter climate. These realities can be partially offset by using improving scientific knowledge of landslides in land-use decisions and by implementing targeted engineering mitigation measures to protect people and property. More fundamental, however, is the need to reduce the risk to the most vulnerable societies through social justice grounded in a more equitable distribution of global resources.

REFERENCES

Agliardi, F., Crosta, G.B. and Frattini, P. (2012). Slow rock slope deformation. In *Landslides: Types, Mechanisms and Modeling*, ed. J.J. Clague and D. Stead. Cambridge, UK: Cambridge University Press, pp. 207–221.

Ale, B.J.M. (2005). Tolerable or acceptable: A comparison of risk regulation in the United Kingdom and the Netherlands. *Risk Analysis*, 25, 231–241.

Arnold, M., Chen, R.S., Deichmann, U. *et al.* (ed.) (2006). *Natural Disaster Hotspots Case Studies*. World Bank Disaster Risk Management Series, 6.

Birkmann, J. (2006). Measuring vulnerability. In *Measuring Vulnerability to Natural Hazards*, ed. J. Birkmann. New York: United Nations University Press, pp. 9–54.

(2007). Risk and vulnerability indicators at different scales: Applicability, usefulness and policy implications. *Environmental Hazards*, 7, 20–31.

Bohle, H.-G. (2001). Vulnerability and criticality: Perspectives from social geography. *Newsletter of the International Human Dimensions Programme on Global Environmental Change*, 2, 1–7.

Chambers, R. (1989). Vulnerability. *Institute of Development Studies Bulletin*, 20(2),1–7.

CRED (Centre for Research on the Epidemiology of Disasters) (2011). *EM-DAT: The OFDA/CRED International Disaster Database*. Université Catholique de Louvain, Centre for Research on the Epidemiology of Disasters [available at www.emdat.be/, accessed December 13, 2011].

Cruden, D.M. and Varnes, D.J. (1996). Landslide types and processes. In *Landslides: Investigation and Mitigation*, ed. A.K. Turner and R.L. Schuster. Transportation Research Board, Special Report 247, pp. 36–75.

Evans, S.G. and Hungr, O. (1993). The assessment of rockfall hazard at the base of talus slopes. *Canadian Geotechnical Journal*, 30, 620–636.

Fell, R., Ho, K.K.S., Lacasse, S. and Leroi, E. (2005). A framework for landslide risk assessment and management. In *Proceedings of the International Conference on Landslide Risk Management*, Vancouver, BC, pp. 3–25.

Friele, P.A., Clague, J.J., Simpson, K. and Stasiuk, M. (2005). Impact of a Quaternary volcano on Holocene sedimentation in Lillooet River valley, British Columbia. *Sedimentary Geology*, 176, 305–322.

Friele, P., Jakob, M. and Clague, J. (2008). Hazard and risk from large landslides from Mount Meager volcano, British Columbia, Canada. *Georisk*, 2, 48–64.

Guthrie, R.H. and Evans, S.G. (2007). Work, persistence, and formative events: The geomorphic impact of landslides. *Geomorphology*, 88, 266–275.

Hungr, O. and Evans, S.G. (2004). The occurrence and classification of massive rock slope failure. *Fachzeitschrift fuer Geomechanik und Ingenieurgeologie im Bauwesen und Bergbau*, 22(2), 16–23.

Hungr, O., McDougall, S. and Bovis, M. (2005). Entrainment of material by debris flows. In *Debris-flow Hazards and Related Phenomena*, ed. M. Jakob and O. Hungr. Berlin: Springer Praxis, pp. 135–158.

International Society for Soil Mechanics and Geotechnical Engineering (2004). *Glossary of Risk Assessment Terms, Version 1*. International Society for Soil Mechanics and Geotechnical Engineering, Technical Committee on Risk Assessment and Management, TC32.

International Strategy for Disaster Reduction (2004). *Living with Risk: A Global Review of Disaster Reduction Initiatives*. Geneva: United Nations, International Strategy for Disaster Reduction.

ISO/TMB/RMWG (2007). *Committee Draft of ISO/IEC Guide 73 'Risk Management Vocabulary'*. Reference No. N48.

Iverson, R.M., Schilling, S.P. and Vallance, J.W. (1998). Objective delineation of lahar-inundation hazard zones. *Geological Society of America Bulletin*, 110, 972–984.

Jakob, M. and Friele, P. (2010). Frequency and magnitude of debris flows on Cheekye River, British Columbia. *Geomorphology*, 114, 382–395.

Leroi, E., Bonnard, C., Fell, R. and McInnes, R. (2005). Risk assessment and management. In *Proceedings of the International Conference on Landslide Risk Management*, Vancouver, BC, pp. 159–198.

McDougall, S., McKinnon, M. and Hungr, O. (2012). Developments in landslide runout prediction. In *Landslides: Types, Mechanisms and Modeling*, ed. J.J. Clague and D. Stead. Cambridge, UK: Cambridge University Press, pp. 187–195.

Moon, A.T., Wilson, R.A. and Flentje, P.N. (2005). Developing and using landslide frequency models. *Proceedings of the International Conference on Landslide Risk Management*, Vancouver, BC, pp. 681–690.

O'Keefe, P., Westgate, K. and Wisner, B. (1976). Taking the naturalness out of natural disasters. *Nature*, 260, 566–567.

Porter, M., Jakob, M., Savigny, K.W., Fougere, S. and Morgenstern, N. (2007). Risk management for urban flow slides in North Vancouver, Canada. In *Proceedings of the 60th Canadian Geotechnical Conference*, Ottawa, ON.

Simpson, K.A., Stasiuk, M., Shimamura, K., Clague, J.J. and Friele, P.A. (2006). Evidence for catastrophic volcanic debris flows in Pemberton Valley, British Columbia. *Canadian Journal of Earth Sciences*, 43, 679–689.

Solomon, S., Qin, D., Manning, M. *et al.* (ed.) (2007). *Climate Change 2007: The Physical Science Basis. Contribution of Working Group*

I to the Fourth Assessment Report of the Intergovernmental Panel on Climate Change. Cambridge, UK: Cambridge University Press.

Stoffel, M. (2006). A review of studies dealing with tree rings and rockfall activity: The role of dendrogeomorphology in natural hazards research. *Natural Hazards*, 39, 51–70.

Stoffel, M., Schneuwly, D., Bollschweiler, M. *et al.* (2005). Analyzing rockfall activity (1600–2002) in a protection forest: A case study using dendrogeomorphology. *Geomorphology*, 68, 224–241.

United Nations Department of Humanitarian Affairs (1992). *Internationally Agreed Glossary of Basic Terms Related to Disaster Management*. Geneva: United Nations Department of Humanitarian Affairs, DNA/93/36.

Varnes, D. J. (1978). Slope movement types and processes. In *Landslides: Analysis and Control*, ed. R. L. Schuster and R. J. Kirzek. US National Academy of Sciences, Transportation Research Board, Special Report 176, 11–33.

Villagrán de León, J.C. (2006). *Vulnerability: A Conceptual and Methodological Review*. Bonn: United Nations University, Institute for Environment and Human Security, 4.

Watts, M.J. and Bohle, H.-G. (1993). The space of vulnerability: The causal structure of hunger and famine. *Progress in Human Geography*, 17, 43–67.

2 Landslides in the Earth system

OLIVER KORUP

ABSTRACT

Landslides convert potential energy into kinetic energy and are thus important agents of topographic change and landscape evolution. They are deformations of Earth's surface that reflect patterns of regional seismic, climatic, and lithospheric stress fields on sloping terrain. Landslides involve fracturing of the lithosphere ranging from microscopic rock fragmentation to giant submarine slope failures, thus spanning more than 26 orders of magnitude in volume. Here I synthesize major rate constraints on landslide distribution, size, and impacts that help gauge their relevance in the Earth system with a focus on the lithosphere, the hydrosphere, and the biosphere. Given sufficient size or frequency, landslides help sculpt local topography, trigger shallow crustal response, limit volcanic edifice growth, modulate bedrock incision as well as water and sediment flux in river systems, trigger far-reaching processes such as tsunamis or catastrophic outburst flows, condition rates of soil production, and alter hillslope and riparian habitats. Most importantly, landslides remain a significant hazard to people, housing, infrastructure, and land use in many parts of the world.

2.1 INTRODUCTION

Landslides are the downhill and outward movement of slope-forming materials under the influence of gravity and also, in most cases, water (Cruden and Varnes, 1996). Mostly triggered by earthquakes, rainstorms, snowmelt, and slope undercutting, they are among the prime producers of sediment and major agents of denudation. Landslides mobilize rock debris, regolith, soil, and biogeochemical constituents in all types of terrain, ranging from the highest peaks in tectonically active mountain belts to the margins of abyssal plains. The growing recognition that landslides play an important role in shifting mass across the Earth's surface, thus helping form and redistribute topography, suggests expanding the classic definition to one that accommodates landslides as deformations of the Earth's surface that reflect patterns of regional seismic, climatic, and lithospheric stress fields on sloping terrain. The objective of this chapter is to synthesize evidence for how the occurrence and consequences of landslides are relevant to Earth as a system, particularly the lithosphere, the hydrosphere, and the biosphere. The intention is to take a deliberate step back from the plethora of detailed landslide case studies and analyses at the hillslope scale and to review landslide impacts within a regional to global context.

2.2 LANDSLIDE DISTRIBUTION AND SIZE

Landslides may initiate almost anywhere within Earth's elevation range, but they abound in tectonically active mountain belts with young, rapidly exhuming, and mechanically weak rocks. There, strong earthquakes and orographically enhanced precipitation fed by monsoonal and cyclonic storms frequently trigger slope instability (Fig. 2.1; Lin *et al.*, 2008). More than half of the largest known terrestrial landslides occur in the steepest 5 percent of Earth's land surface, where the inferred rates of denudation exceed 1 mm per year (Korup *et al.*, 2007). Tectonic fault zones (Strecker and Marrett, 1999; Osmundsen *et al.*, 2009), volcanic arcs (Coombs *et al.*, 2007), rocky coasts (Hapke and Green, 2006), and the edges of continental shelves (Weaver, 2003) are other settings where landslides cluster. Yet even in such highly susceptible terrain, the observed number of landslides per unit area or time ranges through 3–11 orders of magnitude (Fig. 2.2). This variation attests to the broad spectrum of ways in which hillslopes can adjust to external perturbations to their stability through rate changes in landsliding. It

Landslides: Types, Mechanisms and Modeling, ed. John J. Clague and Douglas Stead. Published by Cambridge University Press.

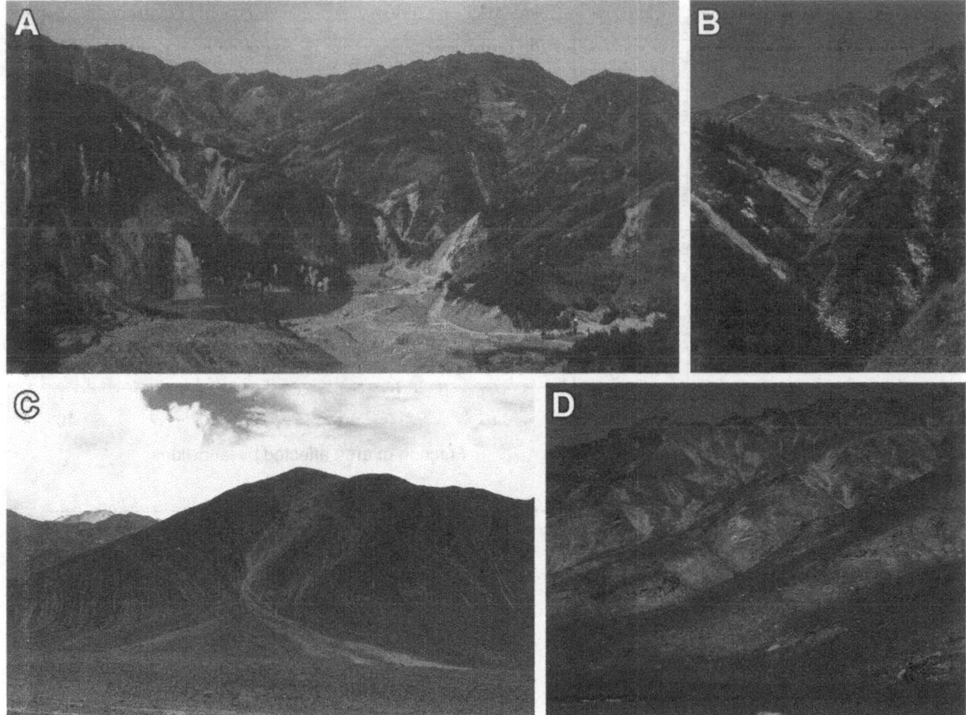

Fig. 2.1. Landslide landscapes. (A) Numerous landslide scars dot subtropical hillslopes of Longmen Shan, southwest China, following the 2008 Wenchuan earthquake. (B) High connectivity between shallow debris slides and headwater channels cut in weak rocks, Swiss Alps. (C) Dozens of small debris flows triggered by heavy rainstorms on semi-arid hillslopes, Ladakh, India. (D) Partly re-vegetated scars of loess flows on soil-mantled hillslopes, Kyrgyz Tien Shan.

also calls for a consistent and comparable measure of landslide frequency or density, as some metrics have higher variance than others for a given area and observation period.

Reported landslide frequencies per unit time range from individual events to landslide episodes that entail the geologically instantaneous occurrence of up to several tens to hundreds of thousands of landslides in the wake of a regional triggering event (Fig. 2.1; Crozier, 2005). Such episodes increase the spatial landslide density per unit study area (km^{-2}), or the proportion of landslide-affected terrain (percent), by an order of magnitude. For example, the 2008 Wenchuan earthquake, China, and the 1999 Chi-Chi earthquake, Taiwan, triggered >10,000 and >22,000 landslides, respectively (Di *et al.*, 2010). Over 50,000 landslides have been documented for individual historic earthquakes (Keefer, 1999; Gorum *et al.*, 2011). These numbers remain estimates where hillslopes stripped completely of vegetation cover make it impossible to discern individual failure scars. Similar detection limits apply to hillslopes with pronounced anthropogenic modification such as terracing or slope-stability mitigation. Rapid vegetation regrowth or dense canopy also commonly obscure smaller slope failures. Depending on study area size and accessibility, as well as the resolution of the remote sensing data used for detection and mapping, reported earthquake-triggered landslide frequencies range from 0.25 to 15 km^{-2} (Barnard *et al.*, 2001; Sato *et al.*, 2007, respectively). Yin *et al.* (2010) estimated a maximum landslide density of 13 km^{-2} for the epicenter region of the 2008 Wenchuan earthquake, based on an analysis of 1 m resolution digital orthophotos. By comparison, rainstorm-triggered landslide densities in small (>10 km^2) catchments have attained values of up to 480 km^{-2} (Crozier, 2005), which is an order of magnitude higher. Indeed, Crozier (2005) lists at least eight rainfall-triggered regional landslide episodes that produced nominally higher landslide densities than those recorded after earthquakes. In small (<1 km^2) mountain catchments subjected to intensive agricultural use, landslide density may soar to highly localized values of 1800 km^{-2} (Thapa and Paudel, 2002). Finally, regional comparison of geological archives suggests that the volumetrically largest landslide episodes in Earth's history may have been triggered by large bolide impacts (Busby *et al.*, 2002).

Distributions of landslide size are largely determined from remote sensing-based measurements of scar or deposit areas. Landslides are commonly identified and measured during regional mapping and compiled into landslide databases or inventories. Landslide area distributions are generally strongly skewed, with large tails that approximate an inverse power law over several orders of magnitude, reflecting – to some degree – invariant geometric scaling over these scales (Figs. 2.3A, B, 2.4). An inventory of 1350 of the Earth's largest documented landslides demonstrates the heavy tail of the landslide size spectrum in terms of volume V_L. While large landslides (V_L > 10^9 m^3) dominate the volumetric distribution on land, some 130 submarine landslides may contain up to 99 percent of the total landslide volume in this inventory (Fig. 2.5). The area covered by these submarine landslides is greater than that of the states of California and Nevada combined. Although heavy tails are robust for landslide areas from 10^3 to 10^6 m^2, landslide size distributions commonly have distinct roll-overs at the smaller end of the size spectrum (Hovius *et al.*, 1997; Malamud *et al.*, 2004), to which several alternative nonlinear distributions may apply

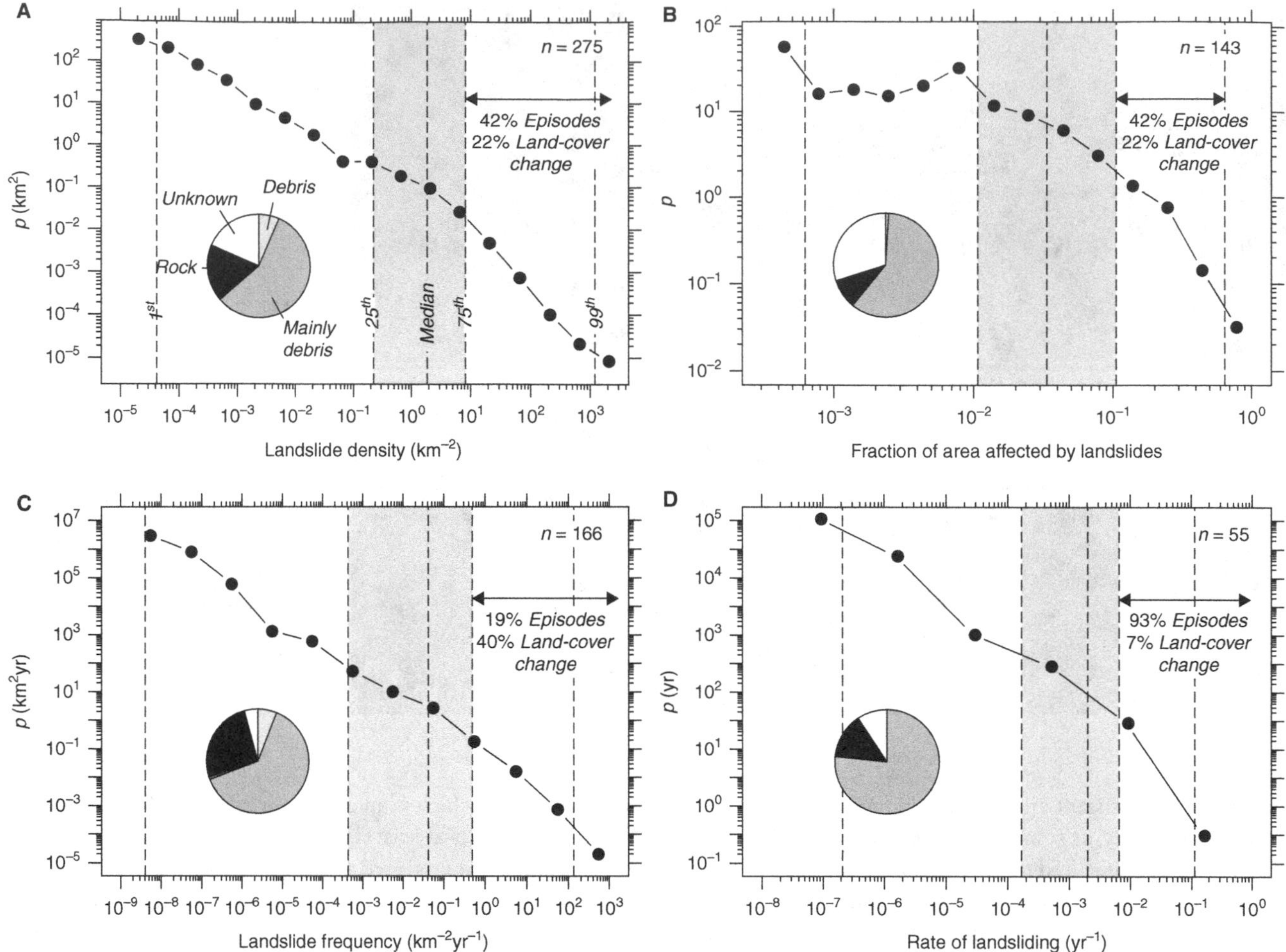

Fig. 2.2. Normalized histograms of measures of landslide occurrence (A, Landslide density (km^{-2}); B, Fraction of area affected by landslides; C, Landslide frequency (km^{-2} per year); D, Rate of landsliding per year) in different mountain belts, compiled from the literature. Vertical lines are percentiles, gray shading is interquartile range, and horizontal arrows encompass the fraction of data derived from historic landslide episodes and human-induced land-cover changes above the 75th percentile in each distribution. Inset pie charts show fraction of dominant landslide material involved. Here "debris" includes soils and regolith.

(ten Brink *et al.*, 2010). Overall, however, smaller landslides are much more frequent than larger ones regardless of the topographic, tectonic, climatic, or lithological setting. Explanatory studies linking landslide size distributions with deterministic slope stability equations argue that the cohesive strength of hillslope materials largely controls the size-frequency scaling properties of soil and bedrock landslides (Van Den Eeckhaut *et al.*, 2007; Stark and Guzzetti, 2009). Similarly, material properties influence the geometry of landslides more than other factors, such as volumetric dilation or entrainment of material during runout. More than half of *ca.* 4200 landslides with field-verified volume and area data have an average thickness of 1 m, a value largely controlled by local soil depth (Larsen *et al.*, 2010; Fig. 2.6). In contrast, large bedrock landslides produce much thicker deposits (10–100 m) than smaller ones.

Whether the different proposed magnitude–frequency distributions are applicable both to the smallest and largest of Earth's landslides awaits further investigation, particularly as scaling exponents vary with trigger mechanism, lithology, and sample size (Fig. 2.4). An inverse gamma distribution, which has been found to successfully model area–frequency data of much smaller terrestrial landslides (Malamud *et al.*, 2004), produces a good fit to five orders of magnitude of Earth's largest landslides with only minor adjustment to the fitting parameters. Eventually, such distributions will help provide estimates of the average frequency of very large, and potentially very destructive, events, which are highly relevant for quantitative hazard and risk assessments.

2.3 LANDSLIDES AND THE LITHOSPHERE

Landslides involve fracturing of the lithosphere on several spatial scales. The lower bound is set by the dynamic fragmentation of rock particles down to sub-millimeter scale during the motion of large bedrock landslides, a process that resembles cataclasis caused by tectonic fault rupture. Thin layers of frictionite,

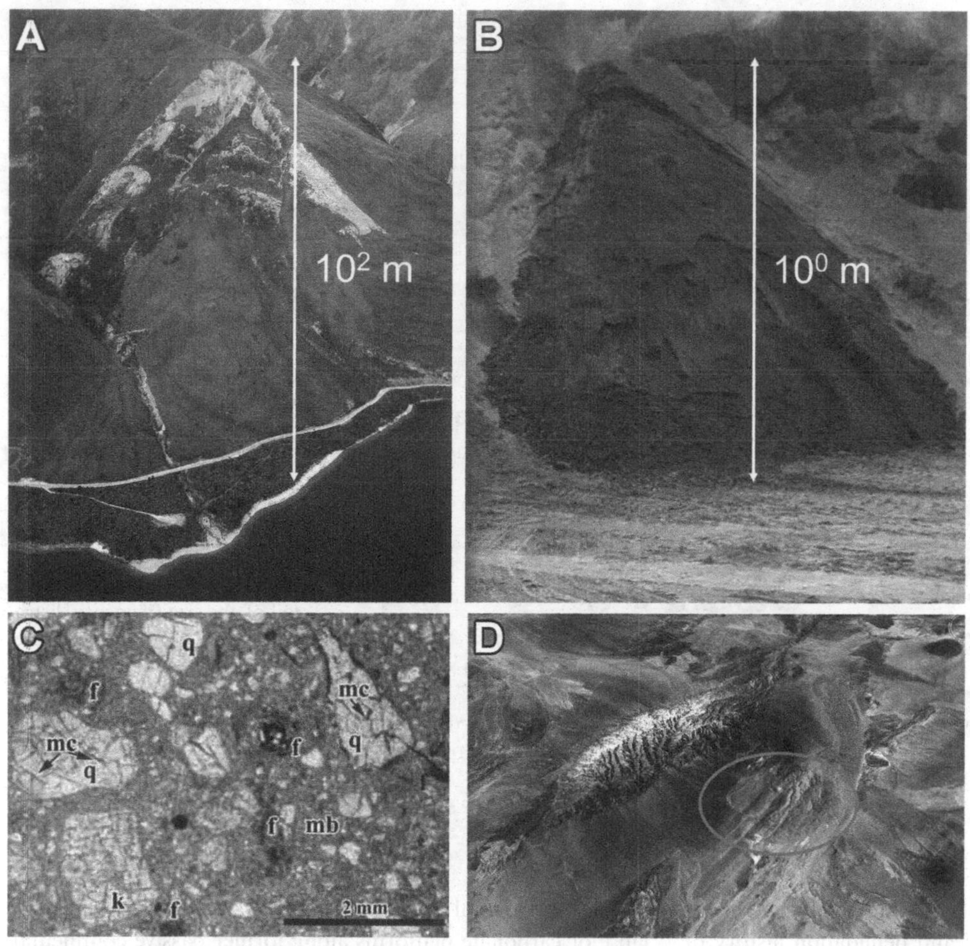

Fig. 2.3. Geometric scaling of landslides. (A) Large deep-seated, slow-moving rockslide. (B) Rotational failure in sandy tailings showing similar surface expression at a scale that is two orders of magnitude smaller. (C) Frictionite produced by dynamic fragmentation of rock particles during catastrophic motion of large rock-slope failures; the particle size ranges down to *ca.* 10^{-15} m^3; f = frictionite, k = K-feldspar q = quartz, mb = micro-breccia, mc = micro-cracks. (D) The 50-km^3 Baga Bogd landslide (gray oval), Mongolia; this landslide approaches a mountain-belt scale in size (Philip and Ritz, 1999).

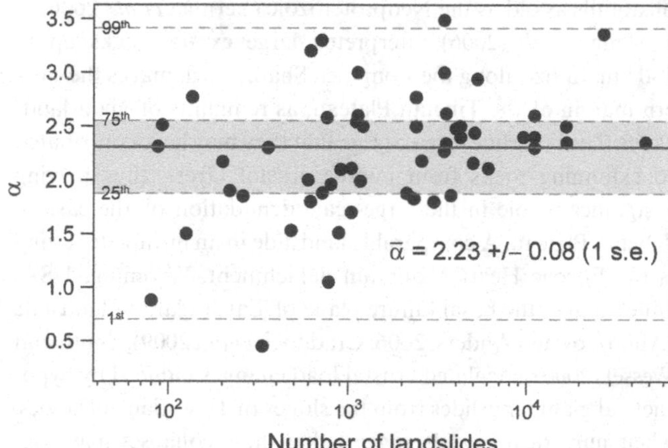

Fig. 2.4. Values of the noncumulative power-law scaling exponent α derived from probability density estimates of landslide area recorded in 60 inventories from different parts of the world, as a function of sample size. Horizontal dashed lines are percentiles of α; mean $\alpha = 2.23$ (s.e. = standard error about the mean), and median $\alpha = 2.31$, which corresponds to the value reported by Van Den Eeckhaut *et al.* (2007).

formed at the base of large landslides, show evidence of incipient mineral melting similar to hyalomylonite along faults (Fig. 2.3C). Frictionite is characterized by physically disintegrated angular particles of *ca.* 10^{-15} m^3 in size with no evidence of chemical weathering (Weidinger and Korup, 2009). The upper bound of lithospheric fracture is defined by the total mobilized volumes of the largest terrestrial landslides ($V_L \approx 10^{10}$ m^3; Fig. 2.3D). These very large landslides are exclusively volcanic debris avalanches linked to catastrophic edifice collapse (Korup *et al.*, 2007). The largest landslides on Earth, however, are submarine slope failures up to 10^{13} m^3 off continental shelves or submarine fans. It follows that the size scales of mechanical breakdown of the lithosphere associated with landsliding span more than 26 orders of magnitude.

From the perspective of long-term landscape evolution, landslides complement the erosional action of rivers and glaciers. The latter, however, increase topographic bedrock relief by deepening valley floors through incision, abrasion, plucking, and scour, whereas landslides reduce bedrock relief through downwasting of interfluves and divides and by systematic lowering of hillslope steepness (Korup, 2006). Especially large failures provide a mechanism for catastrophic shifting of drainage divides, in some cases by >10 km. This limit to local bedrock relief (Schmidt and Montgomery, 1995) has important implications for the height of mountain peaks steep enough to preclude a significant accumulation of snow and hence escape

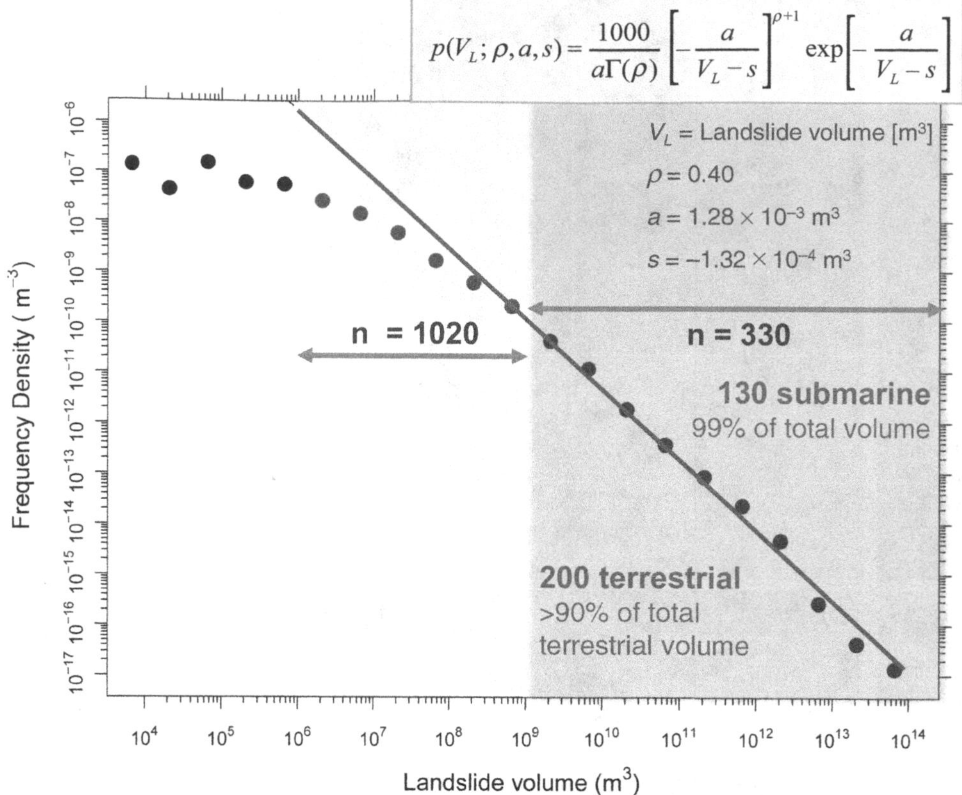

Fig. 2.5. Size-frequency distribution of volumes V_L mobilized by the largest landslides on Earth, based on a global inventory of 1350 landslides with $V_L > 10^6$ m^3 (extending that of Korup *et al.*, 2007). The data are dominated by the largest landslides: 200 landslides represent >90 percent of the total volume contained in the recorded terrestrial landslides, and 130 of the largest submarine landslides include 99 percent of the total inventory volume. A three-parameter inverse gamma distribution $p(V_L, \rho, a, s)$ originally proposed by Malamud *et al.* (2004) for landslide area closely approximates the heavy tail of the size-frequency distribution ($V_L > 10^9$ m^3) over five orders of magnitude, necessitating only minor changes to the original parameter values, such as adjusting for dimensional consistency; Γ is the gamma function.

glacial erosion. Similarly, large catastrophic debris avalanches help constrain the growth, height, and denudation history of volcanic edifices (Coombs *et al.*, 2007; Ponomareva *et al.*, 2006). Arid and submarine environments may preserve evidence of the landscape-forming events for >10^6 years (Wörner *et al.*, 2002; McMurtry *et al.*, 2004; Craddock *et al.*, 2009).

If of sufficient size (>10^9 m^3), landslide detachment and emplacement may also contribute to redistributing local crustal loads and stresses. For example, in the easily eroded rocks of the Penninic nappes of eastern Switzerland, the spatial clustering of many of the largest alpine landslides provides large bodies of intensively crushed debris amenable to rapid fluvial erosion and transport. Simple flexural beam models help estimate the fraction of any resulting erosion-induced uplift. The amount of such uplift largely depends on the value of the flexural rigidity of the crust, its effective elastic thickness, and the wavelength of lithospheric flexure. In the eastern Swiss Alps, erosion-induced unloading may explain up to 40 percent of the measured regional rock uplift rates (Korup and Schlunegger, 2009). Recent sandbox experiments further suggest that the catastrophic emplacement of olistostromes (i.e., large submarine landslides off continental margins) may considerably change both the style and location of deformation of accretionary wedges. Development of such wedges appears to depend on the thickness and extent of landslide loading on 10^2 km scales, resulting in abrupt shifts of deformation to more seaward-located imbricate thrusts (Smit *et al.*, 2010). Similar feedbacks between landslide emplacement and vertical crustal response may apply to the large-scale collapse of carbonate platforms along former passive continental margins, recorded in the geologic archive as olistostromes and olistoliths as old as the Neoproterozoic (Vernhet *et al.*, 2006).

Meng *et al.* (2006) interpreted large exotic blocks up to 150 km^2 in size along the Longmen Shan, which marks the eastern margin of the Tibetan Plateau, as remnants of giant landslides. The landslides are so large that they may have contributed to exhuming rocks from middle crustal layers, thus playing a significant role in the large-scale denudation of the eastern Tibetan Plateau. A comparable landslide in an intraplate setting is the Eocene Heart Mountain detachment, Wyoming, USA, which marks the basal failure plane of Earth's largest landslide (Aharonov and Anders, 2006; Craddock *et al.*, 2009). Smith and Wessel (2000) calculated crustal load changes induced by hypothetical giant landslides from the slopes of Hawaiian volcanoes. Their numerical models suggest that these collapses may trigger crustal uplift and depression of the order of 10^1–10^2 m in the source area, and 10^0 m in the deposition zone, depending on the elastic plate thickness and landslide volume. Such vertical deformation is comparable to other short-term deformation driven by the growth or contraction of magma chambers. Additionally, catastrophic collapse of volcanic island flanks may unroof and decompress magma chambers. The magnitude of decompression depends, among other things, on the depth and elastic properties of the magma chamber and landslide volume, and is estimated to range from a few kPa to several tens of MPa

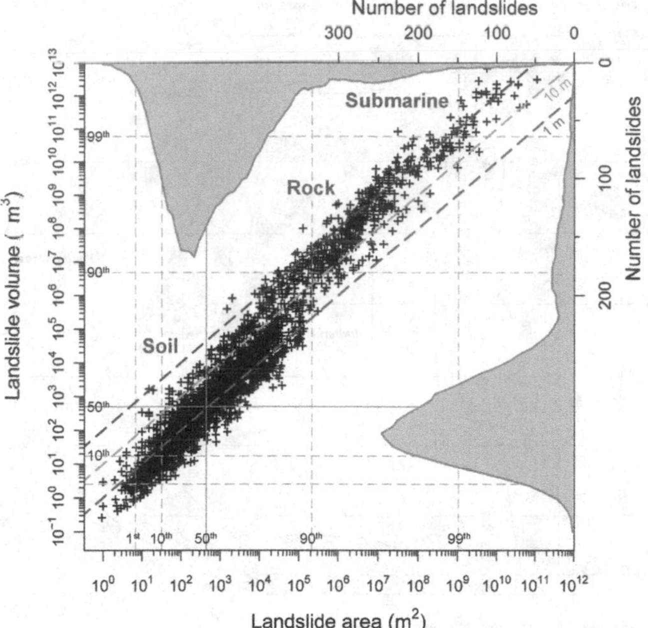

Fig. 2.6. Volume–area scaling for a sample of *ca.* 4200 soil, bedrock, and submarine landslides; based on Larsen *et al.* (2010) and augmented by data on 110 submarine landslides. Dashed lines are mean landslide thicknesses.

(Manconi *et al.*, 2009). Eventually, catastrophic unroofing may further increase volcanic explosive activity and modulate magmatic melt production, differentiation, and flux rates (Longpré *et al.*, 2009). Hence large volcanic debris avalanches are important in the growth and decay of volcanic edifices (Beget and Kienle, 1992). Not surprisingly, significant portions of many volcanoes are composed of landslide debris (McMurtry *et al.*, 2004; Oehler *et al.*, 2008).

2.4 LANDSLIDES AND THE HYDROSPHERE

2.4.1 LANDSLIDES AND THE TERRESTRIAL WATER CYCLE

Syvitski *et al.* (2005) estimate that the contemporary global sediment flux from rivers to oceans is 15.5×10^9 t per year. Fluxes proposed by others are within ±50 percent of this value, depending on the method of interpolation (Beusen *et al.*, 2005). In this context, landslides are significant, although highly localized, prime producers of sediment. For example, the five largest (>10^9 m³) terrestrial landslides mobilized a total volume equal to >1 percent of the global riverine sediment flux during the twentieth century, although they affected only 1.4×10^{-4} percent of the Earth's land surface (Fig. 2.7). At the scale of individual drainage basins, the larger ($V_L > 10^8$ m³) landslides may mobilize –in one single event – volumes that are equivalent to 10^3 years of the mean background catchment erosion (Fig. 2.8).

Landslides of all sizes are important mobilizing agents of soil, debris, and rock, and are a primary mechanism of coupling and modulating mass flux between hillslopes and river channels (Korup, 2005). Rapid delivery of landslide debris to rivers causes sediment pulses or waves (Lisle *et al.*, 1997) that alter channel geometry, fluvial hydraulics, sediment caliber, and flood frequency (Benda and Dunne, 1997; Korup, 2004; Brummer and Montgomery, 2006; Fig. 2.9). Some of the highest reported fluvial sediment yields (>10^5 t km^{-2} per year) have occurred in the wake of regional landslide episodes triggered by earthquakes (Pain and Bowler, 1973; Pearce and Watson, 1986), rainstorms (Page *et al.*, 1994; Trustrum *et al.*, 1999), or a sequence of both (Dadson *et al.* 2004). In Taiwan, for example, the passage of several tropical cyclones has flushed landslide debris produced during the 1999 Chi-Chi earthquake, raising suspended sediment yields in mountain rivers to four times the background level before dropping back to normal in 2005 (Lin *et al.*, 2008).

The hydrologic and geomorphic interaction between landslides and the drainage network is manifold. Debris flows are important transitional phases between landslides and floods and may account for a significant, yet hitherto poorly quantified, fraction of bedrock channel incision in headwater streams (Stock and Dietrich, 2003). Large landslides can have profound impacts on rivers, including changes to channel geometry such as knickpoint creation (Fig. 2.10), avulsions, formation of epigenetic bedrock gorges, and drainage reversals (Korup, 2004, 2006). Landslide debris that overwhelms the ability of a river to remove it creates a blockage that may persist anywhere from a few seconds to 10^4 years (Hewitt, 2010). Such impoundments of water and sediment interrupt and delay the transmission of base-level signals such as headward-migrating bedrock knickpoints (Korup *et al.*, 2010) or sediment pulses. Spatial clustering of landslide dams along a river modulates patterns of fluvial bedrock incision and aggradation, and thus sediment flux from mountain belts to their forelands. Landslide-dammed lakes have spawned some of the most extreme floods and debris flows recorded in the Quaternary (Korup and Tweed, 2007). Such catastrophic outbursts may affect areas of up to 10^3 km. They are both sediment transport events and natural hazards, with a large terrestrial footprint. For example, Dai *et al.* (2005) reported that the sudden dam-break flood from an earthquake-triggered landslide lake on Dadu River, Sichuan, China, in 1786, caused an estimated 100,000 fatalities over a distance of hundreds of kilometers downstream of the blockage site. Observations confirm extreme sediment pulses associated with outburst events, such as the one caused by the failure of the earthquake-triggered Bairaman landslide dam, Papua New Guinea, in 1985 (King *et al.*, 1989). In a period of 3 hours, *ca.* 80×10^6 m³ of sediment was mobilized and flushed downstream in a massive debris flow that had a flow height of *ca.* 100 m just downstream of the dam.

Erosional evidence of landslides, such as detachment scars, is commonly short-lived (Fig. 2.1), but inland water bodies

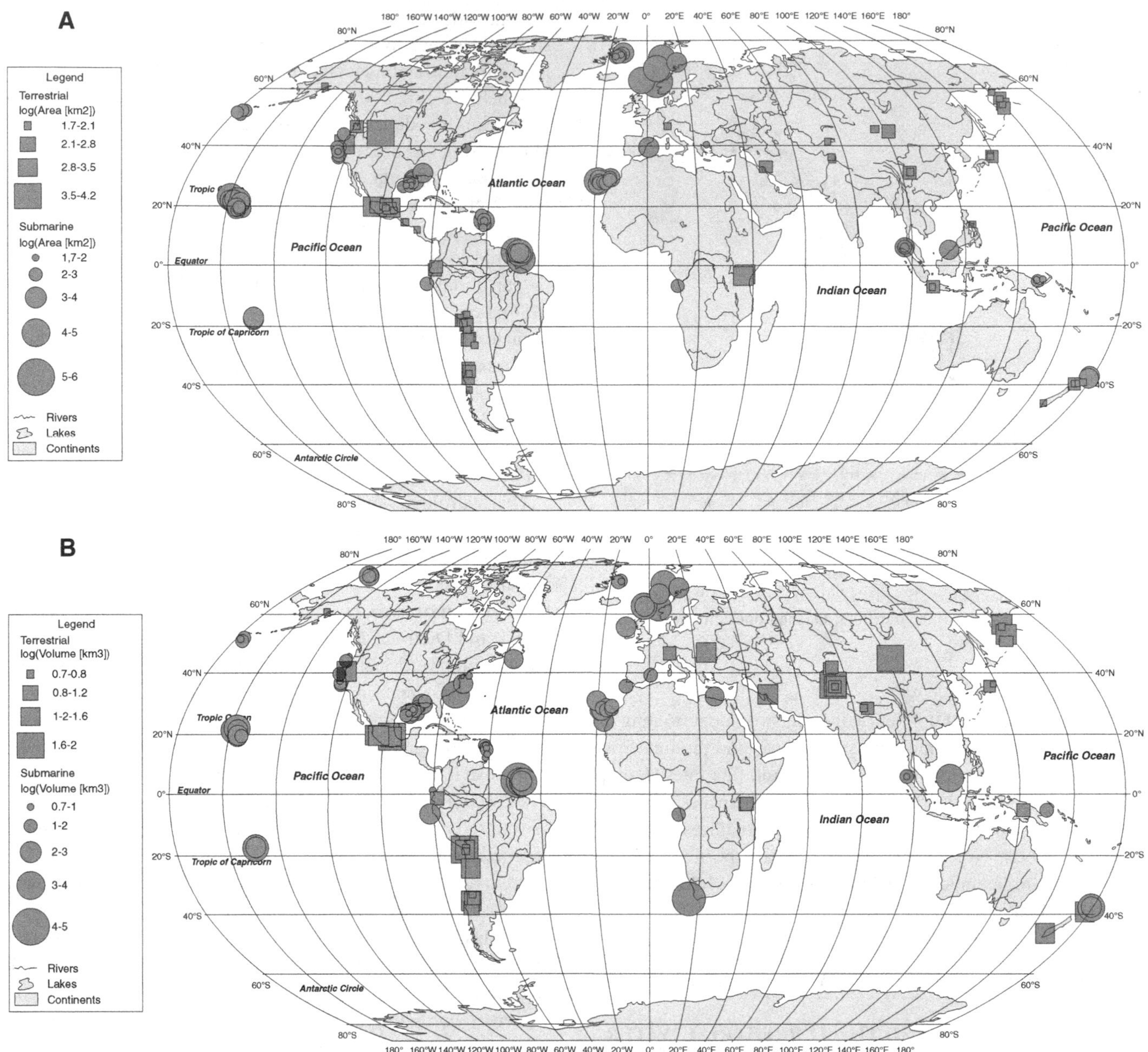

Fig. 2.7. Global distribution of reported landslides. (A) Total affected area >50 km^2. (B) Total mobilized volume >5 km^3. Note that spatial clustering of landslides may not necessarily represent "landslide territories," i.e., areas of pronounced large-scale slope instability (Hampton *et al.*, 1996). Rather, part of the pattern mimics the geographic focus of certain studies, and many more landslide deposits await detection. This map does not include olistostromes or similar stratigraphic, although geomorphically often indistinct, large-scale landslide deposits.

may contain valuable archives of both past slope instability and its triggers. Lake-floor bathymetry reveals lacustrine sediment slumping triggered by earthquake shaking or incoming sediment pulses (Moore *et al.*, 2006; Strasser *et al.*, 2006) and may complement proxies from the fluvial sedimentary record. Appropriately calibrated laminations in lake sediments have the potential to record recurring sediment input in the wake of landslides on millennial timescales (Gomez *et al.*, 2007). If substantially complete, such archives provide important clues about the time series of landsliding in periods preceding detailed documentation through aerial photography or topographic measurements.

2.4.2 LANDSLIDES AND THE OCEANS

Oceans cover two-thirds of Earth's surface and hide most of the submarine landslide processes from direct view. However, high-resolution bathymetric, seismic, and drill-core data have revealed that submarine landslides play a significant role in shaping the seafloor. Indeed, digital bathymetry and seismic

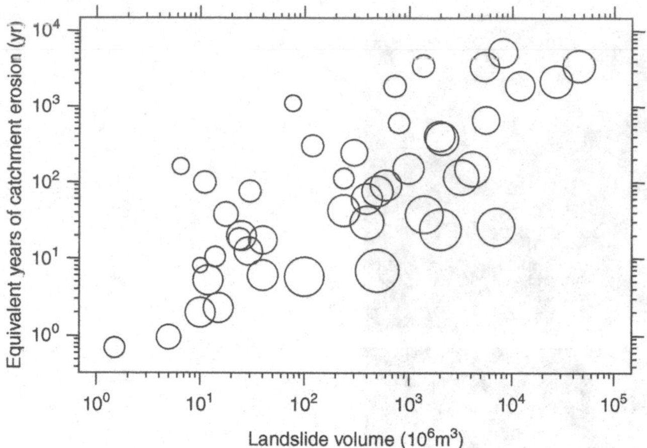

Fig. 2.8. Mobilized volumes of individual large landslides expressed as the equivalent period of time required for background catchment erosion to produce the same volume. Circles are scaled by catchment area and range from 64 to 228,000 km^2.

profiling may provide a more detailed view of giant submarine landslides than is possible for their terrestrial counterparts. Submarine landslides comprise debris avalanches, rotational slides (slumps), and debris or mud flows (turbidites). Motion may be multiphased, with the main failure followed by debris flows that travel up to 10^2 km and terminate at small changes of seafloor-bottom gradients (Talling *et al.*, 2007). The largest submarine landslides are larger than their onshore cousins by up to three orders of magnitude in terms of deposit area and volume (Fig. 2.7). The 1929 Grand Banks submarine landslide mobilized more than 10 times the volume (*ca.* 2×10^{11} m^3; Fine *et al.*, 2005) of the five largest terrestrial landslides in the twentieth century, even adjusting for the high porosity (<60 percent) that is characteristic of many submarine landslide deposits. Silva *et al.* (2010) reported two "megaslides" on the Amazon Fan, each covering >80,000 km^2, or twice the size of Switzerland. Mapping of hundreds of submarine landslide deposits reveals their striking abundance along passive continental margins and chains of volcanic islands (Fig. 2.7). Evidence tends to be sparser along active margins, partly because of the obliteration of landslide deposits by subduction.

In contrast to many large terrestrial landslides (Korup *et al.*, 2007), topographic relief or slope inclination is not a first-order control on the location of submarine landslides. However, changes in regional slope caused by tilting or redistribution of crustal stresses may enhance the susceptibility of the seafloor to failure (Leynaud *et al.*, 2009). Moreover, most of the largest debris avalanches from volcanic island edifices have some of the largest fall heights (>4 km) recorded. However, many other giant submarine landslides have happened in the moderate (<5°) topography of continental shelf margins, largely controlled by changes in bulk sediment strength due to seismicity, storms, loading by incoming sediment, and possibly by slip along surfaces controlled by frozen methane hydrate deposits (Grozic, 2010). The spatial coincidence of many submarine landslide scar areas and methane hydrate layers has spurred a lively "chicken-and-egg" discussion about the sudden release of methane hydrates, which are potent greenhouse gases, into the ocean and atmosphere (Leynaud *et al.*, 2009).

Tsunamis are a major consequence of submarine landslides and those that enter the sea or inland water bodies, translating part of the landslide kinetic energy into wave impact far from the original failure site (Masson *et al.*, 2006; ten Brink *et al.*, 2009; Tappin, 2010). Landslide tsunamis and seiches pose a significant hazard to many marine and inland coastal areas, respectively. The most impressive historic landslide tsunami in terms of wave height was caused by an M_w 8.3 earthquake in southeast Alaska in 1958, when a large rockslide fell into Lituya Bay. The impact created a tsunami with a maximum run-up of 524 m above sea level. The tsunami completely stripped away surficial sediments, soil cover, and vegetation from the fjord wall (Fritz *et al.*, 2009). The largest historic tsunamigenic edifice collapse in 1888 involved *ca.* 5 km^3 – i.e., most of the subaerial portion – of Ritter Island volcano, Papua New Guinea. The resulting >20-km-long debris avalanche caused a tsunami with local wave heights of up to 20 m (Silver *et al.*, 2005). Tsunami deposits that indicate run-up heights locally in excess of 25 m have been documented at numerous locations in the UK and have been attributed to the giant Storegga Slide off the Norwegian continental shelf that occurred around 7900 years ago (Smith *et al.*, 2004).

Sediment mobilization by large submarine landslides along continental escarpments can be enormous. Taylor *et al.* (2002) estimated that half a dozen catastrophic landslides along the formerly glaciated Norwegian Sea margin account for *ca.* 75 percent of the basin sedimentation during the past 30,000 years. Offshore sediment records from abyssal basin turbidite sequences along the northwest African passive margin indicate that landsliding may have reworked up to 20 percent of the continental rise and slope sediments during the past 22 million years (Weaver, 2003). This contribution is limited to organic-rich turbidity currents of volumes $V_L > 10^{10}$ m^3, with average return periods of 10^4 years. In this context, marine drill cores provide, similarly to some terrestrial lake records, detailed sedimentary archives of submarine landslides, glaciogenic debris flows, and stacked turbidite beds. These records have stimulated an enquiry about the possibility of disentangling seismic and climatic triggers from intrinsic instabilities due to variations in sediment supply to continental margins. The cyclic character of many turbidite sequences is an important recorder of recurring phases of slope instability potentially driven by overfilling of continental-shelf sediment repositories, sea-level oscillations, climate change, or strong earthquakes. Bourget *et al.* (2010) reconstructed a decrease in mean turbidite frequency of nearly an order of magnitude following the local sea-level high stand about 8000 years ago at

Fig. 2.9. Thick (10^1-m scale) valley-fill deposits of catastrophic long-runout landslides triggered during the 2008 Wenchuan earthquake.

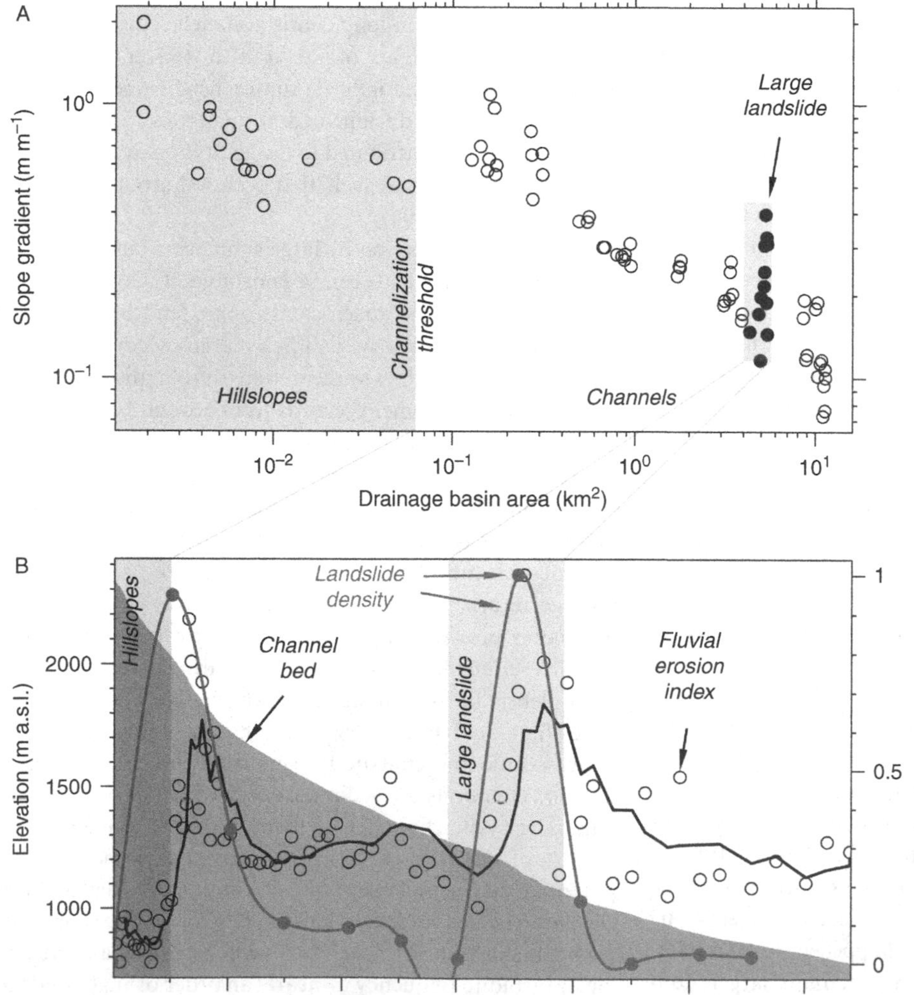

Fig. 2.10. Dynamic feedbacks between landslides and river channel geometry in a small alpine catchment shown by correlation between proxies of fluvial erosion and landslide occurrence. (A) Slope-area plot from 25-m digital elevation model (open circles) shows that large deep-seated bedrock landslide (black circles) spatially coincides with channel knickpoint and a four-fold increase in local bed gradient. (B) Longitudinal river profile with a normalized index of fluvial bedrock erosion potential (open circles; black line is five-point running mean) using unit stream power corrected for discharge effects of regional flood frequency and channel width variations; gray line and filled circles are the spline-interpolated normalized spatial density of 215 shallow landslides. Landslide density was measured incrementally for hillslope portions that contribute to *ca.* 0.5-km-long channel segments. Note the matching of peaks in the fluvial erosion index and shallow landslide density. Although the upstream clustering of landslides may be due to locally enhanced channel incision, the downstream peak in inferred fluvial erosion results from knickpoint formation driven by the large bedrock landslide (gray shading).

the Makran active margin in the Gulf of Oman. They linked these changes to shifts in early Holocene monsoon-driven climate, which favored higher sediment storage near the shelf and thus overburden-induced slope instability. Similarly, Blumberg *et al.* (2008) noted that turbidites deposited in the Chilean forearc were nearly an order of magnitude more frequent during dry and cold glacial stages, when sea level was low and glacial sediments readily available, than during interglacials, when sea level was high. Goldfinger *et al.* (2003) attributed ^{14}C-dated turbidites affecting areas of 10^3 km^2 to giant earthquakes at the Cascadia subduction zone in the US Pacific Northwest. Terrestrial sedimentary archives have rarely been explored with comparable focus or detail, and there remains some debate about whether terrestrial landslide deposits are reliable seismic or climatic proxies.

Other researchers have proposed that global climate change is a driver of catastrophic collapse of volcanic island flanks (Krastel *et al.*, 2001) and deposition of sediment wedges on continental shelves (Vogt and Jung, 2002). They argue that sea-level rise during the transition from glaciations to interglaciations has led to increased pore water pressure and wave action that reduced slope stability in coastal zones (McMurtry *et al.*, 2004; Quidelleur *et al.*, 2008). Such quasi-cyclic models of volcanic edifice growth and decay conditioned by catastrophic flank collapse (Beget and Kienle, 1992) offer important causal linkages between global climate change, large landslides, and volcanism. In nonvolcanic settings, interglacial sea-level rise further increases loading on seafloors. It also promotes bottom-water warming, which contributes to the dissociation of frozen methane hydrates that act as potential weak layers beneath the continental shelf (Vogt and Jung, 2002). Further research is necessary to elucidate whether submarine landslides are a cause or consequence of methane hydrate release.

2.5 LANDSLIDES AND THE BIOSPHERE

Most research on feedbacks between landslides and the biosphere has focused on potentially adverse effects of human activities. Human disturbances such as road cutting, deforestation, and mining are frequent causes and triggers of landslides (Figs. 2.2, 2.11; for an overview see Sidle and Ochiai, 2006). The dynamic interface between landslides and human activities defines the potential for hazard and risk, which are, respectively, the annual probability of adverse impact and the associated expected loss. A large and growing body of literature has focused mainly on assessing landslide susceptibility based on topographic, geologic, hydro-meteorological, and land-use characteristics (Guzzetti *et al.*, 2006), although studies that explicitly quantify landslide hazard and risk in a probabilistic sense are comparatively rare. However, the consequences of landslide occurrence have received increasing attention, including immediate impacts such as tsunamis (ten Brink *et al.*, 2009; Tappin, 2010) and potential long-term feedbacks between landsliding and soil production and productivity (Page *et al.*, 2004; Larsen *et al.*, 2010).

Much research has been concerned with assessing the stabilizing effects of vegetation on hillslopes, particularly with respect to their role in reducing the occurrence of shallow landslides and debris flows. In this context, tree root strength is one of the key parameters for reinforcing soil strength in deterministic force equilibrium equations such as the infinite slope model (Roering *et al.*, 2003; Sidle and Ochiai, 2006). In forested mountainous terrain, riparian slope failures play an important role in recruiting large woody debris (LWD) and particulate organic carbon (POC) to the channel network (May and Gresswell, 2003). In tectonically active mountain belts, landslides, in this manner, contribute significantly to the overall carbon budget, with contemporary yields as high as 10^2 t C km^{-2} per year (Walker *et al.*, 1996; Page *et al.*, 2004; Hilton *et al.*, 2008). Ren *et al.* (2009) estimated that the 2008 Wenchuan earthquake mobilized 235×10^6 t C through biomass erosion – an amount that roughly exceeds the net annual organic carbon sink of the USA. Besides transporting soil, vegetation, and biogeochemical constituents, landslides are also agents of contaminant dispersal. For example, one of the World Bank's natural disaster hotspots is the uranium mining tailings deposits in the Mailuu-Suu Valley, Kyrgyzstan. Maintenance of the tailings deposits has stalled in post-Soviet times, leaving them highly prone to landslides and fluvial entrainment. Short-lived landslide dams promote destabilization of the tailings dams by inundation and sluicing of contaminated fines far downstream into the densely populated reaches of major central Asian rivers (Havenith *et al.*, 2006).

Studies of the ways in which landslides interact with the biosphere have seldom explored potentially beneficial aspects (Hewitt, 2010). For example, landslide-dammed lakes may buffer flashy water and sediment discharge and create important freshwater storage. Also, landslides have long been recognized as modulators and enhancers of biodiversity (Garwood *et al.*, 1979). Landslide-derived LWD significantly alters the hydraulic regime, and hence habitats, of headwater streams, causing log jams, step-pool sequences, and alternating alluvial and bedrock reaches (Massong and Montgomery, 2000). Debris flows in low-order forested mountain rivers promote heterogeneity in channel geometry and habitats by forcing local deposition of LWD and boulder jams (Benda *et al.*, 2003). Debris-flow-derived LWD is also ecologically important in that it creates and partitions riparian and in-channel habitats (Bigelow *et al.*, 2007; Geertsema and Pojar, 2007).

2.6 SUMMARY

Landslides play a profound role in shaping the Earth's surface by mobilizing and re-depositing soil, rock, debris, and biogeochemical constituents. Slope failure occurs on effective length scales that encompass >26 orders of magnitude. The growing amount of quantitative data on their distribution, scaling, and consequences allows systematic first-order synthesis of their effects on the lithosphere, hydrosphere, and biosphere. This

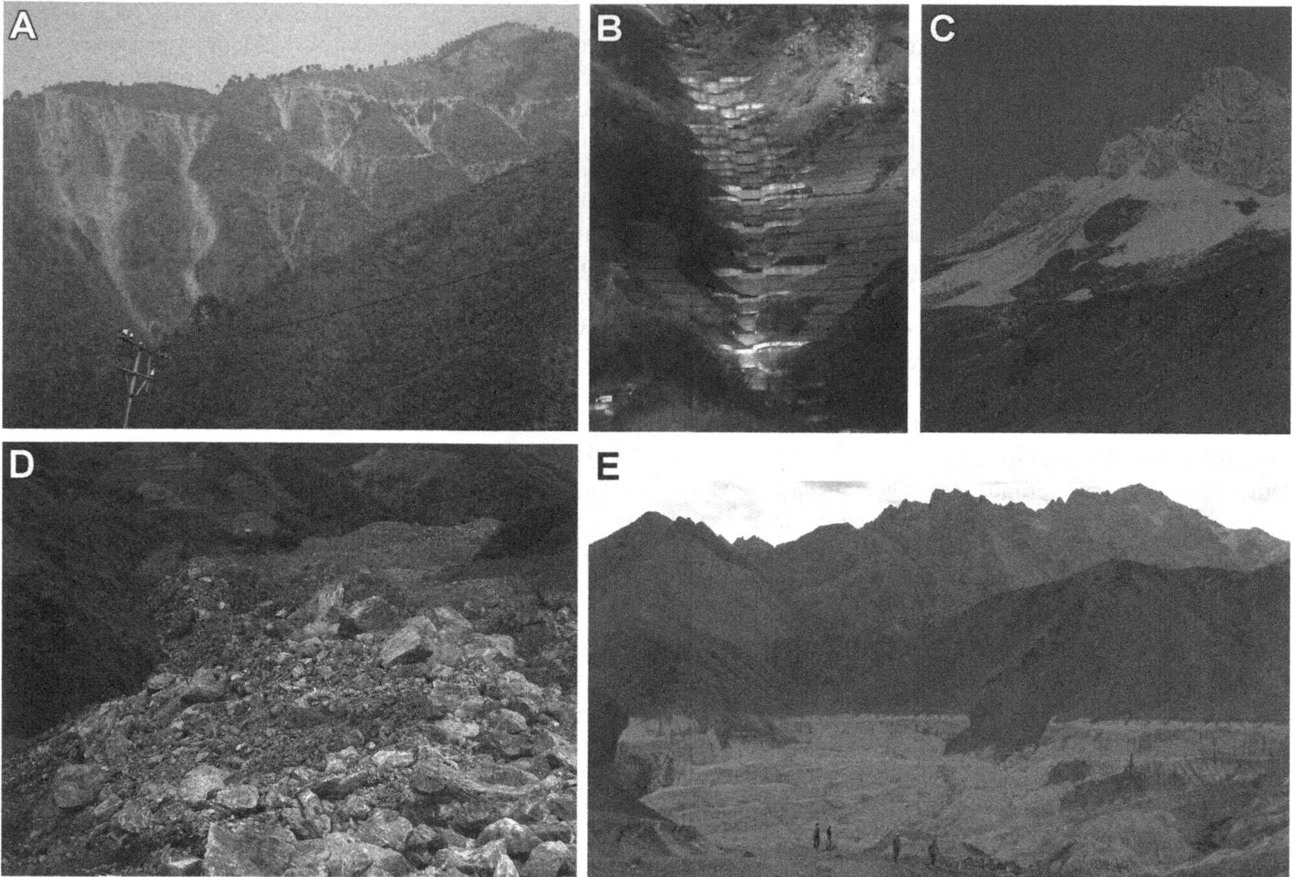

Fig. 2.11. Interactions between landslides and the biosphere. (A) Road-related landslides, Indian Himalaya. (B) A series of debris-flow retention dams, Japan. (C) Rockfall cones and debris flows affecting subalpine vegetation extent and dynamics. (D) Poorly sorted valley-filling landslide deposit, which modulates infiltration and surface runoff. (E) Lake sediments (colored pale ocher with a horizontal top) behind former landslide dam.

chapter shows that, given sufficient size and/or frequency, landslides help to:

- sculpt and limit bedrock topography
- trigger shallow crustal response
- limit volcanic edifice growth
- modulate bedrock incision as well as water and sediment flux in river systems
- trigger potentially far-reaching off-site processes such as tsunamis or catastrophic outburst flows
- condition rates of soil production
- rejuvenate and alter hillslope and riparian habitats
- pose a significant hazard to human lives, housing, infrastructure, and land use in many parts of the world.

Although most landslide research focuses on hazard and risk aspects at the hillslope and regional scale, ample scope remains for further elucidating the effects of landslides on long-term landscape evolution.

ACKNOWLEDGEMENTS

Many colleagues have influenced the development of my research on this topic, but special thanks are due to John Clague and Doug Stead. Part of this work was funded by the German Research Foundation (DFG Heisenberg Program).

REFERENCES

Aharonov, E. and Anders, M.H. (2006). Hot water: A solution to the Heart Mountain detachment problem? *Geology*, 34, 165–168.

Barnard, P.L., Owen, L.A., Sharma, M.C. and Finkel, R.C. (2001). Natural and human-induced landsliding in the Garhwal Himalaya of northern India. *Geomorphology*, 40, 21–35.

Beget, J.E. and Kienle, J. (1992). Cyclic formation of debris avalanches at Mount St Augustine volcano. *Nature*, 356, 701–704.

Benda, L. and Dunne, T. (1997). Stochastic forcing of sediment supply to channel networks from landsliding and debris flow. *Water Resources Research*, 33, 2849–2863.

Benda, L., Veldhuisen, C. and Black, J. (2003). Debris flows as agents of morphological heterogeneity at low-order confluences, Olympic Mountains, Washington. *Geological Society of America Bulletin*, 115, 1110–1121.

Beusen, A.H.W., Dekkers, A.L.M., Bouwman, A.F., Ludwig, W. and Harrison, J. (2005). Estimation of global river transport of sediments and associated C, N, and P. *Global Biogeochemical Cycles*, 19, GB4S05, doi:10.1029/2005GB002453.

Bigelow, P.E., Benda, L.E., Miller, D.J. and Burnett, K.M. (2007). On debris flows, river networks, and the spatial structure of channel morphology. *Forest Science*, 53, 220–238.

Blumberg, S., Lamy, F., Arz, H.W. *et al.* (2008). Turbiditic trench deposits at the South-Chilean active margin: A Pleistocene–Holocene record of climate and tectonics. *Earth and Planetary Science Letters*, 268, 526–539.

Bourget, J., Zaragosi, S., Ellouz-Zimmermann, S. *et al.* (2010). Highstand vs. lowstand turbidite system growth in the Makran active margin: Imprints of high-frequency external controls on sediment delivery mechanisms to deep water systems. *Marine Geology*, 274, 187–208.

Brummer, C.J. and Montgomery, D.R. (2006). Influence of coarse lag formation on the mechanics of sediment pulse dispersion in a mountain stream, Squire Creek, North Cascades, Washington, United States. *Water Resources Research*, 42, W07412, doi:10.1029/2005WR004776.

Busby, C.J., Yip, G., Blikra, L. and Renne, P. (2002). Coastal landsliding and catastrophic sedimentation triggered by Cretaceous–Tertiary bolide impact: A Pacific margin example? *Geology*, 30, 687–690.

Coombs, M.L., White, S.M. and Scholl, D.W. (2007). Massive edifice failure at Aleutian arc volcanoes. *Earth and Planetary Science Letters*, 256, 403–418.

Craddock, J.P., Malone, D.H., Magloughlin, J. *et al.* (2009). Dynamics of the emplacement of the Heart Mountain allochthon at White Mountain: Constraints from calcite twinning strains, anisotropy of magnetic susceptibility, and thermodynamic calculations. *Geological Society of America Bulletin*, 121, 919–938.

Crozier, M.J. (2005). Multiple-occurrence regional landslide events in New Zealand: Hazard management issues. *Landslides*, 2, 247–256.

Cruden, D.M. and Varnes, D.J. (1996). Landslide types and processes. In *Landslides, Investigation and Mitigation*, ed. A.K. Turner and R.L. Schuster. Transportation Research Board, Special Report 247, pp. 36–75.

Dadson, S.D., Hovius, N., Chen, H. *et al.* (2004). Earthquake-triggered increase in sediment delivery from an active mountain belt. *Geology*, 32, 732–736.

Dai, F.C., Lee, C.F., Deng, J.H. and Tham, L.G. (2005). The 1786 earthquake-triggered landslide dam and subsequent dam-break flood on the Dadu River, southwestern China. *Geomorphology*, 65, 205–221.

Di, B., Zeng, H., Zhang, M., Ustin, S.L. *et al.* (2010). Quantifying the spatial distribution of soil mass wasting processes after the 2008 earthquake in Wenchuan, China: A case study of the Longmenshan area. *Remote Sensing of the Environment*, 114, 761–771.

Fine, I.V., Rabinovich, A.B., Bornhold, B.D., Thomson, R.E. and Kulikov, E.A. (2005). The Grand Banks landslide-generated tsunami of November 18, 1929: Preliminary analysis and numerical modeling. *Marine Geology*, 215, 45–57.

Fritz, H.M., Mohammed, F. and Yoo, J. (2009). Lituya Bay landslide impact generated mega-tsunami 50th anniversary. *Pure and Applied Geophysics*, 166, 153–175.

Garwood, N.C., Janos, D.P. and Brokaw, N. (1979). Earthquake-caused landslides: A major disturbance to tropical forests. *Science*, 205, 997–999.

Geertsema, M. and Pojar, J.J. (2007). Influence of landslides on biophysical diversity: A perspective from British Columbia. *Geomorphology*, 89, 55–69.

Goldfinger, C., Nelson, C.H. and Johnson, J.E. (2003). Holocene earthquake records from the Cascadia subduction zone and northern San Andreas fault based on precise dating of offshore turbidites. *Annual Reviews of Earth and Planetary Sciences*, 31, 555–577.

Gomez, B., Carter, L. and Trustrum, N.A. (2007). A 2400 yr record of natural events and anthropogenic impacts in intercorrelated terrestrial and marine sediment cores: Waipaoa sedimentary system, New Zealand. *Geological Society of America Bulletin*, 119, 1415–1432.

Gorum, T., Fan, X.M., van Westen, C.J. *et al.* (2011). Distribution pattern of earthquake-induced landslides triggered by the 12 May 2008 Wenchuan earthquake. *Geomorphology*, 133, 162–167.

Grozic, J.L.H. (2010). Interplay between gas hydrates and submarine slope failure. In *Submarine Mass Movements and their Consequences*, ed. D.C. Mosher, R.C. Shipp, L. Moscardelli *et al.* Dordrecht, Netherlands: Springer, pp. 11–30.

Guzzetti, F., Reichenbach, P., Ardizzone, F., Cardinali, M. and Galli, M. (2006). Estimating the quality of landslide susceptibility maps. *Geomorphology*, 81, 166–184.

Hampton, M.A., Lee, H.J. and Locat, J. (1996). Submarine landslides. *Reviews of Geophysics*, 34, 33– 59.

Hapke, C.J. and Green, K.R. (2006). Coastal landslide material loss rates associated with severe climatic events. *Geology*, 34, 1077–1080.

Havenith, H.-B., Torgoev, I., Meleshko, A. *et al.* (2006). Landslides in the Mailuu-Suu Valley, Kyrgyzstan: Hazards and impacts. *Landslides*, 3, 137–147.

Hewitt, K. (2010). Gifts and perils of landslides. *American Scientist*, 98, 410.

Hilton, R.G., Galy, A. and Hovius, N. (2008). Riverine particulate organic carbon from an active mountain belt: Importance of landslides. *Global Biogeochemical Cycles*, 22, GB1017, doi:10.1029/2006GB002905.

Hovius, N., Stark, C.P. and Allen, P.A. (1997). Sediment flux from a mountain belt derived from landslide mapping. *Geology*, 25, 231–234.

Keefer, D.K. (1999). Earthquake-induced landslides and their effects on alluvial fans. *Journal of Sedimentary Research*, 69, 84–104.

King, J., Loveday, I. and Schuster, R.L. (1989). The 1985 Bairaman landslide dam and resulting debris flow, Papua New Guinea. *Quarterly Journal of Engineering Geology*, 22, 257–270.

Korup, O. (2004). Landslide-induced river channel avulsions in mountain catchments of southwest New Zealand. *Geomorphology*, 63, 57–80.

(2005). Geomorphic imprint of landslides on alpine river systems, southwest New Zealand. *Earth Surface Processes and Landforms*, 30, 783–800.

(2006). Effects of deep-seated bedrock landslides on hillslope morphology, Southern Alps, New Zealand. *Journal of Geophysical Research*, 111, F01018, doi:10.1029/2004JF000242.

Korup, O. and Schlunegger, F. (2009). Rock-type control on erosion-induced uplift, eastern Swiss Alps. *Earth and Planetary Science Letters*, 278, 278–285.

Korup, O. and Tweed, F. (2007). Ice, moraine, and landslide dams in mountainous terrain. *Quaternary Science Reviews*, 26, 3406–3422.

Korup, O., Clague, J.J., Hermanns, R.L. *et al.* (2007). Giant landslides, topography, and erosion. *Earth and Planetary Science Letters*, 261, 578–589.

Korup, O., Montgomery, D.R. and Hewitt, K. (2010). Glacier and landslide feedbacks to topographic relief in the Himalayan syntaxes. *Proceedings of the National Academy of Sciences*, 107, 5317–5322.

Krastel, S., Schmincke, H.U., Jacobs, C.L. *et al.* (2001). Submarine landslides around the Canary Islands. *Journal of Geophysical Research – B3*, 106, 3977–3997.

Larsen, I.J., Montgomery, D.R. and Korup, O. (2010). Landslide erosion controlled by hillslope material. *Nature Geoscience*, 3, 247–251.

Leynaud, D., Mienert, J. and Vanneste, M. (2009). Submarine mass movements on glaciated and non-glaciated European continental margins: A review of triggering mechanisms and preconditions to failure. *Marine and Petroleum Geology*, 26, 618–632.

Lin, G.W., Chen, H., Hovius, N. *et al.* (2008). Effects of earthquake and cyclone sequencing on landsliding and fluvial sediment transfer in a mountain catchment. *Earth Surface Processes and Landforms*, 33, 1354–1373.

Lisle, T.E., Pizzuto, J.E., Ikeda, H., Iseya, F. and Kodama, Y. (1997). Evolution of a sediment wave in an experimental channel. *Water Resources Research*, 33, 1971–1981.

Longpré, M.A., Troll, V.R., Walter, T.R. and Hansteen, T.H. (2009). Volcanic and geochemical evolution of the Teno massif, Tenerife, Canary Islands: Some repercussions of giant landslides on ocean island magmatism. *Geochemistry Geophysics Geosystems*, 10, Q12017, doi:10.1029/2009GC002892.

Malamud, B.D., Turcotte, D.L., Guzzetti, F. and Reichenbach, P. (2004). Landslide inventories and their statistical properties. *Earth Surface Processes and Landforms*, 29, 687–711.

Manconi, A., Longpré, M.A., Walter, T.R., Troll, V.R. and Hansteen, T.H. (2009). The effects of flank collapses on volcano plumbing systems. *Geology*, 37, 1099–1102.

Masson, D.G., Harbitz, C.B., Wynn, R.B., Pedersen, G. and Løvholt, F. (2006). Submarine landslides: Processes, triggers and hazard prediction. *Philosophical Transactions of the Royal Society A*, 364, 2009–2039.

Massong, T.M. and Montgomery, D.R. (2000). Influence of sediment supply, lithology, and wood debris on the distribution of bedrock and alluvial channels. *Geological Society of America Bulletin*, 112, 591–599.

May, C.L. and Gresswell, R.E. (2003). Processes and rates of sediment and wood accumulation in headwater streams of the Oregon Coast Range, USA. *Earth Surface Processes and Landforms*, 28, 409–424.

McMurtry, G.M., Watts, P., Fryer, G.J., Smith, J.R. and Imamura, F. (2004). Giant landslides, mega-tsunamis, and paleo-sea level in the Hawaiian Islands. *Marine Geology*, 203, 219–233.

Meng, Q.R., Hu, J.M., Wang, E. and Qu, H.J. (2006). Late Cenozoic denudation by large-magnitude landslides in the eastern edge of Tibetan Plateau. *Earth and Planetary Science Letters*, 243, 252–267.

Moore, J.G., Schweickert, R.A., Robinson, J.E., Lahren, M.M. and Kitts, C.A. (2006). Tsunami-generated boulder ridges in Lake Tahoe, California-Nevada. *Geology*, 34, 965–968.

Oehler, J.-F., Lenat, J.-F. and Labazuy, P. (2008). Growth and collapse of the Reunion Island volcanoes. *Bulletin of Volcanology*, 70, 717–742.

Osmundsen, P.T., Henderson, I., Lauknes, T.R. *et al.* (2009). Active normal fault control on landscape and rock-slope failure in northern Norway. *Geology*, 37, 135–138.

Page, M.J., Trustrum, N.A., Brackley, H. and Baisden, T. (2004). Erosion-related soil carbon fluxes in a pastoral steepland catchment, New Zealand. *Agriculture Ecosystems Environment*, 103, 561–579.

Page, M.J., Trustrum, N.A. and Dymond, J. (1994). Sediment budget to assess the geomorphic effect of a cyclonic storm, New Zealand. *Geomorphology*, 9, 169–188.

Pain, C.F. and Bowler, J.M. (1973). Denudation following the November 1970 earthquake at Madang, Papua New Guinea. *Zeitschrift für Geomorphologie N.F.*, 18, 92–104.

Pearce, A.J. and Watson, A.J. (1986). Effects of earthquake-induced landslides on sediment budget and transport over a 50-yr period. *Geology*, 14, 52–55.

Philip, H. and Ritz, J.F. (1999). Gigantic paleo-landslide associated with active faulting along the Bogd fault (Gobi-Altay, Mongolia). *Geology*, 27, 211–214.

Ponomareva, V.V., Melekestsev, I.V. and Dirksen, O.V. (2006). Sector collapses and large landslides on Late Pleistocene–Holocene volcanoes, Kamchatka, Russia. *Journal of Volcanology and Geothermal Research*, 158, 117–138.

Quidelleur, X., Hildenbrand, A. and Samper, A. (2008). Causal link between Quaternary paleoclimatic changes and volcanic islands evolution. *Geophysical Research Letters*, 35, L02303.

Ren, D., Wang, J., Fu, R. *et al.* (2009). Mudslide-caused ecosystem degradation following Wenchuan earthquake 2008. *Geophysical Research Letters*, 36, L05401.

Roering, J.J., Schmidt, K.M., Stock, J.D., Dietrich, W.E. and Montgomery, D.R. (2003). Shallow landsliding, root reinforcement, and the spatial distribution of trees in the Oregon Coast Range. *Canadian Geotechnical Journal*, 40, 237–253.

Sato, H.P., Hasegawa, H., Fujiwara, S. *et al.* (2007). Interpretation of landslide distribution triggered by the 2005 Northern Pakistan earthquake using SPOT 5 imagery. *Landslides*, 4, 113–122.

Schmidt, K.M. and Montgomery, D.R. (1995). Limits to relief. *Science*, 270, 617–620.

Sidle, R.C. and Ochiai, H. (2006). *Landslides. Processes, Prediction, and Land Use*. American Geophysical Union Water Resources, Monograph 18, 312 pp.

Silva, C.G., Araujo, E., Reis, A.T. *et al.* (2010). Megaslides in the Foz do Amazonas Basin, Brazilian Equatorial Margin. In *Submarine Mass Movements and their Consequences*, ed. D.C. Mosher *et al.* Dordrecht, Netherlands: Springer, pp. 581–592.

Silver, E., Day, S., Ward, S. *et al.* (2005). Island arc debris avalanches and tsunami generation. *Transactions of the American Geophysical Union*, 86, 485–496.

Smit, J., Burg, J.P., Dolati, A. and Sokoutis, D. (2010). Effects of mass waste events on thrust wedges: Analogue experiments and application to the Makran accretionary wedge. *Tectonics*, 29, TC3003, doi:10.1029/2009TC002526.

Smith, D.E., Shi, S., Cullingford, R.A. *et al.* (2004). The Holocene Storegga Slide tsunami in the United Kingdom. *Quaternary Science Reviews*, 23, 2291–2321.

Smith, J.R. and Wessel, P. (2000). Isostatic consequences of giant landslides on the Hawaiian Ridge. *Pure and Applied Geophysics*, 157, 1097–1114.

Stark, C.P. and Guzzetti, F. (2009). Landslide rupture and the probability distribution of mobilised debris volumes. *Journal of Geophysical Research*, 114, F00A02, doi:10.1029/2008JF001008.

Stock, J. and Dietrich, W.E. (2003). Valley incision by debris flows: Evidence of a topographic signature. *Water Resources Research*, 1089, doi:10.1029/2001WR001057.

Strasser, M., Anselmetti, F.S., Fäh, D., Giardini, D. and Schnellmann, M. (2006). Magnitudes and source areas of large prehistoric northern Alpine earthquakes revealed by slope failures in lakes. *Geology*, 34, 1005–1008.

Strecker, M.R. and Marrett, R. (1999). Kinematic evolution of fault ramps and its role in development of landslides and lakes in the northwestern Argentine Andes. *Geology*, 27, 307–310.

Syvitski, J.P.M., Vörösmarty, C.J., Kettner, A.J. and Green, P. (2005). Impact of humans on the flux of terrestrial sediment to the global coastal ocean. *Science*, 308, 376–380.

Talling, P.J., Wynn, R.B., Masson, D.G. *et al.* (2007). Onset of submarine debris flow deposition far away from original giant submarine landslide. *Nature*, 450, 541–544.

Tappin, D.R. (2010). Submarine mass failures as tsunami sources: Their climate control. *Philosophical Transactions of the Royal Society A*, 368, 2417–2434.

Taylor, J., Dowdeswell, J.A. and Siegert, M.J. (2002). Late Weichselian depositional processes, fluxes, and sediment volumes on the margins of the Norwegian Sea (62–75° N). *Marine Geology*, 188, 61–77.

ten Brink, U.S., Lee, H.J., Geist, E.L. and Twichell, D. (2009). Assessment of tsunami hazard to the U.S. East Coast using relationships between submarine landslides and earthquakes. *Marine Geology*, 264, 65–73.

ten Brink, U.S., Barkan, R., Andrews, B.D. and Chaytor, J.D. (2010). Size distributions and failure initiation of submarine and subaerial landslides. *Earth and Planetary Science Letters*, 287, 31–42.

Thapa, G.B. and Paudel, G.S. (2002). Farmland degradation in the mountains of Nepal: A study of watersheds 'with' and 'without' external intervention. *Land Degradation and Development*, 13, 479–493.

Trustrum, N.A., Gomez, B., Page, M.J., Reid, L.M. and Hicks, D.M. (1999). Sediment production, storage and output: The relative role of large magnitude events in steepland catchments. *Zeitschrift für Geomorphologie N.F.*, 115, 71–86.

Van Den Eeckhaut, M., Poesen, J., Govers, G., Verstraeten, G. and Demoulin, A. (2007). Characteristics of the size distribution of recent and historical landslides in a populated hilly region. *Earth and Planetary Science Letters*, 256, 588–603.

Vernhet, E., Heubeck, C., Zhu, M.Y. and Zhang, J.M. (2006). Large-scale slope instability at the southern margin of the Ediacaran Yangtze platform (Hunan Province, central China). *Precambrian Research*, 148, 32–44.

Vogt, P.R. and Jung, W.Y. (2002). Holocene mass wasting on upper non-Polar continental slopes: Due to post-glacial ocean warming and hydrate dissociation? *Geophysical Research Letters*, 29, 1341, doi:10.1029/2001GL013488.

Walker, L.R., Zarin, D.J., Fetcher, N., Myster, R.W. and Johnson, A.H. (1996). Ecosystem development and plant succession on landslides in the Caribbean. *Biotropica*, 28, 566–576.

Weaver, P.P.E. (2003). Northwest African continental margin: History of sediment accumulation, landslide deposits, and hiatuses as revealed by drilling the Madeira Abyssal Plain. *Paleoceanography*, 18, 1009, doi:10.1029/2002PA000758.

Weidinger, J.T. and Korup, O. (2009). Frictionite as evidence for a large Quaternary rockslide near Kanchenjunga, Sikkim Himalayas, India: Implications for extreme events in mountain relief destruction. *Geomorphology*, 103, 57–65.

Wörner, G., Uhlig, D., Kohler, I. and Seyfried, H. (2002). Evolution of the West Andean Escarpment at 18° S (N. Chile) during the last 25 Ma: Uplift, erosion and collapse through time. *Tectonophysics*, 345, 183–198.

Yin, J., Chen, J., Xu, X.W., Wang, X. and Zheng, Y. (2010). The characteristics of the landslides triggered by the Wenchuan M_s 8.0 earthquake from Anxian to Beichuan. *Journal of Asian Earth Sciences*, 37, 452–459.

3 Earthquake ground motion and patterns of seismically induced landsliding

NIELS HOVIUS AND PATRICK MEUNIER

ABSTRACT

Earthquake strong ground motion changes stresses in hillslopes and reduces the strength of surface materials. This can cause landsliding during earthquakes and enhance rates of slope failure in epicentral areas for longer periods. Rates of earthquake-triggered landsliding are strongly correlated with measured peak ground acceleration. Patterns of landslide density reflect the attenuation of seismic waves and geologic and topographic site effects. Using historic thrust fault ruptures with well-documented ground motion and landslide distributions as examples, we illustrate the links between earthquake mechanisms, seismic wave propagation, and triggered landsliding. The examples have shared geomorphic attributes: a maximum density of triggered landslides where earthquake slip is greatest; a progressive decrease of landslide density away from this maximum; clustering of triggered landslides on topographic ridges and other convex landscape elements; and preferential failure of slopes facing away from the earthquake source. We also show that rates of landsliding can remain high after an earthquake in a geomorphic crisis that fades over a period of years. Continued landsliding adds to the total erosion caused by an earthquake, reducing or possibly canceling seismic surface uplift. The examples underline the potential for the quantitative prediction of patterns of seismically triggered and induced landsliding, use of observed landslide patterns for study of earthquake mechanisms, and inclusion of seismically driven erosion in landscape evolution models.

3.1 INTRODUCTION

Intermediate and large shallow earthquakes can cause landsliding in steep terrain (Oldham, 1899; Mathur, 1953; Pain and Bowler, 1973; Wilson and Keefer, 1979; Harp and Jibson, 1996; Chigira *et al.*, 2010), mainly due to strong ground motion and the associated weakening of the hillslope substrate. Attenuation of seismic waves and topographic site effects give rise to distinct patterns of earthquake strong ground motion. These patterns are reflected in the density of earthquake-triggered landslides on a regional scale, but also within individual topographic features such as hills and mountain ridges. Their effect may persist over time and govern landslide patterns years after an earthquake. Erosion by later, seismically induced, landslides can add substantially to erosion during an earthquake, and affect the topographic change caused by that earthquake. Quantitative evaluations of the patterns of seismically triggered and induced landslides not only help explain the role of earthquakes in mountain building, but also reveal details of the earthquake mechanisms by which they were caused. Moreover, they offer insights into sediment production, landscape evolution, earthquake mechanisms, and seismic hazard risk management and mitigation, with considerable potential for further progress.

3.2 EARTHQUAKES, SEISMIC WAVES, AND GROUND MOTION

Earthquakes occur in solid Earth materials where sufficient elastic strain has accumulated to cause fault rupture. Progressive relative displacement of two geologic blocks separated by a potential or existing fault leads to increased stress and therefore stored strain energy in the rock mass around the fault zone. When the stress exceeds the rock mass strength, sudden failure can occur and stored energy is released. The released energy is used in fracture propagation, permanent displacement of mass, and associated frictional heating, and, importantly, is radiated in elastic strain seismic waves. These seismic waves travel through

Landslides: Types, Mechanisms and Modeling, ed. John J. Clague and Douglas Stead. Published by Cambridge University Press.

the Earth as body waves and along Earth's surface. Body waves radiate outward along curved raypaths that are affected by variations in rock density and stiffness, expanding geometrically until their energy has dissipated. The effect of seismic wave attenuation with distance (*R*) from source is described in its simple form by:

$$A_{(R)} \propto \frac{A_0}{R} e^{\frac{\pi . - f R}{v Q}} \tag{3.1}$$

where *A* is the amplitude of the seismic wave, *f* and *v* are its frequency and velocity, respectively, the subscript 0 denotes the source, and *Q* is a so-called "quality factor" representing energy dissipation due to anelasticity of the rock mass, scattering of waves on geologic structures, and other effects (Taylor *et al.*, 1986; Trifunac, 1994). Surface waves form where sufficient energy reaches the Earth's surface. These waves travel at lower velocities than body waves, but can have considerable amplitudes. The arrival of seismic waves can cause strong motion with directional or orbital shear (Bindi *et al.*, 2010). The combination of these phenomena constitutes an earthquake. Larger earthquakes release more stored energy than smaller ones, causing waves with larger initial amplitudes and attenuation distances. For a given frequency of incoming waves, the greater their amplitude, the larger the resultant ground velocities and accelerations.

The radiation pattern of seismic waves is determined by the focal mechanism of the earthquake (Sommerville *et al.*, 1997, 1999; Anderson *et al.*, 2000). In dip–slip events, motion is perpendicular to the strike of the fault plane. Such fault planes commonly have slopes <60°. In normal faults, the hanging wall moves down with respect to the footwall; and in reverse or thrust faults, it moves up. In both types of dip–slip events, most of the upward projected energy is released into the hanging wall, with significantly less strong ground motion in the footwall (Abrahamson and Somerville, 1996; Allen *et al.*, 1998; Shi *et al.*, 1998; Ogelsby *et al.*, 2000). Because rocks are stronger in compression than in tension, thrust fault earthquakes can be larger than normal fault earthquakes. Moreover, most topographic relief associated with thrust faults is located in the uplifting hanging wall. In contrast, topographic relief associated with normal faults is located in the footwall, the surface of the subsiding hanging wall being evened by deposition. The combination of these factors makes thrust fault earthquakes much more likely to cause significant and widespread landsliding than normal fault earthquakes. Most strike–slip faults are steep (>60°). Earthquakes on such faults commonly result in a more symmetric pattern of strong ground motion, with possible variations along the length of the structure due to directionality of the rupture propagation.

The energy released in an earthquake, expressed as the moment magnitude M_w, is equal to the rigidity of the rock mass multiplied by the product of slip on the fault and the size of the area that slipped. In small earthquakes (M_w <6), relatively small amounts of slip (<1 m) are commonly distributed in a simple way over a slip patch of limited size (length ≤10^1 km). This simple slip distribution makes it possible, for our purposes, to treat such earthquakes as point sources of energy. Larger earthquakes tend to have more extensive and often more complex rupture patterns (Wells and Coppersmith, 1994), with one or multiple slip patches moving along a fault or an array of faults as the earthquake develops. This gives rise to more complex patterns of energy release (Sommerville *et al.*, 1999). Therefore, large earthquakes are better treated as two-dimensional sources of energy, extending over distances of 100 km or more, with segmentation in the most complex cases. For example, the M_w 7.6 Chi-Chi earthquake in central-west Taiwan in 1999 ruptured the north–south trending Chelungpu fault over a distance of about 110 km, with slip increasing from *ca.* 3 m in the south to more than 8 m in the north (Kao and Chen, 2000; Shin and Teng, 2001). Similarly, the M_w 7.9 Wenchuan earthquake in Sichuan Province, China, in 2008, ruptured an array of faults over a distance of *ca.* 200 km, with displacements locally exceeding 10 m and different types of displacement on individual fault segments (Shen *et al.*, 2009). In such cases, treatment of the earthquake as a simple line source, or even a point source of energy, can lead to erroneous conclusions.

A further, relevant complication arises from the fact that, close to faults, peak ground accelerations have been found to saturate with increasing earthquake magnitude (Boore and Atkinson, 2008; Chiou and Youngs, 2008), possibly caused by the dynamics of the earthquake itself or by the geometry of large events (Anderson, 2000). Saturation implies that surface locations close to the rupture may be impervious to earthquake magnitude above a threshold, whereas locations farther afield experience ground motion in proportion to the magnitude of the earthquake. The length scale over which saturation occurs is likely to increase with earthquake size (Schmedes and Archuleta, 2008).

3.3 LANDSLIDING AND EARTHQUAKE STRONG GROUND MOTION

Similar to fault rupture, hillslope failure occurs when the shear stress across a potential failure plane exceeds substrate strength. The addition of earthquake strong ground motion to the ambient gravitational acceleration results in short-lived, cyclic changes of the normal and shear stresses in hillslopes during earthquakes (Newmark, 1965). Strong ground motions may lead to instantaneous failure due to initial or peak ground motion, or because of progressive substrate weakening due to rock mass fracturing (Harp and Jibson, 1996; Lin *et al.*, 2008; Meunier *et al.*, 2008) or breakage of the binding plant root mass over multiple ground motion cycles (Meisling and Sieh, 1980). The greater the amplitude of the incoming seismic waves and the duration of shaking, the greater the likelihood of failure of a steep slope, primed by other erosional processes such as fluvial incision (Kelsey, 1988; Burbank *et al.*, 1996; Densmore and Hovius, 2000). It is therefore expected that relations exist between the magnitude of an earthquake and the extent and intensity of the landsliding it causes.

Global compilations of earthquake and landslide data have been used to explore the effect of earthquake magnitude on the extent of the area affected by landsliding. For example, Keefer (1984) and Rodriguez *et al.* (1999) have tentatively defined maximum affected areas for earthquakes of different magnitudes. Such analyses should take into account the depth and focal mechanism of earthquakes, the frequencies at which seismic waves carry most energy, and the local geology and topography. Source depth is of crucial importance (implicit in the term R in Eq. 3.1), especially in intermediate-sized earthquakes in which the amplitude of seismic waves is likely to attenuate within tens of kilometers to values that do not cause notable landsliding. Therefore, the inclusion of earthquakes with focal depths as large as *ca.* 70 km (Keefer, 1984; Rodriguez *et al.*, 1999) is bound to cause departures from any trend set by shallow events (depth <20 km). Moreover, as low-frequency waves travel farther than high-frequency ones, the dominant frequency of seismic waves (f in Eq. 3.1), which can differ between earthquakes, affects the relevant attenuation length, as do geologic factors (Q in Eq. 3.1). Analyses of global relationships between earthquake moment magnitude and landslide number or concentration (Keefer, 2002), area density, and even volume (Malamud *et al.*, 2004) should therefore be approached with caution.

It is, then, perhaps most productive to consider patterns of landsliding in terms of a geophysical quantity that directly affects slope stability: strong ground motion or, more specifically, ground velocity or acceleration (Luzi and Pergalani, 2000). For many intermediate and large earthquakes that have caused substantial landsliding, insufficient instrumental data exist to evaluate the relation between peak ground velocity (PGV) or peak ground acceleration (PGA) and landsliding. However, where data exist, a strong correlation of ground motion and landslide density (percent area affected by landsliding, A_{ls}) has been found. This relation appears to take a linear form (Meunier *et al.*, 2007):

$$A_{ls} = \alpha\text{PGA} - \beta \tag{3.2}$$

where α is a susceptibility coefficient and $\beta = \alpha\text{PGA}_{cr}$, the minimum ground acceleration required to trigger substantial slope failure (Fig. 3.1). These empirically constrained constants have location-specific values. For >20,000 landslides triggered by the 1999 Chi-Chi earthquake, $\alpha = 2.7\ g^{-1}$ ($5.8\ g^{-1}$) and $\beta = 0.5$ (0.6), where g is the gravitational acceleration, with a regression coefficient $|R| = 0.75$ (0.87) for the horizontal component of PGA (values for the vertical component in parentheses), measured at seven stations in the epicentral area (Dadson *et al.*, 2004). For >11,000 landslides caused by the M_w 6.7 Northridge, California, earthquake in 1994 (Harp and Jibson, 1996), $\alpha = 15.5\ g^{-1}$ ($15.4\ g^{-1}$) and $\beta = 3.3$ (1.4), with $|R| = 0.96$ (0.97). And for >10,000 landslides triggered by the 2004 Chuetsu, Niigata earthquake in Japan (Osanai *et al.*, 2007), $\alpha = 2.4\ g^{-1}$ ($2.6\ g^{-1}$) and $\beta = 1.5$ (0.7), with $|R| = 0.64$ (0.89). In all cases, best-fit regressions have a horizontal acceleration threshold of $\text{PGA}_{cr} \approx 2\ \text{m}\,\text{s}^{-2}$, below which no landslides occurred, despite contrasting climate, relief, and substrate lithology. This threshold applies to failure of the weakest material, regolith, and soil on slopes, the frictional strength of which is relatively uniform over large areas.

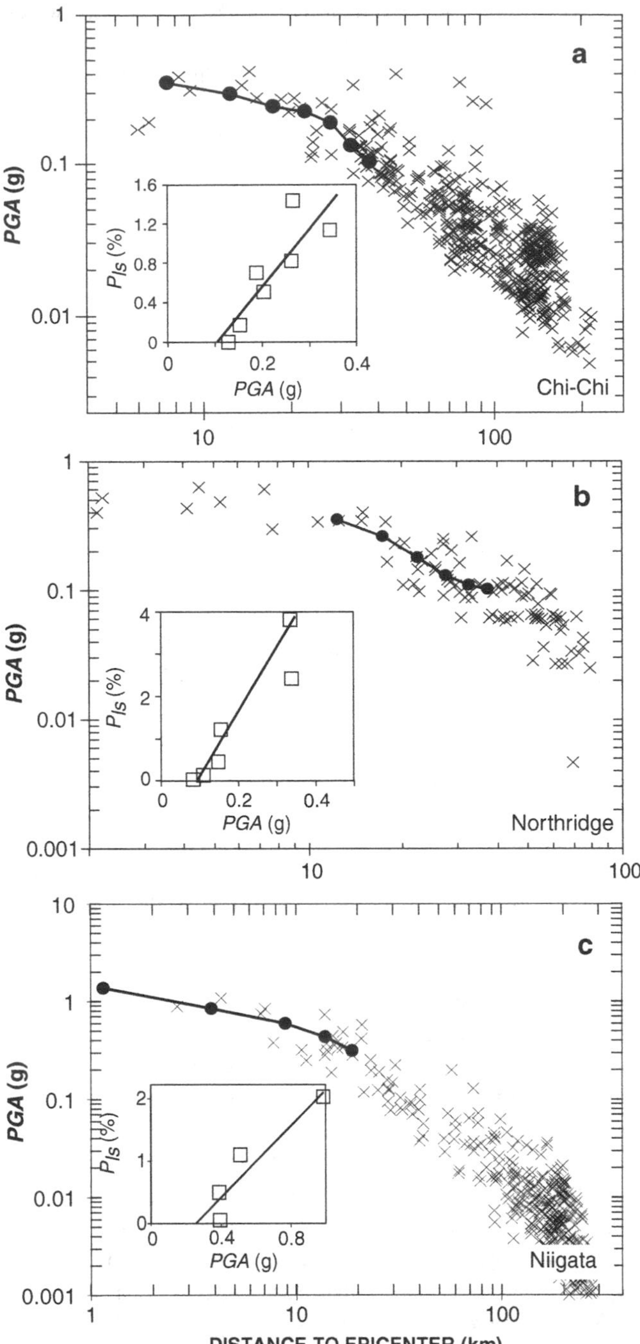

Fig. 3.1. Average landslide density plotted against average vertical PGA for 5-km windows parallel to the fault trend with least squares linear regressions for (a) the 1999 Chi-Chi earthquake (PGA data from Central Weather Bureau, Taipei, Taiwan), (b) the 1994 Northridge earthquake (PGA data from Trifunac and Todorovska, 1996), and (c) the 2004 Niigata earthquake (Honda *et al.*, 2005). (After Meunier *et al.*, 2007; result for the 2004 Niigata earthquake has not been published elsewhere.)

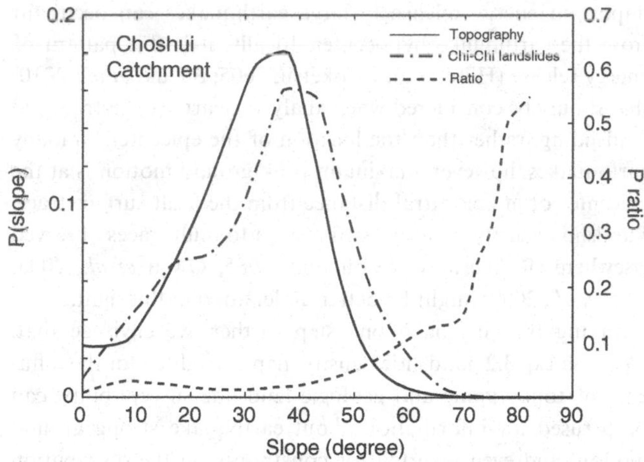

Fig. 3.2. Probability distribution of all topographic slopes (solid line) and slopes affected by earthquake-triggered landsliding (dashed/dotted line; >20,000 landslides) in the Choshui catchment, draining the epicentral area of the 1999 Chi-Chi earthquake. Local topographic slopes have been calculated as the steepest slope within a square of 3 × 3 DEM grid cells. The dashed line is the ratio of the probabilities of all slopes and failed slopes for a given gradient; it is an objective measure of the failure susceptibility. The steepest slopes were most prone to failure during the Chi-Chi earthquake.

Within epicentral areas, rock types can have distinct and different landslide susceptibilities that overprint large-scale patterns of earthquake-triggered landsliding. This effect was shown by Parise and Jibson (2000) for the Northridge earthquake and strongly influenced landslide density patterns of the 2008 Wenchuan earthquake (Yin *et al.*, 2010; Dai *et al.*, 2011) and several recent intermediate-sized earthquakes in Japan (Wang *et al.*, 2007; Kawabata and Bandidas, 2009; Yagi *et al.*, 2009). Moreover, landslide susceptibility is also affected by topography. Topographic gradient is a primary control on slope stability; it partitions gravitational pull into a slope-normal, motion-arresting component and a slope-parallel, shear-inducing component. The result is an increasing likelihood of failure with increasing topographic gradient for a given substrate. This relation can be shown by comparing the probability distribution of local slope values on a given length scale in a landscape and the equivalent probability distribution of failed slopes in that landscape (Fig. 3.2; Lin *et al.*, 2008). The latter normalized by the former is an objective, albeit general, index of the susceptibility of topographic slopes to failure. A landscape with a higher average slope susceptibility index is likely to sustain more landsliding during a given earthquake.

Despite sparse empirical evidence, and mindful of the geologic and topographic controls on landslide susceptibility, we propose that the correlation of peak ground acceleration and landslide density is strong and real and that it is the key to understanding the global attributes of regional and local patterns of earthquake-induced landsliding. Landslide patterns also provide insights into mechanisms of faulting. Further tests of the applicability of Eq. 3.2, and improved understanding of controls on the landslide susceptibility coefficient α and threshold constant β, will depend on studies of future earthquakes in well-instrumented areas and retrospective analysis of well-documented historic cases, and should be considered a priority for research.

3.4 SPATIAL PATTERNS OF EARTHQUAKE-TRIGGERED LANDSLIDES

3.4.1 REGIONAL PATTERNS

Attenuation of seismic waves with increasing distance from the source (Eq. 3.1) and the relation between peak ground acceleration or velocity and landslide rate (Eq. 3.2) give rise to clear patterns of landslide density within an area affected by earthquake strong ground motion. Meunier *et al.* (2007) calculated landslide density, P_{ls}, as the percentage of area with topographic slope >20 percent (an arbitrary threshold) affected by earthquake-triggered landsliding. They plotted P_{ls} against distance from the projected surface trace of the seismogenic fault for three large thrust earthquakes with shallow (<20 km) hypocenters: the 1999 Chi-Chi earthquake; the 1994 Northridge earthquake; and a pair of M_w 6.9 and 6.7 earthquakes on the Ramu-Markham fault in northeast Papua New Guinea; the earthquakes were treated as linear sources of energy (Fig. 3.3). In all three cases, the landslide density peaked in the hanging wall of the seismogenic fault. Moreover, in the case of the 1999 Chi-Chi earthquake and also for the events on the Ramu-Markham fault, maximum landslide densities coincided with the earthquake epicenter. Away from the landslide density maximum, and farther into the hanging wall, P_{ls} decreased quasi-exponentially, mirroring the geometric spreading and attenuation of seismic waves, while P_{ls} dropped more steeply toward the surface trace of the faults. Other researchers have found similar patterns for the concentration of landslides (number per unit area) caused by the 1989 M_w 6.9 Loma Prieta, California, earthquake (Keefer, 2002), and the 2004 M_w 6.6 Chuetsu, Niigata, earthquake (Wang *et al.*, 2007). The Niigata earthquake had a large reverse dip–slip component of movement, like the Chi-Chi and Northridge events (Spudich, 1996; Shin and Teng, 2001), and its landslide pattern shares all the essential characteristics of the other examples in Figure 3.3.

In the case of the 1994 Northridge earthquake, however, the landslide density peak was 8 km north of the epicenter, in the Santa Susanna Mountains. Steep topography is limited near the epicenter, but the landslide density distribution appears to reflect the pattern of energy release during the earthquake. The epicenter of the Northridge earthquake was located above the lower edge of the rupture plane in the flat San Fernando Valley. During the earthquake, as the rupture propagated upward and to the north along the fault plane, recorded ground accelerations remained high and approximately constant up to 10 km north of the epicenter (Todorovska and Trifunac, 1997), which may have set the landslide density pattern. Geomorphically

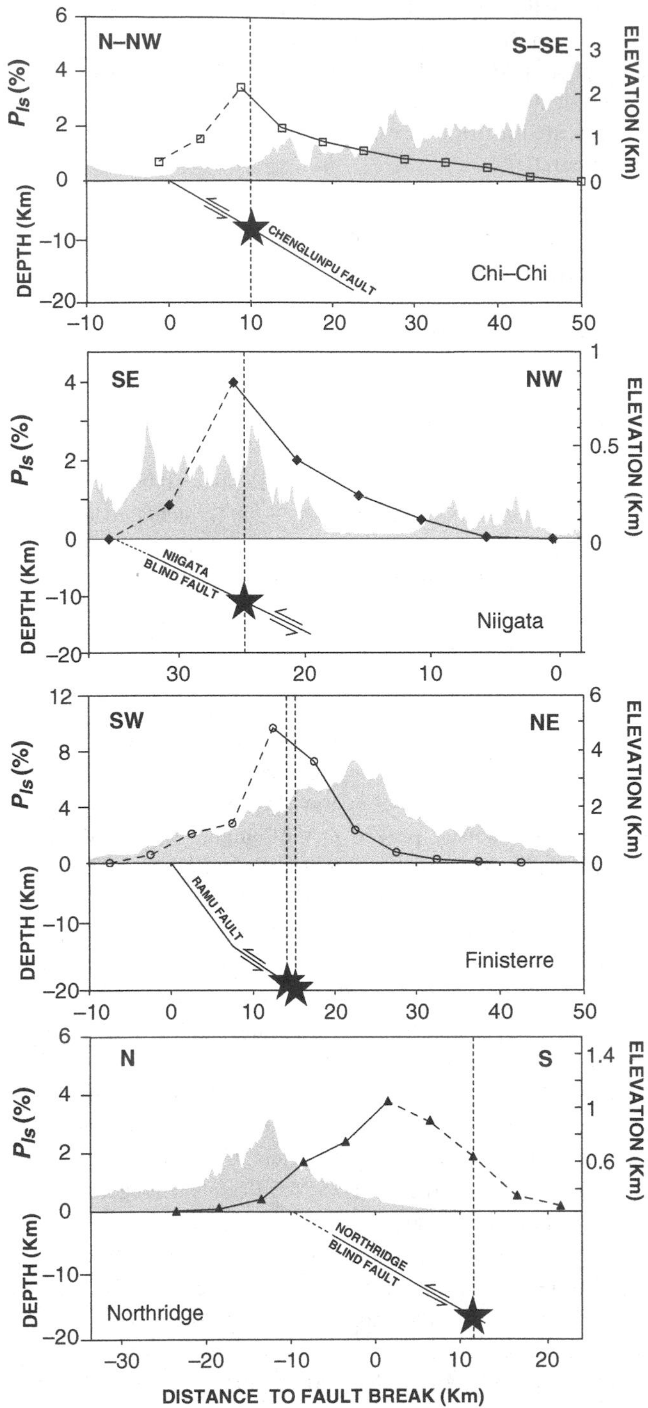

Fig. 3.3. Landslide density (percentage of total area) plotted against distance to the (projected) surface break of the seismogenic fault for the 1999 Chi-Chi, 2004 Niigata, 1993 Finisterre, and 1994 Northridge earthquakes. Fault planes are shown schematically; also shown are positions of earthquake hypocenters (stars). For the Chi-Chi, Finisterre, and Niigata earthquakes, the landslide density peak is located directly above the earthquake hypocenter; for the Northridge earthquake, the landslide density peak is 8 km north of the epicenter at the edge of the Santa Susanna Mountains. Topographic profiles across the epicenter, perpendicular to the fault, are shown in gray. (After Meunier *et al.*, 2007; result for the 2004 Niigata earthquake has not been published elsewhere.)

important energy release in large earthquakes can occur far from the earthquake hypocenter. Ideally, it is this pattern of energy release (Hikima and Koketsu, 2005; Suzuki *et al.*, 2010) that should be considered when analyzing earthquake-triggered landsliding, rather than the location of the epicenter. In many earthquakes, however, maximum peak ground motion is at the epicenter or at epicentral distance from the fault surface trace. Landslide density patterns with respect to fault traces observed elsewhere (Pearce and O'Loughlin, 1985; Owen *et al.*, 2008; Yagi *et al.*, 2009) might be better understood in this light.

Taking this approach one step further, we envisage that, based on Eq. 3.2, landslide density maps, weighted for the influence of topographic and geologic landslide susceptibility, can be perused for information about earthquake strong ground motion, and even inverted for constraints on the distribution of coseismic slip on a fault plane. If robust, this application could complement instrumental records of ground motion that are normally used for this purpose and, where such records are absent, open up a large number of poorly instrumented historic and future earthquakes for further seismological investigation.

3.4.2 TOPOGRAPHIC SITE EFFECTS

Local geologic and topographic features can have important effects on earthquake ground motion. For example, sediment fills of basins or valleys can significantly amplify earthquake ground motion relative to bedrock (Aki, 1993). Such amplification is attributed to the fact that sediments and weak rocks have a lower elastic modulus than strong rocks. Weak geologic materials will undergo a greater displacement for a given force exerted by an incoming seismic wave, although the effect may be nonlinear and may be reduced for large earthquakes (Field *et al.*, 1994). Where hillslopes are formed in weak materials, rates of earthquake-triggered landsliding can be disproportionately high, as was observed in the conglomerate badlands of the 99 Peaks area near the epicenter of the 1999 Chi-Chi earthquake (Hung, 2000). However, on many hillslopes, strong bedrock has only a thin and discontinuous sediment cover. Although this thin cover may locally amplify ground motion (Del Gaudio and Wasowski, 2007), other topographic site effects dominate local landslide patterns in steep uplands.

Large peak vertical accelerations have been recorded at ridge crests during several earthquakes. For example, during the 1987 M_w 5.9 Whittier Narrows, California, earthquake, the amplitude of seismic waves recorded at the crest of 60-m-high Tarzana Hill was 10 times greater than that observed on the surrounding plains (Spudich *et al.*, 1996). On the same hill, instruments recorded a peak horizontal acceleration 1.78 times the gravitational acceleration during the 1994 M_w 6.7 Northridge earthquake (Bouchon and Baker, 1996). Topographic amplification of ground accelerations occurs when seismic waves entering the base of a topographic ridge are partially reflected back into the rock mass and diffracted along the free surface. The seismic waves are progressively focused upward, and the constructive interference of their reflections and the associated diffractions

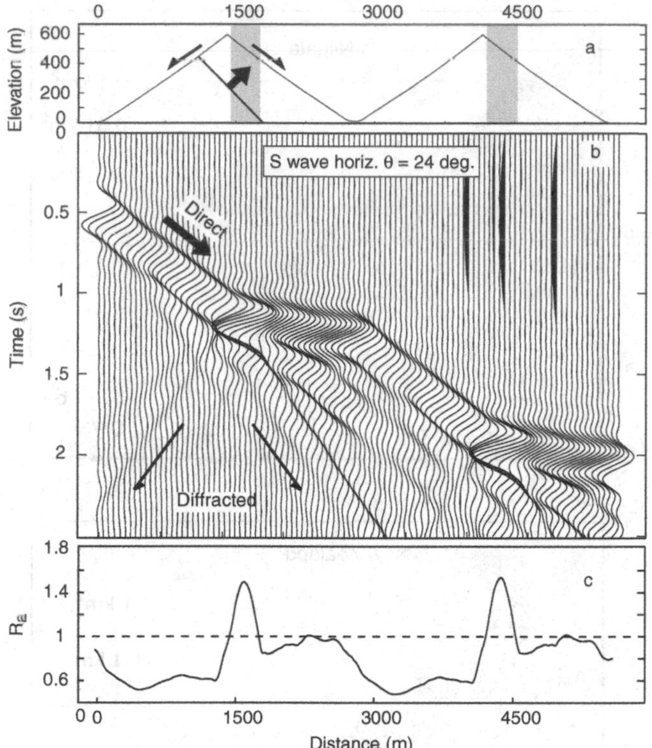

Fig. 3.4. Illustration of topographic site effects in a ridge-and-valley landscape. (a) Topographic profile consisting of two adjacent mountain ridges representing the average cross-profile of ridges in the Finisterre Mountains, Papua New Guinea. Ridge flanks are slightly concave-up and have a length of 1.6 km. (b) Synthetic accelerogram (ground acceleration recorded along the surface with time), generated along the topographic profile. The accelerogram is for the horizontal component of an S wave delta pulse (0–6 Hz) arriving from the left at an angle of 24° from the vertical. (c) Associated ratio, Ra, of local PGA in this model and local PGA in a flat-Earth equivalent. The direct wave interferes constructively with the diffracted wave generated at the ridge crest, causing amplification of the PGA on the ridge flank facing away from the wave source (gray zone in a). (From Meunier *et al.*, 2008.)

increases toward the ridge crest, giving rise to enhanced ground accelerations on topographic highs (Fig. 3.4; Bouchon, 1973; Geli *et al.*, 1988; Meunier *et al.*, 2008). Topographic amplification values are typically small compared with those of sediment fills, but can be >2. Their effect is strongest on S waves, and the exact location of amplification maxima depends on the angle of wave incidence. Seismic waves with wavelengths greater than the base width of a ridge are unaffected by the topography (Meunier *et al.*, 2008). As most of the energy of an earthquake is carried by waves with a relatively low frequency (<1 Hz), larger topographic features tend to be more prone to this effect than smaller features. Oblique incidence of seismic waves causes amplification maxima to shift away from ridge crests or hilltops and into slopes facing away from the earthquake epicenter. Secondary amplification maxima are predicted at smaller, convex-up knickpoints within ridge flanks. Elsewhere in ridge-and-valley landscapes, topographic site effects tend to cause de-amplification of ground motion with respect to the values expected for a "flat-Earth" equivalent.

Topographic site effects can be large enough to significantly affect slope failure during an earthquake (Havenith *et al.*, 2003) and have been shown to govern patterns of landsliding in several large earthquakes (Meunier *et al.*, 2008). Perhaps the clearest example is the 1994 Northridge earthquake, where 56 percent of the area affected by seismically triggered landslides was in the uppermost 25 percent of hillslopes (Fig. 3.5). We have found similar, but somewhat less pronounced, clustering of landslides triggered by the 1993 Ramu-Markham and the 2004 Niigata earthquakes (Fig. 3.5) and the 1999 Chi-Chi earthquake (Meunier *et al.*, 2008). Moreover, secondary landslide clusters occurred near prominent knickpoints above inner gorges along the main valleys of the Finisterre Mountains, in the hanging wall of the Ramu-Markham fault. There, and in the area affected by the Chi-Chi earthquake, landslide rates were highest in slopes facing away from the earthquake epicenter, but – for poorly understood reasons – this effect was not found for the Northridge earthquake (Fig. 3.6).

Due in part to topographic site effects, most earthquake-triggered landslides do not connect directly with river channels and instead deposit debris on hillslopes and in debris-flow channels. As an example, in the Chenyoulan River catchment, south of the epicenter of the Chi-Chi earthquake, 88 percent of earthquake-induced landslides did not reach river channels (Lin *et al.*, 2008). If landslides that did reach channels had the same size distribution as those that did not, then only one-tenth of mobile landslide debris was delivered direct to streams in the catchment. Much of the debris produced by an earthquake may initially remain in a landscape and become entrained in a prolonged downslope cascade of sediment involving multiple episodes of remobilization by mass wasting processes (Dadson *et al.*, 2004). This effect introduces a time dimension to earthquake-induced erosion that will be explored further in the following sections.

3.5 TEMPORAL PATTERNS OF SEISMICALLY INDUCED LANDSLIDING

Substrate weakening through crack propagation and coalescence, and shear damage to vegetation root mass due to rapid cyclic stressing should result in a long-lived increase in slope failures in earthquake-affected areas. These effects are likely to be enhanced by the presence of new landslide debris and colluvium on hillslopes after an earthquake and by the occurrence of earthquake aftershocks. Progressive decay of the seismic moment of aftershocks, closure of cracks due to settling of the shaken rock mass, re-establishment of plant root networks, and erosional removal of debris and weakened materials will act over time to reduce rates of landsliding to background values. These restorative processes are likely to have different time constants, so that their combined effect is not necessarily a simple decay of the landslide rate after an earthquake. However, it is clear that in terms of mass wasting, an earthquake can be the start of a prolonged geomorphic crisis.

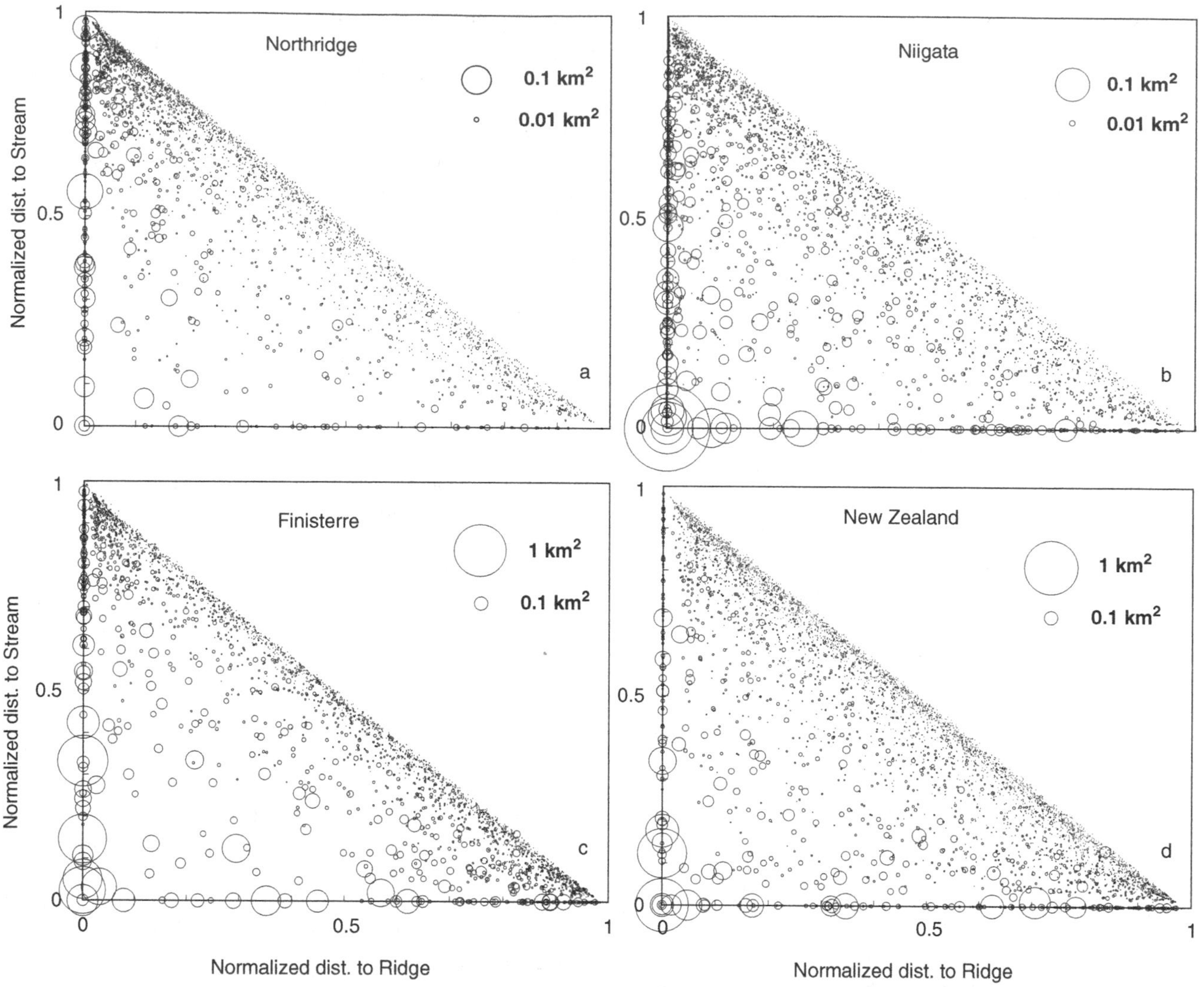

Fig. 3.5. Location of landslides with respect to a ridge crest and stream. Distances from the landslide crown to the nearest ridge and from the lowest point on the landslide lobe to the nearest stream have been measured along the line of steepest descent and normalized for the total length of the slope on which the landslide is located. The surface area of the landslide is indicated with a circle of variable diameter. Landslides triggered by (a) the 1994 Northridge earthquake and (b) the 2004 Niigata earthquake cluster around ridge crests. Landslides triggered by the 1993 earthquakes in the Finisterre Mountains (c) cluster at ridge crests and near the base of slopes. Rainfall-induced landslides in the western Southern Alps, New Zealand (d) are uniformly distributed. (After Meunier *et al.*, 2008; result for the 2004 Niigata earthquake has not been published elsewhere.)

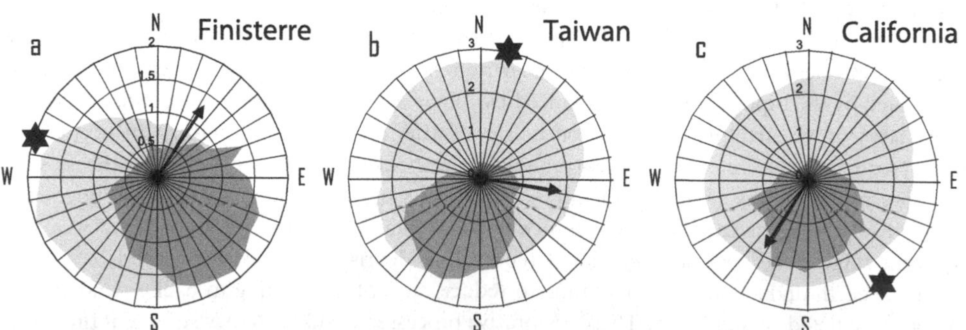

Fig. 3.6. Distribution of slope orientation (in percent; light gray) and normalized distribution of orientation of earthquake-triggered landslides (dark gray) in (a) the Finisterre Mountains, (b) central-west Taiwan, and (c) southern California. Arrows indicate the dip direction of the seismogenic fault. Stars indicate the relative mean position of the earthquake epicenter with respect to the landslides represented in these plots. (From Meunier *et al.*, 2008.)

Studies of the temporal evolution of landsliding in a mountain catchment close to the epicenter of the 1999 Chi-Chi earthquake have confirmed the existence of a progressively decaying seismic perturbation of landslide rates (Lin *et al.*, 2008; Hovius *et al.*, 2011). In the Chenyoulan River catchment (*ca.* 400 km^2), directly south of the earthquake epicenter, the history of mass wasting is known from 16 landslide maps covering the period 1996–2006. A total of 8123 mapped landslides with a combined surface area of 31.5 km^2 occurred prior to the Chi-Chi earthquake, due mainly to typhoon rainfall. About 3800 mapped landslides with a combined area of 16 km^2 were attributed to the earthquake, and a remarkable 48,370 landslides with a total area of 221 km^2 occurred after the earthquake, up to the end of 2006. The post-earthquake landslides, many of which partially occurred at older landslide sites, were triggered by intense, but unexceptional typhoon rainfall. Their rate of occurrence strongly suggests that they were the indirect result of the Chi-Chi earthquake. Prior to 1999, landslides primarily occurred on slope segments adjacent to river channels. Rates of landsliding increased throughout the landscape due to the earthquake (Fig. 3.7), with maximum changes of up to 800 percent at a distance of 300–700 m from streams, and also 1700–2200 m from streams – where many of the principal ridges are located. Due to their convexity, these particular locations in the landscape were predisposed to the largest topographic site effects, confirming that the majority of landslides in this time window were earthquake-induced. Landslide rates have decayed steadily from a maximum in 2001, when the Chenyoulan catchment was first hit by a large typhoon after the earthquake. Moreover, landsliding has migrated to lower positions on hillslopes over this interval. In 2006, landslide rates on lower slopes had returned to pre-earthquake values, but they have remained somewhat elevated on higher slopes (Fig. 3.7b; Hovius *et al.*, 2011).

It is unclear which processes played a role in the progressive restoration of pre-earthquake landslide conditions in the mountain landscape affected by the Chi-Chi earthquake and what determined the rate of this restoration. Restoration was largely unrelated to aftershocks, whose summed moment normalized within 15 months after the Chi-Chi earthquake, when landslide rates were still elevated (Hsu *et al.*, 2009). Moreover, large decreases in groundwater level and pore water pressure were recorded in the epicentral area after the earthquake (Wang *et al.*, 2001), but neither their duration (up to several months) nor the direction of these changes seem to be related to landsliding. Repair of tree root mass can take years or decades after an earthquake (Jacoby, 1997), similar to the timescale of the observed decrease of landslide rate. Finally, the evolution of landslide rates in the Chenyoulan catchment may have been a largely erosional phenomenon, without significant forcing by other factors. Nevertheless, its consequence was that the total amount of erosion due to the Chi-Chi earthquake was much greater than the erosion due to coseismic landsliding alone. It is likely that other large earthquakes have a similar prolonged legacy in landsliding, but quantitative assessments are rare (Koi *et al.*, 2008), in spite of the implications for natural hazard risk, sediment fluxes, and the mass balance of earthquakes.

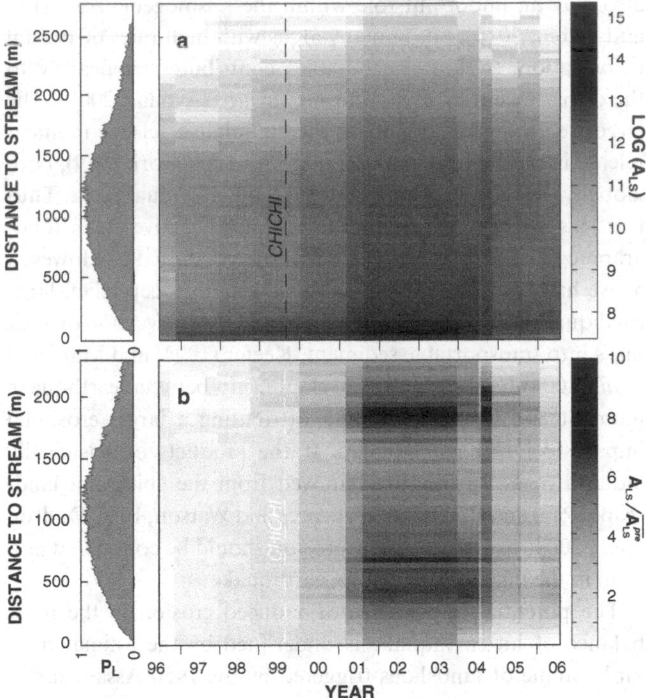

Fig. 3.7. Topographic location and intensity of landsliding in the Chenyoulan catchment, south of the epicenter of the 1999 Chi-Chi earthquake. (a) Time evolution of the area of landsliding, A_{LS} (m^2), reported with flow distance to stream for the period 1996–2006. The stream network is the ensemble of locations with upslope area >1 km^2, a cut-off determined from slope-area scaling. Fifteen time intervals are delimited by landslide maps; the difference between successive maps represents landsliding within the interval. (b) Time evolution of the actual area of landsliding normalized by the area of landsliding prior to the Chi-Chi earthquake (dimensionless), reported with flow distance to stream. (Adapted from Hovius *et al.*, 2011.)

3.6 EARTHQUAKE EROSION AND MASS BALANCE

Slip on faults creates topographic relief. Surface deformation along the fault trace is greatest where dip–slip faults break the Earth's surface (Avouac, 2007). On such faults, uplift and subsidence decrease with distance away from the fault trace in a pattern set by the geometry of the fault plane and the flexural rigidity of the geologic substrate. Pure strike–slip faults do not have much associated topography, other than on fault bends. Most faults have a combination of dip–slip and strike–slip displacement. Deeper earthquakes, for example at subduction zones, do not commonly build local relief, but instead affect surface elevation in a regionally distributed fashion. Finally, below the seismogenic zone, where temperatures and pressures are too high to permit the build-up of elastic strain and brittle failure, deformation is more gradual and is not associated with sudden displacements. Increasingly, it is recognized that gradual deformation, not involving macro-scale seismicity, may also play an important role within the seismogenic zone (Liu and Rubin, 2010), but in many areas with high rates of crustal deformation, shallow earthquakes contribute significantly to the displacement of rocks and the surface (Avouac, 2007). The largest earthquakes dominate the cumulative seismic-moment release in mechanical work (Hanks and Kanamori, 1979), both globally (Bufe and Perkins, 2005) and on individual faults. Thus it is reasonable to expect that in seismically active areas, topographic relief is formed mainly by large earthquakes. However, as we have seen, in uplands with sufficiently steep relief, large earthquakes also cause slope failure and the conversion of rock mass into transportable sediment. Keefer (1994) and Malamud *et al.*, 2004) have proposed a relationship between earthquake moment and landslide volume, attributing a larger erosional impact to larger earthquakes. If the products of seismically induced mass wasting are removed from the epicentral landscape (Pain and Bowler, 1973; Pearce and Watson, 1986, Dadson *et al.*, 2004; Koi *et al.*, 2008), erosion should be considered as a term in the mass balance of an earthquake.

The potential importance of induced erosion in the mass balance of an earthquake is underlined by the estimated 47 km^3 volume of landslides triggered by the 1950 Assam earthquake (M8.6) in the Tibet Himalayas (Mathur, 1953), and the 74–400 mm estimated average surface lowering by landslides caused by a M7.9 earthquake in the Torricelli Mountains of Papua New Guinea in 1935 (Simonett, 1967). These estimates were calculated without detailed constraints on landslide area–volume scaling (Keefer, 1999), and can therefore only serve as general indicators. Guzzetti *et al.* (2009) have shown that, globally, landslide area (A) and volume (V) are related by a power law, $V \propto A^{3/2}$. This relation makes it possible to derive first-order estimates of volumes of material mobilized by populations of landslides induced by an earthquake from comprehensive landslide polygon maps or area–frequency relations (Hovius *et al.*, 1997). Such volume estimates can be paired with constraints on surface displacements associated with earthquakes. However, landslide volumes may differ by as much as two orders of magnitude for a given landslide area (Guzzetti *et al.*, 2009). Moreover, population volume estimates based on landslide area–frequency relations are extremely sensitive to the exact value of the volume–area scaling exponent (Larsen *et al.*, 2010). Thus, detailed knowledge of local conditions is required to reliably use this approach.

Parker *et al.* (2011) examined the potential changes in orogen volume resulting from the M_w 7.9 2008 Wenchuan earthquake in Sichuan, China. The earthquake triggered more than 56,000 landslides, producing an estimated 5–15 km^3 of debris. This amount exceeds the net volume of 2.6 ± 1.2 km^3 added to the orogen by coseismic rock uplift (de Michele *et al.*, 2010), implying that, even if only a fraction of landslide debris is removed from the orogen over the likely 2000–4000-year earthquake return period (Shen *et al.*, 2009), the Wenchuan earthquake will have caused a net material deficit in the Longmen Shan (Fig. 3.8). Similarly, Yanites *et al.* (2010) have estimated that 2–5 km^3 of debris was generated by earthquake-triggered landsliding within the 0.2 *g* contour of the 1999 Chi-Chi earthquake in Taiwan. Complete removal of the landslide material, according to these estimates, would give an average landscape lowering of 0.6–1.7 m (Yanites *et al.*, 2010), far exceeding the average surface uplift induced by the earthquake (Loevenbruck *et al.*, 2004; Yu *et al.*, 2001).

Observing that most landslides do not transfer mass over distances >10^0 km, Hovius *et al.* (2011) argued that it is geophysically more meaningful to consider the export of sediment from an earthquake-affected area, rather than the mass wasting within it. For the Chi-Chi earthquake, they demonstrated that river-suspended sediment concentrations peaked and relaxed in correspondence with landslide rates over a period of *ca.* 7 years. Knowing the expected values of sediment concentrations over the full range of river discharges for aseismic conditions, and assuming a constant proportionality of suspended load and bedload transport, they isolated the seismically induced component of sediment export from the epicentral area of the earthquake. They then distributed this erosion over the earthquake-affected area according to the pattern of earthquake-induced landslide density, subtracting from the co- and post-seismic surface deformation to obtain a conservative estimate of the net topographic effect of the Chi-Chi earthquake (Fig. 3.9). According to their estimates, sediment export removed >30 percent of the added rock mass from the epicentral area, resulting in a reduction of surface uplift by up to 0.25 m, or 35 percent of the local elevation change and a reduction of the area where the Chi-Chi earthquake had built topography. Moreover, they argued that even the largest local surface uplift due to the earthquake could be fully eroded within the earthquake return period.

Both the timescale and the efficiency of removal of earthquake-generated sediment from an epicentral area are likely to differ considerably between cases due to differences in hillslope-channel connectivity, the balance of fluvial transport capacity and sediment supply, and ponding behind landslide dams (Pain and Bowler, 1973; Pearce and Watson, 1986; Korup, 2005; Koi *et al.*, 2008; Yanites *et al.*, 2010). Nevertheless, it is clear that where crustal shortening is accommodated on a shallow detachment, mainly by slip in large earthquakes, average surface elevation is significantly reduced by erosion caused by the same mechanism that constructs topography. Seismically induced landsliding subdues the surface taper of frontal ranges and could hamper the recognition of active, seismogenic faults. This effect may be most pronounced on faults with a significant strike–slip component and reduced surface deformation.

3.7 CONCLUSIONS AND OUTLOOK

The association of landsliding and earthquake strong ground motion is universal in steep uplands. It gives rise to characteristic, and perhaps predictable, patterns of erosion during large earthquakes, the amplitude and geometry of which depend not only on the earthquake mechanism and location, but also on the attenuation and interference of seismic waves due to distance,

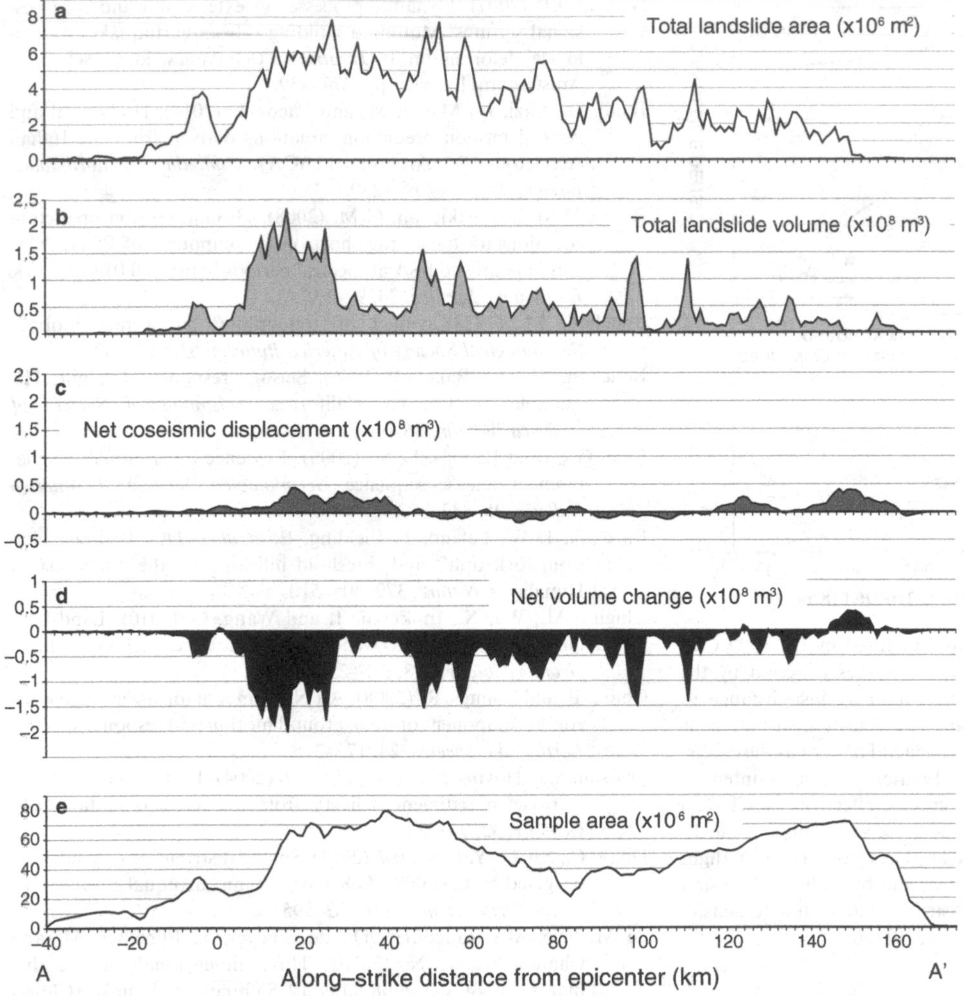

Fig. 3.8. Variations in landslide occurrence and coseismic displacement along the array of faults ruptured in the M_w 7.9 2008 Wenchuan earthquake. All data are projected onto a rupture-parallel line at 1-km intervals. Distance along horizontal axis is given with respect to the perpendicular projection of the epicenter of the main shock onto the fault trend. (a) Total area of landslides within each 1-km-wide strip. (b) Landslide volume derived from a global bedrock landslide-scaling relationship applied to individual landslides within each 1-km-wide strip. (c) Net coseismic displacement (volume change) in each 1-km-wide strip. (d) Potential net volume change determined by subtracting landslide volumes from coseismic volume change. (e) Along-strike distribution of sample area covered by satellite imagery. Local minima in landslide area and volume are not correlated with small sample areas. (From Parker *et al.*, 2011.)

substrate properties, topography, and the propensity to failure of seismically excited locations. These patterns of erosion may persist after a seismic event, dissipating gradually as subsequent drivers mobilize weakened material and shift it through sediment routing systems. In addition, seismically triggered landsliding and subsequent induced erosion can significantly reduce – or even negate – the uplift caused by an earthquake.

These conclusions are derived from quantitative analysis of a small number of geophysically well-instrumented and geomorphologically well-documented cases. Broader underpinning for each of these conclusions will require detailed observations of other historic and future cases of landsliding due to dip–slip and strike–slip earthquakes, with special attention to their seismological dimensions and tectonic setting. The reward will be a deeper insight into the root causes of earthquake-induced landsliding that can be used for scientific gain and societal benefit. With this insight, it will be possible to explore the stochastic nature of seismically induced erosion, and sediment production and routing with topological and temporal precision, thus providing a better understanding of the variability of sediment fluxes from tectonically active landscapes. It will also be possible to introduce earthquakes into landscape evolution models in a way that covers both their constructive and destructive effects, which is likely to result in an improved ability to explain the topography of faulted terrains and will give impetus to the debate on the role of earthquakes in orogenesis. Further, it may increasingly become possible to invert geomorphological observations on landslide populations for the geophysical understanding of faulting and the mechanisms of shallow earthquakes. Finally, it will become possible to quantitatively constrain both site-specific and statistical evaluations of the risk of seismically triggered and induced landsliding, with due regard to the prolonged geomorphological effect of an instantaneous seismic event. These advances could result in a redefinition of natural disaster scenarios for the impact of earthquakes on landslide-prone landscapes and better-informed decisions about the repair and socio-economic regeneration of epicentral areas in steep uplands. Further advances will require

- better constraints on background rates and patterns of landsliding and post-seismic sediment transport
- rapid, comprehensive, and detailed observation of landslides in areas affected by earthquakes

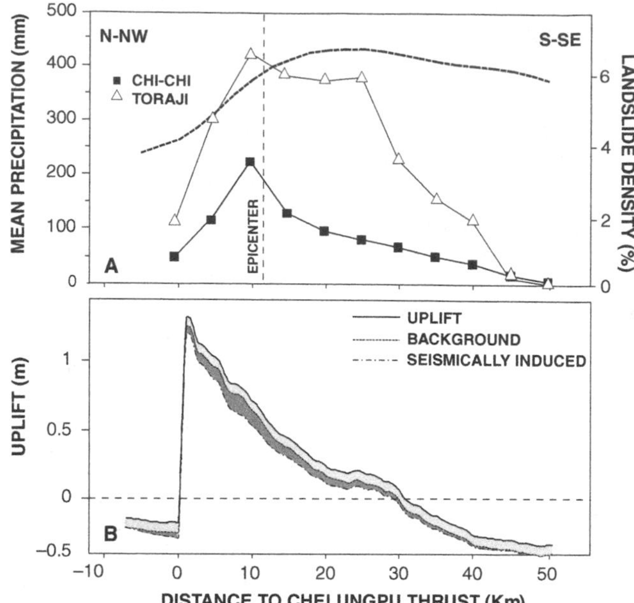

Fig. 3.9. Distribution of uplift and erosion caused by the 1999 Chi-Chi earthquake, Taiwan. (A) Density of landslides triggered by the earthquake and typhoon Toraji (2001), plotted against distance to the surface trace of the Chelungpu fault. The density distribution of typhoon-triggered landslides has the same general pattern as the coseismic landslides, but also reflects the spatial pattern of rainfall intensity shown by the bold black line. (B) Topographic effect of the Chi-Chi earthquake through the Choshui catchment. Seismic uplift of catchment surface (black line) has been reduced by background erosion (light gray), assumed to be uniformly distributed, and by additional erosion caused by the earthquake (dark gray), distributed according to density of coseismic landslides. (From Hovius *et al.*, 2011.)

- instrumental measurements of ground motion in these areas
- a better understanding of geologic and topographic controls on landslide susceptibility
- improved knowledge of the causes of weakening of near-surface materials due to strong ground motion.

REFERENCES

Abrahamson, N.A. and Somerville, P.G. (1996). Effects of the hanging wall and footwall on ground motions recorded during the Northridge earthquake. *Bulletin of the Seismological Society of America*, 86, 93–99.

Aki, K. (1993). Local site effects on weak and strong ground motion. *Tectonophysics*, 218, 93–111.

Allen, C.R., Brune, J.N., Cluff, L.S. and Barrows, A.G., Jr. (1998). Evidence for unusually strong near-field ground motion on the hanging wall of the San Fernando Fault during the 1971 earthquake. *Seismological Research Letters*, 69, 524–531.

Anderson, J.G. (2000). Expected shape of regressions for ground-motion parameters on rock. *Seismological Society of America Bulletin*, 90, 43–52.

Anderson, J.G., Brune, J.N., Anooshehpoor, A. and Ni, S.D. (2000). New ground motion data and concepts in seismic hazard analysis. *Current Science*, 79, 1278–1290.

Avouac, J.P. (2007). Dynamic processes in extensional and compressional settings: Mountain building – from earthquakes to geological deformation. In *Treatise on Geophysics*, ed. G. Schubert. Amsterdam: Elsevier, pp. 377–439.

Bindi, D., Luzi, L., Massa, M. and Pacor, F. (2010). Horizontal and vertical motion prediction equations derived from the Italian Accelerometric Archive (ITACA). *Bulletin of Earthquake Engineering*, 8, 1209–1230.

Boore, D.M. and Atkinson, G.M. (2008). Ground motion prediction equations for the average horizontal component of PGA, PGV, and 5%-damped PSA at spectral periods between 0.01 s and 10 s. *Earthquake Spectra*, 24, 99–138.

Bouchon, M. (1973). Effect of topography on surface motion. *Seismological Society of America Bulletin*, 63, 615–632.

Bouchon, M. and Baker, J. (1996). Seismic response of a hill: The example of Tarzana, California. *Seismological Society of America Bulletin*, 86, 66–72.

Bufe, G.C. and Perkins, D.M. (2005). Evidence for a global seismic-moment release sequence. *Seismological Society of America Bulletin*, 95, 833–853.

Burbank, D.W., Leland, J., Fielding, E. *et al.* (1996). Bedrock incision, rock uplift and threshold hillslopes in the northwestern Himalayas. *Nature*, 379, 505–510.

Chigira, M., Wu, X., Inokuchi, T. and Wang, G. (2010). Landslides induced by the 2008 Wenchuan earthquake, Sichuan, China. *Geomorphology*, 118, 225–238.

Chiou, B. and Youngs, R. (2008). An NGA model for the average horizontal component of peak ground motion and response spectra. *Earthquake Spectra*, 24, 173–215.

Dadson, S.J., Hovius, N., Chen, H. *et al.* (2004). Earthquake-triggered increase in sediment delivery from an active mountain belt. *Geology*, 32, 733–736.

Dai, F.C., Xu, C., Yao, X. *et al.* (2011). Spatial distribution of landslides triggered by the 2008 Ms 8.0 Wenchuan earthquake. *Journal of Asian Earth Sciences*, 40, 883–895.

de Michele, M., Raucoules, D., de Sigoyer, J., Pubellier, M. and Chamot-Rooke, N. (2010). Three-dimensional surface displacement of the 2008 May 12 Sichuan earthquake (China) derived from Synthetic Aperture Radar: Evidence for rupture on a blind thrust. *Geophysics Journal International*, 183, 1097–1103.

Del Gaudio, V. and Wasowski, J. (2007). Directivity of dynamic slope response to seismic shaking. *Geophysical Research Letters*, 34, doi:10.1029/2007GL029842.

Densmore, A.L. and Hovius, N. (2000). Topographic fingerprints of bedrock landslides. *Geology*, 28, 371–374.

Field, E.H., Johnson, P.A., Beresnev, I.A. and Zeng, Y. (1994). Nonlinear ground-motion amplification by sediments during the 1994 Northridge earthquake. *Nature*, 390, 599–602.

Geli, L., Bard, P.-Y. and Jullien, B. (1988). The effect of topography on earthquake ground motion: A review and new results. *Seismological Society of America Bulletin*, 78, 42–63.

Guzzetti, F., Ardizzone, F., Cardinali, M., Rossi, M. and Valigi, D. (2009). Landslide volumes and landslide mobilization rates in Umbria, central Italy. *Earth and Planetary Science Letters*, 279, 222–229.

Hanks, T.C. and Kanamori, H. (1979). Moment magnitude scale. *Journal of Geophysical Research*, 84, 2348–2350.

Harp, E.L. and Jibson, R.W. (1996). Landslides triggered by the 1994 Northridge, California earthquake. *Seismological Society of America Bulletin*, 86, 319–332.

Havenith, H.-B., Vanini, M., Jongmans, D. and Faccioli, E. (2003). Initiation of earthquake-induced slope failure: Influence of topographical and other site-specific amplification effects. *Journal of Seismology*, 7, 397–412.

Hikima, K. and Koketsu, K. (2005). Rupture processes of the 2004 Chuetsu (mid-Niigata prefecture) earthquake, Japan: A series of events in a complex fault system. *Geophysical Research Letters*, 32, L18303.

Honda, R., Aoi, S., Morikawa, N. *et al.* (2005). Ground motion and rupture process of the 2004 Mid Niigata Prefecture earthquake obtained from strong motion data of K-NET and KiK-net. *Earth, Planets and Space*, 57, 527–532.

Hovius, N., Stark, C.P. and Allen, P.A. (1997). Sediment flux from a mountain belt derived by landslide mapping. *Geology*, 25, 231–234.

Hovius, N., Meunier, P., Lin, C.W. *et al.* (2011). Prolonged seismically induced erosion and the mass balance of a large earthquake. *Earth and Planetary Science Letters*, 304, 347–355.

Hsu, Y.J., Avouac, J.P., Yu, S.B. *et al.* (2009). Spatio-temporal slip, and stress level on the faults within the western foothills of Taiwan: Implications for fault frictional properties. *Pure and Applied Geophysics*, 166, 1853–1884.

Hung, J.J. (2000). Chi-Chi earthquake induced landslides in Taiwan. *Earthquake Engineering and Engineering Seismology*, 2, 25–33.

Jacoby, G.C. (1997). Application of tree ring analysis to paleoseismology. *Reviews of Geophysics*, 35, 109–124.

Kao, H. and Chen, W.P. (2000). The Chi-Chi earthquake sequence: Active, out-of-sequence thrust faulting in Taiwan. *Science*, 288, 2346–2349.

Kawabata, D. and Bandidas, J. (2009). Landslide susceptibility mapping using geological data, a DEM from ASTER images and an Artificial Neural Network (ANN). *Geomorphology*, 113, 97–109.

Keefer, D.K. (1984). Landslides caused by earthquakes. *Geological Society of America Bulletin*, 95, 406–421.

(1994). The importance of earthquake-induced landslides to long-term slope erosion and slope-failure hazards in seismically active regions. *Geomorphology*, 10, 265–284.

(1999). Earthquake-induced landslides and their effects on alluvial fans. *Journal of Sedimentary Research*, 69, 84–104.

(2002). Investigating landslides caused by earthquakes: A historical review. *Surveys in Geophysics*, 23, 473–510.

Kelsey, H.M. (1988). Formation of inner gorges. *Catena*, 15, 433–458.

Koi, T., Hotta, N., Ishigaka, I. *et al.* (2008). Prolonged impact of earthquake-induced landslides on sediment yield in a mountain watershed: The Tanzawa region, Japan. *Geomorphology*, 101, 692–702.

Korup, O. (2005). Large landslides and their effect on alpine sediment flux: South Westland, New Zealand. *Earth Surface Processes and Landforms*, 30, 305–323.

Larsen, I.J., Montgomery, D.R. and Korup, O. (2010). Landslide erosion controlled by hillslope material. *Nature Geoscience*, 3, 247–251, doi:10.1038/ngeo776.

Lin, G.W., Chen, H., Hovius, N. *et al.* (2008). Effects of earthquake and cyclone sequencing on landsliding and fluvial sediment transfer in a mountain catchment. *Earth Surface Processes and Landforms*, 33, 1354–1373, doi:10.1002/esp1716.

Liu, Y. and Rubin, A.M. (2010). Role of fault gouge dilatancy on aseismic deformation transients. *Journal of Geophysical Research*, 115, B10414, doi:10.1029/2010JB007522.

Loevenbruck, A., Cattin, R., Le Pichon, X., Dominguez, S. and Michel, R. (2004). Coseismic slip resolution and post-seismic relaxation time of the 1999 Chi-Chi, Taiwan, earthquake as constrained by geological observations, geodetic measurements and seismicity. *Geophysical Journal International*, 158, 310–326.

Luzi, L. and Pergalani, F. (2000). A correlation between slope failures and accelerometric parameters: The 26 September 1997 earthquake (Umbria-Marche, Italy). *Soil Dynamics and Earthquake Engineering*, 20, 301–313.

Malamud, B.D., Turcotte, D.L., Guzzetti, F. and Reichenbach, P. (2004). Landslides, earthquakes, and erosion. *Earth and Planetary Science Letters*, 229, 45–59.

Mathur, L.P. (1953). Assam earthquake of 15th August 1950: A short note on factual observations. In *A Compilation of Papers on the Assam Earthquake of August 15, 1950*, ed. M.B. Ramachandra Rao. National Geographical Research Institute, Hyderabad, Central Board of Geophysics Publication 1, pp. 56–60.

Meisling, K.E. and Sieh, K.E. (1980). Disturbance of trees by the 1857 Fort Tejon earthquake, California. *Journal of Geophysical Research*, 85, 3225–3238.

Meunier, P., Hovius, N. and Haines, A.J. (2007). Regional patterns of earthquake-triggered landslides and their relation to ground motion. *Geophysical Research Letters*, 34, L20408, doi:10.1029/2007GL031337.

(2008). Topographic site effects and the location of earthquake-induced landslides. *Earth and Planetary Science Letters*, 275, 221–232.

Newmark, N.M. (1965). Effects of earthquakes on dams and embankments. *Geotechnique*, 15, 139–160.

Ogelsby, D.D., Archuleta, R.J. and Nielsen, S.B. (2000). The three-dimensional dynamics of dipping faults. *Seismological Society of America Bulletin*, 90, 616–628.

Oldham, R.D. (1899). *Report on the Great Earthquake of 12th June 1897*. Geological Survey of India, Memoir 29.

Osanai, N., Uchida, T., Noro, T. *et al.* (2007). Application of the empirical method of assessing the potential of slope failures to Niigata–ken Chuetsu Earthquake. *Journal of the Japanese Society of Erosion Control Engineering*, 59, 60–65.

Owen, L.A., Kamp, U., Khattak, G.A. *et al.* (2008). Landslides triggered by the 8 October 2005 Kashmir earthquake. *Geomorphology*, 94, 1–9.

Pain, C.F. and Bowler, J.M. (1973). Denudation following the 1970 earthquake at Madang, Papua New Guinea. *Zeitscrift für Geomorphologie*, 18, 92–104.

Parise, M. and Jibson, R.W. (2000). A seismic landslide susceptibility rating of geologic units based on analysis of characteristics of landslides triggered by the 17 January, 1994 Northridge, California earthquake. *Engineering Geology*, 58, 251–270.

Parker, R.N., Densmore, A.L., Rosser, N.J. *et al.* (2011). Mass wasting triggered by the 2008 Wenchuan earthquake exceeds orogenic growth. *Nature Geoscience*, 4, doi:10.1038/ngeo1154.

Pearce, A. and O'Loughlin, C. L. (1985). Landsliding during a M7.7 earthquake: Influence of geology and topography. *Geology*, 13, 855–858.

Pearce, A.J. and Watson, A.J. (1986). Effects of earthquake-induced landslides on sediment budget and transport over 50 years. *Geology*, 14, 52–55.

Rodriguez, C.E., Bommer, J.J. and Chandler, R.J. (1999). Earthquake-induced landslides; 1980–1997. *Soil Dynamics and Earthquake Engineering*, 18, 325–346.

Schmedes, J. and Archuleta, R.J. (2008). Near-source ground motion along strike-slip faults: Insights into magnitude saturation of PGV and PGA. *Seismological Society of America Bulletin*, 98, 2278–2290.

Shen, Z.K., Sun, J., Zhang, P. *et al.* (2009). Slip maxima at fault junctions and rupturing of barriers during the 2008 Wenchuan earthquake. *Nature Geoscience*, 2, 718–724.

Shi, B., Anooshehpoor, A., Brune, J.N. and Zeng, Y. (1998). Dynamics of thrust faulting: 2-D lattice model. *Seismological Society of America Bulletin*, 88, 1484–1494.

Shin, T.C. and Teng, T.L. (2001). An overview of the 1999 Chi-Chi, Taiwan, earthquake. *Seismological Society of America Bulletin*, 91, 895–913.

Simonett, D.S. (1967). Landslide distribution and earthquakes in the Bewani and Torricelli Mountains, New Guinea: A statistical analysis. In *Landform Studies from Australia and New Guinea*, ed. J.N. Jennings and J.A. Mabbutt. Cambridge, UK: Cambridge University Press, pp. 64–84.

Sommerville, P. G., Smith, N., Graves, R. and Abrahamson, N. (1997). Modification of empirical strong ground motion attenuation relations for the amplitude and duration effects of rupture directivity. *Seismological Research Letters*, 68, 199–222.

Sommerville, P. G., Irikura, K., Graves, R. *et al.* (1999). Characterizing earthquake slip models for the prediction of strong ground motion. *Seismological Research Letters*, 70, 59–80.

Spudich, P. (ed.). (1996). *The Loma Prieta, California Earthquake of October 17, 1989: Main Shock Characteristics*. US Geological Survey, Professional Paper 1550-A.

Spudich, P., Hellweg, M. and Lee, W. H. K. (1996). Directional topographic site response at Tarzana observed in aftershocks of the 1994 Northridge, California, earthquake: Implications for mainshock motions. *Seismological Society of America Bulletin*, 86, 193–208.

Suzuki, W., Aoi, S. and Sekigushi, H. (2010). Rupture process of the 2008 Iwate–Miyagi Nairiku, Japan, earthquake derived from near-source strong-motion records. *Geological Society of America Bulletin*, 100, 256–266.

Taylor, S. R., Bonner, B. P. and Zandt, G. (1986). Attenuation and scattering of broadband P and S waves across North America. *Journal of Geophysical Research*, 91, 7309–7325.

Todorovska, M. I. and Trifunac, M. D. (1997). Amplitudes, polarity and time of peaks of strong ground motion during the 1994 Northridge, California, earthquake. *Soil Dynamics and Earthquake Engineering*, 16, 235–258.

Trifunac, M. D. (1994). Q and high frequency strong ground motion spectra. *Soil Dynamics and Earthquake Engineering*, 13, 149–161.

Wang, C. Y., Cheng, L. H., Chin, C. V. and Yu, S. B. (2001). Coseismic hydrologic response of an alluvial fan to the 1999 Chi-Chi earthquake, Taiwan. *Geology*, 29, 831–834.

Wang, H. B., Sassa, K. and Xu, W. Y. (2007). Analysis of a spatial distribution of landslides triggered by the 2004 Chuetsu earthquakes of Niigata Prefecture, Japan. *Natural Hazards*, 41, 43–60.

Wells, D. L. and Coppersmith, K. J. (1994). Analysis of empirical relationships among magnitude, rupture length, rupture area, and surface displacement. *Seismological Society of America Bulletin*, 84, 974–1002.

Wilson, R. C. and Keefer, D. K. (1979). Dynamic analysis of slope failure from the 6 August 1979 Coyote Lake, California, earthquake. *Seismological Society of America Bulletin*, 73, 863–877.

Yagi, H., Sato, G., Hikagi, D., Yamamoto, M. and Yamasaki, T. (2009). Distribution and characteristics of landslides induced by the Iwate–Miyagi Nairiku earthquake in 2008 in Tohoku District, Northeast Japan. *Landslides*, 6, 335–344.

Yanites, B. J., Tucker, G. E., Mueller, K. J. and Chen, Y. G. (2010). How rivers react to large earthquakes: Evidence from central Taiwan. *Geology*, 38, 639–642.

Yin, J., Chen, J., Xu, X.-W., Wang, X. and Zheng, Y. (2010). The characteristics of the landslides triggered by the Wenchuan M_s 8.0 earthquake from Anxian to Beichuan. *Journal of Asian Earth Sciences*, 37, 452–459.

Yu, S. B., Kuo, L. C., Hsu, Y. J. *et al.* (2001). Preseismic deformation and coseismic displacements associated with the 1999 Chi-Chi, Taiwan, earthquake. *Seismological Society of America Bulletin*, 91, 995–1012.

4 Landslides at stratovolcanoes initiated by volcanic unrest

CHRISTOPHER F. WAYTHOMAS

ABSTRACT

Volcanic unrest is related to, and in some cases can cause, large volcanic landslides. Edifice failures accompanied by volcanic landslides are recognized at many stratovolcanoes worldwide in a wide variety of geologic settings. Because of their large volumes (sometimes exceeding 1 km^3) and potential runout distance of several kilometers, volcanic landslides constitute significant hazards. Most volcanic slope failures are the result of a number of factors, and generally no single process acting independently initiates volcanic landslides. Important eruptive processes that lead to, or are intimately associated with, volcanic flank failures and landslides are reviewed. Several examples are discussed in the context of the role that eruptive processes play in triggering or leading to landslides.

4.1 INTRODUCTION

Volcanic eruptions cause a variety of mass-movement phenomena, including pyroclastic flows and surges, lahars, floods, and large landslides. Large eruptions involve sizable quantities of magma that enter the volcanic edifice, move through a conduit system, approach the surface, and vesiculate and fragment, causing explosive eruptions. Displacement of mass, structural changes within the edifice, and hydrothermal activity associated with the rising magma may lead to catastrophic slope failure and large volcanic landslides. Edifice failure accompanied by large volcanic landslides has occurred at many volcanoes in a wide variety of geologic settings. More than 200 volcanoes of Quaternary age worldwide show evidence of volcanic landslides, and multiple landslides may occur on individual volcanoes (Siebert, 1984, 1996; Waythomas *et al.*, 2000). Volcano flank collapse and associated landslides are a significant hazard; this process has caused significant loss of life (Siebert and Simkin, 2002). Stratovolcanoes may have large volumes (several cubic kilometers), steep flanking slopes, and high local relief (thousands of meters), making them susceptible to failure and landslides. A stratovolcano, by definition, is composed of layered, intercalated, or overlapping assemblages of rock and pyroclastic debris of varying strengths, which may make them susceptible to failure.

Oceanic island volcanoes are also prone to structural failures that result in landslides, some of which are among the largest on Earth (Lipman *et al.*, 1988). The stability of oceanic volcanoes is a function of many factors, including volcanic processes such as dike intrusion, cumulate formation, hydrothermal alteration, and elevated edifice pore pressure (Keating and McGuire, 2000). Although important, these processes are only partially responsible for edifice instability at such volcanoes and may or may not constitute driving forces capable of lateral edifice displacement (Iverson, 1995).

In general, landslides on volcanoes characterized by magmatic eruptions are known as "*Bezymianny-type*" events, after the explosive eruption and landslide at Bezymianny volcano in 1956 (Gorshkov, 1962; Siebert *et al.*, 1987). These eruptions and their associated landslides are magmatically driven and result in lateral flank failures and the formation of large debris avalanches with volumes in the range of 0.1–35 km^3 (Ui, 1983; Siebert *et al.*, 1987; Siebert, 1996). Juvenile magmatic products commonly occur within, underlie, or rest on debris-avalanche material, indicating a close temporal association of magmatic activity and flank collapse. In most of the well-studied Bezymianny-type events it is clear that the landslide occurred prior to the main eruptive event and was the primary mechanism causing rapid depressurization of the magmatic system and subsequent explosive activity.

Another class of eruptions that can have a volcanic landslide component is known as "*Bandai-type*" events (Moriya, 1980;

Landslides: Types, Mechanisms and Modeling, ed. John J. Clague and Douglas Stead. Published by Cambridge University Press.

Table 4.1. *Factors leading to eruption-induced volcanic landslides (modified from Voight and Elsworth, 1997).*

Factor	Effect	Examples
Phreatic and phreatomagmatic explosions	Provides explosive driving force capable of laterally displacing volcano flank	Iriga, 1628; Bandai, 1888; Rainier (Holocene)
Magmatic intrusion (cryptodome)	Increases surface slope leading to gravitational instability and failure	Mount St. Helens, 1980; Sheveluch, 1964; Bezymianny, 1956
• Increase in magma pressure within edifice associated with dike intrusion	Increases static load	Augustine, 1883? Jocotitlan? Kilauea
• Changes in internal pore pressure associated with intrusion	Reduces shear strength	Bandai, 1888; Mayuyama, 1792; Papandayan, 1772, 2002; Canary Islands
Magmatic extrusion (lava domes)	Increases surface loading	Soufriere Hill s, 1997
Local ground accelerations associated with volcanic earthquakes and explosions	Increases dynamic load, may induce pore-fluid pressure changes within edifice	Sheveluch (Holocene)

Siebert *et al.*, 1987). These eruptions do not involve direct magmatic activity and are driven by hydrothermal processes (Barberi *et al.*, 1992). They too are temporally associated with large volcanic landslides with volumes in the range of 0.001–35 km^3 (Siebert *et al.*, 1987). Bandai-type events, however, are somewhat smaller and slightly less explosive than their Bezymianny-type counterparts (Siebert *et al.*, 1997).

Prior to the 1980 eruption of Mount St. Helens, landslides caused by eruptions were not widely recognized (Bezymianny being the exception; Gorshkov, 1959), and had not been described in the mainstream scientific literature. The Mount St. Helens eruption catalyzed volcanological research on the topic of volcanic landslides and their causes, and led to the publication of many landmark contributions (Voight *et al.*, 1981, 1983; Ui, 1983; Siebert, 1984; Glicken, 1986; Francis and Self, 1987; Crandell, 1989; Moore *et al.*, 1989; Holcomb and Searle, 1991; Wadge *et al.*, 1995; Jones *et al.*, 1996; Voight and Elsworth, 1997). These studies, and many others not cited here, resulted in the widespread recognition of volcanic landslide deposits at many volcanoes. We now know that many large stratovolcanoes and oceanic volcanoes worldwide have experienced at least one major flank collapse during their history, although many of these are not associated with eruptive activity (Siebert, 1996, 2002).

Following the 1980 Mount St. Helens eruption, studies of volcanic landslides were primarily descriptive and focused on geomorphic features and deposit characteristics (Ui, 1983; Siebert, 1984). Later, however, the focus shifted toward explanations of the relevant processes thought to have caused or initiated the landslides (Elsworth and Voight, 1995; Iverson, 1995; Day, 1996; Voight and Elsworth, 1997, 2000; Elsworth and Day, 1999; Voight, 2000). As a result of these studies, the principal factors that lead to eruption-induced volcanic landslides were described and placed in a conceptual physical framework (Table 4.1). It is emphasized here that most volcanic slope failures are the result of a number of factors and that no single process is responsible for initiating a volcanic flank failure.

In this chapter I review important eruptive processes that lead to, or are intimately associated with, volcanic flank failures and landslides. Several examples are presented and discussed in the context of the role that eruptive processes play in triggering or leading to landslides.

4.2 VOLCANIC LANDSLIDES ASSOCIATED WITH ERUPTIVE ACTIVITY AND UNREST

Large volcanic landslides are almost always temporally associated with eruptive activity of some kind, although it may not be clear whether the landslide was the cause of the eruption or whether the eruption initiated the landslide. The most important considerations have to do with processes that reduce the strength of all or part of the volcanic edifice, increase dynamic or static loads, or both. Whether or not the flank of a volcano will fail and form a large landslide depends on the balance between the forces resisting failure and the forces driving failure (Fig. 4.1). The timescales associated with all of the relevant processes that lead to flank collapse and landslide formation may be thousands, possibly tens of thousands, of years, but the triggering and the event itself may happen in minutes. As will become clear in the discussion that follows, most of what is known about volcanic landslides associated with eruptive activity is either theoretical or inferred from studies of landslide deposits.

4.2.1 VOLCANIC LANDSLIDES ASSOCIATED WITH PHREATIC ERUPTIONS

Phreatic eruptions are steam-driven explosions that occur when ground or surface water is heated by magma, lava, hot rocks, or hot volcanic deposits. Phreatic explosions occur when a confined, high-pressure accumulation of steam is rapidly decompressed. Direct magmatic involvement may lead to what are known as phreatomagmatic eruptions, which are usually inferred from deposits that contain angular, dense, nonvesicular clasts and have a low ratio of juvenile to nonjuvenile material and some evidence of water involvement such as abundant fine ash and accretionary lapilli (Wohletz, 1986; Zimanowski, 1998). In phreatic eruptions, heat causes water to boil and flash

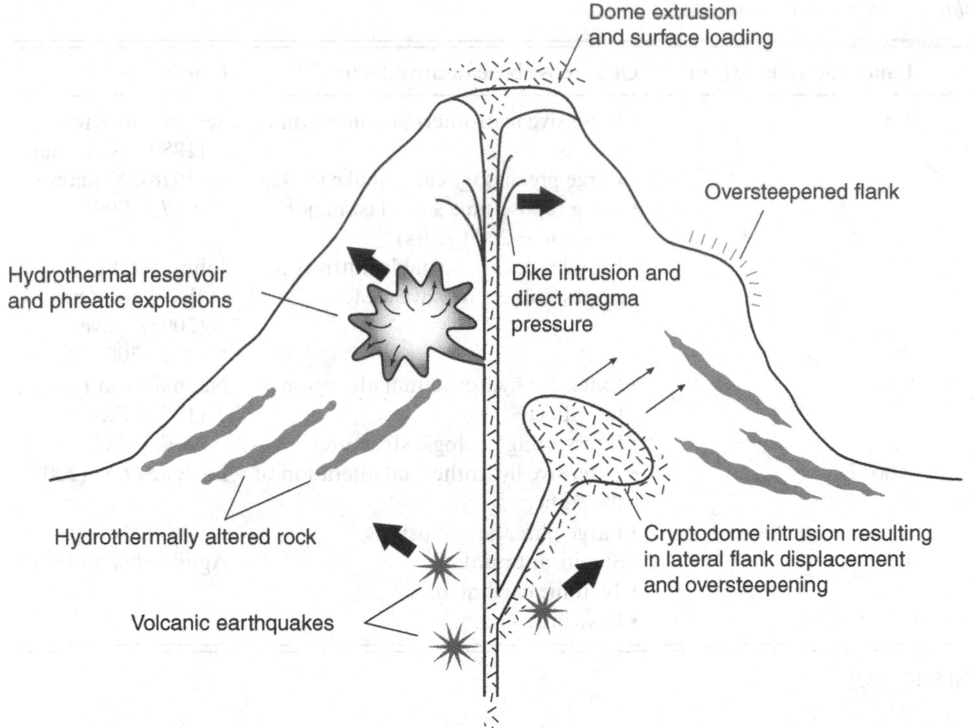

Fig. 4.1. Primary eruptive processes that may bring about volcano flank collapse resulting in landsliding. Bold arrows indicate general direction of driving forces associated with various eruptive processes. The strength of the volcano at the time these processes occur will determine whether or not flank collapse and volcanic landslides develop. Extensive hydrothermal alteration, high internal pore-fluid pressures, and the strengths of the edifice and the substrate are important factors contributing to edifice stability. Modified from McGuire (1996).

to steam, resulting in a violent explosion (Barberi *et al.*, 1992). Phreatic eruptions associated temporally with volcanic landslides are summarized in Table 4.2. In all cases, the presence of phreatic deposits and the lack of unequivocal evidence for coeval juvenile eruptive products is the primary rationale for concluding that the landslides were associated with phreatic eruptive activity.

The most famous example of a volcanic landslide triggered by phreatic activity is the July 15, 1888 eruption of Bandai volcano in Japan (Sekiya and Kikuchi, 1889; Nakamura, 1978). A flank collapse removed most of the north side of the volcano during this eruption and formed a debris avalanche about 1.5 km^3 in volume. The debris avalanche buried several villages and towns and blocked drainages, forming several lakes in the area. Sekiya and Kikuchi (1889) deduced that the eruption was the result of the "sudden expansion of imprisoned steam, unaccompanied by lava flows or pumice ejection." Nakamura (1978) and Yamamoto *et al.* (1999) confirmed the absence of juvenile material in the debris-avalanche deposit, thus confirming that no new magma was involved in the eruption. Although the precise mechanism responsible for the flank collapse and landslide is not known, Yamamoto *et al.* (1999) suggest that local seismic activity led to fracturing of the confining layers of a pressurized hydrothermal system, which caused a significant reduction in the lithostatic load and resulted in rapid steam production and the initiation of several large phreatic explosions. It is not known whether the explosions were sufficient to displace the north flank of the volcano, or whether precursory seismicity also had a role. About one minute prior to the eruption, a strong earthquake occurred, possibly as large as magnitude 5 (Okada, 1983), followed by as many as 20 phreatic explosions. Within 10 minutes after the last explosion, the north flank failed and a pyroclastic density current traveled over the east flank of the volcano, away from the debris avalanche (Yamamoto *et al.*, 1999). Little is known about the composition of the debris-avalanche deposit, although hydrothermally altered rocks were exposed in the head scarp of the avalanche caldera (Yamamoto *et al.*, 1999).

Ritter Island volcano, located in the south Bismark Sea between New Britain and New Guinea, experienced a minor eruption but large flank collapse on March 13, 1888 (Cooke, 1981). The event produced a submarine landslide up to 5 km^3 in volume that initiated a large tsunami in the region (Johnson, 1987; Ward and Day, 2003). The flank collapse removed nearly 90 percent of the subaerial part of the volcano and most of its submarine flank (Johnson, 1987). Eyewitnesses reported that there was only minor explosive activity and little nearby ash fall associated with the flank collapse (Cooke, 1981). The timing of ash emission relative to flank collapse is not known, and juvenile eruptive material associated with the 1888 eruption has not been found. Johnson (1987) concluded that the eruptive activity described by eyewitnesses was probably phreatic in origin.

An explosive eruption, flank collapse, and debris avalanche occurred on the northeast flank of Papandayan volcano in 1772 (Neuman van Padang, 1951). The clay-rich (amount unknown) debris-avalanche deposit covered an area of about 250 km^2 and had a volume of about 0.14 km^3 (Siebert *et al.*, 1987). Studies of the debris-avalanche deposit and eruption have failed to locate

Table 4.2. *Volcanic landslides associated with phreatic eruptions.*

Example	VEI[a]	Landslide volume (km^3)	Other possible causative factors	References
Bandai, 1888	4	1.5	• Extensive hydrothermal alteration of edifice • Large precursory earthquake (~M5) • Long repose time since last major eruption (~2000 years)	Sekiya and Kikuchi (1889), Nakamura (1978), Yamamoto *et al.* (1999)
Ritter Island, 1888	2	4.6	• Island volcano, unstable buttressing slopes, subject to wave attack	Johnson (1987), Ward and Day (2003), Silver *et al.* (2009)
Papandayan, 1772	3	0.14	• Extensive hydrothermal alteration of the edifice • Intersecting geologic structures	Neuman van Padang (1951), Siebert *et al.* (1987)
Papandayan, 2002	2	0.0017	• Extensive hydrothermal alteration of the edifice • Large phreatic explosions	Lavigne *et al.* (2005)
Iriga, 1628	2	1.5	• Significant rainfall • Tectonic earthquakes • Dike intrusion	Aguila *et al.* (1986)

[a] Volcanic explosivity index (Newhall and Self, 1982).

material that is unequivocally juvenile, and it appears that a phreatic explosion had a role in initiating the landslide (Siebert *et al.*, 1987), although a recent paper by Mazot *et al.* (2008) refers to the 1772 Papandayan eruption as magmatically driven. On November 11–15, 2002, several volcanic earthquakes associated with a series of phreatic and phreatomagmatic explosions at Papandayan initiated several small volcanic landslides that transformed into lahars (Lavigne *et al.*, 2005; Hadisantono, 2006). The largest landslide had a volume of 0.0017 km^3 (Lavigne *et al.*, 2005).

Iriga volcano on southeastern Luzon Island in the Philippines experienced a phreatic eruption and ensuing debris avalanche in *ca.* AD 1628 (Aguila *et al.*, 1986). The debris-avalanche deposit has a volume of about 1.5 km^3 and covers an area of 70 km^2 on the southeast flank of the volcano. It contains no juvenile material or evidence of contemporaneous magmatic activity, indicating that a magmatic eruption was probably not involved in the flank collapse and resulting volcanic landslide (Aguila *et al.*, 1986; Lagmay *et al.*, 2000).

4.2.2 HOW DO PHREATIC ERUPTIONS INITIATE VOLCANIC LANDSLIDES?

A pertinent question to address is the physical mechanism by which phreatic eruptions and related hydrothermal processes lead to or initiate volcanic landslides. In general, large-scale flank failure and debris-avalanche formation may occur if the collective mechanical forces generated by heating and hydrothermal pressurization, without direct magmatic intrusion into the edifice, exceed the strength of the volcanic edifice (Reid, 2004).

In the following discussion I consider a typical stratovolcano with a well-established hydrothermal system prone to explosive phreatic activity, whose flanks are weakened internally by hydrothermal alteration and thus possess a lower rock mass strength than the intact edifice. The explosivity of a phreatic eruption and the forces generated depend on the amount of thermal energy associated with the erupting mass, and the efficiency of conversion of this energy to mechanical forces, which can be as high as 10 percent (Wohletz, 1986, 2003; Sato and Taniguchi, 1997). Ejection of tephra and the generation of explosion-related seismic or shock waves are physical expressions of this process. The thermal energy of the eruption mass is a function of several factors, such as its temperature–pressure state, and gas, liquid, and solids content. Pyle (1995) reviewed the thermodynamic processes associated with volcanic explosions, and Mastin (1995) evaluated the theoretical mechanical energy release of steam-driven, nonjuvenile-producing eruptions caused by mixing of water and hot rock, and geyser or boiling-point eruptions. For comparison, gas-driven eruptions, where volcanic gases exsolved from magma bodies become blocked, such as the 1993 Galeras eruption (Stix *et al.*, 1993), result in the maximum mechanical energy release (up to *ca.* 1.3 MJ kg^{-1} of fluid–rock mixture). Phreatic explosions generate less mechanical energy (up to *ca.* 0.4 MJ kg^{-1} of fluid–rock mixture), and boiling-point eruptions the least mechanical energy (up to *ca.* 0.25 MJ kg^{-1} of fluid–rock mixture). For eruptive volumes in the range of 10^6–10^7 m^3, which are plausible for phreatic eruptions (Barberi *et al.*, 1992), and eruption mixture bulk densities of >1000 to <2000 kg m^{-3}, the total mechanical energy release could be in the range of 10^{14}–10^{16} J. Energy release of this magnitude, especially if the release occurs in an explosion or series of closely

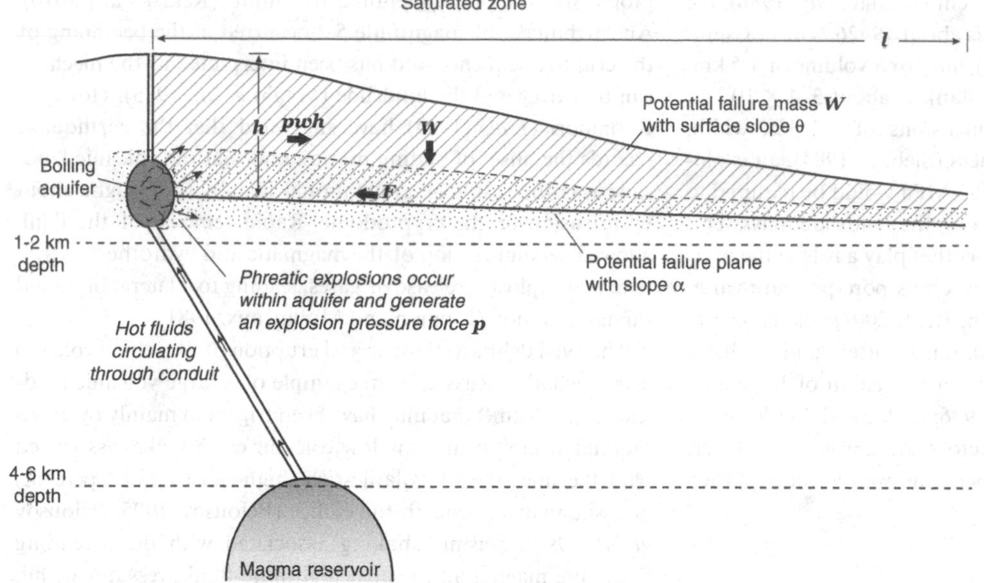

Fig. 4.2. Schematic representation of volcano flank with active hydrothermal system and balance of forces resulting from phreatic explosions within the edifice. The primary driving force is the explosion pressure force ***p*** acting on the volcano flank of width ***w***, length ***l***, and thickness ***h***. The force ***F*** is the resisting force offered by the flank and is a function of its mass and strength.

spaced explosions, may be sufficient to initiate volcanic landslides, although other factors intrinsic to the erupting volcano, such as edifice rock mass strength, also play a role and would need to be evaluated.

Many phreatic explosions occur near the ground surface or at depths of only a few tens of meters (Yokoo *et al.*, 2002). Such explosions may generate significant mechanical energy, but much of it may not effectively modify stresses at sufficient depth to bring about lateral displacement of a volcanic edifice. Phreatic explosions low on the flank, however, could remove material from this area, reduce the volume of the resisting mass, and thereby reduce stability.

Studies of tephra deposits produced by phreatic explosions indicate that nonjuvenile material in the deposits could have been derived from depths of 200–500 m (Nairn and Wiradiradja, 1980; Mastin, 1991), suggesting that phreatic explosions can have deeper focal depths. Retrospective analysis of seismic data collected during phreatic eruptions also confirms that explosions can occur at a depth of several kilometers below the surface (Montalto, 1995; McNutt, 1996; Sudo *et al.*, 1998; Savin *et al.*, 2005).

These observations indicate that phreatic explosions can occur within the upper few kilometers of a typical stratovolcano, provided the hydrostatic pore-fluid pressure is less than the critical pressure for the transition of water to steam (about 22 MPa). The resulting phreatic explosion force could be large enough to displace the edifice and initiate lateral failure of the flank. The balance of forces for a simple example is shown in Figure 4.2. The explosion pressure p acts on a slab-like rigid segment of the edifice of thickness h, width w, and surface slope θ. A slip surface, or potential failure plane, at the base of the rigid segment has slope α. A force F is the total resisting force and is a function of the mass and strength of the edifice. F acts along the failure plane and must be overcome to allow displacement of the rigid segment. The problem can be visualized as a two-dimensional slope stability problem where the stability criterion, or factor of safety (FoS), is the ratio of the total stress that resists displacement to the driving stress associated with the explosion pressure and mass of the edifice segment. The pressure p generated by a phreatic explosion can be expressed as:

$$p = \frac{1}{2}\rho v^2 \tag{4.1}$$

where ρ is fluid–rock mixture density in kg m^{-3} and v is the velocity of the erupting mass in m s^{-2}. Using values for ρ of 1000 kg m^{-3} and v of 400–1000 m s^{-2} (Mastin, 1995), p could be in the range of 80–500 MPa. The ratio of mechanical energy released by a phreatic explosion to phreatic explosion pressure has a dimension of length cubed, which is a volume and can be thought of as the approximate volume of the effective hydrothermal reservoir where explosive phreatic activity takes place. Information about the geometry of active hydrothermal reservoirs in volcanic settings is commonly estimated from geophysical studies (Dzurisin *et al.*, 1994; Furuya, 2004; Hase *et al.*, 2005). Such studies indicate that hydrothermal reservoir volumes at a variety of volcanoes are in the range of 10^6–10^9 m^3, although major geothermal fields may be much larger (Wohletz and Heiken, 1992). Using the values derived above for mechanical energy release (10^{16} J) and maximum phreatic explosion pressure (500 MPa) as rough approximations of phreatic eruption conditions, the theoretical reservoir volume is 20×10^6 m^3. This suggests a possible limiting value for the volume of a hydrothermal reservoir that can produce the driving forces needed to displace a volcanic edifice.

By applying these concepts to a known example of a volcanic landslide associated with phreatic activity, we can evaluate whether plausible explosion pressures generated are sufficient to displace a sector of the volcano and thus initiate a landslide.

In the case of the AD 1628 Iriga event (Aguila *et al.*, 1986), the unit weight of the landslide mass is about 19–26 N m^{-3} (assuming a density of 2000–2700 kg m^{-3}), and for a volume of 1.5 km^3 the force required to displace the flank is about 3–4 × 10^{13} N. Given approximate landslide dimensions of 2.1 km width, 3.5 km length, and 0.2 km thickness (Siebert, 1984), an explosion pressure force of roughly 70–95 MPa would be required to overcome the static resisting force of the landslide mass. This simple analysis ignores other factors that play a role in the overall stability of the volcano flank, such as pore-pressure variations within the edifice (Day, 1996; Reid, 2004), the geometry and extent of hydrothermal alteration (Watters and Delhaut, 1995), and the structure and rock mass strength of the edifice (van Wyk de Vries and Borgia, 1996; Vidal and Merle, 2000; Watters *et al.*, 2000). If these factors are considered, it may be that flank displacements can occur at even lower explosion pressures.

4.2.3 VOLCANIC LANDSLIDES ASSOCIATED WITH MAGMATIC INTRUSIONS

The movement of magma toward the surface and its eventual eruption is another process capable of triggering volcanic landslides. Magmatic intrusion may take the form of a dike or sill, or may involve intrusion of a cryptodome. Regardless of the form of the intrusion, the magmatic driving force displaces the volcano's flank by direct magma pressure, changes in ground surface slope due to injection of magma, or elevated pore-fluid conditions associated with magmatic heat and enhanced hydrothermal fluid circulation (Swanson *et al.*, 1976; Iverson, 1995; Voight and Elsworth, 1997; Elsworth and Day, 1999; Ventura *et al.*, 1999). Although these processes may not necessarily result in an eruption, they can markedly affect edifice stability and increase the likelihood of flank failure.

The influx of magma beneath a volcano may lead to the formation of a cryptodome within the edifice (Voight *et al.*, 1983; Donnadieu *et al.*, 2001), which can increase the surface slope, and thus the gravitational driving force, of the volcano flank. However, the change in surface slope at Mount St. Helens caused by cryptodome intrusion in 1980 only reduced edifice stability by about 3 percent (Reid *et al.*, 2000). Although emplacement of a cryptodome is not necessarily an eruptive process unless it reaches the surface and becomes a lava dome, cryptodome formation can lead to massive flank failure and exhumation of the internal conduit system, rapidly lowering the confining pressure and causing explosive eruptive activity. This mechanism was likely the principal driver for the 1980 Mount St. Helens eruption, the 1964 eruption of Sheveluch volcano, and the 1956 eruption of Bezymianny volcano (Voight *et al.*, 1983; Belousov and Bogoyavlenskaya, 1988; Belousov, 1995).

The 1980 eruption of Mount St. Helens occurred after about 2 months of significant flank deformation associated with the intrusion of a cryptodome (Lipman *et al.*, 1981; Voight *et al.*, 1981). As a result of the intrusion, shear stress within the edifice increased, rock mass strength decreased, and the flank became progressively more susceptible to failure (Reid *et al.*, 2010). An earthquake of magnitude 5.2 occurred at the beginning of the eruptive sequence and has been interpreted as the mechanism that triggered the landslide (Voight *et al.*, 1983). However, Kanamori *et al.* (1984) have suggested that the earthquake records the onset of sliding motion and that the volcanic landslide was the result of gravitational instability brought about by intrusion of the cryptodome. Rapid removal of the flank allowed decompression of the magmatic and hydrothermal system, and explosive release of gases, leading to a lateral blast and Plinian eruption (Lipman and Mullineaux, 1981).

The 1964 debris avalanche and eruption of Shiveluch volcano in Kamchatka, Russia, is an example of a large volcanic landslide (*ca.* 1.5 km^3) that may have been triggered mainly by gravitational instability and shallow volcanic earthquakes associated with the injection of volatile-rich, high-silica (60–62 percent) andesite magma beneath the edifice (Belousov, 1995; Belousov *et al.*, 1999). Seismic shaking associated with the ascending degassing magma and related high pore-fluid pressures within the edifice are thought to have led to massive failure of a lava dome complex that had formed since eruptive activity in 1854 (Belousov *et al.*, 1999). As in the 1980 Mount St. Helens eruption, the volcanic landslide was soon followed by an explosive Plinian eruption, but – unlike Mount St. Helens – no directed blast developed, possibly because magma had not risen high enough in the edifice (Belousov, 1995; Belousov *et al.*, 2007).

The 1956 eruption of Bezymianny volcano in Kamchatka, Russia, is another well-known example of a magmatically driven volcanic landslide and associated directed blast and Plinian eruption (Gorshkov, 1959; Bogoyavlenskaya *et al.*, 1985; Belousov and Bogoyavlenskaya, 1988). The directed blast and subsequent Plinian eruption occurred after about 0.5 km^3 of the southeast flank of the volcano failed catastrophically. Prior to the Plinian eruption, a period of unrest of about 5 months occurred that included numerous vulcanian explosions, and a lava dome began growing in the summit crater in late-November 1955 (Gorshkov, 1959; Gorshkov and Bogoyavlenskaya, 1965). Dome growth was accompanied by as much as 100 m of deformation of the southeast flank, which was likely caused by the emplacement of a cryptodome within the edifice (Gorshkov, 1959; Belousov, 1996). According to Gorshkov (1959), a strong earthquake occurred at "the moment of the eruption," which leaves open the possibility that an earthquake may have triggered the flank collapse or, as suggested by Kanamori *et al.* (1984) in their discussion of the 1980 Mount St. Helens eruption, that the earthquake recorded the first motion of the volcanic landslide.

Although there are other examples of volcanic landslides known or suspected to have been caused by magmatic intrusions (Francis *et al.*, 1985; Siebert *et al.*, 1987; Iverson, 1995; Melekestsev, 2006), the above examples indicate some unresolved questions about the linkage between eruptive activity, flank collapse, and landslide formation. It is possible that large volcanic landslides at these volcanoes would eventually have been initiated by the direct effects of the eruption, by explosion

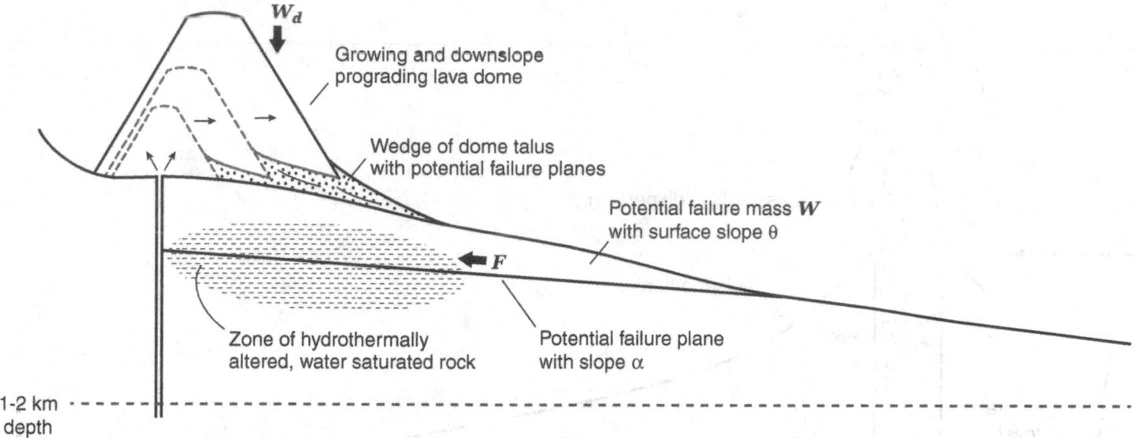

Fig. 4.3. Schematic representation of volcano flank instability developing in response to a growing and downslope-advancing lava dome produced during an effusive dome-building eruption. An unstable shearing lobe may develop at the base of the dome as the mass (W_d) increases and advances downslope over its own talus apron. The volcano is brought even closer to a condition of failure if shallow hydrothermally altered rocks are present beneath the advancing lava dome. This type of situation developed during the 1997 sector collapse and volcanic landslide at Soufriere Hills volcano, Montserrat (Voight *et al.*, 2002).

or pressure forces associated with rapidly ascending and degassing magma reaching the surface, or possibly the collective effects of magmatically related reduction of edifice strength (Day, 1996). It is also possible that the eruptions would have occurred anyway had no large landslide developed, but the eruption magnitude would likely have been less and the overall scope of the event much less spectacular.

4.2.4 VOLCANIC LANDSLIDES ASSOCIATED WITH MAGMA EXTRUSION

Lava flows and domes that develop during eruptive activity can contribute significant mass loads to the flanks of a volcano, which may result in flank failure and landsliding (Le Friant *et al.*, 2003). Exogenous lateral spreading of dome material over a structurally weak substrate, such as talus or hydrothermally altered rock, could bring a growing lava dome to the threshold of failure or cause deeper-seated failure of an already weakened volcano flank. The pertinent forces and processes are illustrated in Figure 4.3.

The December 26, 1997 sector collapse at Soufriere Hills volcano on the Caribbean island of Montserrat is the most recent example of a flank collapse and volcanic landslide associated with external loading of a volcano flank by a growing lava dome (Voight *et al.*, 2002). After more than a year of andesitic dome growth in the summit crater of the volcano, the dome and part of the underlying hydrothermally altered substrate known as Galway's Soufriere collapsed and formed a 5×10^7 m^3 debris avalanche. The sector collapse brought about the explosive decompression of the dome core, resulting in a large pyroclastic flow on the south flank of the volcano (Sparks *et al.*, 2002). Key factors in the flank collapse were a prolonged period of dome growth and related production of a significant volume of talus that was shed off the front of the advancing dome. The thick talus apron was deposited on unstable hydrothermally altered rock. A period of rapid dome growth and exogenous emplacement of a laterally spreading lobe of the lava dome, and localized gravity-driven slip within the shearing, strain-weakened talus and underlying hydrothermally altered rock, triggered the collapse itself (Voight *et al.*, 2002).

Some volcanoes with single conduits that produce summit lava domes may be prone to failure and landslides due to the build-up of unstable dome assemblages. Mount St. Augustine in Alaska is an example of such a volcano, where repeated emplacement of lava domes followed by debris avalanches has been documented (Beget and Kienle, 1992; Siebert *et al.*, 1995). At Mount St. Augustine, dome emplacement during an eruption results in the addition of mass to the summit area that may become unstable during the next eruptive cycle and episode of dome building, as apparently occurred in 1883 (Siebert *et al.*, 1995). A *ca.* 0.3 km^3 debris avalanche occurred during a dome-forming eruption in 1883; since then, lava domes emplaced during eruptions in 1935, 1964, 1986, and 2006 have produced a nested dome cluster within, and now completely obscuring, the original debris-avalanche scar. Although other factors are likely important in promoting flank instability, it is clear that the addition of dome material to the summit area of Mount St. Augustine has increased the likelihood of a volcanic landslide during a future eruptive cycle.

4.2.5 VOLCANIC LANDSLIDES ASSOCIATED WITH EXPLOSIVE CONDUIT PROCESSES

Ascent of volatile-rich magma within a conduit system results in the growth of bubbles within the magma. As the magma continues to rise, the bubbles nucleate and the magma begins to fragment, transforming the bubbly liquid magma into a gas–solid mixture. During the fragmentation process, the potential

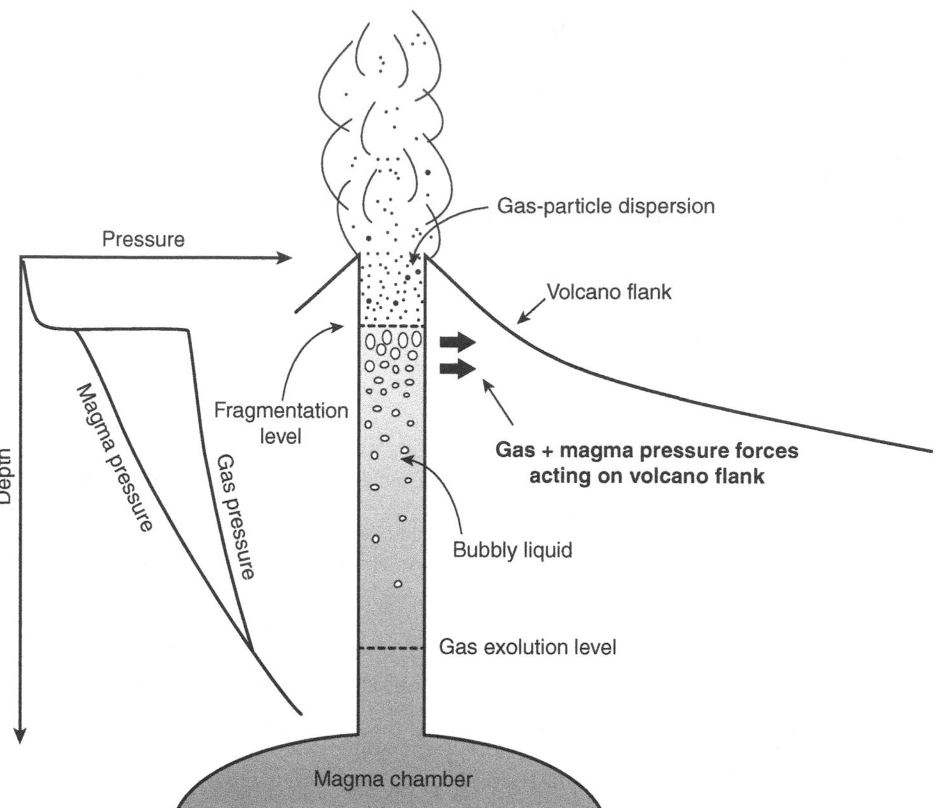

Fig. 4.4. Schematic diagram of volcano flank and simple conduit system during the ascent of viscous gas-rich magma, showing generalized pressure–depth relations for gas and magma. Gas pressure above the fragmentation level is essentially nil. It is plausible that the combined gas and magma pressure forces exerted on the conduit walls below the fragmentation level may be sufficient to initiate lateral displacement of the volcano flank.

energy of the magma is converted to kinetic energy of juvenile particles that constitutes the explosive force of the eruption. Highly viscous magmas, such as most of those of dacitic and rhyolitic composition, offer resistance to bubble growth and can result in high gas overpressures (McBirney, 1973; Sparks, 1978; Navon *et al.*, 1998; Lensky *et al.*, 2001). Because the permeability of viscous magma is generally low, gas overpressures are not able to decrease rapidly enough to inhibit fragmentation, and a large pressure gradient develops across the fragmentation surface (Fig. 4.4). High gas overpressures that develop at or just below the fragmentation level may exceed the local lithostatic pressure and can be greater than the compressive strength of rock, which could induce fracturing in the conduit wall rocks. However, because the pressure gradient in the conduit is primarily vertical, the primary direction of mass flux is also vertical, as is the dominant explosive force vector (Gonnermann and Manga, 2003). If the fragmentation level is shallow (2–3 km) and the eruption mixture flux is high (>10^6 kg s^{-1}), the forces generated during explosive rhyolitic or dacitic eruptions may be at least temporarily large enough to initiate flank collapse (Melnik *et al.*, 2005).

The eruption mixture imparts a shear stress along the conduit walls, although the stress is difficult to quantify (Macedonio *et al.*, 1994). In general, when the pressure of the eruption mixture exceeds the local lithostatic pressure, conduit wall rocks may break. The fractured rocks can then be entrained by the eruption mass, thereby eroding the conduit walls. Highly viscous deforming magmas are most effective at producing the initial fracture network in the conduit wall rocks (Goto, 1999; Tuffen *et al.*, 2003); the high shear stresses develop because of the pressure gradient in the conduit (Tuffen *et al.*, 2002; Gonnerman and Manga, 2003). However, generation of fractures by this process is likely to result in only small amounts of energy release, and the moment magnitude of the seismic signal produced by fracture growth is likely to be <1.0 (Gil Cruz and Chouet, 1997; Miller *et al.*, 1998; Tuffen *et al.*, 2003). The amount of energy released is insufficient for edifice displacement and triggering of volcanic landslides.

Crystal-rich magmas may form viscous caps or plugs in the upper part of the conduit that can retard the upward flow of magma, and they may contain gases exsolved from the magma or water vapor generated by magma interaction with groundwater. Cap formation may lead to the build-up of high pressures at shallow levels within the conduit and cause noteworthy edifice inflation (Diller *et al.*, 2006). Brittle failure of the cap results in discrete, violent explosions, seconds to minutes in duration, known as "vulcanian explosions" (Morrissey and Mastin, 2000). Vulcanian eruptions commonly generate atmospheric shock waves and far-traveled ballistic ejecta, as much as 5 km from source, indicating that conduit pressures can be significantly elevated. Seismic data from the 1998–1999 eruption of Popocatépetl volcano indicate that vulcanian explosions generated forces of 10^9–10^{10} N at a few kilometers depth (Arciniega-Ceballos *et al.*, 1999; Zobin and Martinez, 2010). Although these explosions did not initiate slope failure, it is plausible that at other volcanoes, where the edifice is weaker, such explosions could result in flank collapse.

Collapse calderas that form during major explosive eruptions may initiate large-scale slope failures as a result of caldera subsidence and foundering of the caldera walls. Extensive sequences of landslide debris within calderas that formed by this process are known at many of the large calderas in the western USA (Lipman, 1976).

4.3 DISCUSSION

Volcanic landslides related to unrest and eruptive activity are now a well-known, widely recognized phenomenon at many volcanoes. A review of the literature indicates that at least 200 stratovolcanoes worldwide have experienced volcanic landslides during large-scale flank collapse (Siebert *et al.*, 1987; Siebert, 1996). A review of about 80 case studies where the triggering mechanism of the volcanic landslide was addressed (see Ancillary Materials associated with this volume, www.cambridge.org/9781107002067) indicates that phreatic, phreatomagmatic, or magmatic explosions were temporally associated with about 20 percent of the cases. Magmatic intrusions followed by large eruptions were described in about 15 percent of these examples, and about 8 percent of the cases reviewed were landslide events initiated by mass loading during extrusion of lava domes. However, over half of the cases examined gave no clear explanation or unequivocal evidence linking eruptive activity to landslide formation. This high percentage is, in part, due to the age of some of the landslide deposits. About 80 percent of the volcanic landslides reviewed are more than 300 years old, and many of them are thousands of years old. Because of the antiquity of many volcanic landslide deposits, key field evidence linking eruptive activity to landslide formation has been removed or altered by erosion, or has been covered by the accumulation of post-landslide volcanic deposits. Furthermore, questions about eruptive processes that trigger volcanic landslides are not always a focus of field investigations, and important evidence or relations may not be reported. Correlation of volcanic landslides to magmatically driven eruptive processes (Bezymianny-type events) commonly involves the identification of juvenile material that is closely associated temporally with the landslide deposit. However, in some cases it is not possible to determine whether the volcanic products within the landslide deposit were erupted at the time of the landslide, making it difficult to conclude that a magmatic process triggered the landslide.

It is sometimes possible to document temporal equivalence among volcanic products that can link an eruption with landslide formation. For example, it may be possible to show that tephra or pyroclastic-flow deposits preserved on other sectors of a volcano were erupted at the time of landsliding. Vallance and Scott (1997), for example, showed that the tephrostratigraphy of tephra layer F on Mount Rainier in Washington State was related to shallow magmatic intrusion and explosive activity that likely contributed to flank instability and ultimately the formation of a large volcanic landslide about 5600 years BP, known as the Osceola Mudflow.

Intrusion of magma, and specifically dike emplacement, is commonly invoked as a triggering mechanism for flank collapse (Elsworth and Voight, 1995; Delaney *et al.*, 1998; Delaney and Denlinger, 1999; Elsworth and Day, 1999). However, there are numerous examples of dike emplacement that did not lead to flank collapse and landsliding where such an outcome might have been expected (Roman *et al.*, 2004). Dike emplacement may cause physical displacement of the flank and if the forces resisting displacement are exceeded the volcano flank may approach a condition of failure or may fail outright. Dike emplacement may also lead to increases in pore-fluid pressure within the groundwater system of a volcanic edifice (Reid, 2004), which may promote flank instability and lead to collapse (Day, 1996). Shallow dike emplacement that results in flank collapse could cause an explosive eruption if the dike system were truncated by the collapse scar, causing rapid decompression (Alidibirov and Dingwell, 1996). Dike emplacement is likely to be accompanied by strong seismicity that may also promote flank collapse (Elsworth and Voight, 1995). However, shallow intrusions (dikes and sills) and deep magmatic heat sources can invigorate the hydrothermal system and instigate hydrothermally driven flank collapse (Reid, 2004) without a triggering magmatic eruption.

Volcanic landslides of the Bandai type that are accompanied by phreatic explosions are recognized by the absence of juvenile material in the landslide deposits. Such documentation requires a thorough field investigation of the landslide deposit, as well as knowledge of the volcanic products generated during eruptions and their preservation potential. In many of the case studies of Bandai-type events, well-developed hydrothermal systems and extensive hydrothermal alteration of the edifice have been identified and likely were important factors in bringing the volcanoes closer to failure.

Destabilizing factors such as emplacement of magma, the loading or oversteepening of flanking slopes, and large but rare volcanic explosions may act independently. Such factors may also be linked in a cascading sequence that ultimately results in partial destruction of the volcanic edifice. A flank collapse may cause a sudden decrease of magmatic pressure within the chamber conduit system. The rapid loss of confining pressure resulting from edifice collapse allows rapid bubble growth and magma fragmentation, which nearly always leads to an explosive eruption. Less clear, however, is how (and if) explosive eruptive processes, such as those associated with phreatic, phreatomagmatic, or magmatic eruptions, trigger landslides directly. It appears that, under certain conditions, the forces generated by these explosions may be sufficient to bring about edifice collapse and landslide formation, but historic examples are rare and not well documented.

Large volcanic landslides, regardless of their ultimate trigger mechanism, pose significant hazards to all areas on the flanks of affected volcanoes, and valleys and drainages downslope. It is difficult to know whether a period of volcanic unrest will result in a major volcanic landslide, but there are several factors that make it more likely than not. The rock mass strength of the volcanic edifice and its ability to maintain structural integrity during explosive eruptions is a primary concern and requires thorough investigation. At present, it is very difficult to assess the internal architecture of a volcanic edifice with the degree of certainty necessary to evaluate its stability under various plausible eruptive scenarios. However, geophysical techniques, notably airborne electromagnetic and magnetic methods for assessing the extent of hydrothermal alteration and the distribution of water within a volcanic edifice, are being developed (Finn *et al.*, 2007). The location, degree of alteration, and mineralogy of hydrothermally altered rocks within a volcano play a key role in the stability of the edifice (Lopez and Williams, 1993) and merit careful analysis. Studies of the shear strength of failure surfaces at volcanoes that have experienced historic edifice failures are another possible means of evaluating edifice rock mass strength. Reid *et al.* (2010) presented shear strength test data from rocks sampled along the inferred failure surface of the 1980 Mount St. Helens landslide. They found that reduction in rock mass strength results from progressive shearing. Such strength reduction of older dome rocks within the edifice likely occurred during cryptodome intrusion and gravitational loading. They suggest that slow, but continuous, edifice displacement could be detected through comprehensive deformation monitoring combined with slope stability analysis.

Improved understanding of the three-dimensional structure of volcanoes, better characterization of the rock mass strength of volcanic edifices using high-resolution spatial data such as LIDAR, and high-resolution GPS or InSAR monitoring of flank deformation offer the best new techniques for assessing the likelihood of flank failures and volcanic landslides associated with eruptive activity. In some cases, the difficulties associated with collection of high-resolution field data may be overcome by using laboratory-scale analog models (e.g., Andrade and van Wyk de Vries, 2010), and controlled experiments can be useful in exploring volcanic landslide evolution.

ACKNOWLEDGEMENTS

The author thanks the volume editors for the opportunity to investigate the topic of this chapter. Review comments by B. van Wyk de Vries, M. E. Reid, D. Stead, and J. J. Clague were thorough and helpful, and improved the clarity of several of the concepts discussed.

REFERENCES

Aguila, L., Newhall, C. G., Miller, C. D. and Listanco, E. (1986). Reconnaissance geology of a large debris avalanche from Iriga volcano, Philippines. *Philippine Journal of Volcanology*, 3, 54–72.

Alidibirov, M. and Dingwell, D. B. (1996). Magma fragmentation by rapid decompression. *Nature*, 380, 146–148.

Andrade, S. D. and van Wyk de Vries, B. (2010). Structural analysis of the early stages of catastrophic stratovolcano flank-collapse using analogue models. *Bulletin of Volcanology*, 72, 771–789.

Arciniega-Ceballos, A., Chouet, B. A. and Dawson, P. (1999). Very long-period signals associated with vulcanian explosions at Popocatépetl volcano, México. *Geophysical Research Letters*, 26, 3013–3016.

Barberi, F., Bertagnini, A., Landi, P. and Principe, C. (1992). A review on phreatic eruptions and their precursors. *Journal of Volcanology and Geothermal Research*, 52, 231–246.

Beget, J. E. and Kienle, J. (1992). Cyclic formation of debris avalanches at Mount St Augustine volcano. *Nature*, 356, 701–704.

Belousov, A. B. (1995). The Shiveluch volcanic eruption of 12 November 1964: Explosive eruption provoked by failure of the edifice. *Journal of Volcanology and Geothermal Research*, 66, 357–365.

(1996). Pyroclastic deposits of March 30 1956 directed blast at Bezymianny volcano. *Bulletin of Volcanology*, 57, 649–662.

Belousov, A. B. and Bogoyavlenskaya, G. E. (1988). Debris avalanche of the 1956 Bezymianny eruption. In *Proceedings of the Kagoshima International Conference on Volcanoes*, pp. 460–462.

Belousov, A., Belousova, M. and Voight, B. (1999). Multiple edifice failures, debris avalanches and associated eruptions in the Holocene history of Shiveluch volcano, Kamchatka, Russia. *Bulletin of Volcanology*, 61, 324–342.

Belousov, A., Voight, B. and Belousova, M. (2007). Directed blasts and blast-currents: A comparison of the Bezymianny 1956, Mount St Helens 1980, and Soufriere Hills, Montserrat 1997 eruptions and deposits. *Bulletin of Volcanology*, 69, 701–740.

Bogoyavlenskaya, G. E., Braitseva, O. A., Melekestsev, I. V., Kiriyanov, V. Y. and Miller, C. D. (1985). Catastrophic eruptions of the directed-blast type at Mount St. Helens, Bezymianny and Shiveluch volcanoes. *Journal of Geodynamics*, 3, 189–218.

Cooke, R. J. S. (1981). Eruptive history of the volcano at Ritter Island. In *Cooke-Ravian Volume of Volcanological Papers*, ed. R. W. Johnson. Geological Survey of Papua New Guinea, Memoir 10, pp. 115–123.

Crandell, D. R. (1989). *Gigantic Debris-Avalanche of Pleistocene Age from Ancestral Mount Shasta Volcano, California, and Debris Avalanche Hazard Zonation.* US Geological Survey, Bulletin 1861.

Day, S. J. (1996). Hydrothermal pore fluid pressure and the stability of porous, permeable volcanoes. In *Volcano Instability on the Earth and Other Planets*, ed. W. J. McGuire, A. P. Jones and J. Neuberg. Geological Society Special Publication 110, pp. 77–93.

Delaney, P. T. and Denlinger, R. P. (1999). Stabilization of volcanic flanks by dike intrusion: An example from Kilauea. *Bulletin of Volcanology*, 61, 356–362.

Delaney, P. T., Denlinger, R. P., Lisowski, M. *et al.* (1998). Volcanic spreading at Kilauea. *Journal of Geophysical Research*, 103, 18003–18023.

Diller, K., Clarke, A. B., Voight, B. and Neri, A. (2006). Mechanisms of conduit plug formation: Implications for vulcanian explosions. *Geophysical Research Letters*, 33, L20302.

Donnadieu, F., Merle, O. and Besson, J. (2001). Volcano edifice instability during cryptodome intrusion. *Bulletin of Volcanology*, 63, 61–72.

Dzurisin, D., Yamashita, K. M. and Kleinman, J. W. (1994). Mechanisms of crustal uplift and subsidence at the Yellowstone caldera, Wyoming. *Bulletin of Volcanology*, 56, 261–270.

Elsworth, D. and Day, S. J. (1999). Flank collapse triggered by intrusion: The Canarian and Cape Verde archipelagoes. *Journal of Geophysical Research*, 94, 323–340.

Elsworth, D. and Voight, B. (1995). Dike intrusion as a trigger for large earthquakes and the failure of volcano flanks. *Journal of Geophysical Research*, 100, 6005–6024.

Finn, C.A., Deszcz-Pan, M., Anderson, E.D. and John, D.A. (2007). Three-dimensional geophysical mapping of rock alteration and water content at Mount Adams, Washington: Implications for lahar hazards. *Journal of Geophysical Research*, 112, B10204, doi:10.1029/2006JB004783.

Francis, P.W. and Self, S. (1987). Collapsing volcanoes. *Scientific American*, 255, 90–97.

Francis, P.W., Gardeweg, M., Ramirez, C.F. and Rothery, D.A. (1985). Catastrophic debris avalanche deposit of Socompa volcano, northern Chile. *Geology*, 13, 600–603.

Furuya, M. (2004). Localized deformation at Miyakejima volcano based on JERS-1 radar interferometry: 1992–1998. *Geophysical Research Letters*, 31, L05605.

Gil Cruz, F. and Chouet, B.A. (1997). Long-period events, the most characteristic seismicity accompanying the emplacement and extrusion of a lava dome in Galeras volcano, Colombia, in 1991. *Journal of Volcanology and Geothermal Research*, 77, 121–158.

Glicken, H. (1986). *Rockslide-debris avalanche of May 18, 1980: Mount St. Helens Volcano, Washington*. Ph.D. thesis, University of California, Santa Barbara, CA.

Gonnerman, H.M. and Manga, M. (2003). Explosive volcanism may not be an inevitable consequence of magma fragmentation. *Nature*, 426, 432–435.

Gorshkov, G.S. (1959). Gigantic eruption of the volcano Bezymianny. *Bulletin of Volcanology*, 20, 77–109.

(1962). On the classification and terminology of Pelée and Katmai type eruptions. *Bulletin of Volcanology*, 24, 155–165.

Gorshkov, G.S. and Bogoyavlenskaya, G.E. (1965). *Bezymianny Volcano and the Peculiarities of Its Last Eruption*. Moscow: Nauka (in Russian).

Goto, A. (1999). A new model for volcanic earthquake at Unzen volcano: Melt rupture model. *Geophysical Research Letters*, 26, 2541–2544.

Hadisantono, R.D. (2006). Devastating landslides related to the 2002 Papandayan eruption. *Jurnal Geologi Indonesia*, 1, 83–88.

Hase, H., Hashimoto, T., Sakanaka, S., Kanda, W. and Tanaka, Y. (2005). Hydrothermal system beneath Aso volcano as inferred from self-potential mapping and resistivity structure. *Journal of Volcanology and Geothermal Research*, 142, 259–277.

Holcomb, R.T. and Searle, R.C. (1991). Large landslides from oceanic volcanoes. *Marine Geotechnology*, 10, 19–32.

Iverson, R.M. (1995). Can magma injection and groundwater forces cause massive landslides on Hawaiian volcanoes? *Journal of Volcanology and Geothermal Research*, 66, 295–308.

Johnson, R.W. (1987). Large-scale volcanic cone collapse: The 1888 slope failure of Ritter volcano, and other examples from Papua New Guinea. *Bulletin of Volcanology*, 49, 669–679.

Jones, A.P., McGuire, W.J. and Neuberg, J. (ed.) (1996). *Volcano Stability on the Earth and Other Planets*. Geological Society Special Publication 110.

Kanamori, H., Given, J.W. and Lay, T. (1984). Analysis of seismic body waves excited by the Mount St. Helens eruption of May 18, 1980. *Journal of Geophysical Research*, 89, 1856–1866.

Keating, B.H. and McGuire, W.J. (2000). Island edifice failures and associated tsunami hazards. *Pure and Applied Geophysics*, 157, 899–955.

Lagmay, A.M.F., van Wyk de Vries, B., Kerle, N. and Pyle, D.M. (2000). Volcano instability induced by strike-slip faulting. *Bulletin of Volcanology*, 62, 331–346.

Lavigne, F., Hadisantono, R., Surmayadi, M., Flohic, F. and Geyer, F. (2005). The November 2002 eruption of the Papandayan volcano (Indonesia): Direct and induced hazards, with emphasis on lahars. *Zeitschrift für Geomorphologie (Supplementbande)*, 140, 151–165.

Le Friant, A., Boudon, G., Deplus, C. and Villemant, B. (2003). Large-scale flank collapse events during the activity of Montagne Pelée, Martinique, Lesser Antilles. *Journal of Geophysical Research*, 108, doi:10.1029/2001JB001624.

Lensky, N.G., Lyakhovsky, V. and Navon, O. (2001). Radial variations of melt viscosity around growing bubbles and gas overpressure in vesiculating magmas. *Earth and Planetary Science Letters*, 186, 1–6.

Lipman, P.W. (1976). Caldera-collapse breccias in the western San Juan Mountains, Colorado. *Geological Society of America Bulletin*, 87, 1397–1410.

Lipman, P.W. and Mullineaux, D.R. (1981). *The 1980 Eruptions of Mt. St. Helens, Washington*. US Geological Survey, Professional Paper 1250.

Lipman, P.W., Moore, J.G. and Swanson, D.A. (1981). Bulging of the north flank before the May 18 eruption: Geodetic data. In *The 1980 Eruptions of Mt. St. Helens, Washington*, ed. P.W. Lipman and D.R. Mullineaux. US Geological Survey, Professional Paper 1250, pp. 143–146.

Lipman, P.W., Normark, W.R., Moore, J.G., Wilson, J.B. and Gutmacher, C.E. (1988). The giant Alika debris slide, Mauna Loa, Hawaii. *Journal of Geophysical Research*, 93, 4279–4299.

Lopez, D.L. and Williams, S.T. (1993). Catastrophic volcanic collapse: Relation to hydrothermal processes. *Science*, 260, 1794–1796.

Macedonio, G., Dobran, F. and Neri, A. (1994). Erosion processes in volcanic conduits and an application to the AD 79 eruption of Vesuvius. *Earth and Planetary Science Letters*, 121, 137–152.

Mastin, L.G. (1991). The roles of magma and groundwater in the phreatic eruptions at Inyo Craters, Long Valley Caldera, California. *Bulletin of Volcanology*, 53, 579–596.

(1995). Thermodynamics of gas and steam-blast eruptions. *Bulletin of Volcanology*, 57, 85–98.

Mazot, A., Bernard, A., Fischer, T., Inguaggiato, S. and Sutawidjaja, I.S. (2008). Chemical evolution of thermal waters and changes in the hydrothermal system of Papandayan volcano (West Java, Indonesia) after the November 2002 eruption. *Journal of Volcanology and Geothermal Research*, 178, 276–286.

McBirney, A.R. (1973). Factors governing the intensity of explosive andesitic eruptions. *Bulletin Volcanologique*, 37, 443–453.

McGuire, W.J. (1996). Volcano instability: A review of contemporary themes. In *Volcano Instability on the Earth and Other Planets*, ed. W.J. McGuire, A.P. Jones and J. Neuberg. Geological Society Special Publication 110, pp. 1–23.

McNutt, S.R. (1996). Seismic monitoring and eruption forecasting of volcanoes: A review of the state-of-the-art and case histories. In *Monitoring and Mitigation of Volcanic Hazards*, ed. R. Scarpa and R.I. Tilling. Berlin: Springer, pp. 99–146.

Melekestev, I.V. (2006). Large modern collapses on the active volcanoes of Kamchatka: Causes and mechanisms of formation. In *Landslides from Massive Rock Slope Failure*, ed. S.G. Evans, G. Scarascia Mugnozza, A. Strom and R.L. Hermanns. Amsterdam: Springer, pp. 431–444.

Melnik, O., Barmin, A.A. and Sparks, R.S.J. (2005). Dynamics of magma flow inside volcanic conduits with bubble overpressure buildup and gas loss through permeable magma. *Journal of Volcanology and Geothermal Research*, 143, 53–68.

Miller, A.D., Stewart, R.C., White, R.A. *et al.* (1998). Seismicity associated with dome growth and collapse at the Soufriere Hills volcano, Montserrat. *Geophysical Research Letters*, 25, 3401–3404.

Montalto, A. (1995). Seismic assessment of phreatic-explosion hazard at "La Fossa" volcano (Island of Vulcano, Italy). *Natural Hazards*, 11, 57–73.

Moore, J. G., Clague, D. A., Holcomb, R. T. *et al.* (1989). Prodigious submarine landslides on the Hawaiian ridge. *Journal of Geophysical Research*, 94, 17465–17484.

Moriya, I. (1980). Bandaian eruption and landforms associated with it. In *Collection of Articles in Memory of Retirement of Prof. K. Nishimura, Tohoku University*, pp. 214–219 (in Japanese with English abstract).

Morrissey, M. M. and Mastin, L. G. (2000). Vulcanian eruptions. In *Encyclopedia of Volcanoes*, ed. H. Sigurdsson, B. F. Houghton, S. R. McNutt, H. Rymer and J. Stix. San Diego, CA: Academic Press, pp. 463–475.

Nairn, I. A. and Wiradiradja, S. (1980). Late Quaternary hydrothermal explosion breccias at Kawerau Geothermal Field, New Zealand. *Bulletin of Volcanology*, 43, 1–13.

Nakamura, Y. (1978). Geology and petrology of Bandai and Nekoma volcanoes. *Tokyo University Scientific Report*, 14, 67–119.

Navon, O., Chekhmir, A. and Lyakhovsky, V. (1998). Bubble growth in highly viscous melts: Theory, experiments, and autoexplosivity of dome lavas. *Earth and Planetary Science Letters*, 160, 763–776.

Neuman van Padang, M. (1951). *Indonesia: Catalog of the Active Volcanoes of the World, Volume 1*. Rome: International Association of Volcanology and Chemistry of the Earth's Interior (IAVCEI).

Newhall, C. G. and Self, S. (1982). The volcanic explosivity index (VEI): An estimate of explosive magnitude for historical volcanism. *Journal of Geophysical Research*, 87, 1231–1238.

Okada, H. (1983). Comparative study of earthquake swarms associated with major volcanic activities. In *Arc Volcanism: Physics and Tectonics*, ed. D. Shimozuzuru and I. Yokoyama. Tokyo: TERRAPUB, pp. 43–61.

Pyle, D. M. (1995). Mass and energy budgets of explosive volcanic eruptions. *Geophysical Research Letters*, 5, 563–566.

Reid, M. E. (2004). Massive collapse of volcano edifices triggered by hydrothermal pressurization. *Geology*, 32, 373–376.

Reid, M. E., Christian, S. B. and Bryan, D. L. (2000). Gravitational stability of three-dimensional stratovolcano edifices. *Journal of Geophysical Research*, 105, 6043–6056.

Reid, M. E., Keith, T. E. C., Kayen, R. E. *et al.* (2010). Volcano collapse promoted by progressive strength reduction: New data from Mount St. Helens. *Bulletin of Volcanology*, 72, 761–766.

Roman, D. C., Power, J. A., Moran, S. C. *et al.* (2004). Evidence for dike emplacement beneath Iliamna volcano, Alaska in 1996. *Journal of Volcanology and Geothermal Research*, 130, 265–284.

Sato, H. and Taniguchi, H. (1997). Relationship between crater size and ejecta volume of recent magmatic and phreato-magmatic eruptions: Implications for energy partitioning. *Geophysical Research Letters*, 24, 205–208.

Savin, C., Grasso, J. R. and Bachelery, P. (2005). Seismic signature of a phreatic explosion: Hydrofracturing damage at Karthala volcano, Grande Comore Island, Indian Ocean. *Bulletin of Volcanology*, 67, 717–731.

Sekiya, S. and Kikuchi, Y. (1889). The eruption of Bandai-san. *Journal of the College of Science, Imperial University Japan*, 3, 91–172.

Siebert, L. (1984). Large volcanic debris avalanches: Characteristics of the source areas, deposits and associated eruptions. *Journal of Volcanology and Geothermal Research*, 22, 163–187.

(1996). Hazards of very large volcanic debris avalanches and associated eruptive phenomena. In *Monitoring and Mitigation of Volcano Hazards*, ed. R. Scarpa and R. I. Tilling. New York: Springer-Verlag, pp. 541–572.

(2002). Landslides resulting from structural failure of volcanoes. In *Catastrophic Landslides: Effects, Occurrence and Mechanisms*, ed. S. G. Evans and J. V. DeGraff. Geological Society of America, Reviews in Engineering Geology 15, pp. 209–235.

Siebert, L. and Simkin, T. (2002). *Volcanoes of the World: An Illustrated Catalog of Holocene Volcanoes and Their Eruptions*. Washington, DC: Smithsonian Institution, Global Volcanism Program Digital Information Series, GVP-3.

Siebert, L., Glicken, H. and Ui, T. (1987). Volcanic hazards from Bezymianny- and Bandai-type eruptions. *Bulletin of Volcanology*, 49, 435–459.

Siebert, L., Beget, J. E. and Glicken, H. (1995). The 1883 and late-prehistoric eruptions of Augustine volcano, Alaska. *Journal of Volcanology and Geothermal Research*, 66, 367–395.

Silver, E., Day, S., Ward, S. *et al.* (2009). Volcano collapse and tsunami generation in the Bismarck Volcanic Arc, Papua New Guinea. *Journal of Volcanology and Geothermal Research*, 186, 210–222.

Sparks, R. S. J. (1978). The dynamics of bubble formation and growth in magmas: A review and analysis. *Journal of Volcanology and Geothermal Research*, 3, 1–37.

Sparks, R. S. J., Barclay, J., Calder, E. S. *et al.* (2002). Generation of a debris avalanche and violent pyroclastic density current on 26 December (Boxing Day) 1997 at Soufriere Hills volcano, Montserrat. In *The Eruption of the Soufriere Hills Volcano, Montserrat, from 1995 to 1999*, ed. T. H. Druitt and B. P. Kokelaar. Geological Society Memoir 21, pp. 409–434.

Stix, J., Zapata, G. J. A., Calvache, V. M. *et al.* (1993). A model of degassing at Galeras volcano, Colombia, 1988–1993. *Geology*, 21, 963–967.

Sudo, Y., Tsutsui, T., Ono, H. *et al.* (1998). Seismic activity and ground deformation associated with the 1995 phreatic eruption of Kuju volcano, Kyushu, Japan. *Journal of Volcanology and Geothermal Research*, 81, 245–267.

Swanson, D. A., Duffield, W. A. and Fiske, R. S. (1976). *Displacement of the South Flank of Kilauea Volcano: The Result of Forceful Intrusion of Magma into the Rift Zones*. US Geological Survey, Professional Paper 963.

Tuffen, H., McGarvie, D. W., Gilbert, J. S. and Pinkerton, H. (2002). Physical volcanology of a subglacial-to-emergent rhyolitic tuya at Rauðufossafjöll, Torfajökull, Iceland. *Journal of the Geological Society of London, Special Publication*, 202, 213–236.

Tuffen, H., Dingwell, D. B. and Pinkerton, H. (2003). Repeated fracture and healing of silicic magma generate flow banding and earthquakes? *Geology*, 31, 1089–1092.

Ui, T. (1983). Volcanic dry avalanche deposits: Identification and comparison with non-volcanic debris stream deposits. *Journal of Volcanology and Geothermal Research*, 18, 135–150.

Vallance, J. W. and Scott, K. M. (1997). The Osceola mudflow from Mount Rainier: Sedimentology and hazard implications of a huge clay-rich debris flow. *Geological Society of America Bulletin*, 109, 143–163.

van Wyk de Vries, B. and Borgia, A. (1996). The role of basement in volcano deformation. In *Volcano Instability on the Earth and Other Planets*, ed. W. J. McGuire and A. P. Jones. Geological Society of London, Special Publication 110, pp. 95–110

Ventura, G., Vilardo, G. and Bruno, P. P. (1999). The role of flank failure in modifying the shallow plumbing system of volcanoes: An example from Somma-Vesuvius, Italy. *Geophysical Research Letters*, 26, 3681–3684.

Vidal, N. and Merle, O. (2000). Reactivation of basement faults beneath volcanoes: A new model of flank collapse. *Journal of Volcanology and Geothermal Research*, 99, 9–26.

Voight, B. (2000). Structural stability of andesite volcanoes and lava domes. *Philosophical Transactions of the Royal Society A: Mathematical, Physical and Engineering Sciences*, 358, 1663–1703.

Voight, B. and Elsworth, D. (1997). Failure of volcano slopes. *Geotechnique*, 47, 131.

(2000). Stability and collapse of hazardous gas-pressurized lava domes. *Geophysical Research Letters*, 48, 1–4.

Voight, B., Glicken, H., Janda, R.J. and Douglass, P.M. (1981). Catastrophic rockslide avalanche of May 18. In *The 1980 Eruption of Mount St. Helens, Washington*, ed. P.W. Lipman and D.R. Mullineaux. US Geological Survey, Professional Paper 1250, pp. 347–377.

Voight, B., Janda, R.J., Glicken, H.X. and Douglass, P.M. (1983). Nature and mechanics of the Mount St. Helens rockslide-avalanche of 18 May 1980. *Geotechnique*, 33, 243–273.

Voight, B., Komorowski, J.C., Norton, G.E. *et al.* (2002). The 26 December (Boxing Day) 1997 sector collapse and debris avalanche at Soufriere Hills volcano, Montserrat. In *The Eruption of the Soufriere Hills Volcano, Montserrat, from 1995 to 1999*, ed. T.H. Druitt and B.P. Kokelaar. Geological Society Memoir 21, pp. 363–407.

Wadge, G., Francis, P.W. and Ramirez, C.F. (1995). The Socompa collapse and avalanche event. *Journal of Volcanology and Geothermal Research*, 66, 309–336.

Ward, S.N. and Day, S.J. (2003). Ritter Island volcano: Lateral collapse and the tsunami of 1888. *Geophysical Journal International*, 154, 891–902.

Watters, R.J. and Delhaut, W.D. (1995). Effect of argillic alteration on rock mass stability. In *Clay and Shale Slope Instability*, ed. W.C. Haneberg and S.A. Anderson. Geological Society of America, Reviews in Engineering Geology 10, pp. 139–150.

Watters, R.J., Zimbelman, D.R., Bowman, S.D. and Crowley, J.K. (2000). Rock mass strength assessment and significance to edifice stability Mount Rainier and Mount Hood, Cascade Range volcanoes. *Pure and Applied Geophysics*, 157, 957–976.

Waythomas, C.F., Miller, T.P. and Beget, J.E. (2000). Record of Late Holocene debris avalanches and lahars at Iliamna volcano, Alaska. *Journal of Volcanology and Geothermal Research*, 104, 97–130.

Wohletz, K.H. (1986). Explosive magma–water interactions: Thermodynamics, explosion mechanisms, and field studies. *Bulletin of Volcanology*, 48, 245–264.

(2003). Water/magma interaction: Physical considerations for the deep submarine environment. In *Explosive Subaqueous Volcanism*, ed. J.D.L. White, J.L. Smellie and D. Clague. American Geophysical Monograph 140, pp. 25–40.

Wohletz, K. and Heiken, G. (1992). *Volcanology and Geothermal Energy*. Berkeley, CA: University of California Press.

Yamamoto, T., Nakamura, Y. and Glicken, H. (1999). Pyroclastic density current from the 1888 phreatic eruption of Bandai volcano, NE Japan. *Journal of Volcanology and Geothermal Research*, 90, 191–207.

Yokoo, A., Taniguchi, H., Goto, A. and Oshima, H. (2002). Energy and depth of Usu 2000 phreatic explosions. *Geophysical Research Letters*, 29, 2195.

Zimanowski, B. (1998). Phreatomagmatic explosions. In *From Magma to Tephra*, ed. A. Freundt and M. Rosi. New York: Elsevier, pp. 25–54.

Zobin, V.M. and Martinez, A. (2010). Quantification of the 1998–1999 explosion sequence at Popocatépetl volcano, Mexico. *Journal of Volcanology and Geothermal Research*, 194, 165–173.

5 Mobility of long-runout rock avalanches

TIM DAVIES AND MAURI MCSAVENEY

ABSTRACT

In this chapter we address the conundrum of the surprisingly long runout of large rock avalanches, which has been the subject of many investigations since it was first recognized in the nineteenth century in Switzerland. After describing the nature of the problem quantitatively, we briefly outline the many explanations that have been put forward to explain it; we also describe the wide variety of circumstances in which long runout is known to occur. We then examine the ability of the proposed explanations to apply to this range of circumstances, in order to identify those explanations that can work in all the environments in which long runout occurs. The process of dynamic rock fragmentation appears to be the only mechanism that satisfies this criterion. We outline this mechanism and its energetics in more detail, and summarize its recent success in quantitatively explaining the 40 km runout and the morphological characteristics of the 25 km^3 Socompa debris avalanche deposit in Chile.

5.1 INTRODUCTION

Ever since the well-reported nineteenth-century Elm disaster in Switzerland (Heim, 1882), it has been recognized that the deposits of large rock avalanches can extend much farther from the source, in proportion to their elevation loss, than smaller deposits: in simple terms, large rock avalanches have extraordinarily long runouts. Small granular deposits (10^{-4} m^3 to *ca.* 10^6 m^3) have horizontal runouts given by:

$$R_h/h^* \leq 4 \tag{5.1}$$

where R_h is the runout on a horizontal plane and h^* = (volume)$^{1/3}$ (note that SI units are used throughout). By contrast, larger deposits (>10^6 m^3) have:

$$6 \leq R_h/h^* \leq 10 \tag{5.2}$$

(Fig. 5.1; Davies and McSaveney, 1999).

Early investigators (Scheidegger, 1973; Hsü, 1975) expressed this phenomenon using the ratio H/L (Fig. 5.1; H is the elevation difference between the top of the source area and the distal tip of the deposit, and L the horizontal distance between the same two points). H/L was assumed to equal the internal friction coefficient of the rock avalanche material μ_i, so that for small rock avalanches $H/L = \tan \emptyset_i \approx 0.6$, where $\emptyset_i$ is the internal friction angle, usually about 30–35°. Larger events had $H/L < 0.6$; in an extreme case like the 25 km^3 Socompa deposit in Chile (Kelfoun and Druitt, 2005), $H/L = 0.05$. Many investigators have sought to identify the mechanism by which friction could be reduced in these large events.

The recognition that H/L increases with volume led to the identification of the so-called "size effect" (Scheidegger, 1973). Hsü (1975) proposed that the mobility of such an event could be expressed by the "excess travel distance" L_e:

$$L_e = L - H/\tan 32°. \tag{5.3}$$

This equation implicitly assumes that the fall height H is a controlling variable, which at first sight seems reasonable. However a number of workers (Hsü, 1975; Davies, 1982; Davies and McSaveney, 1999) have shown, using both field and laboratory data, that L does not vary strongly with H. Interestingly, Davies and McSaveney (1999) report a laboratory test in which 1 liter of fine dry sand was allowed to fall about 10 cm down a 35° slope onto a horizontal concrete surface; this experiment gave $H/L = 0.16$, equivalent to $\emptyset_i = 9°$. Because gravity was the only driving force and the friction coefficient of the sand (measured at 0.87; Davies and McSaveney, 1999) represented the only significant resisting force in this test, a small H/L value clearly does not necessarily indicate low friction.

Landslides: Types, Mechanisms and Modeling, ed. John J. Clague and Douglas Stead. Published by Cambridge University Press.

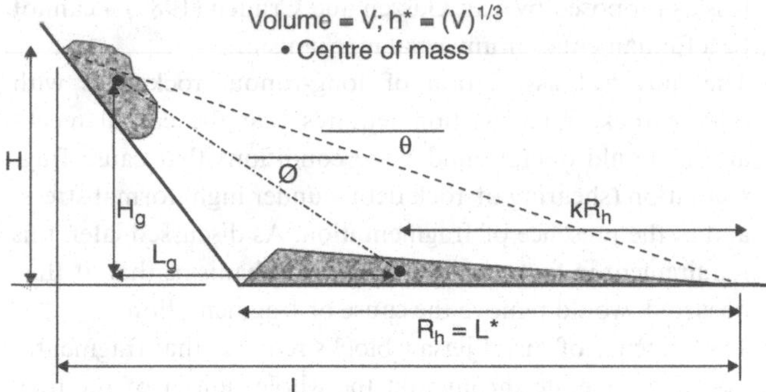

Fig. 5.1. Definition diagram for Eqs. 5.1–5.5.

More recent investigations (e.g., Legros, 2002) have pointed out that the frictional effect on a mass moving down a slope and running out across a less steep surface can be better represented by the reduction in elevation and horizontal translation of the center of mass H_g/L_g than by H/L (Fig. 5.1). Although more difficult to calculate for a natural deposit than H/L, modern surveying and analysis methods now allow this ratio to be determined routinely. The use of H_g/L_g instead of H/L marks a change in perception of the kinematics of the event; whereas H/L applies to a rigid sliding block, the use of H_g/L_g emphasizes that, as well as the center of mass falling and translating, the mass can also spread as it travels, and a realistic representation of the phenomenon needs to incorporate this spreading. Davies (1982) pointed out that the degree of spreading L^* required to match the runout of large field deposits is a well-defined ($R^2 = 0.92$) function of volume (V):

$$L^* = 10V^{1/3} \tag{5.4}$$

which corresponds to $R_h/h^* = 10$ in Eq. 5.1. The picture that emerges is that the total runout can be estimated by adding the horizontal translation of the center of mass to the longitudinal spreading past the center of mass, approximately:

$$L = H/\tan \varnothing_i + k(10V^{1/3}) \tag{5.5}$$

where k is the proportion of horizontal spreading that lies beyond the center of mass of the deposit. The likely imprecision of this estimate can be assessed by considering the spread of data from which Eq. 5.4 is derived; for $V < 10^8$ m^3, the scatter in L^* is about half an order of magnitude; it is much less at larger volumes, suggesting that local topography affects smaller events more than larger ones. The universality of this simple concept has been questioned; in some cases (e.g., the 1980 Mt. St. Helens debris avalanche; Glicken, 1996), the inclination of the line joining the initial and deposit centers of mass is far too low to represent conventional friction, so some other effects must be occurring (Legros, 2002).

While this geometric picture may be adequate for crude estimates of runout knowing H and V, it is far too simplistic for predicting the runout of a rock avalanche through complex three-dimensional terrain. Such a prediction requires a numerical model running over a digital elevation model, which in turn requires that the physical processes leading to Eq. 5.4 be understood. We now examine the mechanisms that have been suggested as causing the unusually large degree of spreading exhibited by falling masses of rock debris greater than about 1 million cubic meters in volume.

5.2 MECHANISMS OF LONG RUNOUT

Many physical explanations for long-runout rock avalanches have been proposed since the phenomenon was first recognized. These include: air-layer lubrication (Shreve, 1968); air-extrusion fluidization (Kent, 1966); fluidization by a dense interstitial dust dispersion (Hsü, 1978); generation of steam by frictional heating of pore water (Goguel, 1978); gaseous pore pressures (Habib, 1975); basal sliding on dissociated or melted rock (Erismann, 1979; Masch *et al.*, 1985; Legros, 2002; De Blasio and Elverhoi, 2008); low-density high-vibration basal layer development (Campbell, 1989); shear of a wet basal zone (Voight and Sousa, 1994); basal pressure wave propagation (Kobayashi, 1997); lubrication by undrained loading of saturated substrates (Abele, 1974; Sassa, 1988; Legros, 2002); mass changes (Van Gassen and Cruden, 1989; Hungr and Evans 2004); mechanical fluidization (McSaveney, 1978; Davies, 1982); acoustic fluidization (Melosh, 1979; Collins and Melosh, 2003); random fragment kinetic energy (Preuth *et al.*, 2010); seismic energy fluidization (Hazlett *et al.*, 1991); oscillation of quasi-rigid plugs (Cagnoli and Quareni, 2009); and dynamic rock fragmentation (Davies and McSaveney, 2002, 2009; Pollet and Schneider, 2004).

A number of summaries and critiques of these mechanisms have also been published (Hungr, 1990; Davies *et al.*, 1999; Erismann and Abele, 1999; Legros, 2002). Our intention here is not to re-evaluate these theories, but instead to first consider the range of situations in which the long-runout phenomenon occurs, over all of which any valid mechanism must be applicable. We adopt the fundamental hypothesis that there is a single causal mechanism for long runout that applies in all situations. This supposition might be incorrect, and different mechanisms might operate in different circumstances. However, if there is indeed a single viable explanation, our hypothesis is supported for the time being, and that explanation has, so far, the advantage of universal applicability and, more significantly, of economy, along the lines of Occam's Razor (a guiding principle of science dating back to Aristotle, which states that a simple explanation is usually preferable to a more complex one). This approach allows many of the already-considered contending mechanisms to be eliminated. We then outline some mechanisms that have not hitherto been reviewed, either because they were previously overlooked or because they are too recent, and consider them within the same framework. Finally we examine in some detail the mechanisms that have not been eliminated.

5.3 ENVIRONMENTS, MATERIALS AND CHARACTERISTICS OF LONG-RUNOUT ROCK AVALANCHES

The long-runout phenomenon has been reported in rock-avalanche deposits in a wide variety of environments:

- subaerial on Earth
- supraglacial on Earth
- subaqueous on Earth (note that we consider here only events that comprise rock-slope failures, not simply sediment flows such as turbidity currents, although the former can give rise to the latter)
- subaerial on Mars
- *in vacuo* on the Moon and other extraterrestrial bodies.

Long-runout rock avalanches also occur in a variety of materials, including crustal rocks (volcanic, metamorphic, sedimentary), and anthropogenic sediments (coal waste tips). They show a variety of common characteristics:

- unusually extensive spreading and runout
- large volume ($>10^6$ m^3)
- substantial increase in total volume between initial failure and deposit ("bulking")
- basal rock fragmentation (in block-slides) or complete rock fragmentation, generally with a carapace of less-fragmented rock, resulting in a fractally distributed mass of angular grains with grain-sizes down to «1 micrometer
- shattered undisaggregated ("jigsaw") clasts at all scales throughout the deposit
- a lack of mixing where the source has more than a single lithology (Hewitt, 2009).

5.4 IMPLICATIONS

The environments, characteristics, and materials of long-runout rock avalanches allow us to formulate a set of implications for the proposed causal mechanisms.

- The existence of long-runout rock-avalanche deposits on the Moon (Guest, 1971; Howard, 1973; Melosh and Ivanov, 1999) requires that the causal mechanism is able to operate in the complete absence of air, water, or other fluids that are capable of generating pore pressure. This requirement eliminates undrained loading and low-friction failure of saturated substrates, high pore water pressures, heating of pore water to generate steam, air-layer lubrication, and air-induced fluidization as potential mechanisms.
- The existence of long-runout rock-avalanche deposits under the ocean (Locat and Lee, 2002) requires that the causal mechanism can operate in water-saturated conditions. This requirement eliminates interstitial dust dispersion as a potential mechanism.
- The restriction of long-runout rock avalanches to volumes $>10^6$ m^3 requires that the causal mechanism either does not operate or is insignificant in effect at smaller volumes.
- The existence of long-runout rock-avalanche deposits on bedrock, subaerial and submarine sediments, soils, and glacier surfaces requires that the phenomenon can occur on any type of substrate. Although some events pick up large volumes of substrate, others do not, so mass change – particularly mass loss as proposed by Van Gassen and Cruden (1989) – cannot be a fundamental or universal mechanism.
- The universal association of long-runout rockslides with intense rock fragmentation requires that the causal mechanism should operate under the conditions that cause fragmentation (shearing of rock debris under high normal stress) and in the presence of fragmentation. As discussed later, this requirement in fact eliminates some mechanisms that, if they occurred, would remove the cause of fragmentation.
- The presence of distal jigsaw blocks requires that fragmentation must operate throughout the whole runout of the rock avalanche, not just at the start. This requirement eliminates all mechanisms that rely on the mechanics of granular flow with nonfragmenting materials and also those that ignore the presence of granular materials (e.g., guided basal waves that require the body of the avalanche to be a nongranular elastic half-space).
- The lack of mixing in rock-avalanche deposits requires that the motion of the rock avalanche mass does not involve turbulence, eddying, or lateral velocity components at anything but a very small scale.
- Although very small quantities of melted rock have been found in a small number of long-runout deposits, its almost universal absence suggests that basal lubrication by molten or dissociated rock is not a fundamental process.
- Although some rock avalanches occur during significant earthquakes, many others do not. Vibration of the moving mass by earthquakes appears to be insufficiently energetic to explain long runout, as are the low-magnitude earthquakes generated by the fall of large rock avalanches.

5.5 OTHER MECHANISMS

New explanations for long-runout rock avalanche mobility continue to appear. For example, acoustic fluidization was applied in detail to long-runout rock avalanches by Collins and Melosh (2003). They recognized that in a nonfragmenting granular shear flow, grains acquire short-term lateral velocity components as they move past one another, and this process takes some longitudinal kinetic energy from the shear flow and transforms it into vibrational energy. It is well known that applied vibration reduces the internal friction of granular flow (Barkan, 1962; Zik *et al.*, 1992; Hori *et al.*, 2008), therefore this effect potentially increases the runout. However, Sornette and Sornette (2000) demonstrated that this process is insufficiently energetic to cause the dramatic increases in shear resistance required to explain low-friction earthquakes. In a similar vein, the proposal that random grain kinetic energy occurs in rock avalanches and explains long runout (Preuth *et al.*, 2010) invokes transverse vibration energy generated by granular shear to reduce friction. It is not clear how this process differs from acoustic fluidization, and it is likely to suffer the same drawbacks.

Cagnoli and Quareni (2009) suggested that the low basal friction required for long runout could be the result of the formation of quasi-rigid plugs of rock-avalanche material developing synchronous vertical oscillations during runout; the reduced time of contact of the plugs with the underlying ground lowers the frictional resistance to motion. The above authors claimed that such motion was reported by Iverson (1997) in debris flows, but what Iverson in fact reported was *uncoordinated* vertical oscillations of *individual grains*, rather than orchestrated oscillations of quasi-rigid plugs. In terms of the criteria developed above, one imagines that it would be difficult for the quasi-rigid plugs to retain their coherence in a submarine environment due to the greatly reduced effective weight of the solid grains involved.

The only mechanism that appears capable of being applied in all the circumstances and environments in which long-runout rock avalanches have been reported is therefore that of dynamic rock fragmentation.

5.6 DYNAMIC ROCK FRAGMENTATION

5.6.1 OUTLINE

We have been studying, for some time, the possibility that the breakage of rocks – universal in long-runout rock avalanches – may hold the key to understanding this phenomenon and have reported a number of stages in the development of the concept (Davies *et al.*, 1999, 2007; Davies and McSaveney, 2002, 2009; McSaveney and Davies, 2007). The basic concept is that, because it requires high stresses to break rock grains, particularly small ones with only microscopic defects, large events might cause pervasive grain breakage and, furthermore, their excess runout might be the result of grain breakage.

This idea has attracted vigorous opposition, mainly on the grounds that breaking a rock consumes kinetic energy and therefore cannot cause increased runout. However, Griffith (1920), in his seminal study of rock breakage, assumed – rather than demonstrated – that the elastic strain energy required to break the rock was consumed as surface energy upon the new surface being generated by breakage. This assumption made no difference to Griffith's rock breakage theory; he was concerned with the energy required to generate a crack, not with what happened to it afterwards. What happens to it after breakage, however, is of great interest to us.

5.6.2 THE ENERGETICS OF BREAKING GRAINS

We recently carried out slow strain-rate, unconfined compression tests on 10-mm-diameter cylinders of greywacke and measured the resulting fragment motions (see Fig. 5.2). The measurements show that the fragments accelerate at about 2×10^4 *g*, implying an outward pressure of about 10 MPa in this particular test. The kinetic energy of fragments can be related to the strain energy stored in the rock prior to failure; tests with greywacke indicated that about 10–40 percent of the total stored energy was released as kinetic energy, whereas with Pyrex cylinders the proportion was 30–95 percent. These data are similar to those obtained by Bergstrom (1963), using rock and Pyrex spheres, and demonstrate that breakage of rock releases as free energy much of the elastic strain energy stored in the rock prior to breakage.

The energy input required to break rock is usually called the fracture energy (U) and is proportional to the new surface area created. Generally $1\ \mathrm{J\,m^{-2}} < U < 10\ \mathrm{J\,m^{-2}}$. If this energy were indeed lost to surface energy, there would exist a limit to the size of grain that could be split in two. The theoretical maximum strain energy that can be stored in unit volume of rock (W_{max}) is given by:

$$W_{max} = Q^2/2E \tag{5.6}$$

(Herget, 1988). Here Q is the ambient compressive strength of the rock material and E its elasticity. In a rock stressed by contact with adjacent grains, the stress distribution is unlikely to be uniform, so $W < W_{max}$. Considering a cubic grain of side d, the energy required to generate a crack of area d^2 is Ud^2 and $W_{max} = Q^2d^3/2E$. Thus, in order to generate a crack splitting the cube:

$$Ud^2 = Q^2d^3/2E. \tag{5.7}$$

Taking a typical strong crustal rock (greywacke sandstone), $Q = 2.5 \times 10^8$ Pa and $E = 7 \times 10^{10}$ Pa (Stewart, 2007), it follows that fragments smaller than 4.5 microns should not exist if $U = 1\ \mathrm{J\,m^{-2}}$ (45 microns if $U = 10\ \mathrm{J\,m^{-2}}$). If the nearby grains that apply the stress to the target grain are also strained in the process, they may also contribute elastic strain energy to the breaking grain; Rice (2006) estimated that the average strong force chain is about 10 grains long, so perhaps 10 times the strain energy in the target grain may be available to generate a new rock surface. This reasoning suggests that the minimum grain size able to be fragmented is *ca.* 1 micron. By contrast, McSaveney and Davies (2007) reported that 90 percent by weight (>99 percent by number) of the grains in the deposit of the 1991 Mt. Cook rock avalanche, New Zealand, are finer than 10 microns. Other rock avalanches have very similar, fractal grain-size distributions to that of the Mt. Cook deposit (Dunning, 2004; Locat *et al.*, 2006; Crosta *et al.*, 2007). Furthermore, recent studies (Reznichenko *et al.*, 2011) have shown that the individual grains in a rock avalanche themselves comprise agglomerations of much smaller grains; thus the Mt. Cook rock-avalanche material is in fact much smaller than reported. In fault gouge, which is also produced by high-stress shearing of granular rock, fractally distributed fragments of 40 nm dimension have been measured (Kuelen *et al.*, 2007), with the smallest recorded grains <10 nm. These data suggest that the assumption that the fracture energy U becomes unavailable as surface energy is incorrect (McSaveney and Davies, 2009). We argue that, instead of consuming fracture energy, the process of fragmentation of solid grains redistributes it in a granular flow; energy is slowly removed from the general shearing motion, stored as elastic strain energy in deforming grains, and released rapidly and locally as grains fail.

The energy released by grain fragmentation in a confined granular flow – that is, in a flow under high confining pressures so that fragments have virtually no free trajectory (Campbell, 2002) – is radiated through the grain mass as acoustic emissions.

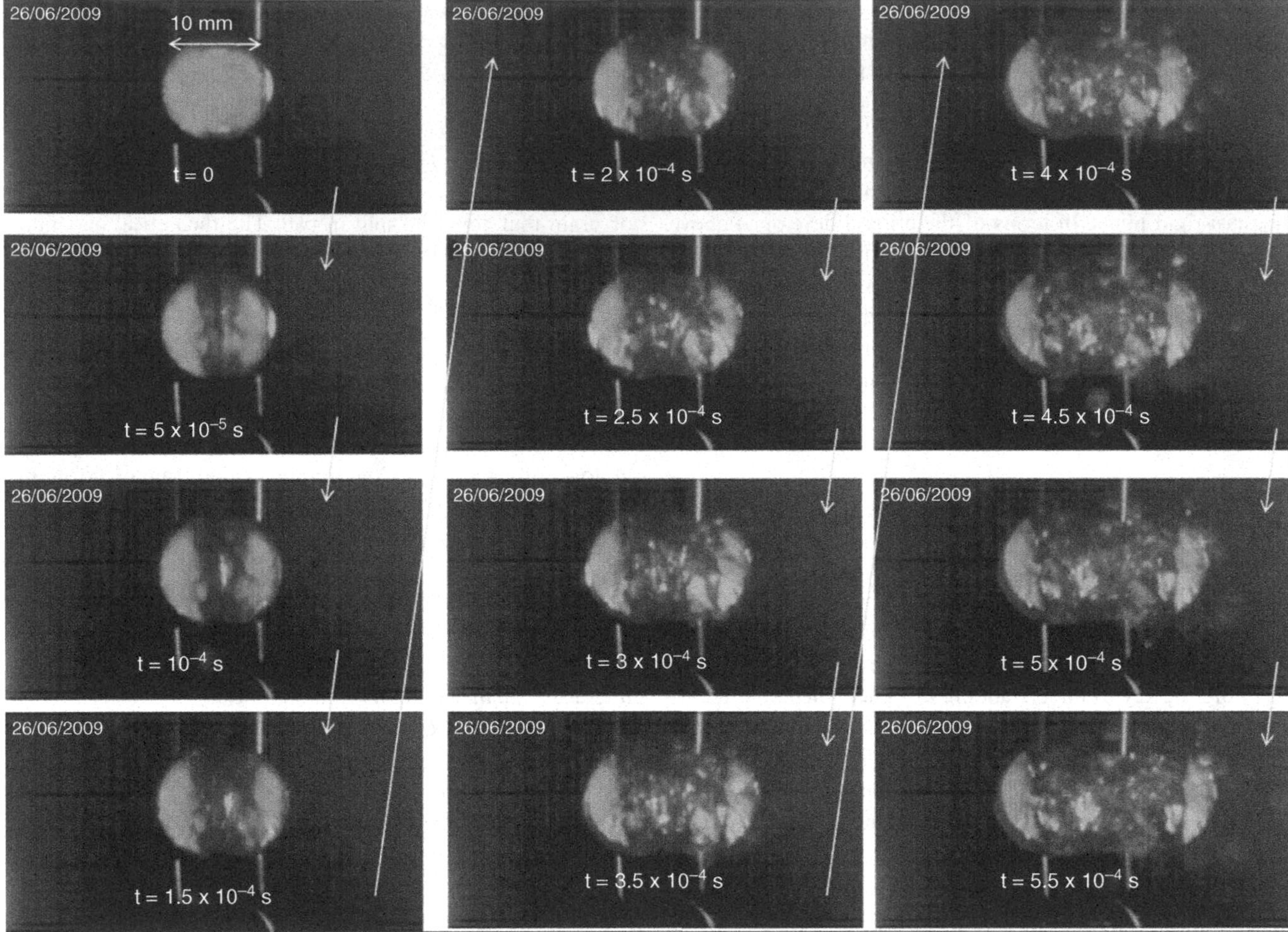

Fig. 5.2. A 10-mm-diameter cylinder of greywacke breaking in slow uniaxial compression; the initial acceleration of the large fragments is 2×10^4 *g* and the stress acting on them is about 10 MPa. Frame rate 20,000 s^{-1}.

These are small-scale, intense seismic P and S waves. Their frequencies correspond to the sizes of grains that fail and are expected to extend into the MHz–GHz range. This seismic energy field will affect grain-contact stresses throughout the grain flow and hence change the flow's resistance to shearing. Experimental data (Barkan, 1962; Zik *et al.*, 1992; Hori *et al.*, 2008) consistently indicate that vibration reduces the shear resistance of a granular flow in proportion to the vibration intensity.

Thus, when rocks are breaking in a granular flow, a source of vibration is present that is not present when rocks are not breaking. Interestingly, the vibration-induced reduction of shear resistance was used by Collins and Melosh (2003) and Preuth *et al.* (2010) to explain rock-avalanche runout, but their source of energy is the lateral velocity components due to intact grains moving past each other. The energy per unit time available from breaking grains is greater than that produced by lateral velocity by a factor of at least 10^3. The reason is that although the strain energy stored temporarily in grains may be almost the same in nonfragmenting as in fragmenting granular flow, a grain can fragment in about 1/1000 the time required for two grains to pass each other in a shear flow. Thus fragmentation is some orders of magnitude more powerful as a source of vibration than that present in nonfragmenting grain flow.

5.6.3 SOCOMPA DEBRIS AVALANCHE

While it seems reasonable, in principle, to explain the mobility of long-runout rock avalanches by grain fragmentation, the true test of any mechanism lies in its ability to reproduce the behavior of a well-documented event in three dimensions. We have recently performed such a test (Davies *et al.*, 2010) using the deposit of the 8000-year-old Socompa debris avalanche (van Wyk de Vries *et al.*, 2001; Kelfoun and Druitt, 2005). This non-eruptive, volcanic debris avalanche of about 25 km^3 volume emplaced a 500 km^2 deposit up to 40 km from the source. Its surface topography is extremely well preserved in the arid desert climate and has been precisely mapped; the pre-event topography is known from groundwater investigations better than for any other event of its size. We used a numerical model ("Volcflow"; Kelfoun and Druitt, 2005) to simulate the motion of the collapsing edifice across the landscape, and utilized an effective stress relationship to calculate the basal resistance to motion at different places at different times. This relationship is based on the concepts of fragmentation outlined above, calibrated with data from other landslides.

We analyzed the effect of fragmentation on a grain flow by assuming that the spatially averaged interstitial pressure generated by the fragmenting grains acts in the same way as a high pore-fluid pressure. It supports some of the overburden stress at the top of the fragmenting layer, so that all grains in the fragmenting layer experience reduced grain-contact stress. We developed a relationship expressing the resistance to granular flow as a function of rock compressive strength and avalanche depth. This relationship made use of the fact that in the Falling Mountain rock avalanche in New Zealand (Davies and McSaveney, 2002) the upper 5 m or so of the deposit was significantly less fragmented than the underlying 70 m of debris, together with a factor derived from our simple simulation of the Waikaremoana block-slide in New Zealand (Davies *et al.*, 2006). Full details of this analysis and simulation are described in Davies *et al.* (2010).

With a slight adjustment of the calibration data, we produced an extremely good representation of the Socompa deposit pattern (Fig. 5.3), as well as the effect of a reflected wave during the emplacement as described by Kelfoun *et al.* (2008). The parameters that drive the model are rock compressive strength, which determines the pressure derived from each fragmentation, and the flow depth; the coefficient of intergranular friction everywhere and at all times has the conventional value of about 0.7. On this basis, the fragmentation concept appears to be capable of explaining the mobility of long-runout rock avalanches without requiring air or water, or weak substrate, or high pore-water pressure, or molten rock.

5.7 DISCUSSION

We have shown that fragmentation

- can operate in all conditions in which long runout occurs
- is compatible with deposit characteristics
- is capable of quantitative representation of the best-described, long-runout rock avalanche for which data are available.

A significant question remains: to what extent is fragmentation compatible with the other mechanisms? In other words, moving away from our basic hypothesis that a single cause is required to explain long runout, how realistic is a mixture of causes?

As noted earlier, several of the proposed explanations are in fact incompatible with fragmentation, because if they occurred, fragmentation would not. The explanations that infer low friction due to high pore-water pressure, large areas of molten rock, weak substrates, and compressed air would result in the collapsed rock from the fall rafting along passively on the low-friction layer, and no mechanism would remain to cause the widespread and intense fragmentation ubiquitous in long-runout rock avalanches. As noted earlier, mechanisms that assume that all fragmentation occurs at the start of the fall are incompatible with the presence of distal shattered undisaggregated clasts, and they also require too much energy to be available at the start of the motion in order to cause the fragmentation.

Fragmentation can clearly occur *in vacuo* and subaerially, but how realistic is it to suggest submarine fragmentation? In principle, a clast can be stressed to fragmentation, storing elastic strain energy, just as well under water as in air or *in vacuo*; the only alteration to the stress conditions of a submarine, as opposed to a subaerial, rock avalanche is the lower overburden stress at any given depth in the avalanche due to the buoyancy effect of water. As a result, we expect that long runout would not occur under water until avalanche thicknesses were perhaps twice those of their subaerial counterparts. If all rock avalanches were geometrically similar, the volume threshold of submarine long runout would be eight times that in the subaerial environment. The available data do not appear to be sufficiently detailed to test this inference. Submarine fragmentation would cause intense water pressure waves to radiate through the grain flow, which would reduce the energy available for a high-frequency seismic wave-field. Intense water pressure variations, however, would tend to reduce intergranular friction, so the end result might not be too different (Davies and McSaveney, 2009). There is no obvious reason for fragmentation to be less effective under water.

In a fragmenting grain flow, the proportion of grains in the shear flow fragmenting at any moment is expected to be small (Davies and McSaveney, 2009). Thus most of the grain shearing that occurs involves nonfragmenting grains and occurs under conventional friction, albeit at low effective stress; thus the mechanisms that generate reduced friction in nonfragmenting grain flow (acoustic fluidization/random kinetic energy) can operate in addition to fragmentation. However, the latter process is several orders of magnitude more powerful than the former, so the former is likely to be negligible in effect. Similarly, the pore-water pressure fluctuations that occur in a saturated grain flow (Iverson, 1997) are intrinsically able to reduce frictional resistance to motion, but are much less effective than fragmentation when the latter is occurring.

5.8 LOOKING FORWARD

There is no doubt that other mechanisms will be proposed in the future to explain the mobility of long-runout rock avalanches. While fragmentation may appear to be a sufficient, and indeed necessary, explanation at present, it has not yet been intensively tested under laboratory conditions and, as with all scientific hypotheses, it requires only one data point to invalidate it. We note, however, that a number of small-scale, high-stress, rock-on-rock friction tests have generated both finely comminuted powder and low friction (Di Toro *et al.*, 2004; Mizoguchi *et al.*, 2009), so our hypothesis does not contradict the existing data. Indeed, our approximate analysis appears to predict the correct

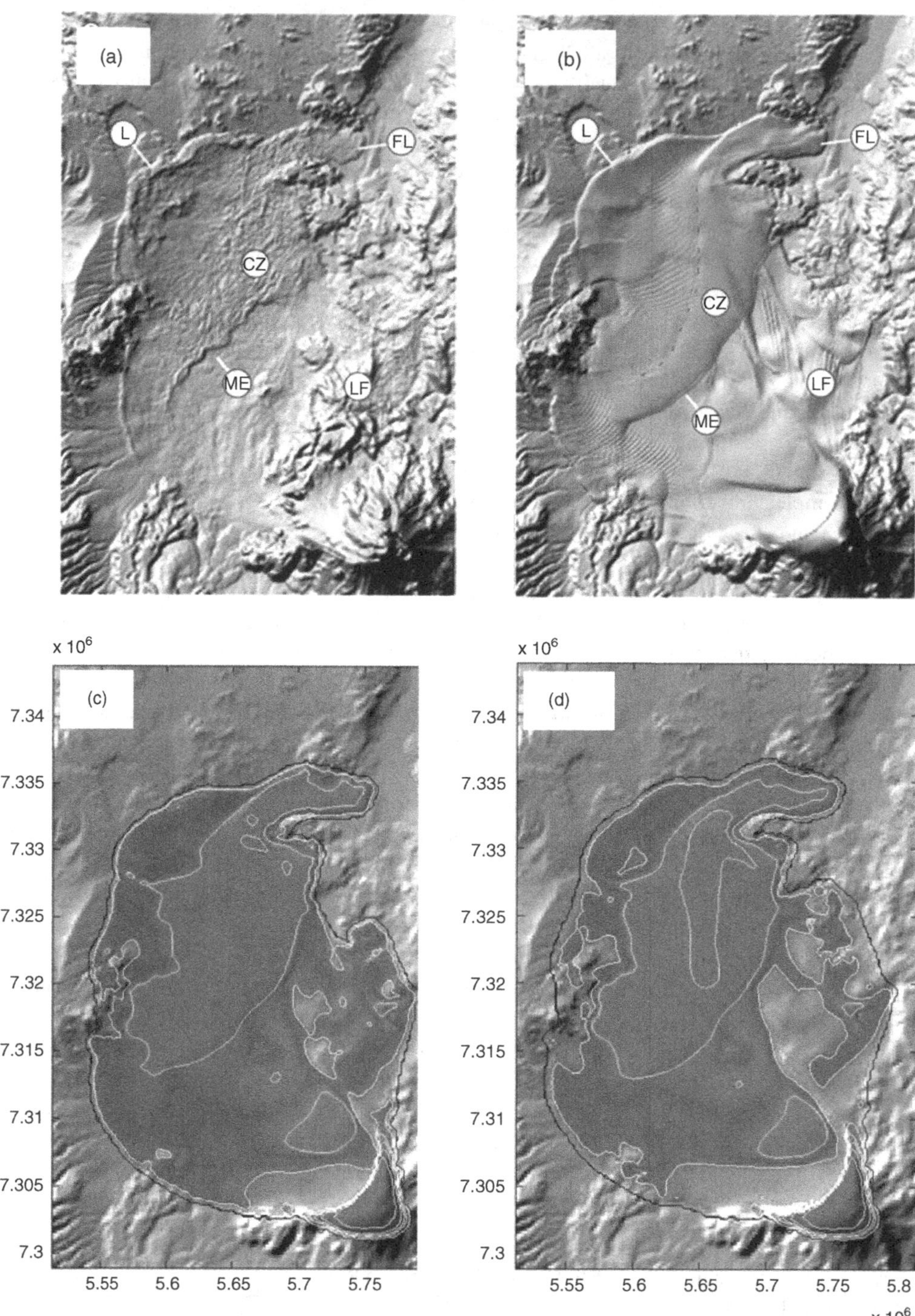

Fig. 5.3. Socompa volcanic debris avalanche. (a) Digital terrain model of deposit; (b) digital terrain model of 52 kPa basal resistance model of Kelfoun and Druitt (2005); (c) deposit thickness map corresponding to (b); (d) thickness map from fragmentation-driven model. L, CZ, FL, ME, and LF in (a) and (b) indicate corresponding features.

degree of friction reduction in the laboratory experiments to which it has been applied (Davies *et al.*, 2007).

Has the long-runout problem been solved? It appears that a solution has been proposed and verified in one case (Socompa; Davies *et al.*, 2010; or three, if its application to the Waikaremoana block-slide (Davies *et al.*, 2006) and to the frictional resistance of the San Andreas fault (Boulton *et al.*, 2009) are included). It also has the attribute of hypothesis parsimony and requires neither unusual environmental conditions nor novel physics apart from the re-utilization of fracture energy. It now needs intensive testing in a variety of circumstances, including detailed laboratory and numerical investigations of the basic rock fracture mechanics and energetics; to this end we are in the process of measuring the complete particle size distribution, including the finest components of the ubiquitous agglomerated microparticles (Reznichenko *et al.*, 2011), and the complete energy budget of fragmenting rocks. At any stage in this testing, our solution may need to be seriously modified, or even discarded. In truth, one can only say that the science is a work in progress.

ACKNOWLEDGEMENTS

We gratefully acknowledge the contributions of many graduate students and colleagues in developing the concepts discussed in this chapter. In particular, we acknowledge many challenging debates with respected, but still unconvinced, colleagues that have forced us to refine our ideas several-fold in the last decade. We hope this process will continue. Comments from John Clague and Doug Stead, and an anonymous reviewer, resulted in significant improvements to the original manuscript.

REFERENCES

Abele, G. (1974). *Bergstürze in den Alpen: ihre Verbreitung, Morphologie und Folgeerscheinungen*. Wissenschaftliche Vereinshefte, 25.

Barkan, D.D. (1962). *Dynamics of Bases and Foundations*. New York: McGraw-Hill.

Bergstrom, C.H. (1963). Energy and size distribution aspects of single particle crushing. In *Rock Mechanics*, ed. C. Fairhurst. New York: Pergamon Press, pp. 155–172.

Boulton, C.J., Davies, T.R.H. and McSaveney, M.J. (2009). The frictional strength of granular fault gouge: Application of theory to the mechanics of low-angle normal faults. In *Extending a Continent: Architecture, Rheology and Heat Budget*. Geological Society Special Publication 321, pp. 9–31.

Cagnoli, B. and Quareni, F. (2009). Oscillation-induced mobility of flows of rock fragments with quasi-rigid plugs in rectangular channels with frictional walls: A hypothesis. *Engineering Geology*, 103, 23–32.

Campbell, C.S. (1989). Self-lubrication for long-runout landslides. *Journal of Geology*, 97, 653–665.

(2002). Granular shear flows at the elastic limit. *Journal of Fluid Mechanics*, 465, 261–291.

Collins, G.S. and Melosh, H.J. (2003). Acoustic fluidization and the extraordinary mobility of sturzstroms. *Journal of Geophysical Research*, 108, 2473, doi:10.1029/2003JB002465.

Crosta, G.B., Frattini, P. and Fusi, N. (2007). Fragmentation in the Val Pola rock avalanche, Italian Alps. *Journal of Geophysical Research*, 112, F01006, doi:10.1029/2005JF000455.

Davies, T.R.H. (1982). Spreading of rock avalanche debris by mechanical fluidisation. *Rock Mechanics*, 15, 9–24.

Davies, T.R.H. and McSaveney, M.J. (1999). Runout of dry granular avalanches. *Canadian Geotechnical Journal*, 36, 313–320.

(2002). Dynamic simulation of the motion of fragmenting rock avalanches. *Canadian Geotechnical Journal*, 39, 789–798.

(2009). The role of dynamic rock fragmentation in reducing frictional resistance to large landslides. *Engineering Geology*, 109, 67–79, doi:10.1016/j.enggeo.2008.11.004.

Davies, T.R.H., McSaveney, M.J. and Hodgson, K.A. (1999). A fragmentation-spreading model for long-runout rock avalanches. *Canadian Geotechnical Journal*, 36, 1096–1110.

Davies, T.R.H., McSaveney, M.J. and Beetham, R.D. (2006). Rapid block glides: Slide-surface fragmentation in New Zealand's Waikaremoana landslide. *Quarterly Journal of Engineering Geology and Hydrogeology*, 39, 115–129.

Davies, T.R.H., McSaveney, M.J. and Deganutti, A.M. (2007). Dynamic rock fragmentation causes low rock-on-rock friction. In *Rock Mechanics: Meeting Society's Challenges and Demands*, ed. E. Eberhardt, D. Stead and T. Morrison. London: Taylor & Francis, pp. 959–966.

Davies, T.R.H., McSaveney, M.J. and Kelfoun, K. (2010). Runout of the Socompa volcanic debris avalanche, Chile: A mechanical explanation for low basal shear resistance. *Bulletin of Volcanology*, 72, 933–944, doi:10.1007/s00445-010-0372-9.

De Blasio, F.V. and Elverhøi, A. (2008). A model for frictional melt production beneath large rock avalanches. *Journal of Geophysical Research, F: Earth Surface*, 113, F02014.

Di Toro, G., Goldsby, G.L. and Tullis, T.E. (2004). Friction falls towards zero in quartz rock as slip velocity approaches seismic rates. *Nature*, 227, 436–439.

Dunning, S.A. (2004). *Rock avalanches in high mountains: A sedimentological approach*. Ph.D. thesis, Luton University, UK.

Erismann, T.H. (1979). Mechanisms of large landslides. *Rock Mechanics*, 12, 15–46.

Erismann, T.H. and Abele, G. (1999). *Dynamics of Rockfalls and Rockslides*. Amsterdam: Springer.

Glicken, H.X. (1996). *Rockslide-Debris Avalanche of May 18, 1980, Mount St. Helens Volcano, Washington*. US Geological Survey, Open File Report 96–677.

Goguel, J. (1978). Scale-dependent rockslide mechanisms. In *Rockslides and Avalanches, Vol. 1*, ed. B. Voight. Amsterdam: Elsevier, pp. 693–705.

Griffith, A.A. (1920). The phenomena of rupture and flow in solids. *Philosophical Transactions of the Royal Society of London, A*, 221, 163–198.

Guest, J.E. (1971). Geology of the farside crater Tsiolkovsky. In *Geology and Physics of the Moon*, ed. G. Fielder. Amsterdam: Elsevier, pp. 93–103.

Habib, P. (1975). Production of gaseous pore pressure during rock slides. *Rock Mechanics*, 7, 193–197.

Hazlett, R.W., Buesch, D., Anderson, J.L., Elan, R. and Scandone, R. (1991). Geology, failure and implications of seismogenic avalanches of the 1944 eruption at Vesuvius, Italy. *Journal of Volcanology and Geothermal Research*, 47, 249–264.

Heim, A. (1882). Der Bergsturz von Elm. *Zeitschrift der Deutschen Geologischen Gesellschaft*, 34, 74–115.

Herget, G. (1988). *Stresses in Rock*. Rotterdam: Balkema.

Hewitt, K. (2009). Catastrophic rock slope failures and late Quaternary developments in the Nanga Parbat-Haramosh Massif, Upper Indus Basin, northern Pakistan. *Quaternary Science Reviews*, 28, 1055–1069.

Hori, T., Sakaguchi, H., Yoshioka, N. and Kaneda, Y. (2008). Shear resistance eduction due to vibration in simulated fault gouge. In *Earthquakes: Radiated Energy and the Physics of Faulting*, ed. R.E. Abercrombie, A. McGarr, G.D. Di Toro and H. Kanamori. American Geophysical Union Geophysical Monograph Series 170, pp. 135–142, doi:10.1029/170GM24.

Howard, K.A. (1973). Avalanche mode of motion: Implications from lunar examples. *Science*, 180, 1052–1055.

Hsü, K.J. (1975). Catastrophic debris streams (sturzstroms) generated by rockfalls. *Geological Society of America Bulletin*, 86, 123–140.

(1978). Albert Heim: Observations on landslides and relevance to modern interpretations. In *Rockslides and Avalanches, Vol. 1*, ed. B. Voight. Amsterdam: Elsevier, pp. 72–93.

Hungr, O. (1990). *Mobility of Rock Avalanches*. Tsukuba, Japan: National Institute for Earth Science and Disaster Prevention, Report 46.

Hungr, O. and Evans, S.G. (2004). Entrainment of debris in rock avalanches: An analysis of a long-runout mechanism. *Geological Society of America Bulletin*, 116, 1240–1252.

Iverson, R. M. (1997). The physics of debris flows. *Reviews in Geophysics*, 35, 245–296.
Kelfoun, K. and Druitt, T.H. (2005). Numerical modeling of the emplacement of Socompa rock avalanche, Chile. *Journal of Geophysical Research*, 110, B12202. doi:10.1029/2005B003758.
Kelfoun, K., Druitt, T.H., van Wyk de Vries, B. and Guilbaud, M.-N. (2008). Topographic reflection of the Socompa debris avalanche, Chile. *Bulletin of Volcanology*, 70, 1169–1187, doi:10.1007/s00445-008-0201-6.
Kent, P.E. (1966). The transport mechanism in catastrophic rockfalls. *Journal of Geology*, 74, 79–83.
Kobayashi, Y. (1997). Long runout landslides riding on guiding basal wave. In *Engineering Geology and the Environment*, ed. P. Marinos, G.C. Koukis, G.C. Tsiambaos and G.C. Stournaras. Rotterdam: Balkema, pp. 761–766.
Kuelen, N., Heilbronner, R., Stünitz, H., Boullier, A.-M. and Ito, H. (2007). Grain size distributions of fault rocks: A comparison between experimentally and naturally deformed granitoids. *Journal of Structural Geology*, 29, 1282–1300.
Legros, F. (2002). The mobility of long-runout landslides. *Engineering Geology*, 63, 301–331.
Locat, J. and Lee, H. (2002). Submarine landslides: Advances and challenges. *Canadian Geotechnical Journal*, 39, 193–212.
Locat, P., Couture, R., Leroueil, S., Locat, J. and Jaboyedoff, M. (2006). Fragmentation energy in rock avalanches. *Canadian Geotechnical Journal*, 43, 830–851.
Masch, L., Wenk, H.R. and Preuss, E. (1985). Electron microscopy study of hyalomylonies: Evidence for frictional melting in landslides. *Tectonophysics*, 115, 131–160.
McSaveney, M.J. (1978). Sherman Glacier rock avalanche. In *Rockslides and Avalanches, Vol. 1*, ed. B. Voight. Amsterdam: Elsevier, pp. 197–258.
McSaveney, M.J. and Davies, T.R.H. (2007). Rockslides and their motion. In *Progress in Landslide Science*, ed. K. Sassa, H. Fukuoka, F. Wang and G. Wang. Berlin: Springer-Verlag, pp. 113–134.
(2009). Surface energy is not one of the energy losses in rock comminution. *Engineering Geology*, 109, 109–113, doi:10.1016/j.enggeo.2008.11.001.
Melosh, H.J. (1979). Acoustic fluidization: A new geologic process? *Journal of Geophysical Research*, 84, 7513–7520.
Melosh, H.J. and Ivanov, B.A. (1999). Impact crater collapse. *Annual Review of Earth and Planetary Science*, 27, 385–415.
Mizoguchi, K., Hirose, T., Shimamoto, T. and Fukuyama, E. (2009). High-velocity frictional behavior and microstructure evolution of fault gouge obtained from Nojima fault, southwest Japan. *Tectonophysics*, 471, 285–296.
Pollet, N. and Schneider, J.-L.M. (2004). Dynamic disintegration processes accompanying transport of the Holocene Flims sturzstrom (Swiss Alps). *Earth and Planetary Science Letters*, 221, 433–448.
Preuth, T., Bartelt, P., Korup, O. and McArdell, B.W. (2010). A random kinetic energy model for rock avalanches: Eight case studies. *Journal of Geophysical Research*, 115, F03036, doi:10.1029/2009JF001640.
Reznichenko, N.V., Davies, T.R.H. and Shulmeister, J. (2011). Diagnostic criteria for the identification of the rock avalanche derived sediments in moraines. In *Abstracts, Snow and Ice Research Group (NZ) Annual Workshop: SIRG 2011*, Fox Glacier, New Zealand, p. 29 [available at www.sirg.org.nz/abstracts/SIRG2011FoxGlacier_Programme.pdf].
Rice, J.R. (2006). Heating and weakening of faults during earthquake slip. *Journal of Geophysical Research*, 111, B05311, doi:10.1029/2005BJ004006.
Sassa, K. (1988). Geotechnical model for the motion of landslides. In *Landslides: Proceedings of the 5th International Symposium on Landslides, 10–15 July, Lausanne*, ed. C. Bonnard. Rotterdam: Balkema, pp. 37–55.
Scheidegger, A.E. (1973). On the prediction of the reach and velocity of catastrophic landslides. *Rock Mechanics*, 5, 231–236.
Shreve, R.L. (1968). *The Blackhawk Landslide*. Geological Society of America, Special Paper 108.
Sornette, D. and Sornette, A. (2000). Acoustic fluidization for earthquakes? *Seismological Society of America Bulletin*, 90, 781–785.
Stewart, S.W. (2007). *Rock mass strength and deformability of unweathered closely jointed New Zealand greywacke*. Ph.D. thesis, University of Canterbury, Christchurch, New Zealand.
Van Gassen, W. and Cruden, D.M. (1989). Momentum transfer and friction in the debris of rock avalanches. *Canadian Geotechnical Journal*, 2, 623–628.
van Wyk de Vries, B., Self, S., Francis, P.W. and Keszthelyi, L. (2001). A gravitational spreading origin for the Socompa debris avalanche. *Journal of Volcanology and Geothermal Research*, 105, 225–247.
Voight, B. and Sousa, J. (1994). Lessons from Ontake-san: A comparative analysis of debris avalanche dynamics. *Engineering Geology*, 38, 261–297.
Zik, O., Stavans, J. and Rabin, Y. (1992). Mobility of a sphere in vibrated granular material. *Europhysics Letters*, 17, 315–319.

6 Rapid rock-slope failures

REGINALD L. HERMANNS AND ODDVAR LONGVA

Our life is short. The memory of mankind as a whole is poor. The few mountain collapses that we experience in our lifetime leave us with the impression that these collapses are very exceptional, extraordinary events. However, that is not the case. Mountain collapses are normal events in the mountains, especially in the high mountains, where they have an important natural role to play in helping to form and shape the mountains; a process that continues relentlessly and steadily. In the mountains we have to expect mountain collapses from time to time, from place to place.

(A. Heim, 1932)

ABSTRACT

Large catastrophic rock-slope failures are a serious threat to humans because

- they cannot be controlled by any physical measures
- they can be highly mobile and reach areas several kilometers from the source
- they can trigger damaging secondary effects.

Nowadays a concerted effort is being made, using advanced remote sensing tools, to identify and monitor slopes that might fail catastrophically. Small accelerations that might be precursors of failure can be used as indicators to help in emergency planning. Characterization of the structure of the rock mass of a potentially unstable slope is an essential step in assessing the likelihood of catastrophic failure. Geologic records indicate that rapid slope failures are not distributed evenly in time due, in part, to climate variability. Analyses of the deposits of large, rapid rock-slope failures in the central Andes indicate that the proximity to active faults and the type of deformation found on these structures control landslide distribution and size. Many catastrophic rock-slope failures are triggered by strong crustal earthquakes. Historic data indicate that subduction earthquakes are less effective in triggering large landslides than crustal earthquakes.

6.1 INTRODUCTION

The topic of this chapter is catastrophic rock-slope failures. We apply the term "catastrophic" to rock-slope failures that involve substantial fragmentation of the rock mass during runout and that impact an area larger than that of a rockfall (shadow angle of *ca.* 28–32° from the source). Most catastrophic rock-slope failures are larger than 10^6 m^3, and failure involves the development of a continuous rupture plane between the underlying rock mass and the sliding rock body. We exclude collapses of volcanic edifices because this topic is covered in Chapter 4.

We first describe the failure process and summarize the conditions that indicate the possibility of an imminent failure. We then describe rock mass structures that control the failure process and depositional features that are characteristic of prehistoric catastrophic rock-slope failures. Next we summarize several chronological studies of large catastrophic landslides in the central Andes and Norway that have implications for climatic conditioning of slopes for failure. Finally we discuss the influence of the tectonic setting based on a systematic analysis of rock-slope failures in the central Andes.

6.2 OVERVIEW OF ROCK-SLOPE FAILURES AND HAZARD

Rock-slope failures of 10^6–10^7 m^3 have occurred nearly every year somewhere on Earth in the past century, and there have been two or more such failures in some years. Rock-slope failures larger than 1 km^3, on the other hand, are much less frequent; the only recent events of this size described in the scientific literature are the 1911 Usoi landslide in Tajikistan (2.2 km^3; Schuster and Alford, 2004) and the 1974 Mayunmarca landslide in Peru (Kojan and Hutchinson, 1978). Even larger events, however, have been documented in the geologic record, some up to several tens of cubic kilometers in size (e.g., Saidmarreh, Iran; Harrison and Falcon, 1934; and Lluta Valley, Chile; Wörner *et al.*, 2002).

Some deposits of large rock-slope failures are today densely populated, for example the Flims landslide deposit, which occurred 8200 ^{14}C years BP (von Poschinger *et al.*, 2006),

Landslides: Types, Mechanisms and Modeling, ed. John J. Clague and Douglas Stead. Published by Cambridge University Press.

and some cities with more than 1 million inhabitants are partially built on young landslide deposits. For example, landslide deposits only about 10,000 years old cover an area of 60 km^2 within the incorporated area of La Paz, Bolivia (Dobrovolny, 1962), and relatively young rockslide deposits are present within Caracas, Venezuela (Ferrer, 1999).

With sufficient relief (150 m, see Keefer, 1984; 400 m, see Hermanns and Strecker, 1999), large catastrophic rock-slope failures achieve high velocities (5–>100 m s^{-1}) in short travel distances; thus evacuation of the runout area, without prior warning of failure, is impossible. Catastrophes involving such landslides include the Huascaran landslide, which destroyed the town of Yungay in Peru in 1970 (Plafker and Ericksen, 1978) and a rock-slope failure that destroyed several villages during the Khait earthquake in Tajikistan in 1949 (Evans *et al.*, 2009a).

Most catastrophic rock-slope failures are "rock avalanches." This term was coined by Hsü (1975), based on Heim's (1932) description of the phenomenon. Heim used the German terms "*Bergsturz*," "*Trümmerstrom*," and "*Sturzstrom*" for streams of rapidly moving debris resulting from the disintegration of a failed large rock mass. The streaming behavior generally develops only when the landslide is larger than 10^6 m^3. Synonyms for rock avalanche include rockfall avalanche, rockfall-generated debris stream, and sturzstrom. Runout distances of rock avalanches commonly exceed several kilometers; their high mobility may be evidenced by high run-up on opposite valley slopes, which is related to the volume of the initial failed rock mass (Scheidegger, 1961), and superelevation of debris at bends in the flow path (Nicoletti and Sorriso-Valvo, 1991). Mobility can be enhanced by the entrainment of saturated soil material, snow, or ice along the flow path (Hungr and Evans, 2004). Flow velocities can be calculated from run-up and superelevation using the equations summarized in Crandell and Fahnestock (1965).

Rock avalanches are not the only landslides that fit our definition of catastrophic rock-slope failures. Others include rock–ice avalanches and rockslides or rockfalls, including those that enter a water body and trigger a tsunami (see Chapter 10). Rock–ice avalanches can involve a range of ratios of rock and ice. For example, a rock–ice avalanche on November 29, 1987 at Estero Parraguirre, Chile, was initiated by the failure of 6 × 10^6 m^3 of rock, but an additional 9 × 10^6 m^3 of debris and ice were entrained from a glacier onto which the rock mass fell (Hauser, 2002). The 1970 Cerro Huascaran rock-slope failure had an initial volume of 6.5 × 10^6 m^3 of rock and ice, but entrained 43 × 10^6 m^3 snow, ice, and debris along its path (Evans *et al.*, 2009b). Both events transformed into debris flows that traveled, respectively, 57 and 180 km. On September 20, 2002, 18.5–27 × 10^6 m^3 of rock and ice dropped from a steep rock slope onto Kolka Glacier in the Russian Caucasus. The rock–ice avalanche removed the lower part of the glacier and traveled down the Genaldon Valley for 20 km, at which point it transformed into a debris flow that traveled to the entrance of the Karmadon Gorge, killing 140 people and causing widespread destruction (Huggel *et al.*, 2005).

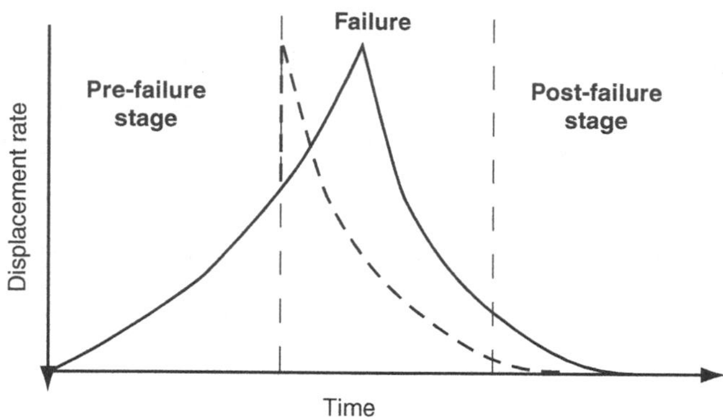

Fig. 6.1. Schematic diagram showing different stages of slope movement leading to catastrophic rock-slope failure. The solid line is for a nonseismic failure; the dashed line is for a seismically triggered failure. The nomenclature follows Leroueil *et al.* (1996).

Secondary effects extend the area of catastrophic rock-slope failures. The most important secondary effects are

- *damming of river valleys*, resulting in upstream flooding behind the debris dam and potential downstream flooding from overtopping and breaching of the dam
- *landslide-triggered displacement waves* (Costa and Schuster, 1988; Clague and Evans, 1994; Tappin, 2010; Evans *et al.*, 2011).

An assessment of the hazard posed by an unstable rock slope must include the possible impacts of secondary phenomena.

The probability of failure is difficult to determine. Regional-scale measures of failure probability include the annual frequency of landslides of a particular size per 10,000 km^2 of mountainous terrain (Hungr, 2006) or the number of events per thousand years per region (see below). Such measures, however, are not helpful at the local scale, where, for example, it may be necessary to assess the probability that a particular slope will fail (Aa *et al.*, 2007; Hermanns *et al.*, 2012). In such situations, it is essential to search for archival and other historic data on the slope of concern (Glastonbury and Fell, 2010), as well as local evidence of past instability such as ground cracks, recent rockfalls, and hydrologic changes on or near the slope. Monitoring of slope deformation may provide some warning of approaching failure (Crosta and Agliardi, 2003 and references therein), although instances where such data have been successfully used to predict catastrophic failure are few. As a rule of a thumb, a protracted acceleration of deformation rates is a clear sign that failure is imminent (Fig. 6.1).

Large catastrophic rock-slope failures can rarely be prevented or mitigated. Risk can be reduced, however, by

- recognizing slopes that potentially might fail suddenly
- slope monitoring and the formulation of warning and emergency evacuation plans.

Two assumptions underlie these measures: first, rock-slope deformation that may lead to failure can be detected and monitored; and second, the deformation accelerates prior to failure. The first assumption is valid, given recent developments in satellite and ground-based remote sensing, although the application of these tools on a continuous basis is expensive. The second assumption may not be true. In seismically active regions, for example, slope deformation accelerates to catastrophic failure within seconds (Fig. 6.1). In these areas, a better option than slope monitoring is zoning and restrictions on land use based on numerical modeling of rock-slope stability and runout (e.g., Welkner *et al.*, 2010).

Regional studies have shown that catastrophic rock-slope failures do not occur uniformly in space (Abele, 1974; Hermanns and Strecker, 1999) or time (Trauth *et al.*, 2000). Rather, they are controlled by lithological and structural conditions, their setting with respect to active faults, and climate. We focus on these controls in the following sections.

6.3 RECOGNITION OF TYPES OF ROCK-SLOPE FAILURES AND THEIR DEPOSITS

Deposits of catastrophic rock-slope failures can be recognized, depending on their climatic setting, for years, decades, centuries, or even millennia. Rock avalanches leave characteristic sheets of debris ranging from several meters to several hundred meters thick with sharp margins. When the landslide is unconstrained by topography, its deposits are relatively thin lobate sheets with lateral levees, frontal rims, and blocks meters to tens of meters in size at the surface. The coarse carapace typically overlies massive angular debris ranging in particle size through many orders of magnitude down to sub-micron size. Many of the clasts are densely fractured and display what have been termed "jigsaw texture" (Yarnold, 1993). Rock-avalanche deposits derived from a single rock type are monolithic. In cases where two or more lithologies are present in the source area, the lowest lithologies in the scarp are concentrated at the outer margin of the deposit, while the highest lithologies in the scarp are concentrated in proximal positions. Failures involving both rock and ice are more difficult to recognize because the initial failed mass commonly transforms into a debris flow or mudflow along its path (Hauser 2002; Huggel *et al.*, 2005; Evans *et al.*, 2009b). Fauqué *et al.* (2009) described deposits of late Pleistocene and early Holocene rock–ice avalanches from the south face of Cerro Aconcagua in South America. The deposits have long been misinterpreted as glacial deposits due to their inclusion of a variety of lithologies, including glacial and alluvial materials entrained along the flow path. The different components in the deposit were not completely mixed; instead, zones consisting mainly of rock-avalanche material are in contact with zones of reworked glacial and alluvial deposits.

Prehistoric catastrophic landslides into a water body are relatively easily identified using bathymetric and seismic data. The blocky deposit of such a landslide will overlie older marine or lacustrine sediments. If the background sedimentation rate is sufficiently high, the deposit will be buried and may be visible only as a layer of chaotic reflectors in a seismic reflection record (Strasser *et al.*, 2006; Longva *et al.*, 2009).

Identifying slopes that might fail catastrophically in the future is a more complex task. Historic data, however, indicate that large rock-slope failures are preceded by slope deformation (Fig. 6.1; Eisbacher and Clague, 1984; Furseth, 2006), which may make it possible to identify a potential rock-slope failure before it happens. A systematic national program to characterize such slopes has been initiated in Norway (Hermanns *et al.*, 2012). Mapping rock slopes that are slowly deforming is the first step in locating the sites of future catastrophic failures. An engineering geologist can then complete a detailed structural analysis of a specific, slowly deforming slope to determine whether catastrophic failure is kinematically possible. Glastonbury and Fell (2010) defined and illustrated, using schematic cross-sections, eight structural environments in which catastrophic slope failures have occurred in the past (Fig. 6.2A–H). In Norway, structural conditions favoring failure in igneous or metamorphic rocks have been identified (Fig. 6.2I, J; Braathen *et al.*, 2004). For example, two sets of joints allow both toppling and sliding (Fig. 6.2J), resulting in a type of landslide referred to as a slide–topple. An investigator should also document how much rock deformation has already taken place and whether and how often catastrophic failures have occurred at the site in the recent past (Guglielmi and Cappa, 2010; Sanchez *et al.*, 2010).

6.4 TEMPORAL ROCKSLIDE DISTRIBUTION AND CLIMATE CHANGE

The impact of climate warming on slope stability, mainly through thawing of alpine permafrost and debuttressing of glacially oversteepened, unstable rock slopes due to glacier retreat, has been the subject of much discussion (Abele, 1974; Evans and Clague, 1994; Noetzli *et al.*, 2007; Fischer *et al.*, 2010; Huggel *et al.*, 2010). Except for the historic period, however, this causative relation can only be demonstrated by dating large numbers of catastrophic rock-slope failures. Based on a compilation of data from the European Alps, Abele (1997) concluded that most large catastrophic rock-slope failures occurred in late-glacial time. Some, however, occurred much later, implying a delay in failure due to progressive rock mass weakening. Most large landslides in the Scottish Highlands appear to be of late-glacial or early-Holocene age and have been attributed to glacial steepening and seismic activity caused by rapid glacio-isostatic rebound (Ballantyne, 1997). Some large landslides in this area, however, happened several thousand years after deglaciation, implying that progressive stress release and joint propagation, and perhaps other time-dependent factors, have played a role (Ballantyne *et al.*, 1998). Cruden and Hu (1993) compiled the ages of large landslides in the Canadian Rocky Mountains and proposed an exhaustion model to explain the decrease in activity through the Holocene. The premises of this model are that there are a finite number of potential failure sites that are conditioned by

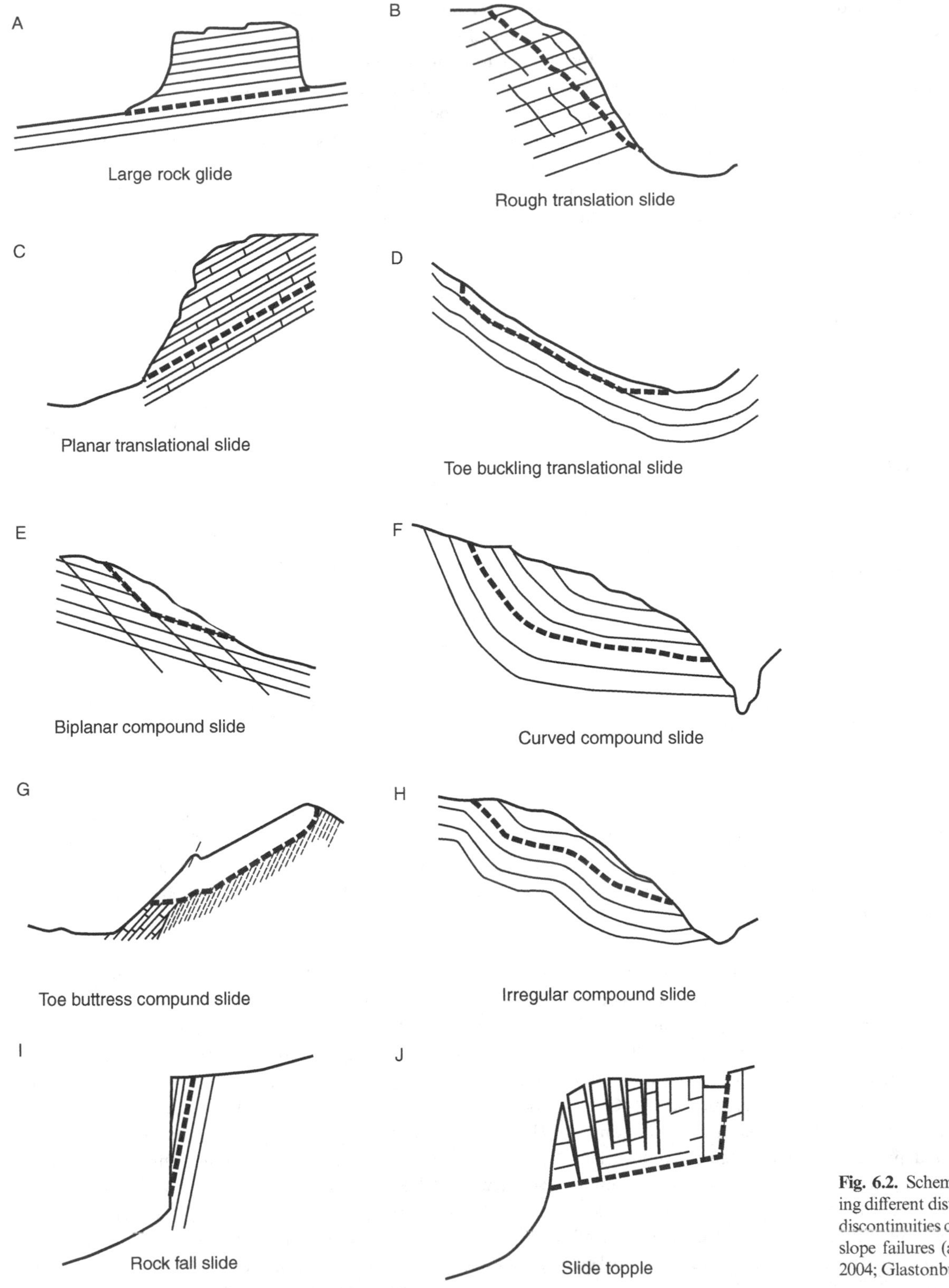

Fig. 6.2. Schematic diagrams showing different distributions of geologic discontinuities controlling large rock-slope failures (after Braathen *et al.*, 2004; Glastonbury and Fell, 2010).

glaciation and that each of these sites fails only once. These premises have been proven to be wrong because more than one slope failure can occur at a single site and because completely new instabilities can be created through the gradual fatigue of rock masses (Hermanns *et al.*, 2006 and references therein; Aa *et al.*, 2007).

Hermanns *et al.* (2000) presented a systematic regional inventory of catastrophic rock-slope failures in northwest

Argentina. The ages of 25 of the 55 mapped deposits were determined through ^{14}C dating, terrestrial cosmogenic nuclide dating, pedological methods, and tephrochronology. The ages of an additional eight deposits were reported later by Hermanns and Schellenberger (2008). None of the 55 landslides in the dataset has a source on glaciated slopes, even though alpine glaciers reached down to 4300 m asl in the easternmost ranges during the late Pleistocene (Haselton *et al.*, 2002). Instead they have sources below 3500 m asl in deeply incised valleys or on slopes several tens of kilometers away from trunk streams. All landslide deposits in the deeply incised valleys date to the late Pleistocene or Holocene, whereas those on the slopes bordering the watershed are all more than 100,000 years old (Hermanns *et al.*, 2000). Landslides in the incised valleys occurred mainly during periods of wetter climate; enhanced runoff and lateral erosion of valley floors were likely the main causes of this landsliding (Trauth *et al.*, 2000; Hermanns and Schellenberger, 2008). However, some large landslides apparently occurred during what are thought to have been dry phases of the Holocene. These exceptions occur near active faults and may have been seismically triggered (Hermanns and Schellenberger, 2008).

Deposits of 22 catastrophic landslides were identified and characterized during a systematic study of a major traffic corridor 800 km south of the region discussed above, within the central Andes (Fig. 6.3; Rosas *et al.*, 2007). Valley glaciers extend down to 3800 m asl in this region today, and reached several hundred meters lower at the maximum of the last glaciation (Fauqué *et al.*, 2009). Twenty of the landslides occurred in ice-free environments; the other two fell onto a valley glacier (Fig. 6.3; Fauqué *et al.*, 2009, Welkner *et al.*, 2010). The ages of 21 of these deposits were determined by terrestrial cosmogenic nuclide dating or radiocarbon dating of plant detritus recovered from lacustrine deposits behind the landslide barriers (Fauqué *et al.*, 2008a, 2008b, 2009; Rosas *et al.*, 2008; Welkner *et al.*, 2010). The landslides occur in two geomorphic settings: (1) fault-controlled valleys that have never been glaciated, and (2) deeply incised valleys that contained glaciers at the Last Glacial Maximum or that drained glaciated valleys. Only one of the 13 dated deposits within the glaciated and trunk valleys dates to a time when glaciers were more extensive than today (Fauqué *et al.*, 2008b, 2009). As shown in Figure 6.3, 10 of the 13 are late-glacial to early-Holocene age, and 2 are younger. The deposits of the two youngest landslides yielded similar ages and are close to one another in a valley with a potentially active fault.

Many catastrophic rock-slope failures have occurred in the deeply incised fjords of Norway. They include disasters in Loen in 1905 and 1936, when large blocks of rock fell into Lake Loenvatnet from the 1493-m-high mountain Ravnefjell (Hermanns *et al.*, 2006 and references therein). The two landslides generated tsunamis with maximum wave heights of 40 and 70 m that, respectively, killed 61 and 74 people. A similar landslide in 1934 in Tafjord triggered a tsunami that killed 40 people (Hermanns *et al.*, 2006 and references therein).

Prehistoric rockslide and rock-avalanche deposits have been found in fjords and valleys throughout Norway; they are best documented in Møre and Romsdal County on the west coast and in Troms County in northern Norway (Blikra *et al.*, 2006; Furseth, 2006). Events in these areas range in age from late-glacial to late-Holocene. However, as seen in Canada, Scotland, and in the Alps, the largest landslides apparently occurred shortly after deglaciation. Detailed mapping and geophysical surveying of Storfjorden revealed deposits of 107 landslides, the largest of which occurred during deglaciation (Table 6.1; Fig. 6.4; Longva *et al.*, 2009). Six of the 107 landslides date to between 12,500 and 11,000 years ago and have an average volume of 59 million m^3; three of these six have volumes of 100–200 million m^3. Landslides were most frequent during the Younger Dryas cold period between 11,000 and 10,000 ^{14}C years BP and at the beginning of the Holocene. Since about 9000 years ago, the frequency of large landslides in Storfjorden has been about five per thousand years, with a slightly increased frequency between 5000 and 1000 years ago. The volumes of the landslides have varied over the Holocene, and there may be both climatic and tectonic signals in the frequency distribution. Norway lies at the western margin of the Baltic Shield, which is thought to be a tectonically passive margin. Some researchers, however, have reported evidence for high tectonic activity at the end of the Pleistocene and into the Holocene in Scandinavia (Mörner, 1996; Bungum *et al.*, 2005). Clusters of rock-slope and soil failures during the late-glacial period have been attributed to earthquakes induced by rapid isostatic rebound (Bøe *et al.*, 2004; Blikra *et al.* 2006). Rapid contemporaneous fluctuations in climate may also have contributed to frequent slope failures at that time. The Holocene climatic optimum in Norway dates to 8000–5000 BP (Hafsten 1986). During that period, the frequency of catastrophic rock-slope failures was the same as earlier and later, but the events – on average – were of smaller size. Climate began to deteriorate about 5000 years ago, and since then the average size of landslides has increased. Systematic regional studies on the temporal distribution of catastrophic slope failures have been carried out in other parts of the world. For example, Bookhagen *et al.* (2005) and Dortch *et al.* (2009) dated 16 catastrophic slope failures in the Himalaya of northern India. Fourteen of these events occurred during two periods, the first a period of increased monsoon activity 40,000 to 30,000 years ago and the second during the most intense monsoon phase of the Holocene, from 8400 to 7200 years ago. On the other hand, Hewitt *et al.* (2011) concluded that earthquakes may have played a greater role than climate in causing the large Holocene rock avalanches in the Karakoram Himalaya that they dated. Soldati *et al.* (2004) documented a cluster of large landslides in the European Alps between about 11,500 and 8500 years ago; many of them were reactivated during the Sub-Boreal period, about 5800–2000 years ago. Prager *et al.* (2008) found that 12 of 14 large (>10^8 m^3) rockslides in the Austrian Alps and surrounding area are Holocene in age, with a minor cluster in the early Holocene and about 4200–3000 years ago in the Sub-Boreal period.

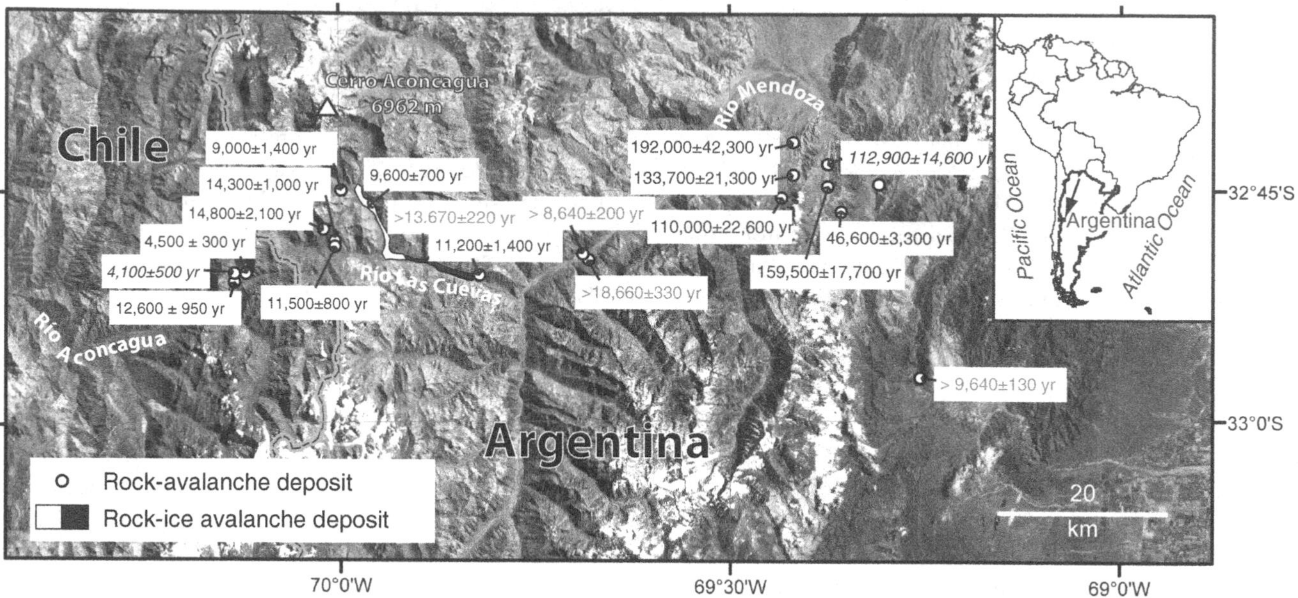

Fig. 6.3. Satellite image of part of the Andes in the vicinity of Cerro Aconcagua, showing the distribution and ages of deposits of catastrophic rock-slope failures. Ages are from Fauqué *et al.* (2008a, 2008b, 2009), Rosas *et al.* (2008), and Welkner *et al.* (2010), and include (1) mean ^{36}Cl exposure ages (black) of 2–6 samples adjusted for an erosion rate of 2.2 mm ka^{-1} (italicized number is an age based on a single sample), and (2) calibrated radiocarbon ages (gray) of plant material recovered from lacustrine deposits behind rockslide barriers. Arrow in inset shows location of the study area.

6.5 CATASTROPHIC LANDSLIDES AND NEOTECTONICS

Abele (1974) discussed the importance of tectonic activity as a preparatory mechanism for landslides. He suggested that intense fracturing of the rock mass adjacent to faults creates conditions conducive to slope failure, and that earthquakes trigger landslides. Many case studies from around the world have demonstrated the importance of tectonic activity both as a conditioning factor and a trigger mechanism.

Tectonic activity contributes to slope instability in three ways. First, it creates zones of weak rock along the fault. All types of faulting break down the rock mass along the fault trace (Brideau *et al.*, 2005, 2009), and folding can produce extension cracks along the hinge zone of anticlines. Second, tectonic activity, operating over long periods, produces relief. Normal and reverse faulting and folding are most efficient in generating important relief (Fig. 6.5). Strike–slip faults, however, can also generate relief in transpressional zones at kinks in the fault trace. Third, tectonic activity can translate inherited structures within the rock mass into positions that are more favorable to failure, for example by producing inclined bedded planes. Folding changes the orientation of discontinuities over the entire structure. In contrast, the effect is more localized in the case of reverse and normal faulting; unfavorably oriented discontinuities in the rock mass can be exposed in the hanging wall of a normal or reverse fault. Rockslides capitalizing on these exposed inherited discontinuities have been reported from around the world (Eisbacher and Clague, 1984; Hermanns and Strecker, 1999; Martino *et al.*, 2004, von Poschinger *et al.*, 2006).

The association of normal faulting and slope collapse has been documented at many sites (Dobrovolny, 1962; Martino *et al.*, 2004; Brideau *et al.*, 2005; Redfield and Osmundsen, 2009), but the number of catastrophic rock-slope failures related to those structures is small. In reverse fault settings (Fig. 6.5B), the situation is generally different, and multiple large landslides have been documented along these structures (Hermanns and Strecker, 1999; Jackson, 2002; Penna *et al.*, 2011). Reverse faulting also produces relief contrasts (Fig. 6.5B) and weakening of rocks along the fault. Thus, in the long term, movement along the fault leads to oversteepening and slope collapse. Some of the world's largest nonvolcanic rockslides have occurred in northern Chile in this tectonic setting (Wörner *et al.*, 2002). Hermanns *et al.* (2001) showed that activity on reverse faults in a mountain range in Argentina caused slope steepening until about 150,000 years ago, when the locus of deformation shifted away from the range front to the foreland. Between about 400,000 and 150,000 years ago, one large rock avalanche occurred, on average, once every 30,000 years. In contrast, there have been no large rock avalanches at the range front during the past 150,000 years. Tectonically inactive mountain fronts built of the same rocks and with the same relief lack any evidence of large rock-slope collapses (Hermanns and

Table 6.1. *Overview of temporal distribution of number and volume of 107 rock avalanches in Storfjorden, Norway.*

Period (^{14}C ka BP)[a]	12.5–11	11–10	10–9	9–8	8–5	5–1	1–0
Number of events	6	19	26	5	16	30	5
Events per 1000 years	4	19	26	5	5	8	5
Total volume (Mm^3)	354	168	31.7	6.7	3.9	15.5	7.6
Average volume (Mm^3)	59	8.8	1.2	1.3	0.2	0.5	1.5
Volume per 1000 years	236	168	31.7	6.7	1.3	3.9	7.6

[a] The ages of events are based on one radiocarbon-dated core and the regional seismic stratigraphy.

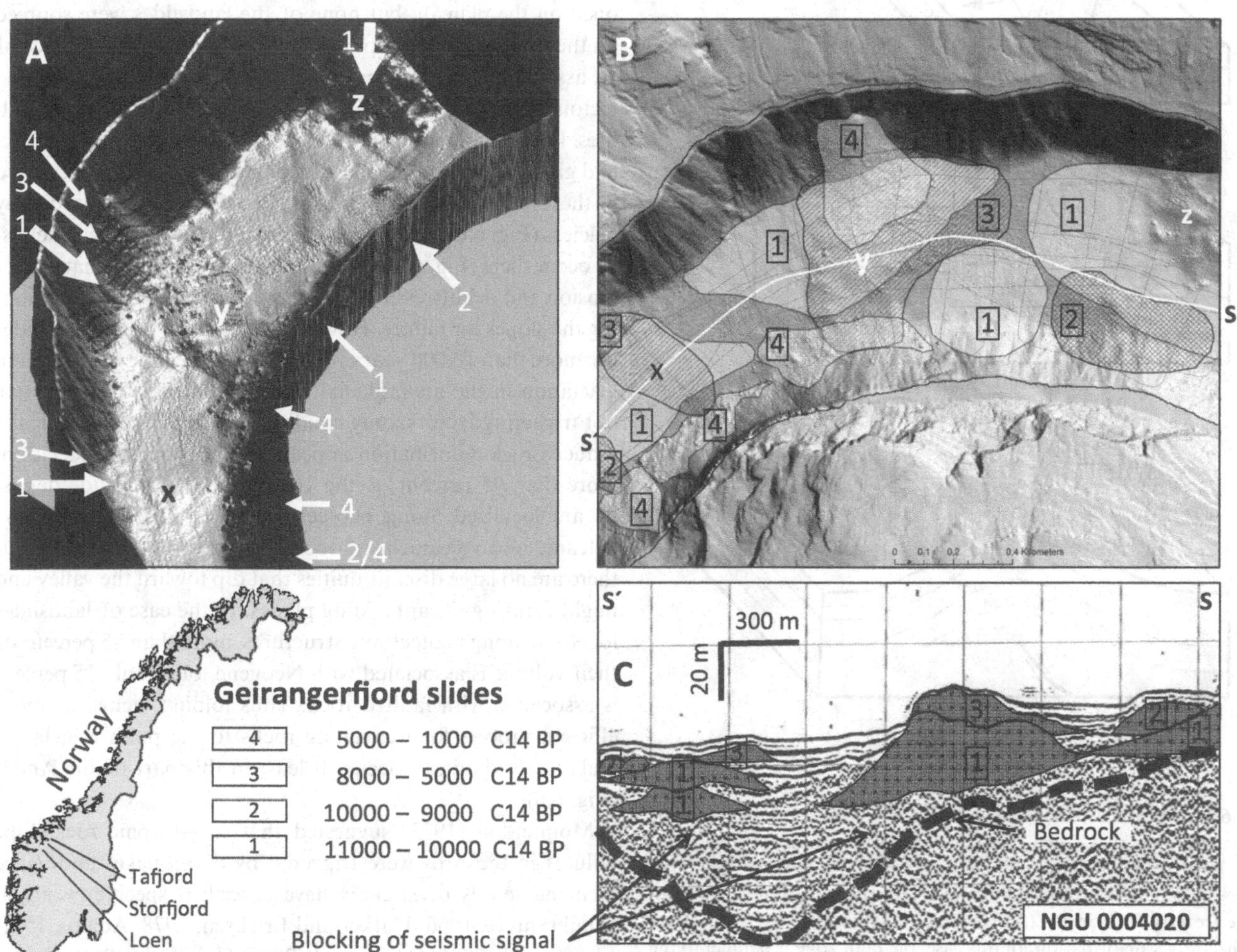

Fig. 6.4. Morphological and seismic interpretation, and estimated age of landslide deposits in Geirangerfjorden, Norway. (A) Shaded relief image; note that deposits of younger landslides have a sharper, more irregular, morphology than those of older, deeper, buried ones. The number and size of each arrow indicates the source, age, and size of the landslide. (B) Seismic lines (white) and depositional areas of landslides. (C) Interpretation of seismic line. Letters X and Z in Parts A and B mark the same locations.

Strecker, 1999). Strecker and Marrett (1999) attribute the high concentration of rock avalanches in this area to a reorganization of tectonic deformation in the Neogene, when former strike–slip faults were reactivated as reverse faults. Catastrophic rock-slope failures occurred also along other strike–slip faults. Sepulveda *et al.* (2010) documented catastrophic rockslides along a strike–slip fault in the Patagonian Andes of Chile, but at those sites the fault intersected deeply incised valleys and the relief required for slope failure was not produced by the fault itself (Fig. 6.5C).

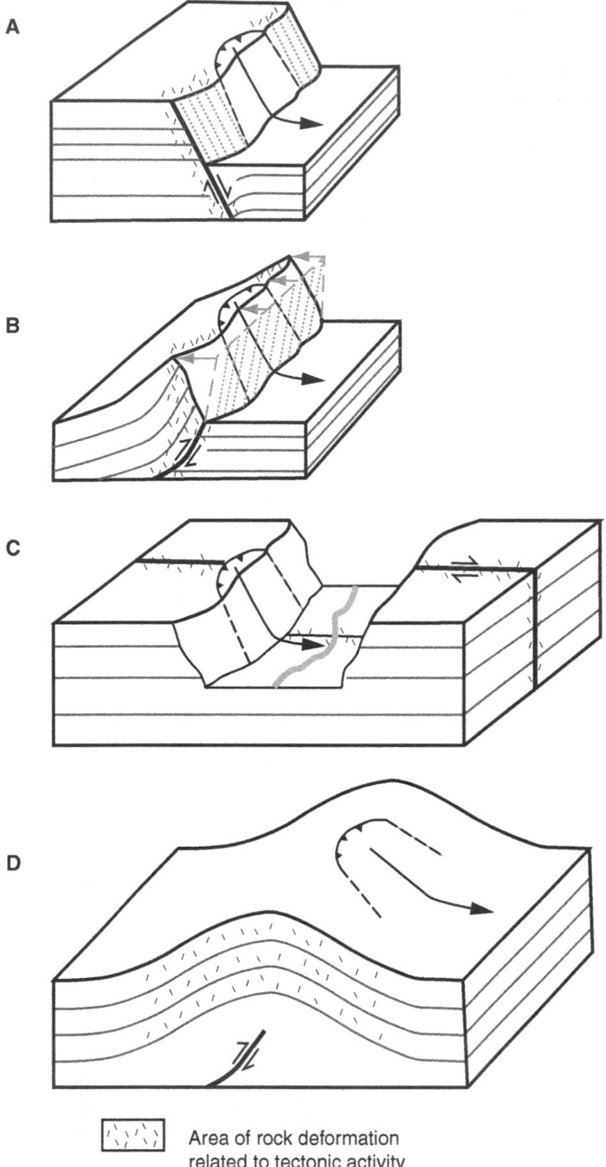

Fig. 6.5. Schematic block diagrams showing the impact of tectonic deformation on rock-slope stability. A simple example is shown here, with horizontally bedded sedimentary rocks. The gray dotted lines represent a fault plane (normal fault in A) or the projection of the fault plane (reverse fault in B). (A) Normal faults produce local relief and localized rock deformation. In this case, the fault zone is parallel to the slope, and sliding can occur along the fault plane. (B) Reverse faults also produce local relief and rock deformation. The area of rock deformation can be especially large in the hanging wall of listric reverse faults, where the rock mass is compressed and tilted. The fault plane is not a possible sliding plane because it dips into the rock mass. Gray dashed line depicts the orientation of the fault in the failed rock mass; arrows indicate slope adjustment by collapse and erosion. (C) Strike–slip faults can enhance local relief by offsetting sloping ground. More importantly, they deform rock along the fault plane; slope failures can occur where the fault crosses a valley. The fault is too steep to be a sliding plane. (D) Folding produces surface relief and can create dipping planes of weakness along which sliding may occur. Rock deformation occurs over a large area and is related to tilting and extension.

In most normal and reverse fault settings, it is difficult to separate the roles of tectonically induced rock deformation, tectonic oversteepening of slopes, and slope steepening by erosion along valleys in conditioning slopes for failure. A region where tectonic oversteepening plays a minor role is the transition between the Central and the Patagonian Andes in Argentina (Penna *et al.*, 2011). The main factors responsible for landslides in this region are erosion along valleys and tectonically induced rock deformation. A total of 19 large landslides have occurred in volcanic and volcaniclastic rocks of Plio-Pleistocene age underlying a plateau that was deformed in the Quaternary by faulting and folding (Fig. 6.6). Local relief of 15–400 m developed on the plateau, but none of the landslides were sourced on the faults or folds responsible for this relief. The absence of an association of the local relief and landslides suggests that tectonic oversteepening of slopes is not responsible for the failures. Local erosional relief of 200–1200 m was created by rivers and glaciers crossing the plateau, and all 19 rockslides occurred in these valleys. Relief is greatest in valley sections eroded by glaciers (Fig. 6.6), and about 80 percent of the landslide deposits occur there (Fig. 6.7). The evidence thus suggests that glacier erosion and debuttressing were important factors in conditioning the slopes for failure. However, because all of the landslides are more than 10,000 years younger than the time of maximum glaciation in the area, glacial erosion and debuttressing were not triggering factors, only conditioning ones. Neotectonically induced rock deformation appears as important as glaciation: more than 95 percent of the volume of the landslide deposits are localized along neotectonic structures. Some of the volcanic and volcaniclastic rocks are intensely fractured, but there are no large discontinuities that dip toward the valley and might form significant sliding planes. In the case of landslides localized along neotectonic structures, more than 85 percent of their volume is associated with Neogene folds; only 15 percent is associated with faulted rock. Thus folding seems the more efficient process for weakening rocks to the point that large-scale landsliding can occur, at least in this part of the Andes (Fig. 6.6).

Montandon (1933) suggested that catastrophic rock-slope failures in the Alps were triggered by earthquakes and, since then, numerous researchers have described specific events in detail (Shreve, 1966; Plafker and Ericksen, 1978; Adams, 1981; Jibson *et al.*, 2006; Owen *et al.*, 2008; Dai *et al.*, 2011). Keefer (1984) and Rodríguez *et al.* (1999) summarized observations on, respectively, 40 and 36 earthquakes that triggered landslides. Most of the earthquakes that triggered large rock avalanches were crustal; only four were great subduction earthquakes. This statistic agrees with evidence gathered in recent years that shallow earthquakes in the continental crust commonly trigger large landslides along or near the surface rupture (Jibson *et al.*, 2006; Sepulveda *et al.*, 2010; Dai *et al.*, 2011), whereas subduction earthquakes only occasionally trigger large landslides. For example, the 1970 Nevado Huascaran rock–ice avalanche (Plafker and Ericksen, 1978) was the only historic catastrophic rock-slope failure in the Cordillera Blanca of Peru to have been triggered by a subduction earthquake. Cerro Huascaran is

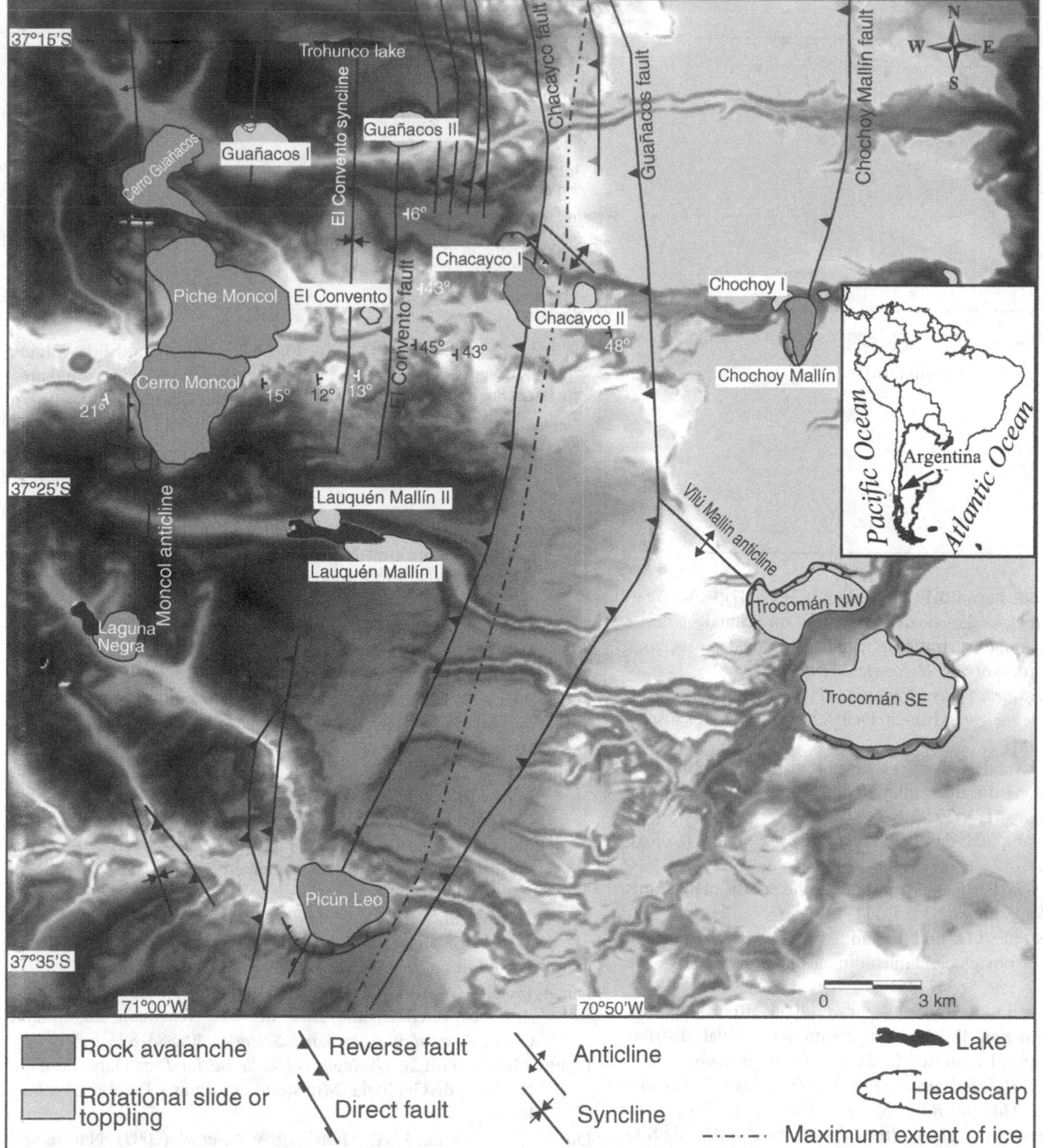

Fig. 6.6. Distribution of neotectonic structures and landslide deposits in plateau basalts in the transitional area between the Central and Patagonian Andes. The map is draped over a digital elevation model; the plateau and high valleys are light gray and white; mountain tops and low valleys are dark gray (modified from Penna *et al.*, 2011). Arrow in inset shows location of the study area.

bordered by a normal fault that had been active in the Holocene (Schwartz, 1988). Only eight years earlier, a similar rock–ice avalanche happened on the same rock face, but without any obvious trigger (Plafker and Ericksen, 1978), and in 1946, a M7.3 crustal earthquake triggered several rock avalanches close to its epicenter, about 60 km from Cerro Huascaran (Heim, 1949; Kampherm *et al.*, 2009). This example illustrates that the coincidence of a catastrophic landslide deposit with a neotectonic fault cannot be taken as proof of seismic triggering. Thus landslide deposits should only be used to supplement other independent evidence for earthquakes, such as fault offsets and seismically induced soft-sediment deformation structures (Hermanns and Niedermann, 2011). Confidence in a seismic trigger may be increased if slope stability models indicate that the failed rock slope could not have been unstable under aseismic conditions (Jibson, 2009).

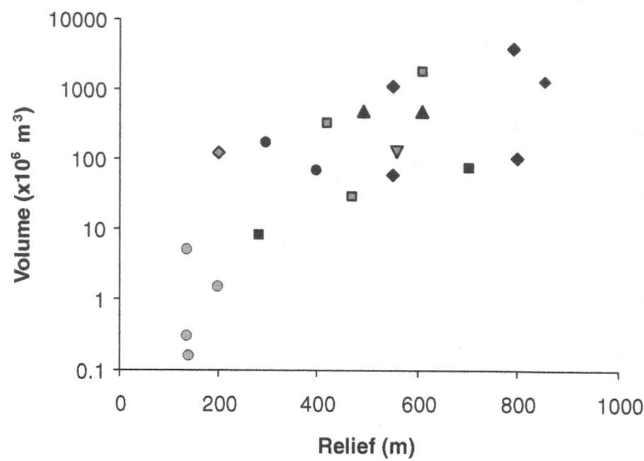

Related to folds
- ◆ Rock avalanche in glaciated valley section
- ◇ Rock avalanche in non glaciated valley section
- ■ Rotational slide or topple in glaciated valley section
- ▣ Rotational slide or topple in non glaciated valley section

Related to faults
- ▲ Rock avalanche in glaciated valley sections
- ▽ Rotational slide in glaciated valley sections

Not related to tectonic structures
- ● Rotational slide or topple in glaciated valley sections
- ○ Rotational slide or topple in non glaciated valley sections

Fig. 6.7. Plot of landslide volume versus relief in the transitional area between the Central and Southern Andes (modified from Penna *et al.*, 2011).

REFERENCES

Aa, A. R., Sjåstad, J., Sønstegaard, E. and Blikra, L. H. (2007). Chronology of Holocene rock-avalanche deposits based on Schmidt-hammer, relative dating and dust stratigraphy in nearby bog deposits, Vora, inner Nordfjord, Norway. *Holocene*, 17, 955–964.

Abele, G. (1974). *Bergstürze in den Alpen: Ihre Verbreitung, Morphologie und Folgeerscheinungen*. Munich: Deutscher und Österreichischer Alpenverein.

Abele, G. (1997). Rockslide movement supported by the mobilization of groundwater-saturated valley floor sediments. *Zeitschrift für Geomorphologie*, 41, 1–20.

Adams, J. (1981). Earthquake-dammed lakes in New Zealand. *Geology*, 9, 215–219.

Ballantyne, C. K. (1997). Periglacial trimlines in the Scottish Highlands. *Quaternary International*, 38–39, 119–136.

Ballantyne, C. K., Stone, J. O. and Fifield, L. K. (1998). Cosmogenic Cl-36 dating of postglacial landsliding at the Storr, Isle of Skye, Scotland. *Holocene*, 8, 347–351.

Blikra, L. H., Longva, O., Braathen, A. *et al.* (2006). Rock slope failures in Norwegian fjord areas: Examples, spatial distribution and temporal patterns. In *Landslides from Massive Rock Slope Failure. Proceedings of the NATO Advanced Research Workshop on Massive Rock Slope Failure: New Models for Hazard Assessment, Celano, Italy, 16–21 June 2002*, ed. S. G. Evans, G. Scarascia Mugnozza, A. Strom and R. L. Hermanns. NATO Science Series IV, Earth and Environmental Sciences 49. Dordrecht, Netherlands: Springer, pp. 475–496.

Bøe, R., Longva, O., Lepland, A. *et al.* (2004). Postglacial mass movements and their causes in fjords and lakes in western Norway. *Norwegian Journal of Geology*, 84, 35–55.

Bookhagen, B., Thiede, R. and Strecker, M. R. (2005). Late Quaternary intensified monsoon phases control landscape evolution in the northwest Himalaya. *Geology*, 33, 149–152.

Braathen, A., Blikra, L. H., Berg, S. S. and Karlsen, F. (2004). Rock-slope failures in Norway: Type, geometry, deformation mechanisms and stability. *Norsk Geologisk Tidsskrift*, 84, 67–88.

Brideau, M., Stead, D., Kinakin, D. and Fecova, K. (2005). Influence of tectonic structures on the Hope Slide, British Columbia, Canada. *Engineering Geology*, 80, 242–259.

Brideau, M., Yan, M. and Stead, D. (2009). The role of tectonic damage and brittle rock fracture in the development of large rock slope failures, *Geomorphology*, 103, 30–49.

Bungum, H., Lindholm, C. and Faleide, J. I. (2005). Postglacial seismicity offshore mid-Norway with emphasis on spatio-temporal-magnitudal variations. *Marine and Petroleum Geology*, 22, 137–148.

Clague, J. J. and Evans, S. G. 1994. *Formation and Failure of Natural Dams in the Canadian Cordillera*. Geological Survey of Canada, Bulletin 464.

Costa, J. E. and Schuster, R. L. (1988). The formation and failure of natural dams. *Geological Society of America Bulletin*, 100, 1054–1068.

Crandell, D. R. and Fahnestock, R. K. (1965). *Rockfalls and Avalanches from Little Tahoma Peak on Mount Rainier, Washington*. US Geological Survey, Bulletin 1221-A.

Crosta, G. B. and Agliardi, F. (2003). Failure forecast for large rock slides by surface displacement measurements. *Canadian Geotechnical Journal*, 40, 176–191.

Cruden, D. M. and Hu, X. Q. (1993). Exhaustion and steady state models for predicting landslide hazards in the Canadian Rocky Mountains. *Geomorphology*, 8, 279–285.

Dai, F. C., Xu, C., Yao, X. *et al.* (2011). Spatial distribution of landslides triggered by the 2008 Ms 8.0 Wenchuan earthquake, China. *Journal of Asian Earth Sciences*, 40, 883–895.

Dobrovolny, E. (1962), *Geologia del Valle de La Paz*. Departamento Nacional de Geología, Ministerio de Minas y Petróleo, La Paz, Bolivia.

Dortch, J. M., Owen, L. A., Haneberg, W. C. *et al.* (2009). Nature and timing of large landslides in the Himalaya and Transhimalaya of northern India. *Quaternary Science Reviews*, 28, 1037–1054.

Eisbacher, G. H. and Clague, J. J. (1984). *Destructive Mass Movements in High Mountains: Hazard and Management*. Geological Survey of Canada, Paper 84–16.

Evans, S. G. and Clague, J. J. (1994). Recent climatic change and catastrophic geomorphic processes in mountain environments. *Geomorphology*, 10, 107–128.

Evans, S. G., Roberts, N. J., Ischuk, A. *et al.* (2009a). Landslides triggered by the 1949 Khait earthquake, Tajikistan, and associated loss of life. *Engineering Geology*, 109, 195–212.

Evans, S. G., Bishop, N. F., Fidel Smoll, L. *et al.* (2009b). A re-examination of the mechanism and human impact of catastrophic mass flows originating on Nevado Huascarán, Cordillera Blanca, Peru in 1962 and 1970. *Engineering Geology*, 108, 96–118.

Evans, S. G., Delaney, K. B., Hermanns, R. L., Strom, A. L. and Scarascia Mugnozza, G. (2011). The formation and behaviour of natural and artificial rockslide dams: Implications for engineering performance and hazard management. In *Natural and Artificial Rock Slide Dams*, ed. S. G. Evans, R. L. Hermanns, A. L. Strom and G. Scarascia Mugnozza. Berlin: Springer, pp. 1–74.

Fauqué, L., Cortés, J. M., Folguera, A. *et al.* (2008a). Edades de las avalanchas de rocas ubicadas en el Río Mendoza aguas abajo de Uspallata. *Actas del XVII Congreso Geológico Argentino, Jujuy*, 1, 282–283.

Fauqué, L., Hermanns, R. L., Wilson, C. *et al.* (2008b). Paleorepresamientos del Río Mendoza entre Polvaredas y Punta de Vacas, Mendoza, Argentina. *Actas del XVII Congreso Geológico Argentino, Jujuy*, 1, 274–275.

Fauqué, L., Hermanns, R. L., Hewitt, K. *et al.* (2009). Megadeslizamientos de la pared sur del Cerro Aconcagua y su relación con depósitos asignados a la glaciación pleistocena. *Revista de la Asociación Geológica Argentina*, 65, 691–712.

Ferrer, C. (1999). Represamientos y rupturas de embalses naturales (lagunas de obstrución) como efectos cosísmicos: Algunos ejemplos en los Andes venezolanos. *Revista Geográfica Venezolana*, 40, 109–121.

Fischer, L., Amann, F., Moore, J. and Huggel, C. (2010). Assessment of periglacial slope stability for the 1988 Tschierva rock avalanche (Piz Morteratsch, Switzerland). *Engineering Geology*, 116, 32–43.

Furseth, A. (2006). *Skredulykker i Norge*. Oslo: Tun Forlag.

Glastonbury, J. and Fell, R. (2010). Geotechnical characteristics of large rapid rock slides. *Canadian Geotechnical Journal*, 47, 116–132.

Guglielmi, Y. and Cappa, F. (2010). Regional-scale relief evolution and large landslides: Insights from geomechanical analyses in the Tinée valley (Southern French Alps). *Geomorphology*, 117, 121–129.

Hafsten, U. (1986). The establishment of spruce forest in Norway, traced by pollen analysis and radiocarbon datings. *Striae*, 24, 101–105.

Harrison, J. V. and Falcon, N. L. (1934). Collapse structures (Kuhgalu district, Persia). *Geological Magazine*, 71, 529–539.

Haselton, K., Hilley, G. and Strecker, M. (2002). Average Pleistocene climatic patterns in the southern Central Andes: Controls on mountain glaciations and paleoclimate implications. *Journal of Geology*, 110, 211–226.

Hauser, A. (2002). Rock avalanche and resulting debris flow in Estero Parraguirre and Río Colorado, region Metropolitana, Chile. In *Catastrophic Landslides*, ed. S. G. Evans and J. V. DeGraff. Geological Society of America, Reviews in Engineering Geology 15, pp. 135–148.

Heim, A. (1932). *Bergsturz und Menschenleben*. Zurich: Beiblatt zur Vierteljahresschrift der Naturforschenden Gesellschaft.

(1949). Observaciones geológicas en la región del terremoto de Ancash de Noviembre de 1946. *Sociedad Geológica del Perú*, 25, 2–21.

Hermanns, R. L. and Niedermann, S. (2011). Late Pleistocene–Early Holocene paleoseismicity deduced from lake sediment deformation and coeval landsliding in the Calchaquíes valleys, NW Argentina. In *Geological Criteria for Evaluating Seismicity Revisited: Forty Years of Paleoseismic Investigations and the Natural Record of Past Earthquakes*, ed. F. A. Audemard, A. Michetti and J. P. McCalpin. Geological Society of America, Special Paper 479, pp. 181–194.

Hermanns, R. L. and Schellenberger, A. (2008). Quaternary tephrochronology helps define conditioning factors and triggering mechanisms of rock avalanches in NW Argentina. *Quaternary International*, 178, 261–275.

Hermanns, R. L. and Strecker, M. R. (1999). Structural and lithological controls on large Quaternary rock avalanches (sturzstroms) in arid northwestern Argentina. *Geological Society of America Bulletin*, 111, 934–948.

Hermanns, R. L., Trauth, M. H., Niedermann, S., McWilliams, M. and Strecker, M. R. (2000). Tephrochronologic constraints on temporal distribution of large landslides in northwest Argentina. *Journal of Geology*, 108, 35–52.

Hermanns, R. L., Niedermann, S., Villanueva Garcia, A., Sosa Gomez, J. and Strecker, M. R. (2001). Neotectonics and catastrophic failure of mountain fronts in the southern intra-Andean Puna Plateau, Argentina. *Geology*, 29, 619–623.

Hermanns, R., Blikra, L., Naumann, M. *et al.* (2006). Examples of multiple rock-slope collapses from Köfels (Ötz Valley, Austria) and western Norway. *Engineering Geology*, 83, 94–108.

Hermanns, R. L., Blikra, L. H., Anda, E. *et al.* (2012). Systematic mapping of large unstable rock slopes in Norway. In *Proceedings of the 2nd World Landslide Forum, Rome*.

Hewitt, K., Gosse, J. and Clague, J. J. (2011). Rock avalanches and the pace of late Quaternary development of river valleys in the Karakoram Himalaya. *Geological Society of America Bulletin*, 123, 1836–1850.

Hsü, K. J. (1975). Catastrophic debris streams (sturzstroms) generated by rockfalls, *Geological Society of America Bulletin*, 86, 129–140.

Huggel, C., Zgraggen-Oswald, S., Haeberli, W. *et al.* (2005). The 2002 rock/ice avalanche at Kolka/Karmadon, Russian Caucasus: Assessment of extraordinary avalanche formation and mobility, and application of QuickBird satellite imagery. *Natural Hazards and Earth System Sciences*, 5, 173–187.

Huggel, C., Fischer, L., Schneider, D. and Haeberli, W. (2010). Research advances on climate-induced slope instability in glacier and permafrost high-mountain environments. *Geographica Helvetica*, 65, 146–156.

Hungr, O. (2006). Rock avalanche occurrence, process and modelling. In *Landslides from Massive Rock Slope Failure. Proceedings of the NATO Advanced Research Workshop on Massive Rock Slope Failure: New Models for Hazard Assessment, Celano, Italy, 16–21 June 2002*, ed. S. G. Evans, G. Scarascia Mugnozza, A. Strom and R. L. Hermanns. NATO Science Series IV, Earth and Environmental Sciences 49. Dordrecht, Netherlands: Springer, pp. 243–266.

Hungr, O. and Evans, S. G. (2004). Entrainment of debris in rock avalanches: An analysis of a long run-out mechanism. *Geological Society of America Bulletin*, 116, 1240–1252.

Jackson, L. E., Jr. (2002). Landslides and landscape evolution in the Rocky Mountains and adjacent foothills area, southwestern Alberta, Canada. In *Catastrophic Landslides*, ed. S. G. Evans and J. V. DeGraff. Geological Society of America, Reviews in Engineering Geology 15, pp. 325–344.

Jibson, R. W. (2009). Using landslides for paleoseismic analysis. In *Paleoseismology*, ed. J. P. McCalpin. Burlington, MA: Academic Press, pp. 565–601.

Jibson, R. W., Harp, E. L., Schulz, W. and Keefer, D. K. (2006). Large rock avalanches triggered by the M7.9 Denali fault, Alaska, earthquake of 3 November 2002. *Engineering Geology*, 83, 144–160.

Kampherm, T. S., Evans, S. G. and Valderrama Murillo, P. (2009). Landslides triggered by the 1946 Ancash earthquake, Peru. *Geophysical Research Abstracts*, 11, EGU2009–13820.

Keefer, D. K. (1984). Landslides caused by earthquakes. *Geological Society of America Bulletin*, 95, 406–421.

Kojan, E. and Hutchinson, J. N. (1978). Mayunmarca rockslide and debris flow, Peru. In *Rockslides and Avalanches. 1. Natural Phenomena*, ed. B. Voight. Amsterdam: Elsevier Scientific Publishing Company, pp. 315–361.

Leroueil, S., Locat, J., Vaunat, J., Picarelli, L. and Faure, R. (1996). Geotechnical characterization of slope movements. In *Proceedings of the International Symposium on Landslides: 1*. Rotterdam: Balkema, pp. 53–74.

Longva, O., Blikra, L. H. and Dehls, J. F. (2009). *Rock avalanches: Distribution and Frequencies in the Inner Part of Storfjorden,*

Møre og Romsdal County, Norway. Norwegian Geotechnical Institute, Report 2009.002.

Martino, S., Moscatelli, M. and Scarascia Mugnozza, G. (2004). Quaternary mass movements controlled by a structurally complex setting in the central Apennines (Italy). *Engineering Geology*, 72, 33–55.

Montandon, F. (1933). Chronologie des grands éboulements alpins du debut de l'ére chrétienne à nos jours. *Matériaux pour i'étude des calamités*, 32, 271–340.

Mörner, N.-A. (1996). Liquefaction and varve deformation as evidence of paleoseismic events and tsunamis: The autumn 10,430 BP case in Sweden. *Quaternary Science Reviews*, 15, 939–948.

Nicoletti, P. G. and Sorriso-Valvo, M. (1991). Geomorphic controls of the shape and mobility of rock avalanches. *Geological Society of America Bulletin*, 103, 1365–1373.

Noetzli, J., Gruber, S., Kohl, T., Salzman, N. and Haeberli, W. (2007). Three-dimensional distribution and evolution of permafrost temperatures in idealized high-mountain topography. *Journal of Geophysical Research*, 112, F02S13. doi:10.1029/2006JF000545.

Owen, L. A., Kamp, U., Khattak, G. A. *et al.* (2008). Landslides triggered by the 8 October 2005 Kashmir earthquake. *Geomorphology*, 94, 1–9.

Penna, I., Hermanns, R. L., Folguera, A. and Niedermann, S. (2011). Multiple slope failures associated with neotectonic activity in the southern central Andes (37°–37°30′S). Patagonia, Argentina. *Geological Society of America Bulletin*, 123, 1880–1895.

Plafker, G. and Ericksen, G. E. (1978). Nevados Huascaran avalanches, Peru. In *Rockslides and Avalanches*, ed. B. Voight. Amsterdam: Elsevier, pp. 277–314.

Prager, C., Zangerl, C., Patzelt, G. and Brandner, R. (2008). Age distribution of fossil landslides in the Tyrol (Austria) and its surrounding areas. *Natural Hazard and Earth System Sciences*, 8, 377–407.

Redfield, T. F. and Osmundsen, P. T. (2009). The Tjellefonna fault system of Western Norway: Linking late-Caledonian extension, post-Caledonian normal faulting, and Tertiary rock column uplift with the landslide-generated tsunami event of 1756. *Tectonophysics*, 474, 106–123.

Rodríguez, C. E., Bommer, J. J. and Chandler, R. J. (1999). Earthquake-induced landslides: 1980–1997. *Soil Dynamics and Earthquake Engineering*, 18, 325–346.

Rosas, M., Baumann, V., Videla, A. *et al.* (2007). *Estudio Geocientífico Aplicado al Ordenamiento Territorial, Puente del Inca, Provincia de Mendoza*. Servicio Geológico Minero Argentino, Informe Final.

Rosas, M., Wilson, C., Hermanns, R. L., Fauqué, L. and Baumann, V. (2008). Avalanchas de rocas de las cuevas una evidencia de la destabilisación de las laderas como consecuencia del cambio climático del Pleistoceno superior. *Actas del XVII Congreso Geológico Argentino, Jujuy*, 1, 313–314.

Sanchez, G., Rolland, Y., Corsini, M. G. *et al.* (2010). Relationships between tectonics, slope instability and climate change: Cosmic ray exposure dating of active faults, landslides and glacial surfaces in the SW Alps. *Geomorphology*, 117, 1–13.

Scheidegger, A. E. (1961). *Theoretical Geomorphology*. Berlin: Springer.

Schwartz, D. (1988). Paleoseismicity and neotectonics of the Cordillera Blanca fault zone, northern Peruvian Andes. *Journal of Geophysical Research*, 93(B5), 4712–4730.

Schuster, R. L. and Alford, D. (2004). Usoi landslide dam and lake Sarez, Pamir Mountains, Tajikistan. *Environmental and Engineering Geoscience*, 10, 151–168.

Sepulveda, S., Serey, A., Lara, M., Pavez, A. and Rebolledo, S. (2010). Landslides induced by the April 2007 Aysén fjord earthquake, Chilean Patagonia. *Landslides*, 7, 483–492.

Shreve, R. L. (1966). Sherman landslide, Alaska. *Science*, 154, 1639–1643.

Soldati, M., Corsini, A. and Pasuto, A. (2004). Landslides and climate change in the Italian Dolomites since the Late Glacial. *Catena*, 55, 141–161.

Strasser, M., Anselmetti, F. S., Fäh, D., Giardini, D. and Schnellmann, M. (2006). Magnitudes and source areas of large prehistoric northern alpine earthquakes revealed by slope failures in lakes. *Geology*, 34, 1005–1008.

Strecker, M. R. and Marrett, R. A. (1999). Kinematic evolution of fault ramps and role in development of landslides and lakes in northwestern Argentine Andes. *Geology*, 27, 307–310.

Tappin, D. R. (2010). Mass transport events and their tsunami hazard. In *Submarine Mass Movements and Their Consequences*, ed. D. C. Mosher, L. Moscardelli, J. D. Chaytor, C. D. P. Baxter, H. J. Lee and R. Urgeles. Berlin: Springer, pp. 667–684.

Trauth, M. H., Alonso, R. A., Haselton, K. R., Hermanns, R. L. and Strecker, M. R. (2000). Climate change and mass movements in the NW Argentine Andes. *Earth and Planetary Science Letters*, 179, 243–256.

von Poschinger, A., Wassmer, P. and Maisch, M. (2006). The Flims rockslide: History of interpretation and new insights. In *Landslides from Massive Rock Slope Failure. Proceedings of the NATO Advanced Research Workshop on Massive Rock Slope Failure: New Models for Hazard Assessment, Celano, Italy, 16–21 June 2002*, ed. S. G. Evans, G. Scarascia Mugnozza, A. Strom and R. L. Hermanns. NATO Science Series IV, Earth and Environmental Sciences 49. Dordrecht, Netherlands: Springer, pp. 329–356.

Welkner, D., Eberhardt, E. and Hermanns, R. L. (2010). Hazard investigation of the Portillo rock avalanche site, central Andes, Chile, using an integrated field mapping and numerical modelling approach. *Engineering Geology*, 114, 278–297.

Wörner, G., Uhlig, D., Kohler, I. and Seyfried, H. (2002). Evolution of the west Andean escarpment at 18°S (N. Chile) during the last 25 ma: Uplift, erosion and collapse through time. *Tectonophysics*, 345, 183–198.

Yarnold, J. C. (1993). Rock-avalanche characteristics in dry climates and the effect of flow into lakes: Insights from mid-Tertiary sedimentary breccias near Artillery Peak, Arizona, *Geological Society of America Bulletin*, 105, 345–360.

7 Risk assessments for debris flows

MATTHIAS JAKOB AND KRIS HOLM

ABSTRACT

Increasingly, debris-flow hazard assessments based on largely arbitrary design return periods are being replaced by risk assessments because the inclusion of consequences is ultimately required for sound management decisions. A full quantitative risk assessment of debris flows requires high-quality and well-researched input parameters in order to obtain defensible results. It adds a degree of sophistication that may not be warranted for all applications because of its higher intensity of field and office effort. For high-consequence cases, however, it appears to be an indispensable tool. This chapter details methods that can be applied to assessing debris-flow risk to fixed developments as well as linear facilities, and highlights the various caveats associated with each method.

7.1 INTRODUCTION

Debris flows are one of the most destructive mass-movement processes worldwide. In urban settings they can damage public and private property, causing economic loss and injury. Outside urban areas, debris flows interrupt roads, highways, railways, pipelines, and electrical transmission and telecommunication lines. Injury or loss of life may also result from debris-flow impact to vehicles, trains, or fixed engineered works such as industrial buildings. Consequently, the many physical and geotechnical aspects of debris flows have been the subject of intensive research for decades (Jakob and Hungr, 2005). Despite this intensive research, however, relatively few papers have addressed methods for assessing debris-flow risk and the levels of such risk that may be tolerated by individuals and society. These topics have important implications for development decisions made in debris-flow hazard areas, including the issuing of permits that allow urban or industrial development to proceed.

Risk is a function of the product of the likelihood that a hazard will occur and its anticipated consequences. Procedures to assess and reduce geohazard risk are available to guide decision makers and the public, and form the first three of four major steps in the larger context of geohazard risk management: identification, analysis, evaluation, and mitigation.

However, it is generally not possible to completely eliminate geohazard risk, and the level of risk considered tolerable or acceptable is commonly not defined in legislation. As a consequence, decision makers are confronted with a difficult choice. In some cases, risk reduction measures are considered and implemented, with some remaining level of residual risk that is not necessarily well defined. In other cases, risk is not explicitly considered, for example within development-permitting legislation based on arbitrary hazard return periods. The difficulties facing a decision maker who is required to make planning decisions based on an assessment of risk are many, and include socio-political resistance to setting risk tolerance levels, technical or budgetary limitations on collecting sufficient data for defensible assessments, and the difficulty of quantifying some losses, such as human suffering. In some jurisdictions, decision makers may resist implementing a risk-based decision-making framework for landslides if they perceive that it might later result in lawsuits or might force expensive capital expenditures.

Nonetheless, an analysis of landslide risk and comparisons with defined thresholds for determining tolerable risk can be a systematic and defensible way to guide decision makers in development planning. This is especially the case for debris flows, because they occur in locations commonly proposed for development, such as fans; because their periodic re-occurrence in an area lends itself particularly well to a risk-based assessment

Landslides: Types, Mechanisms and Modeling, ed. John J. Clague and Douglas Stead. Published by Cambridge University Press.

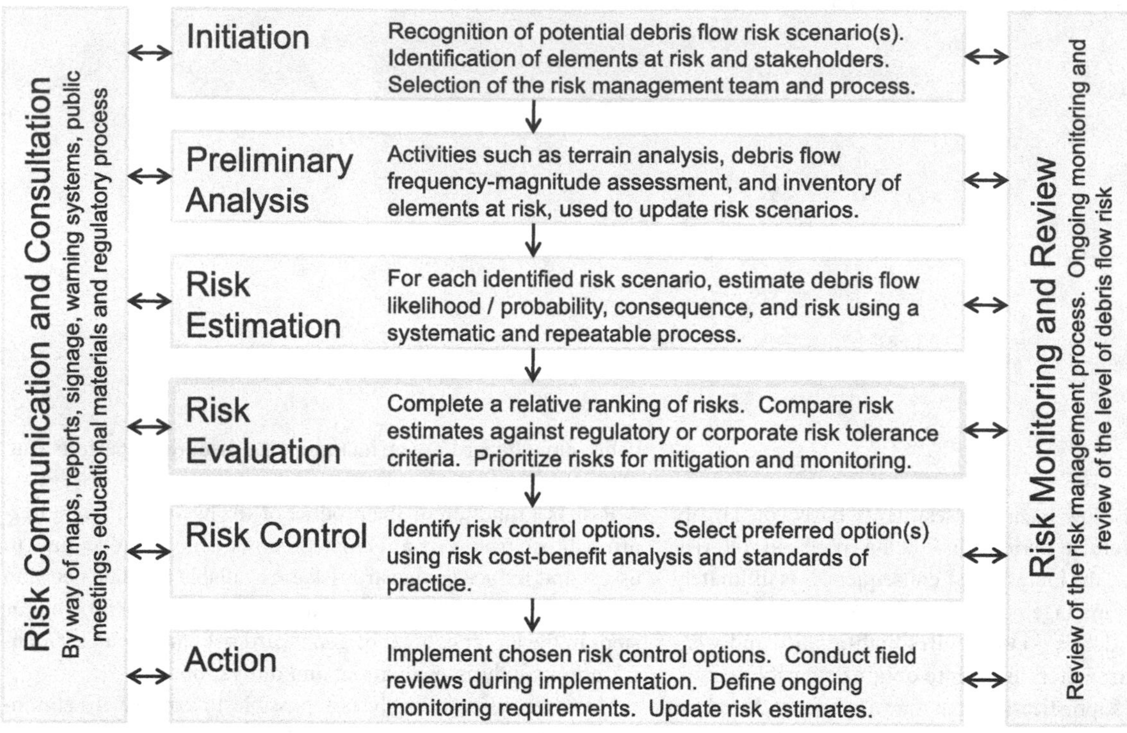

Fig. 7.1. Landslide risk management framework, including debris flows (after CAN/CSA Q850–97, ISO 31000, British Columbia Ministry of Forests, 2004).

method; and because their potential to cause injury and loss of life may be an overriding concern of the approving authorities.

Debris-flow risk assessment involves the first four phases in the larger landslide risk management framework; in sequence the initiation, preliminary analysis, risk estimation, and risk evaluation phases, which all combine to form a risk assessment (Fig. 7.1). This chapter discusses debris-flow risk assessment procedures applicable to development planning and permitting, with emphasis on assessment of risk to life. It provides an overview of the assessment approach, including the definition of a hazard area for study, methods of analysis, and criteria used to define tolerable levels of landslide risk. A highway case study is used to describe the debris-flow risk assessment methodology for nonstationary elements at risk, such as vehicles, and an urban development case study is used to illustrate the methodology for stationary elements at risk, such as buildings. Finally, potential sources of error that limit certainty in the application of results from the assessments are discussed.

7.2 DEBRIS-FLOW RISK ASSESSMENT FRAMEWORK

Debris-flow risks involve the same considerations as any risk assessment (Fig. 7.1). Unacceptable risk is defined as risk that is not accepted by society or decision makers (e.g., government officials) on a voluntary basis. Tolerable risk is defined as risk that society is prepared to live with, where further risk reduction measures are to be implemented when practical and affordable. Broadly acceptable risk is a risk that does not require further reduction.

The first step in a debris-flow risk assessment is to define the area under investigation and to determine the range of debris-flow return periods to be considered in the assessment. The area under investigation is called the "consultation zone" and is discussed in the following section.

7.2.1 CONSULTATION ZONE

In most cases the investigator defines the area that could be affected by the largest credible debris flow. In cases where the study area is on a fan, the consultation zone is commonly the entire surface of the fan, which includes the farthest possible runout. In other cases, coalesced fans might be taken as a single consultation zone for a hazard scenario where debris flows occur coincidentally on two or more surfaces and thus could be regarded as a single event. Debris-flow events in Venezuela (Garcia and Lopez, 2005), or Campania, Italy (Guadagno and Revellino, 2005) are examples. In yet other cases, as – for example – a study area that includes a channel upstream of the fan, additional work including flow modeling may be required to define the consultation zone.

A purely geomorphic definition of the consultation zone is insufficient in many cases, and wrong in others. For example, some fans may be artifacts of a different climatic or geomorphic regime. Uplift or changes in available sediment or climate may have rendered large portions of a fan inactive. On the other hand, debris flows sourced on the flanks of some volcanoes may travel tens or even more than a hundred kilometers downstream without being associated with a fan-shaped landform (Scott, 1988). In such cases, the consultation zone would be the entire downstream area affected by the largest credible volcanic debris flow. It would have to be defined through a first-phase study before a quantitative risk assessment could be conducted. In other cases, changes on a watershed scale, such as might result from landslides that produce large amounts of debris, from extensive logging, or from mining activities, may occur in a watershed. These events have no historic precedent and therefore are not discernible in the stratigraphic record. In these cases, the total potential debris-flow volume, and thus runout, may exceed the largest flow estimated using Quaternary field techniques and dating methods.

A related problem is defining the return period of the "largest credible" debris flow. In areas that were glaciated during the late Pleistocene, the Holocene Epoch – the last 11,600 years – provides a convenient interval for evaluating the frequency and magnitude of debris flows. The largest event that occurred during the Holocene, in many cases, is a reasonable approximation of the "largest credible event." In areas that have not been glaciated, a different approach is required. In many cases, however, the choice of the largest credible debris flow must be based pragmatically on the method that is used to decode past events. Trenching may not be feasible for projects with small budgets, or it may be technically difficult, for example where sediments of interest are below the 5–6 m reach of a standard excavator. Dendrochronological reconstructions of past debris flows can usually only be done for the past few centuries (Jakob, 2010), and studies that employ lake sediments as a proxy for debris-flow activity are only possible in those rare settings where debris flows enter lakes (Kerr Wood Leidal, 2008). Furthermore, many debris-flow fans have been incised by the streams that flow across them and, consequently, lack a complete record of debris-flow activity.

The issues listed above show that it is vital to define an appropriate consultation zone for the purpose of a quantitative risk analysis. This choice affects the outcome and requires input from both qualified practitioners and regulatory authorities.

7.2.2 FREQUENCY–MAGNITUDE ANALYSIS

The next step in a quantitative risk analysis is construction of a frequency–magnitude (F–M) relationship for the range of events that are likely to occur. This range extends from the smallest events that can cause damage or injury to the largest event that can be defended as reasonable given professional best-practice or one that is defined by regulatory authorities. For example, the British Columbia Ministry of Transportation and Infrastructure (2010) specifies that a life-threatening landslide in the context of subdivision permitting requires consideration of the 1:10,000 year partial risk, where partial risk is defined as:

$$R_P = P_H * P_{S:H} \tag{7.1}$$

where P_H is the probability of the event occurring and $P_{S:H}$ is the spatial probability. Based on Eq. 7.1, R_P can be significantly lower than 1:10,000. In practice, this requirement is problematic because of the high cost or impossibility of establishing such a long frequency–magnitude curve. Physical evidence of past debris flows may have been removed by erosion or sediment reworking, and the geologic record has a strong temporal bias, with younger events disproportionately represented compared with older ones.

In the absence of a regulatory framework or requirements, the time window of investigation should at least extend back several hundred years. The time window may approach 10,000 years if the potential consequences of rare events are large, with 1000 or more deaths, or damage to critical infrastructure such as nuclear reactors (Fig. 7.2).

Once a frequency–magnitude relationship has been established for debris flows that occur within the predefined consultation zone, debris-flow hazard scenarios can be defined. Scenarios involve reasonable events or event chains that could have consequences within the consultation zone. In most cases, a hazard scenario may simply be an event of a particular size with a specified return period, for example a hypothetical 100-year debris flow that would damage a proposed development or transportation corridor. An ensemble of hazard scenarios spans the range of event frequencies and magnitudes under consideration. The lowest return period should be that at which the smallest debris flows occur, unless it can be demonstrated that such debris flows pose no risk to present or future development. Such may be the case where a stream has incised so much into its fan that the peak flow associated with the specified return period cannot avulse from the channel.

To illustrate these points, consider a debris-flow risk assessment for a hypothetical small residential subdivision on a fan subject to debris flows. In consultation with the geoscientist, the local government requires a 500-year time window for assessment. A site investigation suggests that even small events pose a hazard, and a 30-year event is determined as the most frequent hazard scenario. Two additional scenario events with 100-year and 500-year average return periods are selected to represent the range of events under consideration.

The number of scenarios considered, in this case three, should balance the need for a reasonable level of study detail against the quality of the data available. In some cases, additional scenarios might be considered. If, for example, the assessment demonstrates that debris flows have avulsed from a channel in

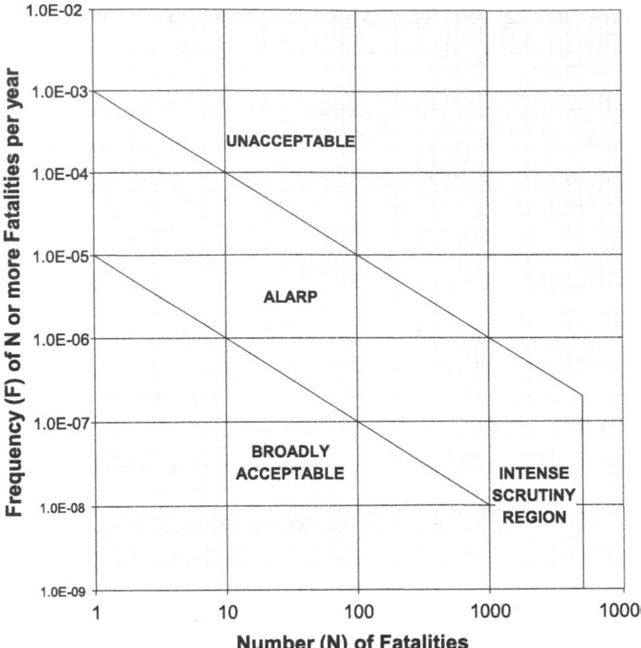

Fig. 7.2. Example of an *F–N* curve for evaluating societal risk.

the past, but may not do so every time, these avulsion scenarios could be included. It is therefore important to carefully define each hazard scenario.

The establishment of a reliable frequency–magnitude relation is a challenging part of the risk assessment. It requires a multidisciplinary approach and, commonly, a variety of methods to reconstruct past events. Friele *et al.* (2005), for example, used borehole stratigraphic and sedimentological data and radiocarbon ages to determine the frequency, flow depth, and runout distance of lahars from Mount Meager, a Pleistocene volcano in southwest British Columbia, and to establish a frequency–magnitude curve spanning the Holocene. Jakob and Friele (2010) combined dendrochronological methods with conventional and AMS (accelerator mass spectrometry) radiocarbon methods to determine debris-flow depths and volumes and peak discharges on the fan of Cheekye River, British Columbia. Similarly, Jakob and Weatherly (2005) used radiometric methods to determine past deposit volumes for debris flows on Jones Creek, Whatcom County, Washington State. Kerr Wood Leidal (2008) used information obtained by Menounos (2006) from lake cores collected from Green Lake to develop debris-flood frequency–volume relationships for Fitzsimmons Creek in southwest British Columbia. Each of these studies has some associated error that needs to be quantified. Because of such unavoidable uncertainties, conservative values may be included as a scenario in the risk assessment.

A frequency–magnitude (F–M) relation is established from a series of debris-flow volumes or peak flows and frequencies. This step is far from being a trivial statistical exercise because the data are discontinuous and are commonly censored (i.e., many small events can no longer be discerned and thus are not included in the analysis). This fact precludes the use of standard frequency analyses such as those applied to hydrologic time series in which annual data are available. The choice of analytical technique and the fitting of a statistical model to the observed frequency distribution are among the most challenging aspects of a debris-flow risk assessment.

7.2.3 DEBRIS-FLOW HAZARD AND INTENSITY MAPS

Once a reliable F–M relationship has been produced, debris-flow volumes or peak flows can be determined numerically and different hazard scenarios quantified. The areas that would be affected by scenario debris flows are then determined using two- or three-dimensional modeling techniques or empirical analysis (Iverson *et al.*, 1998; Simpson *et al.*, 2006). The output values for the risk assessment must correspond to the risk equation (Eq. 7.1). They include debris flow velocity and flow depth, both of which greatly influence debris-flow vulnerability and runout area. The choice of modeling technique is difficult because, at the time of publication of this book, no numerical model can produce a fluid that is equivalent to that of debris flows. Therefore, careful calibration and back-calculation with known events are imperative prior to progressing from hazard scenarios to debris-flow intensity maps.

7.3 DEBRIS-FLOW RISK

For areas with debris-flow hazards, safety provides the technical basis for determining whether the risk to existing development is tolerable, acceptable, or unacceptable, and whether new development should be approved in the absence of hazard mitigation measures. Although it is not the role of the engineer or geoscientist to prescribe an acceptable level of public safety, these professionals need to guide decision makers in the following ways:

- by identifying aspects of risk and conveying the advantages and limitations of different assessment methods
- by demonstrating how hazard and risk acceptance criteria have been applied elsewhere
- by showing how development planning may be affected if risk acceptance criteria are adopted.

The last of these aspects is crucial because, in some cases, the adoption of risk thresholds will mean that developments cannot be increased beyond a certain density on a debris-flow fan unless further mitigation is implemented. Two approaches to estimating and evaluating debris-flow risk levels for development planning are described below.

7.3.1 HAZARD PROBABILITY AND PARTIAL RISK APPROACH

Landslide hazard probability refers to the annual probability of landslide occurrence. In practice, the approving authorities may make decisions based on partial risk, as defined by Eq. 7.1.

In Austria, for example, hazard maps (*Gefahrenzonenpläne*) are based on the 150-year return period event (Austrian Forest Act, 1975), whereas in Switzerland, debris flows with recurrences of up to 300 years must be considered in the design of a structure, with the consideration of residual risk for events of higher return periods (Office Fédéral de l'Aménagement du Territoire, 1997). In British Columbia, hazard acceptability thresholds for development approvals were first proposed by Cave (1991). The thresholds address a range of landslide types, including debris flows, debris avalanches, rockfall, and large rock avalanches. Cave (1991) set different thresholds for different landslide types with the same probability, recognizing that each type imposes a different level of risk. He also distinguished between different types of development, from minor repairs and reconstruction to permitting the construction of new buildings and approval of new subdivisions, because these differences influence the number of people exposed to landslide hazards.

This type of assessment would likely result in a level of safety that meets the expectations of the approving authorities and the public if carried out by qualified professionals and if the results are implemented by the decision makers. However, the approach does not consider the magnitude of loss, that is, the number or value of affected elements at risk. It also does not directly address risk to life, although it is implicit that some threshold likelihood of landslides reaching a housing development poses an unacceptable risk to life.

7.3.2 QUANTITATIVE ASSESSMENTS FOR DEBRIS-FLOW RISK TO LIFE

For debris-flow-prone areas, the potential for loss of life, and thus public safety, may be the primary concern of the approving authorities. Criteria based on the risk of loss of life guide development approvals for landslide-prone areas in Hong Kong and Australia, form part of industrial health and safety regulations in the UK and the Netherlands, and have recently been implemented by one local government in Canada (Australian Geomechanics Society Sub-Committee on Landslide Risk Management, 2000, 2007; Ale, 2005; Leroi *et al.*, 2005; Porter *et al.*, 2007; Whittingham, 2008). They are also occasionally used by developers and some provincial governments in Canada.

Two measures of risk of loss of life are considered in standard quantitative risk analyses: risks to individuals, often expressed as the probability of death of an individual (PDI); and risks to groups, or societal risk. Individual risk addresses the safety of individuals most at risk from an existing or proposed development. It is calculated according to the following equation when dealing with exposure to a single landslide hazard:

$$R = P_H * P_{S:H} * P_{T:S} V * E \quad (7.2)$$

where P_H is the annual probability of the landslide occurring, $P_{S:H}$ is the spatial probability that the landslide will reach the individual most at risk, $P_{T:S}$ is the temporal probability that the individual most at risk will be present when the landslide occurs, V is the vulnerability or probability of loss of life if the individual is impacted, and E is the number of people at risk, which is equal to 1 for determination of individual risk.

Where "risk of loss of life" criteria are used in countries with a Common Law legal system, the maximum tolerable level of risk for new development is typically 10^{-5} per annum for the individual most at risk (Leroi *et al.*, 2005). A distinction is commonly made between new and existing development, with risks as high as 10^{-4} per annum sometimes being tolerated for existing development. This distinction takes into account the chance of hazard avoidance for new development without the necessity of costly risk reduction measures. When the expected loss of life is low (perhaps one or a few people), approval decisions are governed by the estimated level of individual risk. When large groups (usually more than about 10 people) are exposed to a hazard, however, societal risk is more likely to govern whether the proposed development is approved. Societal risk considers the total potential for loss of life for all people exposed to debris flows. It requires calculation of the number of people who would be affected for a range of return periods that are associated with a range of magnitudes and intensities. Societal risk can be estimated using Eq. 7.1 for a single landslide type, with E representing the number of people at risk. If spatial and temporal probabilities and vulnerability vary across the population exposed to the hazard, the group is homogenized by risk exposure and summed for the expected number of fatalities.

Societal risk estimates are graphed by plotting the expected frequency F against the cumulative number of fatalities N (Fig. 7.2). These F–N curves were originally developed for nuclear hazards (Kendall *et al.*, 1977), to reflect societal aversion to multiple fatalities occurring during a single catastrophic event. Subsequently, they have been used in many jurisdictions, including Hong Kong, Australia, the UK, and the District of North Vancouver, British Columbia, as a decision-making tool for evaluating societal risk due to geohazards.

F–N plots are subdivided into (1) unacceptable risk, (2) tolerable risk that should be reduced further if practicable according to the ALARP ("as low as reasonably practicable") principle, and (3) risk that is considered broadly acceptable. In addition, there is a low-probability, high-consequence region that requires intense scientific and regulatory scrutiny, for example dam failures or lahars. From the perspective of potential loss of life, development might be approved conditionally or unconditionally if it can be demonstrated that risks fall in the ALARP or broadly acceptable regions on an F–N plot. If the unmitigated risk falls into the unacceptable region, the development will likely be rejected unless mitigation measures can be implemented that will reduce residual risk to at least the tolerable level. Figure 7.2 provides an example of a F–N curve for evaluating societal risk.

7.4 CASE STUDIES

This section presents two case studies where different methods have been used in debris-flow risk assessments. One case study considers transient elements at risk (vehicles), and the other fixed installations (residential houses). In both cases, assessments are made of individual and group risk. Place names in the first case study have been omitted for reasons of confidentiality. These examples provide distinctly different event scenarios and values of variables in the risk equation. For example, the likelihood that a particular individual within a moving vehicle will be killed by a debris flow is much lower than that of a person living in a debris-flow hazard area, because of the lower proportion of time ($P_{T:S}$) that the former is exposed to the hazard. Societal risk may also be low for a rarely traveled road, but much higher for a heavily traveled one, where vehicles are likely to be in the hazard zone when an event occurs. In both cases, quantitative risk assessments allow estimations of risk that can then be used to prioritize sites where the reduction of risk to tolerable levels is required.

7.4.1 CASE STUDY 1: DEBRIS-FLOW RISK TO HIGHWAY USERS

A major highway crosses a fan subject to size class 4–6 (10^4–10^7 m^3) debris flows (Jakob, 2005), with associated risk to highway users. A highway bridge also crosses the incised main channel crossing the fan.

A frequency–magnitude relationship was established for the hazard area, based on field mapping and dendrochronological and radiocarbon dating of debris-flow deposits. Hazard extents, velocities, and flow depths were then modeled for 20, 50, 100, 200, 500, 2500, and 10,000-year events. These categories represent the range of event magnitudes that are possible, from the smallest events that could potentially cause damage to the largest credible event. Figure 7.3 shows the lengths of road potentially impacted by each size class.

The hazard "consultation zone" for assessment was defined as the length of highway on the fan that could be impacted by a modeled 10,000-year debris flow, including an additional 100 m on either end of the highway section to account for the braking distance of vehicles approaching the hazard zone.

Figure 7.4 shows an event tree summarizing the analytical approach and potential hazard scenarios. These scenarios were selected to encompass a reasonable range of possible outcomes, accounting for uncertainty in the input data: (1) vehicles are impacted by debris; (2) vehicles drive into the gap if the debris flow destroys the highway bridge; and (3) vehicles impact debris already deposited on the highway.

RISK ESTIMATION

For each branch of the event tree, risk was estimated from Eq. 7.1. In this example, E is the number of motorists within the consultation zone, based on vehicle density, velocity, and an assumed number of passengers per vehicle. $P_{T:S}$ is the temporal probability of impact, in this case the likelihood that a particular motorist will be present within the consultation zone at the time of the event (individual risk) or the likelihood that any motorist will be in the consultation zone at the time of the event (group risk). The former is based on assumptions of annual highway travel frequency for an average highway user, and the latter is based on average traffic spacing. $P_{S:H}$ is the spatial probability of impact, corresponding to the percentage of the width of the consultation zone impacted by a debris flow of a particular magnitude.

For the scenario involving risk to persons driving into the void left by the missing bridge, the probability of impact corresponds to the likelihood of bridge destruction and the chance of driving into the void. The likelihood of bridge destruction is estimated by comparing the cross-sectional area under the existing bridge to the estimated debris-flow conveyance for a given return period. The chance of driving into the void is based on an estimated braking distance and field experience with visibility at the site. Similarly, the scenario considering risk to persons crashing into the debris-flow deposit corresponds to the estimated braking distance and field experience with visibility at the site.

INDIVIDUAL RISK SUMMARIES

Individual risk is calculated as the sum of individual risk for all scenarios. The values are reported for highway users who drive through the hazard zone 1, 2, 20, 100, and 500 times per year on average. These travel frequencies range from the one-time tourist to daily commuters. In each return period case, the risk value is a multiple of the number of times the individual passes through the hazard zone.

GROUP RISK SUMMARIES

F–N curves show the cumulative frequency of N or more fatalities. F is calculated as:

$$F = \Sigma f \tag{7.3}$$

where $f = P * P_{S:H} * P_{T:H}$ and is calculated for all scenarios and magnitude categories, ordered from the lowest to highest value of N.

N is calculated as:

$$N = V \times E \tag{7.4}$$

where V is the probability of death in the case of impact, and E is the number of occupants in the vehicle.

RESULTS

Results for individual risk estimates are summarized in Table 7.1. No cases exceed the 10^{-4} threshold for acceptable risk. Results for group risk are shown in Figure 7.5; the majority of the cases lie within the unacceptable risk field on the F–N plot. Although not described in this chapter, these results could be used to determine the effectiveness of particular mitigation options. For example, the analysis could be repeated with lower event magnitudes achieved with an upstream debris barrier, debris basin, or diversion options to determine the most

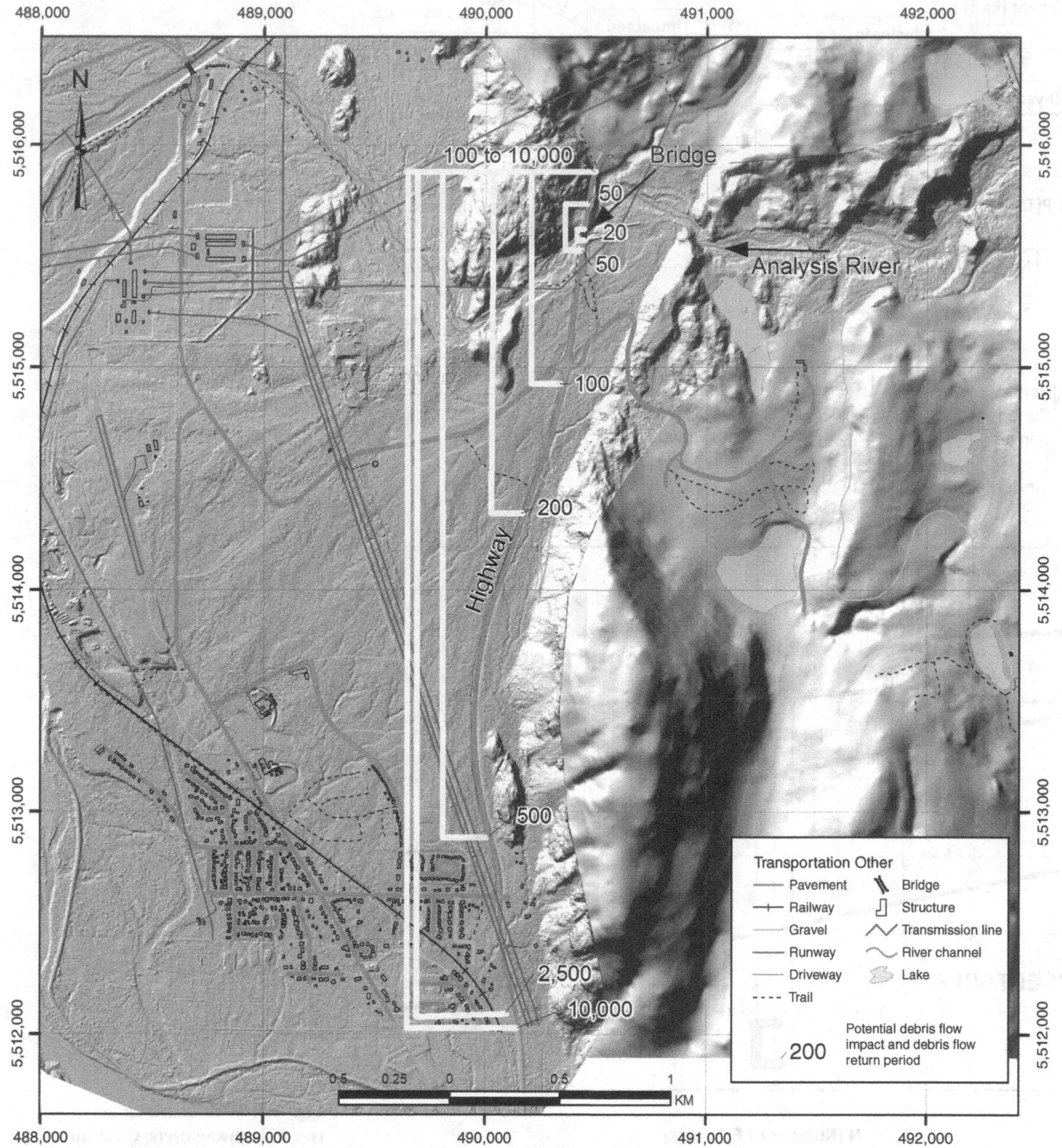

Fig. 7.3. Area of Case study 1 and estimated lengths of road affected by different event magnitudes.

cost-effective strategy to lower group risk to within the ALARP region.

7.4.2 CASE STUDY 2: RISK TO URBAN DEVELOPMENT

This case study considers debris-flow risk to existing development adjacent to Mosquito Creek, a 15.5 km^2 watershed in the District of North Vancouver (DNV), British Columbia. About 4.3 km^2 of the watershed is upstream of the development considered in this example. DNV is located at the foot of the southern flank of the Coast Mountains and receives, on average, about 2400 mm of rainfall annually at lower elevations, mainly during the period from November to February.

Residential development in DNV began in the late 1800s and accelerated from the 1950s to the 1970s, when slopes adjacent to Mosquito Creek were developed. After a damaging flood in the 1950s, a section of Mosquito Creek was routed through a culvert in an attempt to reduce the flood hazard.

More recently, studies have shown that the creek is also subject to debris floods (Kerr Wood Leidal, 2003). Debris floods with return periods ranging from 100 to 2500 years have been modeled based on data derived from empirical relations, dendrochronology, and some professional judgment (BGC

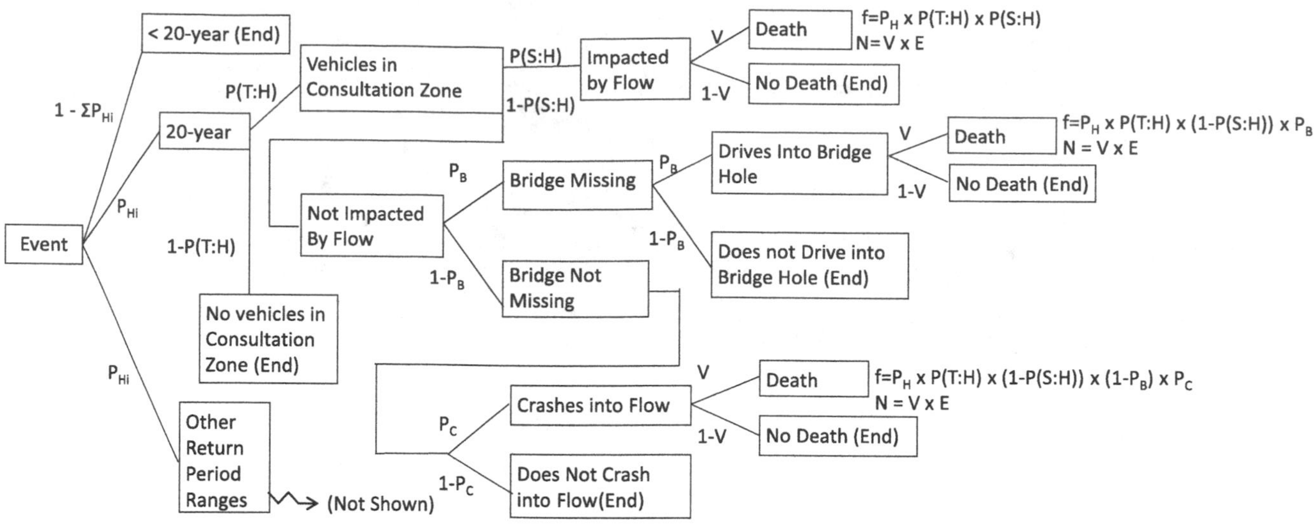

Fig. 7.4. Event tree for determining risk to highway users, Case study 1.

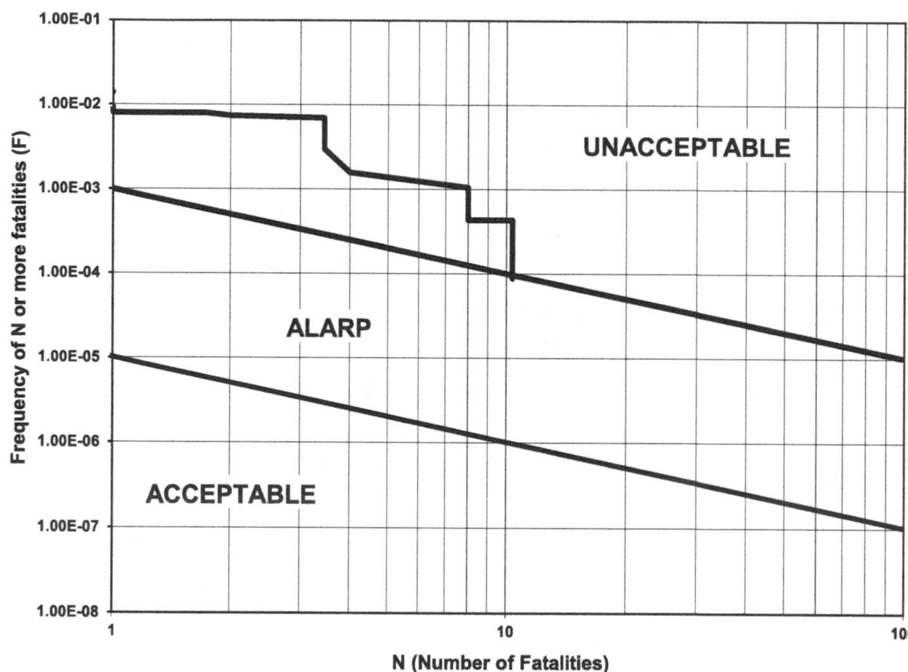

Fig. 7.5. *F–N* curve showing group risk to highway users, Case study 1.

Table 7.1. *Individual risk to highway users, Case study 1.*

Individual driver trips per year	Cumulative risk
1	6.6×10^{-8}
2	1.32×10^{-7}
20	1.32×10^{-6}
100	6.6×10^{-6}
500	3.3×10^{-5}

Engineering, 2010). For return periods exceeding 100 years, the debris flood could overwhelm the intake of the culverted section and discharge through the former channel, impacting a densely populated area.

This case study involved a quantitative assessment of risk to life for existing development along Mosquito Creek. The consultation zone for analysis includes the buildings potentially impacted by a modeled 2500-year debris flow at depths exceeding 0.1 m and with a flow velocity greater than 0.1 m s^{-1}. It was assumed that a flow depth and flow velocity below those values did not constitute a credible loss-of-life risk. Figure 7.6 is an event tree summarizing the analytical approach and the potential scenarios that were considered.

RISK ESTIMATION

For each branch of the event tree, risk was estimated from Eq. 7.1. In this example, elements at risk (E) are the residents

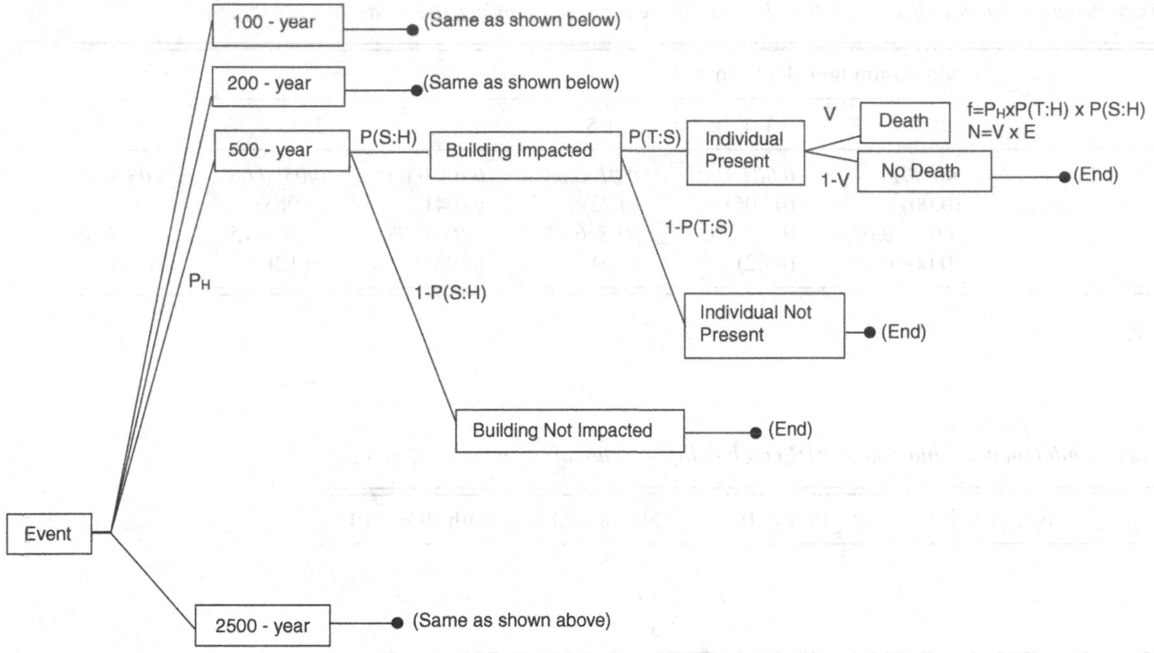

Fig. 7.6. Event tree used for Case study 2 risk analysis (500-year scenario shown).

within buildings, estimated from census data on building locations and occupancy, including residential buildings and a fire hall. The values are approximate, but are reasonable averages.

Spatial probability ($P_{S:H}$) was assigned as 0.9 (very likely) for buildings impacted by modeled flows. Temporal probability ($P_{T:S}$) was assigned an average value of 0.5 for group risk, and a more conservative value of 0.9 for individual risk. The latter case is based on the individual most at risk; that is, someone who spends the most time at home, such as a young child, stay-at-home parent, or an elderly person. For the fire hall, $P_{T:S}$ was assigned an average value of 0.7 for groups, corresponding to the 17 hours per day that the hall is occupied. For risk to individuals, $P_{T:S}$ was assigned an average value of 0.3, corresponding to 12-hour work shifts, 4 days per week.

Vulnerability (V) was estimated based on flow-velocity and flow-depth criteria, as these values can be modeled and are the primary controls on debris-flood intensity (Table 7.2). Values were assigned to each building within the consultation zone. Estimates of vulnerability strongly affect the results of quantitative risk analyses and are also subject to a large degree of uncertainty because factors controlling vulnerability are poorly known. These factors include the location of individuals within and outside of buildings, whether water and debris enter a building, the velocity and depth of water and debris entering the building and how they are distributed inside, the possibility of large floating woody debris, and the ability of individual buildings to withstand impact. Jakob *et al.* (2012) used a large number of case studies worldwide to define a debris-flow vulnerability index ($I_{DF} = dv^2$) where d is flow depth and v is the flow velocity for buildings. Damage class probabilities can be determined from a statistical relationship between damage class and the I_{DF}.

The following four-step approach was taken to provide reasonably defensible estimates of vulnerability.

1. Vulnerability (V) criteria were established based on modeled flow depth and velocity, and were assigned to each building within the debris-flood impact zone.
2. Quantitative risk analysis results were obtained for individuals and groups, based on Eq. 7.1.
3. Numbers of deaths were estimated, based on representative case studies and data compilations (Jakob *et al.*, 2011).
4. Group risk results, calculated in the second and third steps, were compared for consistency.

RISK TO INDIVIDUALS

Risk to individuals (PDI) and for people living in each home were calculated using Eq. 7.1. The total risk is the sum of risk for the 200-, 500-, and 2500-year debris-flood scenarios (Table 7.3). The 100-year return period scenario was not considered because no buildings were impacted by flow depths exceeding 0.1 m.

Minimum PDI values in Table 7.3 are based on the lower end of the vulnerability range and the lower bounds of the probability range. Maximum PDI values are based on the higher end of the vulnerability range and the higher bounds of the probability range. Mean values are the average of the minimum and maximum PDI estimates and are considered a "best estimate."

RISK TO DEVELOPMENT

Figure 7.7 shows group risk summarized on an *F–N* diagram. The bold line represents the mean ("best estimate"), and the

Table 7.2. *Matrix to compute vulnerability (in italics) for loss of life for debris floods, Case study 1.*

Variable		Maximum flow depth (m) <0.3	0.3–1	1–1.5	1.5–2	2–3	3–4
Maximum flow velocity ($m\,s^{-1}$)	<2	*<0.001* (0.001)	*0.001–0.01* (0.006)	*0.01–0.03* (0.02)	*0.03–0.05* (0.04)	*0.05–0.08* (0.065)	*0.08–0.15* (0.12)
	2–5	*0.001–0.01* (0.006)	*0.01–0.03* (0.02)	*0.03–0.05* (0.04)	*0.05–0.08* (0.065)	*0.08–0.15* (0.12)	*0.15–0.30* (0.23)

Table 7.3. *Summary of minimum and maximum PDI (probability of death of an individual) values.*

	No. of buildings with PDI $>10^{-4}$	No. of buildings with PDI $>10^{-5}$
Minimum PDI	1	10
Maximum PDI	10	30
Mean PDI	5	30

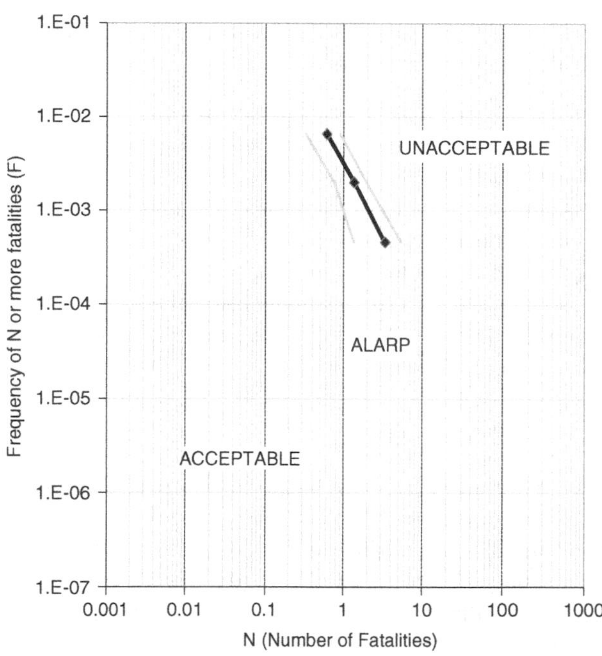

Fig. 7.7. Group risk for the affected population in Case study 2.

gray lines are based on the upper and lower estimates of event probability and vulnerability. The ranges for unacceptable, ALARP, and acceptable risk are consistent with those in Figure 7.2 and represent the tolerance criteria accepted by the local jurisdiction.

Several conclusions can be drawn from an examination of Figure 7.7. Debris-flood risk to life for groups is unacceptable. The risk is highest for the 200-year debris flood because its minimum and maximum value range plots farthest from the line that separates the unacceptable region from the ALARP region; the 100-year event or more frequent events, which are not shown in Figure 7.7, would either plot in the ALARP or the acceptable region; and higher-return-period debris floods would likely plot in the ALARP region.

7.5 ERROR SOURCES

Although the debris-flow risk framework summarized in this chapter is conceptually simple and straightforward, a defensible assessment requires significant effort at both the hazard assessment stage to establish debris-flow frequency–magnitude relationships, and at the risk analysis stage to define credible risk scenarios and to estimate the vulnerability of elements at risk. We have often encountered large knowledge gaps when attempting to define reliable frequency–magnitude relationships and vulnerabilities for debris flows. These knowledge gaps should not prevent practitioners from completing quantitative debris-flow risk assessments, but they do mean that sources of error and study limitations must be documented.

In addition, levels of safety must be considered throughout the design and construction process, and focus should be given to the potential for debris-flow risk to change over time. Changes in risk may result from: slope modification; temporary or long-term changes in the upstream watershed by logging, mining, insect infestations, or forest fires; failure of drainage systems; changes in precipitation due to climate change; development densification; or increased vulnerability due to evolving land use, for example conversion of an industrial area to a residential neighborhood. Some of these factors may be unprecedented: past debris-flow history may not necessarily be representative of future debris-flow activity.

7.6 CONCLUSIONS

This chapter provides the methodology and two examples of debris-flow and debris-flood risk assessments for roads and urban developments. The same methods can be applied to other linear infrastructures or any fixed structures. Quantitative risk assessments enable comparison of observed risk with risk tolerance thresholds, although their successful application is limited when governing jurisdictions refuse to provide formal definitions of thresholds for acceptable, tolerable, and unacceptable levels of risk. In the absence of such regulations, or even guidelines, the practitioner must justify a chosen level of landslide safety. Fortunately, some local governments have taken a leadership role and implemented risk tolerance criteria in their hazard management program. The District of North Vancouver is a leading example in Canada (Porter and Dercole, 2011). It is hoped that other jurisdictions will follow its example.

ACKNOWLEDGEMENT

We acknowledge the District of North Vancouver for permission to publish the data presented in this chapter.

REFERENCES

Ale, B. J. M. (2005). Tolerable or acceptable: A comparison of risk regulation in the United Kingdom and in the Netherlands. *Risk Analysis*, 25, 231–241.

Australian Geomechanics Society Sub-Committee on Landslide Risk Management (2000). Landslide risk management concepts and guidelines. *Australian Geomechanics*, 35(1), 49–92.

Australian Geomechanics Society Sub-Committee on Landslide Risk Management (2007). A national landslide risk management framework for Australia. *Australian Geomechanics*, 42, 1–36.

Austrian Forest Act (1975). Paragraph 11. Gefahrenzonenpläne (Hazard zone maps).

BGC Engineering (2010). *Mosquito Creek Debris Flood Hazard Assessment*. Final Report for Natural Resources Canada. Vancouver: BGC Engineering.

British Columbia Ministry of Forests (2004). *Landslide Risk Case Studies in Forest Development Planning and Operations*. BC Ministry of Forests, Land Management Handbook No. 56. Victoria, BC [available at www.for.gov.bc.ca/hfd/pubs/Docs/Lmh/Lmh56.htm].

British Columbia Ministry of Transportation and Infrastructure (2010). *Subdivision Preliminary Layout Review: Natural Hazard Risk*. Victoria, BC: BC Ministry of Transportation and Infrastructure.

Cave, P. (1991). Hazard acceptability thresholds for development approvals by local governments. In *Proceedings of Geological Hazards Workshop*. BC Geological Survey, Branch Open File 1992–15, pp. 15–26.

Friele, P. A., Clague, J. J., Simpson, K. and Stasiuk, M. (2005). Impact of a Quaternary volcano on Holocene sedimentation in Lillooet River Valley, British Columbia. *Sedimentary Geology*, 176, 305–322.

Garcia, R. and Lopez, J. L. (2005). Debris flows of December 1999 in Venezuela. In *Debris-flow Hazards and Related Phenomena*, ed. M. Jakob and O. Hungr. Berlin: Springer Praxis, pp. 519–538.

Guadagno, F. M. and Revellino, P. (2005). Debris avalanches and debris flows of the Campania region (southern Italy). In *Debris-flow Hazards and Related Phenomena*, ed. M. Jakob and O. Hungr. Berlin: Springer Praxis, pp. 489–508.

Iverson, R. M., Schilling, S. P. and Vallance, J. W. (1998). Objective delineation of lahar-inundation hazard zones. *Geological Survey of America Bulletin*, 110, 972–984.

Jakob, M. (2005). A size classification for debris flows. *Engineering Geology*, 79, 151–161.

(2010). State of the art in debris-flow research: The role of dendrochronology. In *Tree Rings and Natural Hazards: A State-of-the-Art*, ed. M. Stoffel, M. Bollschweiler, D. R. Butler and B. H. Luckman. Springer Advances in Global Change Research 41, pp. 183–192.

Jakob, M. and Friele, P. A. (2010). Frequency and magnitude of debris flows on Cheekye River, British Columbia. *Geomorphology*, 114, 382–385.

Jakob, M. and Hungr, O. (2005). *Debris-flow Hazards and Related Phenomena*. Berlin: Springer Praxis, 739 pp.

Jakob, M. and Weatherly, H. (2005). Debris flow hazard and risk assessment: Jones Creek, Washington. In *Landslide Risk Management*, ed. O. Hungr, R. Fell, R. Couture and E. Eberhardt. London: Taylor & Francis Group, pp. 533–542.

Jakob, M., Holm, K., Porter, M., Weatherly, H. and Dercole, F. (2011). Debris-flood risk assessment on Mosquito Creek, BC. In *5th Canadian Conference on Geotechnique and Natural Hazards*, Kelowna, BC: Canadian Geotechnical Society, Paper No. 212.

Jakob, M., Stein, D. and Ulmi, M. (2012). Vulnerability of buildings to debris flow impact. *Natural Hazards*, 60, 241–261.

Kendall, H. W., Hubbard, R. B., Minor, G. C. and Bryan, W. M. (1977). *The Risks of Nuclear Power Reactors: A Review of the NRC Reactor Safety Study*. Cambridge, MA: Union of Concerned Scientists.

Kerr Wood Leidal (2003). *Debris Flow Study and Risk Mitigation Alternatives for Mosquito Creek*. Report for the District of North Vancouver. Burnaby, BC: Kerr Wood Leidal.

(2008). *Predesign Report for Fitzsimmons Creek, Debris Barrier and Sediment Basin for Resort Municipality of Whistler*. Burnaby, BC: Kerr Wood Leidel, Report 29.173.

Leroi, E., Bonnard, C., Fell, R. and McInnes, R. (2005). Risk assessment and management. In *International Conference on Landslide Risk Management, Vancouver, BC*. State of the Art Paper 6, pp. 159–198.

Menounos, B. (2006). Anomalous early 20th century sedimentation in proglacial Green Lake, British Columbia, Canada. *Canadian Journal of Earth Sciences*, 43, 671–678.

Office Fédéral de L'Aménagement du Territoire (OFAT) (1997). *Prise en Compte des Dangers dus aux Movements de Terrain dans le Cadre des Activités de l'aménagement du Territoire*. Bern: Office Fédéral de l'Aménagement du Territoire.

Porter, M. and Dercole, F. (2011). Urban risk assessments. In *5th Canadian Conference on Geotechnique and Natural Hazards*. Kelowna, BC: Canadian Geotechnical Society, Paper No. 165.

Porter, M., Jakob, M., Savigny, K.W., Fougere, S. and Morgenstern, N. (2007). Risk management for urban flow slides in North Vancouver, Canada. In *Proceedings, Canadian Geotechnical Conference*. Ottawa, ON: Canadian Geotechnical Society, pp. 690–698

Scott, K.M. (1988). *Origins, Behaviour and Sedimentology of Lahars and Lahar-Runout Flows in the Toutle–Cowlitz River System*. US Geological Survey, Paper 1447-A.

Simpson, K.A., Stasiuk, M., Shimamura, K., Clague, J.J. and Friele, P.A. (2006). Evidence for catastrophic volcanic debris flows in Pemberton Valley, British Columbia. *Canadian Journal of Earth Sciences*, 43, 679–689.

Whittingham, R.B. (2008). *Preventing Corporate Accidents: An Ethical Approach*. Oxford: Elsevier.

8 Landslides in quick clay

J. KENNETH TORRANCE

ABSTRACT

Quick clays are sensitive late-glacial and postglacial marine- and brackish-water sediments. Salt removal has decreased their remolded strength, such that, if disturbed, they behave as a liquid. Cementation by nano-sized particles of iron oxides at particle contacts may increase the sensitivity of quick clays in parts of the St. Lawrence Basin. Quick clay landslides are of two main types – stepwise and uninterrupted. In stepwise landslides, sufficient failed material liquefies that the debris from each step flows away from the slide scarp, leaving an unsupported unstable slope that, in turn, fails and flows away. Uninterrupted landslides require advance of the failure plane away from the riverbank at a rate sufficient to produce a continuous series of failing slices and continuous flow of the debris away from the scarp and along the river valley. As debris clogs the valley or loses momentum for any reason, the advance of the failure front slows down and soon stalls. For the end member of the uninterrupted category, the flakeslide, the failure plane advances so rapidly that a large area commences flow as a unit, breaking into subunits as it moves. Because quick clay development involves chemical change, appropriate chemical treatments should be effective and efficient in diminishing landslide risk.

8.1 INTRODUCTION

On May 10, 2010, a quick clay landslide near St. Jude, Quebec took the lives of a family who were watching the Montreal Canadians in a National Hockey League, Stanley Cup playoff game. Unfortunately, almost all the media reports of this tragedy demonstrated that quick clays are widely misunderstood. This chapter, which addresses sediment accumulation, quick clay development, quick clay landslides, failure mechanisms, and possible prevention measures, attempts to provide an insightful overview of the quick clay landslide problem.

8.2 WHAT IS QUICK CLAY?

Quick clay is a naturally occurring, fine-grained sediment, which has high sensitivity (ratio of undisturbed to remolded shear strength) and behaves as a liquid when its structure breaks down. The Norwegian definition that the sensitivity exceeds 30 and the remolded shear strength is <0.5 kPa (Norsk Geoteknisk Forening, 1974) is adopted here. Quick clays are a major hazard because of their susceptibility to large destructive landslides and to major consolidation.

8.3 GEOLOGIC SETTINGS

Quick clays develop within sediments, mostly <12,000 years old, that were deposited during the retreat of the most recent Pleistocene ice sheets from Scandinavia and northern North America. During the glacial period, water storage in the ice sheets lowered sea level to approximately 120 m below their present level (Kenney, 1964), but the continental margins were depressed an even greater amount by their weight. As glacier margins retreated, marine waters inundated the isostatically depressed lowlands of coastal Scandinavia and northern North America.

Initially, the calving margins of the ice sheets delivered poorly sorted, coarse mineral debris directly to the seafloor; later, after direct glacial contact with the sea was lost, the glacially ground rock material was delivered toward the sea by streams and rivers. Gravel and sand accumulated in deltas, and the silt- and clay-sized particles were carried farther offshore where the salt water facilitated their flocculation; deposition of the floccules produced remarkably uniform, fine-grained sediment. These flocculated

Landslides: Types, Mechanisms and Modeling, ed. John J. Clague and Douglas Stead. Published by Cambridge University Press.

Table 8.1. *General model for quick clay development (modified from Torrance, 1983).*

Depositional	Post-depositional
Factors producing high undisturbed strength	
Flocculation [a,b]	Cementation bonds
• Salinity [a]	• Rapidly developed
• Divalent cation adsorption [b]	• Slowly developed
	Slow load increase
• High suspension concentration	• Time for cementation
	Other time-dependent processes
	• Diagenetic changes
Factors producing low remolded strength	
Depositional	Post-depositional
Material properties	Decrease in liquid limit > decrease in water content
• Low-activity minerals dominate [a,b]	• Leaching of salt [a]
	• Dispersants [b]
	Minimal consolidation

[a] Essential in marine clays.
[b] Essential in freshwater clays.

marine sediments have substantially greater water contents than would freshwater sediments of the same mineralogy and texture.

Quick clays occur in the following places: coastal areas of Norway, Sweden, and Finland; the St. Lawrence Basin, the Hudson Bay Lowlands, Quebec, and near Terrace, British Columbia in Canada; the St. Lawrence Basin and Alaska in the USA; and Ariake Bay in Japan. Russia may also have areas of quick clay. No quick clays have been confirmed in the Southern Hemisphere. Quick clays exhibit a wide range of strengths, sensitivities, clay-size contents, relative mineral abundances, and geotechnical behaviors.

8.4 GENERAL MODEL FOR QUICK CLAY DEVELOPMENT

Rosenqvist (1946, 1953) demonstrated that removal of salt from the uplifted marine sediments in Norway was essential to the development of high sensitivity and liquid behavior upon remolding. Penner (1965) confirmed Rosenqvist's conclusions for eastern Canadian quick clays. Quigley (1980) summarized the factors that influence the undisturbed and remolded strengths of these sediments. Torrance (1983) synthesized these findings into a general matrix model for quick clay development (Table 8.1). The table is divided vertically into factors that lead to high undisturbed and low remolded strengths, and horizontally into the requirements that must be met at the time of sediment deposition and those that are post-depositional.

8.4.1 MINERALOGY

The mineralogical requirement for quick clay is domination of the sediment by low-activity, clay-size minerals, where the activity of the clay-size fraction is defined as the sediment's plasticity divided by the percentage of clay-sized (<2 μm) particles. Soils with the same suites and proportions of clay-sized minerals, but different clay-size contents, exhibit the same activity. Most clay-sized minerals have activities less than 1: illite and chlorite from 1 to <0.5; kaolinite <0.5; iron oxides, gibbsite, and clay-sized primary minerals (quartz, feldspars, amphiboles) <0.2. The activities of all these minerals are greater under saline than freshwater conditions and when their exchange sites are dominated by potassium, calcium, or magnesium ions instead of sodium ions. Partial iron oxide coatings on clay particle surfaces slightly increase their liquid limits and thus their activities.

The activities of smectite clay minerals range from 1 to 7. High-swelling commercial smectites (bentonite) in the Na-saturated state at low salinity have activities of approximately 7. Activities decrease to 3–4 at high salinity or when Ca-saturated. Partial iron oxide coatings further reduce smectite activity for each chemical state. Addition of 13 percent high-swelling smectite to quick clay from the 1971 South Nation River landslide (60 percent clay) and 7 percent to sensitive clayey silt (10 percent clay) from Thurso, Quebec, rendered the yield stress of the Na-saturated material nonsusceptible to salinity change (Viljoen, 1989). Viljoen further estimated that 5 percent and 2 percent high-swelling smectite, respectively, would be sufficient to inhibit quick clay development in these sediments. Only traces of smectite have been reported in quick clays from Scandinavia and North America (Berry and Jørgensen, 1971; Torrance, 1988; Berry and Torrance, 1998).

The conclusion that all smectites are incompatible with quick clay development must be qualified, because the most abundant clay mineral in quick clays at Ariake Bay, Japan, which accumulated as sea level rose during melting of the late Pleistocene ice sheets in North America and Scandinavia (Torrance and Ohtsubo, 1995), is a high-iron smectite (Egashira and Ohtsubo, 1982). Clay particles and volcanic ash, eroded from the local soils, flocculated upon entering the marine environment at Ariake Bay. High-iron smectite is a weathering product of volcanic ash in terrestrial environments, and possibly during diagenesis of volcanic ash at shallow depths in marine sediments (Chamley, 1997). In a reducing environment within the sediments, incoming iron is reduced from Fe^{3+} to Fe^{2+} by anaerobic bacteria and the Fe^{2+} is incorporated into the dioctahedral sheets of the smectite to yield smectite with a high negative layer charge that resists the entry of water into its interlayers and does not swell. If the marine pore water is displaced by fresh water, its activity of *ca.* 1 lies at the upper boundary of low activity. On exposure to oxidative weathering, the octahedral Fe^{2+} is oxidized to Fe^{3+}, the interlayer charge decreases, the attractive forces diminish and, as a consequence, the high-iron smectite swells and exhibits higher plasticity and activity (Gates *et al.*, 1998; Torrance, 1999), thereby reducing the sensitivity. Landslides are not a problem in these Japanese quick clays because their surfaces are only a few meters above sea level.

The Scandinavian and eastern Canadian quick clays share the same mineral suite – quartz, feldspars, amphibole, illite, chlorite, calcite, and iron oxides (Moum *et al.*, 1968; Løken,

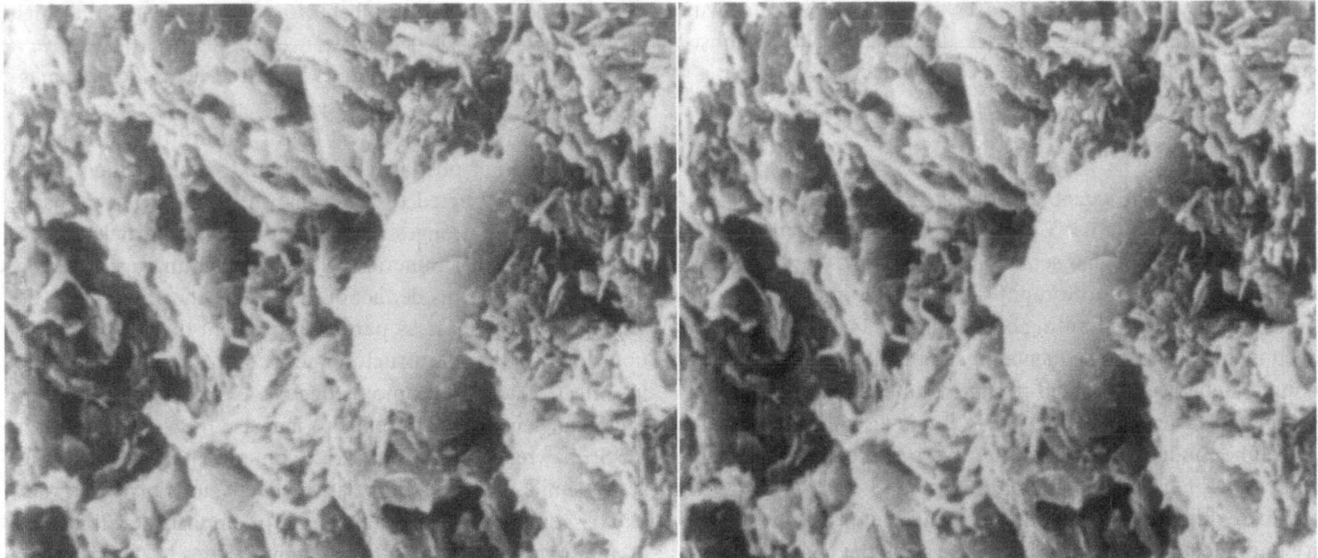

Fig. 8.1. SEM (scanning electron microscope) stereo pair of a fresh dry-fracture surface showing the typical flocculated microstructure of sensitive marine clay. The largest particle is 10 μm long. (Images by Dr. G. Sides.)

1971; Torrance, 1988). The clay-size content ranges from <20 percent to >80 percent, the rest being silt. At most sites the clay-size content varies only slightly with depth. The silt fraction is dominated by quartz and feldspar, with some illite, chlorite, and amphibole in the fine silt fraction. Glacial grinding has been so effective that illite, chlorite, quartz, feldspar, amphibole, and iron oxides are all present in the clay-size fraction. In Norway, and probably in Sweden, phyllosilicates are more abundant than silicates in the clay fraction, whereas in eastern Canada the opposite is the case. The difference probably relates to greater phyllosilicate abundance in the Scandinavian source rocks than in the igneous and metamorphic rocks of the Canadian Shield.

Iron oxides are ubiquitous in quick clays. In eastern Canada, hematite is the dominant iron oxide at the South Nation River landslide site, whereas magnetite dominates at St. Leon, Quebec, 250 km to the east in the Champlain Sea Basin (Torrance *et al.*, 1986). Because the glacial debris had only brief exposure to subaerial weathering and because magnetite is not easily formed in either surface or sedimentary environments, both hematite and magnetite are almost certainly products of glacial grinding of iron oxides present in the rocks and iron ore deposits of the Canadian Shield. The dominance of magnetite in the east points to the iron ore deposits of the eastern Quebec–Labrador border region as the magnetite source.

8.4.2 FLOCCULATION

Flocculation of silt- and clay-size particles upon entering the marine environment provides the accumulated sediment with both high undisturbed strength and high water content. Flocculation does not occur readily in fresh water because, at the near-neutral pH values normally present, the face surfaces of 2:1 phyllosilicates are negatively charged as the result of isomorphous substitution of Al^{3+} for Si^{4+} in the tetrahedral sheets, and the electrical double layers of the particles are well developed. The surfaces of quartz, feldspars, and amphiboles are also negatively charged (Yariv and Cross, 1979). The iron oxide surfaces and the edges of phyllosilicate clay particles are approximately at their isoelectric points. Consequently, interparticle repulsive forces are strong and the individual particles settle through the water at rates that are determined by their size and density rather than as floccules.

High concentrations of salt in solution suppress the double-layer thickness. In saline and brackish water, this suppression of repulsive forces allows the particles to approach closely enough that edge-to-face and edge-to-edge electrostatic bonding occurs between particles. Floccules, consisting of myriad individual particles, form and settle quickly through the water to form sediment with almost random particle orientation and high water content. Even after consolidation due to continuing sediment accumulation, the water content remains high, at approximately the high salinity liquid limit, to depths that can exceed 30 m.

The flocculated structure (Fig. 8.1) is often called a "card-house" structure, invoking images of the very fragile structures that can be built using a standard deck of playing cards. This analogy is misleading. Instead, let us replace the playing cards with custom-made cards that reflect the particle size range – from coarse silt (largest dimension 40 μm) to fine clay (largest dimension <1 μm). The coarse particles are box-like, the phyllosilicate particles are plate-like, and the clay-size quartz and feldspar are flake-like (length and width much greater than their thickness; see Berry, 1974). Over the full range of textures from 10 to 80 percent clay-size material by weight, the minute clay-size particles vastly outnumber the silt particles. Cover the quartz and feldspar surfaces and the planar faces of the phyllosilicates with

Velcro loops and the edge surfaces of the phyllosilicates with Velcro hooks and imagine the intricate and strong "card-house" structures that can be built.

8.4.3 CEMENTATION

Cementation is not a general requirement for quick clay development, but – where it occurs – the greater undisturbed shear strength produces higher sensitivities and allows for higher and steeper slopes. The more rapidly that cementation occurs, the more effective it is in limiting consolidation as further sediment accumulates. The consequence is sediment with higher water content and lower remolded strength than if slow, or no, cementation had occurred. Higher water content also leads to a greater hydraulic conductivity and more rapid post-depositional chemical change, particularly leaching.

The requirements for good cementing agents are low solubility and adequate abundance. Let us assess the commonly proposed candidates – carbonates, iron and aluminum oxides and hydrous oxides, manganese oxides, and amorphous materials – against these criteria. Manganese oxides fail because they occur in only trace amounts and are more soluble under reducing conditions than under oxidizing conditions. Amorphous materials also are present in only small amounts (Torrance, 1990, 1995). Aluminum hydroxide (gibbsite) does not form until the very final stages of subaerial weathering (Jackson and Sherman, 1953); until this extreme stage, Al is restricted to substituting into <15 percent of octahedral sites of hematite and <33 percent in goethite (Murad, 1988). Calcium carbonate is an excellent cementing agent in arid regions because it accumulates at inter-particle contacts, the last place from which water evaporates. In marine clays, most of the carbonate consists of particles of glacially ground carbonate rock that seldom constitute more than 5 percent of the sediment. Despite their abundance, carbonates are not effective cementing agents in water-saturated environments because their modest solubility allows their redistribution by dissolution and re-precipitation; small, high-surface-energy crystals dissolve and the Ca^{2+} and CO_3^{2-} re-precipitate on the surfaces of larger, lower-surface-energy crystals. This process concentrates the carbonate into large concretions (Gadd, 1980), some exceeding 1 kg, with carbonate completely filling the original pore space of the sediment rather than being distributed at inter-particle contacts throughout the sediment.

Iron oxides are the only credible cementation agents in quick clays. They are mostly present as microcrystalline hematite or magnetite (Torrance *et al.*, 1986). Because these iron oxides cannot form under the reducing conditions in the sediment, they must have been present since the time of sediment accumulation. Hematite at the South Nation landslide site in eastern Ontario had a particle size of <20 nm (Torrance *et al.*, 1986). Gonzáles-Lucena (2010) determined that very small particles (<10 nm) of laboratory-prepared iron oxides exhibit much stronger magnetic susceptibilities per unit weight than larger samples; their very limited crystal size results in uncompensated magnetic spins that allow even normally anti-ferromagnetic minerals to exhibit super-paramagnetic behavior. She also noted that high concentrations of suspended iron oxide nanoparticles have a strong tendency to agglomerate. Both the hematite and magnetite in the sediments have high magnetic susceptibility and will be influenced by electrical fields. During the flocculation process, their net positive charge should result in them being preferentially attracted to locations where the negative charge concentration is strongest; namely, at the inter-particle contact points of the other minerals, where a cementing agent will be most effective. If this deduction is correct, the strength of the sediment will increase as particle movement during consolidation slowly makes the structure more compact and the cementing effect of the oxides more effective. Cementation in eastern Canadian sensitive clays tends to increase from west to east in the St. Lawrence Basin. Iron oxide content data (Table 8.2) suggest a similar trend, if the relative effectiveness of the extractants is considered. Borggaard (1988) demonstrated that each of the common extractants has limitations: citrate–dithionite extracts amorphous iron and hematite (suitable at Lemieux and probably Henryville) but not magnetite (underestimates at St. Barnabé, near St. Léon); ammonium oxalate extracts amorphous iron and magnetite (suitable at St. Barnabé), but not hematite; and EDTA extracts amorphous iron but not the crystalline iron oxides (not suitable at any site; magnetite (almost certainly present) is underestimated at the Outardes and Labrador sites). The use of citrate–dithionite and ammonium oxalate procedures, in sequence, is probably the best extraction option for validating or rejecting this inferred iron oxide content trend; Mössbauer spectroscopy studies would be even better.

8.4.4 LIQUID LIMIT

A post-depositional decrease in the liquid limit, while the water content remains essentially constant, is a requirement for quick clay development. Fresh water infiltrating the soil surface or moving upward or laterally through the sediment in response to artesian conditions displaces the original marine pore water (Rosenqvist, 1946). As pore water salinity decreases, double-layer suppression decreases and the inter-particle repulsive forces increase. Nonetheless, the flocculated structure survives, even at very low salinities. However, if the structure is broken down, the strong repulsive forces prevent re-flocculation. The sediment, which had an in-situ water content approximating its high salinity liquid limit before leaching, finds itself at a water content above its now-lower liquid limit and behaves as a liquid if remolded. The liquidity index, after leaching, exceeds 1 and may exceed 4; silty sediments tend to develop higher liquidity indices than clay-rich sediments. Returning to the "Velcro" analogy, the salt removal process has effectively removed the "hooks" from the phyllosilicate edges. Groundwater chemistry profiles indicate that leaching from the surface is responsible for salt removal throughout most of the level expanses of marine sediments in the Ottawa Valley (Torrance, 1988), in areas of the St. Lawrence Lowlands that are not intermingled with rock outcrops, and at the Mink Creek landslide site in British Columbia

Table 8.2. *Fe_2O_3 extracted from sensitive clays in eastern Canada.*

Location	Extraction method	Fe_2O_3 (% by weight)	Reference
Lemieux, ON	Citrate–dithionite	0.4–0.9	Seekings, 1974
Henryville, QC	"	< 1	Torrance, 1990
St. Barnabé, QC	"	0.8–1.8	Torrance and Percival, 2003
St. Barnabé, QC	Ammonium oxalate	1.7–2.5	Torrance and Percival, 2003
Outardes, QC	EDTA	1.34	Loiselle *et al.*, 1971
Labrador, NL	"	1.0	Kenney *et al.*, 1968

(Geertsema and Torrance, 2005). At Drammen, Norway (Moum *et al.*, 1971) and Surte, Sweden (Andersson-Sköld *et al.*, 2005), both of which are located near outcropping rocks, artesian pressures are responsible for upward and laterally directed salt removal. Leaching by artesian pressures is also expected at many other locations, such as St. Liquori (Grondin and Demers, 1996), along the northern contact of the Champlain Sea with the Canadian Shield, and in the landslide-scarred Gouffre River valley, east of Quebec City, where the sediments accumulated in a steep-sided fjord valley (Filion *et al.*, 1991).

Dispersants also decrease the liquid limit. Penner (1965) demonstrated that low concentrations of Na-hexametaphosphate disperse low-salinity Leda clay, but had little effect on high-salinity materials; indeed, at high concentration the "dispersant" acted as a flocculant. Soderblöm (1966) demonstrated the dispersing effects of several organic compounds and argued that organic compounds leached from bogs may account for high sensitivities in some freshwater and brackish-water sediments that underlie bogs.

8.4.5 OTHER FACTORS

Two post-depositional factors – slow load increase and minimal consolidation – are interlinked with flocculation and cementation. Physical, chemical, and mineralogical diagenetic processes also occur in response to increased pressure and temperature as the sediment is buried more deeply. The main processes at the shallow depths relevant to quick clay are bacterial reduction of Fe^{3+} to Fe^{2+} and leaching, which affects the chemical stability of the minerals present.

8.5 DRAINAGE SYSTEM DEVELOPMENT, SUBAERIAL WEATHERING, NODULAR STRUCTURE, AND CUTAN DEVELOPMENT

As isostatic rebound elevated the land out of the sea at the end of the Pleistocene, rivers commenced their incision of the landscape and oxidative subaerial weathering began to develop a surface crust on the marine clays. The weathered surface crust was formed by additions, removals, transformations, and translocations of soil materials (Brady and Weil, 2008, p. 64). Oxygen was introduced, plants grew and added organic matter, aggregated structure developed, and minerals were weathered. Carbonates and iron-bearing silicates were aggressively attacked by, respectively, soil acids and oxidation. Soluble Ca, K, and Mg ions were released and leached downward, changing the pore-water chemistry and adsorbed cation balance. Freezing, thawing, and desiccation caused a network of cracks to form in the near-surface of the weathered soil and on slopes, producing what Eden and Mitchell (1970) termed a "nodular" structure. Iron oxides formed partial coatings on other particles and, along with translocated fine clay-mineral particles and organic materials, formed cutans (oriented coatings) on the surfaces of developing cracks (Fig. 8.2). The presence of cutans indicates that the cracks have remained open more-or-less continuously since they first formed. Crust thickness depends on local conditions: it is greatest at the slope crest and decreases toward the slope base and away from the riverbank. The crust can be several meters thick where the hydraulic gradient is downward, but where artesian pressures are high, upward and lateral water flow will limit the depth of surface oxidative weathering (Rankka *et al.*, 2004). In the long term, downward, upward, and lateral water flows introduce Ca, K, and Mg, produced by weathering in the overlying soil and along rock joints, which, by replacing adsorbed Na on cation exchange sites, slowly increases the remolded strength and reduces the sensitivity (Moum *et al.*, 1971).

8.6 RANGE OF QUICK CLAY PHYSICAL PARAMETERS

Quick clays, although having high sensitivity in common, exhibit great diversity. Approximate ranges of some physical parameters are:

- *texture* – silty clay to clayey silt
- *depth of surface weathering* – 2–5+ m
- *liquidity index of the quick clay zone* – about 1.2–4+
- *thickness of quick clay zone* – about 2–20+ m
- *cementation* – uncemented to strongly cemented
- *pore water pressure distribution* – hydrostatic to artesian
- *cover of other material*, commonly sand – 0–5+ m
- *bedding* – horizontal to gently sloping.

With this diversity, a range in failure modes and other characteristics of quick clay landslides can be expected.

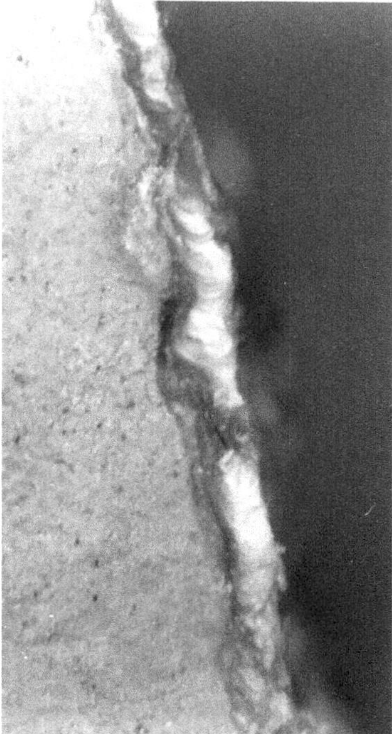

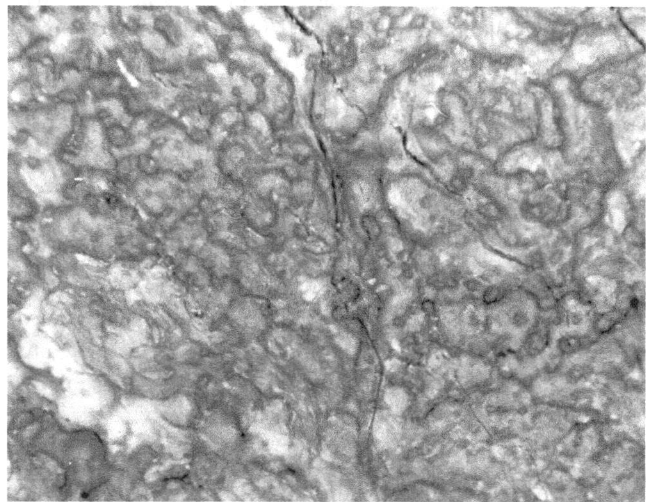

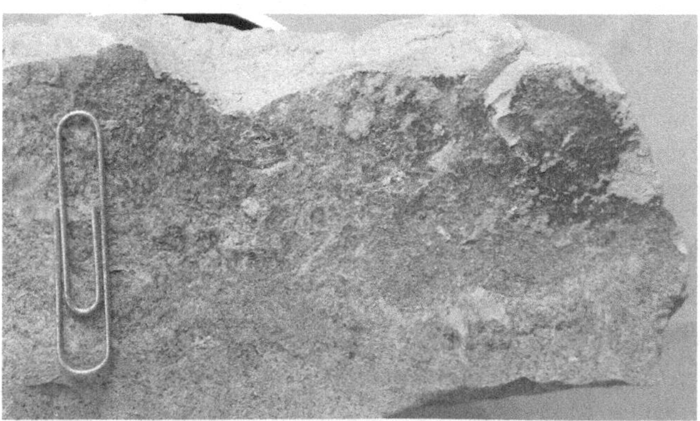

Fig. 8.2. Cutans: (a) cross-section of a clay-dominated cutan (1 mm thick); (b) face view of a clay-dominated cutan (image width *ca.* 1 cm); (c) face view of a cutan consisting of organic matter, iron oxide and clay (paper clip *ca.* 3 cm long). (Photos a and b by J. Germundson; photo c by J. K. Torrance.)

8.7 LANDSLIDE TYPES

The failure process of the Ullensaker (Norway) landslide in 1953 involved a sequence of small failures in which the sediment liquefied and flowed out of the scar through the narrow, bottle-neck-shaped opening and down the valley (Bjerrum, 1955). This mode of failure is described as retrogression, referring to the sequential retreat of the backscarp. Arguably, this nomenclature has influenced people's image of what can also be considered an advance of the failure front into the previously stable landscape.

Bjerrum (1967) identified the requirements for progressive failure in overconsolidated clays as being, in summarized form: (1) local shear stresses exceed the peak shear strength of the clay; (2) an advance of the failure surface is accompanied by sufficient differential strain to take the clay beyond failure; and (3) the post-failure shear strength is low enough that sufficient differential strain causes the zone of stress concentration to move into the neighboring unfailed clay. These constraints also apply to failures in normally consolidated and cemented sensitive marine clays, including quick clays. However, the consequences of these processes in overconsolidated clays and sensitive marine clays are different. Deformation accompanying failure in overconsolidated clays generates negative pore water pressures that, until relieved, inhibit progress of the landslide, whereas deformation accompanying failure in sensitive marine clays causes microstructural collapse, which generates positive pore water pressure that exacerbates the situation and facilitates rapid undrained progress of the failure. We gain new insights into these landslides by explicitly recognizing that retrogressive failures in quick clays occur in an aggressive manner.

Carson and Lajoie (1981) identified five major landslide types, based on topographic constraints, in sensitive marine sediments:

1. two-dimensional spreading failures, with size controlled by topography
2. aborted retrogression
3. excess retrogression
4. multidirectional retrogression
5. flakeslides.

Torrance (1987) added a type 0 category to designate the simple rotational failures that occur within the weathered crusts along river valley slopes (and in some river terrace slopes) in which none of the debris becomes liquid when remolded. The debris remaining in flow bowls of types 1, 2, and sometimes types 3 and 5 landslides has a ribbed morphology, with ridges of displaced nonliquefied debris in the scar oriented perpendicular to the direction of debris flow. Landslides cease to retrogress because some factor (such as sediment properties, gradual decrease in the failure plane depth, or the valley becoming filled with debris) becomes unfavorable to their continuation.

In terms of the failure process involved, type 1–5 landslides can be grouped into two broad categories: *stepwise landslides*, which can proceed stepwise because the debris from each failure step almost completely remolds and exits the landslide scar; and *uninterrupted landslides* in which the thoroughly remolded debris is insufficient for the failure process to continue unless the momentum of the moving debris is maintained from start to finish by an uninterrupted supply of new debris of adequate liquidity.

8.8 LANDSLIDE INITIATION

Human actions and earthquakes (Aylesworth *et al.*, 2000) can initiate landslides in sensitive marine sediments, but riverbank erosion that heightens and steepens the slope, in combination with saturation of the weathered crust on the slope by rainfall or melting snow, is the most common cause. Saturation of the slope creates excess pore water pressure, and lowers effective stress shear strength; artesian pressures may also be a factor. Eden and Mitchell (1970) described the "nodular" weathered crust of the eastern Canadian clays as failing by a dilatant mechanism in which the individual nodules remain largely intact, and frictional forces are dominant. In landslide-prone areas of Norway, cracks at the top of slopes and small failures of the weathered slope material are recognized by local farm families as indicators of imminent landsliding (Kenney, 1967; Drury, 1968; Løken *et al.*, 1970). Such lore is not common in eastern Canada; a substantial 50 × 150 m riverbank failure that preceded the tragic St. Jean Vianney landslide by 10 days raised no alarm locally (Tavenas *et al.*, 1971). Most quick clay landslides are surprises; no one lives adjacent to the slope and no regular inspection or other monitoring occurs. Eyewitnesses have provided anecdotal information about some landslides, but most assessments of the landslide process are based on after-the-fact examination of the scar and debris. In eastern Canada, Mitchell and Markell (1974) noted that the initial landslide at a site may "develop over long periods of time" and that subsequent "retrogression may take several hours or even days to develop," but, once major earthflows in the sensitive soils start, they "take place in a relatively short period of time." Not all riverbank failures in sensitive clays are followed by a quick clay landslide, but in areas known to be prone to large quick clay landslides, any riverbank failure should trigger an immediate geotechnical assessment, particularly where human infrastructure is nearby.

8.9 LANDSLIDE DEVELOPMENT

Several factors influence the development of the failure and the final state of the debris.

- First, the undisturbed shear strength profile, as affected by consolidation, cementation, and weathering, determines the slope height and angle that are stable, the depth where the initial failure of quick clay occurs, the ease of structural breakdown during failure, and the extent to which quick clay above the initial failure zone experiences post-failure remolding.
- Second, the remolded shear strength, related to the liquidity index (Locat and Demers, 1988), determines the flow properties of the debris.
- Third, the thickness of the quick clay layer that initially fails and the proportion of the depth profile above the initial failure zone that liquefies influence how rapidly debris can exit the site and how far it can travel.
- Fourth, the thickness of the surface crust and other overburden at the site influences the difficulty of transporting the debris away from the landslide scar.
- Finally, bedding planes that dip toward the river facilitate debris transport.

Let us now consider what predisposes a site to each landslide type, starting with the type with the most demanding requirements. A geomorphological perspective will be taken.

8.9.1 STEPWISE LANDSLIDES

MULTIDIRECTIONAL RETROGRESSION

Multidirectional retrogression produces landslide scars that are either bottle-necked or progress in directions other than directly into the slope. These landslides are facilitated by the presence of a thick zone of quick clay with a high liquidity index and by a setting where the available gravitational potential energy is sufficient to thoroughly remold most of the quick clay that lies above the failure plane. Cementation does not prevent this type of landslide.

The extent of the quick clay that becomes liquid must be sufficient for the overlying sediment that does not liquefy to be transported out of the landslide scar and along the valley. Also, the material that does not liquefy must break into chunks that are small enough that they do not become grounded in the landslide scar. Multidirectional retrogression is common in the soft marine sediments of Scandinavia, for example at Ullensaker and Hekseberg (Drury, 1968); it occurred during the first stage of the Rissa landslide, where quick clay may have commenced at depths as shallow as 2–3 m and extended to depths of 10–20+ m (Aas, 1967; Gregersen, 1981). Similar landslides occur in the St. Lawrence Basin of eastern Canada: the St. Jean Vianney landslide (La Salle and Chagnon, 1968) and many of the landslides in the Yamachiche (Karrow, 1972) and Maskinongé (Lajoie, 1978) river basins.

8.9.2 UNINTERRUPTED LANDSLIDES

Aborted, 2-dimensional, excess retrogression, and flake landslides represent a continuum in which failure involves an uninterrupted sequence of failures. Differences along this continuum are related to the extent and rate of advance of the failure zone, the rate and degree to which the debris is carried away, and changing conditions along the failure zone. Uninterrupted landslides are common in the clays of the St. Lawrence Basin: South Nation (1971), Lemieux (1993), and many others. They appear to be rare in Scandinavia, but have occurred: e.g., Selnes in 1965 (Kenney, 1967).

Fig. 8.3. (a) The St. Liquori landslide (1989) and (b) the South Nation River landslide (1971), showing typical ribbed morphology shortly after their occurrence. (Photos by J. K. Torrance.)

FLAKESLIDES

Flakeslides in quick clays are initiated when a relatively thin layer of a quick clay zone collapses and turns liquid. The overlying material ruptures at the ultimate location of the landslide backscarp and the detached flake commences movement as an intact unit. The flake may break up into large chunks, as in the strongly cemented sediments at Toulnustouc (Conlon, 1966), into long narrow ridges perpendicular to the flow direction and separated by down-faulted grassed surfaces (Fig. 8.3a; St. Liquori), or rapidly disintegrate during transport, as in the flakeslide stages at Rissa, Norway (Gregersen, 1981). The proportion of the debris remaining within the landslide scar, or transported away, depends on the thickness, undisturbed strength, degree of cementation, ease of remolding, and liquidity index of the flake material and the capacity of the valley.

THE RISSA LANDSLIDE

The Rissa landslide (Gregersen, 1981; Norwegian Geotechnical Institute, 2006), which was initiated by an earthfill that was placed along a lake shore, was witnessed from start to finish. It began when 80 m of shoreline slid away. Over the next 40 minutes it slowly progressed slightly up and across the slope as a mildly multidirectional stepwise series of small slides in which the debris liquefied and flowed away to produce a narrow, 450-m-long slide scar. Then came the surprise. A 150 × 200 m flakeslide commenced and liquefied as it flowed toward the lake along the natural slope, not through the long narrow slide scar. Further flakesliding followed immediately and, in an uninterrupted sequence, extended the scar 1 km along the mountainside in about 5 minutes. That almost all the landslide debris liquefied and almost none remained within the scar is a measure of the low soil strength, the high liquidity index of the remolded material, and the thinness of the weathered crust. The flakeslides constituted over 90 percent of the landslide area. Probably the most remarkable aspect is that during the uninterrupted flakeslide sequence, the failure progressed 1 km in 5 minutes – representing a rate of over 10 km per hour!

ABORTED, TWO-DIMENSIONAL, AND EXCESS RETROGRESSION

Aborted, two-dimensional, and excess retrogression landslide types illustrate differences in landslide propagation into a riverbank relative to the adjacent valley's capacity to contain

Table 8.3. *Effect of various factors on the probability of different retrogressive failures.*

	Aborted	Two-dimensional	Excess
Variable soil	increase	increase	decrease
Thick weathered crust	increase	increase	decrease
Thick quick clay zone	decrease	decrease	increase
Thick failure zone	decrease	decrease	increase
Thorough remolding	decrease	decrease	increase
High liquidity index	decrease	decrease	increase
Dipping bedding plane	decrease	increase	increase
High debris momentum	decrease	decrease	increase

the landslide debris. Respectively, they stop before the adjacent valley is filled, spread out until debris reaches the opposite side, and exceed the capacity of the adjacent valley, with debris flowing downstream and upstream. These landslides are widest at their outlet and exhibit a wide range of width-to-retrogression ratios. They differ in the ability of the failure zone to propagate into the slope, which depends on site characteristics and the amount and characteristics of the debris. The latter is determined mainly by the proportion and the liquidity index of the debris that is thoroughly remolded. All three processes can produce landslides with ribbed surfaces. Possible site and debris characteristic scenarios are summarized in Table 8.3 in an attempt to assess the effect of potential factors influencing landslide propagation.

Possible reasons for *aborted retrogression* are:

- higher remolded strength and lower sensitivity at the failure plane depth decreases debris mobility
- the presence of only a thin zone of quick clay limits debris transport.

These landslides are rarely a serious problem.

Potential reasons for *two-dimensional retrogression* include:

- the debris is unable to change direction and flow along the valley because the valley is too narrow, an insufficient proportion of the sediment is quick clay, or the unremolded overburden resists major remolding and breaks up into large blocks that are difficult to transport
- increasing soil strength at the depth of the failure inhibits propagation of the failure zone.

These landslides may be a serious problem, as at St. Jude.

In the case of *excess retrogression*, none of these restrictions applies and the failure zone continues to propagate away from the riverbank. If the quick clay layer above the failure zone is thick and is remolded to the liquid state as the overlying sediments collapse, the residual blocks of unremolded weathered surface crust and other unremolded sediment are readily transported out of the flow bowl. Until the valley is choked with debris, the momentum of the debris and, hence, its ability to travel far along the valley, is sustained by rapid continuous retrogression of the backscarp. Excess retrogression landslides have a large associated risk.

8.9.3 DEVELOPMENT OF RIBBED LANDSLIDES

The 1964 earthquake-triggered Turnagain Heights landslide in Anchorage, Alaska is the most thoroughly investigated of the ribbed landslides in sensitive clay sediments (sensitivity 5–35). Seed and Wilson (1967) concluded that the earthquake's strong and extended vibrations caused liquefaction of a sand layer underlying the sensitive clay. Liquefaction led to a 2D spreading failure of the overlying clay in which the clay broke into two types of elongated ridges, both oriented perpendicular to the direction of movement: (1) those that were the consequence of rotational failure; and (2) those that consisted of sharp-peaked, inverted-triangular wedges formed by down-dropping of blocks as the sensitive clay extended laterally due to the spreading movement. Carson (1977, 1979) drew attention to the presence of these same types of ridges in the ribbed landslides of the St. Lawrence Basin and proposed spreading as the most credible mechanism for producing inverted-wedge-shaped ridges in the flow bowls. Mitchell (1978) emphasized the dynamic nature of the processes that produce large ribbed landslides: an acceleration stage is followed by a rapid stage and then a deceleration stage.

The 1971 South Nation River landslide (Fig. 8.3b), which extended approximately 650 m along the river and 500 m away from the river, produced a substantial number of bare, inverted-wedge-shaped ridges that had moved into and along the river valley or came to rest in the scar. When landsliding ceased, the back portion of the scar was dominated by long narrow rotated ridges that were grass-covered on the trailing side. The grass-covered ridges developed as long slices only a few meters wide that slumped from the backscarp in such a rapid sequence that the base of each succeeding slice lay under the upper portion of the preceding slice and no liquefied debris had splashed onto the grassy surfaces. This underlap is consistent with the observation by Eden and Mitchell (1970) that, in multi-slice rotational failures in quick clay, the failure plane of the initial landslide in the weathered crust may develop very slowly but subsequent rotational failures occur rapidly, and effectively continuously, until landsliding ceases.

The 1989 St. Liquori landslide debris (Fig. 8.3a) was dominated by inverted-wedge-shaped ridges with nearly horizontal, down-faulted, grassed surfaces between ridges (Grondin and Demers, 1996). These ridges developed either sequentially or nearly simultaneously. Sequential formation of inverted wedges and down-faulted grassed surfaces appears to involve a series of steps.

1. The leading side of the first inverted wedge is created by a rotational failure.
2. The failure zone at depth in the quick clay rapidly advances so that it underlies the entire base of the unit that will calve off.
3. The trailing side is created by tension failure as the inverted wedge tips forward in response to the flow of remolded clay from below.
4. The inverted wedge starts to slide forward as tension failure on the trailing side reaches the depth of the failure zone, leaving behind unsupported overhanging sediment that, in turn, tips forward and initiates another tension failure of the opposing slope, thereby creating a wedge-shaped unit.
5. The lower portion of the wedge-shaped unit remolds, liquefies, and flows, causing this wedge to both slump and move forward, lowering its grassy surface and leaving behind a slope that will become the leading edge of the next inverted wedge.

While all this is happening, the failure zone continues to advance and sets the stage for the next inverted wedge in the sequence. Inverted-wedge formation can continue as long as the failure zone advances rapidly; when the advance of the failure zone slows sufficiently, rotational failures may again occur. Nearly simultaneous formation of inverted wedges appears to require a rapid advance of the failure zone, so that it underlies an area sufficiently large to fail as a small or large flakeslide, which then breaks up by the down-faulting mechanism described above. The minimum requirement for formation of inverted wedges is that the failure zone must extend beyond the wedge base before the wedge can break away.

In ribbed landslides, inverted-wedge formation should dominate only at times when the failure zone at depth is rapidly advancing. Rotational landsliding should dominate during the acceleration and deceleration phases when the failure zone at depth is advancing slowly. Eventually, continuous advance of the landslide will cease when conditions become unfavorable, because the depth to the failure zone becomes too shallow, soil properties change, or the river channel becomes blocked and the remaining debris produced accumulates in the scar. The process of landsliding can be envisaged as an "advancing conveyor belt" – the failing quick clay zone at depth continuously extends into the sediment and carries away successive rotational failures, inverted wedges, and wedges of the overlying material. The formation of ribbed landslides is favored where the quick clay above the failure zone is sufficiently strong, relative to the gravitational potential energy, that post-failure remolding is moderate in the debris produced by rotational failure, almost nonexistent in the inverted wedges as they advance on their broad bases, but nearly complete in the lower parts of the down-faulted wedges.

Quinn *et al.* (2011) proposed a new model for large landslides in sensitive clay, based on a fracture mechanics approach. Briefly, they adopt a theoretical progressive failure model, similar to that of Bjerrum (1967), in which, by one sudden failure or a series of small localized failures over time, "the entire failure surface for the landslide develops prior to the first instance of significant, noticeable movement." This mechanism is the same as described earlier for flakeslides, the end member of the uninterrupted landslide continuum. This new theoretical model appears to be a significant advance in analyzing the large landslide problem. It should be possible to adapt this fracture mechanics model to the situation of the rapidly advancing failure plane and sequential failure mechanism proposed for the other members of the uninterrupted landslide continuum described in Section 8.9.2.

8.10 HAZARD ASSESSMENT AND PROSPECTIVE METHODS FOR PREVENTION OF QUICK CLAY LANDSLIDES

Quick clay hazard assessment and risk management are being actively pursued in Norway, Sweden, and Quebec. The initial activity in each jurisdiction involved acquiring a complete inventory of large landslides in marine sediments. With this information, weight-of-evidence statistical methods can be applied to identify areas of high landslide risk (Quinn *et al.*, 2010), and probability risk assessment can be applied at individual sites (Cassidy *et al.*, 2008). Areas of low pore-water salinity, where quick clay may be present, can be identified within larger areas of marine sediments through electrical resistance surveys (Carson, 1981; Hunter *et al.*, 2010), However, because not all the low-conductivity sediment is quick clay, the presence of quick clay should be confirmed with cone penetration probes (Rømoen *et al.*, 2010). In the context of landslide prevention, the four major factors involved in quick clay formation – flocculation, cementation, low-activity minerals, and leaching – are chemical in nature. The main methods that are commonly used to prevent landslides – installation of berms to limit fluvial and wave undercutting of the slope, soil removal to reduce the slope angle, installation of horizontal drains, and pumping to control pore pressures in slopes – are physical in nature. The amount of earth material and disturbance involved, slope height and steepness, and the valley narrowness eliminate these methods as practical solutions. It is apparent to this observer that long-term solutions to the quick clay landslide problem will involve chemical changes that increase the liquid limit of the material or strengthen critical zones within the soil. A decrease in water content by electro-osmosis (Bjerrum *et al.*, 1968) and the addition of salt in wells (Moum *et al.*, 1968) have both been successfully used in Norway to increase the strength of sensitive quick clays, and lime columns have been used in Norway, Sweden, and Canada to address consolidation problems in these materials. For landslide control, the most promising approach appears to be the introduction of a saturated solution of an appropriate chemical into small-diameter boreholes, from which it diffuses into the quick clay zone (Moum *et al.*, 1968). The depth and distance from the slope that will need to be treated can be determined through numerical modeling. Sodium chloride (NaCl) could be used, but quick clay would eventually return when this salt leaches out, leaving a Na-dominated, low-salinity soil. Potassium chloride (KCl) is superior because it will diffuse in the sediment more rapidly (Moum *et al.*, 1968) and, when the potassium salt has leached out, it will leave behind a K-dominated soil with a sufficiently high liquid limit that the quick clay condition is unlikely to recur. Enlargement of riverbank failures would still occur, but their enlargement into large quick clay landslides would be prevented or greatly inhibited. The major unanswered questions about such chemical remediation relate to its environmental acceptability in areas where people rely on local groundwater.

REFERENCES

Aas, G. (1967). Vane tests for investigation of anisotropy of undrained shear strength of clays. In *Proceedings of the Geotechnical Conference, Oslo, Vol. 1*, pp. 3–8.

Andersson-Sköld, Y., Torrance, J.K., Lind, B. *et al.* (2005). Quick clay: A case study of chemical perspective in southwest Sweden. *Engineering Geology*, 82, 107–118.

Aylesworth, J.M., Lawrence, D.E. and Guertin, J. (2000). Did two massive earthquakes in the Holocene induce widespread landsliding and near-surface deformation in part of the Ottawa Valley, Canada? *Geology*, 28, 903–906.

Berry, R.W. (1974). Quartz cleavage and quick clays. *Science*, 184, 183.

Berry, R.W. and Jørgensen, P. (1971). Grain size, mineralogy and chemistry of a quick clay sample from the Ullensaker slide, Norway. *Engineering Geology*, 5, 73–84.

Berry, R.W. and Torrance, J.K. (1998). Mineralogy, grain-size distribution and geotechnical behavior of Champlain clay core samples, Quebec. *Canadian Mineralogist*, 36, 1625–1636.

Bjerrum, L. (1955). Stability of natural slopes in quick clay. *Geotechnique*, 5, 101–119.

(1967). Progressive failure in slopes of overconsolidated plastic clay and clay shales. *Journal of the Soil Mechanics and Foundation Division, ASCE*, 93, SM5, 1–49.

Bjerrum, L., Moum, J. and Eide, O. (1968). *Application of Electro-osmosis to a Foundation Problem in a Norwegian Quick Clay*. Norwegian Geotechnical Institute, Publication 74.

Borggaard, O.K. (1988). Phase identification by selective extraction techniques. In *Iron in Soils and Clay Minerals*, ed. J.W. Stucki, B.A. Goodman and U. Schwertmann. Dordrecht, Netherlands: D. Reidel Publishing Company, pp. 83–98.

Brady, N.C. and Weil, R.R. (2008). *The Nature and Properties of Soils*. Upper Saddle River, NJ: Pearson Education.

Carson, M.A. (1977). On the retrogression of landslides in sensitive muddy sediments. *Canadian Geotechnical Journal*, 14, 582–602.

(1979). On the retrogression of landslides in sensitive muddy sediments: Reply. *Canadian Geotechnical Journal*, 16, 431–444.

(1981). Influence of porefluid salinity on instability of sensitive marine clays: A new approach to an old problem. *Earth Surface Processes and Landforms*, 6, 499–515.

Carson, M.A. and Lajoie, G. (1981). Some constraints on the severity of landslide penetration in sensitive deposits. *Géographie Physique et Quaternaire*, XXXV, 301–316.

Cassidy, M.J., Uzielli, M. and Lacasse, S. (2008). Probability risk assessment of landslides: A case study at Finneidfjord. *Canadian Geotechnical Journal*, 45, 1250–1267.

Chamley, H. (1997). Clay mineral sedimentation in the ocean. In *Soils and Sediments: Mineralogy and Geochemistry*, ed. H. Paquet and N. Clauer. Berlin: Springer-Verlag, pp. 269–302.

Conlon, R.J. (1966). Landslide on the Toulnustouc River, Quebec. *Canadian Geotechnical Journal*, 3, 113–144.

Drury, P. (1968). *The Hekseberg Landslide, March 1967*. Norwegian Geotechnical Institute, Publication 75, pp. 27–31.

Eden, W.J. and Mitchell, R.J. (1970). The mechanics of landslides in Leda clay. *Canadian Geotechnical Journal*, 7, 285–296.

Egashira, K. and Ohtsubo, M. (1982). Smectite in marine quick-clays of Japan. *Clays and Clay Minerals*, 30, 275–280.

Filion, L., Quinty, F. and Bégin, C. (1991). A chronology of landslide activity in the valley of Rivière du Gouffre, Charlevoix, Quebec. *Canadian Journal of Earth Sciences*, 28, 250–256.

Gadd, N.R. (1980). Maximum age for a concretion at Green Creek, Ottawa. *Géographie Physique et Quaternaire*, XXXIV, 229–238.

Gates, W.P., Jaunet, A.-M., Tessier, D. *et al.* (1998). Swelling and texture of iron-bearing smectites reduced by bacteria. *Clays and Clay Minerals*, 49, 487–497.

Geertsema, M. and Torrance, J.K. (2005). Quick clay from the Mink Creek landslide near Terrace, British Columbia: Geotechnical properties, mineralogy, and geochemistry. *Canadian Geotechnical Journal*, 42, 907–918.

González-Lucena, F. (2010). *Mineral magnetism of environmental reference material: Iron oxyhydroxide nanoparticles*. Ph.D. thesis, University of Ottawa, Ottawa, ON.

Gregersen, O. (1981). *The Quick Clay Landslide at Rissa, Norway*. Norwegian Geotechnical Institute, Publication 135, pp. 1–6.

Grondin, G. and Demers, D. (1996). The 1989 Saint-Liquori flakeslide: Characterization and remedial works. In *Landslides: Proceedings of the Seventh International Symposium on Landslides, Oslo, Vol. 2*, ed. K. Senneset. Rotterdam: Balkema, pp. 743–748.

Hunter, J.A., Burns, R.L., Good, R.L. *et al.* (2010). Near-surface geophysical techniques for geohazards investigations: Some Canadian examples. *Leading Edge*, August 2010, 964–977.

Jackson, M.L. and Sherman, G.D. (1953). Chemical weathering of minerals in soils. *Advances in Agronomy*, 5, 211–317.

Karrow, P.F. (1972). Earthflows in the Grondines and Trois Rivières areas, Québec. *Canadian Journal of Earth Sciences*, 9, 561–573.

Kenney, T.C. (1964). Sea-level movements and the geological histories of the post-glacial marine soils at Boston, Nicolet, Ottawa and Oslo. *Géotechnique*, 14, 203–210.

(1967). Slide behaviour and shear resistance of a quick clay determined from a study of the landslide at Selnes, Norway. In *Proceedings of the Geotechnical Conference, Oslo, Vol. 1*, pp. 57–64.

Kenney, T.C., Moum, J. and Berre, T. (1968). An experimental study of bonds in a natural clay. In *Proceedings of the Geotechnical Conference, Oslo, Vol. 1*, pp. 65–69.

La Salle, P. and Chagnon, J.-Y. (1968). An ancient landslide along the Saguenay River, Quebec. *Canadian Journal of Earth Sciences*, 5, 548–549.

Lajoie, G. (1978). *La présence et la distribution des coulées argileuses dans la vallée de la Maskinongé*. M.Sc. thesis, McGill University, Montréal, PQ.

Locat, J. and Demers, D. (1988). Viscosity, yield stress, remolded strength, and liquidity index relationships for sensitive clays. *Canadian Geotechnical Journal*, 25, 799–806.

Loiselle, A., Massieera, M. and Sainani, U. R. (1971). A study of cementation bonds of the sensitive clay of the Outardes River region. *Canadian Geotechnical Journal*, 8, 479–498.

Løken, T. (1971). Recent research at the Norwegian Geotechnical Institute concerning the influence of chemical additions on quick clay. *Geologiska Föreningen i Stockholm Förhandlingar*, 92, 133–147.

Løken, T., Jørstad, F. and Heiberg, S. (1970). *Gamle Leirskred på Romerike*. Sæavtrykk av Romerike Historielags Årbok VII (in Norwegian).

Mitchell, R. J. (1978). On the retrogression of landslides in sensitive muddy sediments: Discussion. *Canadian Geotechnical Journal*, 15, 446–450.

Mitchell, R. J. and Markell, A. R. (1974). Flowsliding in sensitive soils. *Canadian Geotechnical Journal*, 11, 11–31.

Moum, J., Sopp, O. I. and Løken, T. (1968). *Stabilization of Undisturbed Quick Clay by Salt Wells*. Norwegian Geotechnical Institute, Publication 81, pp. 1–7.

Moum, J., Løken, T. and Torrance, J. K. (1971). A geochemical investigation of the sensitivity of normally consolidated clay from Drammen, Norway. *Geotechnique*, 21, 329–340.

Murad, E. (1988). Properties and behaviour of iron oxides as determined by Mössbauer spectroscopy. In *Iron in Soils and Clay Minerals*, ed. J. W. Stucki, B. A. Goodman and U. Schwertmann. Dordrecht, Netherlands: D. Reidel Publishing Company, pp. 309–350.

Norsk Geoteknisk Forening (1974). *Retlingslinjer for Presentasjon av Geotekniske Undersökelser*. Norsk Geoteknisk Forening, Oslo, Norway (in Norwegian).

Norwegian Geotechnical Institute (2006). *The Rissa Landslide*. DVD, Norwegian Geotechnical Institute, Oslo, Norway.

Penner, E. (1965). A study of sensitivity in Leda clay. *Canadian Journal of Earth Sciences*, 2, 425–441.

Quigley, R. M. (1980). Geology, mineralogy and geochemistry of Canadian soft soils: A geotechnical perspective. *Canadian Geotechnical Journal*, 9, 261–285.

Quinn, P. E., Hutchinson, D. J., Diederichs, M. S. and Rowe, R. K. (2010). Regional-scale landslide susceptibility mapping using weight-of-evidence method: An example applied to linear infrastructure. *Canadian Geotechnical Journal*, 48, 1151–1162.

Quinn, P. E., Diederichs, M. S., Rowe, R. K. and Hutchinson, D. J. (2011). A new model for large landslides in sensitive clay using a fracture mechanics approach. *Canadian Geotechnical Journal*, 48, 1151–1162.

Rankka, K., Andersson-Sköld, Y., Hulten, C. *et al.* (2004). *Quick Clay in Sweden*. Swedish Geotechnical Institute, Report 65.

Rømoen, M., Pfaffhuber, A. A., Karlsrud, K. and Helle, T. E. (2010). Resistivity on marine sediments retrieved from RCPTU-soundings: A Norwegian case study. In *Proceedings of the International Symposium on Cone Penetration Testing, Huntington Beach, CA, Vol. 2*, pp. 289–304.

Rosenqvist, I. T. (1946). Om leirers kvikkaktighet. *Statens Vegvesen, Veglaboratoriet Meddelelsen*, 4, 5–12.

Rosenqvist, I. T. (1953). Considerations on the sensitivity of Norwegian quick-clays. *Géotechnique*, 3, 195–200.

Seed, H. B. and Wilson, S. D. (1967). The Turnagain Heights landslide, Anchorage, Alaska. *Proceedings of the American Society of Civil Engineers*, 93, 325–353.

Seekings, D. R. (1974). *A geotechnical investigation of a borehole from the site of the South Nation River landslide*. M.A. thesis, Carleton University, Ottawa, ON.

Söderblom, R. (1966). Chemical aspects of quick clay formation. *Engineering Geology*, 1, 415–431.

Tavenas, F., Chagnon, J.-Y. and LaRochelle, P. (1971). The Saint-Jean-Vianney landslide: Observations and eye-witness accounts. *Canadian Geotechnical Journal*, 8, 463–478.

Torrance, J. K. (1983). Towards a general model of quick clay development. *Sedimentology*, 30, 547–555.

(1987). Quick clays. In *Slope Stability*, ed. M. G. Anderson and K. S. Richards. Chichester, UK: J. Wiley and Sons, pp. 447–473.

(1988). Mineralogy, pore-water chemistry and geotechnical behaviour of Champlain Sea and related sediments. In *The Late Quaternary History of the Champlain Sea Basin*, ed. N. R. Gadd. Geological Association of Canada, Special Paper 35, pp. 259–275.

(1990). Oxide minerals in the sensitive post-glacial marine clays. *Applied Clay Science*, 5, 307–323.

(1995). On the paucity of amorphous minerals in the sensitive marine clays. *Canadian Geotechnical Journal*, 32, 535–538.

(1999). Mineralogy, chemistry and the structure and microstructure of sediments. In *Characterization of Soft Marine Clays*, ed. T. Tsuchida and A. Nakase. Rotterdam: Balkema, pp. 203–211.

Torrance, J. K. and Ohtsubo, M. (1995). Ariake Bay quick clays: A comparison to the general model. *Soils and Foundations*, 35, 11–19.

Torrance, J. K. and Percival, J. B. (2003). Experiences with selective extraction procedures for iron oxides. In *2001: A Clay Odyssey*, ed. E. A. Domínquez, G. R. Mas and F. Cravero. Amsterdam: Elsevier, pp. 219–226.

Torrance, J. K., Hedges, S. W. and Bowen, L. H. (1986). Mössbauer spectroscopic study of the iron mineralogy of post-glacial marine clays. *Clays and Clay Minerals*, 34, 314–322.

Viljoen, D. W. (1989). *The influence of smectite concentration on the rheological properties of low-activity marine clays: A quantitative analysis*. B.Sc. Honours paper, Department of Geography, Carleton University, Ottawa, ON.

Yariv, S. and Cross, H. (1979). *Geochemistry of Colloid Systems*. Berlin: Springer-Verlag.

9 Controls on the distribution of major types of submarine landslides

DAVID J. W. PIPER, DAVID C. MOSHER, AND D. CALVIN CAMPBELL

ABSTRACT

This study uses the eastern Canadian continental margin as a type example to assess controls on the distribution of different types of submarine landslides. A brief summary is provided of the major styles of submarine landslides recognized globally, their transport mechanisms, and the factors responsible for both preconditioning and triggering failure. The eastern Canadian continental margin differs along its length as a result of the northward-decreasing age of oceanic rifting and resulting depth of the ocean floor. It has been strongly modified by Quaternary glaciation. Landslides on the margin are recognized from 2D and 3D seismic data, multibeam bathymetry, and 10-m-long cores. The ages of landslides are well defined from regional stratigraphic studies. The distribution of different types of landslides on the eastern Canadian margin depends on whether the margin is progradational or erosional. Progradational slopes have unconsolidated sediment available for retrogressive slumps, whereas on more erosional slopes more consolidated sediment is available to form blocky disaggregated landslides. Most large landslides appear to result from regional failure during large, but rare, passive-margin earthquakes. Preconditioning factors such as sedimentation rate or flux of basinal fluids do not seem to have had a major impact on the distribution of larger failures, but could be locally significant for small landslides.

9.1 INTRODUCTION

Submarine landslides are widespread on the steep slopes of the ocean floor, particularly on continental slopes, where sedimentation rates from the adjacent continent are high. They are also common on mid-ocean ridges, transform faults, and the flanks of seamounts and volcanic islands. Tectonic deformation on accretionary prisms at subduction zones also promotes submarine landsliding. Mass wasting by submarine landslides is volumetrically the most important form of submarine erosion (Mosher *et al.*, 2010a). The range of failure types known on land – from rockfalls to rotational slumps to debris flows – is also recognized in the ocean.

Although the types of failures under the ocean are similar to those on land, there are two important differences between terrestrial and submarine landslides. First, most of the seafloor is an area of net deposition, with thick successions of sediment, some of which may be underconsolidated as a result of rapid burial or release of deep fluids. Second, failures on the seafloor, particularly those involving only weakly consolidated sediment, can mix with water. Thus as a submarine landslide travels downslope, it may break up and transform first to a debris flow, then with further inmixing of water to a turbulent turbidity current (Hampton, 1972; Piper *et al.*, 1999). Furthermore, movement of the landslide may be facilitated by the process of hydroplaning (Mohrig *et al.*, 1998). Turbidites – the deposits of turbidity currents – are volumetrically the most abundant type of deep-sea deposit. Turbidity currents are a threat to deep-sea telecommunication cables (Hsu *et al.*, 2008), and buried turbidites are the predominant deep-water type of petroleum reservoir (Stow and Mayall, 2000).

Submarine landslides are known principally from their seabed bathymetric expression and from seismic reflection profiles of the resulting deposits. The tops of many landslide deposits have been sampled with 10-m-long cores, and a few deposits have been penetrated by Ocean Drilling Program cored boreholes or by other drilling techniques. The high costs of comprehensive in-situ geotechnical investigations have generally been justified only for major petroleum installations. Because submarine landslides are episodic, they are difficult to monitor on the ocean floor.

Our understanding of submarine landslides has advanced greatly in the past two decades, largely as a result of technological

Landslides: Types, Mechanisms and Modeling, ed. John J. Clague and Douglas Stead. Published by Cambridge University Press.

advances. Precise navigation using global positioning systems (GPS), coupled with inertial navigation, both developed for military purposes, has allowed the acquisition of precise multibeam bathymetry along swaths of the seafloor. 3D seismic imaging techniques developed by the petroleum industry provide relatively high-resolution images of the subsurface structure of submarine landslides, particularly on petroliferous continental margins. The interest of the petroleum industry in seabed stability, particularly in the Gulf of Mexico and in the Ormen Lange field at the head of the Storegga Slide off Norway, has resulted in new in-situ technologies for geomechanical measurements, which can be monitored through the developing technology of seafloor cabled observatories. The recognition that many tsunamis are generated by submarine landslides, following studies of the 1998 Papua New Guinea tsunami (Tappin *et al.*, 2001), gave new impetus to the assessment of geohazards resulting from submarine landslides (Mosher *et al.*, 2010a).

The first part of this chapter describes the range of transport processes involved in submarine landslides and the resulting architecture of the deposit. It is followed by an assessment of the principal causes of submarine failures. The emphasis is on features that differ in the marine realm from those of landslides on land. These concepts are then applied in order to analyze the factors responsible for differences in the style of submarine landslides along the glaciated continental margin of eastern Canada. Based on this analysis, we attempt to derive some general thoughts on the factors that control the distribution of submarine landslides.

9.2 SEDIMENT TRANSPORT PROCESSES IN SUBMARINE LANDSLIDES

Submarine landslides have rarely, if ever, been observed directly. The interpretation of sediment transport processes is thus principally based on the overall architecture of the landslide (Bull *et al.*, 2009), and to a lesser extent the character of deposits recovered in cores (Tripsanas *et al.*, 2008b) and comparisons with better-known processes on land (Nardin *et al.*, 1979; Mulder and Cochonat, 1996). Sediment transport processes are also informed by tank experiments (Mohrig *et al.*, 1998) and numerical modeling (Mulder and Moran, 1995).

Creep is slow, long-term permanent deformation of sediment under constant load. Creep in the submarine environment results in folded stratified sediments above a decollement surface (Fig. 9.1), which on steep slopes may result in sediment failure (Lee and Chough, 2001).

Debris or **rock falls** involve the failure and movement of fragmented bedrock or indurated sediment on a steep slope, such as the wall of a submarine canyon (Blikra and Nemec, 1998). The clasts tumble freely downslope and accumulate at the base of slope, producing a deposit that ranges from isolated clasts to gravels with no matrix (Prior and Doyle, 1985; Blikra and Nemec, 1998).

Debris or **rock avalanches** are sourced in a similar manner to debris or rock falls, but involve large volumes of failed material, in which the interacting clasts collide and share their momentum in a manner similar to that prevailing in a grain flow. In the marine realm, debris avalanches are derived either from land (e.g., the steep flanks of volcanoes or fjords; Fig. 9.2) or from deep (>100 m) rotational failures on steep slopes (>10°) (Mulder and Cochonat, 1996; Masson *et al.*, 2002), the latter transforming into debris avalanches through shearing, fragmentation, and dilation (Pollet and Schneider, 2004). Debris and rock avalanches produce an ungraded to normally graded breccia or conglomerate (Blikra and Nemec, 1998). Submarine debris avalanches derived from failures in sedimentary rocks and highly consolidated sediment have been interpreted in the marine realm (Piper *et al.*, 1997; Collot *et al.*, 2001; Bohannon and Gardner, 2004; Normark *et al.*, 2004). Most examples lack direct evidence for debris avalanche processes because samples of the deposit were not available, but piston cores from the Gulf of Mexico show blocks with a jigsaw texture, which is diagnostic of debris-avalanche processes (Tripsanas *et al.*, 2008b).

Slides and **slumps** are movements of coherent sediment masses over discrete basal shear planes. They terminate upslope in headscarps (Fig. 9.3), with relief of a few to hundreds of meters (Coleman and Prior, 1988; Mulder and Cochonat, 1996). The marine community has generally used the criteria of Nardin *et al.* (1979) and Coleman and Prior (1988) to differentiate slumps and slides. Slumps are rotational slides, with a Skempton ratio $h\,l^{-1}$ (h: depth of landslide; l: length of slide) >0.33, and evidence of displacement of relatively intact blocks over curved or spoon-like slip surfaces with limited downdip transport (e.g., rotational blocks abut headscarps). Slides have $h\,l^{-1}$ ratios <0.15 and basal shear or sliding zones parallel to the surface slope, over which slabs of sediments are displaced downslope. Deformation in slide deposits varies from low (inclined and microfaulted bedding) to moderate (convolute/folded and lenticular bedding), with the initial sedimentary structures and facies of the failed sediment only partially recognizable (Mosher *et al.*, 1994; Coussot and Meunier, 1996; Tripsanas *et al.*, 2008b).

If the deformation is high enough, slumps or slides transform into **debris flows**. Three processes are involved:

1. widespread Coulomb failure within a sloping sediment mass
2. partial or complete liquefaction of the mass by high pore-fluid pressures
3. conversion of landslide translational energy to internal vibrational energy (Iverson *et al.*, 1997).

Debris flows are (pseudo-)plastic, poorly sorted flows, in which clasts float in a fine-grained matrix (mud to sand) with finite shear strength. In cohesive debris flows (Mulder and Alexander, 2001), the matrix contains sufficient clay to reduce the settling rate of coarser sediment and entrainment of ambient water (Ilstad *et al.*, 2004), so that it behaves as a Bingham flow. The upper plug zone, in which the yield stress is not exceeded, overlies a basal shear zone in which the shear stress exceeds the yield stress of the matrix (Johnson, 1970; Iverson *et al.*, 1997; De Blasio *et al.*, 2004). Deposition from cohesive debris flows

A

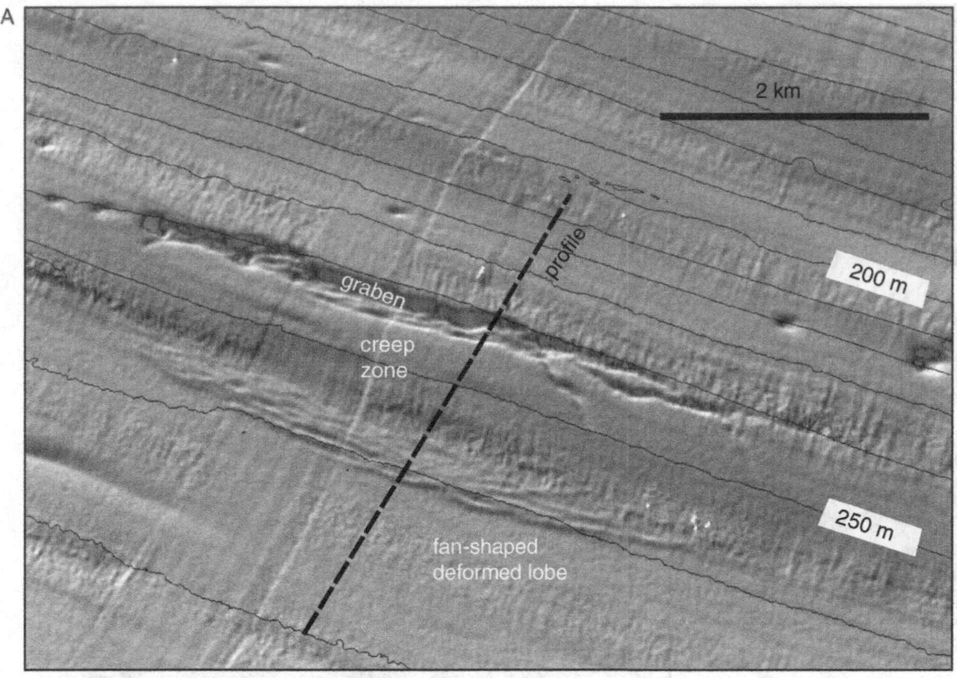

B

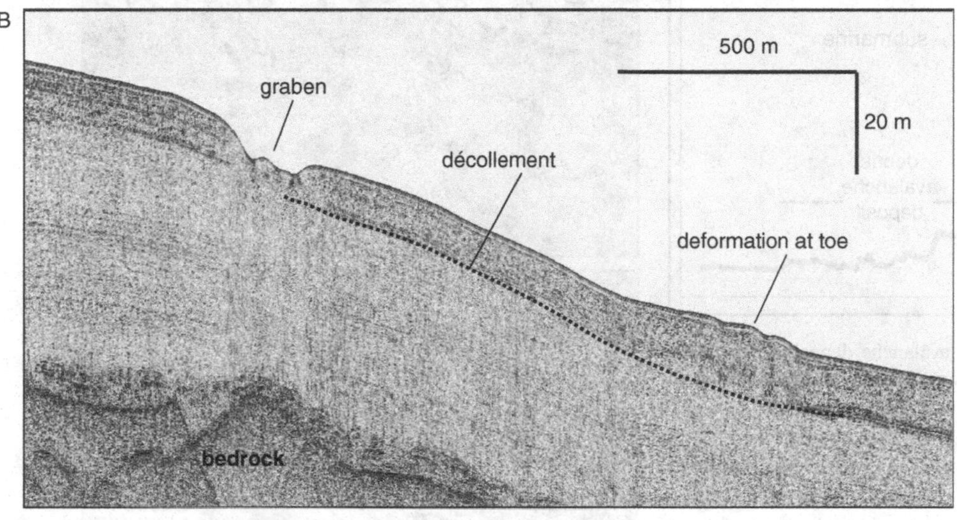

Fig. 9.1. Bathymetric image (A) and seismic profile (B) of sediment creep on the flank of the St. Lawrence estuary (modified from Campbell *et al.*, 2008).

occurs during sudden cessation of flow, when the applied shear stress falls below a threshold value (the yield stress). Rapid deceleration results in *en masse* deposition, initiated at the front of the flow (Prior *et al.*, 1984; Huang and García, 1999).

9.3 ARCHITECTURAL TYPES OF SUBMARINE SLIDES

9.3.1 INTRODUCTION

The architecture of submarine landslides can be inferred from their seafloor morphological expression and geophysical imaging, in many cases even when an old landslide deposit has been covered by tens of meters of younger sediment (Longva *et al.*, 2008; Mosher and Campbell, 2011). The information from seismic reflection profiling is limited by the basic physics of sound and the geometry of the deposits. No coherent reflection is returned if there is no impedance contrast from laterally continuous interfaces that dip <15°. In the older literature, deposits of submarine landslides were recognized by their incoherent seismic reflection character and commonly a positive seafloor expression; they were generally termed "debris flows" (Damuth, 1980). The term "mass-transport deposit" is now generally used for deposits such as these, for which a more precise architecture cannot be determined. Where available, 3D seismic-reflection profiling provides enhanced imagery of submarine landslides, particularly through the use of layer-parallel slices.

9.3.2 FRONTALLY EMERGENT AND FRONTALLY CONFINED SLIDES

The overall morphology of the downslope part of landslides depends on whether they are frontally emergent, flowing over

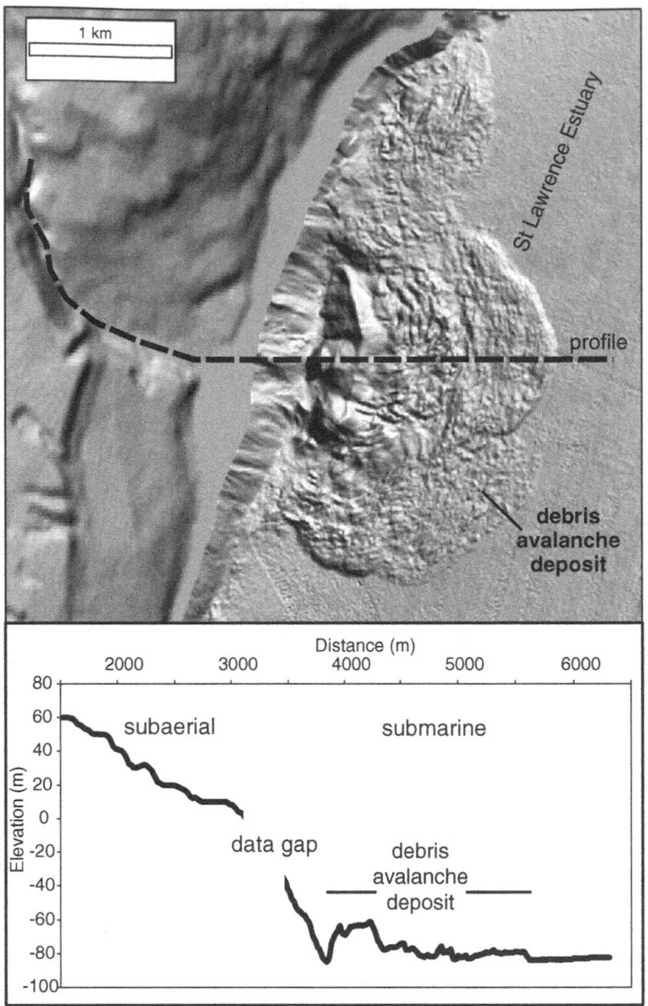

Fig. 9.2. Bathymetric image of debris avalanche deposit on the steep glaciated flank of the St Lawrence estuary (modified from Poncet *et al.*, 2010).

the seabed and terminating at a snout (Fig. 9.4), or whether they are frontally confined, showing progressive frontal thrusting in the landslide deposit and in the autochthonous sediment beyond (Frey-Martinez *et al.*, 2006; Moscardelli *et al.*, 2006). Frontally confined landslides undergo a restricted downdip translation and do not overrun the undeformed downslope strata. In frontally emergent slides, the downdip translation is much larger and the landslide is able to ramp up from its original basal shear surface and spread in an unconfined manner over the seafloor. Which type of slide develops depends principally on the depth of the basal shear plane, but also on the nature of the sediment and the regional gradient (Moernaut and De Batist, 2011).

9.3.3 SIMPLE SLUMPS

Simple slumps involve rotational sliding on a basal shear plane, with a limited degree of retrogression, leaving a headscarp that is amphitheatre-shaped. The slump deposit consists

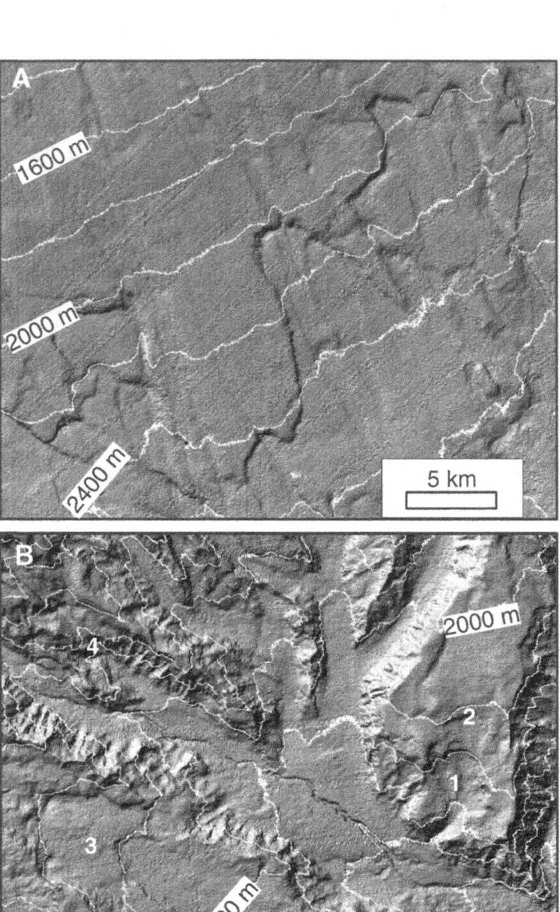

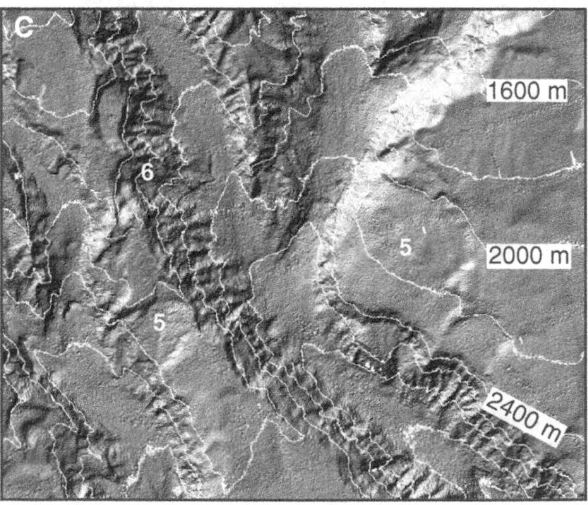

Fig. 9.3. Multibeam bathymetry images of the central Scotian Slope at about 2000 m water depth. (A) Stepped bathymetry with scarps <20 m high resulting from evacuation of sediment by retrogressive slumping. (B) Part of Logan Canyon showing (1) amphitheatre-like scarps resulting from slump conditioned by erosional steepening, (2) retrogressive slump above a simple slump, (3) retrogressive slumping and evacuation in inter-canyon area, and (4) apparent deep retrogressive slumping away from the main valley. (C) Part of Bonnecamps Canyon showing (5) stepped retrogressive slumping and evacuation of sediment away from canyon walls and (6) amphitheatre-like scarps on the canyon wall.

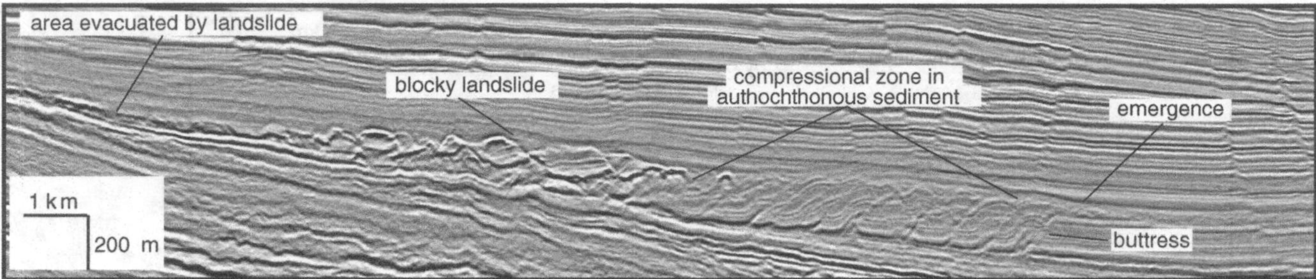

Fig. 9.4. Buried Miocene frontally emergent slide, Torbrook prospect, Scotian Slope, showing the general features of a frontally emergent slide. Reproduced with permission from Encana.

of a series of rotated blocks, giving a characteristic transverse-lineated morphology (e.g., Afen Slide: Jackson *et al.*, 2004; Eivissa slides: Lastras *et al.*, 2004). Such slumps are common in oversteepened areas such as canyon walls (Mosher *et al.*, 2004), in tectonically steepened accretionary prisms (López *et al.*, 2010), and in areas of active salt tectonics (Tripsanas *et al.*, 2008b). The downslope behavior of the slump differs according to setting: some are frontally confined, some frontally emergent, and some are subsequently eroded by canyon-floor processes. On steep slopes, failed blocky material may transform into a debris flow or a turbidity current, as on the Scotian Slope (Piper *et al.*, 1985) and in the 1979 Nice Airport failure (Dan *et al.*, 2007).

9.3.4 MAJOR RETROGRESSIVE SLUMPS

Large retrogressive slumps generally involve failure over hundreds of square kilometers on the continental slope of passive continental margins. They show complex retrogression, typically in the upper few tens of meters of sediment, along one or more weak horizons. This style of failure is characteristic of some of the best-known submarine failures, such as the 1929 Grand Banks landslide off eastern Canada (Piper *et al.*, 1999; Mosher and Piper, 2007) and the *ca.* 8000-year-old Storegga landslide off Norway (Haflidason *et al.*, 2004). Retrogression is demonstrated by seabed morphology in areas of evacuation (Fig. 9.3) and by the sequence of sediment types in the resulting deposits (Tripsanas *et al.*, 2008a). Upslope retreat continues until stronger upper slope or shelf sediments are encountered, either because of a change in sediment type, degree of consolidation, or change in fluid-related overpressure; an irregular headscarp is thus formed.

Such failures result in a wide range of deposits. Rotational slump blocks may be preserved below the headscarp at the upslope limit of retrogression. These may pass downslope into blocky slide deposits if the depth of failure was sufficient to mobilize well-consolidated sediment. Shallower, less consolidated sediment may mobilize into a debris flow that will transform into a turbidity current if it flows over steep pre-existing topography such as submarine canyon walls (Piper *et al.*, 1999). In some cases, evacuation of sediment is sufficient to leave a stepped seafloor morphology with little or no overlying mass-transport deposit (Fig. 9.3).

9.3.5 LARGE COMPLEX LANDSLIDES

Large complex landslides are characterized by the translation of large blocks (10^2–10^3 m in length) of well-consolidated sediment that break up into smaller blocks of a range of sizes. The internal structure of such slides is best known from 3D seismic studies (Gee *et al.*, 2005; Mosher and Campbell, 2011; Fig. 9.5). These landslides may show evidence of some basal erosion, including weakly developed linear grooves (Mosher and Campbell, 2011). Ancient examples, now exposed in sedimentary sequences on land (Lucente and Pini, 2003), provide information on the deformation within such slide deposits. Because of their sediment strength, large complex landslides of pelagic and hemipelagic sediments with a high content of biogenic silica are particularly susceptible to internal deformation (Davies and Clark, 2006).

9.3.6 BLOCKY DISAGGREGATED LANDSLIDES

Large slides consisting of blocks of indurated material, apparently lacking an organized structure, are widespread on many continental margins (Dahlgren *et al.*, 2002; Joanne *et al.*, 2010). 3D seismic studies show that many have prominent erosional grooves at the base of landslide deposits (Fig. 9.6), on a scale of 10–100 m below the autochthonous seabed (Posamentier and Kolla, 2003; Gee *et al.*, 2005); the grooves indicate non-turbulent parallel flow spreading slightly downslope. Such landslides involve well-consolidated sediment, so that blocks maintain their coherence during transport. The basal linear scours suggest a debris avalanche process for some deposits, whereas others may be predominantly slides in which blocks have broken up more than in the large complex slides described above.

9.4 THE CAUSES OF FAILURES

In general, failures require a load on a slope, and shear failure takes place on the weakest plane. Submarine failures may result

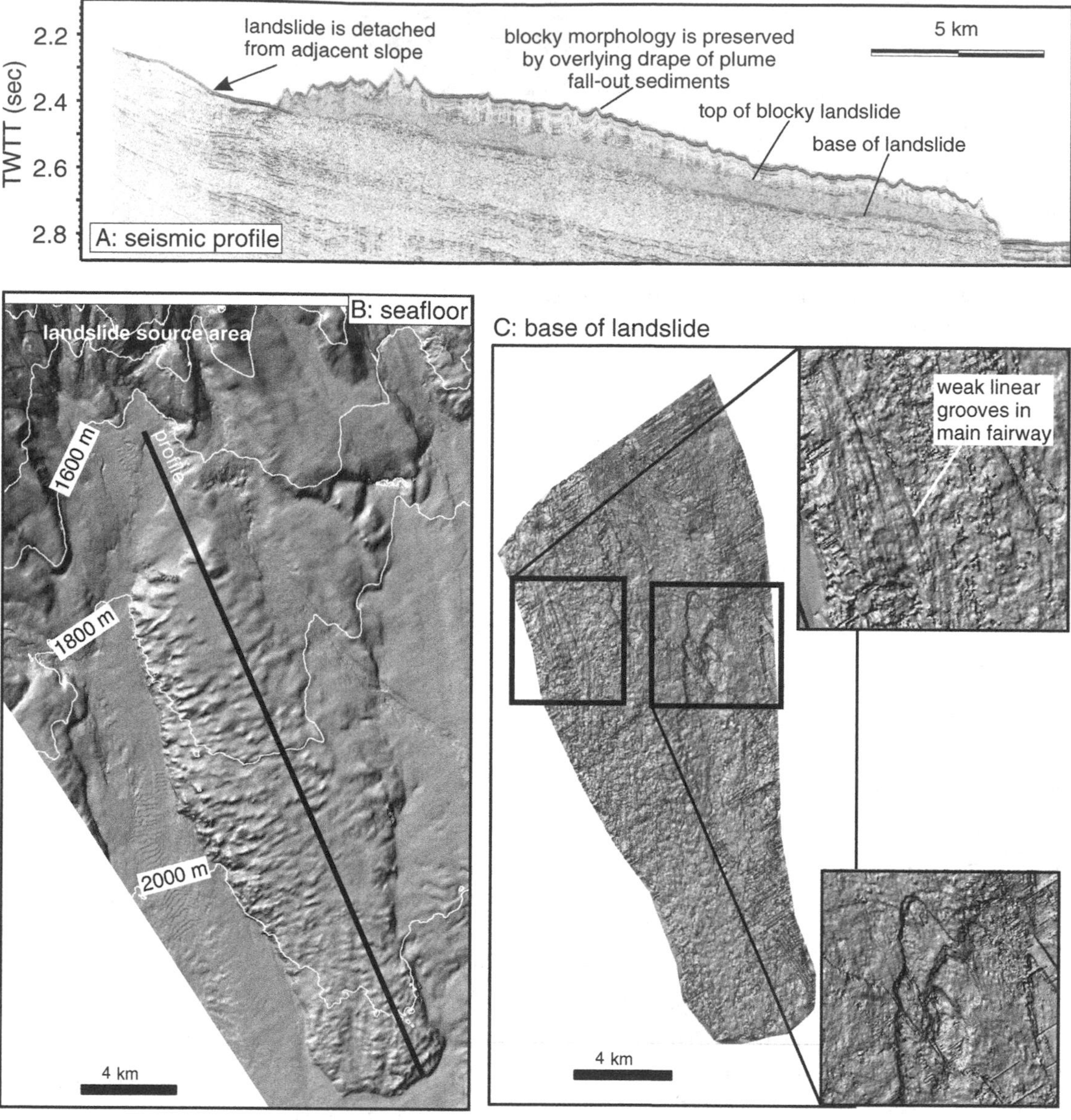

Fig. 9.5. Barrington blocky disaggregated landslide on the western Scotian Slope (modified from Mosher and Campbell, 2011). (A) Longitudinal seismic profile. (B) Bathymetry derived from 3D seismic data. (C) 3D seismic images of the base of the landslide. Reproduced with permssion from Encana.

from steepening or loading the seabed, or reducing the strength of critical weak layers. It is useful to consider two stages in the development of failures: preconditioning and triggering. Many factors that precondition sediments for failure may also act as a trigger, but clear evidence for triggers other than earthquakes is rare.

Earthquakes act to trigger failures in two ways. First, the ground motions resulting from the passage of seismic waves may lead to the yield strength in weak layers being exceeded. Second, the cyclic loading of fine-grained granular sediments will lead to structural breakdown, creating excess pore pressure and thus weakening the sediment. Both of these effects can take place over a large area. The occurrence of coincident failures at multiple locations on a continental slope, including different canyon systems, is evidence of such widespread effects (Adams, 1990; Goldfinger *et al.*, 2000, 2003). Most preconditioning processes are unlikely to cause synchronous failure over a large area, except perhaps in shallow water where tsunami waves, ice loading, or heavy rain and groundwater effects can have a regional impact.

Preconditioning due to steepening of the seabed can result from a range of tectonic processes, from depositional progradation (as in deltas) to erosion by turbidity currents and enhanced oceanographic flows, particularly in submarine canyons. That steepening may also be a trigger is shown by the presence of blocks of canyon-wall sediment in canyon-fill turbidites.

Preconditioning through loading is favored by high sedimentation rates, which also result in lower sediment strength through incomplete draining during compaction, and thus underconsolidation. Loading by rising sea level or by grounded ice (Mulder and Moran, 1995) may also precondition sediment for failure.

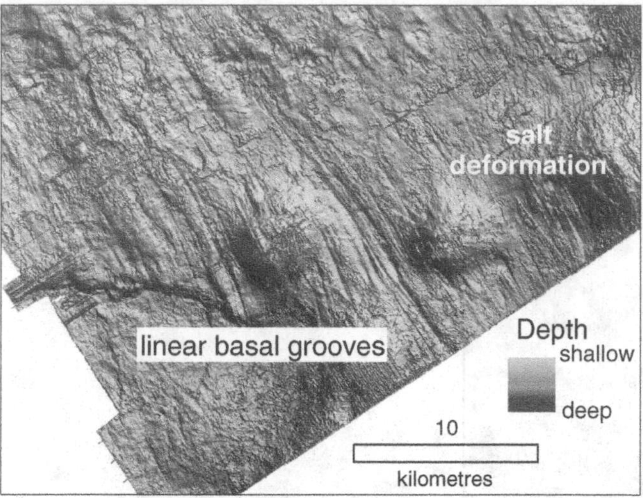

Fig. 9.6. 3D seismic image of basal linear grooves beneath a Miocene landslide on the western Scotian Slope (modified from Campbell and Mosher, 2010). Reproduced with permission from Encana.

Preconditioning by reducing the strength of weak layers may result from basinal liquid or gaseous fluids rising through the sediment column. Sediment failures are commonly associated with seafloor pockmarks, subsurface fluid chimneys, and subsurface gas hydrates. Strength properties of sediments in shallow water are also influenced by terrestrial groundwater flow into deltaic sediments (Stegmann *et al.*, 2011). The important role of gas hydrates has been inferred from relations between failure frequency and falling sea level (Maslin *et al.*, 2004) and warming bottom waters (Kennett and Fackler-Adams, 2000). These conditions favor the release of free gas from hydrates and thus precondition sediment by reducing sediment strength. However, clear evidence that these processes trigger anything other than local failures is lacking.

9.5 THE GLACIATED EASTERN CANADIAN CONTINENTAL MARGIN

9.5.1 DATA AVAILABILITY

Individual landslides on the eastern Canadian continental margin are known from 2D and 3D seismic data, multibeam bathymetry, and 10-m-long cores (Mosher, 2009). The data are insufficient to systematically map landslides, but sufficient to recognize most large near-seabed landslides south of Davis Strait (Fig. 9.7). The paucity of 3D seismic data from landslides limits interpretation of transport mechanisms. Landslide ages are well defined from regional stratigraphic studies, in which the chronology is based on upper-slope till tongues and regional Heinrich layers that form prominent seismic reflections because of their high content of detrital carbonate (Mosher *et al.*, 2004; Piper 2005). Many individual landslides have been described in papers and theses. Papers by Mosher *et al.* (2004) and Piper (2005) provide a regional view of margin architecture. This chapter emphasizes controls on failure based on published studies.

9.5.2 PROGRADATIONAL AND EROSIONAL CONTINENTAL MARGINS

The overall morphology of the eastern Canadian continental margin (Fig. 9.7) is controlled by asymmetric rifting and sequential seafloor spreading history of the North Atlantic Ocean, Labrador Sea, and Baffin Bay. The oldest continental margin, with the deepest adjacent ocean basin, but also some of the lowest continental slope gradients, is the Scotian margin, where seafloor spreading began in the Early Jurassic. Transform margins, such as the Southwest Grand Banks Slope, are steeper. Iberia separated from the Grand Banks in the early Cretaceous, the Labrador Sea opened in the late Cretaceous, and Baffin Bay in the Paleocene, with a corresponding northward decrease in ocean water depth. Complexities in rifting, transform faulting, and volcanism are responsible for second-order features such as Orphan Knoll (a horst of continental crust), Davis Strait (transform volcanism), and seamount chains.

Quaternary glaciation, more intense in the mid to late Pleistocene than earlier, had a profound effect on margin morphology. At maximum ice extent, most or all of the continental shelf was covered by ice, with a grounding line at a present water depth of 500–600 m on the continental slope (Piper *et al.*, 2012). Ice streams excavated deep troughs on the continental shelf, separated by shallow banks, on some of which ice domes developed. In places, such as off Hudson Strait and the major transverse troughs of the Labrador Shelf, the continental shelf prograded more than 10 km during the Quaternary and corresponding slope deposits are hundreds of meters thick. Progradation was greatest off northern-latitude ice streams (type A margin in Fig. 9.7), where ocean depths are less and margin gradients lower, allowing the accumulation of glacigenic debris flows (Tripsanas and Piper, 2008; Li *et al.*, 2011). Progradation also took place on smooth slope segments seaward of shallow shelf banks (type B margin in Fig. 9.7), where abundant sediment was supplied by fallout from meltwater plumes that advected southward and westward in the Labrador Current. Directly down-flow from major ice outlets, sedimentation rates exceeded 5 m per thousand years at glacial maxima.

Erosion of the continental slope took place during the Quaternary in several settings. Glacigenic debris flows transformed into turbidity currents that accelerated until they reached the continental rise seaward of ice streams at lower latitudes (type C margin in Fig. 9.7), where ocean depth is greater and the continental margin steeper. Erosion was facilitated by subglacial meltwater discharge, with hyperpycnal flows eroding deep and wide submarine canyons; for example, seaward of Laurentian Channel. Some areas seaward of shelf banks have been deeply dissected by submarine canyons (type D margin in Fig. 9.7), related to erosion when sea level was low and discharge of subglacial meltwater through tunnel valleys. Other areas seaward of banks have shallow canyons to *ca.* 1500 m water depth, cutting a steep slope with a base-of-slope ramp having abundant mass-transport deposits (type E margin in Fig. 9.7). A final type of margin is composite (type F margin). This resembles the erosional types in its distribution of canyons

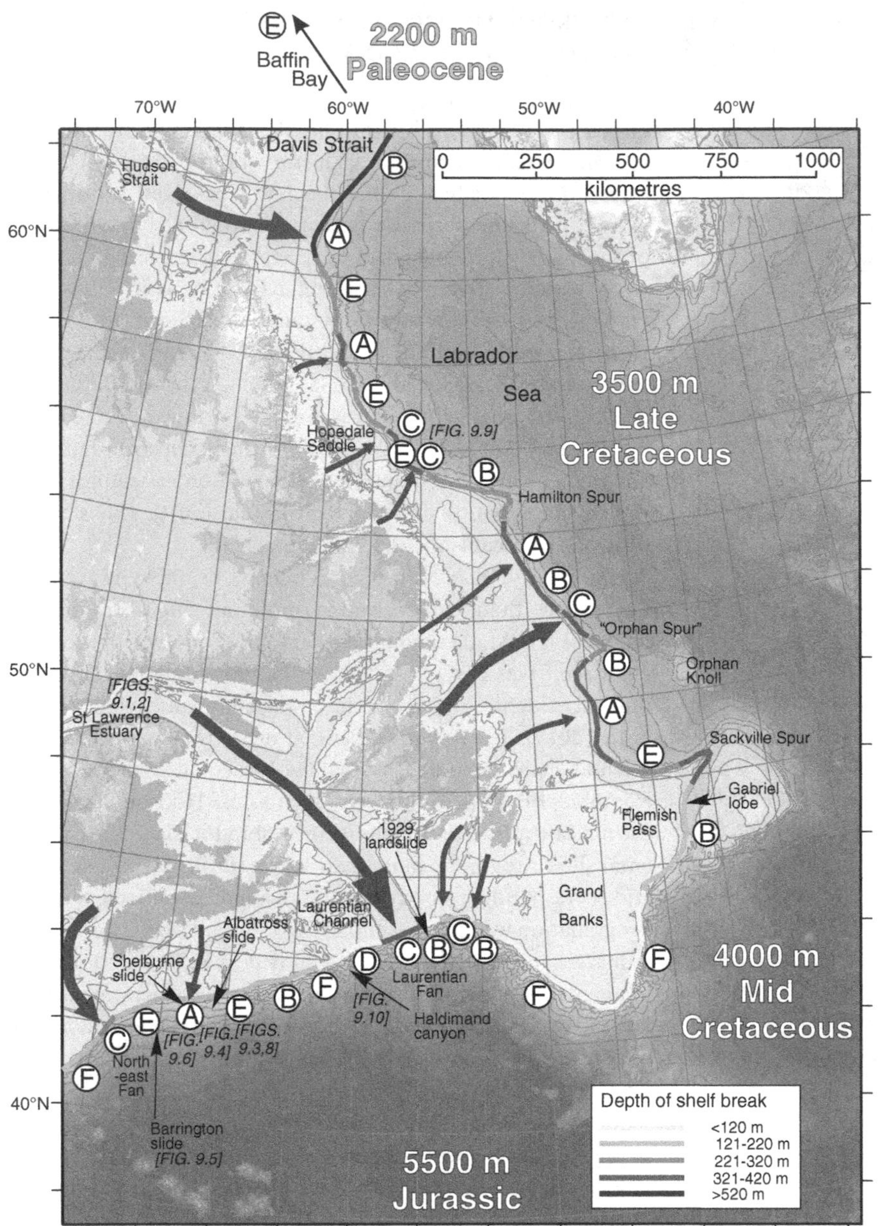

Fig. 9.7. Bathymetric map of the eastern Canadian margin, showing margin types (A) to (F). Line at shelf break indicates present depth of shelf break. Large arrows schematically show major Pleistocene ice streams on the shelf. Large text on ocean basins shows time of onset of seafloor spreading and depth of ocean basin floor.

and depositional facies, but has prograded overall through the Quaternary by deposition of shelf-edge till.

9.5.3 PROGRADATIONAL SMOOTH SLOPES AND SEDIMENT DRIFTS

Progradational smooth slopes (type B) have gradients of less than 4°. They are characterized by shallow failures initiated on local steep gradients that retrogress up to the limit of till on the upper slope, producing thin debris-flow deposits and turbidites. The classic example of such a landslide is the 1929 Grand Banks failure, triggered by a M7.2 earthquake (Piper *et al.* 1988, 1999; Mosher and Piper, 2007). In the case of this landslide, debris-flow deposits and thin blocky slides are preserved on much of the continental slope. Elsewhere, similar failures have almost completely evacuated the source area and transport path, leaving numerous step-like unconformities in the overall progradational slope (Fig. 9.3). These unconformities are draped by proglacial plume fallout deposits that may fail in a similar manner (Fig. 9.8). The basal shear planes of such failures, where penetrated by cores, consist of thin beds of turbidite sand (Campbell, 2000). The failures occur in multiple submarine canyons and gullies and are thus interpreted to be seismically triggered. Failure may be preconditioned by the high sedimentation rates, but sediments on slopes of <6° are essentially stable in the absence of earthquake shaking (Mosher *et al.*, 1994). They are probably also preconditioned by the release of basinal fluids; chimneys and pockmarks are widespread in areas of failure in 1929 and on the central Scotian Slope. Gas hydrates are also present in these areas (Mosher, 2011) but the times of failures do not correlate with periods of significant sea-level change. Rather,

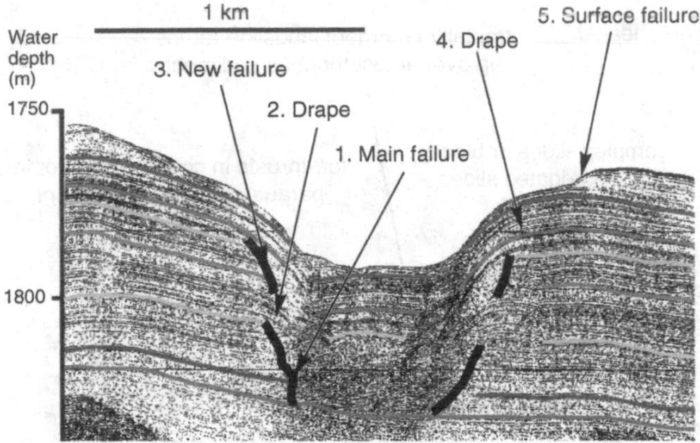

Fig. 9.8. Sequence of repeated retrogressive failure and deposition of plume fallout sediments at the head of a slump scarp on the central Scotian Slope.

the magnitude–frequency relationship of failures is similar to that for earthquakes, with large failures that extend more than 100 km along the slope having a recurrence interval of about 10,000 years.

Large sediment drifts have built regional mounds on the continental slope and rise, termed "spurs," notably in the southern Labrador Sea (Fig. 9.7). Sediment is transported to these spurs by the Labrador Current, which acts to at least 1500 m water depth, and by the Western Boundary Undercurrent at *ca.* 3000 m water depth. Sedimentation rates and slope gradients are high, the latter up to 4°, on the lee side of spurs, whereas spur crest sediments are commonly winnowed and accumulated slowly, resulting in considerable consolidation. The spurs tend to host gas hydrates due to increased porosity (Mosher, 2011). Blocky disaggregated slides, apparently retrogressive, are well preserved on the south side of Sackville Spur (Piper and Campbell, 2005).

9.5.4 EROSIONAL MARGIN SEGMENTS

Erosional margin segments (types C–E) and the composite margin type (F) have similar types of submarine landslides, but in different proportions. All erosional margin segments have rare, but large, blocky disaggregated slides or complex slides. These may have sources on the outer shelf, as in the case of the large complex frontally confined landslide off Hopedale Saddle in southern Labrador (Fig. 9.9; Deptuck *et al.*, 2007). More commonly, however, they are derived from shallow canyons on the steep upper slope, as is the case for the *ca.* 14,000-year-old Albatross Slide on the central Scotian Slope (Shor and Piper, 1989) and the larger *ca.* 60,000-year-old Barrington Slide on the western Scotian Slope (Fig. 9.5; Mosher and Campbell, 2011). Other large landslides have sources on high canyon walls, such as the *ca.* 150,000-year-old mass-transport deposit on the eastern Scotian Rise (Fig. 9.10; Piper and Ingram, 2003) and the large slides on Northeast Fan, where the upper slope has been deeply incised by subglacial meltwater flows (Piper *et al.*, 2012). The large Shelburne complex slide of Pliocene age (Mosher *et al.*, 2010b), with a volume exceeding 800 km^3, also appears to have an origin on the upper slope. In every case, at least 50 m of sediment appears to have failed and, where sampled, the blocks are well indurated (Mulder *et al.*, 1997). These large failures are infrequent, with approximately one failure per 200 km length of erosional margin in 10,000 years and, as a result, the triggering mechanism is uncertain and the nature of any weak layer is unknown. In some cases, the weak layer may be a zone of interstratal slip or decollement, observed locally in inter-canyon areas at sub-bottom depths of 50–200 m (Piper, 2005). Where deep canyons are present, blocky disaggregated landslides are partly trapped within canyon floors, but larger slides have spread out on the rise and have frontally emergent terminations. In some cases, such blocky mass-transport deposits are present in multiple canyons, suggesting a seismic trigger (Fig. 9.10). Where the outer shelf has failed, a trigger by ice loading or fresh pore water seems probable (Mulder and Moran, 1995), but has yet to be demonstrated.

Smaller failures on the erosional margins include complex landslides and blocky disaggregated landslides derived from canyon walls (Jenner *et al.*, 2007), most of which accumulate on canyon floors. In cases where canyons are short, blocky disaggregated landslide deposits extend to a frontally emergent, steeply terminating lobe beyond the end of the canyon, as is the case of the Gabriel lobe in Flemish Pass (Piper and Campbell, 2005). Simple rotational slumps are common on canyon walls (Fig. 9.3B, C), presumably preconditioned and possibly triggered by erosional oversteepening. On composite margins (type F), retrogressive slumps occur in surficial material on inter-canyon ridges (Fig. 9.3C), similar to those in margin type B.

9.6 DISCUSSION

A first-order control on the type of landslides on the eastern Canadian margin is regional gradient (Fig. 9.11), which in turn is a result of large-scale tectonics, including the opening history and age of the passive margin. The regional slope gradient has been modified by Quaternary processes, with a reduction in gradient at high-latitude glacier outlets and a steepening of gradient at lower-latitude glacier outlets. Gradient influences whether landslides break up and accelerate as debris avalanches or turbidity currents, or whether they come to rest near their source.

The other first-order control on the type of landslide is the strength of the failed sediment (Fig. 9.11). Very weak sediment, such as that deposited by glacigenic debris flows, readily transforms to turbidity currents on slopes >2.5° (Piper and Normark, 2009). The surficial sediments involved in large retrogressive failures break up easily, a process probably assisted by the presence of thin sand and silt turbidites that readily shear during earthquake loading and downslope movement. Debris flows depend on the availability of cohesive sediment and will transform to turbidity currents on steep valley walls (Piper *et al.*, 1999). Disaggregated blocky landslides are favored where sediments consolidated by depths of 50 m or more are available on the steep upper slope or on canyon walls; complex slides may

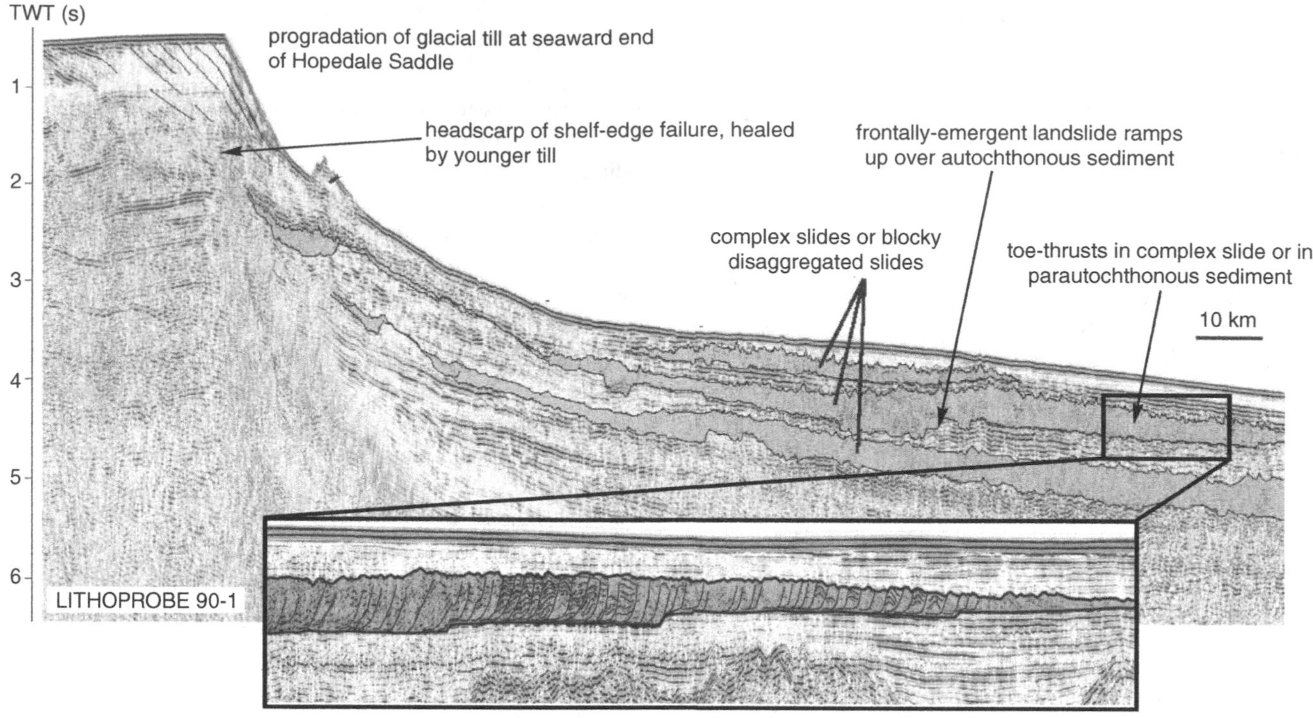

Fig. 9.9. Shelf-edge collapse at Hopedale Saddle that resulted in a major complex landslide with toe thrusts (modified from Deptuck *et al.* 2007).

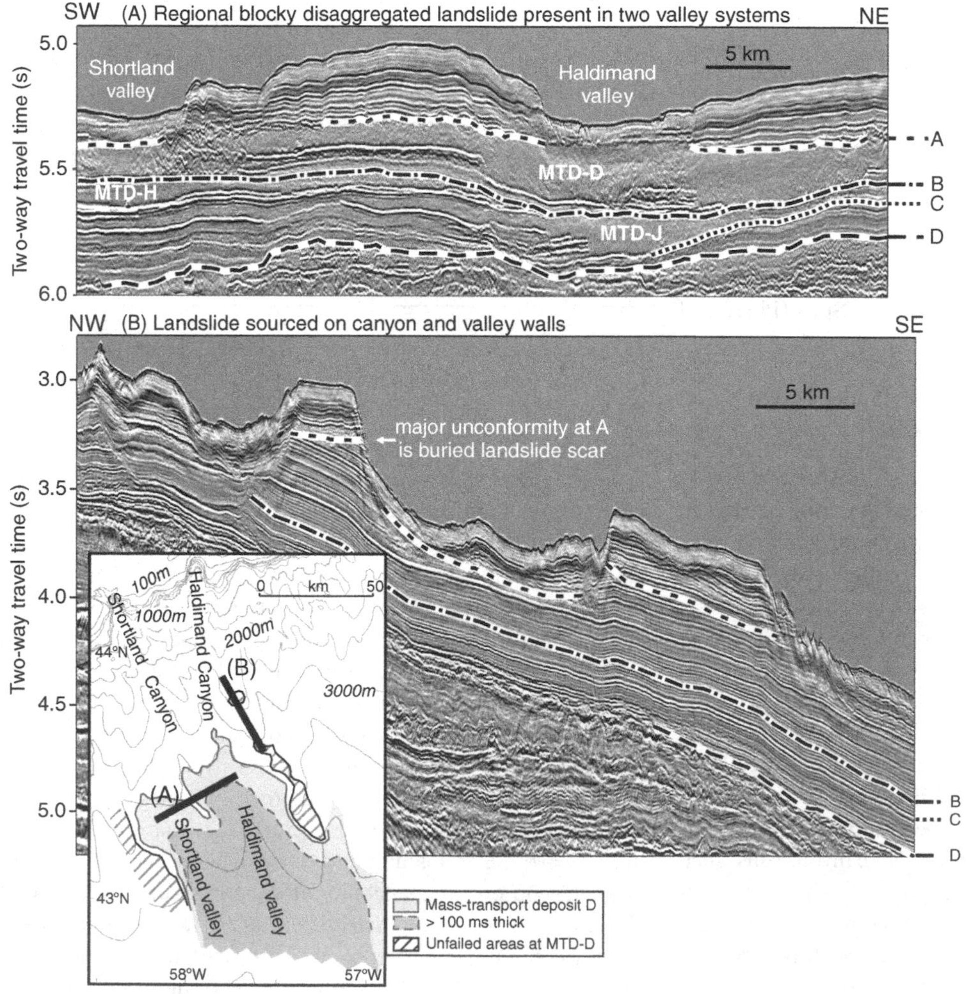

Fig. 9.10. (A) Strike seismic reflection profile showing a blocky disaggregated landslide deposit on the eastern Scotian Rise. (B) Dip seismic reflection profile showing an unconformity on the adjacent valley wall that marks the healed scar of the landslide. Downslope termination of slide is unknown. Modified from Piper and Ingram (2003). Reproduced with permission from TGS-NOPEC.

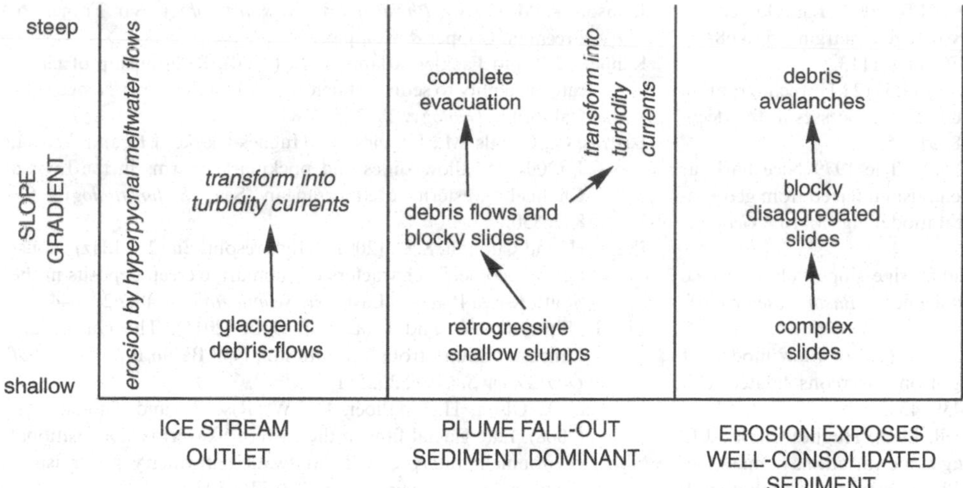

Fig. 9.11. Summary diagram showing the effect of gradient of the continental slope on different types of submarine landslides on the eastern Canadian continental margin.

occur on lower gradients, as is the case of the Hopedale Saddle Slide (Fig. 9.9).

The distribution of landslides in time and space suggests that most were triggered by rare earthquakes. The distribution does not support an important role for gas hydrates. Nor is there evidence that regional variation in basinal fluid flow, which might precondition failures, has an important role in controlling where large landslides have occurred. Loading of the outer shelf by glacier ice may have triggered some large failures. Others may have resulted from progressive deformation of decollement surfaces, where sediment is unsupported on eroded canyon walls.

9.7 CONCLUSIONS

Four principal types of submarine landslides are distinguished on the glaciated continental margin of eastern Canada: simple slumps, retrogressive slumps, large complex landslides, and blocky disaggregated landslides. Landslides transitional between these types also occur. The transport mechanism for each landslide type is reasonably well understood except for blocky disaggregated landslides, where the roles of simple sliding and debris avalanching are difficult to distinguish using the available data. Well-developed systems of basal erosional grooves and cored blocks with jigsaw texture, however, favor a debris avalanche process. Break-up of landslides and their transformation into debris flows and turbidity currents are important mechanisms in marine environments, particularly in failures involving weak surficial sediments.

The distribution of different types of landslides on the eastern Canadian continental margin depends on whether the margin is progradational (weak sediment is available for retrogressive slumps) or erosional (more consolidated sediment is available to form blocky disaggregated slides). This gross margin style is strongly influenced by the tectonic history of opening of the adjacent ocean and by Quaternary glaciation. Most large slides appear to have resulted from regional failure during large, but rare, earthquakes. Preconditioning factors such as sedimentation rate or flux of basinal fluids do not seem to have had a major impact on the distribution of larger failures, but could be locally significant for small landslides.

REFERENCES

Adams, J. (1990). Paleoseismicity of the Cascadia subduction zone: Evidence from turbidites off the Oregon-Washington margin. *Tectonics*, 9, 569–583.

Blikra, L.H. and Nemec, W. (1998). Postglacial colluvium in western Norway: Depositional processes, facies and paleoclimatic record. *Sedimentology*, 45, 909–959.

Bohannon, R.G. and Gardner, J.V. (2004). Submarine landslides of San Pedro Escarpment, southwest of Long Beach, California. *Marine Geology*, 203, 261–268.

Bull, S., Cartwright, J. and Huuse, M. (2009). A review of kinematic indicators from mass-transport deposits using 3D seismic data. *Marine and Petroleum Geology*, 26, 1132–1151.

Campbell, D.C. (2000). *Relationship of Sediment Properties to Failure Horizons for a Small Area of the Scotian Slope*. Geological Survey of Canada, Current Research 2000-D08.

Campbell, D.C. and Mosher, D.C. (2010). Middle to Late Miocene slope failure and the generation of a regional unconformity beneath the western Scotian Slope, eastern Canada. In *Submarine Mass Movements and Their Consequences: IV*, ed. D.C. Mosher, C. Shipp, L. Moscardelli, J. Chaytor, C. Baxter, H. Lee and R. Urgeles. Dordrecht, Netherlands: Springer, pp. 645–755.

Campbell, D.C., Duchesne, M. and Bolduc, A. (2008). Geomorphological and geophysical evidence of Holocene seafloor instability on the southern slope of the Lower St. Lawrence Estuary, Québec. In *Fourth Canadian Conference on Geohazards: From Causes to Management*, ed. J. Locat, D. Perret, D. Turmel, D. Demers and S. Leroueil. Québec: Presse de la Université Laval, pp. 367–374.

Coleman, J.M. and Prior, D.B. (1988). Mass wasting on continental margins. *Annual Reviews of Earth and Planetary Science*, 16, 101–119.

Collot, J.-Y., Lewis, K., Lamarche, G. and Lallemand, S. (2001). The giant Ruatoria avalanche on the northern Hikurangi margin, New Zealand: Result of oblique seamount subduction. *Journal of Geophysical Research*, 106, 19271–19297.

Coussot, P. and Meunier, M. (1996). Recognition, classification and mechanical description of debris flows. *Earth Science Reviews*, 40, 209–227.

Dahlgren, K. I. T., Vorren, T. O. and Laberg, J. S. (2002). Late Quaternary glacial development of the mid-Norwegian margin: 65 to 68° N. *Marine and Petroleum Geology*, 19, 1089–1113.

Damuth, J. E. (1980). Use of high-frequency (3.5–12 kHz) echograms in the study of near-bottom sedimentation processes in the deep-sea: A review. *Marine Geology*, 38, 51–75.

Dan, G., Sultan, N. and Savoye, B. (2007). The 1979 Nice harbour catastrophe revisited: Trigger mechanism inferred from geotechnical measurements and numerical modelling. *Marine Geology*, 245, 40–64.

Davies, R. J. and Clark, I. R. (2006). Submarine slope failure primed and triggered by silica and its diagenesis. *Basin Research*, 18, 339–350.

De Blasio, F. V., Elverhoi, A., Issler, D. *et al.* (2004). Flow models of natural debris flows originating from overconsolidated clay materials. *Marine Geology*, 213, 439–455.

Deptuck, M. E., Mosher, D. C., Campbell, D. C., Hughes-Clarke, J. E. and Noseworthy, D. (2007). Along slope variations in mass failures and relationships to major Plio-Pleistocene morphological elements, SW Labrador Sea. In *Submarine Mass Movements and Their Consequences: III*, ed. V. Lykousis, D. Sakellariou and J. Locat. Dordrecht, Netherlands: Springer, pp. 37–45.

Frey-Martinez, J., Cartwright, J. and James, D. (2006). Frontally confined versus frontally emergent submarine landslides: A 3D seismic characterization. *Marine and Petroleum Geology*, 23, 585–604.

Gee, M. J. R., Gawthorpe, R. L. and Friedmann, J. S. (2005). Giant striations at the base of a submarine landslide. *Marine Geology*, 214, 287–294.

Goldfinger, C., Kulm, L. D., McNeill, L. C. and Watts, P. (2000). Super-scale failure of the southern Oregon Cascadia margin. *Pure and Applied Geophysics*, 157, 1189–1226.

Goldfinger, C., Nelson, C. H. and Johnson, J. (2003). Deep-water turbidities as Holocene earthquake proxies: The Cascadia subduction zone and northern San Andreas fault systems. *Annali Geofisica*, 46, 1169–1194.

Haflidason, H., Sejrup, H. P., Nygård, A. *et al.* (2004). The Storegga Slide: Architecture, geometry and slide development. *Marine Geology*, 213, 201–234.

Hampton, M. A. (1972). The role of subaqueous debris flow in generating turbidity currents. *Journal of Sedimentary Petrology*, 42, 775–793.

Hsu, S.-K., Kuo, J., Lo, C.-L. *et al.* (2008). Turbidity currents, submarine landslides and the 2006 Pingtung earthquake off SW Taiwan. *Terra Atmosphere Oceanological Science*, 19, 767–772.

Huang, X. and García, M. H. (1999). Modeling of non-hydroplaning mudflows on continental slopes. *Marine Geology*, 154, 131–142.

Ilstad, T., Elverhøi, A., Issler, D. and Marr, J. G. (2004). Subaqueous debris flow behaviour and its dependence on the sand/clay ratio: A laboratory study using particle tracking. *Marine Geology*, 213, 415–438.

Iverson, R. M., Reid, M. E. and LaHusen, R. G. (1997). Debris-flow mobilization from landslides. *Annual Review of Earth and Planetary Science*, 25, 85–138.

Jackson, P. D., Gunn, D. A. and Long, D. (2004). Predicting variability in the stability of slope sediments due to earthquake ground motion in the AFEN area of the western UK continental shelf. *Marine Geology*, 213, 363–378.

Jenner, K. A., Piper, D. J. W., Campbell, D. C. and Mosher, D. C. (2007). Lithofacies and origin of late Quaternary mass transport deposits in submarine canyons, central Scotian Slope, Canada. *Sedimentology*, 54, 19–38.

Joanne, C., Collot, J.-Y., Lamarche, G. and Migeon, S. (2010). Continental slope reconstruction after a giant mass failure, the example of the Matakaoa Margin, New Zealand. *Marine Geology*, 268, 67–84.

Johnson, A. M. (1970). *Physical Processes in Geology*. San Francisco: Freeman, Cooper & Company.

Kennett, J. P. and Fackler-Adams, B. N. (2000). Relationship of clathrate instability to sediment deformation in the upper Neogene of California. *Geology*, 28, 215–218.

Lastras, G., Canals, M., Urgeles, R., Hughes-Clarke, J. E. and Acosta, J. (2004). Shallow slides and pockmark swarms in the Eivissa Channel, western Mediterranean Sea. *Sedimentology*, 51, 837–850.

Lee, S. H. and Chough, S. K. (2001). High-resolution (2–7 kHz) acoustic and geometric characters of submarine creep deposits in the South Korea Plateau, East Sea. *Sedimentology*, 48, 629–644.

Li, G., Piper, D. J. W. and Campbell, D. C. (2011). The Quaternary Lancaster Sound trough-mouth fan, NW Baffin Bay. *Journal of Quaternary Science*, 26, 511–522.

Longva, O., Olsen, H. A., Piper, D. J. W., Rise, L. and Thorsnes, T. (2008). Late glacial fans in the eastern Skagerrak: Depositional environment interpreted from swath bathymetry and seismostratigraphy. *Marine Geology*, 251, 110–123.

López, C., Spence, G., Hyndman, R. and Kelley, D. (2010). Frontal ridge slope failure at the northern Cascadia margin: Margin-normal fault and gas hydrate control. *Geology*, 38, 967–970.

Lucente, C. C. and Pini, G. A. (2003). Anatomy and emplacement mechanism of a large submarine slide within a Miocene foredeep in the northern Apennines, Italy: A field perspective. *American Journal of Science*, 303, 565–602.

Maslin, M., Owen, M., Day, S. and Long, D. (2004). Linking continental-slope failures and climate change: Testing the clathrate gun hypothesis. *Geology*, 32, 53–56.

Masson, D. G., Watts, A. B., Gee, M. J. R. *et al.* (2002). Slope failures on the flanks of the western Canary Islands. *Earth Science Reviews*, 57, 1–35.

Moernaut, J. and De Batist, M. (2011). Frontal emplacement and mobility of sublacustrine landslides: Results from morphometric and seismostratigraphic analysis. *Marine Geology*, 285, 29–45.

Mohrig, D., Whipple, K. X., Hondzo, M., Ellis, C. and Parker, G. (1998). Hydroplaning of subaqueous debris flows. *Geological Society of America Bulletin*, 110, 387–394.

Moscardelli, L., Wood, L. and Mann, P. (2006). Mass-transport complexes and associated processes in the offshore area of Trinidad and Venezuela. *American Association of Petroleum Geologists Bulletin*, 90, 1059–1088.

Mosher, D. C. (2009). Submarine landslides and consequent tsunamis in Canada. *Geoscience Canada*, 36, 179–190.

(2011). A margin-wide BSR gas hydrate assessment: Canada's Atlantic margin. *Marine and Petroleum Geology*, 28, 1540–1553.

Mosher, D. C. and Campbell, D. C. (2011). The Barrington submarine landslide, western Scotian Slope. In *Mass-Transport Deposits in Deepwater Settings*, ed. C. Shipp, P. Weimer and H. Posamentier. Society of Economic Paleontologists and Mineralogists, Special Publication 96, pp. 151–160.

Mosher, D. C. and Piper, D. J. W. (2007). Analysis of multibeam seafloor imagery of the Laurentian Fan and the 1929 Grand Banks landslide area. In *Submarine Mass Movements and Their Consequences: III*, ed. V. Lykousis, D. Sakellariou and J. Locat. Dordrecht, Netherlands: Springer, pp. 77–88.

Mosher, D. C., Moran, K. and Hiscott, R. N. (1994). Late Quaternary sediment, sediment mass-flow processes and slope stability on the Scotian Slope. *Sedimentology*, 41, 1039–1061.

Mosher, D. C., Piper, D. J. W., Campbell, D. C. and Jenner, K. (2004). Near-surface geology and sediment-failure geohazards of the central Scotian Slope. *American Association of Petroleum Geologists Bulletin*, 88, 703–723.

Mosher, D. C., Moscardelli, L., Shipp, R. *et al.* (2010a). Submarine mass movements and their consequences: Introduction. In *Submarine Mass Movements and Their Consequences: IV*, ed. D. C. Mosher,

C. Shipp, L. Moscardelli, J. Chaytor, C. Baxter, H. Lee and R. Urgeles. Dordrecht, Netherlands: Springer, pp. 1–8.

Mosher, D.C., Xu, Z. and Shimeld, J. (2010b). The Pliocene Shelburne mass-movement and consequent tsunami, western Scotian Slope. In *Submarine Mass Movements and Their Consequences: IV*, ed. D.C. Mosher, C. Shipp, L. Moscardelli, J. Chaytor, C. Baxter, H. Lee and R. Urgeles. Dordrecht, Netherlands: Springer, pp. 765–775.

Mulder, T. and Alexander, J. (2001). The physical character of subaqueous sedimentary density flows and their deposits. *Sedimentology*, 48, 269–299.

Mulder, T. and Cochonat, P. (1996). Classification of offshore mass movements. *Journal of Sedimentary Research*, 66, 43–57.

Mulder, T. and Moran, K. (1995). Relationship among submarine instabilities, sea-level variations and the presence of an ice sheet on the continental shelf: An example from the Verrill Canyon area, Scotian Shelf. *Paleoceanography*, 10, 137–154.

Mulder, T., Berry, J.A. and Piper, D.J.W. (1997). Links between morphology and geotechnical characteristics of large debris flow deposits in the Albatross area on the Scotian Slope (SE Canada). *Marine Geotechnique*, 15, 253–281.

Nardin, T.R., Hein, F.J., Gorsline, D.S. and Edwards, B.D. (1979). A review of mass movement processes, sediment and acoustic characteristics, and contrasts in slope and base-of-slope systems versus canyon-fan-basin floor systems. In *Sedimentation on Continental Slopes*, ed. L.J. Doyle and O.H. Pilkey. Society of Economic Paleontologists and Mineralogists, Special Publication 27, pp. 61–73.

Normark, W.R., McGann, M. and Sliter, R. (2004). Age of Palos Verdes submarine debris avalanche, southern California. *Marine Geology*, 203, 247–259.

Piper, D.J.W. (2005). Late Cenozoic evolution of the continental margin of eastern Canada. *Norwegian Journal of Geology*, 85, 305–318.

Piper, D.J.W. and Campbell, D.C. (2005). Quaternary geology of Flemish Pass and its application to geohazard evaluation for hydrocarbon development. In *Petroleum Resources and Reservoirs of the Grand Banks, Eastern Canadian Margin*, ed. R.N. Hiscott and A.J. Pulham. Geological Association of Canada, Special Paper 43, pp. 29–43.

Piper, D.J.W. and Ingram, S. (2003). *Major Quaternary Sediment Failures on the East Scotian Rise*. Geological Survey of Canada, Current Research 2003-D01.

Piper, D.J.W. and Normark, W.R. (2009). The processes that initiate turbidity currents and their influence on turbidites: A marine geology perspective. *Journal of Sedimentary Research*, 79, 347–362.

Piper, D.J.W., Farre, J.A. and Shor, A.N. (1985). Late Quaternary slumps and debris flows on the Scotian Slope. *Geological Society of America Bulletin*, 96, 1508–1517.

Piper, D.J.W., Shor, A.N. and Hughes Clarke, J.E. (1988). The 1929 Grand Banks earthquake, slump and turbidity current. In *Sedimentologic Consequences of Convulsive Geologic Events*, ed. H.E. Clifton. Geological Society of America, Special Paper 229, pp. 77–92.

Piper, D.J.W., Pirmez, C., Manley, P.L. *et al.* (1997). Mass transport deposits of Amazon Fan. *Proceedings of the Ocean Drilling Program, Scientific Results*, 155, 109–146.

Piper, D.J.W., Cochonat, P. and Morrison, M.L. (1999). Sidescan sonar evidence for progressive evolution of submarine failure into a turbidity current: The 1929 Grand Banks event. *Sedimentology*, 46, 79–97.

Piper, D.J.W., Deptuck, M.E., Mosher, D.C., Hughes-Clarke, J.E. and Migeon, S. (2012). Erosional and depositional features of glacial meltwater discharges on the eastern Canadian continental margin. In *Application of Seismic Geomorphology Principles to Continental Slope and Base-of-slope Systems: Case Studies from Seafloor and Near-Seafloor Analogues*, ed. B.E. Prather, M.E. Deptuck, D.C. Mohrig, B. van Hoorn and R.B. Wynn. Society of Economic Mineralogists and Paleontologists, Special Publication 99, 61–80.

Pollet, N. and Schneider, J.-L.M. (2004). Dynamic disintegration processes accompanying transport of Holocene Flims sturzstrom (Swiss Alps). *Earth and Planetary Science Letters*, 221, 433–448.

Poncet, R., Campbell, C., Dias, F., Locat, J. and Mosher, D. (2010). A study of the tsunami effects of two landslides in the St. Lawrence estuary. In *Submarine Mass Movements and Their Consequences: IV*, ed. D.C. Mosher, C. Shipp, L. Moscardelli, J. Chaytor, C. Baxter, H. Lee and R. Urgeles. Dordrecht, Netherlands: Springer, pp. 755–764.

Posamentier, H.W. and Kolla, V. (2003). Seismic geomorphology and stratigraphy of depositional elements in deep-water settings. *Journal of Sedimentary Research*, 73, 367–388.

Prior, D.B. and Doyle, E.H. (1985). Intra-slope canyon morphology and its modification by rockfall processes, U.S. Atlantic continental margin. *Marine Geology*, 67, 177–196.

Prior, D.B., Bornhold, B.D. and Johns, M. (1984). Depositional characteristics of a submarine debris flow. *Journal of Geology*, 92, 707–727.

Shor, A.N. and Piper, D.J.W. (1989). A large Late Pleistocene blocky debris flow on the central Scotian Slope. *Geomarine Letters*, 9, 153–160.

Stegmann, S., Sultan, N., Kopf, A., Apprioual, R. and Pelleau, P. (2011). Hydrogeology and its effect on slope stability along the coastal aquifer of Nice, France. *Marine Geology*, 280, 168–181.

Stow, D.A.V. and Mayall, M. (2000). Deep-water sedimentary systems: New models for the 21st century. *Marine and Petroleum Geology*, 17, 125–135.

Tappin, D.R., Watts, P., McMurtry, G.M., Lafoy, Y. and Matsumoto, T. (2001). The Sissano, Papua New Guinea tsunami of July 1998: Offshore evidence on the source mechanism. *Marine Geology*, 175, 1–23.

Tripsanas, E.K. and Piper, D.J.W. (2008). Late Quaternary glacigenic debris flows in Orphan Basin. *Journal of Sedimentary Research*, 78, 724–744.

Tripsanas, E., Piper, D.J.W. and Campbell, D.C. (2008a). Evolution and depositional structure of earthquake-induced mass movements and gravity flows, southwest Orphan Basin, Labrador Sea. *Marine and Petroleum Geology*, 25, 645–662.

Tripsanas, E., Piper, D.J.W., Jenner, K.A. and Bryant, W.R. (2008b). Sedimentary characteristics of submarine mass-transport deposits: New perspectives from a core-based facies classification. *Sedimentology*, 55, 97–136.

10 Tsunami hazard assessment related to slope failures in coastal waters

BRIAN D. BORNHOLD AND RICHARD E. THOMSON

ABSTRACT

Although subaerial and subaqueous landslides have been responsible for many tsunamis in high-relief coastal areas around the world, routine assessments of these hazards are rarely undertaken. Assessment must draw on the expertise of geoscientists, engineers, and hydrodynamicists, and requires analyses of both the landslide and the resulting waves. Landslide tsunami assessments aim to determine:

- occurrences of past events
- likelihood of future occurrences
- magnitudes of past events
- locations experiencing greatest impact
- conditions and triggers that led to failure
- wave characteristics and coastal run-up.

Key assessment considerations include the geologic evidence of past failures, both subaerial and subaqueous, and the written or oral history of past events. These can aid in determining whether further assessment studies are warranted. The general paucity of observations of past events, however, makes empirical assessment difficult. As a consequence, physical and numerical modeling are critical tools in characterizing the phenomena. Because modern numerical models are fast to run and relatively inexpensive, they are now widely used for specific case studies. Much can be learned before failures occur in areas prone to tsunamigenic landslides. Hydrodynamic modeling, combined with geologic and geotechnical evidence, can be used to assess the tsunamigenic potential of landslides. Although definitive estimates of the frequency of occurrence and magnitude of tsunamigenic events are difficult to make, analyses can place valuable constraints on the siting and design of coastal facilities.

10.1 INTRODUCTION

Subaerial and subaqueous landslides have been responsible for many tsunamis in high-relief coastal areas around the world. Unlike earthquake-generated tsunamis, landslide tsunamis are most dangerous when generated in shallow water. Norway has had the largest number of destructive landslide-generated tsunami events over recent centuries because it has a relatively large population living along its fjord coastline (Jørstad, 1968). However, other regions such as Alaska, British Columbia, Greenland, and Chile have also experienced landslide-generated waves that have caused major destruction and many deaths. The highest wave ever documented (524 m) was produced in Lituya Bay, Alaska on July 9, 1958 by a 5–10 × 10^6 m^3 subaerial rockslide (Fig. 10.1), triggered by a magnitude (M) 7.9 earthquake (Miller, 1960; Lander, 1996). Although landslide-generated tsunamis are more localized than those generated by earthquakes, they can produce remarkably high waves and significant run-up, particularly when trapped within narrow inlets or semi-enclosed bays. Such waves are often the result of landslides triggered by earthquakes, even moderate ones such as the M6.2 earthquake in Aisén Fjord, Chile in April 2007 (Naranja and Arenas, 2009; Sepulveda and Serey, 2009). Many landslide-generated tsunamis also occur in the absence of earthquakes, due to a variety of conditioning factors and triggers.

Subaerial landslides typically produce larger tsunamis than subaqueous landslides. The collapse of a huge cliff on the flank of Mount Campalla near Scilla in Calabria, Italy in 1783 generated a tsunami that killed 1500 people who had gathered on a beach following a devastating earthquake the day before; the maximum wave run-up was 8.3 m and inland inundation reached 200 m (Graziani *et al.*, 2006; Tappin, 2010). A tsunami generated by a 2–3 × 10^6 m^3 subaerial failure in Tafjord, Norway in

Landslides: Types, Mechanisms and Modeling, ed. John J. Clague and Douglas Stead. Published by Cambridge University Press.

Fig. 10.1. View east along Lituya Bay in southeast Alaska showing the trimline (left of image) produced by the tsunami triggered by the tsunami in July 1958. (Photo courtesy of Bill Eichenlaub.)

1934 reached 62 m above sea level and killed 40 people (Harbitz *et al.*, 1993; Blikra *et al.*, 2005). More than 130 additional people have been killed by landslide-generated tsunamis in Norway. A landslide at Paatuut in coastal Greenland in 2000 produced a 50-m-high run-up near the source and a 28-m-high run-up 20 km away (Dahl-Jensen *et al.*, 2004). No one was killed in this incident, but all buildings below 30 m elevation in the abandoned coal mining town of Qullisat, 20 km away, were destroyed. Sometime around AD 1570, about 3–4 × 10^6 m^3 of rock fell into Knight Inlet, British Columbia and generated a tsunami that completely destroyed the First Nations community of Kwalate, killing at least 100 people (Bornhold *et al.*, 2007). The site has never been reoccupied. On July 30 and August 6, 2011, rock avalanches in Lysefjoren, near Lysebotn, Norway (near the 1934 Tafjord event) resulted in tsunamis, one of which caused damage to nearby boats. The failures plummeted from a height of 250–300 m above sea level and resulted in 1.5 m tsunamis (I. Didenkulova, personal communication, 2011).

Tsunamis produced by subaqueous failures can also be devastating, even though they are typically smaller than those produced by subaerial landslides. Furthermore, the hazard associated with such events is less likely to be recognized. In 1975, a subaqueous failure of roughly 25 × 10^6 m^3 near Kitimat, British Columbia generated tsunami waves over 8 m high that damaged coastal facilities and several vessels (Murty, 1979; Prior *et al.*, 1984; Skvortsov and Bornhold, 2007). A similar subaqueous failure in 1974 from the opposite side of the fjord also produced a damaging tsunami. The M9.2 Alaska earthquake of 1964 set off many subaqueous slope failures that spawned devastating local tsunamis that inundated and damaged Seward and Valdez (Lee *et al.*, 2006; Suleimani *et al.*, 2009a). In 1994, a subaqueous failure occurred at Skagway, Alaska during replacement of a cruise-ship dock; it caused a tsunami estimated to have been up to 11 m high that killed one worker and caused extensive damage to a ferry terminal, other port facilities, small vessels, and the dock under construction (Cornforth and Lowell, 1996; Kulikov *et al.*, 1996; Rabinovich *et al.*, 1999; Thomson *et al.*, 2001). In 1946, a M7.3 earthquake on Vancouver Island created a coastal slope failure and tsunami; the wave swamped a small boat and its occupant drowned (Rogers, 1980; Mosher *et al.*, 2001).

We focus in this chapter on some aspects of landslide-generated tsunamis that occur in confined, high-relief, coastal marine areas. Similar events have also occurred in lakes and reservoirs in mountainous regions. We begin by briefly examining the factors responsible for destructive, local landslide-generated tsunamis, and then suggest a methodology for assessing the hazard when developing coastal infrastructure. Finally, we recommend a set of activities for these assessments and identify their uncertainties and limitations.

Previous studies have identified some of the relevant aspects of assessing tsunami risk from landslides; few, however, have explicitly addressed assessment approaches for both subaerial and subaqueous failures in coastal areas. Such hazard assessments require analysis of two distinct physical aspects of the problem – the landslide and the wave – and must draw on expertise in several professional communities, including geoscience, engineering, and hydrodynamics. Leroueil *et al.* (2003) present a procedure for risk assessment associated with submarine landslides, and Masson *et al.* (2006) discuss hazard prediction related to submarine landslides. Blikra *et al.* (2005) quantify, at a general level, the tsunami hazard associated with subaerial landslides and tsunamis in Norway. Similarly, van Zeyl (2009) assesses the hazards presented by subaerial failures and tsunamis in coastal British Columbia.

10.2 COMMON CONDITIONING FACTORS AND TRIGGERS

The causes of landslides are summarized in other chapters in this volume and need to be dealt with only briefly here. The conditioning factors and triggers that affect subaqueous and subaerial failures are similar and include steep slopes, sensitive soils, unfavorable stratification, artesian conditions, toe erosion, rapid loading, and earthquakes (Leroueil *et al.*, 2003). In addition, human activities, such as emplacement of fill in shallow

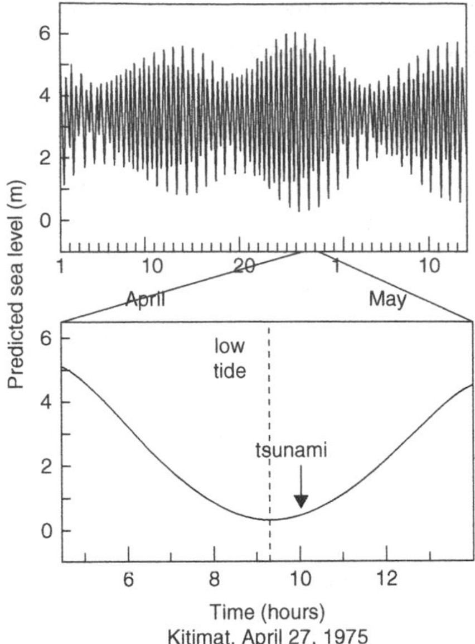

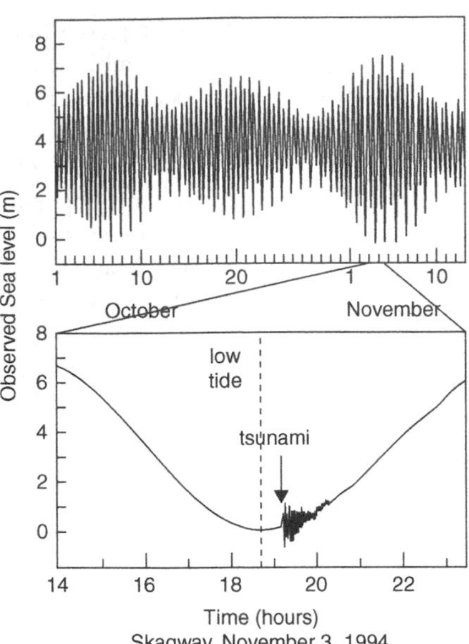

Fig. 10.2. Tide gauge records from Kitimat, British Columbia (April 1975), and Skagway, Alaska (November 1994) showing two failure-related tsunamis shortly after low tide. (Adapted from Kulikov *et al.*, 1996.)

subaqueous zones, have been implicated in many failure-related tsunamis.

Toe erosion of slopes, most commonly caused by nearby failures, has been documented as an important trigger of subsequent failures. Karlsrud and Edgers (1980), for example, documented a failure in Orkdalsfjord, Norway that triggered two additional failures a short time later. Similarly, Suleimani *et al.* (2009a) concluded that a tsunamigenic failure near Seward, Alaska during the 1964 Alaska earthquake caused a second landslide that produced a tsunami.

Several studies (e.g., Kulikov *et al.*, 1998; Thomson *et al.*, 2001) demonstrate an important relationship between times of very low tides and subaqueous slope failures. Examples include the 1956 failure in Howe Sound, British Columbia, which occurred a few minutes after low tide and destroyed part of a local mill; the 1975 landslide and tsunami near Kitimat, British Columbia, which occurred 50 minutes after a very low tide (Fig. 10.2; Prior *et al.*, 1984); and a similar event in Skagway Harbor in 1994, which occurred about 25 minutes after the lowest tide in many weeks (Fig. 10.2; Kulikov *et al.*, 1998). As with similar cases in Norway, construction activity in the coastal zone was implicated in these failures.

One important trigger of subaqueous slope failures that is often overlooked is the rapid sea-level drawdown that accompanies the receding waves during major tsunamis, whatever their source. The rapid fluctuations in the level of the sea during the tsunami may expose intertidal and shallow subtidal areas and place exceptional loads on adjacent steep slopes.

Along fjord coastlines, such as those in Alaska, Norway, British Columbia, and Chile, many of the conditions that lead to tsunamis may occur together. The high-relief coastal areas of western North and South America, for example, are seismically active, locally have groundwater under pressure, and experience high rainfall and high sediment fluxes. In some regions, isostatically uplifted, sensitive, marine clays can also contribute to failure.

10.3 PROGRESSIVE TSUNAMI HAZARD ASSESSMENT APPROACH

It is prudent, when planning coastal development in an area of high subaerial or subaqueous relief, to examine the potential hazard and risk from landslide-generated tsunamis. Van Zeyl (2009) presents some useful guidelines related to subaerial failures and tsunamis on the British Columbia coast, building on those reported by Blikra *et al.* (2005) for a coastal area in Norway. In this study, we present a staged approach for both subaerial and subaqueous failures and tsunamis. The approach commences with activities that are least costly, and then proceeds, if necessary, to more expensive activities such as geophysical and bathymetric field surveys, geotechnical assessments, and geotechnical and hydrodynamical numerical modeling. The aims are to determine:

- the frequency of occurrence of past tsunamis in the region
- the likelihood of future tsunamis in the region
- magnitudes of past events near the proposed development site
- locations that experienced the greatest past impact
- the conditions and triggers that led to past subaqueous or subaerial failure
- wave amplitudes, wavelengths, coastal run-up, and wave-induced currents likely to accompany future tsunamis at the site.

10.3.1 PAST ACCOUNTS OF EVENTS

A review of existing information about local tsunamis is the easiest and least expensive method of gaining an appreciation of the tsunami hazard. In areas that have been inhabited for decades to centuries, this information can include local and regional written accounts, government records related to damage and reconstruction efforts, and anecdotal accounts from eyewitnesses or local industries impacted by the failures or tsunamis. While eyewitness reports are commonly imprecise, they can provide valuable information about past events.

Much of the industrial activity in coastal British Columbia and Alaska is connected with the forestry industry. Several tsunamis that damaged log booms and other coastal facilities have been reported by personnel involved in these activities. On March 3, 1989, a 250-m-long causeway near the head of Bute Inlet, British Columbia failed, along with adjacent sediments, triggering a tsunami of unknown amplitude (B. Thomson, personal communication, 1992). In 1999, a large subaerial landslide in Knight Inlet, British Columbia caused a tsunami that disrupted log-booming activities in the fjord (R. Guthrie, personal communication, 2005). In Chile, landslides into inlets have seriously damaged fish-farming operations, such as the event near Puerto Aisén in April 2007 (Naranja and Arenas, 2009).

Determination of the locations, frequency of occurrence, and magnitudes of landslide-generated tsunamis in sparsely populated coastal areas is difficult, owing to the infrequency of large events, the remoteness of many of the sites, and the absence of written records prior to the arrival of Europeans. In western Canada, oral history can contribute evidence of past events or corroborate other lines of evidence, but it lacks the temporal fidelity or descriptive detail required to establish accurate occurrence frequencies or wave magnitudes. The example of the Kwalate failure in Knight Inlet, British Columbia, in which a subaerial rock avalanche created a large tsunami that destroyed a First Nations village, is indicative of the problems faced when using these sources to decipher details of the event such as the date of occurrence. Barrow (1935) concluded that the event occurred about three generations ago, or possibly earlier. Oral history provided to us by the current Hereditary Chief of the Da'naxda'xw/A'wa'etlala First Nation indicates that the event occurred in the middle of the nineteenth century. Both radiocarbon dating and archeological evidence, however, show that the event took place around AD 1570 and thus must have occurred prior to the arrival of Europeans (Bornhold *et al.*, 2007).

10.3.2 GEOLOGIC INVESTIGATIONS

Geologic investigations are recommended if: (1) historic reports indicate that locally generated tsunamis have occurred in the past; (2) the area is too remote to have had historic documentation; or (3) the history of human occupation has been too short to provide confidence that such events do not occur.

Van Zeyl (2009) emphasizes the importance of the relationships between steep slopes, their proximity to the shoreline, and the geologic factors that predispose slopes to failure when evaluating tsunami hazards in coastal environments. He compared two British Columbia fjords: Knight Inlet, which has experienced many local tsunamis, and Howe Sound, where there have been very few documented events even though it is more populated than Knight Inlet. He used indices such as the average subaerial slope within 5 km of the shoreline, the total area of slopes >70° within 5 km of the shoreline, and total area of slopes with relief greater than 800 m. Knight Inlet scored significantly higher than Howe Sound on all of these criteria.

Fig. 10.3. Photograph of the cliff at Adeane Point, Knight Inlet. A major rock avalanche at this site is presumed to have been the source of the tsunami that destroyed the First Nations village of Kwalate in the late 1500s (Bornhold *et al.*, 2007).

Aerial photographs and digital elevation models have been used to identify subaerial landslides that reached coastal waters and to suggest the possibility of past occurrences of tsunamis from this type of event (Blikra *et al.*, 2005). If available, several generations of aerial photos may help determine the frequency of slope failures in the region of interest. The 840-m-high mountain slope opposite the former Kwalate village site at Knight Inlet, for example, shows clear evidence of past failure into the fjord below (Fig. 10.3). Although not all subaerial landslides entering coastal waters create significant tsunamis, identification of such events points to the need for further detailed investigation; absence of such events could indicate that this hazard is not present in the area.

Detailed existing bathymetric surveys, especially multibeam echosounding, can also provide evidence of possible tsunamigenic landslides, both subaqueous and subaerial (Blikra *et al.*, 2005; Lee *et al.*, 2006; Bornhold *et al.*, 2007; Suleimani *et al.*, 2009a). The evidence includes failure scarps on steep submarine slopes, and debris accumulations on the basin floor. As with terrestrial landslides along the coast, such evidence does not prove that tsunamis were generated (the slide movement could have been slow or the debris could have accumulated over time), but it does point to the need for further investigation. If detailed bathymetric data do not exist, but other evidence indicates a significant hazard, multibeam echosounding field programs should be considered. Ideally, the multibeam survey should be coupled with high-resolution sub-bottom profiling to elucidate sediment thicknesses in potential failure zones and areas of debris accumulation.

Tsunami waves commonly erode sediments from the offshore environment and deposit them in the inundation zone (Atwater, 1987; Clague and Bobrowsky, 1994; Dawson, 1999; Dawson and Shi, 2000). The deposits typically consist of sheets of thin sand or silty sand, in some cases with gravel, deposited in coastal wetlands, adjacent lakes, and on nearby soils. Bornhold *et al.* (2007) reported a layer of clean sand about 1.5 cm thick lying on a shell midden and beneath a forest soil throughout the area of the Kwalate village.

10.3.3 GEOTECHNICAL STUDIES

If documentary evidence and geologic studies indicate that landslide-triggered tsunamis are of concern, a geotechnical program should be undertaken to define more precisely the frequency and magnitude of the hazard. For subaerial landslides, field investigations can range from a visual examination of existing slope failures to determine the failure mechanisms and causes, to drilling programs. Drilling may be done within areas of instability to determine the geometry and geotechnical character of the potential failed rock or soil mass, as well as in the surrounding area to determine the extent of any adverse conditions that could lead to future failures. Based on these studies, an engineering geologist can assess the likelihood of failure under different forcing factors, such as earthquakes or elevated pore pressures. Landslide velocities and fragmentation rates following failure, which are important inputs to the hydrodynamic modeling of tsunamis, can also be estimated. For rocky coastal areas, the structural geology of potentially unstable rock masses should also be assessed to determine the degree and style of fracturing (Bjerrum and Jørstad, 1968; Braathen *et al.*, 2004; van Zeyl, 2009). Discontinuity mapping, for example, has been undertaken for kinematic studies and to identify potential rock-slope failure mechanisms. Such studies, when combined with discontinuum numerical modeling, have significant potential for identifying possible rock-slope failure volumes, as well as the likelihood of frequent small failures. A major initiative has been undertaken in Norway at Åknes, the site of a potentially devastating rock-slope failure and tsunami, in which a variety of innovative and promising monitoring and slope characterization techniques have been used to assess the landslide hazard and risk (Harbitz *et al.*, 2009 and references therein; Blikra, 2012, Chapter 26, this volume).

For subaqueous failures, cores or boreholes in both failed and nonfailed areas can help to elucidate the geometries and characteristics of past landslides and aid in determining the likelihood of future failures. Dynamic modeling of subaqueous failures, using known or estimated geotechnical properties to determine the rates of anticipated seabed deformation, is also important, as it provides key inputs to hydrodynamical tsunami models. Such an approach was recently undertaken, for example, in investigations of subaqueous slope failures in Kitimat Arm, British Columbia, using LSDYNA and DAN-W modeling approaches (AMEC Environment and Infrastructure, 2011). As discussed in the modeling section (Section 10.3.5), a block failure, which remains intact with little friction on the plane of failure, creates a larger tsunami than the same volume of material failing as a fragmented slide or debris flow (Rabinovich *et al.*, 2003).

10.3.4 FREQUENCY OF TSUNAMIGENIC LANDSLIDES

In the absence of long historical records it is generally difficult to determine the frequency of terrestrial or subaqueous tsunamigenic landslides. On the other hand, if an analysis indicates that events in a particular region are always triggered by seismic shaking, it is possible to estimate probabilities of future events based on the recurrence intervals of a particular threshold of ground acceleration. Where there is not a clear relationship between seismicity and landslides, most studies to date have attempted to estimate the number of failures and their associated volumes, and to make an expert judgment as to whether they would be likely to produce tsunamis. The number of landslides is divided by the time over which the failures occurred to obtain a sense of the hazard. Blikra *et al.* (2005), for example, identified 59 events with volumes greater than 0.5 million m^3 in the Storfjorden area, of which half had volumes greater than 1 million m^3. For each sector of the fjord, they divided the number of large landslides by 10,000–12,000 years, the elapsed time since deglaciation, to obtain frequency ranges of "more than 1 event per 1000 years," "1 event per 2500–5000 years," or "more than 1 event per 5000 years." These occurrence rates were placed within the context of Norwegian building codes, which require protection from events with frequencies of 1 per 1000 years or greater for homes, and protection from events with lower average recurrence intervals for public buildings such as schools, hotels, and industrial complexes. Blikra *et al.* (2005) concede that such an approach is only a "first approximation," but it provides the basis for undertaking additional geologic and geotechnical investigations at key locations.

Tappin (2010) has suggested that the risk of future tsunamigenic landslides could be less than that indicated by the number of observed landslides. Many events occurred during or shortly after late-Pleistocene glaciation, and the present risk under today's climate could be quite different. Conversely, land-use changes may have increased the hazard; as an example, two major tsunamigenic failures at different sites at the head of the fjord near Kitimat, British Columbia occurred within a period of 7 months (1974–1975). Bathymetric data collected by the Canadian Hydrographic Service show that there is little other evidence over the past 10,000 years for other similarly large failures. The region has, however, seen significant coastal development, and the surrounding watersheds have been extensively modified by logging.

A refined assessment of hazard requires identification of the conditions or triggers that can cause failure, including – for example – large floods or periods of very high rainfall or of increased erosion resulting from deforestation. Estimates of recurrence can be made using dating techniques such as tree-ring analysis (Carrara and O'Neill, 2003), radiocarbon dating (Xu *et al.*, 2003), or optically stimulated luminescence (Gartia, 2008), but the number of events that can be dated is generally insufficient to confidently determine landslide frequency. Such dating is rarely possible for subaqueous failures because relevant datable materials are commonly lacking, sediments overlying failed deposits may be thin, and underlying sediments are infrequently sampled and may not provide a closely limiting age of the failure.

10.3.5 LANDSLIDE-GENERATED TSUNAMI MODELING

The paucity of observations of landslide-generated tsunamis and the large range of possible conditions under which they are generated make it difficult to use historic information to predict the effects of future landslides in coastal waters. As a consequence, physical (laboratory) and numerical modeling of such events are important tools in characterizing and predicting these phenomena, and they are most effective when used in combination (Chaudhry *et al.*, 1983; Heinrich, 1992; Assier-Rzadkiewicz *et al.*, 1996, 1997; Fritz *et al.*, 2004, 2009; Shigihara *et al.*, 2006). Physical modeling is conducted under controlled conditions to improve our understanding of specific physical processes, whereas numerical modeling is used to study specific events in natural settings with complex geometry. Physical models provide preliminary qualitative estimates of tsunami characteristics and behavior (Fritz, 2002); numerical models provide more exact quantitative estimates that can account for the complexity of the waves in realistic marine basins.

Landslide-generated tsunamis depend on the volume, velocity, and coherence of the slide mass. The local water depth and the fraction of the slide volume that is subaerial are also important factors. A failure model is determined, in part, by the constraints of the problem being addressed: if a high degree of spatial and temporal accuracy is required, a more complex – and presumably realistic – failure model is required. Six types of failure models, from simple to complex, are listed and described below.

1. *Empirical models*. Estimates of the initial potential energy of the failed mass can be used to generate theoretical estimates of the leading tsunami wave height, assuming a certain proportion of potential energy is converted into the total (kinetic plus potential) energy of the waves (Murty, 1979; Bornhold *et al.*, 2007). Typical conversion percentages range from 0.01 to 0.1 percent. Fritz (2002) presented results of laboratory studies of conversion of landslide kinetic energy to total wave energy.
2. *Rigid-body (block-slide) models*. The entire volume of failed material is assumed to move downslope as a rigid block (Fig. 10.4a). The block does not deform or spread as it moves under the forces of gravity and bottom friction. Rigid bodies are used to investigate impulse tsunami generation (Fritz, 2002).
3. *Viscous flow models*. The failed material rapidly deforms into a viscous (Newtonian) fluid with gravity, kinematic viscosity $\mu = \tau/(\partial u/\partial z)$, and bottom drag determining how the material moves down the slope (Fig. 10.4b). Here τ is the shear stress, u is the slide velocity, and z is the vertical coordinate normal to the flow direction. In their study of viscous slides, Jiang and LeBlond (1992, 1994) assumed that the landslide mass rapidly acquires a parabolic profile due to bottom friction and viscous forces. The governing equations for the flow and the displaced water are treated separately, except for non-linear interaction terms that allow for direct surface wave generation by the advancing landslide.
4. *Bingham plastic slide models*. This model is similar to the viscous flow model except that the failed mass has a non-Newtonian fluid viscosity of the form $\mu = (\tau - \tau_o)/(\partial u/\partial z)$. In this case, the material behaves as a solid body until the shear stress exceeds a certain threshold, or yield stress, τ_o (Fig. 10.5). The model typically requires a more detailed understanding of the rheology of the failed material than the viscous model does.
5. *Impulse wave models*. The failed material may enter a lake or the sea as a rockfall or steep rockslide (Bornhold *et al.*, 2007). It forms an impact crater as it plummets into the water, producing an impulse wave similar to that created by a small meteorite or asteroid (Mader and Gittings, 2002). Fritz (2002) and Fritz *et al.* 2009) provide a thorough treatment of this topic.
6. *Three-dimensional wave modeling*. Few attempts have been made to simulate tsunamis in three dimensions, due to the numerical modeling complexities. Exceptions include the studies of Grilli and Watts (1999) and Grilli *et al.* (2001, 2002), which are based on potential flow theory and which, in effect, use a boundary element method (BEM) to decrease the initial three-dimensional problem to two dimensions.

Wiegel (1955) pioneered the use of physical modeling of submarine landslides and associated tsunamis. Many aspects of the problem, including the initial stages of tsunami generation, have been studied using physical models (Fritz, 2002; Fritz *et al.*, 2004), and physical parameters that are crucial to numerical simulations were originally derived through laboratory modeling (Sasitharan *et al.*, 1993; Watts, 1998, 2000; Fritz, 2002). Most physical studies of landslide-generated tsunamis have been conducted using rigid-body slides (Wiegel, 1955; Heinrich, 1992;

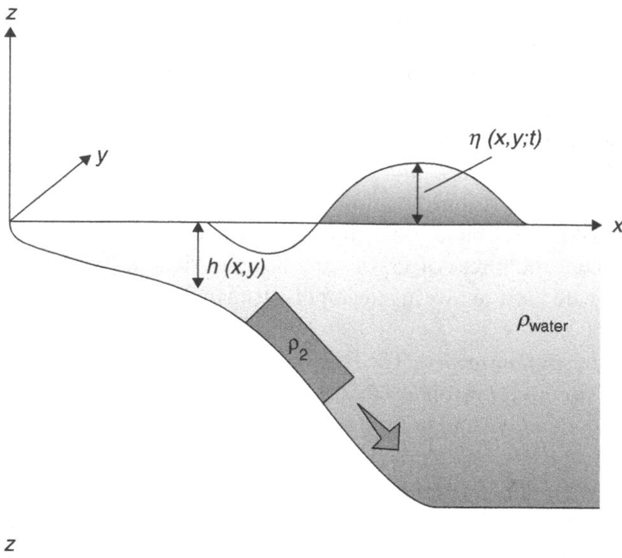

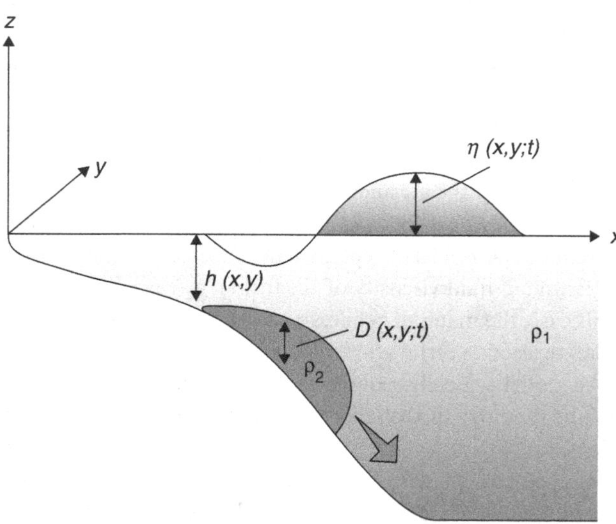

Fig. 10.4. Sketches showing tsunami waves of amplitude η generated by (a) a rigid-body, or block-slide, and (b) a viscous slide of density ρ_2 in water depth $h(x, y)$. ρ_{water} is the water density and x, y, z are coordinates in the cross-shore, alongshore, and vertical directions, respectively.

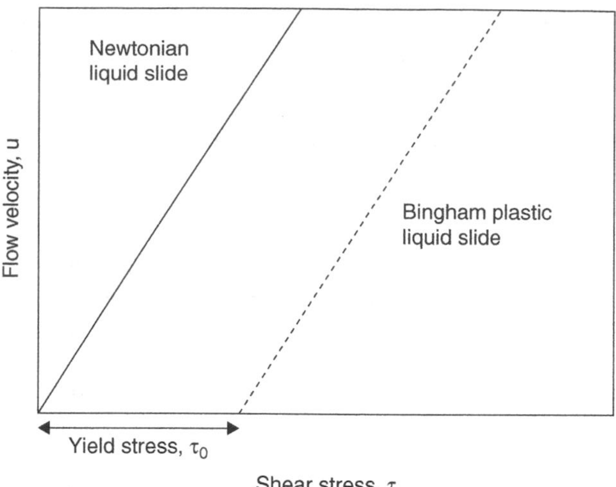

Fig. 10.5. Differences in the slide volume flow velocity, u, as a function of internal viscous shear stress, τ, for a slide behaving as a Newtonian fluid (solid line), and a Bingham plastic fluid (dashed line).

Watts, 1998, 2000; Watts *et al.*, 2000), but a few have considered fluid (viscous and Bingham) failures (Assier-Rzadkiewicz *et al.*, 1997; Shigihara *et al.*, 2006; Fritz *et al.*, 2009). Almost all physical models of submarine landslides and landslide-generated tsunamis assume uniform linear slopes (Heinrich, 1992; Assier-Rzadkiewicz *et al.*, 1996, 1997; Watts, 1998, 2000; Watts *et al.*, 2000; Fritz *et al.*, 2004, 2009; Shigihara *et al.*, 2006). However, natural slopes are more complex and, consequently, give rise to much more complicated motions.

Because today's numerical models run quickly and produce results that compare favorably with observation, numerical simulation is now widely used for specific case studies. Comparisons of rigid-body and viscous-slide models (Fine *et al.*, 2003; Rabinovich *et al.*, 2003) show that the former generate much higher tsunami waves than the latter for the same input parameters. There are also significant differences in the propagation, frequency content, and decay of tsunami waves generated by the two models. Differences also exist between viscous and rheologically more complex, Bingham-type models. The three-dimensional, shallow-water, viscous-slide model developed by Jiang and LeBlond (1992, 1994) provides a reasonable compromise between the rigid-body and Bingham slide models. It appears to be the first model to account for all major submarine landslide effects, including the coupling of the landslide and associated surface waves.

Fine *et al.* (2003) corrected errors in the governing equations of the Jiang and LeBlond model and generalized it to include arbitrary bottom topography. They also made similar modifications to the commonly used rigid-body, shallow-water slide model and examined the effect of different landslide parameters such as slope angle, slide position, water depth, and friction on surface wave generation. Versions of this model were used to simulate the 1999 Papua New Guinea tsunami (Heinrich *et al.*, 2000; Titov and González, 2001; Imamura *et al.*, 2001) and the 1979 tsunami caused by slumping of the Nice harbor extension (Assier-Rzadkiewicz *et al.*, 2000). The Fine *et al.* (2003) model allows for a subaerial component to the landslide but, as with most models, does not address the problem that wet and dry areas change during the tsunami, thus creating a variable boundary between the two areas. Only a few papers (Heinrich, 1992; Heinrich *et al.*, 1998) deal with this issue.

The numerical simulation of landslide-generated tsunamis is typically based on the shallow-water approximation (Harbitz, 1992), which has three main assumptions.

1. The surface waves and landslide satisfy the long-wave (hydrostatic) approximation, implying that the wavelength, λ, of the water waves is much greater than the water depth, and that the width and length of the landslide (in the horizontal x, y directions) is much greater than the thickness of the sliding mass. In the case of the surface waves, the wavelength is

typically assumed to be more than 25 times the water depth, h (specifically, kh « 1, where k is the wavenumber, $k = 2\pi/\lambda$).

2. In the case of rigid-body landslides, the material moves as a nondeformable body with specified bottom friction. For viscous landslides, the moving mass is treated as an incompressible, isotropic, laminar fluid that rapidly reaches a steady state.
3. Seawater is treated as an incompressible inviscid fluid.

The lower layer consists of viscous sediments or a rigid body, with density ρ, dynamic viscosity μ (or friction coefficient k in case of a rigid body), and horizontal velocity **U** with components U and V. The landslide is bounded by an upper surface $z = -h(x,y;t)$ and the seabed surface $z = -h_s(x,y)$, which gives a thickness of the landslide mass (D) as a function of time, t, of $D(x,y;t) = h_s(x,y) - h(x,y;t)$. A schematic of the computational domain for a landslide with a subaerial component is shown in Figure 10.6. The domain consists of four zones: (1) the dry coastal area, D; (2) the dry portion of the slide, S_D, corresponding to the subaerial part of the slide; (3) the wet portion of the slide, S_W, corresponding to the submarine part of the slide; and (4) the water, W. The numerical model must account for the time-varying changes in the areas and locations of these zones. Both the slope and the slide have small angles, so the motion is essentially horizontal. A study by Assier-Rzadkiewicz *et al.* (1997) indicates that the long-wave assumption becomes inaccurate for slopes steeper than about 10°. A study by Rabinovich *et al.* (2003) suggests that the long-wave approximation is in error by about 8 percent for a slope of 16° and 15 percent for a slope of 23°. Based on this analysis, Suleimani *et al.* (2009a) argued that for the average pre-event slopes of 10–20° off Seward, Alaska, the possible error introduced by a slide moving down these higher gradients slopes could be around 10 percent. Bottom friction (drag on the bottom currents) is a major energy dissipation term in the numerical simulations; the component of wave energy flux normal to the open boundaries of the model is transmitted through the boundary without reflection. Without friction and the outward energy flux to adjoining oceanic regions, waves and currents in the model would not attenuate with time.

Numerical and laboratory modeling studies indicate that purely submarine landslides are less effective tsunami generators than subaerial landslides. In the case of a subaerial rigid-body landslide, wave height is proportional to the height of the initial slide mass above the sea or lake. For viscous subaerial landslides, however, there is an optimal height of fall that produces the largest waves. In both cases, an increase in slide volume, density, or slope angle increases the energy of the generated waves. The rapid speed at which subaerial landslides plummet into the water is one of the reasons that subaerial landslides can be much more effective tsunami generators than slower-moving submarine landslides. Wave generation is most efficient at a state of resonance, where the slide Froude number $F_r = v_s / \sqrt{gh}$ (the ratio of the slide speed to the phase speed of the surface gravity waves) = 1.0. For purely submarine landslides with density

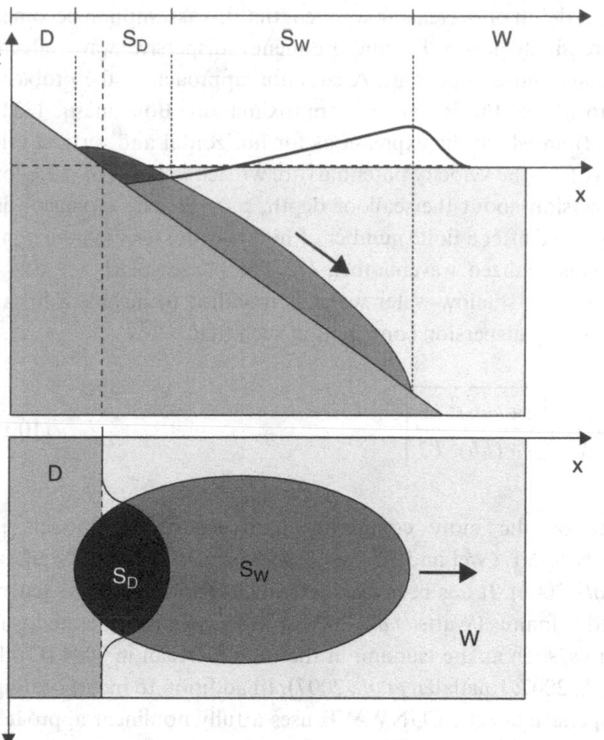

Fig. 10.6. Sketch of a combined subaerial and submarine landslide immediately after failure near the coast. The top panel is a side view and the bottom panel a plan view of the landslide. The domain consists of four zones: (1) the dry coastal area, D; (2) the dry, or subaerial, portion of the landslide, S_D; (3) the wet, or submarine, portion of the landslide, S_W; and (4) water, W (adapted from Fine *et al.*, 2003).

$\rho_s \leq 2.0$ g cm^{-3}, the Froude number is always less than 1 and resonance coupling of the landslide mass and surface waves is physically impossible. For subaerial landslides, resonance coupling is possible, with a significant transfer of energy from the landslide mass into surface waves. Subaerial landslides such as that at Lituya Bay in 1958 and Skagway, Alaska in 1994 displace a large volume of water at high speed as they enter the sea. Because the efficiency of tsunami generation is inversely proportional to water depth through $\sqrt{gh}$ in the Froude number expression, subaerial slides are particularly effective wave generators. Raichlen *et al.* (1996) demonstrated that the subaerial component of the Skagway landslide in 1994 significantly increased the wave amplitude of the tsunami.

Numerical modeling of tsunamis generated by submarine landslides is much more complicated than modeling of seismically generated tsunamis. The duration of the landslide is sufficiently long that it affects the characteristics of the surface waves. As a consequence, coupling between the slide mass and the surface waves must be considered. Moreover, the shape and internal characteristics of the slide mass change significantly during movement, altering the surface waves over the course of the event.

Most tsunami modeling studies can reliably use the shallow-water approximation and equations. However, for increasing water depth or decreasing wavelengths, this assumption becomes increasingly less valid and frequency dispersive wave effects become more important. A common approach to the problem is to invoke the Boussinesq approximation (Boussinesq, 1871, 1872), in which the expressions for horizontal and vertical velocity (i.e., the velocity potential) are written in terms of a Taylor expansion about the seafloor depth, $z = -h$. The expansion is truncated after a finite number of higher-order terms in the non-dimensionalized wavenumber, kh. The phase speed, $c = \sqrt{gh}$, for strictly shallow-water waves is modified to include a linear frequency dispersion contribution such that:

$$c = \sqrt{gh\left[\frac{1+(kh)^2/6}{1+(kh)^2/2}\right]}. \qquad (10.1)$$

One of the more commonly used Boussinesq models is FUNWAVE (Wei and Kirby, 1995; Kirby *et al.*, 1998; Madsen *et al.*, 2006). It has been used to simulate both landslide-generated tsunamis (Watts *et al.*, 2000) and seismically generated tsunamis, such as the tsunami in the Indian Ocean in 2004 (Grilli *et al.*, 2007; Ioualalen *et al.*, 2007). In addition to incorporating dispersive effects, FUNWAVE uses a fully nonlinear approach and allows the modeler to take into account wetting–drying effects (i.e., run-up and drain-down) along the coast. The model has been tested and verified against laboratory modeling and exact analytical solutions.

When waves strike the shore at an angle, their energy can become trapped in the near-coastal area, giving rise to edge waves. Edge waves are part of the physics of numerical wave models. Nearshore bottom topography and coastline geometry determine the frequency, wavelength, phase speed, amplitude, and velocity structure of these waves. Because tsunamis are most effectively generated by the shallow subaqueous or subaerial components of a landslide, the resulting wave energy can be trapped along the coast as edge waves (Rabinovich, 2009).

10.3.6 THINGS WE USUALLY DON'T KNOW

Reconstruction of past landslide-generated tsunamis or an assessment of future risk must deal with several poorly known factors, most of which are geologic or geotechnical. In contrast, the hydrodynamics of landslide-generated tsunamis are now well understood due to the pioneering work of Mader (1988), Heinrich (1992), Jiang and LeBlond (1992), and others, although subtleties, such as edge waves, are only now being adequately addressed (Bricker *et al.*, 2007).

A major limitation to studies of past subaqueous or subaerial landslides is the inevitable uncertainty about the volume, shape, and internal character of the landslide mass. Even if the pre- and post-failure slope morphologies are known from detailed bathymetric or topographic data, which is rarely the case, the fraction of the failed volume that generated the tsunami is generally uncertain; secondary retrogressive failures may contribute little to tsunami generation. Often one must consider a "worst-case" scenario in which the entire mass failed at one time, although such assumptions can result in unrealistically conservative scenarios. Similarly, large uncertainties exist around the velocity of the failed subaerial mass as it enters the water, which is a critical factor in determining the magnitude of the resultant tsunami. Also important are the physical characteristics of the failed mass: Was it a single block, a highly fragmented rock mass, or did it evolve into a debris flow? Uncertainty can be reduced through combined kinematic and mechanistic modeling of rock slopes; for example by determining potential maximum volumes of failure wedges, as has been done in Knight Inlet, British Columbia by van Zeyl (2009).

Murty (2003) stressed the importance of landslide volume over other factors, such as bottom slope, water depth, density of slide material, and duration of the landslide. Volume is clearly an important determinant of tsunami wave height, but a broad range of tsunami amplitudes is possible with the same landslide volume, suggesting that volume alone is a poor indicator of wave height (Fig. 10.7). The dataset in Figure 10.7 includes the most reliable published estimates of volume (V) and observed wave heights (H_{max}), but these estimates have large uncertainties. The relationship $H_{max} = 1.15 + 5.39 \times V$ explains about 78 percent of the variance for landslides smaller than 1 million m^3; however, only about half the variance is explained by the relationship $H_{max} = 3.61 + 0.10 \times V$ for landslides larger than 1 million m^3. Forcing the linear regressions to pass through the origin considerably weakens the relationships.

There is also considerable uncertainty about the proportion of landslide potential and kinetic energy that is converted to tsunami wave energy. Different values have been proposed, but they range over several orders of magnitude (Fritz, 2002; Harbitz *et al.*, 2006.

Eyewitness accounts or field evidence, such as observed elevations of tsunami deposits, constrain uncertainties about the amplitude of landslide-generated tsunamis, but they are rarely definitive. Eyewitnesses are notoriously inaccurate in estimating the amplitudes and periods of waves; they may lack a good vantage point or may be in danger from the event. In addition, local infrastructure that may aid in estimating amplitudes is commonly destroyed during the tsunami. Local bathymetric and topographic effects strongly control tsunami run-up and can result in significant differences over distances as small as tens of meters, even for relatively uniform coastlines.

When making an assessment of future tsunami hazards in a particular region, it may not be possible to define important parameters without extensive field work, involving, for example, geophysical surveys and geotechnical investigations. What volume is likely to fail? How will a rock mass behave after it fails? How do possible contributory factors such as groundwater, joints, faults, and lithology differ within an area of possible future failure?

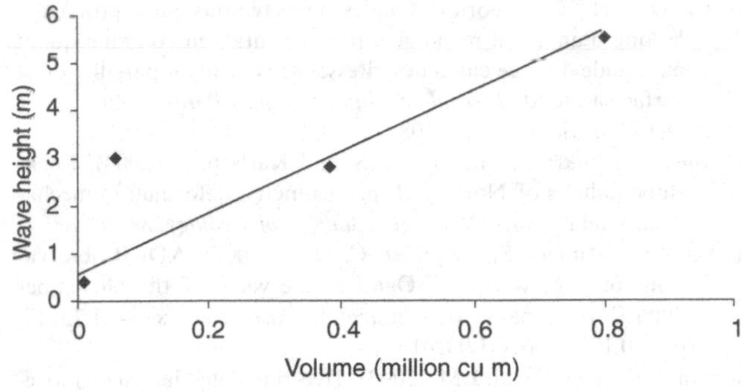

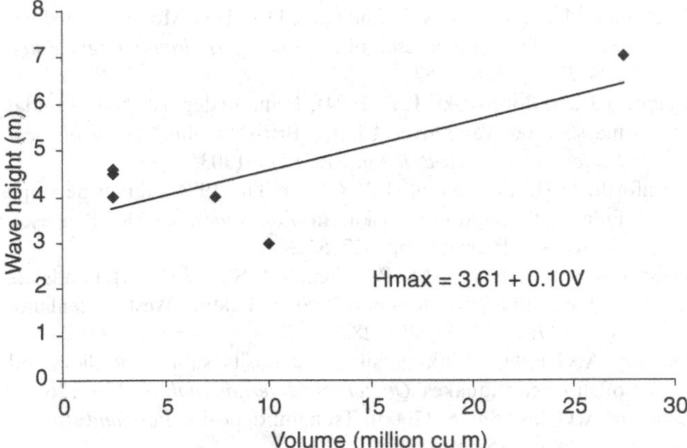

Fig. 10.7. The relationship between subaqueous landslide volume and maximum tsunami wave height for landslides less than 1 million m^3 (upper panel) and greater than 1 million m^3 (lower panel). Regression values are based on reliable data published by Jørstad (1968), Assier-Rzadkiewicz *et al.* (2000), Thomson *et al.* (2001), Papadopoulos and Kortekaas (2003), Strasser *et al.* (2007), Schnellmann *et al.* (2006), and Suleimani *et al.* (2009b).

10.3.7 SUMMARY OF LANDSLIDE TSUNAMI HAZARD ASSESSMENT APPROACH

It is difficult to produce a universal flowchart for tsunami hazard assessment programs, but there are several key elements that should be addressed.

- Are there newspaper accounts, anecdotes, and other local government or industry documents that indicate that local tsunamis have occurred? If so, what information do they provide on possible triggers, wave heights, locations of maximum drawdown or inundation, wave period, and failure recurrence?
- How long has the area been settled? Is this period of time sufficient to document these relatively rare events?
- If written records exist for less than a century or so, are there stories or other accounts from Indigenous peoples that might indicate that tsunamis can occur? What information can be gleaned from these accounts?
- Is there photographic or geologic evidence of landslides having entered coastal waters from adjacent steep slopes?
- Is there evidence of subaqueous failures from detailed bathymetry, sidescan sonar images, or high-resolution seismic data? This evidence includes failure zones on delta fronts or steep fjord sidewalls, including headscarps, downslope troughs, debris accumulation at the base of the slope, and extensive accumulations of debris flows or turbidites in the deeper parts of the basin.
- Is there any monitoring evidence of past or ongoing failure provided by, for example, repeat LIDAR or InSAR surveys or terrestrial InSAR continuous monitoring?

Based on the answers to these questions, an investigative team would then decide what further activities, such as field investigations or numerical modeling, should be undertaken. The decision to proceed with increasingly more costly phases of a study should be based on traditional risk assessment approaches. Because of the many uncertainties in such investigations, the decision to proceed will generally be based on the qualitative expert judgment of the investigative team.

10.4 CONCLUSIONS

Our understanding of landslide-generated tsunamis has improved greatly over the past few decades as devastating events around the world have been documented. Many uncertainties, largely geologic and geotechnical, remain and preclude definitive hazard assessments. Current models of slope failures and their resultant tsunamis are, on the other hand, well developed. What is typically lacking in hazard assessments is a good understanding of the failure characteristics, which are important for determining the nature of the tsunami. Such uncertainties, which are mostly related to a lack of knowledge of geologic conditions and are frequently hampered by difficult access to potential failure sites, will likely persist into the foreseeable future, with only modest improvements in estimates of probabilities of tsunami occurrence. Recent efforts to merge geotechnical and hydrodynamical models offer hope for a better understanding of the tsunami wave regime. Initiatives in Norway, especially at Åknes, have pioneered the use of a variety of monitoring and assessment techniques, such as rod extensometers, instrumented boreholes, laser and radar technologies, and a geophone network. Other promising technologies that may help reduce some of these uncertainties include airborne InSAR, terrestrial InSAR (TinSAR) slope monitoring, and the use of mobile LIDAR. We have presented a stepwise approach to hazard assessment for tsunamis caused by subaerial and subaqueous slope failures. Each phase of the assessment can be critically reviewed to determine whether it is necessary to proceed to the next phase. If, for example, there is no anecdotal, historic, or geologic evidence of past landslide-generated tsunamis in an area with generally low slopes within 5 km of the coastline in an aseismic region, it generally would not be necessary to proceed with the assessment. On the other hand, an area with modest slopes and intermediate seismicity, and with evidence for one or two historic events, would probably warrant continued assessment.

We echo the counsel of Bardet *et al.* (2003) that there is much that can be learned before failures actually occur in areas prone to underwater landslides, and that hydrodynamic modeling can be used effectively to assess the tsunamigenic potential of such landslides. We would extend their recommendation to subaerial failures. Landslide-generated tsunami hazard assessment should be a critical part of any infrastructure project in a moderate- to high-relief coastal environment. Although definitive estimates of the frequency and magnitude of these events can rarely be made, the analyses can place valuable constraints on the siting and design of coastal facilities.

ACKNOWLEDGEMENTS

We acknowledge the considerable scientific input we received from Isaac Fine and Alexander Rabinovich. Sheri Ward and Wendy Seward assisted with the preparation of several of the figures. John Clague, Doug Stead, and Andrée Blais-Stevens provided valuable comments on earlier drafts of this chapter.

REFERENCES

AMEC Environment and Infrastructure (2011). *Landslide-Generated Wave Hazard Analysis, Kitimat Arm, Enbridge Northern Gateway Project*. Report submitted to the National Energy Board of Canada [available at https://www.neb-one.gc.ca/ll-eng/livelink.exe?func=ll&objId=716802&objAction=browse&redirect=3].

Assier-Rzadkiewicz, S., Mariotti, C. and Heinrich, P. (1996). Modeling of submarine landslides and generated water waves. *Physics and Chemistry of the Earth*, 21, 7–12.

(1997). Numerical simulation of submarine landslides and their hydraulic effects. *Journal of Waterway, Port, Coastal and Ocean Engineering, ASCE*, 123, 149–157.

Assier-Rzadkiewicz, S., Heinrich, P. Sabatier, P.C., Savoye, B. and Bourillet, J.F. (2000). Numerical modelling of landslide-generated tsunami: The 1979 Nice event, *Pure and Applied Geophysics*, 157, 1707–1727.

Atwater, B.F. (1987). Evidence for great Holocene earthquakes along the outer coast of Washington. *Science*, 236, 942–944.

Bardet, J.-P., Synolakis, C.E., Davies, H.L., Imamura, F. and Okal, E.A. (2003). Landslide tsunamis: Recent findings and research directions. *Pure and Applied Geophysics*, 160, 1793–1809.

Barrow, F.J. (1935). *Barrow Catalogues/Pictographs (Complete), Book No. 2, No. EgSj-DiRx1*. Report on file at the Royal British Columbia Museum, Victoria, BC.

Bjerrum, L. and Jørstad, E. (1968). *Stability of Rock Slopes in Norway*. Norwegian Geotechnical Institute, Publication 79.

Blikra, L.H. (2012). The Åknes rockslide, Norway. In *Landslides: Types, Mechanisms and Modeling*, ed. J.J. Clague and D. Stead. Cambridge, UK: Cambridge University Press, pp. 323–335.

Blikra, L.H., Longva, O., Harbitz, C. and Løvholt, F. (2005). Quantification of rock avalanche and tsunami hazard in Storfjorden, western Norway. In *Landslides and Avalanches*, ed. K. Senneset, K. Flaate and J.O. Larsen. London: Taylor and Francis Group, pp. 7–63.

Bornhold, B.D., Harper, J.R., McLaren, D. and Thomson, R.E. (2007). Destruction of the First Nations village of Kwalate by a rock avalanche-generated tsunami. *Atmosphere–Ocean*, 45, 123–128.

Boussinesq, J. (1871). Théorie de l'intumescence liquide, appellée onde solitaire ou de translation, se propagent dans un canal rectangulaire. *Comptes Rendus de l'Académie des Sciences*, 72, 755–759.

Boussinesq, J. (1872). Théorie des ondes et des remous qui se propagent le long d'un canal rectangulaire horizontal, en communiquant au liquide dans ce canal des vitesses sensiblement pareilles de la surface au fond. *Journal de Mathématiques Pures et Appliquées, Deuxième Série*, 17, 55–108.

Braathen, A., Blikra, L.H., Berg, S.S. and Karlsen, F. (2004). Rock-slope failures of Norway: Type, geometry, deformation mechanisms and stability. *Norwegian Journal of Geology*, 84, 67–88.

Bricker, J.D., Munger, S., Pequinet, C. *et al.* (2007). ADCP observations of edge waves off Oahu in the wake of the November 2006 Kuril tsunami. *Geophysical Research Letters*, 34, L23617. doi:10.1029/2007GL032015.

Carrara, P.E. and O'Neill, J.M. (2003). Tree-ring dated landslide movements and their relationship to seismic events in southwestern Montana. *Quaternary Research*, 59, 25–35.

Chaudhry, M.H., Mercer, A.G. and Cass, D. (1983). Modeling of slide-generated waves in a reservoir. *Journal of Hydraulic Engineering, ASCE*, 109, 1505–1520.

Clague, J.J. and Bobrowsky, P.T. (1994). Tsunami deposits beneath tidal marshes on Vancouver Island, British Columbia. *Geological Society of America Bulletin*, 106, 1293–1303.

Cornforth, D.H. and Lowell, J.A. (1996). The 1994 submarine slope failure at Skagway, Alaska. in *Landslides*, ed. K. Senneset. Rotterdam: Balkema, pp. 527–532.

Dahl-Jensen, T., Larsen, L.M., Pedersen, S.A.S. *et al.* (2004). Landslide and tsunami 21 November 2000 in Paatut, West Greenland. *Natural Hazards*, 31, 277–287.

Dawson, A.G. (1999). Linking tsunami deposits, submarine slides and offshore earthquakes. *Quaternary International*, 60, 119–126.

Dawson, A.G. and Shi, S. (2000). Tsunami deposits. *Pure and Applied Geophysics*, 157, 875–897.

Fine, I.V., Rabinovich, A.B., Thomson, R.E. and Kulikov, E.A. (2003). Numerical modeling of tsunami generation by submarine and subaerial landslides. In *Submarine Landslides and Tsunamis*, ed. A.C. Yalciner, E.N. Pelinovsky, C.E. Synolakis and E. Okal. Dordrecht, Netherlands: Kluwer Academic Publishers, pp. 69–88.

Fritz, H.M. (2002). *Initial phase of landslide generated impulse waves*. Dipl. Bau-Ing. dissertation. Swiss Federal Institute of Technology, Zurich, Switzerland.

Fritz, H.M., Hager, W.G. and Minor, H.E. (2004). Near-field characteristics of landslide generated impulse waves. *Journal of Waterway, Port, Coastal and Ocean Engineering, ASCE*, 130, 287–302.

Fritz, H.M., Mohammed, F. and Jeseon, Y. (2009). Lituya Bay landslide impact generated mega-tsunami 50th anniversary. *Pure and Applied Geophysics*, 166, 153–175.

Gartia, R.K. (2008). Luminescence dating of some historical/prehistorical natural hazards of India. In *Proceedings of the Third International Conference on Luminescence and Its Applications*. Bangalore: MacMillan India, pp. 47–54.

Graziani, L., Maramai, A. and Tinti, A. (2006). A revision of the 1783–1784 Calabrian (southern Italy) tsunamis. *Natural Hazards and Earth System Sciences*, 6, 1053–1060.

Grilli, S.T. and Watts, P. (1999). Modelling of waves generated by a moving submerged body: Application to underwater landslides. *Engineering Analysis with Boundary Elements*, 23, 645–656.

Grilli, S.T., Guyenne, P. and Dias, F. (2001). A fully nonlinear model for three-dimensional overturning waves over arbitrary bottom. *International Journal of Numerical Methods in Fluids*, 35, 829–867.

Grilli, S.T., Vogelmann, S. and Watts, P. (2002). Development of a 3D numerical wave tank for modeling tsunami generation by underwater landslides. *Engineering Analysis with Boundary Elements*, 26, 301–313.

Grilli, S.T., Ioualalen, M., Asavanant, J. *et al.* (2007). Source constraints and model simulation of the December 26, 2004, Indian

Ocean tsunami. *Journal of Waterway, Port, Coastal and Ocean Engineering, ASCE*, 133, 414–428.

Harbitz, C.B. (1992). Model simulations of tsunamis generated by the Storegga slides. *Marine Geology*, 105, 1–21.

Harbitz, C., Pedersen, G. and Gjevik, B. (1993). Numerical simulations of large water waves due to landslides. *Journal of Hydraulic Engineering*, 119, 1325–1342.

Harbitz, C.B., Løvholt, F., Pedersen, G. and Masson, D.G. (2006). Mechanisms of tsunami generation by submarine landslides: A short review. *Norwegian Journal of Geology*, 86, 255–264.

Harbitz, C.B., Glimsdal, S., Løvholt, F. *et al.* (2009). Tsunami hazard assessment and early warning systems for the North East Atlantic. In *Proceedings of the DEWS Mid-term Conference 2009: Lessons Learned from Concept to Demonstrator.* Potsdam, Germany.

Heinrich, P. (1992). Nonlinear water waves generated by submarine and aerial landslides, *Journal of Waterway, Port, Coastal and Ocean Engineering, ASCE*, 118, 249–266.

Heinrich, P., Mangeney, A., Guibourg, S. *et al.* (1998). Simulation of water waves generated by a potential debris avalanche in Montserrat, Lesser Antilles. *Geophysical Research Letters*, 25, 3697–3700.

Heinrich, P., Piatensi, A., Okal, E. and Hébert, H. (2000). Near-field modeling of the July 17, 1998 tsunami in Papua New Guinea. *Geophysical Research Letters*, 27, 3037–3040.

Imamura, F., Hashi, K. and Imteaz, M.M.A. (2001). Modeling for tsunamis generated by landsliding and debris flow. In *Tsunami Research at the End of a Critical Decade*, ed. G. Hebenstreit. Dordrecht, Netherlands: Kluwer, pp. 209–228.

Ioualalen, M., Asavanant, J., Kaewbanjak, N. *et al.* (2007). Modeling the 26th December 2004 Indian Ocean tsunami: Case study of impact in Thailand. *Journal of Geophysical Research*, 112, C07024, doi:10.1029/2006JC003850.

Jiang, L. and LeBlond, P.H. (1992). The coupling of a submarine slide and the surface waves which it generates. *Journal of Geophysical Research*, 97, 12731–12744.

(1994). Three-dimensional modeling of tsunami generation due to a submarine mudslide, *Journal of Physical Oceanography*, 24, 559–572.

Jørstad, F.A. (1968). Waves generated by landslides in Norwegian fjords and lakes. *Norwegian Geotechnical Institute*, 79, 13–32.

Karlsrud, K. and Edgers, L. (1980). Some aspects of submarine slope stability. In *Marine Slides and other Mass Movements*, ed. S.E. Saxov and J.K. Nieuwenhuis. New York: Plenum in cooperation with NATO Scientific Affairs Division, pp. 61–81.

Kirby, J.T., Wei, G., Chen, Q., Kennedy, A.B. and Dalrymple, R.A. (1998). FUNWAVE 1.0. *Fully Nonlinear Boussinesq Wave Model: Documentation and User's Manual.* Report CACR-98-06. Newark, DE: University of Delaware, Center for Applied Coastal Research (CACR), Department of Civil and Environmental Engineering.

Kulikov, E.A., Rabinovich, A.B., Thomson, R.E. and Bornhold, B.D. (1996). The landslide tsunami of November 3, 1994, Skagway Harbor, Alaska. *Journal of Geophysical Research*, 101, 6609–6615.

Kulikov, E.A., Rabinovich, A.B., Fine, I.V., Bornhold, B.D. and Thomson, R.E. (1998). Tsunami generation by landslides at the Pacific coast of North America and the role of tides. *Oceanology*, 38, 323–328.

Lander, J.F. (1996). *Tsunamis Affecting Alaska, 1737–1996*. Boulder, CO: US Department of Commerce.

Lee, H.J., Ryan, H.F., Keyen, R.E. *et al.* (2006). Varieties of submarine failure morphologies of seismically induced landslides in Alaskan fjords. *Norwegian Journal of Geology*, 86, 221–230.

Leroueil, S., Locat, J., Levesque, C. and Lee, H.J. (2003). Towards an approach for the assessment of risk associated with submarine mass movements. In *Submarine Mass Movements and Their Consequences*, ed. J. Locat and J. Mienert. Dordrecht, Netherlands: Kluwer, pp. 59–67.

Mader, C.L. (1988). *Numerical Modeling of Water Waves*. Berkeley, CA: University of California Press.

Mader, C.L. and Gittings, M.L. (2002). Modeling the Lituya Bay mega-tsunami II. *Science of Tsunami Hazards*, 20, 241–250.

Madsen, P.A., Fuhrman, D.R. and Wang, B. (2006). A Boussinesq-type method for fully nonlinear waves interacting with a rapidly varying bathymetry. *Coastal Engineering*, 53, 487–504.

Masson, D.G., Harbitz, C.B., Wynn, R.B., Pedersen, G. and Lovholt, F. (2006). Submarine landslides: Processes, triggers and hazard prediction. *Philosophical Transactions of the Royal Society (A)*, 364, 2009–2039.

Miller, D.J. (1960). The Alaska earthquake on July 10, 1958: Giant wave in Lituya Bay. *Bulletin of the Seismological Society of America*, 50, 253–266.

Mosher, D.C., Monahan, P. A, and Barrie, J.V. (2001). Submarine failures in the Strait of Georgia, British Columbia: Landslides of the 1946 Vancouver Island earthquake. In *2001 An Earth Odyssey: Proceedings of the 54th Canadian Geotechnical Conference, Calgary, Alberta*. Richmond, BC: Bitech Publishers, pp. 744–751.

Murty, T.S. (1979). Submarine slide-generated water waves in Kitimat, British Columbia. *Journal of Geophysical Research*, 84, 7777–7779.

(2003). Tsunami wave height dependence on landslide volume. *Pure and Applied Geophysics*, 160, 2147–2153.

Naranja, J.A. and Arenas, M. (2009). Mass movement-induced tsunamis: main effects during the Patagonian Fjordland seismic crisis in Aisén (45° 25′ S), Chile. *Andean Geology*, 36, 137–145.

Papadopoulos, G.A. and Kortekaas, S. (2003). Characteristics of landslide-generated tsunamis from observational data. In *Submarine Mass Movements and Their Consequences*, ed. J. Locat and J. Mienert. Dordrecht, Netherlands: Kluwer, pp. 367–374.

Prior, D.B., Bornhold, B.D. and Johns, M. (1984). Depositional characteristics of a submarine debris flow. *Journal of Geology*, 92, 707–727.

Rabinovich, A.B. (2009). Seiches and harbor oscillations. In *Handbook of Coastal and Ocean Engineering*, ed. Y.C. Kim. Singapore: World Scientific Publications, pp. 193–236.

Rabinovich, A.B., Thomson, R.E., Kulikov, E.A., Bornhold, B.D. and Fine, I.V. (1999). The landslide-generated tsunami of November 3, 1994 in Skagway Harbor, Alaska: A case study. *Geophysical Research Letters*, 26, 3009–3012.

Rabinovich, A.B., Thomson, R.E., Bornhold, B.D., Fine, I.V. and Kulikov, E.A. (2003). Numerical estimation of tsunami risk associated with landslide-generated tsunamis with application to the Strait of Georgia, British Columbia. *Pure and Applied Geophysics*, 160, 1273–1313.

Raichlen, F., Lee, J.J., Petroff, C. and Watts, P. (1996). The generation of waves by a landslide: Skagway, Alaska: A case study. In *Coastal Engineering 1996: Proceedings of the 25th International Coastal Engineering Conference, Orlando, Florida*, ed. B.L. Edge. ASCE, pp. 1478–1490.

Rogers, G.C. (1980). A documentation of soil failure during the British Columbia earthquake of 23 June 1946. *Canadian Geotechnical Journal*, 35, 122–127.

Sasitharan, S., Robertson, P.K., Sego, D.C. and Morgenstern, N.R. (1993). Collapse behaviour of sand. *Canadian Geotechnical Journal*, 30, 569–577.

Schnellmann, M., Anselmetti, F.S., Giardini, D. and McKenzie, J.A. (2006). 15,000 years of mass-movement history in Lake Lucerne: Implications for seismic and tsunami hazards. *Eclogae Geologicae Helvetiae*, 99, 409–428.

Sepulveda, S.A. and Serey, A. (2009). Tsunamigenic, earthquake-triggered rock failures during the April 21, 2007 Aisén earthquake, southern Chile (45.5° S). *Andean Geology*, 36, 131–136.

Shigihara, Y., Goto, D., Imamura, F. *et al.* (2006). Hydraulic and numerical study on the generation of subaqueous landslide-induced tsunami along the coast. *Natural Hazards*, 39, 159–177.

Skvortsov, A. and Bornhold, B.D. (2007). Numerical simulation of landslide-generated tsunami in Kitimat Arm, British Columbia, Canada, April 27, 2005. *Journal of Geophysical Research*, 112, F02028.

Strasser, M., Stegmann, S., Bussmann, F. *et al.* (2007). Quantifying subaqueous slope instability during seismic shaking: Lake Lucerne as model for ocean margins. *Marine Geology*, 240, 77–97.

Suleimani, E., Nicolsky, D.J., Haeussler, P. and Hansen, R. (2009a). Combined effects of tectonic and landslide-generated tsunami runup at Seward, Alaska, during the M9.2 1964 earthquake. *Pure and Applied Geophysics*, 166, 131–152.

Suleimani, E., Hansen, R. and Haeussler, P. (2009b). Numerical study of tsunami generated by multiple slope failures in Resurrection Bay, Alaska during the M_w 9.2 1964 earthquake. *Pure and Applied Geophysics*, 166, 131–152.

Tappin, D.R. (2010). Mass transport events and their tsunami hazard. In *Submarine Mass Movements and their Consequences*, ed. D.C. Mosher, R.C. Shipp, L. Moscardelli *et al.* Dordrecht, Netherlands: Springer, pp. 667–684.

Thomson, R.E., Rabinovich, A.B., Kulikov, E.A., Fine, I.V. and Bornhold, B.D. (2001). On numerical simulation of the landslide-generated tsunami of November 3, 1994 in Skagway Harbor, Alaska. In *Tsunami Mitigation*, ed. G. Hebenstreit. Dordrecht, Netherlands: Kluwer, pp. 243–282.

Titov, V.V. and González, F. (2001). Numerical study of the source of the July 1998 PNG tsunami. In *Tsunami Research at the End of a Critical Decade*, ed. G. Hebenstreit. Dordrecht, Netherlands: Kluwer, pp. 197–207.

van Zeyl, D. (2009). *Evaluation of subaerial landslide hazards in Knight Inlet and Howe Sound, British Columbia.* M.Sc. thesis, Simon Fraser University, Burnaby, BC.

Watts, P. (1998). Wavemaker curves for tsunamis generated by underwater landslides. *Journal of Waterway, Port, Coastal and Ocean Engineering, ASCE*, 124, 127–137.

(2000). Tsunami features of solid block underwater landslides. *Journal of Waterway, Port, Coastal and Ocean Engineering, ASCE*, 126, 144–152.

Watts, P., Imamura, F. and Grilli, S. (2000). Comparing model simulations of three benchmark tsunami generation cases. *Science of Tsunami Hazards*, 18, 107–123.

Wei, G. and Kirby, J.T. (1995). Time-dependent numerical code for extended Boussinesq equations. *Journal of Waterway, Port, Coastal and Ocean Engineering, ASCE*, 121, 251–261.

Wiegel, R.L. (1955). Laboratory studies of gravity waves generated by movement of a submerged body. *Transactions of the American Geophysical Union*, 36, 759–774.

Xu, S., Zheng, G.D. and Lang, Y.H. (2003). Radiocarbon dating and Mössbauer spectroscopic study of the Mukaiyama landslide, Japan. *Journal of Radioanalytical and Nuclear Chemistry*, 258, 307–314.

11 Physical impacts of climate change on landslide occurrence and related adaptation

CHRISTIAN HUGGEL, NIKOLAY KHABAROV, OLIVER KORUP, AND MICHAEL OBERSTEINER

ABSTRACT

We review current understanding of the effects of climate change on the occurrence of landslides and debris flows in cold, temperate, and tropical mountains. We start with a summary of observed impacts of climate change on shallow landslides and debris flows, followed by discussions of rock-slope failures and the physical processes that make climate an important cause and trigger of landslides. While an increase in extreme precipitation has been observed in many regions worldwide over the past decades, changes in frequency and magnitude of landslides are more difficult to identify. In high mountain regions with snow, glaciers, and permafrost, slope stability is sensitive not only to changes in precipitation but also to changes in temperature. In the European Alps, the number of high alpine rock-slope failures has increased over the past few decades, coincident with an increase in mean air temperature. Model-based projections of future climate indicate that extreme precipitation events are likely to increase, causing more landslides. Seasonal variations in precipitation and earlier snowmelt imply changes in the seasonality of landslide occurrence. In addition, changes in sediment supply can strongly condition debris-flow frequency and magnitude. We conclude with a case study that outlines the potential and limitations of adaptation to future changes in precipitation.

11.1 INTRODUCTION

Economic losses from landslides have been rising over recent decades (Guzzetti, 2000; Petley *et al.*, 2005; ISDR, 2009), mainly because of increasing development and investment in landslide-prone areas (Schuster and Highland, 2001; Petley *et al.*, 2007). Although changes in land use, including deforestation, have affected slope stability in developed and populated areas, the causes of the increasing impact of landslides are not well understood (Petley *et al.*, 2007). The anthropogenic overprint makes it difficult to detect the potential impacts of climate change on landslide magnitude and frequency, which determine landslide hazard (Lateltin *et al.*, 2005). The magnitude/frequency concept and its application in risk appraisals form a useful framework for quantifying the potential impacts of climate change on slope stability, but detecting systematic changes in landslide occurrence, as well as accurately forecasting future impacts of climate change, remain difficult. Few sufficiently detailed landslide inventories exist to accurately quantify changes in the magnitude or frequency of landslides. The period of historically documented landslide events is too short and the record is too incomplete to reveal statistically significant trends. In addition, there is a strong documentation bias, with many more events documented for recent years than for earlier periods. This bias exists for all areas, but is a particular problem for high mountain regions and many developing countries. Yet many high mountain areas have the advantages of being less affected by human activity and of being more sensitive to climate change. In these areas, landslide occurrence is controlled by a variety of factors, including geology, hydrology, topography, and climate, which are all linked through feedback mechanisms. For example, atmospheric warming drives the retreat of glaciers and the exposure of fresh glacial sediments, thus potentially affecting the frequency and magnitude of debris flows (Evans and Clague, 1994; Hewitt *et al.*, 2008).

This chapter focuses on landslides and debris flows in cold, temperate, and tropical mountains. We start with the observed impacts of climate change on shallow landslides and debris flows, followed by discussions of rock-slope failures and the physical processes that make climate an important cause and trigger of landslides. We then consider landslide activity in the context

Landslides: Types, Mechanisms and Modeling, ed. John J. Clague and Douglas Stead. Published by Cambridge University Press.

Fig. 11.1. Two shallow landslides in the Combeima Valley, Colombia that occurred in 2006 (right panel) and 2009 (left panel). The image on the left shows the source zone of a shallow, unchannelized landslide that initiated on a steep (*ca.* 35°), cultivated slope. The landslide on the right is much larger and was channelized, with deposit thicknesses of 4–5 m (people and car included for scale). (Photos by C. Huggel.)

of anticipated climate change. We review the latest studies on extreme meteorological events and the development of conceptual approaches for evaluating the role of climate change in future landslide activity. We conclude with a model and a case study illustrating the effect of changes in rainfall on landslide activity and related losses. This study illustrates a possible avenue for adaptation within the framework of an early-warning system.

11.2 OBSERVED IMPACTS OF CLIMATE CHANGE ON LANDSLIDE ACTIVITY

11.2.1 EFFECTS ON SHALLOW LANDSLIDES AND DEBRIS FLOWS

Precipitation is an important landslide trigger. In mountain regions with seasonal or perennial snow cover, glaciers, and permafrost, temperature is an additional factor affecting slope stability. Intense and prolonged rainfall saturates soils and produces high transient pore pressures (Fig. 11.1; Iverson, 2000). Other factors include total rainfall, rainfall intensity and duration, and antecedent rainfall (Wieczorek and Glade, 2005; Sidle and Ochiai, 2006). Rapid snowmelt may further reduce the stability of shallow soils (Kim *et al.*, 2004). Numerous empirical models relate observed occurrences of shallow landslides and debris flows to rainfall intensity and duration (Caine, 1980; Larsen and Simon, 1993; Jakob and Weatherly, 2003) to derive critical intensity–duration thresholds (Fig. 11.2). These relationships provide first-order metrics that can be used for early warning (Keefer *et al.*, 1987; Guzzetti *et al.*, 2007).

Records of rainfall intensity and duration in high mountain regions are limited, reflecting the scarcity of climate stations, except for parts of the European Alps. In the Swiss Alps, Zimmerman *et al.* (1997) reported debris flow-triggering rainfall intensities of *ca.* 30–60 mm per hour for durations of *ca.* 1 hour. Dahal and Hasegawa (2008) found that rainfall intensity thresholds for shallow landslides in Nepal, a country with a limited meteorological network, are substantially different (Fig 11.2).

Antecedent rainfall modulates critical rainfall intensities (Kim *et al.*, 1991; Glade, 1998), although many rainfall records have insufficient temporal resolution to test this assertion (Huggel *et al.*, 2010a). Records of landslide occurrence and corresponding rainfall parameters often suffer from uncertainties regarding exact location and timing. Furthermore, orographic effects on precipitation may not be captured adequately by a single (or few) rain gauges. In the Alps, for example, mid- to high-elevation areas generally receive the most precipitation, whereas in the tropical Colombian Andes, precipitation is lowest in high elevated glacierized areas. In the latter area, rainfall-triggered landslides are most common at lower to intermediate elevations (Huggel *et al.*, 2010a). Reviews of intensity–duration thresholds of rainfall-triggered landslides have shown order-of-magnitude variability that depends, among other things, on local climate, geology, soil characteristics, and land use. The reliability of individual precipitation thresholds also depends on the quality of records relating to rainfall-triggered landslides.

Soil and rock mechanics provide a deterministic framework for slope stability under both dry and wet conditions. The rate of infiltration is controlled by the physical properties of the soil: notably porosity, hydraulic conductivity, pore size distribution, preferential flow networks, vegetation cover, topography, freezing, and land use (Sidle and Ochiai, 2006). Several models of

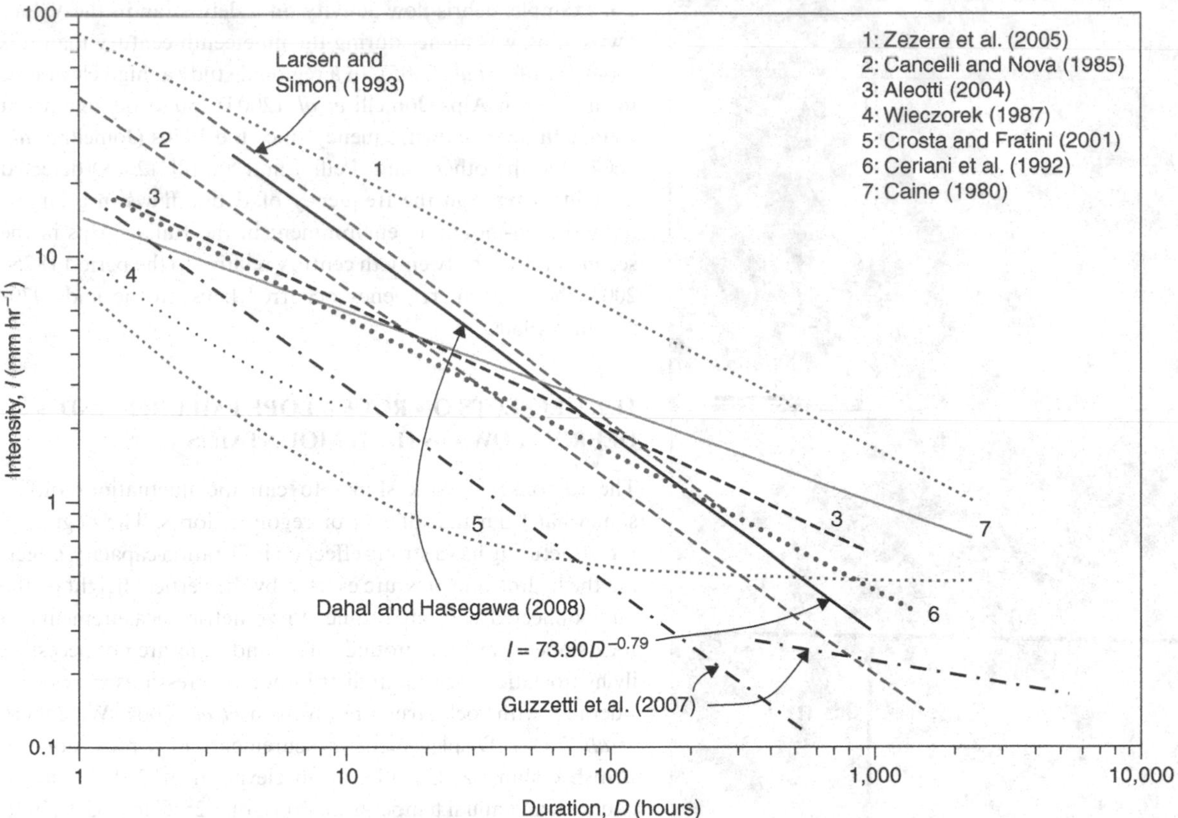

Fig. 11.2. Published landslide-triggering thresholds of rainfall intensity and duration. The thresholds can differ by an order of magnitude depending on climate, hydrology, topography, geology, land cover, land use, and other factors (modified from Dahal and Hasegawa, 2008).

subsurface flow at different scales have been developed, based in most cases on Darcy's law for unsaturated and saturated flow (Montgomery *et al.*, 1997; Iverson, 2000; Uchida *et al.*, 2001). In mountainous terrain, variations in transient pore pressure may trigger shallow landslides and debris flows, commonly at or near the soil– or regolith–bedrock interface (Rickenmann and Zimmermann, 1993). Permafrost may act as a hydraulic barrier that promotes elevated pore pressures in the active layer (Fig. 11.3; Zimmermann and Haeberli, 1992; Rist and Phillips, 2005).

Piezometer and tensiometer measurements in natural and artificial soils have demonstrated the large spatial and temporal variability in pore pressures (Iverson and LaHusen, 1989; Harp *et al.*, 1990; Montgomery *et al.*, 2002; Simoni *et al.*, 2004). Pierson (1980) found that 77–91 percent of the variation in maximum pore pressure could be explained by the amount of 24-hour rainfall, exclusive of the effects of antecedent rainfall. Increases in pore pressure may follow precipitation peaks by up to several hours, depending on local soil characteristics (Montgomery *et al.*, 2002). In contrast, Harp *et al.* (1990) observed that pore pressures may also decrease just before slope failure.

An increase in extreme precipitation may cause more frequent landslides, all other factors being equal. Extreme precipitation may be defined by:

- relative thresholds such as percentiles of a statistical reference distribution
- absolute thresholds (in millimeters)
- return periods of precipitation events
- the severity of damage (Beniston *et al.*, 2007; Trenberth *et al.*, 2007).

The 95th or 99th percentiles are commonly used to quantify precipitation extremes or heavy precipitation (Trenberth *et al.*, 2007). The frequency of precipitation extremes has increased in many parts of the world over the past few decades, although the increases differ seasonally and spatially in different regions (Trenberth *et al.*, 2007), and large uncertainties remain in mountain belts. For example, Kyselý (2009) found an increase in heavy winter precipitation in the Czech Republic from 1961 to 2005, whereas Pavan *et al.* (2008) argued the opposite for summer precipitation in the mountains of the Emilia-Romagna region of Italy between 1951 and 2004. Marengo *et al.* (2010) concluded that precipitation extremes increased in parts of Brazil, Argentina, northwest Peru, and Ecuador during the second half of the twentieth century. Increases in rainfall have also been reported in many parts of North America and some areas of Asia; for example, India and western and northern China (Krishnamurthy *et al.*, 2009; Pryor *et al.*, 2009). On the other hand, no significant change has been observed in other regions of these continents (Choi *et al.*, 2009).

Rainstorms during the summers of 1987 and 2005 in the Swiss Alps and adjacent areas resulted in widespread debris-

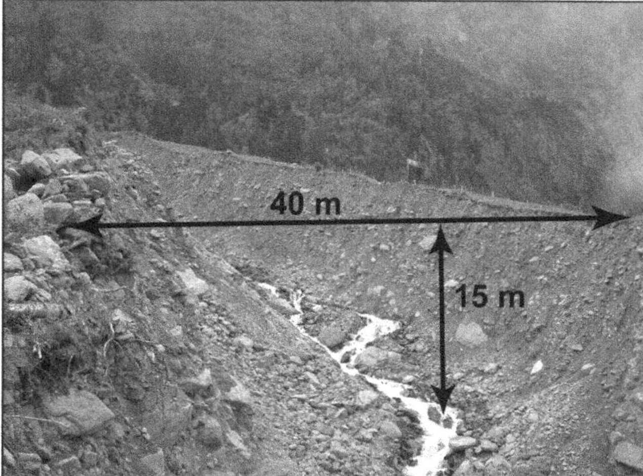

Fig. 11.3. Source (upper image) and eroded channel (lower image) of the August 22, 2005 Rotlaui debris flow near Guttannen in the central Swiss Alps. The dashed white line on the aerial image delineates a body of permanently frozen sediment, formerly covered by Homad Glacier. The arrow indicates the initiation zone of the debris flow (*ca.* 2400 m asl). A significant percentage of the debris-flow volume was entrained on the Holocene debris fan between *ca.* 1100 and 1300 m asl (lower image). The debris flow was the largest (*ca.* 500,000 m^3) in Switzerland in at least 20 years.

flow damage (Bezzola and Hegg, 2007). The largest debris flows – Varuna in 1987 and Guttannen in 2005 – originated from periglacial areas and had volumes up to 0.5 million m^3 (Rickenmann and Zimmermann, 1993; Scheuner *et al.*, 2009). The Guttannen debris flow occurred during a storm that dropped about 170 mm of rainfall in 48 hours, a 100-year event and the largest single storm precipitation since 1876 in that area (Scheuner *et al.*, 2009). The effect of the storm was exacerbated by a rise in the 0°C-isotherm to well above 3000 m asl, resulting in rainfall and runoff at very high elevations. In 1987, intense rainfall of up to 260 mm, together with antecedent precipitation, triggered numerous debris flows (Zimmermann, 1990).

How trends in landslide occurrence are linked to these precipitation changes awaits further study and the integration of different time series. Landslide frequency data are becoming more widely available, although mostly at the catchment scale. For example, debris-flow activity on a debris fan in the Valais, Swiss Alps, was higher during the nineteenth century than it is today (Stoffel *et al.*, 2005). In a regional study at high elevations in the French Alps, Jomelli *et al.* (2004) found no significant change in debris-flow frequency since the 1950s (Jomelli *et al.*, 2004). On the other hand, Pelfini and Santilli (2008) detected a slight increase in the frequency of debris flows in a largely natural high-mountain environment in the Italian Alps in the second half of the twentieth century relative to the period 1875–2003; the peak in frequency occurred between the mid-1970s and mid-1980s.

11.2.2 EFFECTS ON ROCK-SLOPE FAILURES AND DEBRIS FLOWS IN HIGH MOUNTAINS

The response of rock slopes to climatic fluctuations differs somewhat from that of soil or regolith slopes. The degree of rock fracturing has a strong effect on infiltration capacity, affecting the hydrostatic pressure exerted by the vertical height of the interconnected saturated zone. Piezometric measurements in rock slopes show that groundwater conditions are not necessarily hydrostatic. The spatial distribution of pressures varies considerably with rock structure (Watson *et al.*, 2004; Willenberg *et al.*, 2008). Displacement measurements in a rock slope in British Columbia, Canada, at an elevation of 500–800 m asl and with an annual temperature range of −25°C to +35°C, indicate a possible relation between transient pore-pressure variations and infiltration (Watson *et al.*, 2004), which overprint the annual temperature cycle. Cooling introduces deviatoric stresses and a reduction in effective normal stress, resulting in cracking of rock bridges, continuous loss of cohesion in the rock mass, and – eventually – slip (Krähenbühl, 2004; Watson *et al.*, 2004).

Few measurements of displacement have been made on steep frozen rock slopes in high mountains. However, observations at Jungfraujoch (3580 m asl) in the Swiss Alps show extension during periods of cooling and contraction in otherwise warm periods (Wegmann and Gudmundsson, 1999). Precipitation at such high-elevation sites is predominantly snow, and snowmelt may be the main contributor to infiltration, depending on local topography, aspect, and rock fracturing. In addition, liquid water generated by permafrost thaw modulates hydrostatic pressure, loss of bonding, and reduction in shear strength (Gruber and Haeberli, 2007). Recent field measurements at the Matterhorn in the Swiss Alps have contributed to an improved understanding of the role of meltwater circulation in fracture systems, permafrost thaw, and related deformation of alpine rock slopes (Hasler, 2011).

Only in a few cases, such as the 1988 Tschierva rock avalanche (*ca.* 3100 m asl) in the eastern Swiss Alps, does the proximity of a rain gauge allow a reliable reconstruction of antecedent rainfall. The record at this site indicates exceptionally high precipitation (>100 mm) on two consecutive days, 2 weeks before failure (Fischer *et al.*, 2010). Temperatures were around freezing during this time, and therefore some of the precipitation was

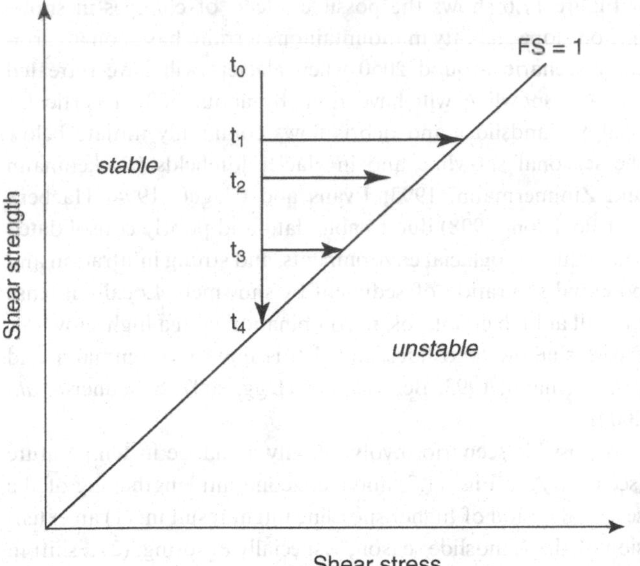

Fig. 11.4. Simplified scheme illustrating a theoretical slope history with a long-term gradual decrease in slope strength due, for example, to permafrost degradation (vertical axis) and short-term increases in shear stress due, for example, to an earthquake or heavy rainfall (horizontal axis). At different times, short-term events may increase shear stress to the threshold of instability (t_1 and t_3), whereas in others the threshold is not reached (t_2). The factor of safety (FS) distinguishes between stable and unstable slope conditions.

snow. Subsequent melting and infiltration may have compromised the stability of the slope. In high mountain regions with permanent snow, glaciers, and permafrost, both a long-term gradual increase in temperature and brief temperature extremes may destabilize slopes (Haeberli and Beniston, 1998; Gruber and Haeberli, 2007; Huggel *et al.*, 2008). Transient temperature increases and associated meltwater production may trigger landslides without precipitation. Analysis of a number of large landslides and ice avalanches in Alaska, the European Alps, and the Southern Alps of New Zealand has shown a pattern of unusually high temperatures in the days and weeks before many failures (Huggel *et al.*, 2010b). Although the specific processes are not understood in detail in every case, it is reasonable to assume that meltwater due to high temperatures infiltrated rock slopes along fractures and joints, thawing frozen clefts by heat advection, possibly increasing hydrostatic pressures and thus reducing shear strength. Meltwater can also penetrate to the base of steep glaciers and reduce their resistance to failure (Fig. 11.4). Also, a gradual increase in mean temperature can reduce shear strength through permafrost thaw.

Recent studies indicate that both small and large rock-slope failures have increased significantly in number in the central European Alps in recent decades (Fischer, 2010; Ravanel and Deline, 2011). This increase coincides with a marked temperature increase and also with an increase in temperature extremes. Most of the Earth's land regions have experienced an increase in high temperatures, expressed as the 90th or 95th percentiles of the long-term record, over the past 50–100 years (Trenberth *et al.*, 2007). The frequency of hot days in Europe has almost tripled over the period 1880–2005 (Della-Marta *et al.*, 2007), and Ding *et al.* (2007) and Kunkel *et al.* (2008) found a marked increase in heat waves since the 1960s in, respectively, China and the USA. However, studies specifically targeted at temperature trends in kilometer-long regions are rare, and any causal linkage to landslide occurrence requires further study.

11.3 PROJECTED IMPACTS OF CLIMATE CHANGE ON LANDSLIDE ACTIVITY

Precipitation and temperature extremes may contribute to destabilizing slopes and triggering landslides (Fig. 11.4), but they also affect boundary conditions such as topography. Local topography can change significantly over years to decades due to shrinkage of glaciers and landslides (Holm *et al.*, 2004; Geertsema *et al.*, 2006; Paul and Haeberli, 2008; Hewitt *et al.*, 2008). Temperature and precipitation, however, change on much shorter timescales. Extreme precipitation and temperature events will likely increase in frequency in the twenty-first century. Uncertainties, however, are large, especially for precipitation (Meehl *et al.*, 2007), although efforts have been made to reduce these uncertainties with improved general circulation models (GCM) and regional climate models (RCM) (Hawkins and Sutton, 2011). At the global scale, Kharin *et al.* (2007) have projected that extreme temperature and precipitation events will be twice as frequent by the middle of the twenty-first century than they were at the end of the twentieth century.

Several studies have applied statistical and dynamical downscaling to project extreme precipitation in areas of complex topography. Downscaling of GCMs for North America indicates a strong increase in heavy precipitation over the southern and central USA during the second half of the twenty-first century, but a decrease in extreme precipitation over the Canadian prairies (Wang and Zhang, 2008). An increase in precipitation extremes is also indicated for northern and central Europe, with a possible decrease in southern Europe (Beniston *et al.*, 2007; Schmidli *et al.*, 2007). These trends are consistent with the results of RCMs for areas in Europe (Kyselý and Beranová, 2009). Similarly, an increase in extreme precipitation is projected for the period 2071–2100 for most of southeast South America and western Amazonia (Marengo *et al.*, 2009). The seasonal distribution of heavy precipitation can affect slope stability in mountain areas. Snow melt in spring or early summer may enhance the effects of heavy precipitation and destabilize slopes. Several models show a projected increase in extreme precipitation in winter for northern and some parts of central Europe (Christensen *et al.*, 2007; Kyselý and Beranová, 2009), as well as increases in extreme precipitation in winter, spring, and fall for the UK (Buonomo *et al.*, 2007; Fowler and Ekström, 2009).

Periods with temperatures in the uppermost percentiles will be more intense, more frequent, and of longer duration in the future (Meehl *et al.*, 2007). Daily minimum temperatures are projected to increase faster than daily maximum temperatures, which would lead to a decrease in the diurnal temperature range, with consequent effects on weathering processes in mountains. Important progress in regional climate modeling in Europe has been made through the ENSEMBLES project, which used a large number of RCMs with identical boundary conditions over Europe (van der Linden and Mitchell, 2009). The RCMs were run with 25- or 50-km horizontal resolution for the period 1951–2050 (some until 2100), assuming a greenhouse gas emission scenario of rapid economic growth (SRES A1B) (Nakicenovic and Swart, 2000). In one of the few studies focusing on temperature extremes in high mountains, Huggel *et al.* (2010b) used ENSEMBLE climate model data to analyze the frequency of 5-day to 30-day positive temperature anomalies with pronounced melting (above 5°C air temperature) at high-elevation sites. They found that such events will be between 1.5 and 4 times more frequent, and in some models up to 10 times more frequent, by 2050 as compared with 1951–2000. These events have implications for rock slope and glacier stability, because glaciers also produce large amounts of meltwater.

Projections of long-term increases in mean temperature are generally more robust than attempts to forecast future temperature extremes, although in the former case the magnitude of the temperature increase is uncertain because different climate models and emission scenarios are used. In Europe, mean temperatures are likely to increase more than the global average, with warming strongest in winter over northern Europe and in summer over southern Europe (Christensen *et al.*, 2007). Over the Alps, the warming may be +2°C ± 1°C by 2050, and +2.5–3°C ± 1.5°C by 2070, in comparison with 1990 (Organe Consultatif sur les Changements Climatique, 2007). In the Andes, temperatures may rise by 3.5°C ± 1.5°C by 2071–2100 with respect to 1961–1990 (Urrutia and Vuille, 2009). Similar projections have been made for North America, with a mean warming of about 3–4°C by 2080–2099 relative to 1980–1999, but with potential extreme winter warming over Alaska of up to 10°C (Christensen *et al.*, 2007).

The potential effects of warming of this magnitude on landslide activity are difficult to assess, as they involve multiple feedback processes. Some studies have attempted to link semi-empirical and physically based slope stability models with downscaled climate model output (Buma, 2000; Collison *et al.*, 2000; Schmidt and Glade, 2003; Bathurst *et al.*, 2005), but uncertainties about future rainfall intensities (Parry *et al.*, 2007) and site-specific conditions reduce confidence in forecasts of future landslide frequency. Malet *et al.* (2007) highlighted the role of reduced snow cover on slope stability in the French Alps near the end of the twenty-first century, using downscaled climate models and a hydrologic and slope stability model. Jakob and Lambert (2009) projected changes in rainfall in the Coast Mountains of British Columbia using a number of climate models, and applied a statistical rainfall intensity–duration curve for landslide-triggering storms to analyze changes in landslide frequency (Fig. 11.5).

Figure 11.6 shows the possible effects of changes in snowline on slope stability in mountainous terrain, based on an arbitrary scenario around 2050 when glaciers will have retreated and the snowline will have risen by about 300 m vertically. Shallow landslides and debris flows commonly initiate below the seasonal snowline and in glacier forefields (Rickenmann and Zimmermann, 1993; Evans and Clague, 1994; Haeberli and Beniston, 1998) due to abundant and poorly consolidated sediment in proglacial environments, and strong infiltration and potential saturation of sediment by snowmelt. Locally intense rainfall at high elevations, in combination with a high snowline, favors shallow landslides and debris flows (Rickenmann and Zimmermann, 1993; Bezzola and Hegg, 2007; Scheuner *et al.*, 2009).

A possible scenario involving only a change in temperature (scenario A in Fig. 11.7) and a concomitant lengthening of the seasonal period of higher snowline might result in: (1) an extension of the "landslide season," especially in spring; (2) a shift in the peak of landslide activity to early summer; and (3) a slightly higher landslide frequency in winter, depending on snowline and site elevation. The cumulative frequency of landslides could remain about the same, but system feedbacks make projections difficult. For a scenario involving increases in both mean temperature and precipitation intensity (scenario B in Fig. 11.7), one could envision a similar seasonal landslide frequency distribution as in scenario A, but with a higher frequency of landslide occurrence. More reliable future assessments of extreme precipitation variability require improvements in climate models (Kyselý and Beranová, 2009).

In addition to temperature and rainfall intensity, sediment supply and land use are important determinants of landslide frequency and magnitude. Observations at several sites in the Swiss Alps indicate that sediment supply can change significantly due to permafrost thaw in rock and scree slopes. In the Valais in the southern Swiss Alps, rock glaciers delivered up to five times more sediment to debris-flow channels between about 1950 and 1990 than earlier in the century (Lugon and Stoffel, 2010). The average velocity of rock glaciers is typically <1 m per year, but ground-based monitoring and remote sensing since 2002 have revealed an acceleration of rock glaciers to 4 m per year and, in exceptional cases, to as high as 15 m per year (Delaloye *et al.*, 2008; Roer *et al.*, 2008). Similar increases in the rate of movement of some unstable slopes have been reported from various areas of the Alps and are well correlated with increases in mean annual air temperature (Kääb *et al.*, 2007; Roer *et al.*, 2008). As a consequence, debris-flow magnitude and frequency may change in these areas because of increased sediment availability.

A possible model for a warmer future could be the situation at Ritzlihorn–Guttannen in the central Swiss Alps. Strong rockfall activity commenced on the frozen north face of Ritzlihorn in 2009, probably caused by warming and degrading permafrost. Rockfall and avalanche debris accumulates at the apex of the large Holocene debris fan on which the community of

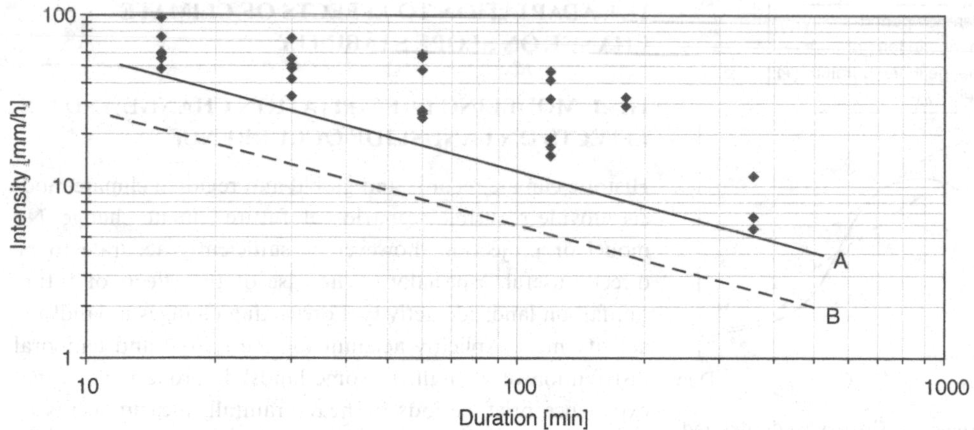

Fig. 11.5. Rainfall intensity–duration plot for a number of landslides in the Cordillera Central in Colombia. Line A is a threshold function for conditions of the past few years. Line B is a scenario threshold function for possible future conditions with more frequent landslides but the same rainfall intensities. Currently it is not clear whether more intense or more frequent rainfall would shift the threshold from A to B (or any similar shift). Other processes may have a more important effect, for example a decrease in vegetation cover.

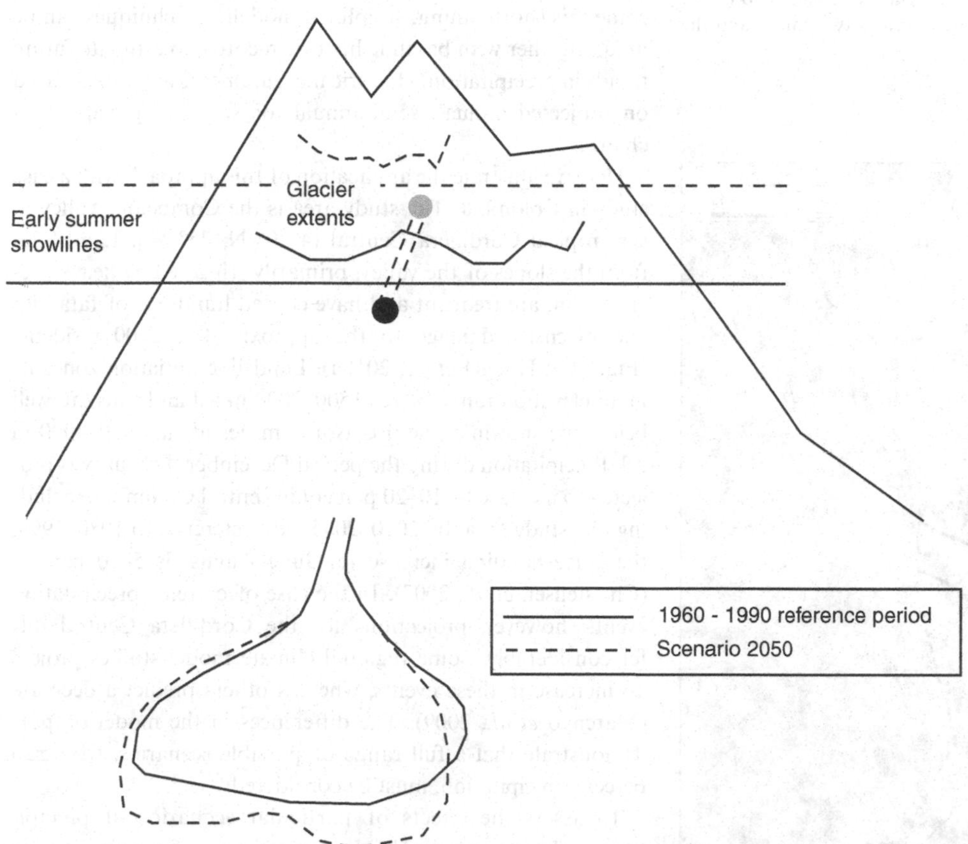

Fig. 11.6. Sketch showing changes in conditions favoring early summer debris flows on a snow- and glacier-clad mountain between a reference period (in this example 1960–1990) and the future (2050). Black and gray dots indicate the points of debris-flow initiation for, respectively, the reference period and the future. The limit of glacier and seasonal snow cover shifts to higher elevations, potentially elevating debris-flow initiation locations. The corresponding increase in potential energy of debris flows may lead to longer runouts, as shown in the sketch.

Guttannen is located (Fig. 11.8). Debris flows have begun to initiate on the apex of the fan during rainstorms, due in part to the saturation of sediments by melting of avalanched snow (D. Tobler, personal communication, 2011). Debris flows entrain material as they move down the channel draining the fan and, on occasion, obstruct the Aare River. 'Just before reaching the river, debris flows cross avalanche sheds that protect an important highway and a transnational gas pipeline. Both of these infrastructure lifelines were interrupted and severely damaged by debris flows in 2009 and 2010. Unprecedented rockfall activity from steep thawing slopes, which in turn trigger equally unprecedented debris flows at Ritzlihorn–Guttannen may be a forerunner of future conditions in similar steep high-mountain watersheds.

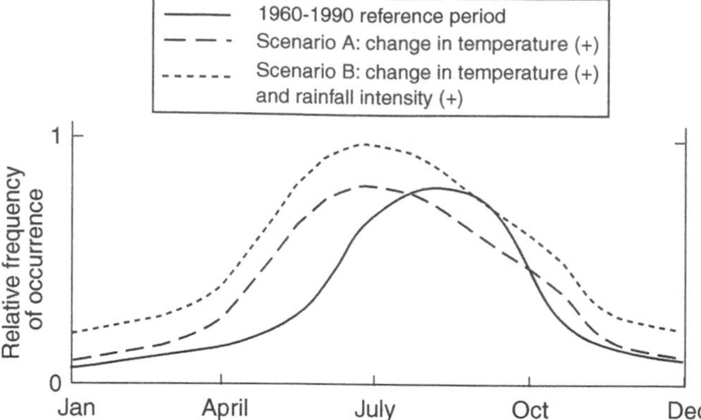

Fig. 11.7. Relative frequency of occurrence of shallow landslides and debris flows in mountain areas with seasonal snow cover. For the reference period (1960–1990, solid line), landslides are most frequent in early summer when the snowline recedes and meltwater runoff is high. In Scenario A (long dashes), with a temperature increase, snowmelt begins earlier in the year and, accordingly, shallow landslides and debris flows are more likely to occur during the spring. In addition, the landslide season is extended due to later onset of snowfall. In Scenario B (short dashes), warming is accompanied by higher intensity rainfall, and the frequency of landslides increases.

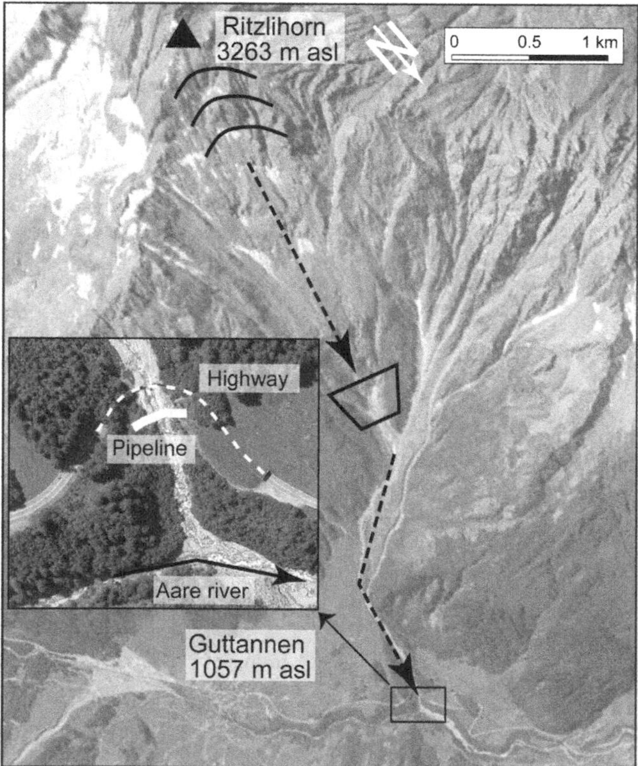

Fig. 11.8. The situation at Ritzlihorn–Guttannen in the central Swiss Alps. Half circles indicate source zones of rockfall from permafrost areas on the north face of Ritzlihorn. Rockfall (upper dashed arrow) accumulates at the apex of the Holocene fan and is remobilized by debris flows (lower dashed arrow) that flow across the fan to the Aare River. The debris flows cross a highway gallery and a transnational gas pipeline (inset). (Imagery retrieved from Google Earth).

11.4 ADAPTATION TO EFFECTS OF CLIMATE CHANGE ON SLOPE STABILITY

11.4.1 MODELING PRECIPITATION CHANGE AND ITS EFFECTS ON LANDSLIDE OCCURRENCE

Historic climate records and global and regional climate models provide plausible scenarios of future climate change. No model or projection, however, is sufficiently accurate to be directly useful, especially in the case of the effects of future rainfall on landslide activity. Forecasting changes in landslide activity must explicitly account for the spatial and temporal distribution of rainfall. In some landslide-prone regions that experience brief periods of heavy rainfall, measurements of rainfall intensity over 10–15 minutes are required to derive adequate threshold values for triggering landslides. Generally, such detail is either unavailable or not sufficiently reliable for long-term projections, even when boundary conditions established by a GCM are refined in a regional climate model that is specifically calibrated for a particular region or area. To overcome this shortcoming, simplified modeling techniques can be used, together with baseline historic records, to estimate future trends in precipitation. Historic measurements are scaled based on projected annual, semi-annual, or seasonal precipitation change.

Here we illustrate the application of this approach with a case study in Colombia. The study area is the Combeima Valley in the tropical Cordillera Central (4°30′ N, 75° W). Landslides from the slopes of the valley, primarily triggered by heavy precipitation, are frequent and have caused hundreds of fatalities and extensive damage to the approximately 5500 residents (Fig. 11.1; Huggel *et al.*, 2010a). Landslide initiation zones lie in an elevation range of *ca.* 1500–3000 m asl and thus are well below the snowline: the 0°C-isotherm lies at *ca.* 4800–4900 m asl. Precipitation during the period December–February is projected to increase by 10–20 percent in central Colombia, including our study area, by 2090–2095 with reference to 1980–1999; the corresponding increase for June–August is 5–10 percent (Christensen *et al.*, 2007). In the case of extreme precipitation events, however, projections for the Cordillera Central differ considerably. Some regional climate model studies project an increase in these events, whereas others predict a decrease (Marengo *et al.*, 2009). The differences in the model outputs demonstrate that a full range of possible scenarios, with ±20 percent precipitation, must be considered.

To assess the effects of particular scenarios of precipitation change on landslide occurrence and damage, we applied a recently developed landslide early-warning system (LS-EWS) model (Huggel *et al.*, 2010a). For the baseline, we used ERA-40 reanalysis precipitation data with 6-hour resolution (Uppala *et al.*, 2005) for the period 1991–2000 from the grid cell closest to the Combeima Valley. The model applies an empirical rainfall intensity–duration threshold to simulate landslide occurrence and subsequent evacuation when a landslide is simulated. Monte Carlo simulations were used to model a rainfall measurement error. The error, in turn, translates into a range of evacuation and damage scenarios within a LS-EWS scheme derived from local conditions in the Combeima Valley (see Huggel *et al.*, 2010a, for a more detailed description of the model). Precipitation change scenarios were run on the ERA-40 reference data. A single value for the precipitation change is likely to produce a biased damage assessment, because it does not provide information on the change in magnitude and frequency of heavy rainfalls. Thus, we defined scenarios with emphasis on heavy rainfalls, together with simple scenarios where all the rainfall intensities were multiplied by the same scaling coefficient, independent of the intensity.

We established a scaling parameter δ and compared two scenarios: one in which all rainfalls were scaled equally by a parameter (1 + δ); and the other in which heavy rainfalls were scaled by one parameter (1 + δ) and lighter rainfalls were scaled by another parameter (1 − δ), so that heavy rainfalls have even higher intensity and light rainfalls are even less intense, with the total amount of yearly precipitation remaining constant. For our dataset, the two scenarios deliver the same results in terms of the number of landslides and expected damage (Fig. 11.9). This outcome is not surprising because landslides are triggered by extreme precipitation and both scenarios modify extreme rainfall in the initial dataset identically: a relatively small positive value of δ does not, in either case, transform small rainfall events into large events. The model is dimensionless and spatially not explicit, such that atmospheric moisture dynamics do not affect model performance.

11.4.2 ADAPTATION OF A LANDSLIDE EARLY-WARNING SYSTEM TO CHANGING PRECIPITATION

We applied the LS-EWS model to assess damage under different precipitation scenarios. The purpose of an EWS is to issue reliable warnings to allow people in a hazardous area to evacuate to safety without concern for damage to buildings. To evaluate EWS performance in a changing climate, it is necessary to discriminate between losses that can be prevented by such a system and those that cannot. For that reason, the losses estimated here, including those shown in Figure 11.9, are restricted to evacuation costs, injuries, and fatalities, and exclude damage to infrastructure. The capacity of an EWS to adjust to changes in precipitation has internal and external elements. Internal elements involve rules relating to warning and evacuation, for example the circumstances that lead to issuing of an evacuation order. External elements involve changes in the key functional components of the EWS, for example improvements in spatio-temporal resolution of rainfall measurements. Figure 11.10 illustrates the internal and external adaptation capacities of the EWS in our case study. The negative values of δ represent a reduction in precipitation and lead to a smaller number of landslides, and consequently reduced losses (down to zero for δ = −0.2; Fig. 11.10, panel (a)). For a fixed rainfall measurement error (RME), only the evacuation threshold is adjusted in

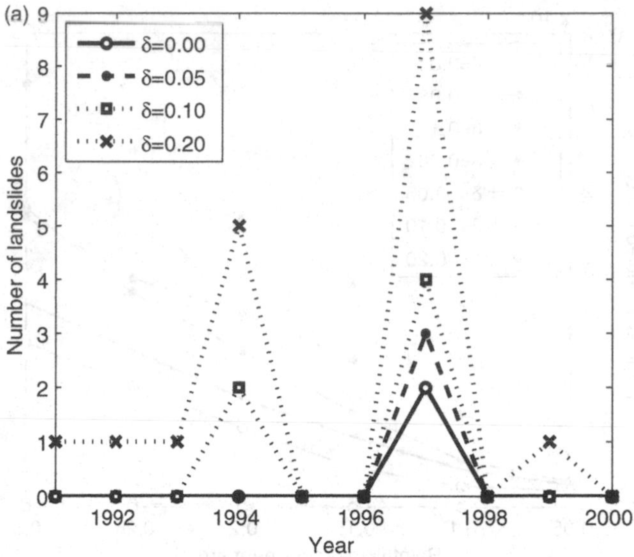

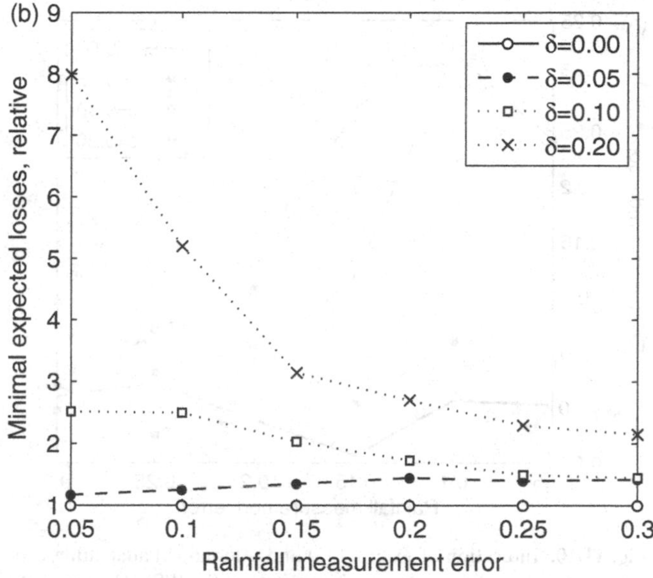

Fig. 11.9. (a) Modeled number of landslides for four precipitation scenarios. All rainfall intensities were scaled equally by (1 + δ), independently of the initial rainfall magnitude. (b) Expected damage (evacuation costs and fatalities) caused by landslides for different changes in precipitation, δ, relative to a baseline.

response to a change in precipitation. The sensitivity of the relevant losses to the evacuation threshold for the EWS is shown in Figure 11.10 (panel (b)). The relative loss reduction of the adjusted EWS versus the unadjusted EWS is expressed as:

$$(L_u - L_a) / L_u \quad (11.1)$$

where L_u are expected losses for unadjusted EWS and L_a are those for adjusted EWS. The relative loss reduction is moderate to fairly high (10–25 percent) for a combination of moderate precipitation change (δ = 0.05, δ = 0.1) and a high-precision EWS (RME <0.15). For a less precise EWS (RME >0.2), the efficiency gain from the adjusted evacuation threshold is

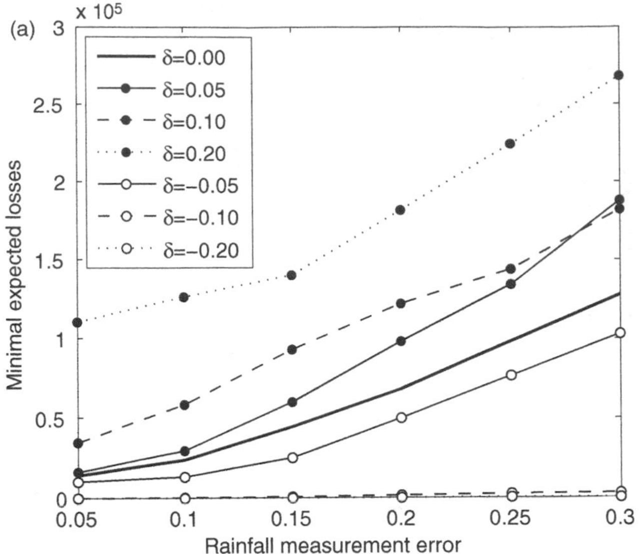

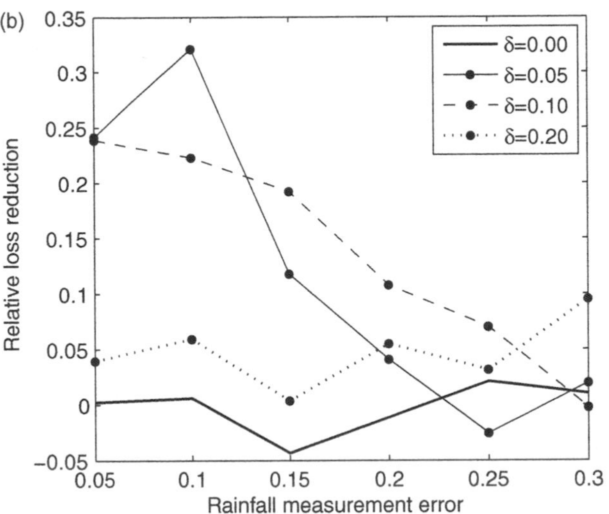

Fig. 11.10. Indication of external (a) and internal (b) adaptation capacities of a landslide early-warning system (LS-EWS). Here external adaptation capacity refers to the reduction of the rainfall measurement error (RME). The internal adaptation capacity refers to the evacuation threshold adjustment for a fixed RME.

negligible and comparable to the numerical variations of *ca.* 5 percent that are inherent in the Monte Carlo simulation procedures of the model. The efficiency gain is also *ca.* 5 percent for a larger increase in precipitation (δ = 0.2) for all RMEs, indicating little or no sensitivity of the EWS to the evacuation threshold adjustment for high precipitation changes.

Figure 11.10 (panel (a)) shows the potential and limitations of RME improvement for EWS adaptation purposes. For a baseline scenario under present conditions (δ = 0) with a high rainfall measurement error (RME = 0.3), losses can be slightly reduced with substantially improved rainfall monitoring capabilities (RME = 0.05) for a climate change scenario with strong precipitation changes (δ = 0.2). Our modeling results indicate that a reduction in RME, representing improved external adaptation capacity, has more potential for successfully improving an EWS than an adjustment in the evacuation threshold alone. Figure 11.10 (panel (a)) further shows that the relative loss reduction of an RME-improved EWS (RME = 0.05) over the initial EWS (RME = 0.3), even for an extreme precipitation increase (δ = 0.2), may be as large as (2.7 − 1.1)/2.7 ≈ 60 percent, which is substantially larger than the loss reduction achieved with an adjustment in the threshold alone (Figure 11.10, panel (b)).

Several considerations of LS-EWS in general, and this approach of assessing EWS adaptation capacity in particular, should be taken into account. Increased precipitation will most likely result in a greater number of landslides under conditions that are otherwise unchanged. Structural mitigation measures such as dams, or passive avoidance such as relocation of endangered assets, including population, may be alternative or complementary options to EWS. Furthermore, the assumption that the system is stationary – for example, the triggering threshold does not change – is an oversimplification. A major difficulty with a LS-EWS is that rainfall events with intensity–duration relations exceeding the threshold may not trigger landslides, but do require an alarm to be issued. Recent studies using different rainfall exceedance probabilities are promising but do not fully address the problem of non-landslide rainfall events (Brunetti *et al.*, 2010). More sophisticated soil mechanics models may be difficult to implement in an operational LS-EWS, given the high spatial variability of soil and rainfall conditions. Finally, human response to warnings remains a critical issue, and the willingness of people to evacuate may decrease after several false alarms (Dash and Gladwin, 2007). Communication about the early-warning procedures and uncertainties is crucial (Dow and Cutter, 1998), although clearly more research is needed to reliably incorporate such aspects into a numerical model.

11.5 CONCLUSIONS

Different types of landslides respond differently to changes in precipitation and temperature. Despite substantial growth in the amount of data relevant to this issue, it is unclear whether an increase in extreme rainfall events is likely to increase the occurrence of shallow landslides and debris flows, assuming otherwise constant conditions. Even though extreme precipitation events have become more frequent in many regions of the world over the past decades, changes in landslide magnitude–frequency distributions have not yet been conclusively demonstrated. Reasons include incomplete and biased documentation of events, which limit the extraction of climatic signals, and overprinting by land-use changes. In some high mountain regions such as the central European Alps, observed large rock-slope failures have increased during the past 100 years. Although the mechanisms are not completely understood, glacier shrinkage and permafrost degradation are likely to have had a considerable influence.

Reliable forecasts of future climatically driven changes in landslide activity must overcome a number of challenges. Hydrologic and slope stability modeling demonstrates the strong influence of local site conditions, resulting in scenarios in which landslides may become more or less frequent. Most climate models, however, indicate an increase in the frequency of extreme precipitation and temperature events during the twenty-first century. It is likely that landslide frequency and magnitude will be influenced by such trends, especially in high mountains, where changes in sediment supply can be significant.

The need for landslide early-warning systems is rising. Changing rainfall patterns, however, pose an additional challenge to the development of robust and reliable warning systems. Studies indicate that damage from landslides may rise considerably with higher rainfall in the future. Adaptation involving improved rainfall observations may be a better option than internally adjusting evacuation procedures, such as changing the evacuation thresholds for rainfall. On a more general level, our study supports the call for improved environmental monitoring as one necessary and effective adaptation strategy for future climate change.

ACKNOWLEDGEMENTS

The authors thank the editors – John J. Clague and Douglas Stead, and reviewer Philip Deline for their constructive comments that helped to improve the manuscript.

REFERENCES

Bathurst, J.C., Moretti, G., El-Hames, A., Moaven-Hashemi, A. and Burton, A. (2005). Scenario modelling of basin-scale, shallow landslide sediment yield, Valsassina, Italian Southern Alps. *Natural Hazards and Earth System Sciences*, 5, 189–202.

Beniston, M., Stephenson, D.B., Christensen, O.B. *et al.* (2007). Future extreme events in European climate: An exploration of regional climate model projections. *Climatic Change*, 81, 71–95.

Bezzola, G.R. and Hegg, C. (2007). *Ereignisanalyse Hochwasser 2005, Teil 1: Prozesse, Schäden und erste Einordnung*. Bundesamt für Umwelt BAFU, Eidgenössische Forschungsanstalt WSL. Umwelt-Wissen, 707.

Brunetti, M.T., Peruccacci, S., Rossi, M. *et al.* (2010). Rainfall thresholds for the possible occurrence of landslides in Italy. *Natural Hazards and Earth System Sciences*, 10, 447–458.

Buma, J. (2000). Finding the most suitable slope stability model for the assessment of the impact of climate change on a landslide in southeast France. *Earth Surface Processes and Landforms*, 25, 565–582.

Buonomo, E., Jones, R., Huntingford, C. and Hannaford, J. (2007). On the robustness of changes in extreme precipitation over Europe from two high resolution climate change simulations. *Quarterly Journal of the Royal Meteorological Society*, 133, 65–81.

Caine, N. (1980). The rainfall intensity: Duration control of shallow landslides and debris flows. *Geografiska Annaler Series A: Physical Geography*, 62, 23–27.

Choi, G., Collins, D., Ren, G. *et al.* (2009). Changes in means and extreme events of temperature and precipitation in the Asia-Pacific Network region, 1955–2007. *International Journal of Climatology*, 29, 1906–1925.

Christensen, J., Hewitson, B., Busuioc, A. *et al.* (2007). Regional climate projections. In *Climate Change 2007: The Physical Science Basis. Contribution of Working Group I to the Fourth Assessment Report of the Intergovernmental Panel on Climate Change*, ed. S. Solomon, D. Qin, M. Manning *et al.* Cambridge and New York: Cambridge University Press, pp. 847–940.

Collison, A., Wade, S., Griffiths, J. and Dehn, M. (2000). Modelling the impact of predicted climate change on landslide frequency and magnitude in SE England. *Engineering Geology*, 55, 205–218.

Dahal, R.K. and Hasegawa, S. (2008). Representative rainfall thresholds for landslides in the Nepal Himalaya. *Geomorphology*, 100, 429–443.

Dash, N. and Gladwin, H. (2007). Evacuation decision making and behavioral responses: Individual and household. *Natural Hazards Review*, 8, 69–77.

Delaloye, R., Strozzi, T., Lambiel, C., Perruchoud, E. and Raetzo, H. (2008). Landslide-like development of rockglaciers detected with ERS-1/2 SAR interferometry. In *Proceedings of the ESA FRINGE Symposium 2007, Frascati, Italy*. ESA SP-649.

Della-Marta, P.M., Haylock, M.R., Luterbacher, J. and Wanner, H. (2007). Doubled length of Western European summer heat waves since 1880. *Journal of Geophysical Research – Atmospheres*, 112, D15103.

Ding, T., Qian, W. and Yan, Z. (2007). Changes in hot days and heat waves in China during 1961–2007. *International Journal of Climatology*, 30, 1452–1462.

Dow, K. and Cutter, S.L. (1998). Crying wolf: Repeat responses to hurricane evacuation orders. *Coastal Management*, 26, 237–252.

Evans, S.G. and Clague, J.J. (1994). Recent climatic change and catastrophic geomorphic processes in mountain environments. *Geomorphology*, 10, 107–128.

Fischer, L. (2010). *Slope instabilities on perennially frozen and glacierized rock walls: Multi-scale observations, analysis and modelling*. Ph.D. thesis, University of Zurich, Switzerland.

Fischer, L., Amann, F., Moore, J. and Huggel, C. (2010). Assessment of periglacial slope stability for the 1988 Tschierva rock avalanche (Piz Morteratsch, Switzerland). *Engineering Geology*, 116, 32–43.

Fowler, H.J. and Ekström, M. (2009). Multi-model ensemble estimates of climate change impacts on UK seasonal precipitation extremes. *International Journal of Climatology*, 29, 385–416.

Geertsema, M., Clague, J.J., Schwab, J.W. and Evans, S.G. (2006). An overview of recent large catastrophic landslides in northern British Columbia, Canada. *Engineering Geology*, 83, 120–143.

Glade, T. (1998). Establishing the frequency and magnitude of landslide-triggering rainstorm events in New Zealand. *Environmental Geology*, 35, 160–174.

Gruber, S. and Haeberli, W. (2007). Permafrost in steep bedrock slopes and its temperature-related destabilization following climate change. *Journal of Geophysical Research*, 112, F02S18.

Guzzetti, F. (2000). Landslide fatalities and the evaluation of landslide risk in Italy. *Engineering Geology*, 58, 89–107.

Guzzetti, F., Peruccacci, S., Rossi, M. and Stark, C.P. (2007). Rainfall thresholds for the initiation of landslides in central and southern Europe. *Meteorology and Atmospheric Physics*, 98, 239–267.

Haeberli, W. and Beniston, M. (1998). Climate change and its impacts on glaciers and permafrost in the Alps. *Ambio*, 27, 258–265.

Harp, E.L., Wells, W.G. and Sarmiento, J.G. (1990). Pore pressure response during failure in soils. *Geological Society of America Bulletin*, 102, 428–438.

Hasler, A. (2011). *Thermal conditions and kinematics of steep bedrock permafrost*. Ph.D. thesis, University of Zurich, Switzerland.

Hawkins, E. and Sutton, R.T. (2011). The potential to narrow uncertainty in projections of regional precipitation change. *Climate Dynamics*, 37, 407–418.

Hewitt, K., Clague, J.J. and Orwin, J.F. (2008). Legacies of catastrophic rock slope failures in mountain landscapes. *Earth-Science Reviews*, 87, 1–38.

Holm, K., Bovis, M. and Jakob, M. (2004). The landslide response of alpine basins to post-Little Ice Age glacial thinning and retreat in southwestern British Columbia. *Geomorphology*, 57, 201–216.

Huggel, C., Caplan-Auerbach, J. and Wessels, R. (2008). Recent extreme avalanches: Triggered by climate change? *EOS, Transactions of the American Geophysical Union*, 89, 469–470.

Huggel, C., Khabarov, N., Obersteiner, M. and Ramírez, J. (2010a). Implementation and integrated numerical modeling of a landslide early warning system: A pilot study in Colombia. *Natural Hazards*, 52, 501–518.

Huggel, C., Salzmann, N., Allen, S. *et al.* (2010b). Recent and future warm extreme events and high-mountain slope stability. *Philosophical Transactions of the Royal Society A*, 368, 2435–2459.

ISDR (International Strategy for Disaster Reduction) (2009). *Global Assessment Report on Disaster Risk Reduction: Risk and Poverty in a Changing Climate*. Geneva, Switzerland: United Nations.

Iverson, R.M. (2000). Landslide triggering by rain infiltration. *Water Resources Research*, 36, 1897–1910.

Iverson, R.M. and LaHusen, R.G. (1989). Dynamic pore-pressure fluctuations in rapidly shearing granular materials. *Science*, 246, 796–799.

Jakob, M. and Lambert, S. (2009). Climate change effects on landslides along the southwest coast of British Columbia. *Geomorphology*, 107, 275–284.

Jakob, M. and Weatherly, H. (2003). A hydroclimatic threshold for landslide initiation on the North Shore Mountains of Vancouver, British Columbia. *Geomorphology*, 54, 137–156.

Jomelli, V., Pech, V.P., Chochillon, C. and Brunstein, D. (2004). Geomorphic variations of debris flows and recent climatic change in the French Alps. *Climatic Change*, 64, 77–102.

Kääb, A., Frauenfelder, R. and Roer, I. (2007). On the response of rockglacier creep to surface temperature increase. *Global and Planetary Change*, 56, 172–187.

Keefer, D.K., Wilson, R.C., Mark, R.K. *et al.* (1987). Real-time landslide warning during heavy rainfall. *Science*, 238, 921–925.

Kharin, V.V., Zwiers, F.W., Zhang, X. and Hegerl, G.C. (2007). Changes in temperature and precipitation extremes in the IPCC ensemble of global coupled model simulations. *Journal of Climate*, 20, 1419–1444.

Kim, H.J., Sidle, R.C., Moore, R.D. and Hudson, R. (2004). Throughflow variability during snowmelt in a forested mountain catchment, coastal British Columbia, Canada. *Hydrological Processes*, 18, 1219–1236.

Kim, S.K., Hong, W.P. and Kim, Y.M. (1991). Prediction of rainfall-triggered landslides in Korea. In *Proceedings of the 6th International Symposium on Landslides*, ed. D.H. Bell. Rotterdam: Balkema, pp. 989–994.

Krähenbühl, R. (2004). Temperatur und Kluftwasser als Ursachen von Felssturz. *Bulletin für Angewandte Geologie*, 9, 19–35.

Krishnamurthy, C.K., Lall, U. and Kwon, H.H. (2009). Changing frequency and intensity of rainfall extremes over India from 1951 to 2003. *Journal of Climate*, 22, 4737–4746.

Kunkel, K.E., Bromirski, P.D., Brooks, H.E. *et al.* (2008). Observed changes in weather and climate extremes. In *Weather and Climate Extremes in a Changing Climate. Regions of Focus: North America, Hawaii, Caribbean, and US Pacific Islands*, ed. T.S. Karl *et al.* Washington, DC: US Climate Change Science Program and the Subcommittee on Global Change Research, pp. 35–80.

Kyselý, J. (2009). Trends in heavy precipitation in the Czech Republic over 1961–2005. *International Journal of Climatology*, 29, 1745–1758.

Kyselý, J. and Beranová, R. (2009). Climate-change effects on extreme precipitation in central Europe: Uncertainties of scenarios based on regional climate models. *Theoretical and Applied Climatology*, 95, 361–374.

Larsen, M.C. and Simon, A. (1993). A rainfall intensity-duration threshold for landslides in a humid-tropical environment, Puerto Rico. *Geografiska Annaler, Series A: Physical Geography*, 75A, 13–23.

Lateltin, O., Haemmig, C., Raetzo, H. and Bonnard, C. (2005). Landslide risk management in Switzerland. *Landslides*, 2, 313–320.

Lugon, R. and Stoffel, M. (2010). Rock-glacier dynamics and magnitude–frequency relations of debris flows in a high-elevation watershed: Ritigraben, Swiss Alps. *Global and Planetary Change*, 73, 202–210.

Malet, J., Remaître, A., Maquaire, O. *et al.* (2007). Assessing the influence of climate change on the activity of landslides in the Ubaye Valley. In *Proceedings of the International Conference on Landslides and Climate Change: Challenges and Solutions*, ed. R. McInnes, J. Jakeways, H. Fairbank and E. Mathie. London: Taylor & Francis, pp. 195–205.

Marengo, J.A., Jones, R., Alves, L.M. and Valverde, M.C. (2009). Future change of temperature and precipitation extremes in South America as derived from the PRECIS regional climate modeling system. *International Journal of Climatology*, 29, 2241–2255.

Marengo, J., Rusticucci, M., Penalba, O. and Renom, M. (2010). An intercomparison of observed and simulated extreme rainfall and temperature events during the last half of the twentieth century. Part 2. Historical trends. *Climatic Change*, 98, 509–529.

Meehl, G., Stocker, T., Collins, W. *et al.* (2007). Global climate projections. In *Climate Change 2007: The Physical Science Basis. Contribution of Working Group I to the Fourth Assessment Report of the Intergovernmental Panel on Climate Change*, ed. S. Solomon, D. Qin, M. Manning *et al.* Cambridge and New York: Cambridge University Press, pp. 747–845.

Montgomery, D.R., Dietrich, W.E., Torres, R. *et al.* (1997). Hydrologic response of a steep, unchanneled valley to natural and applied rainfall. *Water Resources Research*, 33, 91–109.

Montgomery, D.R., Dietrich, W.E. and Heffner, J.T. (2002). Piezometric response in shallow bedrock at CB1: Implications for runoff generation and landsliding. *Water Resources Research*, 38, 1274, doi:10.1029/2002WR001429.

Nakicenovic, N. and Swart, R. (2000). *Special Report on Emissions Scenarios: A Special Report of Working Group III of the Intergovernmental Panel on Climate Change*. Cambridge, UK: Cambridge University Press.

Organe Consultatif sur les Changements Climatique (2007). *Klimaänderung und die Schweiz 2050*. Bern: OcCC and ProClim.

Parry, M.L., Canziani, O.F., Palutikof, J.F., van der Linden, P.J. and Hanson, C.E. (ed.) (2007). *Climate Change 2007: Impacts, Adaptation and Vulnerability. Contribution of Working Group II to the Fourth Assessment Report of the Intergovernmental Panel on Climate Change*. Cambridge, UK: Cambridge University Press.

Paul, F. and Haeberli, W. (2008). Spatial variability of glacier elevation changes in the Swiss Alps obtained from two digital elevation models. *Geophysical Research Letters*, 35, L21502.

Pavan, V., Tomozeiu, R., Cacciamani, C. and Di Lorenzo, M. (2008). Daily precipitation observations over Emilia-Romagna: Mean values and extremes. *International Journal of Climatology*, 28, 2065–2079.

Pelfini, M. and Santilli, M. (2008). Frequency of debris flows and their relation with precipitation: A case study in the Central Alps, Italy. *Geomorphology*, 101, 721–730.

Petley, D.N., Dunning, S.A. and Rosser, N.J. (2005). The analysis of global landslide risk through the creation of a database of worldwide landslide fatalities. In *Landslide Risk Management*, ed. O. Hungr, R. Fell, R. Couture and E. Eberhardt. Amsterdam: A.A. Balkema, pp. 367–374.

Petley, D.N., Hearn, G., Hart, A. *et al.* (2007). Trends in landslide occurrence in Nepal. *Natural Hazards*, 43, 23–44.

Pierson, T.C. (1980). Piezometric response to rainstorms in forested hillslope drainage depressions. *Journal of Hydrology New Zealand*, 19, 1–10.

Pryor, S.C., Howe, J.A. and Kunkel, K.E. (2009). How spatially coherent and statistically robust are temporal changes in extreme precipitation in the contiguous USA? *International Journal of Climatology*, 29, 31–45.

Ravanel, L. and Deline, P. (2011). Climate influence on rockfalls in high-alpine steep rockwalls: The north side of the Aiguilles de Chamonix (Mont Blanc massif) since the end of the 'Little Ice Age'. *Holocene*, 21, 357–365.

Rickenmann, D. and Zimmermann, M. (1993). The 1987 debris flows in Switzerland: Documentation and analysis. *Geomorphology*, 8, 175–189.

Rist, A. and Phillips, M. (2005). First results of investigations on hydrothermal processes within the active layer above alpine permafrost in steep terrain. *Norsk Geografisk Tidsskrift*, 59, 177–183.

Roer, I., Haeberli, W., Avian, M. *et al.* (2008). Observations and considerations on destabilizing active rock glaciers in the European Alps. In *Proceedings of the 9th International Conference on Permafrost*, ed. D.L. Kane and K.M. Hinkel. Fairbanks, AK: University of Alaska, pp. 1505–1510.

Scheuner, T., Keusen, H., McArdell, B. and Huggel, C. (2009). Murgangmodellierung mit dynamisch-physikalischem und GIS-basiertem Fliessmodell. *Wasser Energie Luft*, 101, 15–21.

Schmidli, J., Goodess, C.M., Frei, C. *et al.* (2007). Statistical and dynamical downscaling of precipitation: An evaluation and comparison of scenarios for the European Alps. *Journal of Geophysical Research*, 112, D04105.

Schmidt, M. and Glade, T. (2003). Linking global circulation model outputs to regional geomorphic models: A case study of landslide activity in New Zealand. *Climate Research*, 25, 135–150.

Schuster, R.L. and Highland, L. (2001). *Socioeconomic and Environmental Landslides in the Western Hemisphere*. US Geological Survey, Open File Report 01-276.

Sidle, R.C. and Ochiai, H. (2006). *Landslides: Processes, Prediction, and Land Use*. Washington, DC: American Geophysical Union.

Simoni, A., Berti, M., Generali, M., Elmi, C. and Ghirotti, M. (2004). Preliminary result from pore pressure monitoring on an unstable clay slope. *Engineering Geology*, 73, 117–128.

Stoffel, M., Lièvre, I., Conus, D. *et al.* (2005). 400 years of debris-flow activity and triggering weather conditions: Ritigraben, Valais, Switzerland. *Arctic, Antarctic, and Alpine Research*, 37, 387–395.

Trenberth, K., Jones, P., Ambenje, P. *et al.* (2007). Observations: Surface and atmospheric climate change. In *Climate Change 2007: The Physical Science Basis. Contribution of Working Group I to the Fourth Assessment Report of the Intergovernmental Panel on Climate Change*, ed. S. Solomon, D. Qin, M. Manning *et al.* Cambridge and New York: Cambridge University Press, pp. 235–336.

Uchida, T., Kosugi, K. and Mizuyama, T. (2001). Effects of pipeflow on hydrological process and its relation to landslide: A review of pipeflow studies in forested headwater catchments. *Hydrological Processes*, 15, 2151–2174.

Uppala, S., Kallberg, P., Simmons, A. *et al.* (2005). The ERA-40 re-analysis. *Quarterly Journal of the Royal Meteorological Society*, 131, 2961–3012.

Urrutia, R. and Vuille, M. (2009). Climate change projections for the tropical Andes using a regional climate model: Temperature and precipitation simulations for the end of the 21st century. *Journal of Geophysical Research*, 114, D02108.

van der Linden, P. and Mitchell, J.F.B. (2009). *ENSEMBLES: Climate Change and Its Impacts: Summary of Research and Results from the ENSEMBLES Project*. Exeter, UK: Met Office Hadley Centre.

Wang, J. and Zhang, X. (2008). Downscaling and projection of winter extreme daily precipitation over North America. *Journal of Climate*, 21, 923–937.

Watson, A.D., Moore, D.P. and Stewart, T.W. (2004). Temperature influence on rock slope movements at Checkerboard Creek. In *Proceedings of the 9th International Symposium on Landslides. Rio de Janeiro*. Rotterdam: Balkema, pp. 1293–1298.

Wegmann, M. and Gudmundsson, G.H. (1999). Thermally induced temporal strain variations in rock walls observed at subzero temperatures. In *Advances in Cold-Region Thermal Engineering and Sciences*. Lecture Notes in Physics. Berlin/Heidelberg: Springer, pp. 511–518.

Wieczorek, G.F. and Glade, T. (2005). Climatic factors influencing occurrence of debris flows. In *Debris-flow Hazards and Related Phenomena*, ed. M. Jakob and O. Hungr. Berlin: Springer Praxis, pp. 325–362.

Willenberg, H., Evans, K.F., Eberhardt, E., Spillmann, T. and Loew, S. (2008). Internal structure and deformation of an unstable crystalline rock mass above Randa (Switzerland). Part II. Three-dimensional deformation patterns. *Engineering Geology*, 101, 15–32.

Zimmermann, M. (1990). Debris flows 1987 in Switzerland: Geomorphological and meteorological aspects. In *Hydrology of Mountainous Regions. II. Artificial Reservoirs, Water and Slopes*. International Association of Hydrological Sciences, Publication 194, pp. 386–393.

Zimmermann, M. and Haeberli, W. (1992). Climatic change and debris flow activity in high-mountain areas: A case study in the Swiss Alps. *Catena Supplement*, 22, 59–70.

Zimmermann, M., Mani, P. and Romang, H. (1997). Magnitude–frequency aspects of alpine debris flows. *Eclogae Geologicae Helvetiae*, 90, 415–420.

12 Landslides and geologic environments

ROBIN FELL, DAVID STAPLEDON, AND PATRICK MACGREGOR

ABSTRACT

The geologic environment has a major influence on the likelihood, classification, and mechanisms of landsliding, the hydrogeology as it affects landsliding, and the strength of potential rupture surfaces in rock and soil. The geologic environment includes the type of rock (igneous, sedimentary, or metamorphic), stratigraphy, structure (bedding, folding, and faulting), defects (joints and sheared and crush zones), and weathering. The soils developed in these environments and their susceptibility to landsliding are related to the underlying rocks. Other factors that affect the likelihood, classification, and magnitude of landsliding include topography, climate, seismicity, and geologic history – for example whether the area has been glaciated.

12.1 INTRODUCTION

This chapter discusses some common geologic environments and their influence on the classification and mechanisms of landslides, hydrogeology as it affects landsliding, and the strength of potential rupture surfaces in rock and soil. It is based on the authors' experience and is not meant to be an exhaustive coverage of the topic. More details on some aspects are provided by Fell (1992), Fell *et al.* (1992, 2005), and Stapledon (1995).

12.2 THE INFLUENCE OF ROCK TYPE, STRATIGRAPHY, WEATHERING, AND STRUCTURE ON LANDSLIDING

12.2.1 GRANITIC ROCKS

Included under this heading are granite and other medium- to coarse-grained igneous rocks. In fresh unweathered exposures, granitic rocks are generally highly durable, strong to extremely strong, and contain widely spaced (>2 m) joints in a roughly rectangular pattern. Sheet joints are common, due to stress relief resulting from erosion of the overlying rock. The joints are typically parallel to the ground surface and are preferentially weathered, forming exfoliating sheets parallel to the slope.

Fig. 12.1. Granite core stones below granitic soil on Granite Island, Victor Harbor, South Australia (Fell *et al.*, 2005).

Chemical weathering is initiated at, and proceeds from, the ground surface and from sheet joints, tectonic joints, and faults, ultimately reducing the size of roughly rectangular joint blocks to smaller, more rounded forms that are increasingly separated by weathered sandy material (Figs. 12.1, 12.2, and 12.3). The profile grades from residual granitic soil near the surface down to fresh rock at depth, with residual "boulders" of fresh or partly weathered rock at any level. Fresh outcrops or large fresh boulders at the ground surface may or may not be underlain

Landslides: Types, Mechanisms and Modeling, ed. John J. Clague and Douglas Stead. Published by Cambridge University Press.

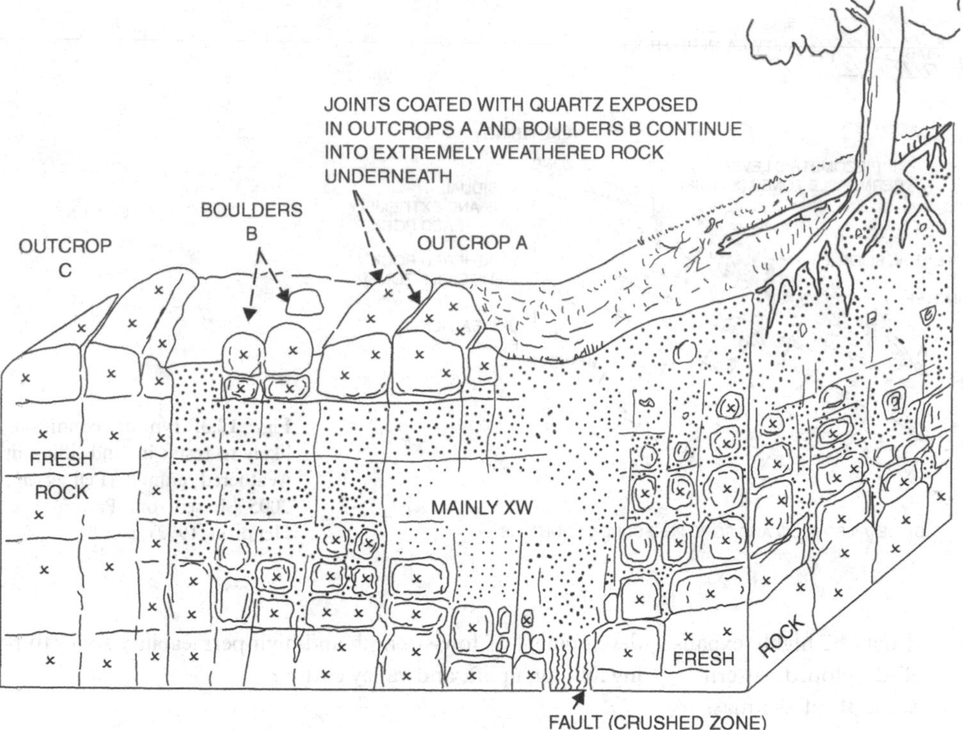

Fig. 12.2. Features characteristic of weathered masses of granitic and other igneous rocks (Fell *et al.*, 2005).

by fresh rock. It is not uncommon to find that weathering has occurred beneath such outcrops, along sheet joints or gently dipping tectonic joints, or within previously altered granitic rock. Understanding this potential variability in weathered granite profiles is important when considering the stability of slopes in these materials. Although some landslides in weathered granitic rocks undoubtedly occur by failure through the fabric of extremely weathered material, failure in many cases is localized wholly or partly along relict joints or other defects in the weathered rock.

Figure 12.3 shows typical conditions that may lead to landsliding in granitic environments. Wedge ABD is bounded by relict joints in the low-permeability zone of a residual soil and extremely weathered (XW) rock. It may be subject to excess water pressures from the underlying more permeable zone, causing it to fail and slide downslope. A larger, deeper slide is possible if a continuous thin seam of extremely weathered granite parallel to surface AB is present within the rockier zone IIA and daylights at the lower line of springs S (Fig. 12.3). The contrast in permeability between the upper, more-weathered, low-permeability soil and XW rock to the underlying, less-weathered rock with open joints is critical to understanding the hydrogeology and pore pressures in the slope.

Active landsliding, or evidence of past landsliding, is relatively common on steep slopes underlain by weathered granitic rocks, particularly in areas with high rainfall, for example Hong Kong. However, many of these landslides are the result of poor construction practices involving cut and fill rather than an inherent instability of the slope materials.

Large landslides are not common in these materials and, if present, may be bounded by intrusions such as dolerite dikes and sills that weather to lower strengths than the granitic rocks. Lacerda (2004) provides insight into the detailed geotechnical conditions present in deeply weathered granitic soils.

12.2.2 BASALT

Basalt flows can overlie a rock surface, residual soil, or sediment. In many instances a basalt deposit consists of multiple layers from successive flows. Depending on the mineralogical composition of the basalt, the thickness of the flows, and whether they were deposited subaerially or under water, the flows may be massive and of low permeability, or jointed (e.g., columnar joints) or vesicular and of higher permeability.

It is common for basalts to be interbedded with pyroclastic materials such as ash and lava fragments or tuff and agglomerate produced by volcanic explosions. Also flows may be separated by alluvial or other sediments. The result may be interlayered high and lower-permeability strata as shown in Figure 12.4. The interlayering of basalts and higher-permeability sediments has a significant effect on the pore pressures and, potentially, the shear strength in the slope and must be understood to properly assess slope stability.

Basalts are susceptible to chemical weathering. Extremely weathered basalts and the surface residual soils developed on

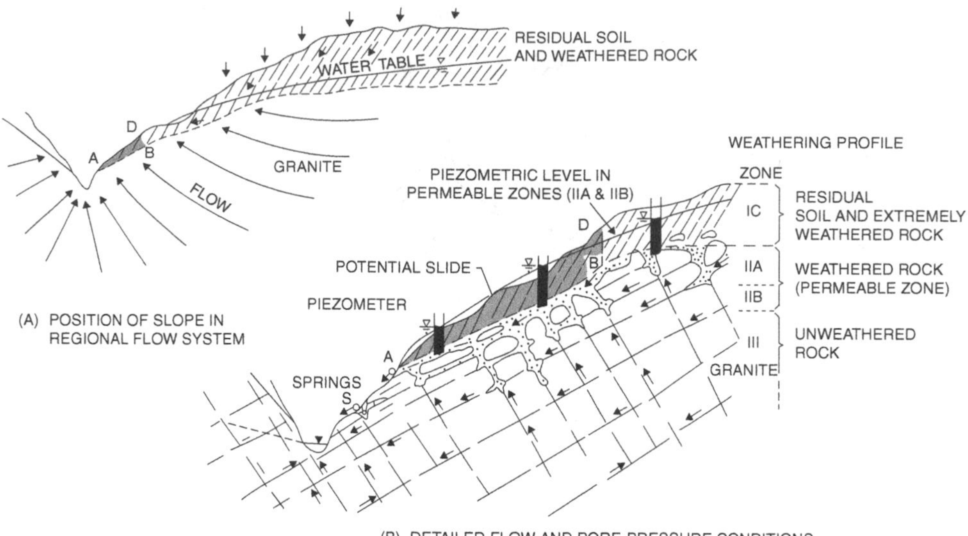

Fig. 12.3. Typical conditions likely to result in landsliding in weathered granite (Fell *et al.*, 2005, based on Patton and Hendron, 1972).

them typically have high plasticity and may be highly expansive and fissured. The presence of a well-developed pattern of slickensided fissures causes the shear strength of the mass to be significantly lower than that of the intact material.

Many slopes steeper than about 10° that are underlain by such materials show geomorphological evidence of past or current landsliding. Such landsliding occurs commonly at the steep margins of plateaus or hills capped by basalt flows that overlie old weathered land surfaces (Figs. 12.4 and 12.5). The sliding occurs, in some cases, simply as a result of oversteepening of the hillside by erosion, and in other cases from the pressure exerted by groundwater exiting from a permeable zone beneath the extremely weathered basalt. The permeable zone may be jointed, less-weathered basalt, as in Figure 12.5, or alluvial sands or gravels on the old buried land surface. Examples are given in Fell (1992).

12.2.3 SEDIMENTARY ROCKS

Mudrocks are sedimentary rocks formed by the consolidation and cementation of mainly clay, silt, or clay–silt admixtures. The principal rock types are claystone (predominantly clay-sized grains), siltstone (predominantly silt-sized grains), mudstone (clay–silt admixtures), and shale (mudstone with well-developed fissility). The sandstone group is defined here to include sandstones and conglomerates. Carbonate rocks are those that contain significant amounts of the soluble minerals calcite, aragonite, or dolomite; the most common carbonate rocks are limestone and dolomite. Weakly metamorphosed equivalents of the above rocks – for example quartzite, metasiltstone, meta-conglomerate, and marble – are grouped with nonmetamorphosed sedimentary rocks for the purpose of this discussion. These rocks are commonly interbedded, which may have a significant effect on the stability of slopes due to the development of low-strength and high-permeability zones during folding, uplift, and valley cutting.

12.2.4 THE EFFECTS OF FOLDING OF SEDIMENTARY STRATA

During folding, thin seams of crushed rock develop within the mudrocks, typically at their boundaries with interbedded sandstone, conglomerate, or carbonate rocks. These seams develop due to interbed slip during folding. Mudrocks have a higher stiffness than the interbedded, higher-strength sandstones, conglomerates, or carbonates (Fig. 12.6). Seams of crushed rock within mudrocks are known as bedding-surface faults, bedding-surface shears, or bedding-plane shears. Even gentle folding may produce bedding-surface faults. These consist mainly of clay, are nearly planar, and commonly have slickensided surfaces both within them and at their boundaries.

Although they commonly extend over large areas, bedding-surface faults may be only a few millimeters thick. Such defects are difficult to recover and recognize in diamond drill cores. As discussed below, bedding-surface faults may have low strength and have a significant influence on the stability of slopes.

12.2.5 THE EFFECTS OF STRESS RELIEF FROM FORMATION OF VALLEYS

When valleys are formed, the horizontal stresses in the Earth's crust are relieved, which may result in the formation of valley stress-relief features such as sheet joints in massive granitic rocks, joints in bedded sedimentary rocks, and bedding-surface faults. This may also result in valley bulge features. Figures 12.7 and 12.8 show these features schematically in flat-lying sedimentary rocks.

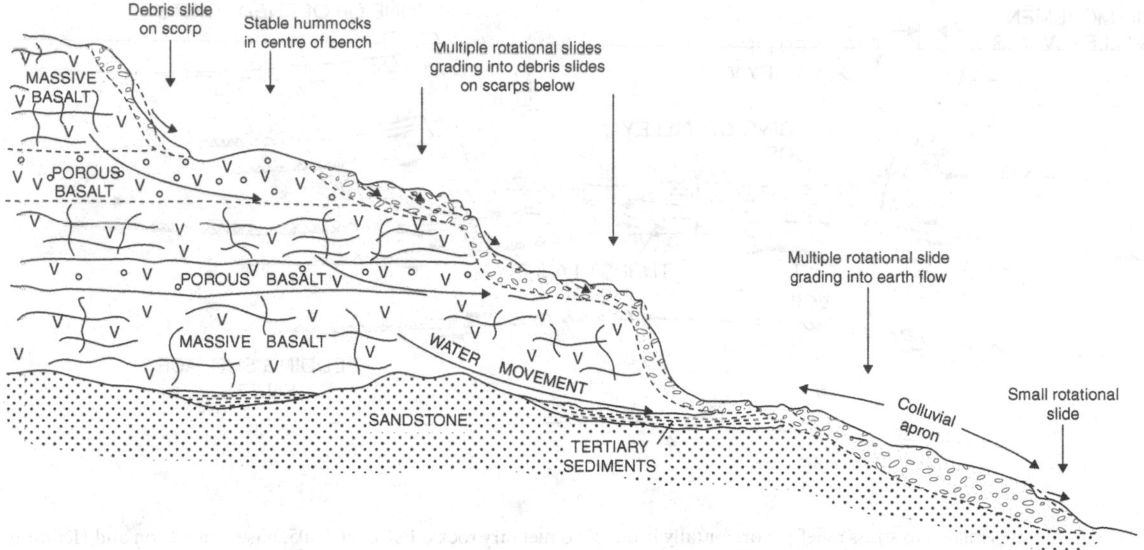

Fig. 12.4. Interpretive hydrogeologic conditions involved in landsliding in Tertiary basalts in Queensland (Willmott, 1983).

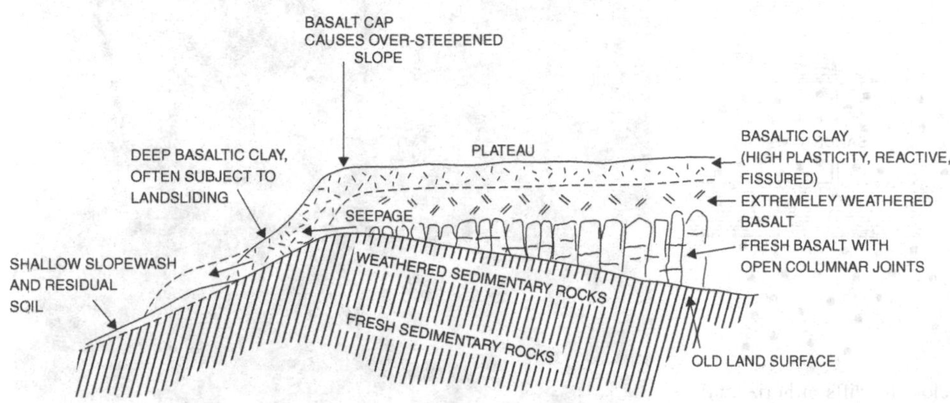

Fig. 12.5. Typical profile through the margin of a basalt plateau, showing conditions that lead to slope instability (Fell, 1992).

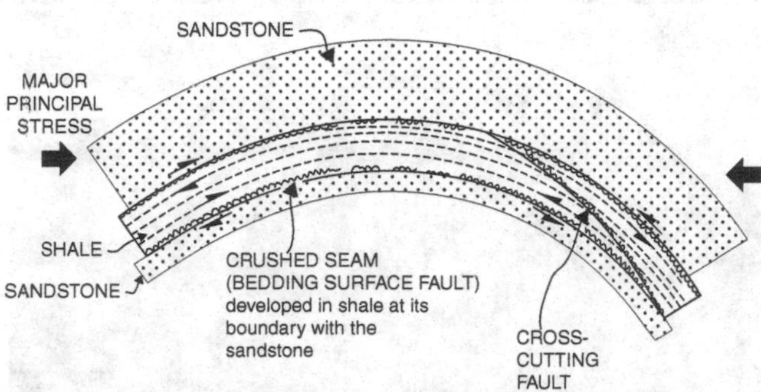

Fig. 12.6. Formation of bedding-surface faults in mudrocks (Fell *et al.*, 2005).

Gentle anticlines, in some cases with associated thrust faults, have been recorded in many river valleys cutting through near-horizontal sedimentary rocks of moderate to low strength (Fig. 12.7). The phenomenon is referred to as valley bulging or valley rebound, and has been described by many authors including Patton and Hendron (1972), Hutchinson (1988), Fell *et al.* (1992, 2005), and Jones and Lee (1994). Most of the features in the valley floor shown in Figure 12.7 have developed as a result of buckling and shear failure under high horizontal compressive stresses and low vertical stresses. Stresses were concentrated beneath the valley floor as a result of load transfer when the valley was excavated. During incision, lateral support was removed from the rocks above the valley floor and vertical load was removed from the rock below the valley floor.

Joints in the valley sides open because of differential movements in the rock strata. The stiff sandstone bed in Figure 12.8, for example, expands less than the weaker shale as the locked-in stresses are relieved during formation of the valley. The differential movement results in the opening of joints in the stiffer sandstone and in the formation of bedding-surface faults. Figure 12.9 shows an example of stress-relief joints that have been infilled with clay.

The opening of joints results in higher rock mass permeability, particularly in the stiffer sandstone beds; the joints become

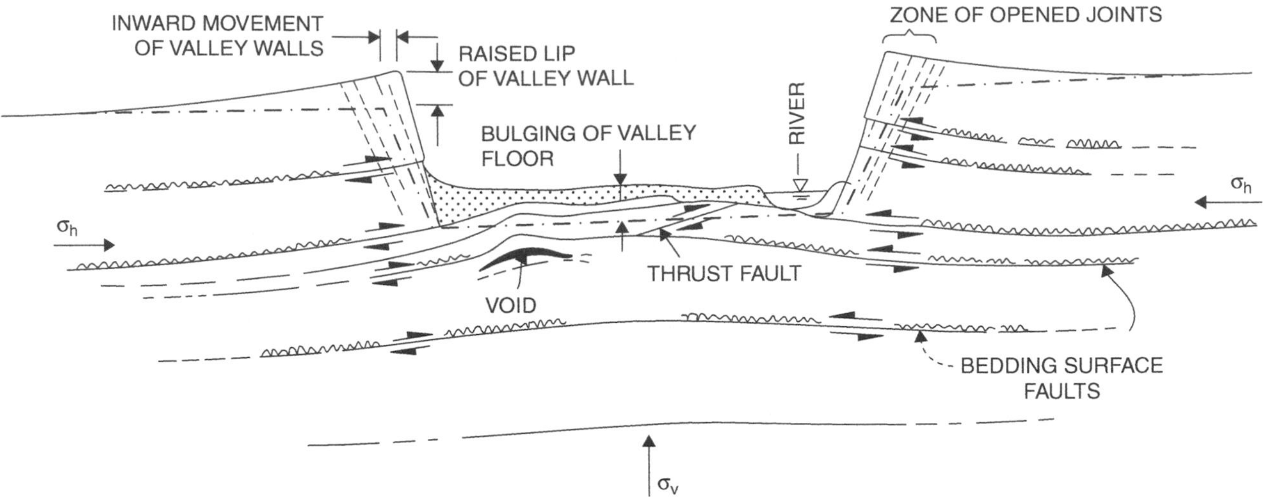

Fig. 12.7. Complex valley structures related to stress relief in horizontally bedded sedimentary rocks (Fell *et al.* 2005, based on Patton and Hendron 1971).

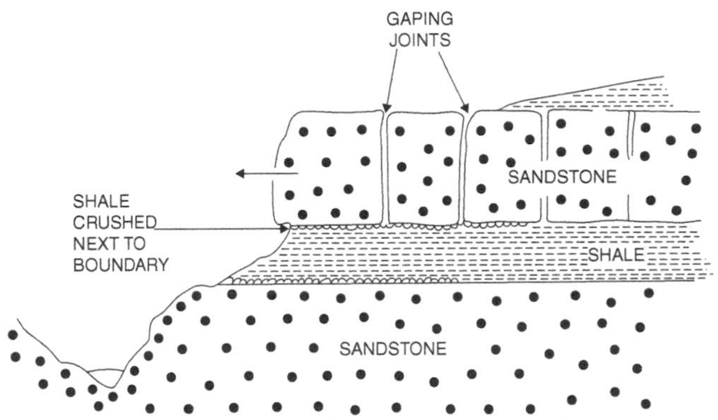

Fig. 12.8. Commonly observed features close to cliffs in horizontally bedded sandstones with shale or siltstone interbeds (Fell *et al.*, 2005, based partly on Deere and Patton, 1971).

a preferential path for groundwater. As shown in Figure 12.10, groundwater may be confined by lower-permeability colluvium deposits on the valley slopes, leading to high water pressures and instability in the colluvium (slides A and B in Fig. 12.10). Stress relief may also open joints and bedding surfaces in the finer-grained, less-stiff shale, leading to increased pore pressures and the potential for deeper sliding along a bedding-surface fault at the base of the shale (slide C).

Fell *et al.* (1987) and Fell (2006) give examples of bedding-surface faults and landsliding on slopes inclined as little as 6° in gently folded and incised claystone and shale. Figure 12.11 shows typical characteristics of such landslides. Investigation of a number of these landslides show that the rupture surface is well defined. The back-analyzed effective shear strength of the rupture surfaces is 10–12°. Test pits excavated through landslide deposits have encountered significant water flows at the upper surface of the in-situ weathered rock. These flows

Fig. 12.9. Photographs of the abutment of Shannon Creek Dam, showing infilled stress-relief joints. The joints are up to 80 mm wide and terminate on shale beds. Photographs courtesy of L McDonald and Clarence Valley Council.

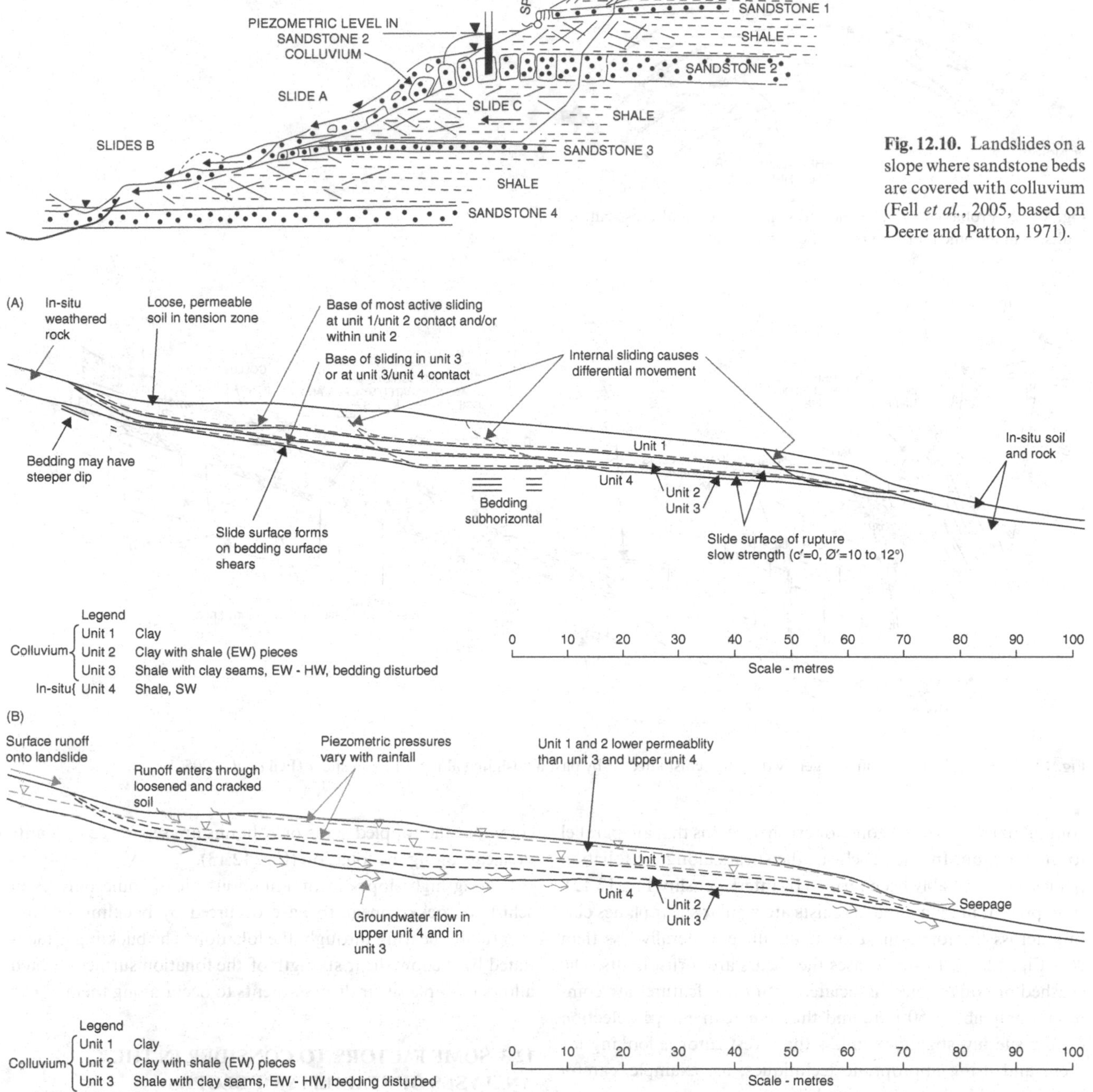

Fig. 12.10. Landslides on a slope where sandstone beds are covered with colluvium (Fell *et al.*, 2005, based on Deere and Patton, 1971).

Fig. 12.11. Geotechnical and hydrologic characteristics of slides on shale in the Sydney area. (A) Geotechnical characteristics of the landslides. (B) Some hydrogeologic characteristics of the landslides (Fell, 2006).

are confined by the lower-permeability landslide debris and control pore pressures and the stability of the slopes. The debris under tension in the upper part of the landslide commonly has higher permeability than the debris lower on the slope, allowing surface water to enter and build up in the landslide. On a larger scale, multiple landslides are common where mudrocks are interbedded with other sedimentary rocks (Fell, 1992, 1995).

12.2.6 FOLIATED METAMORPHIC ROCKS

This group includes slate, phyllite, and schist, all of which have a pronounced cleavage or planar foliation arising from the parallel arrangement of platy minerals. Slate, phyllite, and most mica schists have formed by the regional metamorphism of mudstone or siltstone, whereas greenschists have formed by the regional metamorphism of basalt and gabbro. Schists commonly

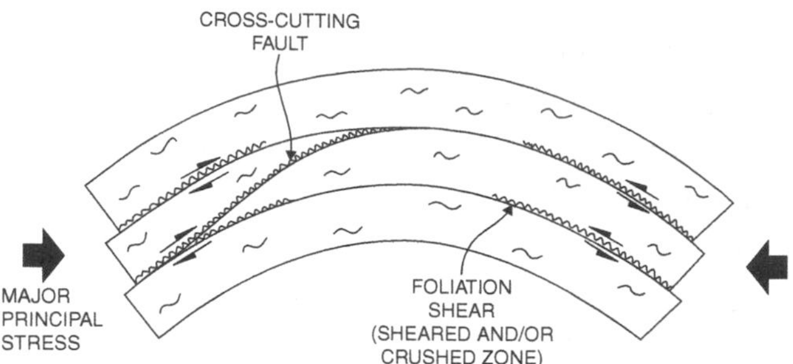

Fig. 12.12. Probable way in which foliation shears and cross-cutting faults form in some schistose rocks (Fell *et al.*, 2005).

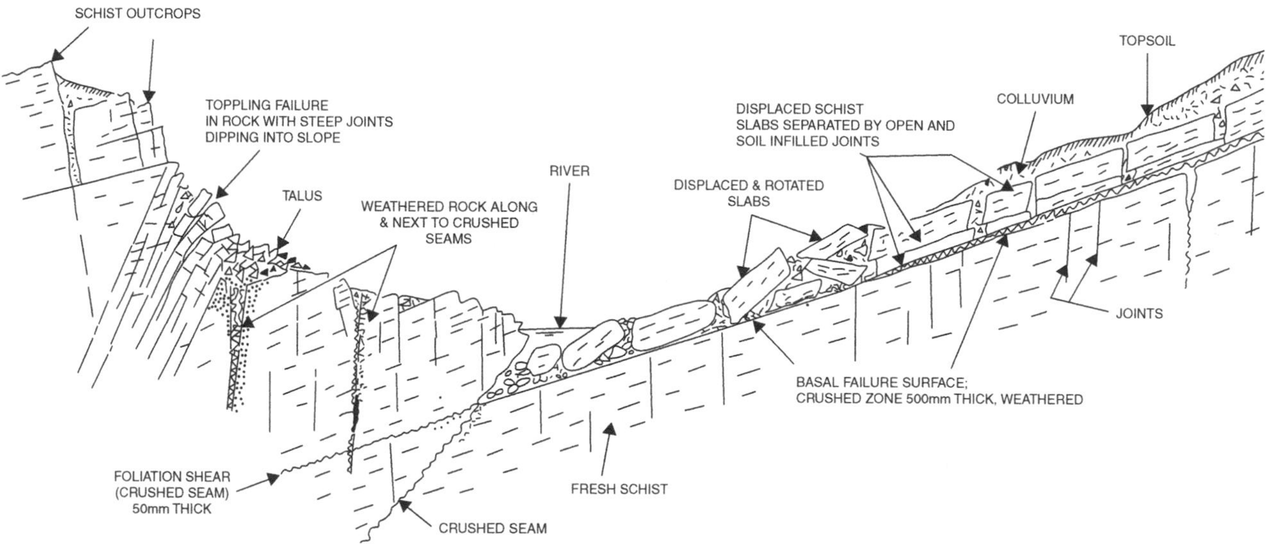

Fig. 12.13. Typical valley profile in gently dipping schist affected by past landsliding along foliation shear (Fell *et al.*, 2005).

contain narrow sheared zones or crushed seams that are parallel to the foliation. In folded schists, the shears along the foliation planes have probably been formed by interlayer slip (Fig. 12.12). Also present in many folded schists are similar shear planes cutting across the foliation at an acute angle, generally less than 20° (Fig. 12.12). In some cases the shears are thrust faults. The crushed or gouge zones associated with these features are commonly only about 50 mm, and thus they can escape detection during site investigations unless the investigator is looking for them and using appropriate techniques; for example, careful trenches and high-quality core drilling.

Such defects are commonly the initiating failure surfaces of landslides in schistose rocks (Fig. 12.13). As shown in Figure 12.12, however, failure can occur at an acute angle to the foliation. In the case of larger landslides, such as those shown in Figure 12.14, several levels of sliding are common, with different degrees of disturbance of the landslide mass and compartmentalization of groundwater. These effects can result in large differences in pore pressure within the landslide mass. In steeply dipping schistose rocks, failure by toppling is also relatively common; the toppled slabs or columns are separated by joints or shears along the foliation (Fig. 12.13).

In long high slopes in mountainous areas, some failures in schistose rocks appear to have occurred by buckling of, and eventually shearing through, the foliation. The buckling is facilitated by the low shear strength of the foliation surfaces, which allows multiple shear displacements to occur along them.

12.3 SOME FACTORS TO CONSIDER IN THE ANALYSES OF STABILITY OF SLOPES

12.3.1 THE MECHANICS OF SLIDING

It is essential to consider carefully the mechanics of sliding when assessing the method of analysis and the strengths that should be assigned to potential rupture surfaces. Figures 12.15 and 12.16 show examples of translational and compound slides in rock. For planar translational slides, the surface of rupture will be a bedding surface or foliation. The shear strength of the surface may be between peak and residual and will depend on the

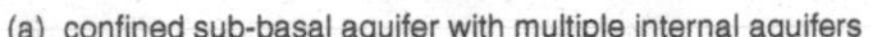

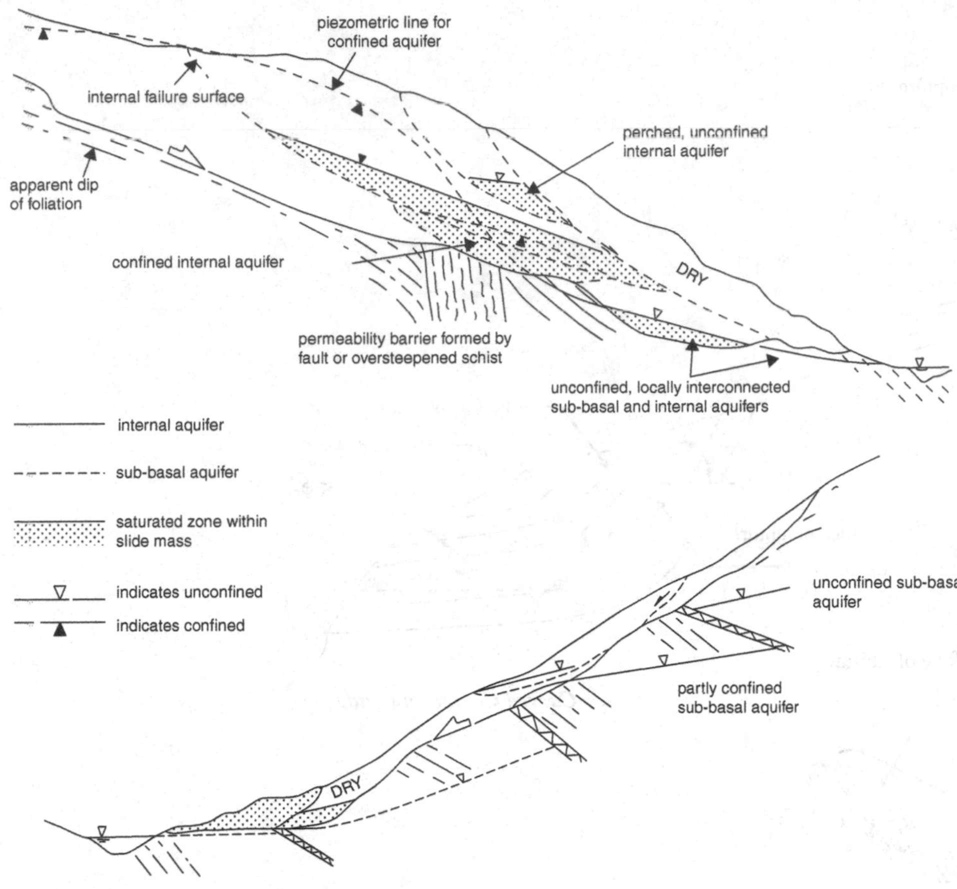

Fig. 12.14. Cromwell Gorge landslides, showing geometry, relation to foliation, and groundwater systems (Macfarlane *et al.*, 1991).

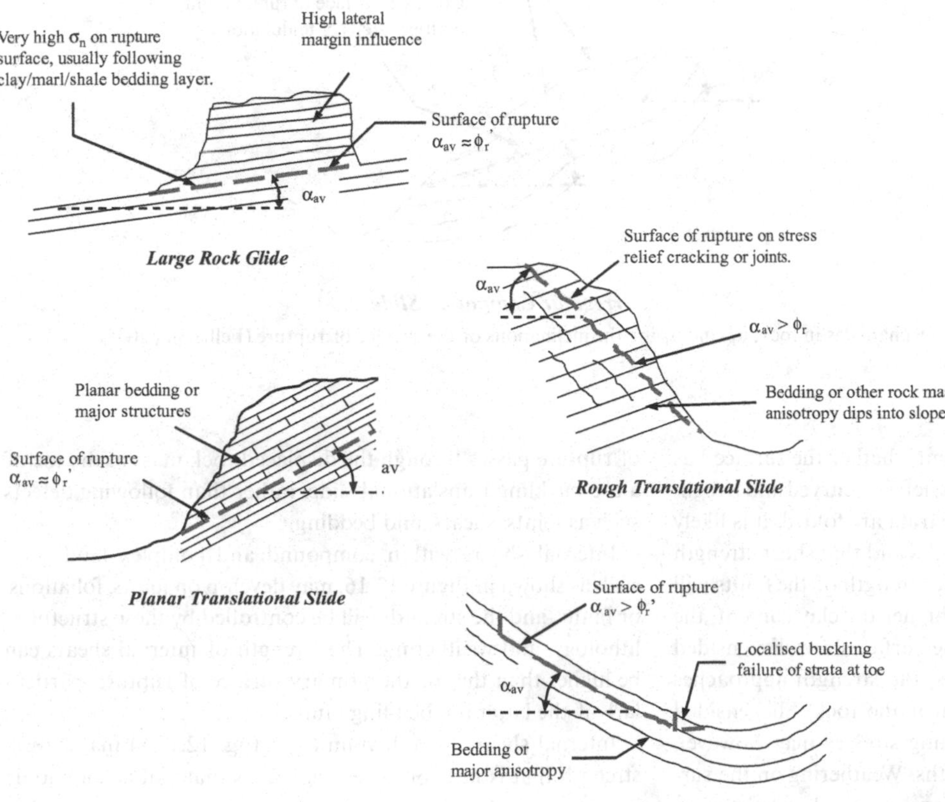

Fig. 12.15. Examples of translational slide mechanisms in rock; α_{av} is the inclination of the surface of rupture (Fell *et al.*, 2007).

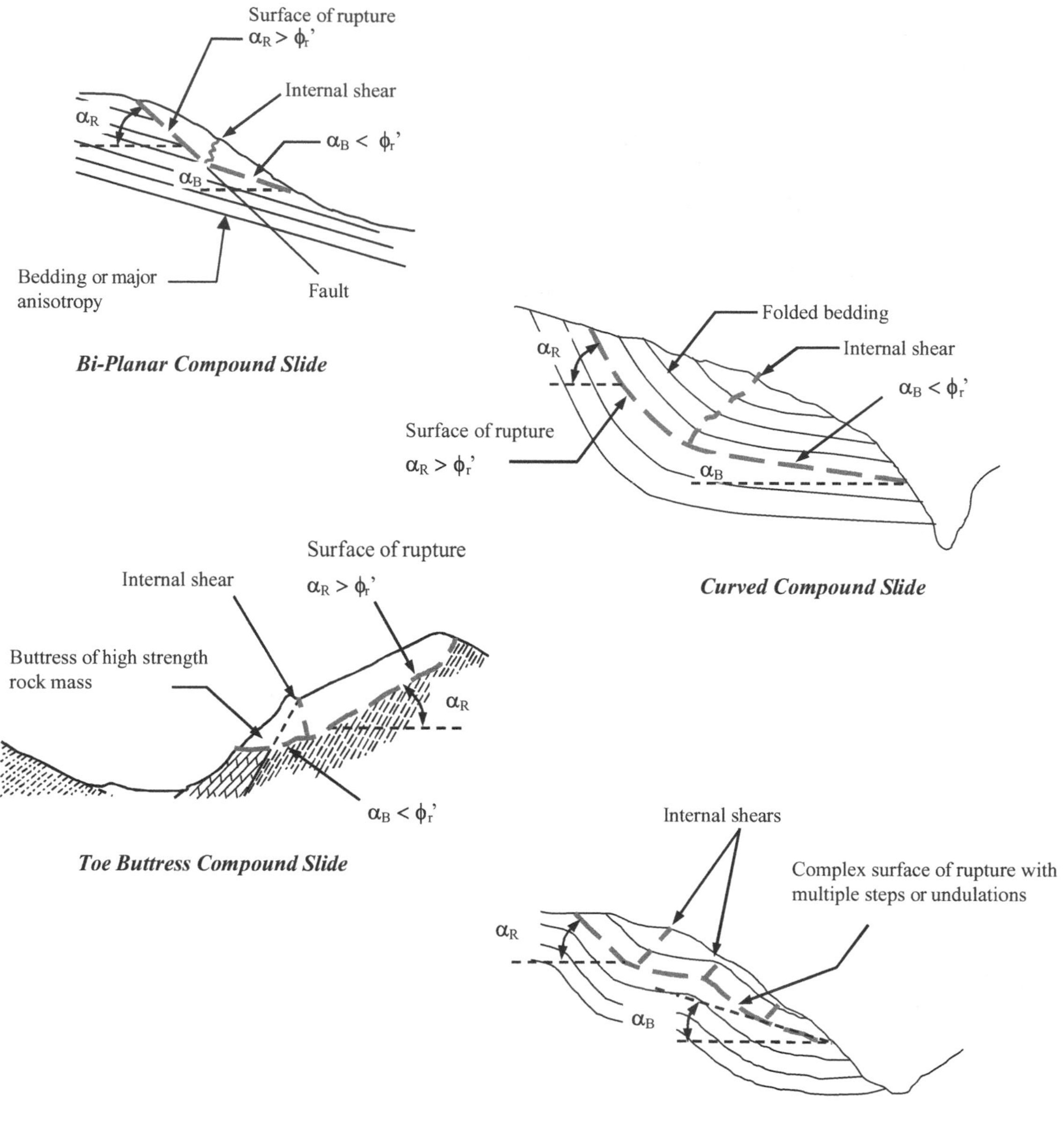

Fig. 12.16. Examples of compound slide mechanisms in rock; α_R and α_B are the inclinations of the surface of rupture (Fell *et al.*, 2007).

lithology, planarity of the surface, and whether the surface has been subject to shearing from stress relief. For curved and irregular compound landslides, where the strata are folded, it is likely that there will be bedding-surface faults and that shear strength will be controlled by those faults. The strength of the faults will be determined by the lithology (the higher the clay content, the lower the strength), and whether the surfaces are slickensided or only crushed. For crushed seams, the strength approaches the large strain or residual strength of the rock. Slickensided surfaces formed under larger confining stresses may, however, have significantly lower shear strengths. Weathering on the surface of rupture may also affect the rock strength. The surface of rupture passes through the displaced rock mass at the toe of a toe-buckling translational slide, rather than following defects such as joints, shears, and bedding.

Internal shears within compound and complex landslides, such as shown in Figure 12.16, may develop on joints, foliations, or faults, and the strength will be controlled by these structures, lithology, and weathering. The strength of internal shears can be higher than that of the primary surface of rupture, particularly if the latter is a bedding fault.

Internal shears in colluvium (e.g., Fig. 12.11A) may have a strength that is near or at residual. This material is commonly a mixture and may include particles of weathered rock; it may have a significantly higher strength than a basal rupture surface that follows bedding-surface shears.

12.3.2 ASSESSMENT OF SHEAR STRENGTH PARAMETERS IN ROCK

Shear strength parameters should always be assessed by people who are experienced in rock and soil mechanics. Furthermore, these people should be familiar with the site and the sliding mechanism.

Hoek and Bray (1981) reported representative strengths of a range of rocks, as well as the peak and residual strength of infilled discontinuities and shear surfaces. Slickensided shear surfaces in rock commonly have very low strengths. For example, such surfaces in tuffaceous claystone in Sydney, Australia have a shear strength of around $c' = 0$, $\phi' = 10$–$12°$ (Fell, 1995). It is important to note, however, that the critical strengths are those determined by testing shear surfaces. Residual strength tests on the adjacent rock may give misleadingly high values. In a case described by Fell *et al.* (1987), for example, claystone samples taken from adjacent to the surface of rupture had residual strengths of $c' = 0$, $\phi' = 15$–$16°$, whereas the slide surface had shear strengths of $c' = 0$, $\phi' = 10$–$12°$. This difference was due, not to problems in testing, but rather to tectonic folding that had created lower-strength bedding planes than could have formed by shearing the adjacent rock. Sliding had occurred on these surfaces. Faults in granitic rocks can have similar strengths to silty sand, with $c' = 0$, $\phi' = 20$–$25°$ or more, which is much lower than the residual strength of extremely weathered granite.

12.3.3 ASSESSMENT OF PORE PRESSURES IN SLOPES

As discussed earlier, pore pressures in slopes are controlled by the relative permeability of the different strata, which in turn is controlled by stratigraphy, stress-relief features, and weathering. These factors must be taken into account when attempting to predict pore pressures and when installing piezometers to measure pressures. It must be recognized that seasonal and annual variations in climate can control pore pressures, and that it may take several years of monitoring with short intervals between readings to properly understand groundwater conditions. Pore pressures in slopes such as shown in Figure 12.11 may vary significantly over a period of hours during heavy rainfall, due to direct rainfall infiltration and surface flows onto the slopes.

REFERENCES

Deere, D. U. and Patton, F. D. (1971). Slope stability in residual soils. In *Proceedings of the Pan-American Conference on Soil Mechanics and Foundation Engineering, Vol. 1*, pp. 87–170.

Fell, R. (1992). Theme address: Landslides in Australia. *Proceedings of the Sixth International Symposium on Landslides, Vol. 3*, ed. D. H. Bell. Rotterdam: Balkema, pp. 2059–2100.

(1995). Landsliding in the Gosford-Lake Macquarie-Newcastle Area. In *Engineering Geology of the Newcastle-Gosford Region*, ed. S. W. Sloan and M. A. Allman. Sydney: Australian Geomechanics Society, pp. 265–291.

(2006). Landslides in the Wainamatta Group, Baulkham Hills Shire, Sydney. *Australian Geomechanics*, 41, 53–62.

Fell, R., MacGregor, J. P., Williams, J. and Searle, P. (1987). A landslide in Patonga claystone on the Sydney–Newcastle Freeway. *Geotechnique*, 37, 255–270.

Fell, R., MacGregor, J. P. and Stapledon, D. H. (1992). *Geotechnical Engineering of Embankment Dams*. Rotterdam: Balkema.

Fell, R., Macgregor, J. P., Stapledon, D. and Bell, G. (2005). *Geotechnical Engineering of Dams*. Leiden: Balkema.

Fell, R., Glastonbury, J. and Hunter, G. (2007). Rapid landslides: The importance of understanding mechanisms and rupture surface mechanics. *Quarterly Journal of Engineering Geology and Hydrology*, 40, 9–27.

Hoek, E. and Bray, J. W. (1981). *Rock Slope Engineering*. London: Institute of Mining and Metallurgy.

Hutchinson, J. N. (1988). Morphological and geotechnical parameters of landslides in relation to geology and hydrogeology. In *Proceedings of the Fifth International Symposium on Landslides*, ed. C. Bonnard. Rotterdam: Balkema, pp. 3–36.

Jones, D. K. C. and Lee, E. M. (1994). *Landsliding in Great Britain*. London: Her Majesty's Stationery Office.

Lacerda, W. A. (2004). The behavior of colluvial slopes in a tropical environment. In *Landslides: Evaluation and Stabilization*, ed. W. A. Lacerda, M. Ehrlich, S. A. B. Fontoura and A. S. F. Sayao. Leiden: Balkema, pp. 1315–1342.

Macfarlane, D. F., Pattle, A. D. and Salt, G. (1991). Nature and identification of Cromwell Gorge landslides groundwater systems. In *Landslides, Vol. 1*, ed. D. H. Bell. Rotterdam: Balkema, pp. 509–517.

Patton, F. D. and Hendron, A. J. (1972). General report on mass movements. In *Proceedings of the Second Congress of the International Association of Engineering Geology*, VGR pp. 1–57.

Stapledon, D. H. (1995). Geological modelling in landslide investigation. In *Proceedings of the Sixth International Symposium on Landslides*, ed. D. H. Bell. Rotterdam, Balkema, pp. 199–523.

Willmott, W. F. (1983). *Slope Stability and Its Constraints on Closer Settlement on the Mapleton–Maleny Plateau, South East Queensland*. Geological Survey of Queensland, Record 1983/9.

13 Numerical modeling of rock-slope instability

DOUGLAS STEAD AND JOHN COGGAN

ABSTRACT

The last decade has seen considerable advances in the application of numerical modeling to rock slopes. In this chapter we present the current state of the art, beginning with simple kinematic and limit equilibrium approaches and then moving on to continuum and discontinuum modeling techniques. We review the available brittle fracture modeling methods and finally discuss some of the future challenges facing the successful application of numerical models.

13.1 INTRODUCTION

Applications of numerical modeling to the study of rock slopes and landslides have increased considerably over the past decade. An investigation of a soil or rock landslide is now almost always accompanied by some form of slope analysis, whether a simple kinematic analysis of rock slopes, initial limit equilibrium analysis, or the use of a sophisticated numerical code. Stead *et al.* (2006) discuss developments in the numerical analysis of rock slopes, emphasizing the use of methods that are appropriate to the requisite level of stability analysis. Figure 13.1 shows a toolbox of slope analytical methods that are dependent on the complexity of the slope failure mechanism being investigated or analyzed. In this chapter we provide a state-of-the-art review of the methods that are available for both two- and three-dimensional analysis of rock-slope instability.

We stress at the outset that slope analysis methods are applied in a wide variety of environments and in situations where different amounts of geotechnical data are available. For example, some slope analyses are conducted in simple geologic environments or where there are large amounts of geotechnical data, whereas others are performed on high mountainous slopes where there are limited data and the mechanism of failure is not well constrained. Numerical modeling is valuable in both situations, either for predicting future slope behavior or exploring a range of slope failure mechanisms. We have found preliminary modeling of landslides with limited data to be useful prior to in-depth geotechnical mapping. It can lead the investigator to the critical parameters to investigate, for example the continuity, roughness, and orientation of critical joint sets for kinematic failure.

Stead *et al.* (2006) provide a detailed discussion of the applications and limitations of the different methods of slope analysis. In this chapter, we review new developments since this publication and their application to landslide studies.

13.2 KINEMATIC ANALYSIS

Stereographic analysis is the first step in most investigations of rock-slope stability, and helps in assessing different kinematic failure mechanisms, including planar, wedge, and toppling instability. Preliminary stereographic techniques using concepts such as the daylight envelope, toppling envelope, and friction cones have been invaluable in the investigation of road-cut instabilities, bench-scale instabilities in quarries and mines, and low-height landslides on rock slopes. These techniques must, however, be used with caution as the size of the slope increases. Kinematic stereographic analysis is relevant for persistent structures such as faults on high mountain slopes, but is less useful for assessing the failure potential of complex rock mass failures involving multiple joint sets or step-paths. Non-daylighting wedge failures in high-stress, large slopes or in weak rock masses can occur by the failure of intact rock bridges. Biplanar, buckling, and multiplanar translational failure mechanisms are not generally amenable to simple kinematic stereographic analysis.

Landslides: Types, Mechanisms and Modeling, ed. John J. Clague and Douglas Stead. Published by Cambridge University Press.

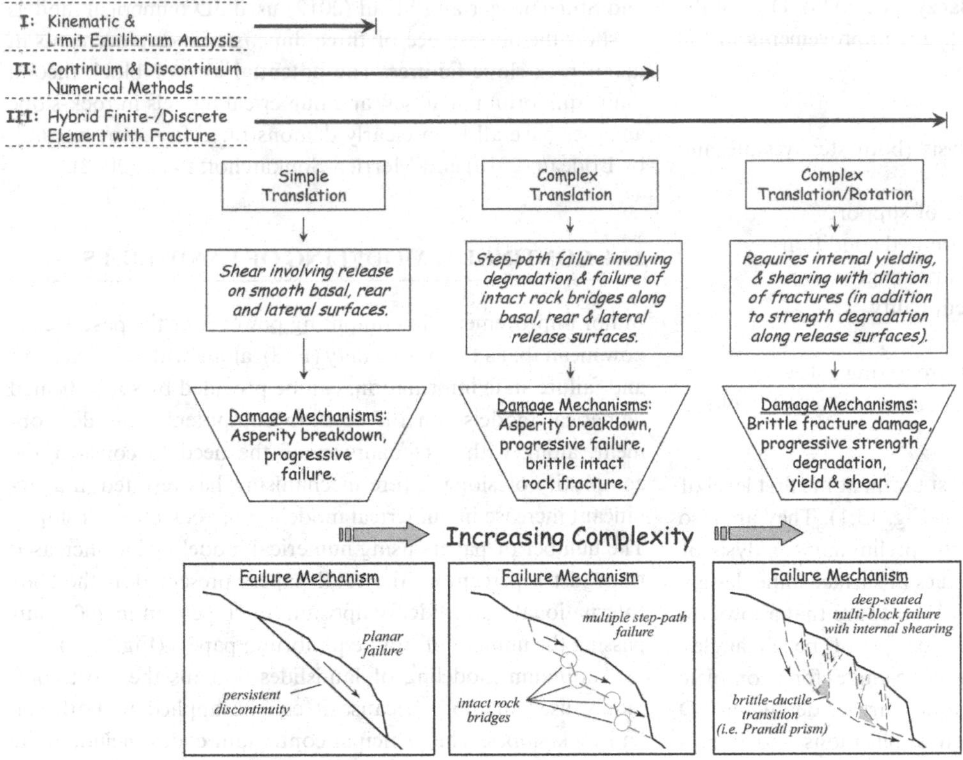

Fig. 13.1. Levels of landslide analysis and mechanisms of failure (modified from Stead *et al.*, 2006).

Currently available stereographic codes such as DIPS (Rocscience, 2011) offer a wide range of methods to consider the attributes of discontinuities measured during a discontinuity survey (orientation, persistence, spacing, infill, roughness, water). It is essential that full use of these techniques be made and overreliance on orientation data alone be avoided. The design of appropriate support for rock slopes makes consideration of data other than orientation such as spacing, persistence, roughness and infill mandatory. The essence of block theory (Goodman and Shi, 1985) is to determine critical keyblocks that may control the stability of a rock slope. Block theory has traditionally been used with limit equilibrium analysis to assess stability, but few studies combining keyblock analysis and numerical modeling of landslides have been published. An important area of future research in landslide modeling is the use of block theory in combination with discrete fracture networks (DFNs). Consideration of not only the orientation of blocks, but also their shape and size and the presence of rock bridges, is fundamental to understanding the three-dimensional characteristics of rock-slope failures at all scales.

13.3 ADVANCES IN LIMIT EQUILIBRIUM SLOPE ANALYSIS

Limit equilibrium methods remain the most common form of slope analysis not only for soil slopes but also for rock slopes, where failure is assumed to be either planar, multiplanar, or wedge in nature. The percentage of papers presented at International Landslide Symposiums in 2000, 2004, and 2008 that use limited

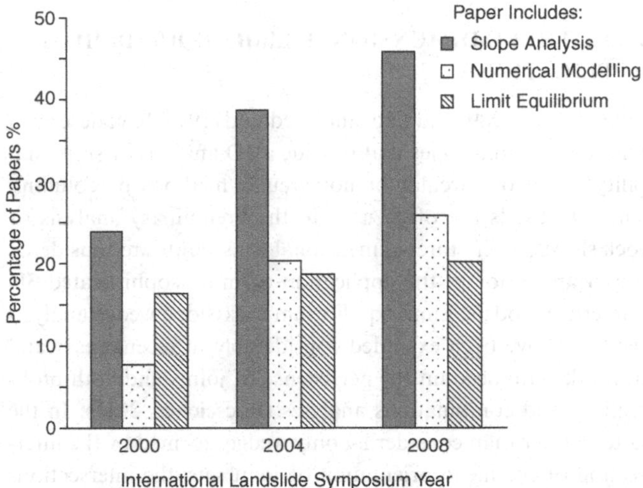

Fig. 13.2. Trends in methods of landslide analysis, 2000–2008.

equilibrium techniques ranges from 16 to 20 percent (Fig. 13.2; Bromhead *et al.*, 2000; Lacerda *et al.*, 2004; Chen *et al.*, 2008). Using the same dataset, we found that, more generally, the percentage of papers incorporating slope analysis techniques has risen from 24 to 47 percent over this 8-year period.

13.3.1 TWO-DIMENSIONAL LIMIT EQUILIBRIUM APPLICATIONS

Many commercial 2D limit equilibrium codes are available, ranging from simple infinite slope analysis software to more complex step-path failure codes (Baczynski, 2008). During the past decade, there have been significant improvements in 2D limit equilibrium methods including:

- more choice in failure criteria
- integration of groundwater analysis (both steady state and transient)
- incorporation of a variety of types of support
- application to saturated and unsaturated conditions
- ability to consider anisotropic shear strength
- improved sensitivity analysis functionality
- improved probabilistic analysis
- improved graphical pre- and post-processing software
- optimized searching routines (circular, composite, block, hybrid simulated annealing).

Limit equilibrium techniques are most useful at the first level of rock-slope stability analyses (Level 1, Fig. 13.1). They are also the current method of choice for the preliminary analysis of landslides and for pre-feasibility studies in surface mine design. Many companies use in-house flexible software that allows for rapid batch-mode limit equilibrium design of bench angles, berm widths, and inter-ramp overall slope angles for a complete range of open-pit mine design sectors. Further details on 2D limit equilibrium methods and their applications and limitations are provided by Stead *et al.* (2006).

13.3.2 THREE-DIMENSIONAL LIMIT EQUILIBRIUM APPLICATIONS

Hungr *et al.* (1989) and Lam and Fredlund (1993) have described limit equilibrium codes that provide a 3D analysis of slope stability based on circular or noncircular methods of columns. These methods are often used in the preliminary analysis of rock slopes, where three-dimensional constraints are considered important, prior to the application of more sophisticated 3D numerical models. Limit equilibrium rock-slope wedge analysis methods have been extended considerably in recent years and now take into account the persistence of joints and both probabilistic and combinations analyses (Rocscience, 2011). In the latter, rather than considering only wedges formed by the intersection of the mean orientation of joint sets, the intersections between all the joint poles on a stereonet are considered and factors of safety (FoS) for each intersection are derived. Modern methods also allow the assessment of wedge weights, which is an important consideration in designing remedial measures. Recent work by Hungr and Amann (2011), using 3D analytical and limit equilibrium methods constrained by 3D numerical models, shows the importance of wedge asymmetry and rotation about a vertical axis. These authors showed that conventional wedge analyses yield overly conservative factors of safety where wedge rotation and induced bending moments produce tensile stresses and failure along the sub-vertical release plane (what Hungr and Amann, 2011, term the "constraint surface" to asymmetric wedges). Brideau (2010), Sturzenegger (2010) and Sturzenegger and Stead (2012) used 3D numerical models to show the importance of three-dimensional displacements in major rock-slope failures. The potential uses of block theory, limit equilibrium analyses, and numerical models in rock-slope analyses have all been clearly demonstrated in recent research by Brideau (2010) and Merrien-Soukatchoff *et al.* (2012).

13.4 CONTINUUM MODELING OF LANDSLIDES

Major improvements in computing power over the past decade now mean that a factor of safety (FoS), along with displacement and failure state information, can be provided by sophisticated numerical models run on desktop computers. This development, along with a recognition of the need to consider the complexity of slope failure mechanisms, has resulted in a significant increase in numerical modeling of rock and soil slopes. The number of papers using numerical modeling has increased from just 8 percent of the total papers presented at the 2004 International Landslide Symposium to 26 percent in 2008, surpassing the number of limit equilibrium papers (Fig. 13.2).

Continuum modeling of landslides remains the most commonly used method, because it can be applied to both soil and rock slopes. The principal continuum codes include finite element models (FEM) and finite difference models (FDM). A wide range of in-house and commercial FEM codes, including Phase2 (Rocscience, 2011) and Abaqus (Simulia, 2011) are being used in both two- and three-dimensional analyses of soil and rock slopes. In contrast, FDM codes, although widely used, are few in number: they include FLAC and FLAC3D (Itasca, 2011). Figure 13.3 shows the use of continuum codes in modeling different types of rock-slope failures, including a 2D analysis of the Frank Slide (Fig. 13.3a; Benko and Stead, 1998) and 3D analyses of a china clay quarry slope (Fig. 13.3c; Coggan *et al.*, 2000) and the Åknes rock slope (Fig. 13.3d; Grøneng *et al.*, 2010).

An important application of continuum codes has been to assess the importance of topography on in-situ stress distribution in high mountain slopes. Kinakin and Stead (2005), using digital elevation models and FLAC, characterized deep-seated gravitational slopes in British Columbia, Canada into three types (hogsback, dogtooth, and castellated). They used a FLAC sensitivity analyses approach to characterize the slope stress distributions for each geometry type. Ambrosi and Crosta (2011) have extended this work using FLAC3D and the Mohr ubiquitous constitutive criterion to investigate the influence of varying rock anisotropy, slope geometry, in-situ stresses, and water pressures. This work has significantly improved our understanding of the importance of postglacial valley evolution on the propagation of plastic strain and its relation to slope model parameters. An important aspect of rock-slope stability is estimating the influence of the disturbance or rock mass damage associated with slope formation. In nature, this may be due to geomorphic processes, whereas in engineering practice it can result from poor-quality blasting. A disturbance factor, *D*, is included

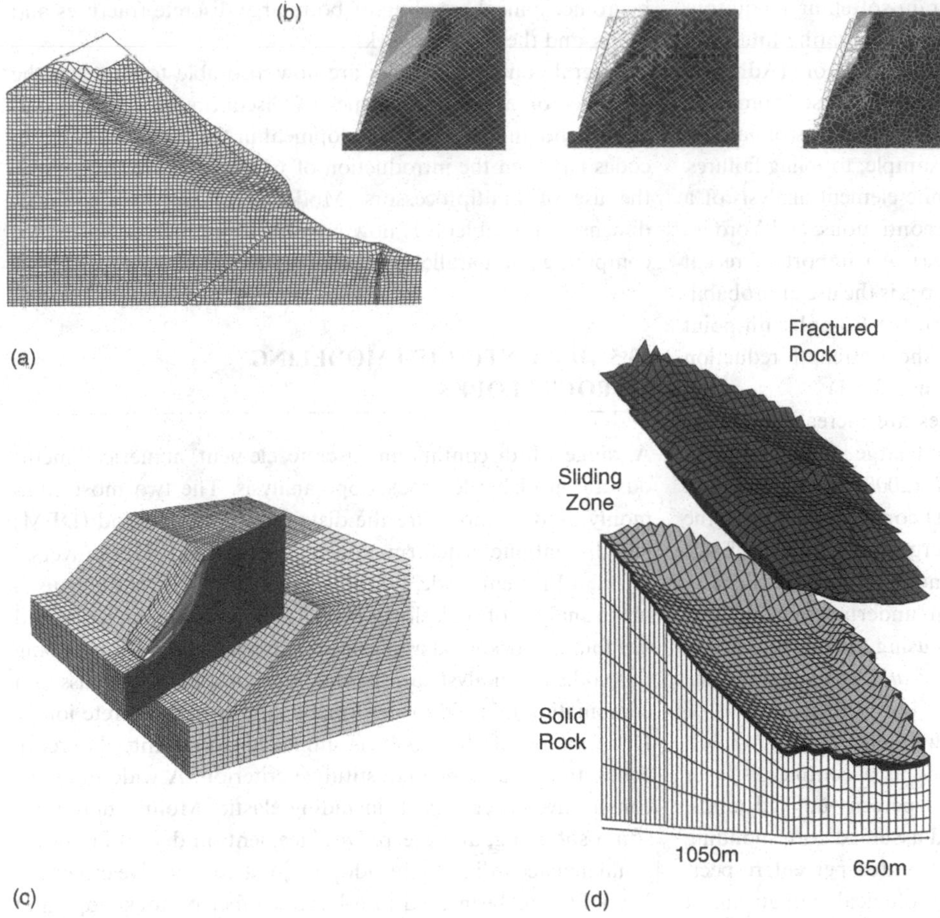

Fig. 13.3. Continuum analysis of rock slopes. (a) Continuum model of the Frank Slide (modified from Benko and Stead, 1998). (b) Incorporation of joints in a Phase-2 finite element analysis of a 60-m-high rock slope; from left to right: continuous, nonpersistent, and Voronoi jointing (modified from Hammah *et al.*, 2009a). (c) FLAC3D analysis of an altered china clay slope. (d) FLAC3D analysis of the Åknes rock slope, Norway (modified from Grøneng *et al.*, 2010).

within the Hoek–Brown failure criterion to represent the associated reduction in rock mass strength, but little research has been done to determine the extent of such damage zones within a rock slope. Li *et al.* (2011), however, provide an interesting study of the influence of the assumed disturbance factor on calculated open-pit slope stability using a combined 2D limit equilibrium and finite element approach.

The main development in continuum models in the past decade has been the increased use of shear strength reduction (SSR) techniques (Dawson *et al.*, 1999; Griffiths and Lane, 1999). These techniques determine the factor by which shear strength must be reduced for a slope to reach a critical state of equilibrium. The approach is analogous to determining the FoS in conventional limit equilibrium codes. In FEM codes, the method involves finding the reduction in shear strength at which the analysis just fails to converge, whereas in FDM codes it is related to the maximum unbalanced forces or nodal displacement within the model. Diederichs *et al.* (2007) provide a useful discussion of the merits of SSR techniques.

An additional recent development has been to combine dynamic programming and finite element analysis (Pham and Fredlund, 2003). Dynamic programming finite element methods offer the advantage that assumptions related to the choice of the interslice force function in limit equilibrium analysis are unnecessary. Pham and Fredlund (2003) note that combined use of the dynamic programming method and FEM analysis eliminates the need for the user to define the shape of the slip surface; instead, the choice is optimized by the program. They suggest that their results "tend to point towards a movement away from a complete reliance upon limit equilibrium methods of slope stability analysis." We agree with this observation, because SSR techniques are now available in many 2D and 3D continuum and discontinuum modeling codes. However, there remains an important role for limit equilibrium methods for the preliminary analysis of structurally controlled failure mechanisms in rock slopes where data are limited.

Although perhaps best suited to landslides in soils, continuum codes have nevertheless found increasing use in the analysis of landslides in high mountainous slopes characterized by discontinuous rock masses, and in large open-pit mines. The choice of the constitutive criterion is important if a continuum model is to realistically capture the landslide mechanism. The most commonly adopted criterion is the Mohr plasticity criterion, which is appropriate for landslide analysis. In many recent studies, however, researchers have attempted to capture anisotropy resulting from preferential planes of weakness within plasticity models by using a strain-softening ubiquitous criterion (Sainsbury *et al.*, 2008), by incorporating interfaces (Hammah *et al.*, 2008), or by using Cosserat theory (Adhikary and Dyskin, 2007; Riahi and Curran, 2008). These approaches have enabled more realistic continuum simulations of failures that are dominated by foliation; for example, toppling failures. Figure 13.3b shows an example of finite element analysis of a 70-m-high slope using continuous, discontinuous, and Voronoi fractures, after Hammah *et al.* (2009a). An important recent development in the analysis of rock slopes is the use of probabilistic finite element techniques based on the Rosenbleuth point estimate method, integrated with the shear strength reduction factor (Hammah *et al.*, 2009b; Rocscience, 2011)

Three-dimensional continuum codes are increasingly being used in the analysis of landslides and large open-pit slopes. Francois *et al.* (2007), Aringoli *et al.* (2008), and Chemenda *et al.* (2009) demonstrate the use of 3D continuum codes in the modeling, respectively, of the Triesenberg (Liechtenstein; FEM), Montelparo (central Italy; FDM) and La Clapière (France; FDM) landslides. Research has been undertaken to address time-dependent deformation in slopes using the creep constitutive criterion. For example, Grøneng *et al.* (2010) describe the use of Burger's viscoplastic creep criterion in their analysis of the Åknes landslide in Norway using FLAC3D (Fig. 13.3d). They modeled the present sliding zone as a continuous layer with 3 percent intact rock bridges, 64 percent unfilled joints, and 35 percent gouge-filled joints, and assumed corresponding future degraded characteristics of 1, 62, and 35 percent, respectively. They further assumed that the geological strength index (GSI) of the rock mass above the sliding zone would decrease from 62 to 37. To estimate the creep properties of the sliding zone, they calibrated the FLAC3D model against the average recorded annual displacement rate of the central area of the landslide of 26 mm per year. They then forecast the creep displacement over a 100-year period, based on their assumptions about the strengths of unfilled and gouge-filled joints, and noted the instability associated with a reduction in the percentage of intact rock bridges. Grøneng *et al.* (2010) emphasize the preliminary nature of their work and call attention to the need for improved groundwater modeling in view of the importance of fracture permeability and optimization of the model block element or mesh size (and hence runtime).

Finite element models of steady-state and transient groundwater conditions can be used as input to limit equilibrium models. Increasingly, geomechanical continuum codes incorporate coupled or uncoupled groundwater-mechanical analyses of multiphase fluids. Guglielmi *et al.* (2008) describe coupled hydromechanical modeling of an extensively instrumented, porous, fractured carbonate rock slope, at the Coaraze laboratory in southern France, in response to unconfined, free-water-surface movement. Instrumentation included extensometers, pressure gauges, and surface tiltmeters. Coupled modeling was undertaken using FLAC3D and a 2D-discontinuum code showing close agreement between observed and simulated displacements. Parametric studies indicated the importance of hydromechanical coupling of both large discrete fractures and faults, and the joint network.

Several continuum codes are now also able to simulate the influence of a variety of types of discontinuity fracture networks, and an important development in the use of continuum codes has been the introduction of 64-bit software that allows the use of multiprocessors. Modeling of large-scale, three-dimensional problems is now possible using high-performance computing and parallel-processing architecture.

13.5 DISCONTINUUM MODELING OF ROCK SLOPES

A range of discontinuum discrete element numerical methods are available for rock-slope analysis. The two most commonly used methods are the distinct element method (DEM) and discontinuous deformation analysis (DDA). The Universal Distinct Element Code, UDEC (Itasca, 2011), is routinely used in the analysis of rock slopes in civil and mining engineering and has found widespread application in landslide modeling. Using this code, the analyst can simulate the landslide rock mass as a combination of rigid or deformable blocks and discrete joints. Deformation of the blocks is simulated using finite difference zones and an assumed constitutive criterion. A wide range of constitutive criteria exist, including elastic, Mohr, ubiquitous, strain-softening, and creep. Displacement on discontinuities is simulated according to the adopted joint constitutive criteria.

UDEC has been used in a large number of rock-slope and landslide studies to investigate the factors influencing translational failures (Hencher *et al.*, 1996), buckling (Stead and Eberhardt, 1997; Tommasi *et al.*, 2009) and, in particular, toppling failures (Pritchard and Savigny, 1990; Coggan and Pine, 1996; Nichol *et al.*, 2002; Baczynski *et al.*, 2008). Many authors have used UDEC to simulate deep-seated gravitational slope displacements (e.g., Bovis and Stewart, 1998; Kinakin, 2004; Hensold, 2011). Several major rockslides have also been analyzed using complex two-dimensional UDEC models. In many cases, these analyses have complemented modeling with limit equilibrium, finite element, and other discrete element codes. The use of UDEC in modeling of a quarry slope involving a toppling or sliding failure in slates (Coggan and Pine, 1996) is illustrated in Figure 13.4a. Figure 13.4b illustrates UDEC modeling of the Frank Slide, Canada by Benko and Stead (1998). This exercise involved a complex two-dimensional fold geometry, with slip on bedding planes and induced tensile fracturing in the fold hinge area. UDEC has been used extensively in the recent modeling of complex landslide geometries, including Åknes, Norway (Kveldsvik *et al.*, 2009), Randa, Switzerland (Eberhardt *et al.*, 2004, Gischig, 2011), and La Clapière, France (Cappa *et al.*, 2004).

Rose and Scholtz (2009) describe a methodology for simulating complex deformation in large open pits using UDEC and a strain-softening criterion. They simulated sequential bench mining and used survey data to calibrate the models at

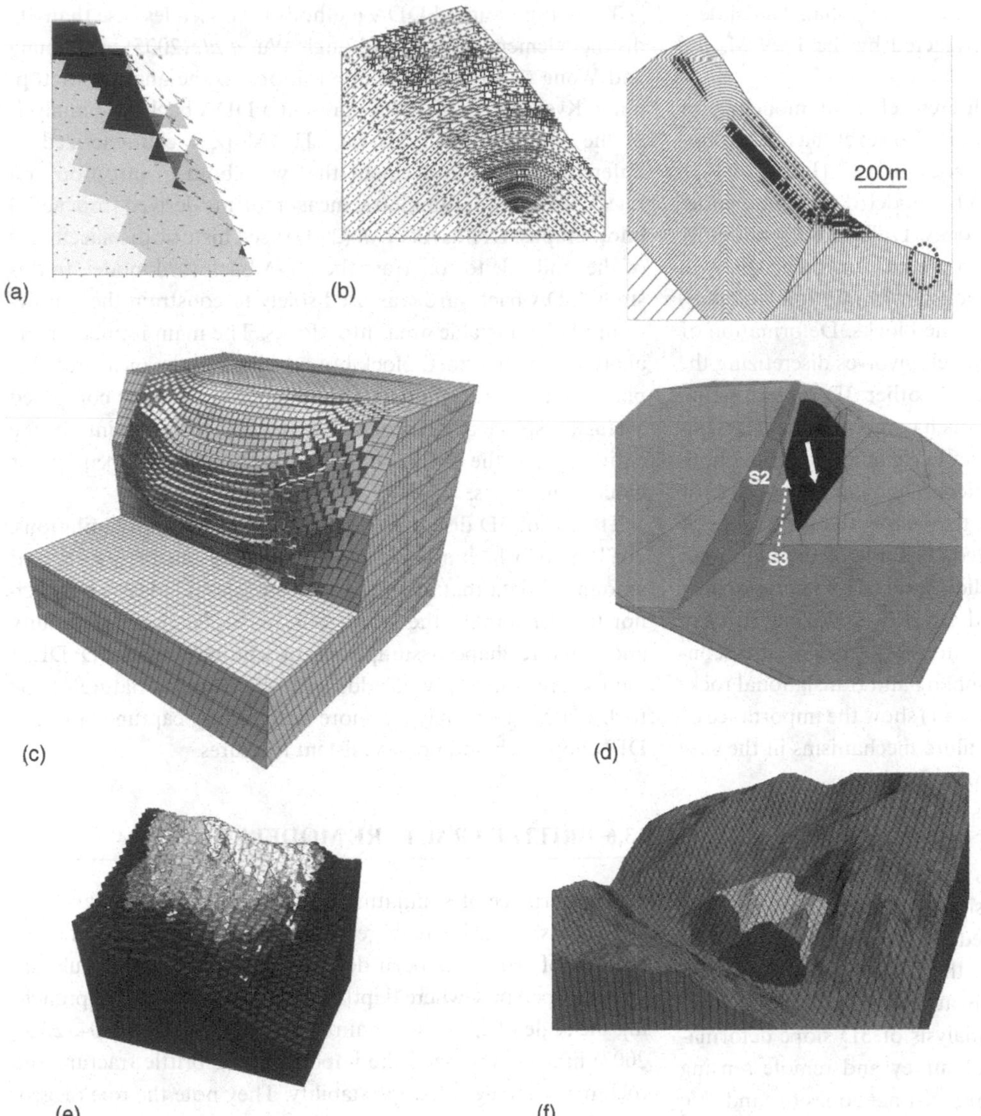

Fig. 13.4. Discontinuum analysis of rock slopes. (a) UDEC 2D distinct element analysis of the Delabole slate quarry failure, Cornwall, UK. (b) UDEC 2D distinct element analysis of the Frank Slide, showing shear failure along bedding planes (dark bands) and tensile failure in the fold hinge (T) (modified from Benko and Stead, 1998). (c) 3DEC 3D distinct element model of conceptual toppling failure (modified from Brideau and Stead, 2010). (d) 3DEC distinct element model of the Palliser landslide (modified from Sturzenegger and Stead, 2012). (e) 3DEC block model of Turtle Mountain; areas of higher displacement are lighter gray. (f) Complex 3DEC topographic model of the Beauregard landslide, created using SURFER™ and geostatistical fits to data (modified from Kalenchuk, 2010).

each excavation stage. Their models closely followed the mining stress path; thus they were able to evaluate the effects of mining-induced stresses on progressive slope deformation. The simulated and measured slope displacements in their four case studies were in close agreement. Armstrong and Rose (2009), using a similar UDEC analysis, noted that a strain-softening UDEC approach can provide valuable information on the slope disturbance parameter, *D*, used in the Hoek–Brown failure criterion. The parameter is estimated from the relative percentage of at-yield to non-yielded zones and its variation with depth.

The applicability of distinct element codes to dynamic analysis allows the effects of seismicity to be explored (Eberhardt and Stead, 1998; Havenith *et al.*, 2003; Bhasin and Kaynia, 2004). Recent studies have also considered the influence of thermal stresses on rock-slope deformation using UDEC, with constraints provided by field instrumentation (Watson *et al.*, 2004; Gischig, 2011). Thermally induced slope deformation may affect displacements along discontinuities at depth within a rock slope and are an important topic for further research.

Discontinuous deformation analysis (DDA) was originally developed by Goodman and Shi (1985) as a back-analysis method to fit a deformed block system to field-measured displacements or observations. It was later modified to allow a complete analysis of deformation of blocky systems. Unlike the distinct element methods, which use an explicit solution scheme, DDA uses an implicit solution to solve for unknown displacements. Interaction between blocks is simulated by mechanical springs or penalty functions. Khan (2010) provides a useful comparison of the theory and application of DEM and DDA, and many other authors have verified DDA codes against distinct element codes. The use of DDA in rockslide modeling has been demonstrated by numerous researchers, including Sitar and MacLaughlin (1997), Hatzor *et al.* (2004), and Sitar *et al.* (2005). Wu *et al.* (2009) recently undertook a dynamic DDA analysis of the central Taiwan Chiu-fen-erh-shan landslide – a large 36 million m^3 landslide, triggered by the 1999 M_w 7.6 Taiwan Chi-Chi earthquake.

To date, relatively little 3D discrete element modeling of landslides has been done. Some work, however, has been done using the 3D distinct element codes 3DEC (Itasca, 2011), which allows a blocky medium to be modeled assuming either deformable or non-deformable blocks. Different discontinuity constitutive criteria (including area contact elastic/plastic with Coulomb slip failure and continuously yielding joint models) are used to model displacement of the blocks. Deformation of the blocks, as in the 2D UDEC model, involves discretizing the blocks into finite difference zones. Another 3D discontinuum method is the 3D DDA method, which combines discrete element representation of joints with finite elements to model both discontinuity and block deformation. The major advantage of 3D discrete element codes is that they allow the kinematics of the landslide to be fully appreciated. The role of different joint sets or other structures such as sliding surfaces, lateral or rear release surfaces can be investigated. Brideau (2010) and Brideau and Stead (2010, 2011) used 3DEC to study the role of discontinuity set orientations on both toppling and translational rock-slope failures. Brideau and Stead (2011) show the importance of kinematic release on rock-slope failure mechanisms in the case of both conceptual models and major landslides.

Three-dimensional DEM codes can provide a realistic simulation of the complex nature of landslides, in which several failure mechanisms may evolve as the failure progresses or may occur in different parts of the slope. Kalenchuk (2010) performed two of the best-constrained 3D discontinuum landslide modeling studies to date: one on the Downie Slide in British Columbia and the other on the Beauregard landslide in Italy. Both case studies allowed back analysis of 3D slope deformations constrained by conventional survey and remote sensing data and borehole instrumentation. Distinct zones of landslide movement were reproduced in the 3D DEM model, providing significant advances in our understanding of the development of landslides. Kalenchuk (2010) also demonstrated the value of geostatistical surface-fitting algorithms for making optimal use of site data in 3D models. It is widely accepted that many, if not most, large-scale rockslides have three-dimensional complexities that may reflect variations in geologic structure, geotechnical properties, groundwater pressures, and slope geometry. Figure 13.4c and 13.4d illustrate the wide range of applications of 3DEC, from a conceptual kinematic investigation of toppling failure (Brideau and Stead, 2010) to a bedding-controlled translational failure (Sturzenegger, 2010). Figure 13.4e and 13.4f are applications of 3DEC in modeling more complex, 3D landslide geometries and structures. Figure 13.4e shows part of an investigation to assess the influence of discontinuity sets on slope deformation at Turtle Mountain, Alberta (Brideau *et al.*, 2011). Figure 13.4f illustrates Kalenchuk's (2010) 3DEC model of the Beauregard Dam abutment slope, where geostatistical software was used to investigate the optimal shear zone geometry from borehole and geophysical data, modeling utilized constraints imposed by InSAR monitoring data.

Three-dimensional DDA methods have seen less use than 3D distinct element models, although Wu *et al.* (2005) and Yeung and Wong (2007) describe applications to the analysis of toppling. Kveldsvik *et al.* (2009) present a DDA backward analysis of the Åknes slope in Norway. This slope is characterized by different areas of movement that vary both in direction and rate. Using photogrammetric measurements derived from aerial photographs, Kveldsvik *et al.* (2009) used historic displacements of the landslide to constrain the DDA backward model. In this study DDA backward was used solely to constrain the subdivision of the unstable area into blocks. The main inputs for the analysis were assumed block boundaries and the measured displacements. The outputs from the analysis were the computed average displacement error of all the measuring points in the entire model, the strains in each block, and the openings of block boundaries.

The use of 3D discrete element codes has several limitations, not least of which are the demands placed on the quality and amount of data that are required. Care should always be taken not to oversimplify the slope rock mass. The joint-continuity and fracture-shape assumptions in routinely used 3D DEM models are commonly at odds with the complex nature of the rock mass, which may be more realistically captured using a DFN approach and non-persistent fractures.

13.6 BRITTLE FRACTURE MODELING

The importance of simulating brittle fracture in the analysis of large landslides is being increasingly recognized. A considerable amount of work has been done on brittle fracture simulation in large open pits, where depths of 1 km or more are approaching the scale of high mountain slopes. Stead *et al.* (2004, 2006, 2007) have summarized the importance of brittle fracture and rock mass damage in slope stability. They note the role of geomorphic and geologic processes in promoting damage within a rock mass, as well as the role that brittle fracture plays in sliding release and internal deformation within complex rockslides.

13.6.1 FEM/DEM

Hybrid finite–discrete element codes that incorporate a fracture mechanics algorithm are finding increasing use in the analysis of rock slopes. These codes are particularly useful in modeling brittle fracture mechanisms associated with high mountain slopes and large open pits, where the high stress concentrations make intact rock fracture an important component of failure. Stead and Coggan (2006) describe the use of the ELFEN code (Rockfield, 2011) in modeling rockslides in what they term a "total slope analysis" approach. The ELFEN code allows the progressive transition of an intact blocky rock mass into rock debris during runout and deposition to be simulated. Each newly created particle is discretized and the software checks for further fracturing as the simulation continues. Stead and Coggan (2006) demonstrate the importance of brittle fracture in the Elm landslide in Switzerland. They also show the importance

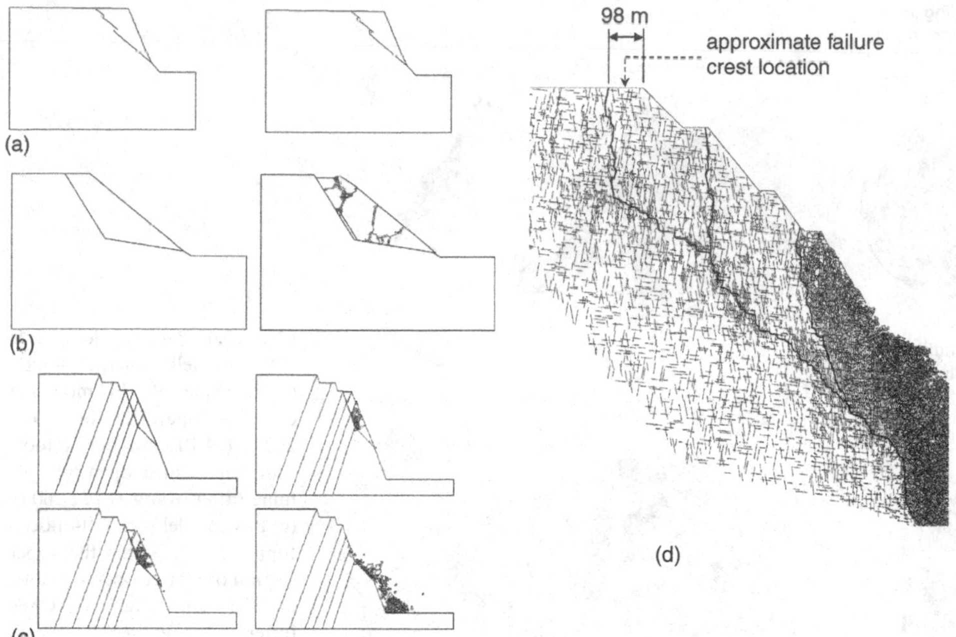

Fig. 13.5. Brittle fracture modeling of rock slopes using FEM/DEM. (a) Simulation of rock bridge failure using a step-path. (b) Simulation of brittle fracture during active-passive failure of a 50-m-high slope in a weak rock mass. (c) Total slope analysis of a failure in the Delabole slate quarry (see Fig. 13.4a). (d) Simulation of 800-m-high brittle fracture slope failure at the Palabora open-pit mine, South Africa, induced by underground block caving.

of brittle fracture associated with other failure mechanisms, including planar, biplanar, buckling, and toppling. Eberhardt *et al.* (2004) used the FEM/DEM approach in a toolbox of methods, progressing from continuum, through discontinuum, to FEM/DEM modeling, to illustrate the importance of progressive fracture in the Randa rockslide in Switzerland.

The importance of step-path failure and fracture mechanics in rock-slope failure has been recognized for several decades, but only since the advent of FEM/DEM modeling has it been possible to realistically model brittle fracture processes associated with step-path failure generation (Stead *et al.*, 2007; Yan, 2008; Vyazmensky *et al.*, 2010). Recently, discrete fracture networks and FEM/DEM codes have been integrated into brittle fracture modeling of more realistic, fractured rock masses containing multiple discontinuity sets. Figure 13.5 shows a range of FEM/DEM slope applications to rock-slope modeling, ranging from a simple step-path jointed rock mass (Fig. 13.5a) to a combined FEM/DEM–DFN model of an 800-m-high slope failure induced by underground block cave mining (Fig. 13.5d; Vyazmensky *et al.*, 2010). Figure 13.5b and 13.5c show, respectively, a simulation of fracture during a 50-m-high active–passive rock-slope failure and a combined sliding–toppling failure. The computing requirements of FEM/DEM brittle fracture codes are high, and currently this method has been limited in application to 2D slope models. With the growth of computing power and parallelization of the FEM/DEM codes, we suggest that these methods will soon allow even more realistic simulation of complex 3D slope failures.

13.6.2 PARTICLE FLOW CODES

An alternative method of modeling both brittle fracture in rock slopes and runout is to use particle flow codes. Both 2D and 3D particle flow codes are commercially available and have been applied in rock-slope modeling. In particle flow codes, circular particles (2D) or spheres (3D) are bonded together. For landslide applications, the bonds between particles are broken when the induced stresses exceed the bond strength, and the deformation is then controlled by inter-particle friction. Figure 13.6a illustrates the principles of a 2D particle flow model (PFC2D) that incorporates intact fracture simulation between bonded particles and sliding along defined joint sets (Lorig *et al.*, 2009). Particle flow codes such as PFC2D and PFC3D (Itasca, 2011) allow initial failure and debris transport and runout to be modeled, and have considerable potential for characterizing the different stages of slope failure (Stead *et al.*, 2007). Wang *et al.* (2003) successfully used PFC2D to model footwall slope plowing and buckling failures (Fig. 13.6b). PFC2D has also been used to model large 1000-m-high slopes at the Chucquimata open-pit mine in Chile, where toppling failure mechanisms have been simulated and found to be in close agreement with observations (Fig. 13.6c; Lorig *et al.*, 2009). Detailed modeling of the initiation and progression of brittle fracture in 3D models of landslides is currently limited by the size of the model that can be constructed. High mountain slopes and large open pits cannot be modeled without the assumption of unrealistically high sphere size, which results in an increase in model uncertainty. Poisel and Preh (2008) describe the application of PFC3D in runout modeling of large landslides, including the Frank Slide (Fig. 13.6d), and Bazgard *et al.* (2011) demonstrate the potential use of PFC3D in the dynamic analysis of high mountain slopes.

13.6.3 UDEC DAMAGE MODELS

UDEC damage models have been proposed by Alzo'ubi (2009) for a wide range of instabilities, from physical models to high open-pit slopes and engineered highway cuts. Damage models

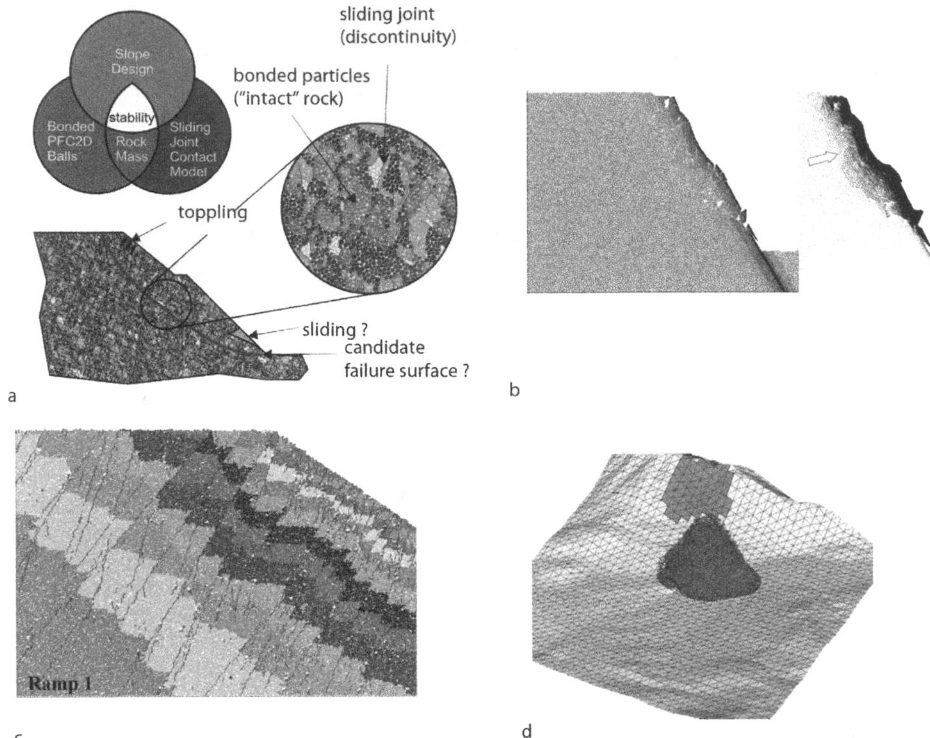

Fig. 13.6. Particle flow code (PFC) modeling of rock slopes. (a) Principles of PFC modeling of a rock slope (after Lorig *et al.*, 2009). (b) PFC model of footwall failure in a open-pit coal mine (after Wang *et al.*, 2003). (c) PFC model of a 130-m-deep toppling zone within the upper section of a large open-pit slope at the Chucquimata, Chile (after Lorig *et al.*, 2009). (d) Modeling of rock-slope failure and runout of the Frank Slide using PFC3D (after Poisel and Preh, 2008).

are based on the use of the UDEC Voronoi code, where the block rock mass in a UDEC model is discretized into polygonal zones. The boundaries between these zones are assigned intact rock properties, such that when the strength of a boundary is exceeded by induced stresses in a slope, the rock fractures along the polygon contacts. Voronoi models have been available for some time, but only recently has this technique been used within UDEC to model rock-slope fracture. Care must be taken when using this method to ensure that the Voronoi contact properties are calibrated and the sizes of the polygons realistically capture the slope fracture process.

Alzo'ubi (2009) successfully used the UDEC Voronoi method to model the Checkerboard Creek rock slope and a toppling failure at the Highland Valley open-pit mine, both in British Columbia. He also showed that rock support measures could be included in models of rock slopes; that changes in tensile strength due to weathering are an important factor in brittle failure; and that the UDEC Voronoi method was better able to capture toppling failure in physical models than conventional UDEC.

13.6.4 THE SYNTHETIC ROCK MASS (SRM) APPROACH

The effect of scale on rock mass strength is a major issue in rock-slope modeling. Conventional approaches use rock mass classification and characterization schemes and the Hoek–Brown criterion, but their appropriateness to high mountain slopes and large open pits has been questioned. It is not possible to test large rock masses in compression or tension, and therefore researchers have used virtual numerical testing of large synthetic rock masses (SRMs) which consist of PFC3D models incorporating discrete fracture or joint networks (DFNs). Large cylindrical SRM samples ranging in length from a few meters to >80 m can be numerically tested in uniaxial compression, uniaxial tension, and triaxial compression. Samples are loaded at a variety of orientations to the joint networks, allowing both the influence of scale and anisotropy on rock mass strength to be assessed. Once this strength has been determined in SRM (PFC3D + DFN) models, an equivalent anisotropic strain-softening constitutive model can then be incorporated into a continuum code such as FLAC3D. Figure 13.7 shows the principles of the SRM technique; a DFN is incorporated into a particle flow code model for testing under different stress regimes in order to provide a synthetic rock mass at the desired scale (see Mas Ivars *et al.*, 2011, for further details). This approach has been used successfully in modeling large open-pit slopes (Sainsbury *et al.*, 2008). The information required to derive the synthetic rock mass is not trivial and the method may only be appropriate for the most intensively investigated landslides. A similar SRM approach, but with FEM/DEM codes, is also being used in 2D modeling of open pits and engineered rock slopes. In this case, the SRM is an ELFEN model incorporating a DFN.

An important recent development in the SRM approach has been the use of the lattice-spring method in the 3D analysis of rock slopes. The method involves the construction of models

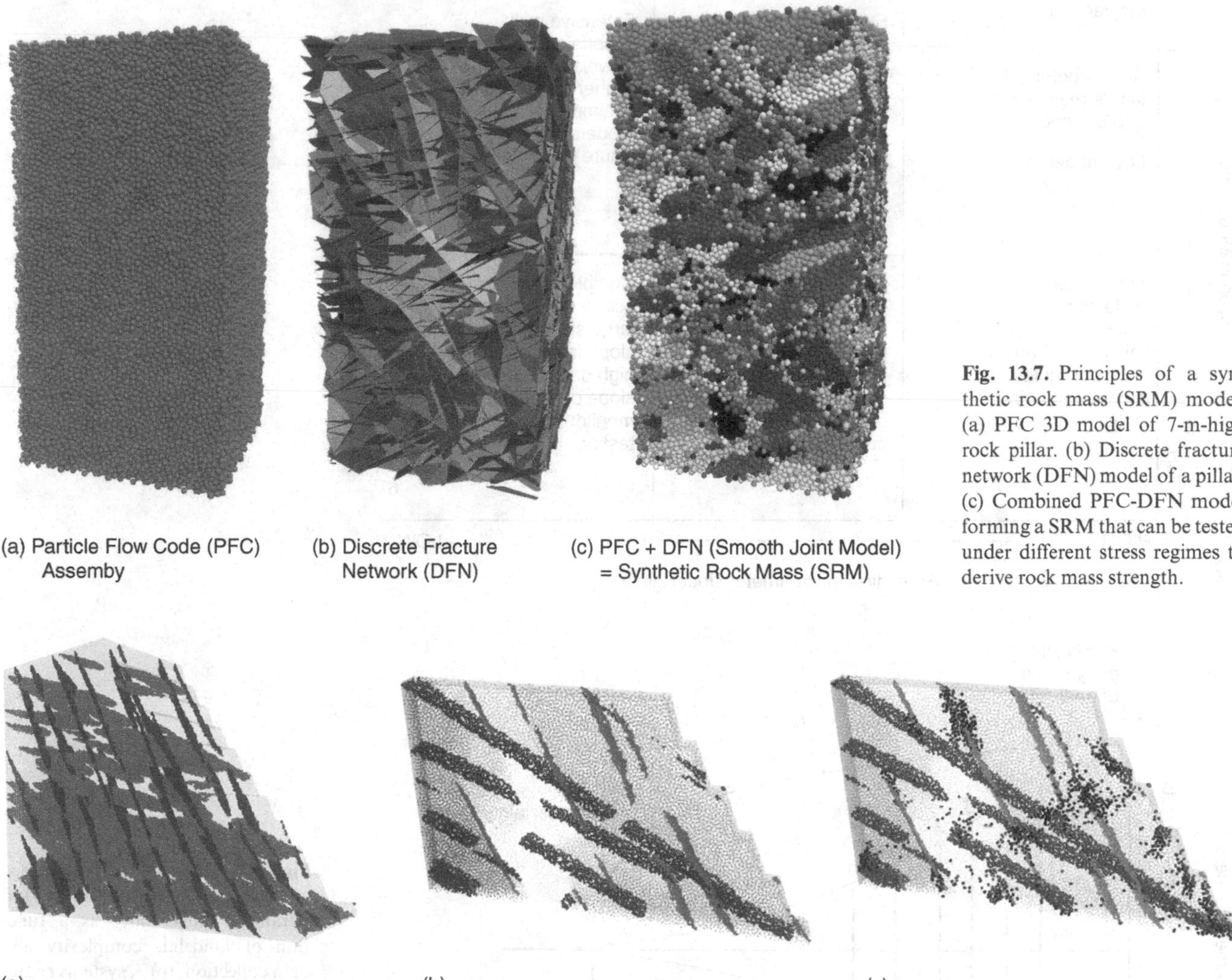

Fig. 13.7. Principles of a synthetic rock mass (SRM) model. (a) PFC 3D model of 7-m-high rock pillar. (b) Discrete fracture network (DFN) model of a pillar. (c) Combined PFC-DFN model forming a SRM that can be tested under different stress regimes to derive rock mass strength.

Fig. 13.8. Lattice-spring modeling of an open-pit slope: (a) 500-m sector of 1000-m-high slope with 10 benches showing DFN with two discontinuous joint sets; (b) Cross-section slice through the upper part of the 1000-m slope showing the joint traces within the lattice-spring model; (c) Stress-induced failure of intact rock bridges between joints (black circles), new fractures can act as conduits for groundwater. (Modified from Cundall and Damjanac, 2009.)

consisting of a lattice of point masses connected by springs. Cundall and Damjanac (2009) successfully demonstrate the use of a 3D lattice-spring code (slope model) in simulating rock bridge failure in a 500-m-long section of a 1000-m-high rock slope consisting of 10 benches that are cut by 2 non-persistent discontinuity sets (Fig. 13.8). They comment that lattice-spring methods are versatile and combine the strengths of traditional distinct element methods to simulate block and joint displacements with the ability to simulate intact rock fracture. Slope model can accept DFNs and simulate both mechanical and coupled groundwater mechanics behavior. It is similar to PFC3D, but with much greater efficiency because sphere and ball contacts are replaced, respectively, by point masses (nodes) and springs. The 3D lattice-spring modeling method, if combined with DFN models, has significant potential for improving our understanding of landslide failure mechanisms.

13.7 FUTURE CHALLENGES IN THE NUMERICAL ANALYSIS OF ROCK SLOPES

13.7.1 CONSIDERATION OF SPATIAL VARIABILITY IN NUMERICAL MODELING

Most continuum modeling of rock slopes is undertaken using a deterministic approach, where the sensitivity of the model response to varying input is investigated. The current state of practice in the numerical modeling of rock slopes, large landslides, and open-pit mines is to divide the slope into geomechanical domains or design sectors based on the geological strength

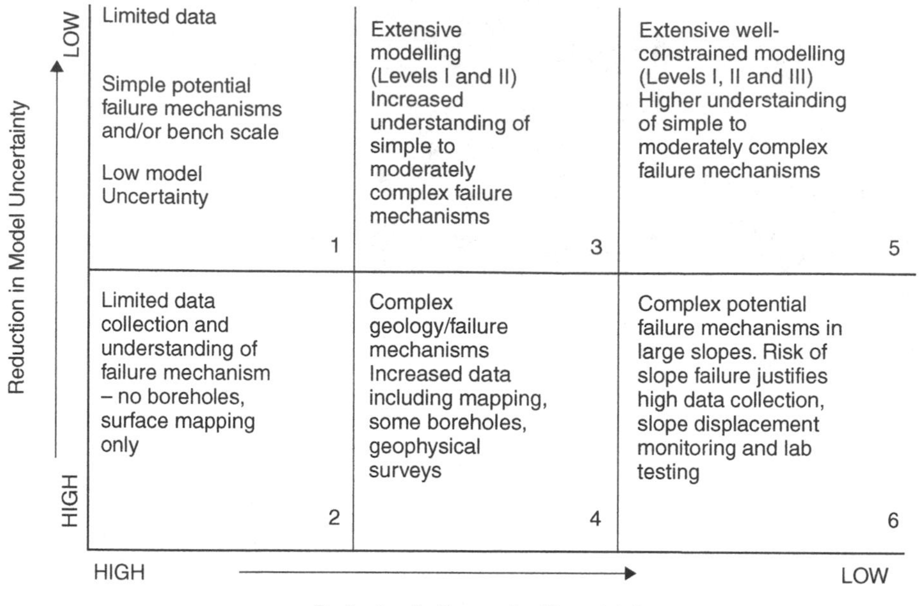

(a)

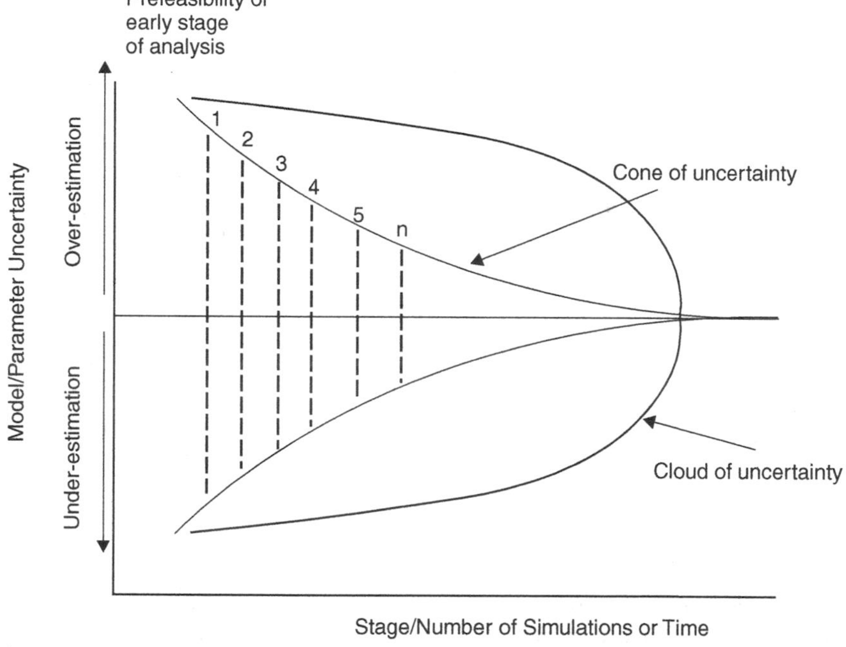

(b)

Fig. 13.9. Uncertainty in rock-slope analysis. (a) Changes in model and parameter uncertainty as a function of landslide complexity and data collection. (b) A systems engineering approach to uncertainty in rock-slope modeling using the cone of uncertainty and the cloud of uncertainty. The width of the cone of uncertainty narrows with increasing model experience (during stages 1 to n) and with carefully planned monitoring/ laboratory testing constraints. A cloud of uncertainty exists where models/additional data are not optimized to reduce uncertainty.

index (GSI), joint set characteristics, and slope orientation. Constant GSI-derived rock mass strengths are generally used to represent geomechanical domains. Where data are available, a probabilistic approach can be used to incorporate differences in the rock mass properties of the units. Spatial variation of rock mass strength is rarely considered, but it is generally inherent in the slope being considered. Jefferies *et al.* (2008) discuss the influence of rock strength spatial variability on slope stability. They described the initial development of a stochastic model for slope failure in rock, based on a measured distribution of GSI from an open-pit mine. They used FLAC and a Hoek–Brown failure criterion to model a 500-m-high, 45° slope. Using a custom-programmed subroutine, they populated the FLAC zones with values of cohesion and friction randomly chosen from normally distributed values of GSI and with stress values calculated during an initial elastic model state. Incorporation of the inherent variability of stochastic mechanical properties and geology into models allows more reliable analyses. Jefferies *et al.* (2008) compared the results of the stochastic model with a deterministic model that assumes a mean value of GSI. The factor of safety for the deterministic mean was 1.17, yet approximately 20 percent of the stochastic simulations resulted in failure. Jefferies *et al.* (2008) concluded that the stochastic model allowed "the failure mechanisms to seek the weaker ground."

Griffiths and Marquez (2007) demonstrated the importance of considering the 3D spatial variability of geomechanical properties in slope modeling. They developed a random-variable, finite element method to consider spatial variability in a 3D FEM model. They showed that as the spatial variability of strength increased and, depending on the length–depth ratio of the slope, the 3D factor of safety of the slope could be less than that predicted by both FEM models with no variability and 2D limit equilibrium methods. This finding is in contrast to the normal assumption that the 2D factor of safety is always less than the 3D factor of safety. We conclude from this study that, unless the spatial variability of 3D geomechanical properties is included in a continuum model, an apparently more sophisticated analysis may lead to incorrect results. For large-scale slopes, large-memory, multiprocessor platforms may be required to undertake reliable, probabilistic-geostatistical modeling, but promising results have been reported using mine-scale 3D FEM models and the alternate point estimate method (Reusch *et al.*, 2008).

Pasculli *et al.* (2006) discuss the effects of the spatial variability of mechanical properties in a 3D study of a landslide. They conducted more than 50 FLAC3D simulations of the same landslide geometry, with different mechanical parameters assigned to the model grid based on assumed Gaussian distributions. Adopting a shear strength reduction approach, they characterized the result from each spatial property mesh model run by its proximity to the Mohr–Coulomb failure envelope, using what they termed a strength failure indicator (SFI). By statistically analyzing the SFI parameters for each sector of the FLAC3D model, they developed a numerical microzonation concept and defined areas that had undergone plastic damage using a degradation index. Their results emphasize the need to consider the effect of spatial variability in rock strength on slope stability. We believe that the increased use of geostatistical methods to characterize input into both 2D and 3D models of rock slopes is essential, and see this as an important area for further research in order to reduce uncertainty in current numerical modeling practice.

13.7.2 TREATMENT OF UNCERTAINTY IN NUMERICAL MODELING

Uncertainty has been recognized as an important consideration in risk assessment for a long time. Consideration of model and data uncertainty is a key, but often inadequately considered, operation when undertaking the numerical modeling of landslides. Lacasse and Nadim (1996) discuss uncertainty in characterizing soil properties in terms of aleatoric and epistemic uncertainty: the former a function of spatial parameter variability and the latter including measurement technique variability, data quantity, and modeling errors. To date, however, there has been little consideration of the importance of aleatory and epistemic uncertainty in the modeling of rock slopes. Numerous techniques have been proposed to evaluate the role of uncertainty in rock engineering, including fuzzy analysis, value of additional information methods, and Bayesian updating. The objectives are to understand which parameters have a major influence on model results, to reduce parameter and model uncertainty, and to quantify the reliability of a slope model. Figure 13.9 shows a systems engineering approach to visualizing uncertainty in the numerical modeling of rock slopes based on the "cone of uncertainty" and the "cloud of uncertainty." The often recommended "toolbox approach" or the "start simple and build in complexity approach" represent two ways of reducing the dimensions of the cone of uncertainty in slope modeling. Uncertainty is greatest at the beginning of a numerical modeling project or in the pre-feasibility stage of mine slope design. As different levels of modeling (levels I –III) are applied, constrained by additional data collection and instrumentation, the width of the cone of uncertainty is reduced. If analyses are poorly planned or do not add information, then a cloud of uncertainty may exist where additional analyses fail to reduce the uncertainty. We suggest that a systems engineering approach to rock-slope modeling may have potential in recognizing, and thereby reducing, model and parameter uncertainty.

13.7.3 IMPROVED MODELING CONSTRAINTS

Increasing use has been made of instrumentation to constrain models in recent years, but much remains to be done in this area. Most models include numerical time steps; only a few use a creep criterion and provide simulated displacements over real time. The calibration of numerical time steps and simulated displacement against measured survey displacement, both surface and subsurface, has allowed a significant increase in confidence in the results of numerical modeling to rock slopes. Researchers have also begun to use nonconventional instrumentation measurements to constrain model simulations, including displacements determined using ground- and satellite-based InSAR, slope stability radar, and ground-based LIDAR. The increased use of geophysical instruments (ground-penetrating radar; borehole radar; electrical, seismic, thermal, and microseismic tools; and acoustic emission) has provided not only improved characterization of rock masses and failure surfaces, but also of slope kinematics. We anticipate that, in the future, numerical modeling will increasingly be complemented by remotely sensed displacement measurements and the use of a wide range of geophysical instruments to constrain movement and changes in rock mass quality within a slope. The greater availability of 3D geophysical datasets and borehole data will require the use of efficient 3D visualization methods, such as virtual and augmented reality.

13.8 CONCLUSIONS

Computing power has increased tremendously over the past 50 years, and the engineer now has the ability to run highly sophisticated numerical models in both 2D and 3D on a desktop computer. This growth will continue as multiprocessor PCs and parallel processing software become more widely available. An increase in the ability to model slopes more realistically carries an obligation to use the available three-dimensional data in the most appropriate way. The engineer must be conversant with a wide range of numerical methods in what is commonly referred to as the "numerical modeling toolbox." Simple limit equilibrium analysis will continue to be important in preliminary slope parametric studies, but increasing complexity or risk requires the use of more sophisticated numerical models. It is particularly important that the engineer and geoscientist recognize and appreciate model and parameter uncertainty when using these sophisticated numerical techniques. The development of equally sophisticated data collection methods and the integration of data with numerical modeling are prerequisites for realistic and rigorous modeling. Important concepts that must be further researched in order to ensure that our modeling capability is matched by the required model input data include the characterization of discontinuity persistence and rock bridges, scale effects on rock mass strength, hydromechanical effects in fractured rock, and time-related geologic and geomorphic processes.

REFERENCES

Adhikary, D. P. and Dyskin, A. V. (2007). Modelling of progressive and instantaneous failures of foliated rock slopes. *Rock Mechanics and Rock Engineering*, 40, 349–362.

Alzo'ubi, A. M. (2009). *The effect of tensile strength on the stability of rock slopes*. Ph.D. thesis, University of Alberta, Edmonton, AB.

Ambrosi, C. and Crosta, G. B. (2011). Valley shape influence on deformation mechanisms of rock slopes. In *Slope Tectonics*, ed. M. Jaboyedoff. Geological Society of London, Special Publication 351, pp. 215–233.

Aringoli, D., Calista, M., Gentili, B., Pambianchi, G. and Sciarra, N. (2008). Geomorphological features and 3D modelling of Montelparo mass movement (central Italy). *Engineering Geology*, 99, 70–84.

Armstrong, J. and Rose, N. D. (2009). Mine operation and management of progressive slope deformation on the south wall of the Barrick Goldstrike Betze-Post open pit. In *Proceedings of the Slope Stability 2009 Conference, Santiago, Chile*.

Baczynski, N. R. P. (2008). STEPSIM4: Revisited network analysis methodology for critical paths in rock mass slopes. In *Proceedings of SHIRMS, Vol. 2*, ed. Y. Potvin, J. Carter, A. Dyskin and R. Jeffrey. Perth: Australian Centre for Geomechanics, pp. 405–418.

Baczynski, N. R. P., Sheppard, I. K., Smith, K. J., Simbina, P. and Sakail, R. (2008). Toppling slope failure: Predicted versus actual, Ok Tedi, Papua New Guinea. In *Proceedings of SHIRMS, Vol. 2*, ed. Y. Potvin, J. Carter, A. Dyskin and R. Jeffrey. Perth: Australian Centre for Geomechanics, pp. 419–431.

Bazgard, A., Davies, T. R. and Cundall, P. A. (2011). Dynamic analysis of mountain response using PFC3D and FLAC3D. In *Proceedings of the 2nd International FLAC/DEM Symposium*, ed. D. Sainsbury, R. Hart, C. Detournay and M. Nelson. Minneapolis, MN: Itasca International Inc., pp. 577–583.

Benko, B. and Stead, D. (1998). The Frank slide: A re-examination of the failure mechanism. *Canadian Geotechnical Journal*, 35, 299–311.

Bhasin, R. and Kaynia, A. M. (2004). Static and dynamic simulation of a 700-m-high rock slope in western Norway. *Engineering Geology*, 71, 213–226.

Bovis, M. J. and Stewart, T. W. (1998). Long-term deformation of a glacially undercut rock slope, southwest British Columbia. In *Proceedings of the 8th International Congress of the International Association for Engineering Geology and the Environment*, pp. 1267–1276.

Brideau, M.-A. (2010). *Three-dimensional kinematic controls on rock slope stability conditions*. Ph.D. thesis, Simon Fraser University, Burnaby, BC.

Brideau, M.-A. and Stead, D. (2010). Controls on block toppling using a three-dimensional distinct element approach. *Rock Mechanics and Rock Engineering*, 43, 241–260.

(2011). The influence of three-dimensional kinematic controls on rock slope stability. In *Proceedings of the 2nd International FLAC/DEM Symposium*, ed. D. Sainsbury, R. Hart, C. Detournay and M. Nelson. Minneapolis, MN: Itasca International Inc., pp. 213–220.

Brideau, M.-A., Pedrazzini, A., Stead, D. *et al.* (2011). Three-dimensional slope stability analysis of South Peak, Crowsnest Pass, Alberta, Canada. *Landslides*, 8, 139–158.

Bromhead, E., Dixon, N. and Ibsen, M.-L. (ed.) (2000). *Landslides in Research, Theory and Practice*. London: Thomas Telford.

Cappa, F., Guglielmi, Y., Soukatchoff, V. M. *et al.* (2004). Hydromechanical modeling of a large moving rock slope inferred from slope levelling coupled to spring long-term hydrochemical monitoring: Example of the La Clapière landslide (Southern Alps, France). *Journal of Hydrology*, 291, 67–90.

Chemenda, A. I., Bois, T., Bouissou, S. and Tric, E. (2009). Numerical modelling of the gravity-induced destabilization of a slope: The example of the La Clapière landslide, southern France. *Geomorphology*, 109, 86–93.

Chen, Z., Zhang, J., Li, Z., Wu, F. and Ho, K. (2008). Landslides and engineered slopes, from the past to the future. In *Proceedings of the 10th International Symposium on Landslides, Xi'an, China*. London: Taylor & Francis Group.

Coggan, J. S. and Pine, R. J. (1996). Application of distinct-element modelling to assess slope stability at Delabole slate quarry, Cornwall, England. *Transactions of the Institute of Mining and Metallurgy, Section A*, 105, A22–A30.

Coggan, J. S., Stead, D. and Howe, J. H. (2000). Characterisation of a structurally controlled flowside in a kaolinised granitic slope. In *Landslides in Research: Theory and Practice*, ed. E. Bromhead, N. Dixon and M.-L. Ibsen. London: Thomas Telford, pp. 299–304.

Cundall, P. A. and Damjanac, B. (2009). A comprehensive 3D model for rock slopes based on micromechanics. In *Proceedings of the Slope Stability 2009 Conference, Santiago, Chile*.

Dawson, E. M., Roth, W. H. and Drescher, A. (1999). Slope stability analysis by strength reduction. *Geotechnique*, 49, 835–840.

Diederichs, M. S., Lato, M., Hammah, R. and Quinn, P. (2007). Shear strength reduction approach for slope stability analyses. In *Proceedings of the First Canadian–US Rock Mechanics Symposium*, ed. E. Eberhardt, D. Stead and T. Morrison. Rotterdam: A. A. Balkema, pp. 319–327.

Eberhardt, E. and Stead, D. (1998). Numerical analysis of slope instability in thinly bedded weak rock. In *Proceedings of the*

8th International Association of Engineering Geology Congress, Vancouver, Canada, September 1998, pp. 3011–3018.

Eberhardt, E., Stead, D. and Coggan, J.S. (2004). Numerical analysis of initiation and progressive failure in natural rock slopes: The 1991 Randa rockslide. *International Journal of Rock Mechanics and Mining Science*, 41, 69–87.

Francois, B., Tacher, L., Bonnard, C., Laloui, L. and Triguero, V. (2007). Numerical modelling of the hydrogeological and geomechanical behaviour of a large slope movement: The Triesenberg landslide (Liechtenstein). *Canadian Geotechnical Journal*, 44, 840–857.

Gischig, V.S. (2011). *Kinematics and failure mechanisms of the Randa rock slope instability (Switzerland)*. Ph.D. thesis, ETH Zurich, Switzerland.

Goodman, R.E. and Shi, G. (1985). *Block Theory and Its Application to Rock Engineering*. Englewood Cliffs, NJ: Prentice Hall.

Griffiths, D.V. and Lane, P.A. (1999). Slope stability analysis by finite elements. *Geotechnique*, 49, 387–403.

Griffiths, D.V. and Marquez, R.M. (2007). Three-dimensional slope stability analysis by elasto-plastic finite elements. *Geotechnique*, 57, 537–546.

Grøneng, G., Lu, M., Nilsen, B. and Jenssen, A.K. (2010). Modelling of time-dependent behavior of the basal sliding surface of the Åknes rockslide area in western Norway. *Engineering Geology*, 114, 442–422.

Guglielmi, Y., Cappa, F.J., Rutqvist, J., Tsang, C.-F. and Thoraval, A. (2008). Mesoscale characterization of coupled hydromechanical behavior of a fractured-porous slope in response to free water-surface movement. *International Journal of Rock Mechanics and Mining Science*, 45, 862–878.

Hammah, R.E., Yacoub, T., Corkum, B. and Curran, J.H. (2008). The practical modelling of discontinuous rock masses with finite element analysis. In *Proceedings of the 42nd U.S. Rock Mechanics Symposium*, ARMA 08–180.

Hammah, R.E., Yacoub, T. and Curran, J.H. (2009a). Variation of failure mechanisms of slopes in jointed rock masses with changing scale, ROCKENG09. In *Proceedings of the 3rd CANUS Rock Mechanics Symposium*, ed. M. Diederichs and G. Grasselli. Toronto, ON: Canadian Association of Rock Mechanics, Paper 3956.

Hammah, R.E., Yacoub, T. and Curran, J.H. (2009b). Probabilistic slope analysis with the finite element method. In *U.S. Rock Mechanics Symposium and 4th U.S.–Canada Rock Mechanics Symposium*, ARMA 09–149.

Hatzor, Y.H., Arzi, A.A., Zaslavsky, Y. and Shapira, A. (2004). Dynamic stability analysis of jointed rock slopes using the DDA method: King Herod's Palace, Masada, Israel. *International Journal of Rock Mechanics and Mining Science*, 41, 813–832.

Havenith, H.-B., Strom, A., Calvetti, A.F. and Jongmans, D. (2003). Seismic triggering of landslides. Part B. Simulation of dynamic failure processes. *Natural Hazards and Earth System Sciences*, 3, 663–682.

Hencher, S.R., Liao, Q.H. and Monaghan, B.G. (1996). Modelling slope behavior for open pits. *Transactions of the Institute of Mining and Metallurgy, Section A: Mining Industry*, 105, 37–47.

Hensold, G. (2011). *An integrated study of deep-seated gravitational slope deformations at Handcar Peak, southwestern British Columbia*. M.Sc. thesis, Simon Fraser University, Burnaby, BC.

Hungr, O. and Amann, F. (2011). Limit equilibrium of asymmetric laterally constrained rockslides. *International Journal of Rock Mechanics and Mining Science*, 48, 748–758.

Hungr, O., Salgado, F.M. and Byrne, P.M. (1989). Evaluation of a three-dimensional method of slope stability analysis. *Canadian Geotechnical Journal*, 26, 679–686.

Itasca (2011). *FLAC 7.0, FLAC3D V4, UDEC 5.0, 3DEC 4.0, PFC 4.0, PFC3D 4.0*. Minneapolis, MN: Itasca Consulting Group Inc. [available at www.itascacg.com/].

Jefferies, M., Lorig, L. and Alvarez, C. (2008). Influence of rock-strength spatial variability on slope stability. In *Continuum and Distinct Element Numerical Modeling in Geo-Engineering 2008*, ed. R. Hart, C. Detournay and P. Cundall. Minneapolis, MN: Itasca Consulting Group, Paper 01–05.

Kalenchuk, K.S. (2010). *Multi-dimensional analysis of large, complex slope instability*. Ph.D. thesis, Queen's University, Kingston, ON.

Khan, M.S. (2010). *Investigation of discontinuous deformation analysis for application in jointed rock masses*. Ph.D. thesis, University of Toronto, Toronto, ON.

Kinakin, D. (2004). *Occurrence and genesis of alpine linears due to gravitational deformation in southwestern British Columbia*. M.Sc. thesis, Simon Fraser University, Burnaby, BC.

Kinakin, D. and Stead, D. (2005). Analysis of the distributions of stress in natural ridge forms: Implications for the deformation mechanisms of rock slopes and the formation of sackung. *Geomorphology*, 65, 85–100.

Kveldsvik, V., Einstein, H.H., Nilsen, B. and Blikra, L. (2009). Numerical analysis of the 650,000 m^2 Åknes rock slope based on measured displacements and geotechnical data. *Rock Mechanics and Rock Engineering*, 42, 689–728.

Lacasse, S. and Nadim, F. (1996). Uncertainties in characterizing soil properties. In *Uncertainty in the Geologic Environment: From Theory to Practice: Proceedings of Uncertainty '96, Madison, Wisconsin*, ed. C.D. Shackelford, P.P Nelson, and M.J.S. Roth. Geotechnical Special Publication 58. New York: ASCE, pp. 49–75.

Lacerda, W.A., Erlich, M., Fontoura, S.A.B. and Sayao, A.S.F. (ed.) (2004). *Landslides: Evaluation and Stabilization*. Rotterdam: A.A. Balkema. pp. 1293–1298.

Lam, L. and Fredlund, D.G. (1993). A general limit equilibrium model of three-dimensional slope stability analysis. *Canadian Geotechnical Journal*, 30, 905–919.

Li, A.J., Merifield, R.S. and Lyamin, A.V. (2011). Effect of rock mass disturbance on the stability of rock slopes using the Hoek–Brown failure criterion. *Computers and Geotechnics*, 38, 546–558.

Lorig, L., Stacey, P. and Read, J. (2009). Slope design methods. In *Guidelines for Open Pit Slope Design*, ed. J. Read and P. Stacey. Collingwood, Australia: CSIRO Publishing, pp. 237–264.

Mas Ivars, D., Pierce, M.E., Darcel, C. *et al.* (2011). The synthetic rock mass approach for jointed rock mass modelling. *International Journal of Rock Mechanics and Mining Science*, 48, 219–244.

Merrien-Soukatchoff, V., Korini, T. and Thoraval, A. (2012). Use of an integrated discrete fracture network code for stochastic stability analyses of fractured rock masses. *Rock Mechanics and Rock Engineering*, 45, 159–181.

Nichol, S.L., Hungr, O. and Evans, S.G. (2002). Large-scale brittle and ductile toppling of rock slopes. *Canadian Geotechnical Journal*, 39, 773–788.

Pasculli, A., Calista, M. and Mangifesta, M. (2006). The effects of spatial variability of mechanical parameters on a 3D landslide study. In *4th International FLAC Symposium on Numerical Modeling in Geomechanics*, ed. R. Hart and P. Varona Minneapolis, MN: Itasca Consulting Group, Paper 01–05.

Pham, H.T.V. and Fredlund, D.G. (2003). The application of dynamic programming to slope stability analysis. *Canadian Geotechnical Journal*, 40, 830–847.

Poisel, R. and Preh, A. (2008). Modifications of PFC3D for rock mass fall modeling. In *Continuum and Distinct Element Numerical Modeling in Geo-Engineering 2008*, ed. R. Hart, C. Detournay and P. Cundall. Minneapolis, MN: Itasca Consulting Group, Paper 01–04.

Pritchard, M.A. and Savigny, K.W. (1990). Numerical modelling of toppling. *Canadian Geotechnical Journal*, 27, 823–834.

Reusch, F., Beck, D. and Tyler, D. (2008). Quantitative forecasting of sidewall stability and dilution in sub-level caves. In *Massmin2008*, ed. H. Schunnesson and E. Nordlund. Lulea, Sweden: Lulea Technology Press.

Riahi, A. and Curran, J. H. (2008). Application of Cosserat continuum approach in the finite element shear strength reduction analysis of jointed rock slopes. In *Proceedings of the 12th International Conference of International Association for Computer Methods and Advances in Geomechanics (IACMAG), Goa, India*. New York: ASCE, pp. 110–118.

Rockfield (2011). *ELFEN 2D/3D Numerical Modelling Package*. Swansea, UK: Rockfield Software Ltd. [available at www.rockfield.co.uk/].

Rocscience (2011). *Rocscience Software Products: DIPS V. 5.1, Rocplane V.2, Swedge V. 5, Slide V.6, Phase2 V.8*. Toronto, ON: Rocscience Inc. [available at www.rocscience.com].

Rose, N. D. and Scholz, M. F. (2009). Analysis of complex deformation behaviour in large open pit mine slopes using the Universal Distinct Element Code (UDEC). In *Proceedings of Slope Stability 2009, Santiago, Chile*.

Sainsbury, B., Pierce, M. E. and Mas Ivars, D. (2008). Analysis of caving behaviour using a synthetic rock mass–ubiquitous joint rock mass modelling technique. In *Proceedings of First Southern Hemisphere International Rock Mechanics Symposium (SHIRMS 2008), Perth, Australia*. International Society for Rock Mechanics (ISRM), pp. 243–253.

Simulia (2011). *Abaqus V 6.11*. Providence, RI: Simulia [available at www.simulia.com].

Sitar, N. and MacLaughlin, M. M. (1997). Kinematics and discontinuous deformation analysis of landslide movement. In *2nd Pan-American Symposium on Landslides, Rio de Janeiro*. International Society for Soil Mechanics and Geotechnical Engineering.

Sitar, N., MacLaughlin, M. M. and Doolin, D. M. (2005). Influence of kinematics on landslide mobility and failure mode. *Journal of Geotechnical and Geoenvironmental Engineering*, 131, 716–728.

Stead, D. and Coggan, J. S. (2006). Numerical modelling of rock slopes using a total slope failure approach. In *Landslides from Massive Rock Slope Failures*, ed. S. G. Evans, R. Hermanns and A. Strom. Dordrecht, Netherlands: Springer, pp. 131–142.

Stead, D. and Eberhardt, E. (1997). Developments in the analysis of footwall slopes in surface coal mining. *Engineering Geology*, 46, 41–61.

Stead, D., Coggan, J. S. and Eberhardt, E. (2004). Realistic simulation of rock slope failure mechanisms: The need to incorporate principles of fracture mechanics. *International Journal of Rock Mechanics and Mining Science*, 41, CD-ROM.

Stead, D., Eberhardt, E. and Coggan, J. S. (2006). Developments in the characterization of complex rock slope deformation and failure using numerical modelling techniques. *Engineering Geology*, 83, 217–235.

Stead, D., Coggan, J. S., Elmo, D. and Yan, M. (2007). Modelling brittle fracture in rock slopes: Experience gained and lessons learned. In *International Symposium on Rock Slope Stability in Open Pit and Civil Engineering*, ed. Y. Potvin. Australian Centre for Geomechanics, pp. 239–252.

Sturzenegger, M. (2010). *Multi-scale characterization of rock mass discontinuities and rock slope geometry using terrestrial remote sensing techniques*. Ph.D. thesis, Simon Fraser University, Burnaby, BC.

Sturzenegger, M. and Stead, D. (2012). The Palliser rockslide: Characterization and modelling of a stepped failure surface. *Geomorphology*, 138, 145–161.

Tommasi, P., Verrucci, L. Campedel, P., Veronese, L. Pettinelli, E. and Ribacchi, R. (2009). Buckling of high natural slopes: The case of Lavini di Marco (Trento-Italy). *Engineering Geology*, 109, 93–108.

Vyazmensky, A., Stead, D., Elmo. D. and Moss, A. (2010). Numerical analysis of block caving induced instability in large open pit slopes: A finite element / discrete element approach. *Rock Mechanics and Rock Engineering*, 43, 21–39.

Wang, C., Tannant, D. D. and Lilly, P. A. (2003). Numerical analysis of the stability of heavily jointed rock slopes using PFC2D. *International Journal of Rock Mechanics and Mining Sciences*, 40, 415–424.

Watson, A. D., Moore, D. P. and Stewart, T. W. (2004). Temperature influence on rock slope movements at Checkerboard Creek. In *Landslides: Evaluation and Stabilization*, ed. W. A. Lacerda, M. Erlich, S. A. B. Fontoura and A. S. F. Sayao. Rotterdam: A. A. Balkema, pp. 1293–1298.

Wu, J.-H., Ohnishi, Y., Shi, G.-H. and Nishiyama, S. (2005). Theory of three-dimensional discontinuous deformation analysis and its application to a slope toppling at Amatoribashi, Japan. *International Journal of Geomechanics*, 5, 179–195.

Wu, J.-H., Shang Lin, J.-S. and Chen, C.-S. (2009). Dynamic discrete analysis of an earthquake-induced large-scale landslide. *International Journal of Rock Mechanics and Mining Science*, 46, 397–407.

Yan, M. (2008). *Numerical modelling of brittle fracture and step-path failure: from laboratory to rock slope scale*. Ph.D. thesis, Department of Earth Sciences, Simon Fraser University, Burnaby, Canada.

Yeung, M. R. and Wong, K. L. (2007). Three-dimensional kinematic conditions for toppling. In *Proceedings of the First Canadian–US Rock Mechanics Symposium*, ed. E. Eberhardt, D. Stead and T. Morrison. Rotterdam: A. A. Balkema, pp. 335–340.

14 Remote sensing techniques and landslides

DAVID PETLEY

ABSTRACT

Over the past three decades, geophysical techniques, most notably remote sensing, have become a key component of landslide investigation. Key developments have been: (1) a move from the use of analog aerial and oblique photography through satellite image interpretation to the application of radar and laser technologies; and (2) the increased use of photogrammetry to allow remote collection of rock-slope property data. Throughout this period there has been a progressive improvement in the spatial resolution of the data available, cost, and geographic coverage. The result has been a continuous and often underappreciated improvement in our capacity to identify and analyze landslides. Some improvements have been progressive and gradual, such as in the capacity of optical satellite remote sensing, while others have occurred almost instantaneously, and with little fanfare, such as the appearance of the Google Earth tool. As a consequence, this field is vast and rapidly developing. This chapter aims to examine recent key developments in the use and application of these techniques in landslide investigation, focusing primarily on remote sensing. Likely future developments in this area are outlined in the final section, although with the caveat that predicting likely advances in technology and techniques is difficult and not without risk.

14.1 INTRODUCTION

A wide range of applied geophysical techniques are used in landslide investigations. In general, they can be divided into two main classes: remote sensing techniques, which examine the characteristics of the Earth's surface from a distance, and subsurface techniques, which probe the near-surface geologic environment without the need for excavation. This chapter provides a brief review of recent advances in these geophysical techniques in the context of landslide investigations, without seeking to provide a comprehensive review of this vast topic. Inevitably the coverage of the topic is subjective and reflects the experience of the author.

What is beyond doubt is that remote sensing remains the more widely used of the two approaches, and it is probably also true to say that it is the more rapidly evolving of the two. For that reason alone, the balance of this review is tipped toward remote sensing, although a comparatively brief consideration of engineering geophysical approaches is also provided. It should be noted that in this context the term "landslide investigation" encompasses identification, classification, characterization, mapping, and monitoring, whether for research or for practical and/or commercial purposes.

Over the past two decades, two key reviews of remote sensing techniques in landslide investigation have been published (Mantovani *et al.*, 1996; Metternicht *et al.*, 2005). Perhaps surprisingly, only 15 years ago Mantovani *et al.* (1996, p. 202) concluded that "aerial photos are the most important remote sensing tools in landslide studies. The application of presently available satellite remote sensing is limited as far as it refers to the direct mapping of slope instability features."

They noted, however, that rapid developments in remote sensing technologies, both optical and microwave, held great potential for landslide investigation. Nine years later, Metternicht *et al.* (2005, p. 301) stated that

> the literature reviewed shows that the contribution of remote sensing to the mapping, monitoring, spatial analysis and hazard prediction of mass movements (e.g., landslides, debris flows) has largely been in the form of stereo air photo and satellite image interpretations of landslide characteristics (e.g., distribution

Landslides: Types, Mechanisms and Modeling, ed. John J. Clague and Douglas Stead. Published by Cambridge University Press.

and classification) and factors (e.g., slope, lithology, geostructure, land use/land cover, rock anomalies).

Since then, significant progress has been made in the application of satellite imagery, but work has remained focused primarily on landslide identification and factor-based hazard mapping. Although there had been some progress in the use of satellite-based remote sensing techniques, both of the reviews indicate a sense of frustration in that the application of these techniques was failing to exploit their full potential. This chapter was written six years after the study of Metternicht *et al.* (2005), and thus a key consideration here is whether improvements in capability have led to better application of the tools for landslide investigation.

This chapter does not provide a comprehensive overview of the application of remote sensing techniques to landslides – the two earlier reviews cover the majority of the approaches used, and there is little point in replicating them here. Nor does it attempt to explain the physics behind remote sensing; there are textbooks that achieve that in more detail than is possible here (Lillesand *et al.*, 2008, for example). Instead, it seeks to explore the key, and in fact substantial, advances that have been made in the use of remote sensing since the publication of the Metternicht *et al.* (2005) review.

Fig. 14.1. TOPSAT image of the Hattian Bala landslide in Pakistan. High-resolution satellite data allow detailed mapping of landslide features (image courtesy of Qinetiq / British National Space Centre).

14.2 KEY RECENT TECHNOLOGICAL ADVANCES

Before looking at the key advances in the use of remote sensing in landslide investigation, it is worth reviewing the important improvements in sensor capability and data processing methods, that have made these advances possible over the past decade:

- **Digital image capture.** Digital photogrammetry has now displaced conventional film-based photography in almost every application. During this time there has been a huge advance in the spatial resolution of digital cameras, matched with a dramatic reduction in costs. Furthermore, image processing techniques have evolved rapidly, thanks both to better software and, more importantly, improved computer processing power and speed.
- **Optical satellite sensors.** The spatial resolution of optical satellites has improved substantially over the past decade. A number of commercial satellite sensors now provide spatial resolutions of better than 5 m. In particular, this period has seen the widespread availability of GeoEye-1 (0.4 m optimum spatial resolution), QuickBird (0.6 m), Ikonos (0.8 m), Formosat-2 (2.0 m), and SPOT-5 (2.5 m) datasets, albeit at substantial cost. These instruments allow high-quality imaging, even of small landslide features, in some cases in stereo. Dunning *et al.* (2007) demonstrated that the instruments with the highest resolution have allowed mapping of very small-scale landslide features, such as tension cracks (Fig. 14.1).
- **Google Earth.** Although the development and release of the Google Earth tool in 2004 and 2005 might not seem, at first sight to be as important as the launch of a new satellite system, the availability of this free tool is revolutionizing spatial analysis in many areas of the earth sciences. The availability of georeferenced, high-quality imagery, commonly in multiple epochs, using software that is freely available, has permitted the landslide community to use remote sensing in new and innovative ways. Its importance continues to grow, and the tool is likely to be critically important in the next few years.
- **Radar satellite sensors.** A further major advance has been in the development of radar satellite instruments and, more particularly, tools that allow the interpretation of radar datasets. Although radar in itself is not a good tool for identifying or classifying landslides, radar techniques allow small landslide movements to be accurately determined. This technology is far from mature at present, but the next generation of sensors, which will collect data in a wider range of spectral bands, at higher spatial resolutions, and with improved temporal coverage, is likely to be an important area of progress.
- **Terrestrial laser scanners.** The rapid development, in recent years, of terrestrial laser scanners (TLS), commonly referred to as ground-based LIDAR, has revolutionized field data collection, in particular on steep unstable slopes. The ability to remotely generate 3D models of slopes and to use repeat scans to image rockfall activity and coastal retreat has led to a much better understanding of mass movements. Again, the technology, and in particular the software needed to make best use of the data, is far from mature, and so will be an area of important development in the future.
- **Airborne LIDAR.** Although airborne LIDAR data have been available for some time, the past decade has seen substantial improvements in the quality and availability of the data, and in the capability of processing systems. The major remaining challenges are the cost of data collection and the ability to process the huge resultant datasets, but recent technological developments are encouraging.

- **Ground-based radar.** A few studies have now been undertaken, in particular in Italy (Casagli *et al.*, 2010), Norway (Norland, 2006), and Switzerland (Gischig *et al.*, 2009), into the use of radar systems located on the ground to monitor landslide movement. These systems are proving to be particularly helpful in (1) routine monitoring of quarry walls to provide advanced warning of potential collapse events, and (2) emergency situations in which ongoing landslide movement threatens communities or key infrastructure. This technology is likely to be important in the future, but remains immature at present for most applications.
- **Engineering geophysics.** The use of engineering geophysical techniques in landslide investigation also remains in its infancy, but key developments are occurring, primarily led by the landslide research community in France. Notable among these are techniques to evaluate the subsurface environment of landslides in three dimensions and in real time. It is likely that the fusion of these data with those obtained using conventional approaches will be a key research frontier over the next few years.

14.3 PHOTOGRAMMETRY

Photogrammetry is the earliest application of remote sensing to the study of landslides, and aerial photograph analysis remains a key tool in both the commercial and the research domains. Apart from the aforementioned increasing availability of high-quality digital imagery, which provides better resolution and easier data transfer, a key development has been the increasing use of images draped over digital elevation models (DEMs). Photo-draped DEMs provide perspective views, greatly easing the mapping process (Pesci *et al.*, 2004). The most important development, however, has been the use of photogrammetry to examine landslide movement patterns and thus to elucidate the kinematics of the landslide system (e.g., Kääb, 2002). Interestingly, in most studies the images used have been conventional analog (film) products, scanned to a high resolution and then orthorectified with control points. Mapping accuracy of 0.2–0.3 m can be achieved, allowing movement fields to be generated both for areas of permafrost creep and for areas of rockslides in high mountains. The result is a dense displacement field that, with high temporal resolution imagery, can provide both the pattern of strain on the ground and the potential to analyze changes in strain in the context of environmental and other forcings. Displacement fields can also permit analysis of the deeper structure of landslides, especially those with a component of rotation. It is frustrating that this powerful new tool has been used so little in building a proper four-dimensional understanding of landslide systems. It is hoped that this will become a focus in the near future.

Photogrammetry remains a key technique for producing inventories for multiple synchronous landslides, such as those generated by large earthquakes. The most notable examples to date are the huge number of landslides triggered by the 2008 Wenchuan earthquake in China. The affected area is well suited to the use of remote sensing techniques because most of the landslides changed the land cover from forest to bare soil or rock. Such changes are easy to detect with automatic algorithms. A notable study is that of Dai *et al.* (2011), who mapped 56,000 landslides triggered by the earthquake, mainly from aerial photographs. This dataset provided a firm, consistent foundation upon which to examine the spatial distribution of the landslides in relation to possible causative factors such as lithology, slope angle, and aspect. However, the necessary investment of time is very large, which is likely to limit the future application of the technique in favor of satellite-based approaches (see below).

Improvements in the resolution of calibrated digital cameras have also stimulated interest in using this approach to examine the evolution of rock cliffs. Lim *et al.* (2005), for example, explored the use of the technique to investigate the evolution of near-vertical coastal rock cliffs. The ability to resolve rockfall locations of even small blocks from one image to the next, in principle allows monitoring for individual detachments across a large, inaccessible rock face. Serious difficulties, however, were encountered with the view angle, and in particular the oblique view that is provided when the camera is located at the base of even a modest cliff if it is near-vertical. Large numbers of tie-points were required to generate the requisite data quality, and – even then – substantial errors were seen in the data. The conclusion was that the technique offers some potential for high-resolution monitoring of rockfall events in lower cliffs, with appropriate levels of correction, but is of limited use for monitoring rockfall from tall, steep cliffs.

Another use of the technique is measuring cliff retreat, where individual detachment events are less important (Wangensteen *et al.*, 2007). Two images collected some time apart can provide a large volume of data, especially where older imagery is available, allowing an early baseline to be established. Despite its potential, the technique has failed to gain traction for this application, due primarily to the development of terrestrial laser scanning (see below), which can provide better-quality rockfall detachment data at higher speeds and with less processing. As a consequence, this application of photogrammetry is likely to be remembered as a technique that matured just as it was superseded, and future applications probably remain limited. Notwithstanding these limitations, digital photogrammetry is proving to be extremely useful for characterizing and analyzing rock slopes, in particular in relation to discontinuity – and thus stability – analysis. Sturzenegger and Stead (2009) reviewed the application of close-range digital photogrammetry for rock-slope characterization and showed that widely available techniques are adequate for engineering purposes. The methods allow cost-effective quantification of discontinuity location, orientation, and curvature.

Compared with digital photogrammetric systems, the instruments required for terrestrial laser scanning, at the moment at least, are comparatively large, sensitive to environmental conditions, and very expensive, while digital cameras are inexpensive. Thus for cliff monitoring in real time, especially in difficult environments, digital photogrammetry may offer an advantage over terrestrial laser scanning, as long as processing times can be kept short. Further research in this area is justified, making use of the ongoing development of software that can rapidly generate automatic ortho-images and detect changes between them. Developing technologies also allow cameras to be mounted on unmanned aerial vehicles (UAVs), including remotely piloted helicopters. The development of small aircraft with substantive lifting capabilities and pre-programmed flight paths is allowing high-quality photogrammetric imagery to be collected at low cost. This technology in its infancy with respect to slopes, although Gong *et al.* (2010) demonstrated its use in their examination of landslides triggered by the 2008 Wenchuan earthquake.

14.4 OPTICAL SATELLITE IMAGERY, INCLUDING THE APPLICATION OF THE GOOGLE EARTH TOOL

The highest-resolution optical satellites are now able to generate imagery that has a spatial resolution comparable to, or even better than, that of conventional aerial photographs. The two main drawbacks are cost and the lack of stereo capability, although some sensors can generate stereo imagery, albeit at prodigious expense. Although the costs of specially acquired, high-resolution satellite imagery are generally high, they compare favorably with the cost of procuring a dedicated aerial survey over a comparable area. The use of aerial imagery, however, is often based on the acquisition of existing image sets that are available at comparatively low cost. Such images are available for most developed countries and many developing ones, and are often available in multitemporal sets. Unfortunately, this is rarely the case for high-resolution satellite imagery and, even when it is available, the vendors still charge a high price. In addition, optical satellite data are dependent on cloudless conditions, whereas aerial imagery can generally be collected under high clouds. QuickBird imagery was available soon after the 2005 Kashmir (Pakistan and India) earthquake, which occurred in late fall, when the skies were almost cloudless. In contrast, the 2008 Wenchuan (China) earthquake occurred in May, the start of the rainy season, severely limiting the use of optical satellite tools for many weeks.

Three key applications have emerged from the availability of new high-resolution optical satellite imagery. The first stems from earlier work aimed at creating landslide susceptibility and hazard maps using factor data (e.g., vegetation cover) derived from the imagery in combination with other datasets, such as DEMs and lithology maps. For example, Lee (2005) evaluated landslide hazard in Malaysia using a factor-based approach, including an assessment of land use from Landsat data and a vegetation index value from SPOT imagery. Other factors included a range of topographic and geologic indices, all of which were combined in a GIS using a logistic regression model. Numerous similar studies have been completed (Sarkar and Kanungo, 2004; Castellanos Abella and van Westen, 2008). In addition the improved imagery can be used to benchmark the success of the algorithm by comparing the areas identified as being of high susceptibility/hazard with the locations of actual landslides (Guzzetti *et al.*, 2005). A comprehensive review of data needs, and the role that imagery can play in providing data, for this approach is provided by van Westen *et al.* (2008), who concluded that "it is the spatial data availability that is the limiting factor in landslide hazard and risk assessment." Thus, there is much more work to do in this area in the coming decade.

The second application is also a continuation of earlier work, in this case creating inventories of landslides after large events such as earthquakes (Harp *et al.*, 2002; Saba *et al.*, 2010; Parker *et al.*, 2011). The main advantage of the high spatial resolution that is now available from satellite imagery is that the inventory is more likely to be complete. The disadvantage is that mapping landslides by hand is even more time-consuming, given the large area covered. For this reason, several attempts have been made to develop automated algorithms to generate such maps. Initial attempts have focused on comparatively simple tools such as texture, but recently technology has permitted more sophisticated analyses, for example using object shape. None of these approaches is mature at this time, but there is little doubt that it will eventually be a powerful technique. Borghuis *et al.* (2007) concluded that an algorithm-based approach successfully mapped 63 percent of the landslides in Taiwan, and Yang and Chen (2010) showed that in a test area in Sichuan Province in China, an algorithm was able to detect about 90 percent of the landslides. The difference between the two results may represent improvements in technology between the two studies. However, Martha *et al.* (2010) found that automated techniques were only able to detect 69–77 percent of the landslides in a test area in the Himalayas of India, which may imply that the success of the technique continues to be dependent on the environment, the quality of images available, and the types of landslide present. In all cases, these approaches rely on detecting landslides by mapping loss of vegetation cover. Thus, they remain limited to areas with forests or similar dense vegetation. Where landslides occur in regions without dense vegetation, automatic detection remains exceptionally challenging, to the point of not being viable. Ramli and Petley (2006) examined combinations of multiple spectral bands for landslide detection in temperate environments and concluded that automated landslide mapping using this approach remains fraught with difficulties. On the other hand, Parker *et al.* (2011) generated a very detailed inventory of landslides triggered by the 2008 Wenchuan earthquake and used these data to estimate the volume loss from the mountain range as a result of mass movements during the event. They further compared this volume with estimates of the volume of material advected into the mountain range, allowing a complete sediment budget to be generated for the earthquake. Thus, considerable progress has been made with automated mapping, although this is likely to be an important area for future development.

Fig. 14.2. Google Earth image of the Knipe Point landslide in North Yorkshire, UK. Upper panel dates from 2002 and lower panel from 2007. See Figure 14.3 for an interpretation of the change between the images.

The third application has been the use of Google Earth imagery to examine landslides. Since 2008, Google Earth has been obtaining and providing free GeoEye-1 data with a spatial resolution of about 0.5 m. In addition, Google has increasingly uploaded older imagery, usually a combination of satellite images and aerial photographs, in some places dating back to the 1940s. The availability of these data has provided researchers with an opportunity to map the temporal development of slope failures at no cost. Shroder and Weihs (2010) mapped the Lake Shewa ancient landslide dam in Afghanistan using Google Earth imagery, and Sato and Harp (2009) tested the use of Google Earth as a base map for imported imagery to map landslides triggered by the 2008 Wenchuan earthquake. They concluded that "Google Earth is an effective and rapid reconnaissance tool."

Figure 14.2 compares two Google Earth images of the Knipe Point landslide in North Yorkshire, UK. The first (upper panel) was collected on March 5, 2002 and the second (lower panel) on May 7, 2007. This coastal landslide reactivated between these two dates. The annotated image shown in Figure 14.3 demarcates the major changes, which include substantial movement in the central portion of the landslide, most notably retrogression of the landslide crown, causing the loss of a property and some loss of land close to the beach. Although the techniques used to evaluate the imagery are not in any way new, the availability of several epochs of images at no cost is revolutionary. Interpretations can be aided using the simple terrain model available in Google Earth, although it is not very useful in high mountain areas due to its 30-m height resolution. A further disadvantage is that the resolution of Google Earth coverage, especially in remote terrain where landslides may be most frequent, is too low to map most landslides.

In recent years, Google has made imagery available quickly in areas affected by natural disasters. Thus, for example, Geoeye

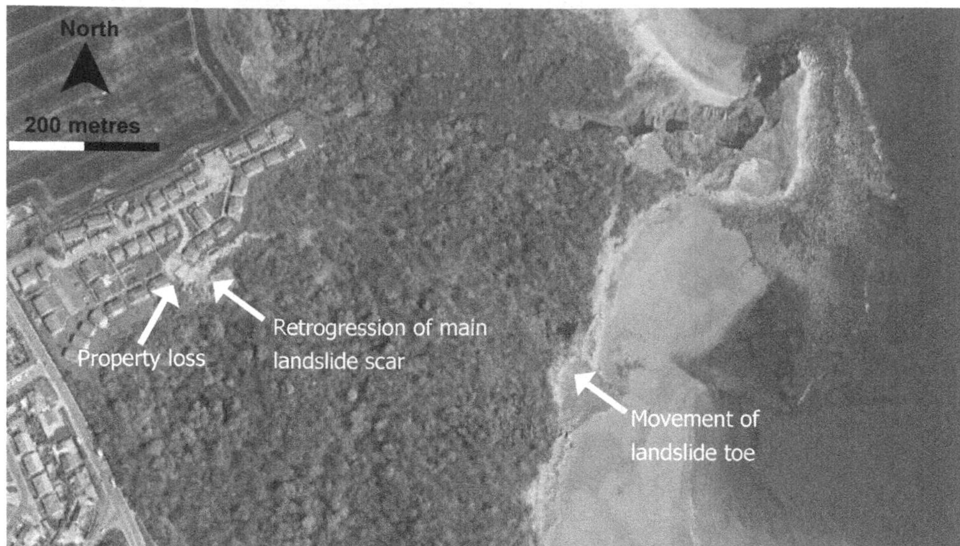

Fig. 14.3. Annotated Google Earth image of the Knipe Point landslide, illustrating changes between 2002 and 2007.

data were released for the areas of Brazil worst affected by landslides in January 2011, and imagery was also quickly released for the area impacted by the Haiti earthquake. The rapid availability of high-resolution satellite imagery is facilitating community mapping efforts in disaster areas. For example the GEO-CAN community coordinated mapping of building damage in Port-au-Prince in the days after the earthquake (Bevington *et al.*, 2010). This effort was undertaken in two phases using 50-cm GeoEye-1 images that were obtained the day after the earthquake. Each of 50 volunteers was allocated a 500 × 500 m image, and within 48 hours, more than 5000 collapsed buildings had been mapped in an area of 133 km^2 (Bevington *et al.*, 2010). The mapping exercise was subsequently expanded to cover a greater area using a larger team and higher-resolution imagery. Within a week of the disaster approximately 30,000 buildings had been identified as heavily damaged or destroyed, solely based on remotely sensed imagery. No such community mapping effort has been undertaken for a landslide disaster to date, but the potential is self-evident. Assuming that Google continues to maintain and improve the tool, the best is yet to come.

14.5 RADAR SATELLITE INSTRUMENTATION

When satellite radar imagery first became available, its applications to landslides appeared to be limited. Radar instruments provide an all-weather capability, as clouds are essentially opaque to electromagnetic radiation in these wavelengths, but the low spatial resolution and comparatively poor image quality, combined with early problems of using radar imagery in areas with steep slopes, meant that the tool was used sparingly at first. In the mid-1990s, however, a new application of radar for landslide investigation emerged – radar interferometry, a technique that is frequently referred to as InSAR (Fruneau *et al.*, 1996). This technique compares two satellite images taken at different times from approximately the same sensor position. Small shifts in the terrain distance along the direction of the radar beam cause a change in the phase of the returning radar wave. As the wavelength is known, and the phase change can be precisely measured, the deformation of the ground can be determined. Thus, the movement of landslides can, in principle, be determined.

The practical application of InSAR to landslides has, however, been exceptionally challenging for a number of reasons.

- The technique detects any major change in the surface along the direction in which the radar wave travels, regardless of cause. Unfortunately there are many other causes of such changes in addition to landslide movement, for example the growth of vegetation.
- The InSAR technique relies on the measurement of movement in the direction of the radar beam. Thus, the landslide movement vector and the radar track need to be reasonably closely aligned.
- Radar data are collected at oblique angles and thus are compromised in areas of high relief, in particular on slopes facing away from the direction in which the radar waves travel.
- Larger deformations are often difficult to resolve, especially where the boundary of the deforming surface is sharp, such as at a lateral shear.
- High-quality measurement relies on the ability to identify points that have not moved, which in some cases can be difficult.

In the case of landslides, many of these problems have been resolved using a technique known as permanent scatter interferometry (PS InSAR), in which points that reflect similar amounts of radiation in several images across a longer epoch of time are selected and compared (Colesanti *et al.*, 2003). This technique provides a much lower spatial coverage of deformation than InSAR, but the quality of each data point is much better. PS InSAR techniques have proven to be particularly useful in areas with a substantial number of buildings and other human-made structures, as these generally provide consistent radar returns. The technique has also been applied to rural regions, but generally with a lower point concentration. In some cases metal point reflectors have been constructed to increase the number of permanent scatterers.

These approaches have been widely applied (Carnec *et al.*, 1996; Kimura and Yamaguchi, 2000; Bulmer *et al.*, 2006; Colesanti and Wasowki, 2006), but are still being refined. A number of problems remain.

- Long-term deformation measurements generally correspond well with data collected from other, more conventional, tools, but the InSAR displacement time plot is frequently less accurate, with a pattern that is too linear. Improvements, however, are being made, and good results have been achieved for some large landslides (see the review in Carlo and Wasowski, 2005).
- Processing times are long, due to the complexity of the technique. Again, improvements are being made, but the method remains far from straightforward.
- Image availability is limited. Most areas of the Earth are imaged rarely at best, leading to long temporal separations between images. However, the launch of new radar instruments is improving this situation rapidly.
- Ensuring that data is delivered in a timely enough manner to allow processing to be undertaken rapidly enough to meet end-user needs, especially in disaster situations, has been challenging.

Many of these issues will be resolved over the next few years. Importantly, there are plans to launch several new instruments over the next decade (Fig. 14.4), which will reduce the temporal base length and coverage issues. Furthermore, the availability of instruments in new radar bands, most notably L-band, will improve the capability of instruments to resolve surface deformation. There is no doubt that InSAR will become an important method in the toolbox of landslide scientists.

Finally, some success is being achieved in the use of airborne radar instrumentation. Wang *et al.* (2009) used this technique to document landslides in the aftermath of the 2008 Wenchuan earthquake. They found that, with a spatial resolution of about 1 m, the technique was reasonably effective in locating landslides, although it did not allow quantification of their volumes. The technique identified valley-blocking landslides due to the clear backscatter properties of the water surface of the impounded lakes. Nonetheless, terrain shadowing remains a major limitation of all radar techniques in high mountain areas, and some landslides will inevitably be missed.

14.6 AIRBORNE LIDAR

LIDAR uses a laser beam to scan the Earth's surface. The cloud of points can be displayed as a digital elevation model. In the case of airborne LIDAR, the instrument is mounted on an aircraft that flies across the area of interest while the laser beam scans the surface. The distance from the aircraft platform to the surface is then determined by measuring either the time of flight of the laser beam or a phase change in the laser signal; the location is determined by measuring the orientation and inclination of the beam relative to the known location of the sensor. Thus, the position and elevation of the platform are recorded, typically using differential GPS and inertial navigation units, supplemented with carefully surveyed ground control points. In many cases aerial photographs are acquired at the same time as the LIDAR dataset. The photographs can then be orthorectified and draped over the LIDAR imagery, creating extremely high-quality, three-dimensional virtual models of the terrain. Manipulation of these images can simplify and improve landslide mapping.

Jaboyedoff *et al.* (2012) examine in depth the use of LIDAR data in landslide investigation; thus the topic will only briefly be considered here. Airborne LIDAR analyses of landslides have served several purposes. First, the data provide extremely accurate digital elevation models, which can be used to produce landslide susceptibility and hazard maps (McKean and Roering, 2004). Repeat surveys allow analysis of surface change and permit, for example, quantification of erosion or volumes of displaced materials. Vectors of movement across a landslide over the period between the two datasets can be determined. In addition, the quality of the dataset allows the precise determination of slope angles across the landslide; where appropriate, profiles can be created and input into slope modeling software (Jones, 2006). In some cases, landslide mapping has even been undertaken without the use of aerial photographs. The ability to exaggerate the vertical scale allows the researcher to see subtle features that may not be evident from aerial photographs (Glenn *et al.*, 2006).

The most impressive recent application of airborne LIDAR is analysis of the terrain beneath forest land (Van Den Eeckhaut *et al.*, 2007). Mapping slope features in such environments is exceptionally difficult. Conventional airborne photography does not penetrate the forest canopy, and radar data lacks the required resolution. Field mapping is possible, but is extremely difficult and can be expensive. Accurately pinpointing one's location in the field is challenging, and GPS does not function beneath dense tree cover. Furthermore, identifying features correctly is also difficult where the ground is disturbed by tree roots and fallen branches, and where lighting is poor. Airborne LIDAR can overcome many of these problems. In all but the most densely forested terrain, a small number of radar pulses reach the ground surface, and these can easily be selected from the total dataset. Although the point density is much lower than that of the full dataset, these "last return" data points can be

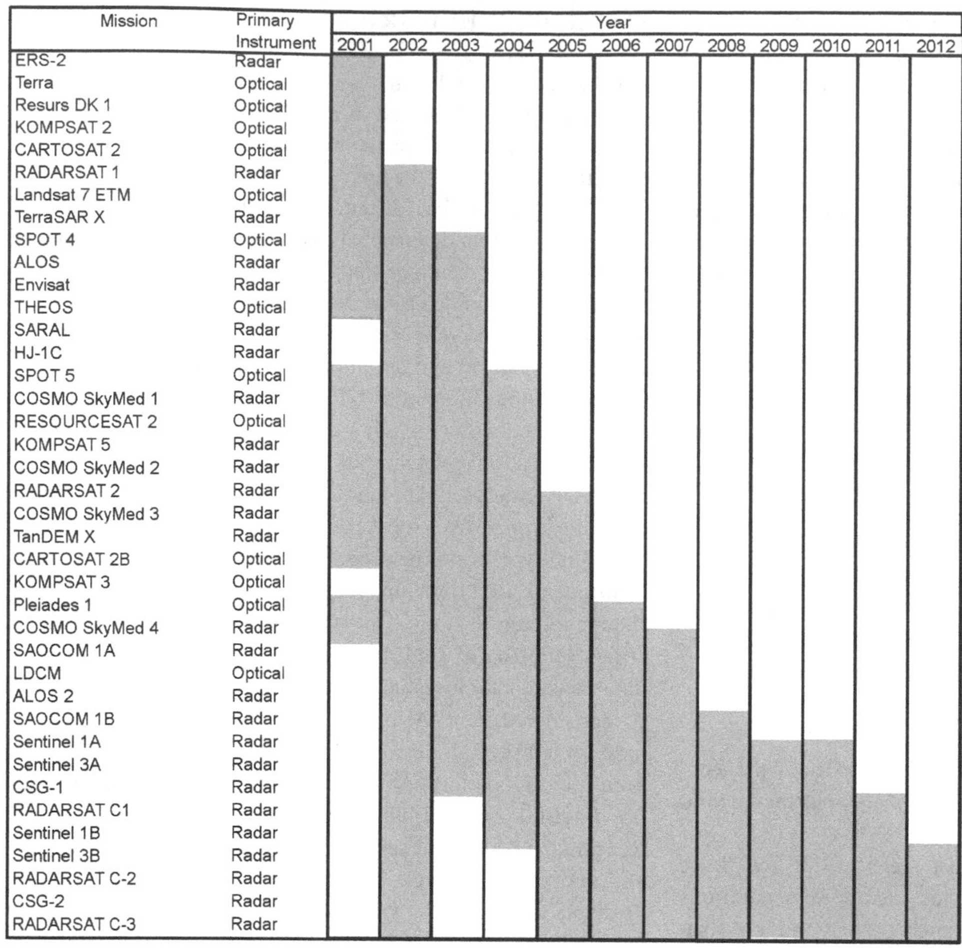

Fig. 14.4. Existing and planned radar Earth observation satellites; anticipated lifespans are indicated in gray. It is evident that radar capability will be substantially enhanced over the next decade.

sufficient to generate a good-quality digital elevation model of the ground surface. From these data, it may be possible to map landslides.

The technique does not work in areas of thick vegetation cover in which no radar pulses penetrate the canopy, such as in tropical forests. In more temperate areas, data can often be improved substantially by timing the flight for the middle of winter, when leaf cover is at a minimum.

The disadvantages of airborne LIDAR are data availability, processing complexity, and cost. Only a relatively small part of the Earth's surface has been mapped by LIDAR. Data are rarely free, and data acquisition is extremely expensive and beyond the reach of most landslide investigators. At present, LIDAR surveys generally require the use of a fixed-wing or rotary aircraft, expensive equipment, and considerable manpower. Some groups are working on reducing the size of LIDAR instruments so that they can be flown on UAVs, which has the potential to reduce the cost substantially. At present, however, this remains a distant prospect.

A final consideration is the large quantity of data that LIDAR surveys generate. In many cases the volume of data is so large that end-users resort to using LIDAR output for little more than a high-quality digital elevation model. The high demands of handling and processing LIDAR data are often forgotten when planning such campaigns.

14.7 TERRESTRIAL LASER SCANNING (GROUND-BASED LIDAR)

In the past seven years, terrestrial LIDAR, more commonly known as terrestrial laser scanning (TLS), has probably been the most important tool used in mass-movement research. TLS is a fundamental development in our capability to investigate mass movements on very steep slopes.

TLS is similar to airborne LIDAR in that the scanner uses a laser to generate a three-dimensional representation of the surface being studied. In most cases the instrument is located at a fixed point, although mobile scanners mounted on a vehicle or boat are also available. The latest instruments are capable of collecting more than 200,000 data points per second over ranges of up to about 4 km. The instruments are

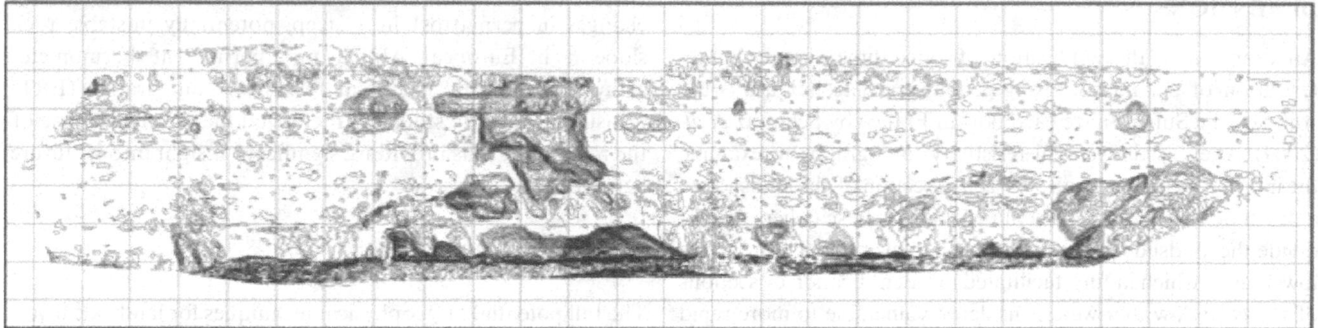

Fig. 14.5. Example of TLS data, showing a 70-m-high cliff in northeast England. The top image shows the digital terrain model; the bottom image shows contours of change over time, highlighting the location of major rockfalls.

expensive, costing $100,000 or more – although lower-capability instruments are much cheaper – but running costs are low and utilization rates can be high. With the software provided with the instrument, the raw data can rapidly be converted into a DEM.

A important advantage of TLS is its capability of generating repeat surveys of steep or vertical cliffs (Fig. 14.5). The surfaces can then be compared and changes detected. These changes have been used to identify key aspects of the behavior of steep slopes, including (1) inventories of rockfalls and the establishment of frequency–magnitude relations (Abellán *et al.*, 2006), (2) quantification of erosion rates, for example of shorelines (Lim *et al.*, 2009), precursory phenomena (Rosser *et al.*, 2007), or rockfalls associated with permafrost degradation (Oppikofer *et al.*, 2008), and (3) the relation between landslides and environmental forcing factors such as precipitation and temperature (Rosser *et al.*, 2005).

Monitoring of steep slopes has previously been difficult. Direct measurements have not been possible due to the combination of the large surface areas involved and the physical danger of working in areas of possible rockfall. Until recently, photogrammetric methods could not be used because of geometric distortions; however, as discussed earlier, many of these problems have now been overcome. TLS also circumvents these problems and is rapidly becoming the technique of choice in rockfall research. The widespread application of this technique will lead to a major change in our understanding of the processes operating on steep cliffs over the next decade.

TLS is a reliable technique for collecting discontinuity data, particularly in inaccessible or large areas. It thus facilitates kinematic analysis of stability, modeling of the likely behavior of slopes, and detailed engineering design for the mitigation of mass movement (Sturzenegger and Stead, 2009; Brideau *et al.*, 2011).

14.8 GEOPHYSICAL TECHNIQUES

14.8.1 SUBSURFACE ENGINEERING GEOPHYSICAL INVESTIGATION

Jongmans and Garanbois (2007) note that three factors limit the use of geophysical techniques in landslide investigation.

1. Resolution decreases with depth.
2. In many cases the solution obtained from a set of data is not unique, requiring independent data.
3. The information obtained is typically in the form of physical parameters rather than the geologic or geotechnical properties with which the landslide community is more familiar.

Add to these factors the difficult and complex settings of many landslides, with their rough terrain, steep slopes, clay-rich or disrupted sediments, and the presence of large volumes of water, and the relatively limited use of geophysical techniques is understandable. Jongmans and Garanbois (2007), however, highlighted two important developments that have the potential to change the use of engineering geophysics in the investigation of landslides. First, in recent years 3D geophysical imaging techniques have been applied to landslides, providing a detailed understanding of subsurface conditions. Second, permanent arrays of geophysical sensors are now being installed as a part of landslide monitoring systems.

14.8.2 IMAGING OF LANDSLIDES IN THREE DIMENSIONS

An example of the application of three-dimensional characterization of a landslide using geophysicical techniques is the study of the Super Sauze earthflow in France by Schmutz *et al.* (2009). They used electrical resistivity tomography to examine the internal structure of a 150 × 200 m section of the landslide to a depth of about 10 m. The data enabled the researchers to subdivide the landslide into a flowing upper layer and a more stable lower layer, which in turn facilitated the identification of sections of the earthflow that were considered vulnerable to more rapid flow. It should be noted, however, that this high-quality geophysical characterization was done on only a part of the landslide and reached a depth of just 10 m, serving to highlight the challenges associated with the routine use of these techniques.

Three-dimensional geophysical data are also proving to be helpful in the analysis of the hydrology of slopes. Heincke *et al.* (2010), for example, used 3D electric and seismic tomography methods to identify near-surface tension fractures of the Åknes rockslide in western Norway, and thus to evaluate the role they play in the hydrology of the landslide. The fractures were imaged using a 3D seismic survey, and the water distribution within the rock mass was interpreted using intersecting 2D geoelectric profiles, which were then integrated with the 3D seismic data. The results clearly demonstrate that landslide-induced tension fractures were responsible for changing the hydrology of the landslide and that "replumbing" with increasing strain may be a important feature in landslide dynamics (Jomard *et al.*, 2007).

A particularly interesting and potentially important study is that of Renalier *et al.* (2010), who used ambient seismic noise collected at 13 stations over a 15-day period to generate a 3D model of the Avignonet landslide in France. The study was designed only to compare the effectiveness of the technique with a more conventional method, but it provides a noninvasive method of characterizing the internal structure of a landslide and thus has much potential.

14.8.3 PERMANENT ARRAYS

Few long-term or permanent arrays for geophysical data collection have been reported in the literature. Norman *et al.* (2010), however, described the use of an array of seismometers to investigate the delivery of wave energy to a steep coastal cliff, demonstrating that an understanding of the manner of energy dissipation on a cliff can aid in the understanding of rates of coastal erosion through rockfalls. On the other hand, real-time geophysical monitoring during a water injection test on the lower part of the La Clapière landslide in France provided insights into patterns of water drainage in space and time. The study demonstrated that 30 percent of the injected water drained quickly along a complex shear plane 10 m below the surface. The study changed the model of the geometry of the landslide and thus its likely behavior. Finally, Krautblatter and Hauck (2007) and Krautblatter *et al.* (2010) monitored changes in permafrost in a steep, potentially unstable rock slope in the European Alps using a permanent thermometer array and monthly electrical resistivity tomography (ERT) measurements. The study characterized the thermal regime of the rock mass, thus providing significant insight into its future behavior.

14.8.4 DATA FUSION

The full potential of geophysical techniques for landslide investigation will be realized when the datasets are fully integrated with those acquired during conventional investigations to generate a four-dimensional understanding of landslide behavior. Naudet *et al.* (2008) used a combination of electrical resistivity tomography (ERT) and self-potential (SP) measurements, together with borehole data and conventional multi-epoch airphoto interpretation and geomorphological surveys, to determine the geometry of a landslide. They used the results of their study to analyze the likely effectiveness of a planned drainage scheme.

The problems of evaluating processes on very steep slopes have been highlighted previously in this chapter. Deparis *et al.* (2008) combined TLS data with geophysical information to evaluate the fracture pattern on steep rock cliffs. In this case, electrical resistivity tomography and ground-penetrating radar (GPR) surveys were undertaken on the plateau at the top of the cliff, allowing fractures to be identified and mapped. By combining the GPR and TLS datasets, the authors were able to determine the characteristics of the main joint sets within the slope, which ultimately determine the stability of the cliff. Most importantly, this approach allowed a hitherto unknown, but important, master fracture determining the future stability of the cliff to be identified.

14.9 CONCLUSIONS

At the start of this chapter I noted the observations of Mantovani *et al.* (1996) and Metternicht *et al.* (2005) that the potential of remote sensing had not been matched by the application of technologies to landslide studies. I also posed the question of whether the situation had improved in the six years between the publication of the more recent of these two studies and the writing of this chapter. My review shows that the answer to this question is yes. There are a number of reasons for this improvement, notably the availability of:

- high-resolution satellite imagery, in particular that with submeter spatial resolution, which has permitted proper observation of landslide features
- high-quality digital terrain models, notably those generated by SRT
- the Google Earth tool, which places high-quality satellite data on every desktop computer with a fast internet connection
- satellite radar data and the development of new techniques for detecting millimeter-scale displacements
- TLS instruments and associated software, which permit detailed characterization of steep slopes for the first time.

As a consequence we have, perhaps without realizing it, entered a golden age of landslide investigation using remote sensing and geophysical techniques – the future looks exciting. Five key developments are likely in the near future.

1. *Better-quality satellite radar data will become available from new satellite platforms.* It will eventually be possible to automatically generate deformation maps for large areas with a high temporal frequency. At the moment, this is possible for small areas using ground-based radars; upscaling this technology to large geographic areas (e.g., a mountain chain) offers real potential for enhanced forecasting of rock-slope failure.
2. *We can expect greater access to high-resolution optical data.* Although it seems unlikely that spatial resolution will improve dramatically (in part because of governmental restrictions on the distribution of these datasets for security reasons), increased competition between vendors and the provision of Geoeye data through the Google Earth tool are likely to improve data availability.
3. *LIDAR datasets will become increasingly available*, perhaps through the mounting of instruments on small, low-cost, unmanned aerial platforms, and repeat surveys for increasingly large areas of the Earth's surface. Some regions, and even countries, are undertaking surveys of large areas that will in time prove highly beneficial.
4. *There will be improvements in TLS instruments.* A key development is that of fixed scanners that repeatedly collect data for the same area, allowing the evolution of rockfalls to be examined properly and even to be used as rockfall warning systems.
5. *We can expect an increase in 3D subsurface data collection on landslides using geophysical tools*, perhaps ultimately combined with real-time data analysis.

It is likely that geophysical tools will be increasingly used in landslide investigations in the years to come. The challenge, but also the key opportunity, will be to use the geophysical data to develop a full understanding of landslide processes. There is little doubt that we have come a long way since the reviews of Mantovani *et al.* (1996) and Metternicht *et al.* (2005), but it also seems likely that substantial progress will be made in the next 10 years.

REFERENCES

Abellán, A., Vilaplana, J.M. and Martinez, J. (2006). Application of a long-range Terrestrial Laser Scanner to a detailed rockfall study at Vall de Núria (Eastern Pyrenees, Spain). *Engineering Geology*, 88, 136–148.

Bevington, J., Adams, B. and Eguchi, R. (2010). GEO-CAN debuts. *Imaging Notes*, 25, 26–32.

Borghuis, A.M., Chang, K. and Lee, H.Y. (2007). Comparison between automated and manual mapping of typhoon-triggered landslides from SPOT-5 imagery. *International Journal of Remote Sensing*, 28, 1843–1856.

Brideau, M.A., Pedrazzini, A., Stead, D. *et al.* (2011). Three-dimensional slope stability analysis of South Peak, Crowsnest Pass, Alberta, Canada. *Landslides*, 8, 139–158.

Bulmer, M.H., Petley, D.N., Murphy, W. and Mantovani, F. (2006). Detecting landslide deformation using two-pass differential interferometry: Implications for planetary studies. *Journal of Geophysical Research*, 111, E06S16, doi:10.1029/2005JE002593.

Carlo, C. and Wasowski, J. (2005). Investigating landslides with space-borne synthetic aperture radar (SAR) interferometry. *Engineering Geology*, 88, 173–199.

Carnec, C., Massonnet, D. and King, C. (1996). Two examples of the use of SAR interferometry on displacement fields of small spatial extent. *Geophysical Research Letters*, 23, 3579–3582.

Casagli, N., Catani, F., Del Ventisette, C. and Luzi, G. (2010). Monitoring, prediction, and early warning using ground-based radar interferometry. *Landslides*, 7, 291–301.

Castellanos Abella, E.A. and van Westen, C.J. (2008). Qualitative landslide susceptibility assessment by multicriteria analysis: A case study from San Antonio del Sur, Guantánamo, Cuba. *Geomorphology*, 94, 453–466.

Colesanti, C. and Wasowski, J. (2006). Investigating landslides with space-borne Synthetic Aperture Radar (SAR) interferometry. *Engineering Geology*, 88, 173–199.

Colesanti, C., Ferretti, A., Prati, C. and Rocca, F. (2003). Monitoring landslides and tectonic motions with the Permanent Scatterers Technique. *Engineering Geology*, 68, 3–14.

Dai, F.C., Xua, C., Yao, X. *et al.* (2011). Spatial distribution of landslides triggered by the 2008 Ms 8.0 Wenchuan earthquake, China. *Journal of Asian Earth Sciences*, 40, 883–895.

Deparis, J., Fricout, B., Jongmans, D. *et al.* (2008). Combined use of geophysical methods and remote techniques for characterizing the fracture network of a potentially unstable cliff site (the 'Roche du Midi', Vercors massif, France). *Journal of Geophysics and Engineering*, 5, 147–157.

Dunning, S.A., Mitchell, W.A., Rosser, N.J. and Petley, D.N. (2007). The Hattian Bala rock avalanche and associated landslides triggered by the Kashmir earthquake of 8 October 2005. *Engineering Geology*, 93, 30–44.

Fruneau, B., Achache, J. and Delacourt, C. (1996). Observation and modelling of the Saint-Etienne-de-Tinee landslide using SAR interferometry. *Tectonophysics*, 265, 181–190.

Gischig, V., Loew, S., Kos, A. *et al.* (2009). Identification of active release planes using ground-based differential InSAR at the Randa rock slope instability, Switzerland. *Natural Hazards and Earth System Sciences*, 9, 2027–2038.

Glenn, N.F., Streutker, D.R., Chadwick, D.J., Thackray, G.D. and Dorsch, S.J. (2006). Analysis of LiDAR-derived topographic information for characterizing and differentiating landslide morphology and activity. *Geomorphology*, 73, 131–148.

Gong, J.H., Wang, D.C., Li, Y. *et al.* (2010). Earthquake-induced geological hazards detection under hierarchical stripping classification framework in the Beichuan area. *Landslides*, 7, 181–189.

Guzzetti, F., Reichenbach, P., Cardinali, M., Galli, M. and Ardizzone, F. (2005). Probabilistic landslide hazard assessment at the basin scale. *Geomorphology*, 72, 272–299.

Harp, E.L., Castañeda, M.R. and Held, M.D. (2002). *Landslides Triggered by Hurricane Mitch in Tegucigalpa, Honduras*. US Geological Survey, Open File Report 02–33.

Heincke, B., Günther, T., Dalsegg, E. *et al.* (2010). Combined three-dimensional electric and seismic tomography study on the Åknes rockslide in western Norway. *Journal of Applied Geophysics*, 70, 292–306.

Jaboyedoff, M., Oppikofer, T., Abellán, A. *et al.* (2012). Use of LIDAR in landslide investigations: A review. *Natural Hazards*, 61, 5–28.

Jomard, H., Lebourg, T., Binet, S., Tric, E. and Hernandez, M. (2007). Characterization of an internal slope movement structure by hydrogeophysical surveying. *Terra Nova*, 19, 48–57.

Jones, L.D. (2006). Monitoring landslides in hazardous terrain using terrestrial LiDAR: An example from Montserrat. *Quarterly Journal of Engineering Geology and Hydrogeology*, 39, 371–373.

Jongmans, D. and Garanbois, S. (2007). Geophysical investigation of landslides: A review. *Bulletin Société Géologique de France*, 178, 101–112.

Kääb, A. (2002). Monitoring high-mountain terrain deformation from repeated air- and spaceborne optical data: Examples using digital aerial imagery and ASTER data. *Journal of Photogrammetry and Remote Sensing*, 57, 39–52.

Kimura, H. and Yamaguchi, Y. (2000). Detection of landslide areas using satellite radar interferometry. *Photogrammetric Engineering and Remote Sensing*, 66, 337–344.

Krautblatter, M. and Hauck, C. (2007). Electrical resistivity tomography monitoring of permafrost in solid rock walls. *Journal of Geophysical Research – Earth Surface*, 112, F02S20.

Krautblatter, M., Verleysdonk, S., Fores-Orozco, A. and Kemna, A. (2010). Temperature-calibrated imaging of seasonal changes in permafrost rock walls by quantitative electrical resistivity tomography (Zugspitze, German/Austrian Alps). *Journal of Geophysical Research – Earth Surface*, 115, F02003.

Lee, S. (2005). Application of logistic regression model and its validation for landslide susceptibility mapping using GIS. *International Journal of Remote Sensing*, 26, 1477–1491.

Lillesand, T., Kiefer, R.W. and Chipman, J. (2008). *Remote Sensing and Image Interpretation*, 6th edn. Chichester, UK: Wiley.

Lim, M., Petley, D.N., Rosser, N.J., Allison, R.J. and Long, A.J. (2005). Digital photogrammetry and time-of-flight laser scanning as an integrated approach to monitoring cliff evolution. *Photogrammetric Record*, 20, 109–129.

Lim, M., Rosser, N., Allison, R. and Petley, D. (2009). Coastal rock cliff geomorphology: Spatial patterns and erosional processes. *Geomorphology*, 114, 12–21.

Mantovani, F., Soeters, R. and van Westen, C.J. (1996). Remote sensing techniques for landslide studies and hazard zonation in Europe. *Geomorphology*, 15, 213–225.

Martha, T.R., Kerle, N., Jetten, V., van Westen, C.J. and Kumar, K.V. (2010). Characterising spectral, spatial and morphometric properties of landslides for semi-automatic detection using object-oriented methods. *Geomorphology*, 116, 24–36.

McKean, J. and Roering, J. (2004). Objective landslide detection and surface morphology mapping using high-resolution airborne laser altimetry. *Geomorphology*, 57, 331–351.

Metternicht, G., Hurni, L. and Gogu, R. (2005). Remote sensing of landslides: An analysis of the potential contribution to geo-spatial systems for hazard assessment in mountainous environments. *Remote Sensing of Environment*, 98, 284–303.

Naudet, V., Lazzari, M., Perrone, A. *et al.* (2008). Integrated geophysical and geomorphological approach to investigate the snow-melt-triggered landslide of Bosco Piccolo. *Engineering Geology*, 98, 156–167.

Norland, R. (2006). Differential interferometric radar for mountain rock slide hazard monitoring. In *IEEE International Symposium on Geoscience and Remote Sensing (IGARSS 2006)*. Denver, CO: IEEE, pp. 3293–3296.

Norman, E., Rosser, N., Petley, D., Barlow, J. and Brain, M. (2010). Exploring the relationship between tidal duration and energy delivery to a coastal cliff. In *Geologically Active*, ed. A.L. Williams, G.M. Pinches, C.Y. Chin, T.J. McMorran and C.I. Massey. London: CRC Press (CD).

Oppikofer, T., Jaboyedoff, M. and Keusen, H.R. (2008). Collapse at the eastern Eiger flank in the Swiss Alps. *Nature Geoscience*, 1, 531–535.

Parker, R.N., Densmore, A.L., Rosser, N.J. *et al.* (2011). Mass wasting triggered by the 2008 Wenchuan earthquake is greater than orogenic growth. *Nature Geoscience*, 4, 449–452.

Pesci, A., Baldi, P., Bedin, A. *et al.* (2004). Digital elevation models for landslide evolution monitoring: Application on two areas located in the Reno River Valley (Italy). *Annals of Geophysics*, 47, 1339–1353.

Ramli, M.F. and Petley, D.N. (2006). Best band combination for landslide studies in temperate environments. *International Journal of Remote Sensing*, 27, 1219–1231.

Renalier, F., Jongmans, D., Campillo, M. and Bard, P.Y. (2010). Shear wave velocity imaging of the Avignonet landslide (France) using ambient noise cross correlation. *Journal of Geophysical Research – Earth Surface*, 115, F03032.

Rosser, N.J., Petley, D.N., Lim, M., Dunning, S.A. and Allison, R.J. (2005). Terrestrial laser scanning for monitoring the process of hard rock coastal cliff erosion. *Quarterly Journal of Engineering Geology and Hydrogeology*, 38, 363–376.

Rosser, N.J., Lim, M., Petley, D.N. and Dunning, S.A. (2007). Patterns of precursory rockfall prior to slope failure. *Journal of Geophysical Research – Earth Surface*, 112, F04014, doi:10.1029/2006JF000642.

Saba, S.B., van der Meijde, M. and van der Werff, H. (2010). Spatiotemporal landslide detection for the 2005 Kashmir earthquake region. *Geomorphology*, 124, 17–25.

Sarkar, S. and Kanungo, D.P. (2004). An integrated approach for landslide susceptibility mapping using remote sensing and GIS. *Photogrammetric Engineering and Remote Sensing*, 70, 617–625.

Sato, H.P. and Harp, E.L. (2009). Interpretation of earthquake-induced landslides triggered by the 12 May 2008, M7.9 Wenchuan earthquake in the Beichuan area, Sichuan Province, China using satellite imagery and Google Earth. *Landslides*, 6, 153–159.

Schmutz, M., Guerin, R., Andrieux, P. and Maquaire, O. (2009). Determination of the 3D structure of an earthflow by geophysical methods: The case of Super Sauze, in the French southern Alps. *Journal of Applied Geophysics*, 68, 500–507.

Shroder, J.F. and Weihs, B.J. (2010). Geomorphology of the Lake Shewa landslide dam, Badakhshan, Afghanistan, using remote sensing data. *Geografiska Annaler, Series A – Physical Geography*, 92A, 469–483.

Sturzenegger, M. and Stead, D. (2009). Close-range terrestrial digital photogrammetry and terrestrial laser scanning for discontinuity characterization on rock cuts. *Engineering Geology*, 106, 163–182.

Van Den Eeckhaut, M., Poesen, J., Verstraeten, G. *et al.* (2007). Use of LIDAR-derived images for mapping old landslides under forest. *Earth Surface Processes and Landforms*, 32, 754–769.

van Westen, C.J., Castellanos, E. and Kuriakose, S.L. (2008). Spatial data for landslide susceptibility, hazard, and vulnerability assessment: An overview. *Engineering Geology*, 102, 112–131.

Wang, C., Zhang, H., Wu, F. *et al.* (2009). Disaster phenomena of Wenchuan earthquake in high resolution airborne synthetic aperture radar images. *Journal of Applied Remote Sensing*, 3, 031690.

Wangensteen, B., Eiken, T., Odegard, R.S. and Sollid, J.L. (2007). Measuring coastal cliff retreat in the Kongsfjorden area, Svalbard, using terrestrial photogrammetry. *Polar Research*, 26, 14–21.

Yang, X.J. and Chen, L.D. (2010). Using multi-temporal remote sensor imagery to detect earthquake-triggered landslides. *International Journal of Applied Earth Observation and Geoinformation*, 12, 487–495.

15 Engineering geomorphology of landslides

JAMES S. GRIFFITHS AND MALCOLM WHITWORTH

ABSTRACT

Both natural and human-made landscapes change through time as a result of geomorphological processes, including landsliding. Geomorphologists regard landsliding as a process–response system that moves soil and rock downslope by materials falling, sliding and flowing in response to changes in the extant conditions. In studying landslides, geomorphologists seek to place the site-specific processes of downslope material movement within a framework of the overall landscape situation; a perspective which has important implications for all engineering investigations of ground movement. The geomorphological methods of studying landslides are the same as those employed by engineering geologists and geotechnical engineers, but with greater emphasis on field mapping and the use of remote sensing data (imagery, geophysics, LIDAR etc.). The overall aim is to provide a four-dimensional model of ground conditions which can feed directly into engineering investigations and any project design.

15.1 INTRODUCTION

Landsliding is a geomorphological process and integral to the natural evolution of the landscape. As such, landslides have been studied by geomorphologists in terrestrial (e.g., Dikau *et al.*, 1996; Turner and Schuster, 1996), submarine (e.g., Hampton *et al.*, 1996), and extraterrestrial (e.g., Forster, 2004) environments. Geomorphological studies of landslides are similar to studies of all natural landforms and require an understanding of the processes that created them, the materials they are composed of, their evolution through time, their potential for future changes, and how they fit within an overall model of landscape development. Highland and Bobrowsky (2008) provide a general introduction to landslides that brings out the importance of studying the surface form of the failure. Matthews *et al* (1997) presented the case for rapid mass movement (i.e., landslides) being a source of evidence for Holocene climate change in Europe. Cooper (2007) recognized that these periods of significantly increased landslide activity left many relict features in the contemporary landscape. Korup (2009) reviewed the linkages between landslides, hillslope erosion, and landscape evolution, and highlighted the breadth of techniques employed to quantify landslide processes. He also underscored the importance of using detailed field observations, high-resolution digital data, which increasingly means LIDAR (Schulz, 2007), and geochronological methods (e.g., Ballantyne *et al.*, 1998) in landslide studies.

Landslides, therefore, have a long history of investigation from an academic geomorphological perspective. However, studying the engineering geomorphology of landslides requires that investigations be targeted specifically at the geomorphological components of the landslide or landslide-prone terrain that may have engineering significance. Some aspects of the engineering geomorphology of landslides are reviewed by Skempton and Hutchinson (1969), Hutchinson (2001), Griffiths (2005), Fookes *et al.* (2007), and notably Hutchinson (2008), which provides a record of his unique and important contribution to the subject. This chapter examines landslides within the framework of landscape-forming process–response systems and demonstrates that landslides cannot be investigated without reference to the overall geomorphological environment within which they are located. This approach is based on the long-established geographic concept of the more general environmental conditions associated with the "situation" as opposed to the localized conditions at a specific "site" (e.g., Wilder, 1985). To do this requires an understanding of both contemporary dynamic processes and long-term landscape evolution, which includes investigating relict landforms resulting from processes

Landslides: Types, Mechanisms and Modeling, ed. John J. Clague and Douglas Stead. Published by Cambridge University Press.

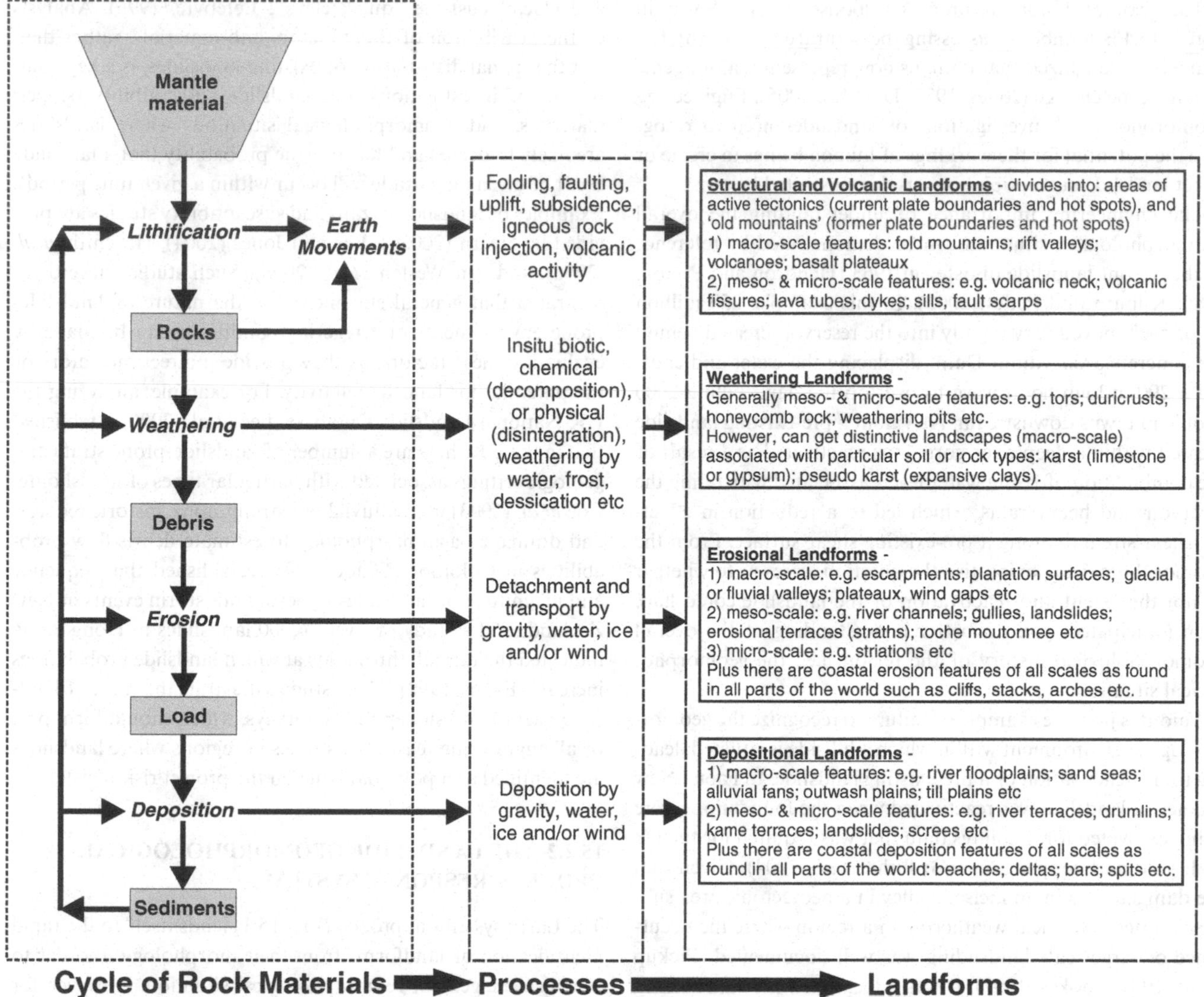

Fig. 15.1. Earth systems landscape model based on the rock cycle and four main landform categories (from Griffiths and Stokes, 2008; reproduced with the permission of the Geological Society of London).

that are no longer present or active in an area. The chapter then looks at the engineering geomorphological investigation of landslides using desk studies, remote sensing, digital data analysis, and field mapping techniques. Finally a geomorphological classification of landslides is presented that recognizes the complexity of the landscape forms produced, which has an important bearing on the engineering investigation of landslide-prone or failed slopes.

15.2 LANDSLIDES: SITUATION AND SITE

15.2.1 GEOMORPHOLOGICAL PROCESS–RESPONSE SYSTEMS

Chorley *et al.* (1984, p. 5) defined a geomorphological system as "a structure of interacting processes and landforms that function individually and jointly to form a landscape complex." Such systems represent conceptual models that require inputs, throughputs, and outputs of mass and energy (Fookes *et al.*, 2007). The value of these models is that they provide a framework for assessing the way in which the transporting agents of wind, water, ice, and gravity create landforms that combine to form "landscapes." Griffiths and Stokes (2008) suggested using the geologic rock cycle as a basis for establishing an overall "Earth systems" approach to define contemporary landscapes (Fig. 15.1), within which geomorphological processes are seen to create landforms of four main categories (structural and volcanic, weathering, erosional, and depositional). The Earth systems approach establishes that landslides, along with all landforms, are part of a larger geomorphological system. Thus any site-specific investigation of a geomorphological feature such as an alluvial fan, delta, river terrace, or landslide must take account of the overall geomorphological environment or "situation" in which the feature is located.

The geomorphological process–response model shown in Figure 15.1 is suitable for assessing the natural environment, but it must be recognized that humans now represent major agents of landscape change (Jones, 1977; Douglas, 2005). Engineering geomorphological investigations of landslides need to recognize the potential for the workings of human beings to create or affect landslide-susceptible sites, as demonstrated below.

The engineering importance of understanding the overall geomorphological situation can be demonstrated by reference to the Vaiont landslide disaster in 1963 (Hendron and Patton, 1986; Kilburn and Petley, 2003). In this example, 250 million m^3 of rock moved very rapidly into the reservoir created behind the concrete arch Vaiont Dam, displacing the water and causing a 200-m-high flood wave to overtop the dam, killing 2000 people in towns downstream. At Vaiont a pre-existing landslide located within a deep river gorge was reactivated as a result of the combination of rising water levels associated with filling the reservoir and heavy rains, which led to a reduction in effective shear strength along a pre-existing shear surface. From the geomorphological perspective the question arises as to whether or not the extent and reactivation of the landslide could have been anticipated, given a better understanding of the overall geomorphological history of the region, i.e., the geomorphological situation.

Vaiont is just one example of failure to recognize the geomorphological environment within which a hillside is situated, leading to the reactivation of a pre-existing landslide. In Papua New Guinea, a landslide was reactivated when the foundations were being excavated for a tailings dam that was required for the Ok Tedi gold and copper mine (Griffiths *et al.*, 2004). Given that the dam site was in an incised valley in a neotectonic area subject to intense tropical weathering in a region where the occurrence of large-scale landsliding was well documented (Pickup *et al.*, 1981; Fookes, 1997), it was a location where landsliding definitely could have been anticipated (Griffiths, 1999). Another example is Carsington in the UK, where an embankment dam failed after being built over Pleistocene periglacial solifluction deposits. The deposits were subsequently found to be at residual strength, and the material had not been recognized during the ground investigations (Skempton and Vaughan, 1993). Any study of the geomorphological history of the Carsington area would have established the widespread occurrence of solifluction deposits, which are a relict mass-movement landform feature and are known to contain relict shear surfaces at residual strength (Skempton and Weeks, 1976).

The examples cited above suggest that landslides occur in particular geomorphological settings, and landslide-prone slopes tend to possess certain distinctive features. Cooke and Doornkamp (1990) suggested that landslides are associated with the following factors: steep slopes; high slopes; high pore water pressures; thick, deeply weathered soil cover; undercutting of the base of the slope; weak, incoherent material outcropping below stronger, more coherent material; and intense rainfall or earthquakes. In addition, particular materials are recognized as being particularly landslide-susceptible (e.g., glaciomarine and glaciolacustrine "quick clays"; Lefebvre, 1996). Analysis of the distribution of these factors and materials, rather than just the spatial distribution of existing landslides, is a key component of investigations into landslide susceptibility (slopes, materials, and geomorphological situations where landslides are likely to occur) and hazard (the probability that a landslide event of given magnitude will occur within a given time period). Examples of landslide hazard and susceptibility studies are provided by Hearn (2002a), Lee and Jones (2004), Havenith *et al.* (2006), and van Westen *et al.* (2008). Such studies have demonstrated that general statements on the nature of landslide-prone environments or triggering conditions can be made by evaluating these factors, as they provide indirect indicators of the potential for landslide activity. For example, analyzing the UK National Landslide Database, Lee *et al.* (2000) established that in the UK there are a number of landslide-prone strata and geologic settings associated with particular types of landsliding. Coe *et al.* (2003) used alluvial fan stratigraphy, historic records, and drainage basin morphology to estimate debris-flow probabilities in Colorado. Glade (1998) established the frequency and magnitude of landslide-triggering rainstorm events in New Zealand, and a study of over 26,000 landslides in Hong Kong indicated the rainfall thresholds at which landslide probabilities increase (Evans, 1998). These studies illustrate the value of landslide hazard and susceptibility surveys, which should form part of all engineering feasibility studies in regions where landslides are identified as a potential issue on the project risk register.

15.2.2 THE LANDSLIDE GEOMORPHOLOGICAL PROCESS–RESPONSE SYSTEM

The Earth systems approach (Fig. 15.1) lends itself to the rapid identification of landforms from their morphology in order to establish the general geomorphological situation. However, for detailed engineering investigation of features such as landslides, it becomes necessary to focus on the geomorphological processes themselves and the factors controlling them at a specific site. While geomorphological process–response systems can be defined at a range of scales from a whole river catchment down to a single hillside, they all contain the following elements:

- Individual morphological components combined in a landform assemblage, which can be established using the approach outlined in Figure 15.1
- Transfers or cascades of energy, such as the kinetic energy in moving ice or flowing water, and mass (i.e., sediment).

The full scope of a detailed process–response model for a coastal cliff site where a slope failure or landslide has occurred is presented in Figure 15.2 (Brunsden, 2002; Brunsden and Lee, 2004). This illustrates the type of site-specific geomorphological data that must be obtained for an investigation of a coastal landslide area. Such a process–response model could form the basis of a hazard and risk assessment, or a full-slope stability evaluation if all relevant parameters can be quantified. Figure 15.2 shows the following:

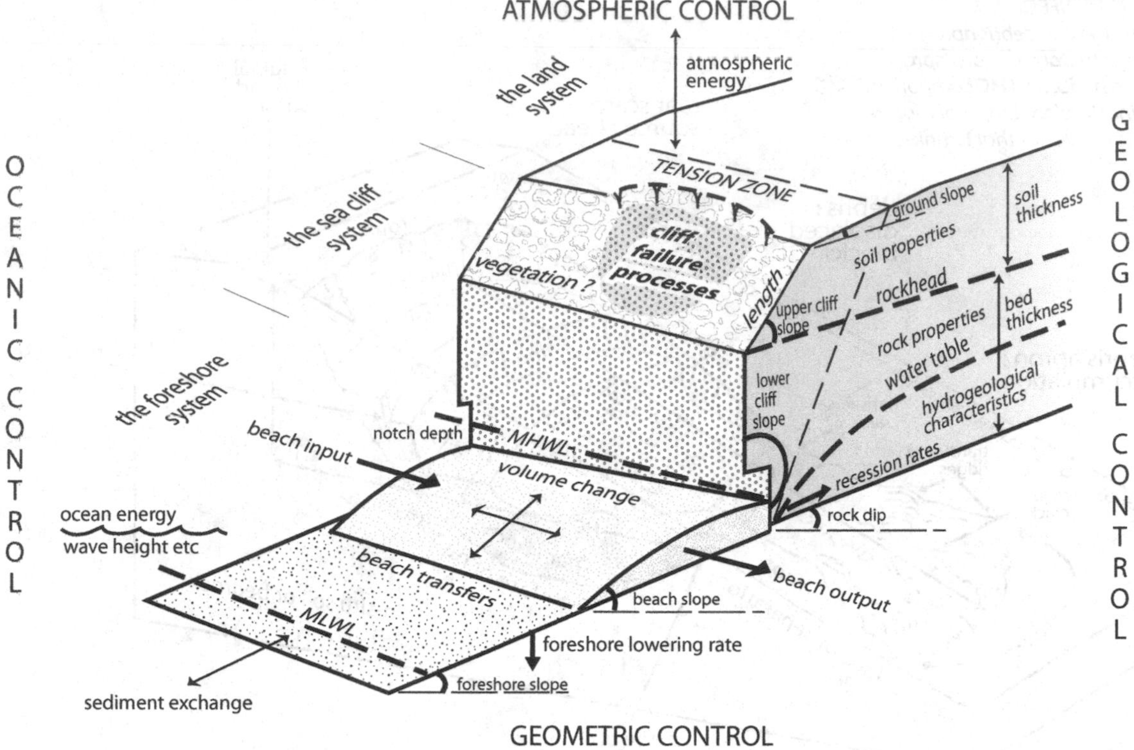

Fig. 15.2. The sea cliff geomorphological process–response system showing the factors that control a cliff failure (modified from Brunsden, 2002; reproduced with the permission of the Geological Society of London).

- Three geomorphological process–response systems can be identified in the model, and these have created the features on the foreshore, the cliff, and the land behind. Each of these process–response systems has its own inputs, throughputs, and outputs.
- The sea cliff system includes the "cliff failure processes," which can be regarded as a landslide process–response subsystem.
- The foreshore, sea cliff, and land process–response systems are subject to control by energy from the ocean and the atmosphere. The first-order factors that control the nature, shape, and – critically – the stability of the cliff are the atmosphere, geology, ocean, the pre-existing geometry of the landforms, and how all of these have changed over time.
- The atmospheric controls include: precipitation type, magnitude, and frequency; solar energy; evapotranspiration; and temperature range, magnitude and frequency.
- The geologic controls are primarily a function of the effective rock mass and shear strength of the materials/discontinuities forming the landforms. Data are required on: discontinuity and strata dip and dip direction, soil or unconsolidated sediment properties, rock properties, soil and rock thickness, weathering profiles, relevant hydrologic conditions, and hydrogeologic characteristics of all the materials.
- The oceanic controls, which are partly responsible for causing changes in the stress on the cliff, include: wave height and length, wave frequency, and mean low and high tide heights.
- The geometric controls are a function of the geomorphological history of the development of the cliff and include: height, length, and slope of the foreshore, beach, lower cliff, upper cliff, and the sloping ground above the cliff.
- Sediment movement is controlled dominantly by the ocean and gravity, and can be regarded as part of the foreshore process–response system. It requires evaluation of: offshore sediment exchange, beach sediment transfer along both the shore and nearshore, beach volume change, and volume of any cliff failure.
- The processes active in the foreshore systems directly affect the stresses and stress distribution in the cliff and lead to slope failure through wave notch erosion, beach erosion, and sedimentation.
- The processes that created the land surface behind the cliff (e.g, long-term fluvial erosion, weathering, sea-level fluctuations, periglacial or glacial processes, earlier mass movements) will have a bearing on the nature of the soils and rocks that now make up the cliff system.

The cliff failure processes, or landslide "subsystem," shown in Figure 15.2 must be assessed with respect to the inputs and outputs of the process–response systems on the foreshore, the whole of the cliff, and the land behind. Other factors could be built into the model, such as vegetation and any human interference, including the construction of sea defenses. In theory, all the parameters, or factors, shown in Figure 15.2

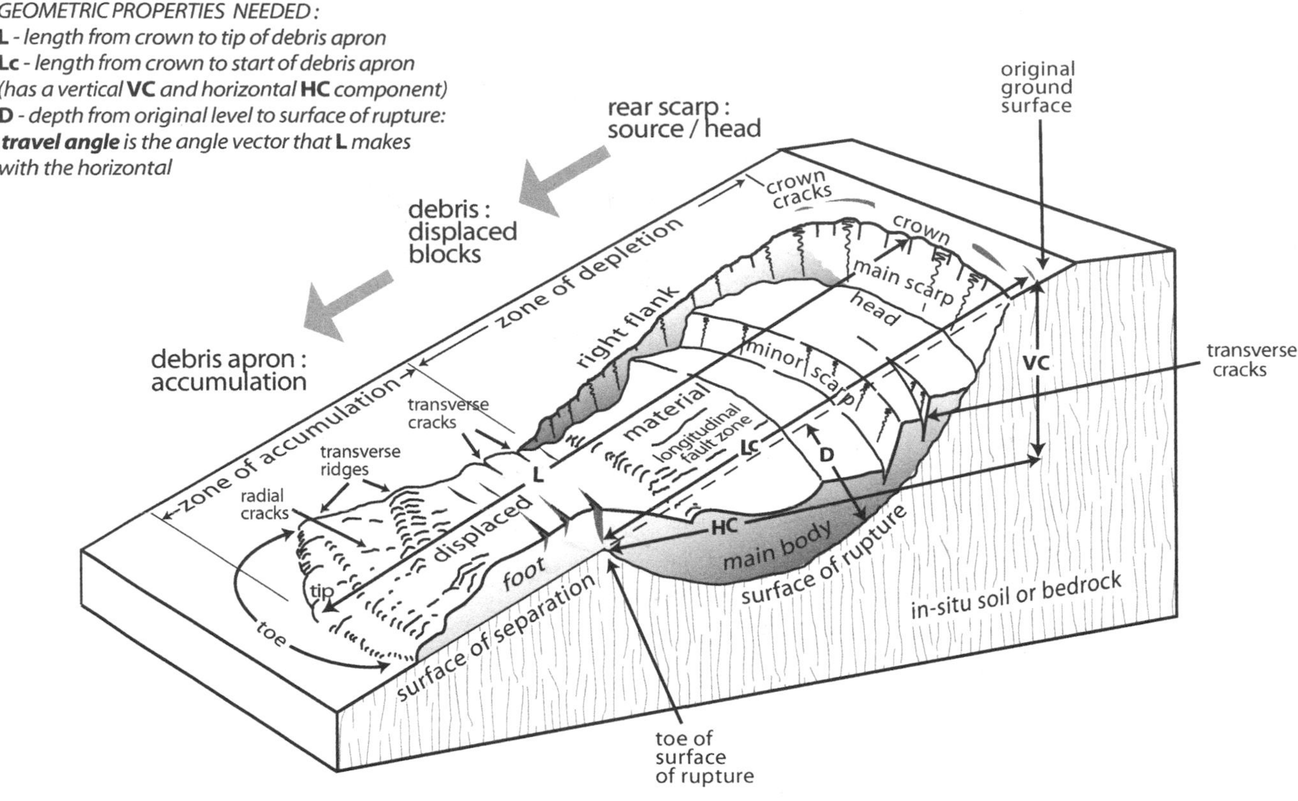

Fig. 15.3. Three-dimensional sketch of the components of a landslide system (adapted from an original drawing by Varnes, 1978).

could be quantified for a particular cliff location, in which case the geomorphological systems at the site level would be fully understood. A full understanding would allow the landslide subsystem to be modeled in a deterministic way, enabling any changes in the stability to be linked to the main controlling factors (e.g., storm magnitude, rates of weathering causing changes to the properties of soils and rocks, or interference by humans). The modeling could then form the basis for establishing the likelihood of coastal cliff failures or forecasting the effects of, for example, climate change on the stability of the cliff. In practice it is not possible for all these parameters to be quantified; thus probabilistic approaches or simple judgment-based models to assess the risk from erosion, including landsliding, are often more appropriate (Lee *et al.*, 2001; Lee, 2005).

Figure 15.2 shows the cliff failure processes, and hence landsliding, as a process–response subsystem within the overall conceptual model of the coastal cliff. However, it is possible to examine an individual landslide system in more detail using the same approach. An example of how this might be undertaken for a pre-existing failure is presented in Figure 15.3 (based on Varnes, 1978; WP/WLI, 1990, 1995). As shown in this figure, the geomorphological process–response system comprises three components: (1) *the source or head* defined by the rear scarp; (2) *the main body* of the landslide or, in the case of slides and flows (see Ancillary Materials associated with this volume, www.cambridge.org/9781107002067), the landslide track; and (3) *the toe*, where the debris apron of most of the displaced material accumulates. To define the overall process–response system, one must measure the geometry of the landslide, establish the engineering characteristics, and calculate the volumes of materials involved. By compiling the data for the landslide system in this way, it is possible to establish the interaction between the individual components that influence the overall landslide behavior and thus focus site-specific investigations and monitoring of movements (Lee *et al.*, 2004).

The approach presented above is appropriate for investigating pre-existing landslides. In order to examine a site where the potential for landsliding exists, it is necessary to return to establishing the susceptibility of the site based on the indirect

Table 15.1. *Hierarchical terrain systems approach to landslide mapping.*

Mapping scale	Topographic DEM data available	Image data available	Map data available	Application
Land system >1:60,000	SRTM DEM ASTER DEM	ASTER imagery Landsat 7 ETM	Pre-existing topographic and geologic maps	Evaluation of possible constraints due to landsliding using regional assessment of the terrain and mapping of environment factors such as drainage, slope, geology, land use and vegetation land cover.
Land facet 1:10,000–1:60,000	Satellite-derived DEM data Airborne LIDAR	High-resolution satellite imagery Stereo airphotos	Pre-existing topographic and geologic maps	Evaluation of potential landsliding using more detailed aerial photographic reconnaissance and desk study data. Should be undertaken in conjunction with a site visit.
Land element 1:500– 1:10,000	Satellite-derived DEM data Airborne LIDAR Ground LIDAR	High-resolution satellite imagery Stereo airphotos Digital orthophotos	Site survey maps Topographic maps Orthophoto basemaps	Detailed characterization of landslides within and adjacent to study site. Detailed geomorphological mapping of landforms associated with landslides using high-resolution data and field reconnaissance.

geomorphological factors. However, in both situations a spatially correct geomorphological map of the landslide complex must be produced in order to evaluate an existing landslide system, to establish the potential for landsliding, and to plan engineering site investigations.

15.3 ENGINEERING GEOMORPHOLOGICAL INVESTIGATION OF LANDSLIDES

15.3.1 THE NATURE AND SCOPE OF THE INVESTIGATION

The geomorphological investigation of landslides provides a framework for describing and mapping surface landslide processes and predicting future process behavior. Such investigations include the application of geomorphological techniques to map surface form, describe the nature of surface materials present, and record evidence of processes (Embleton and Verstappen, 1988; Cooke and Doornkamp, 1990; Hutchinson, 2001; Smith *et al.*, 2011). These techniques were initially developed for geological maps and evolved through the work of Waters (1958) and Savigear (1965) into a method of morphometric and morphological mapping (Brunsden and Jones, 1972; Mitchell, 1991). This is now a widely accepted part of conventional terrain evaluation and plays an important role in slope stability assessments, especially where it is supported by complementary methods such as desk study, historical analysis, remote sensing interpretation, subsurface investigation, and monitoring (Lee, 2001). Applications of geomorphological surveys in the UK include: the assessment of the Channel Tunnel portal and high-speed rail link (Griffiths *et al.*, 1995; Birch and Griffiths, 1996; Waller and Phipps, 1996; Griffiths, 2001a), landslide mapping along the Cotswolds escarpment (Whitworth *et al.*, 2005), coastal studies in Dorset (Brunsden and Jones, 1972; Griffiths and Abraham, 2008), and landslide investigations at Whitehaven (Boggett *et al.*, 2000), the Taff Valley (Brunsden *et al.*, 1975), northeast Derbyshire (Al-Dabbagh and Cripps, 1987), and along the Perth–Inverness trunk road (Chaplow, 1983). Notable studies outside the UK include geomorphological assessments of road design in Nepal (Hearn, 2002b), the Central Cross Island Highway in Taiwan (Petley, 1998, 2001), pipeline corridor assessment in Georgia (Fookes *et al.*, 2001a), the extensive work undertaken for all forms of development in Hong Kong (King, 1996; Parry and Ng, 2010), landslide mapping in the Polish Carpathians (Wojcik and Mrozek, 2002), and the range of examples compiled for the European Commission LIFE project on Coastal Change, Climate and Instability (McInnes and Jakeways, 2000).

When undertaking engineering geomorphological assessments, Waller and Phipps (1996) and Phipps (2003) recommended a terrain systems approach (Table 15.1). This approach involves separating terrain into a series of hierarchical units from the regional level (*terrain system*) down to the site level (*terrain elements*). The approach fits perfectly within the framework provided by the Earth systems methodology presented in Figure 15.1. At a regional scale, assessment is reserved for initial reconnaissance of the site to place it within its overall

Table 15.2. *Remote sensing tools for the geomorphological investigation of landslides.*

Data type	Data processing methods	Applications
Stereo aerial photography	Stereo interpretation of aerial photography. Low sun angle required to enhance surface morphological features. Photogrammetric processing to generate digital elevation models and orthophotography.	Common technique for landslide investigation due to availability of current and historic air photo coverage. Landslide monitoring using repeat airphoto coverage.
Orthorectified photography	Overlay onto digital elevation models for three-dimensional interpretations. Integration with other digital data within GIS environment.	Planimetric and georeferenced orthorectified imagery can be used during field mapping alongside topographic base maps.
Airborne and ground- based LIDAR	Shaded relief images generated using multiple illumination azimuth and elevation angles. Plan and profile slope curvature can be used to identify extent and types of geomorphological slope units. Slope angle can be used to enhance individual slope units and landforms such as scarps and rotated landslide blocks.	High-resolution LIDAR topographic data Vegetation penetration capabilities can provide bare earth elevation models for landslide mapping beneath tree cover. Shaded relief, slope and curvature images can be used during field mapping alongside orthophotography and topographic base maps.

geomorphological situation, and is typically used for route selection and feasibility studies (Phipps, 2001). Mapping at this scale relies on existing desk study data such as topographic and geologic maps, low-cost moderate-resolution satellite imagery, and limited field reconnaissance. Most landslide investigations, however, typically involve mapping at the site-specific scale, where stereo aerial photography and airborne LIDAR replace satellite imagery as the tools of choice, and the data are integrated with a desk study prior to a more detailed walk-over survey or geomorphological assessment of the site. The advantage of this technique is that it emulates the typical engineering approach to site investigation, whereby initial desk study and photo interpretation are followed by more detailed field reconnaissance (Lawrance *et al.*, 1993; Griffiths, 2001b; Phipps, 2003).

15.3.2 DESK STUDIES AND DIGITAL DATA ANALYSIS

Aerial photography remains the most popular medium for the preliminary investigation of landslides. Its continued use reflects the wide availability of historic and current imagery, general ease of interpretation, and stereo viewing capabilities. Consequently, aerial photography forms an integral component of the desk study stage of most engineering investigations (Hart *et al.*, 2009). Stereo photographs are preferable for landslide mapping, and those that are acquired with low sun angle enhance some of the subtle terrain features associated with slope instability (Lawrance *et al.*, 1993). However, orthorectified photographs are now widely available and, although they do not have stereo capability unless viewed with a digital elevation model, they are planimetric and georeferenced images that can be used as base maps during field mapping alongside standard topographic maps and hand-held GPS (Griffiths and Abraham, 2008). The general procedures for aerial photographic interpretation for landslide investigations have been discussed by Ho *et al.* (2006) and Lawrance *et al.* (1993), and are summarized in Table 15.2. It is important to note that the scale of the photography will affect the size of the landforms that are visible, and therefore a field survey should always be undertaken as part of the site interpretation to verify any aerial photographic interpretation (Hutchinson, 2001; Hart *et al.*, 2009).

Where historic aerial photographs are available, it is possible to study the temporal development of landslide systems (van Westen and Lulie Getahun, 2003). Taken further, quantitative data can be extracted using digital photogrammetry to derive digital elevation models and thereby calculate vertical ground displacements and volumetric changes (Walstra *et al.*, 2007; Dewitte *et al.*, 2008; Prokesova *et al.*, 2010) and to quantify three-dimensional landslide displacements (LePrince *et al.*, 2007).

These photographic techniques are rapidly being superseded by LIDAR, which is becoming more widely available, is accurate, and can be rapidly processed. Airborne laser scanning systems (LIDAR) are now routinely used to capture information for terrain analyses and landslide investigations (McKean and Roering, 2003; Glenn *et al.*, 2006). The main reason for the increased use of LIDAR is the high spatial resolution and precision of the digital elevation data and the ability to remove the masking effects of vegetation (Van Den Eeckhaut *et al.*, 2007). LIDAR data are an important source of morphometric indices for landslide susceptibility assessment, including elevation, slope gradient, and slope curvature – all of which are important intrinsic factors that influence the distribution of landslides in the terrain (Jaboyedoff *et al.*, 2012).

Given that landslides disrupt the land surface, it is possible to use LIDAR data to identify slope instability by enhancing

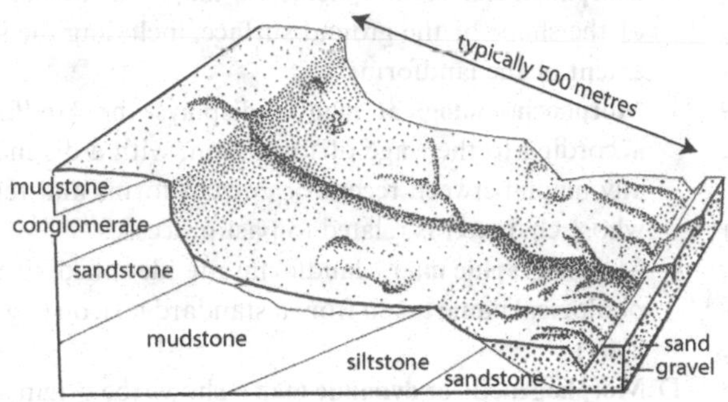

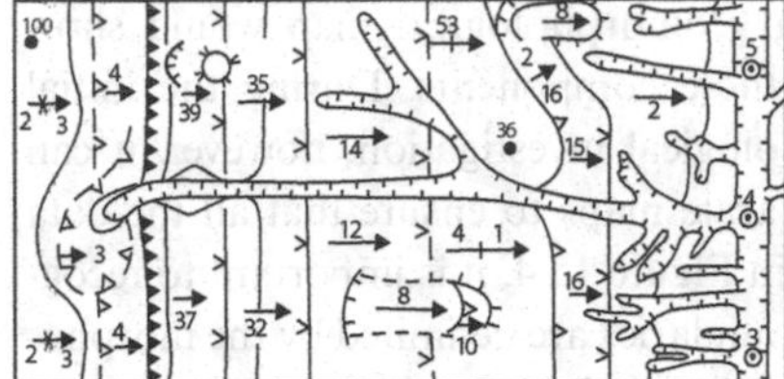

B Morphochronological map

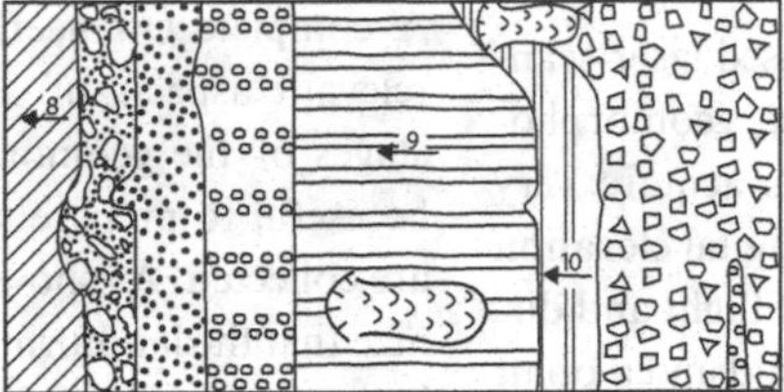

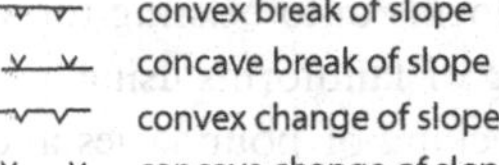

C Morphographic map

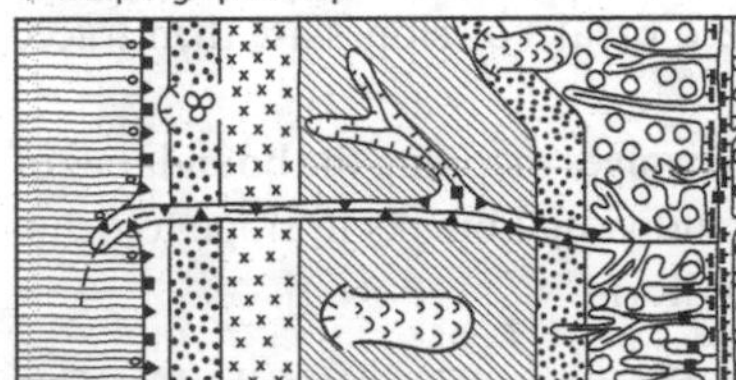

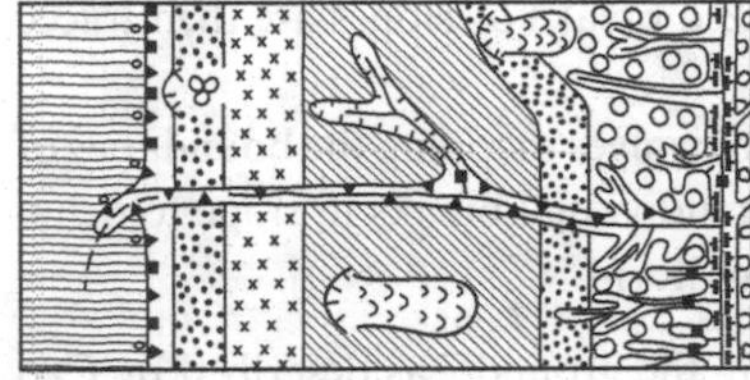

D Morphogenetic / Dynamic map

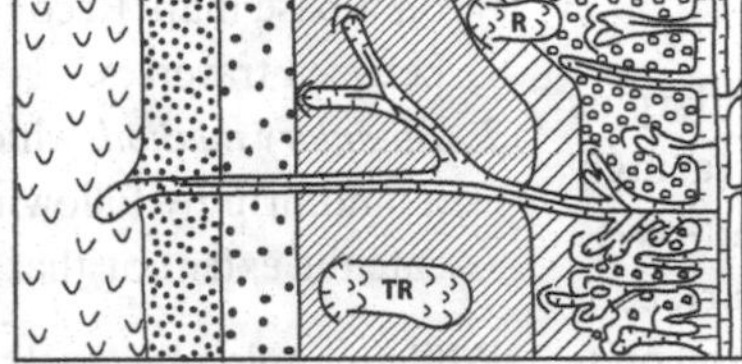

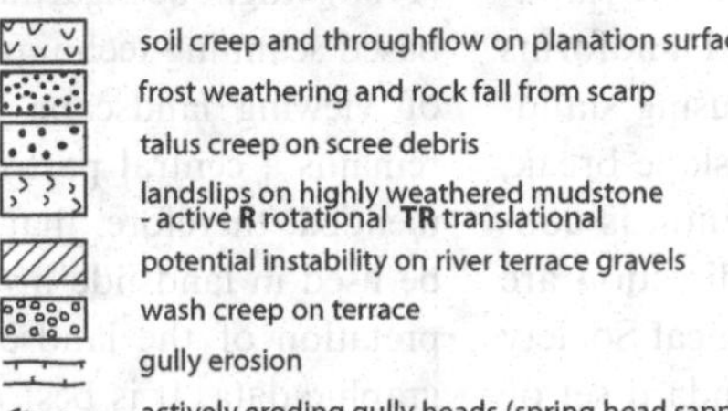

Fig. 15.4. Geomorphological mapping symbols and map types: (A) morphological or morphometric map; (B) morphochronological; (C) morphographical; (D) morphogenetic or dynamic map (from Cooke and Doornkamp, 1990; reprinted with the permission of Oxford University Press).

the topographic signatures of landslides in the data. The topographic signature of a landslide can be enhanced either visually using shaded relief maps or by interpreting slope angle, curvature, and roughness (Table 15.2). Indeed, Schulz (2007) demonstrated that LIDAR was a more effective tool for identifying landslides than conventional aerial photography.

Where LIDAR data are unavailable, topographic data can be derived from high-resolution satellites such as SPOT, QuickBird, and Ikonos; each is able to produce accurate elevation models suitable for landslide mapping (van Westen *et al.*, 2008). Alternatively, unmanned aerial vehicles can be used to acquire either optical imagery using digital camera technology or high-resolution topographic data using LIDAR systems (Niethammer *et al.*, 2011).

Recent developments in the use of ground-based laser-scanning systems present new opportunities for the geomorphological assessment of landslides. These systems acquire very high-density digital elevation data, from which digital elevation models can be built. The technology uses time of flight and direction between a laser scanner mounted on a tripod and a point on a reflective surface to determine the distance and location of a point with 15-mm accuracy (Jaboyedoff *et al.*, 2012). The use of these systems, however, requires careful planning when dealing with complex landslide terrains. Furthermore, multiple scan locations are often required when the extent of the area exceeds the scanner's range or where the ground surface is obstructed by trees or buildings. Despite these limitations, the technology has been used successfully for direct landslide mapping (Rowlands *et al.*, 2003), calculation of landslide volumes (Corsini *et al.*, 2007), the temporal study of landslide movement (Glenn *et al.*, 2006), and the assessment of hazards along transportation routes (Lato *et al.*, 2009).

15.3.3 FIELD-BASED GEOMORPHOLOGICAL MAPPING

Field-based geomorphological mapping commonly underpins large-scale landslide assessment, although it is increasingly being integrated with the results obtained from ground-based laser systems. The morphology of the topographic surface is documented during mapping, allowing subsequent interpretation and identification of the surface processes and landforms. The boundaries of different slope units and landforms are represented on a base map in plan position using standard symbols indicating the location and type of slope break, whether sharp or gentle, and whether this transition is convex or concave in section. Slope angle and dip direction are also commonly recorded (Lee, 2001). The Geological Society Working Party (Anonymous, 1982) defined a standard set of symbols to be used on morphological maps (see map A in Fig. 15.4) that are essentially an extension of the notation initially devised by Waters (1958) and Savigear (1965). Four separate map types are shown in Figure 15.4 (from Cooke and Doornkamp, 1990):

- A **Morphological or morphometric map** – a factual portrayal of the shape of the ground surface, including the shape and extent of the landforms
- B **Morphochronological map** – displays the landform units according to their age of formation, with a distinction usually made between recent or current forms and relict forms whose character is related to past processes
- C **Morphographic map** – landforms are identified by a geomorphological name taken from a standard lexicon (e.g., Goudie, 2004)
- D **Morphogenetic or dynamic map** – shows the origin and development of the landscape, including details of the processes that are presently active.

A complete engineering geomorphological map would show relevant aspects of all these components. During the initial stages of the geomorphological investigation, however, it can be useful to prepare separate maps to ensure that all the data are collected. As shown in Figure 15.4, it is important to recognize that the landform boundaries are delimited by the morphological units identified in Figure 15.4A.

The most common method of geomorphological assessment involves either field mapping of landforms onto existing topographic maps or office-based mapping of landforms using aerial photographs, followed by field checking of boundaries and unit characteristics, termed "ground-truthing." Hand-held GPS devices are now used to supplement these techniques, thereby allowing digital mapping directly into a GIS system and the use of orthophotographic base maps, which provide a greater level of surface detail than topographic maps alone (Griffiths and Abraham, 2008).

In the past, the technique of field geomorphological mapping was commonly based on two approaches (Catt, 1988).

1. *Traverse method*, which involves walking parallel traverses across the main landforms and recording breaks and changes on a base map. Recognized breaks of slopes are then traced between traverses.
2. *Boundary method*, which involves mapping the boundaries of a landform by following individual breaks in slope to delineate the extent of that landform.

While mapping by "walking the ground," as described by Catt (1988), might be regarded as having been superseded by ground-based scanning techniques (see Section 15.3.2), the importance of viewing landscape features from all possible viewpoints remains a central part of interpreting landforms. It is recommended, therefore, that the practice of "walking the ground" be used in landslide investigations, even when an initial interpretation of the landscape is derived from digital or photographic data. It is best to adopt the transverse method with a GPS and slope inclinometer when investigating a landslide. The locations of the breaks or changes in slopes are identified along each traverse, and the type of slope change (convex or concave) and the slope angle are recorded. The topographic base map is populated with morphological information by joining breaks or

Table 15.3. *Geomorphological and hydrologic features characteristic of slope instability (based on Soeters and van Westen (1996).*

Geomorphological features	Relation to slope instability
Semicircular break in slope	Head of landslide scarp with exposed failure plane. Arcuate scarp to rotational landslide. Head source zone to mudflow or debris flow.
Step-like morphology	Individual retrogressive landslides or multiple units within the same landslide.
Flat or back-tilted slope units (tilted upslope)	Remains of rotated landslide block.
Linear ridge running parallel to contours	Downslope edge of rotated landslide block.
Lobate features on lower slopes	Accumulation zones from landslide in the form of flows style movements.
Irregular slope morphology	Shallow flows, small retrogressive landslides or zone of complex landslide activity. Landslide amalgamation may have occurred through repeated slope movement.
Back-tilted trees (tilted upslope).	Rotational landslide movement and presence of rotated landslide blocks.
Forward-tilted trees (tilted downslope)	Area of uplift or heave in landslide zone. Downslope creep processes.
Hydrologic features	**Relation to slope instability**
Stagnated drainage	Water accumulation within back-tilted block and areas of irregular ground with the landslide.
Disrupted drainage patterns	River re-routed around frontal lobes of landslide or down the lateral sides of the landslide.
Anomalous seepage or drainage patterns	Disruption of surface and subsurface water flow by landslide. Emergence of water from failure surface and the front of accumulation zones (lobes).

changes in slope with the same features identified on previous traverses. The final morphological map is generated through a number of traverses across the whole site (Lee, 2001). In many cases, this traverse method is supplemented by using the boundary mapping technique in order to accurately define the boundary and extent of landforms between traverses.

Morphological information should be supplemented with observations on the nature of any observed slope instability, including information on the size, form, and frequency of slope movements. Evidence of slope instability can be derived from terrain elements such as slope morphology, drainage, and vegetation, using a number of published field checklists (Table 15.3; Crozier, 1984; Bromhead, 1992; Soeters and van Westen, 1996; Dikau *et al.*, 1996). Where slope instability has been identified, particular effort should be made to determine its age (Crozier, 1984). Sharp boundaries and scarps are indicative of recent slope movement; over time, degradation results in increasingly subdued landslide forms (Doornkamp, 1994), therefore relict landslides will display rounded and subdued morphology (Table 15.4).

Geomorphological field mapping can be particularly important in areas where dense vegetation obscures surface landforms, making interpretation using aerial photographs difficult. This is particularly significant for engineering projects because woodlands may conceal evidence of past or present landslide activity (Hutchinson, 2001). Field mapping can provide vital insights into the nature of movement and the evolution of landslides; it is also a noninvasive method for determining the nature of the underlying failure surface (Hutchinson, 1983; Carter and Bentley, 1985).

Production of a final engineering geomorphological map requires an integration of field information on surface morphology and landslide distribution with an interpretation of subsurface deposits. The result will depend on the end-user requirements (Griffiths and Hearn, 1990), but the map should, at the very least, include engineering geomorphological, geological, hydrological, and hydrogeological information (Hutchinson, 2001), and it should be accompanied by an interpretative report that outlines the main findings of the survey and conveys the nature of the landslide hazards in the study area (Griffiths and Hearn, 1990).

15.4 GEOMORPHOLOGICAL CLASSIFICATION OF LANDSLIDES

Integral to the geomorphological investigation of landslides is a common landslide terminology. The geomorphological classification of landslides relies on the description of surface morphological characteristics and landforms to define the type of landslide and thereby determine the mechanism of movement. A number of different classifications have been proposed, including the early schemes of Zaruba and Mencl (1969) and Nemčok *et al.* (1972) and more recent classifications by Varnes (1978), Zaruba and Mencl (1982), Hutchinson (1988), Cruden (1991), and Cruden and Varnes (1996). Many of these schemes use little or no geomorphological information, although those suggested by Hutchinson (1988), Dikau *et al.* (1996), and Soeters and van Westen (1996) are based on surface morphology and type of material and can differentiate landslide types. A summary of

Table 15.4. *Activity classification for landslides (based on Cruden and Varnes, 1996; after Mather* et al., *2003).*

Activity state	Identification of causes of movement	Condition of main scarp	Condition of lateral margins	Internal morphology	Estimated age (years)[a]
Active – currently moving (includes inactive landslides that have been reactivated)	Causes of movement identifiable and active	Sharp; unvegetated	Sharp; unvegetated; streams at edge	Hummocky; fresh scarps; reverse slopes; undrained depressions; fresh tension cracks; active movement	< 100 (historic)
Suspended – moved within the last 12 months and likely to become active again	Causes of movement identifiable and likely to re-occur	Sharp; unvegetated	Sharp; unvegetated; streams at edge	Hummocky; fresh scarps; reverse slopes; undrained depressions; identifiable tension cracks	< 100 (historic)
Dormant – young (inactive)	Cause of movement still identifiable and could re-occur	Relatively sharp; partially vegetated	Relatively sharp; partially vegetated; lateral streams fed by small tributaries flowing off the main body of the slide	Hummocky; relatively sharp and fresh scarps; reverse slopes; undrained depressions; tension cracks closed and vegetated but marked by small depressions	100–5000 (late Holocene)
Dormant – mature (inactive); described as abandoned by Dikau *et al.* (1996)	Cause of movement still identifiable but not likely to re-occur	Smooth; vegetated	Smooth vegetated; lateral streams fed by tributaries flowing off the main body of the slide	Smooth rolling topography; disrupted and disjointed internal drainage network	5000–10,000 (early Holocene)
Dormant – old (inactive) or relict[b]	Causes of movement may be inferred but associated with different climatic or geomorphological conditions	Dissected; vegetated	Vague lateral margins; no lateral drainage	Smooth, undulating topography; normal stream pattern	10,000–100,000 (Late Pleistocene)
Fossil (inactive) or ancient[b]	Causes of movement unknown but associated with different climatic and geomorphological conditions	May not be identifiable; likely to be at least partially, if not completely, removed by erosion	May not be identifiable; likely to be at least partially, if not completely, removed by erosion	Fully integrated into the existing topography and very little indication of the former landslide morphology remains	> 100,000

[a] Caution is required in the interpretation of timescales, as weathering rates differ considerably in different environmental settings.
[b] Relict or fossil landslides may also be covered by younger deposits. Rib and Liang (1978) described these as "hidden" or "buried" landslides, which could potentially be exhumed and reactivated. This situation has been described by Schultz and Harper (1996) for the Late Carboniferous paleo-landslides reactivated in building excavations in Pennsylvania, USA.

these landslide classification schemes, combined with key surface morphological indicators that can be used to identify and classify each landslide type based upon geomorphological criteria, can be found in the Ancillary Materials associated with this volume, www.cambridge.org/9781107002067).

A geomorphological classification of landslides is inherently difficult because of the complex mechanical behavior of landslide systems and the fact that many landslides exhibit more than one type of movement. However, it is still possible to identify and map each of the individual movement forms even though the overall movement may be complicated. Slides form along well-defined failure surfaces or shear planes that are commonly semicircular or planar in shape. In some cases, more complex shear surfaces may develop in heterogeneous materials, forming compound slides. Successive rotational landslides typically form a series of separate shallow failures, each arranged head to toe. Like many types of landslide, successive landslides can be *retrogressive* (i.e., a series of failures start at the base of a slope and retrogress upward), *progressive* (a failure starts at the top of a slope and progresses downward), or a *combination* of the two. This situation contrasts with multiple rotational slides, in which several shear surfaces develop to form separate distinct rotational units within a single landslide complex. Planar translational slides move roughly parallel to the slope and occur in weathered or sheared soils and rock along pre-existing discontinuities that daylight out of the slope. Compound landslides typically involve two types of movement that occur concurrently within the same landslide system, whereas complex landslides involve two or more types of movement through a change in behavior as the landslide progresses downslope (Dikau, 1999). Landslides in rock masses typically involve movement or detachment along pre-existing discontinuities and can result in either planar slides, rockfall, or toppling. A fall denotes freefall or a bounding movement from a steep rock face. A topple is a pivoting movement of a block from a rock face; it generally involves forward rotation, although backward rotation is also possible. Flow-type landslides behave as a fluidized mass in which water plays a significant role; they are subdivided according to the nature and grain size of the material involved in the flow. Three-dimensional block diagrams prepared by the British Geological Survey (2011) are useful aids for distinguishing landslide types.

The derivation of a landslide classification from field geomorphological information involves the following stages:

- **Morphological** or **morphometric mapping** of surface features, including the locations or any changes in slope and measurement of slope angles for each individual slope unit (Fig. 15.4A).
- **Field observations** of slope features, vegetation, geology, and surface hydrologic features.
- **Interpretation of field data** to generate a morphographic map (Fig. 15.4C) that includes a description and classification of each landform.

The first two stages of this procedure have been discussed in the previous section. The final classification of the landslide should be based on the nature of the underlying geology (soil, debris, or rock) and the landforms present in each of the three landslide zones (scarp, track/main body, and toe/debris apron; Fig. 15.3). The required important observations include: the shape of the landslide and scarp in plan view; the nature of landforms within the body of the landslide, such as grabens or back-tilted slopes; preservation of vegetation and structures on the landslide and any sense of movement these may provide, such as forward or back rotation; the behavior of the landslide at the edge of the sliding zone; and the nature and extent of the accumulation zone at the toe of the landslide.

A landslide can be further described according to how it develops. It may be retrogressive, progressive, or *enlarging*: in the last case, instability propagates sideways parallel to the slope contours. Most landslides that retrogress also become larger as the instability propagates upslope. It is also important to consider whether the landslide is a *first-time failure* in previously unsheared ground or a *reactivation of a pre-existing landslide* along existing shears in the subsurface (Hutchinson, 1988, 1995).

15.5 CONCLUSIONS

Engineering geomorphological assessments of landslides and landslide-prone terrain provide crucial insights into the nature and extent of the landslide hazard. Process–response landslide models derived from geomorphological investigations and interpretation are based on the integration of information provided by desk studies, remote sensing, digital data analysis, and field mapping. The models provide important spatial information on the extent, boundary conditions, types, and age of landslides in an area. They also place landslides within the overall geomorphological situation. The models should guide engineering site investigations and are cost-effective in determining locations for any engineered work (Brunsden *et al.*, 1975; Griffiths and Marsh, 1986; Fookes, 1997). In addition, the overall geomorphological model creates a spatial framework for interpreting borehole and trial pit geology and monitoring data, and for developing a fully integrated, ground behavior model (Cooke and Doornkamp, 1990; Griffiths *et al.*, 2011). This spatial framework, in turn, provides a basis for interpolating and extrapolating subsurface information from single-point data sources, such as trial pits and boreholes, to form a three-dimensional model of ground conditions, thereby creating the total geology model suggested as being necessary for all engineering projects (Fookes *et al.*, 2000, 2001b).

REFERENCES

Al-Dabbagh, T.H. and Cripps, J.C. (1987). Data sources for planning: Geomorphological mapping of landslides in north-east Derbyshire. In *Planning and Engineering Geology*, ed. M.G. Culshaw, F.G. Bell, J.C. Cripps and M. O'Hara. Geological Society, Engineering Geology Special Publication 4, pp. 101–114.

Anonymous (1982). Geological Society Working Party on Land Surface Evaluation for Engineering Practice: Report by a working party under the auspices of the Geological Society. *Quarterly Journal of Engineering Geology*, 55, 265–316.

Ballantyne, C.K., Stone, J.O. and Fifield, L.K. (1998). Cosmogenic Cl-36 dating of postglacial landsliding at The Storr, Isle of Skye, Scotland. *Holocene*, 8, 347–351.

Birch, G.P. and Griffiths, J.S. (1996). Engineering geomorphology. In *Engineering Geology of the Channel Tunnel*, ed. C.S. Harris, M.B. Hart, P.M. Varley and C.D. Warren. London: Thomas Telford, pp. 64–75.

Boggett, A.D., Mapplebeck, N.J. and Cullen, R.J. (2000). South Shore Cliffs, Whitehaven: Geomorphological survey and emergency cliff stabilization works. *Quarterly Journal of Engineering Geology and Hydrogeology*, 33, 213–226.

British Geological Survey (2011). *How Does the BGS Classify Landslides?* [available at www.bgs.ac.uk/landslides/how_does_BGS_classify_landslides.html, accessed June 14, 2011].

Bromhead, E. (1992). *The Stability of Slopes*, 2nd ed. London: Blackie.

Brunsden, D. (2002). Geomorphological roulette for engineers and planners: Some insights into an old game. *Quarterly Journal of Engineering Geology and Hydrogeology*, 35, 101–142.

Brunsden, D. and Jones, D.K.C. (1972). The morphology of degraded landslide slopes in south west Dorset. *Quarterly Journal of Engineering*, 5, 205–222.

Brunsden, D. and Lee, E.M. (2004). *Behaviour of Coastal Landslide Systems: An Interdisciplinary View*. Special issue of *Zeitschrift für Geomorphologie*, 134.

Brunsden, D., Doornkamp, J.C., Fookes, P.G., Jones, D.K.C. and Kelly, J.M.H. (1975). Large scale geomorphological mapping and highway design. *Quarterly Journal of Engineering Geology*, 8, 227–253.

Carter, M. and Bentley, S. (1985). The geometry of slip surfaces beneath landslides. *Canadian Geotechnical Journal*, 22, 234–238.

Catt, J.A. (1988). *Quaternary Geology for Scientists and Engineers*. Chichester, UK: Wiley.

Chaplow, R. (1983). Engineering geomorphological investigations of a possible landslide, Killiecrankie Pass, Scotland. *Quarterly Journal of Engineering Geology*, 16, 301–308.

Chorley, R., Schumm, S.A. and Sugden, R. (1984). *Geomorphology*. London: Methuen.

Coe, J.A., Godt, J.W., Parise, M. and Moscariello, A. (2003). Estimating debris-flow probability using fan stratigraphy, historic records, and drainage-basin morphology, Interstate 70 highway corridor, central Colorado, U.S.A. In *Debris-Flow Hazards Mitigation: Mitigation, Prediction and Assessment*, ed. D. Rickenmann and C.-L. Chen. Rotterdam: Millpress, pp. 1085–1096.

Cooke, R.U. and Doornkamp. J.C. (1990). *Geomorphology in Environmental Management*. 2nd ed. Oxford: Oxford University Press.

Cooper, R.G. (2007). *Mass Movement in Great Britain*. Peterborough, UK: Joint Nature Conservation Committee, Geological Conservation Series 33.

Corsini, A., Borgatti, L., Coren, F. and Vellico, M. (2007). Use of multi-temporal airborne LiDAR surveys to analyze post-failure behaviour of earthslides. *Canadian Journal of Remote Sensing*, 33, 116–120.

Crozier, M.J. (1984). Field assessment of slope instability. In *Slope Instability*, ed. D. Brunsden and D.B. Prior. Chichester, UK: Wiley, pp. 103–142.

Cruden, D.M. (1991). A simple definition of a landslide. *Bulletin of the International Association for Engineering Geology and the Environment*, 43, 27–29.

Cruden, D. and Varnes, D.J. (1996). Landslide types and processes. In *Landslides: Investigation and Mitigation*, ed. A.K. Turner and R.L. Schuster. Washington, DC: National Academy Press, pp. 36–75.

Dewitte, O., Jasselette, J., Cornet, Y. *et al.* (2008). Tracking landslide displacements by multi-temporal DEMs: A combined aerial stereophotogrammetric and LIDAR approach in western Belgium. *Engineering Geology*, 9, 11–22.

Dikau, R. (1999). The recognition of landslides. In *Floods and Landslides: Integrated Risk Assessment*, ed. R. Casale and C. Margottini. Düsseldorf: Springer, pp. 39–44.

Dikau, R., Brunsden, D., Schrott, L. and Ibsen, M.-L. (1996). *Landslide Recognition: Identification, Movement and Causes*. Chichester, UK: Wiley.

Doornkamp, J.C. (1994). Imaging the geomorphological environment: Mapping in order to understand processes. *East Midland Geographer*, 17(1,2), 51–57.

Douglas, I. (2005). Urban geomorphology. In *Geomorphology for Engineers*, ed. P.G. Fookes, E.M. Lee and G. Milligan. Caithness, UK: Whittles, pp. 757–779.

Embleton, C. and Verstappen, H.T. (1988). The nature and objectives of applied geomorphological mapping. *Zeitschrift für Geomorphologie Supplementband*, 68, 1–8.

Evans, N.C. (1998). The natural terrain landslide study. In *Slope Engineering in Hong Kong*, ed. K.S. Li, J.N. Kay and K.K.S. Ho. Rotterdam: Balkema, pp. 137–144.

Fookes, P.G. (1997). First Glossop Lecture. Geology for engineers: The geological model, prediction and performance. *Quarterly Journal of Engineering Geology*, 30, 290–424.

Fookes, P.G., Baynes, F.J. and Hutchinson, J.N. (2000). Total geological history: A model approach to the anticipation, observation and understanding of site conditions. In *GeoEng2000: An International Conference on Geotechnical and Geological Engineering*, Melbourne, Australia.

Fookes, P.G., Lee, E.M. and Sweeney, M. (2001a). Pipeline route selection and ground characterization, Algeria. In *Land Surface Evaluation for Engineering Practice*, ed. J.S. Griffiths. Geological Society of London, Engineering Geology Special Publication 18, pp. 129–133.

Fookes, P.G., Baynes, F.J. and Hutchinson, J.N. (2001b). Total geological history: A model approach to understanding site conditions. *Ground Engineering*, 34(3), 42–47.

Fookes, P.G., Lee, E.M. and Griffiths, J.S. (2007). *Engineering Geomorphology: Theory and Practice*. Caithness, UK: Whittles.

Forster, A. (2004). Is there a role for engineering geology on Mars? *Quarterly Journal of Engineering Geology and Hydrogeology*, 37, 5–6.

Glade, T. (1998). Establishing the frequency and magnitude of landslide-triggering rainstorm events in New Zealand. *Environmental Geology*, 35, 160–174.

Glenn, N.F., Streutker, D.R., Chadwick, D.J., Thackray, G.D. and Dorsch, S.J. (2006). Analysis of LiDAR-derived topographic information for characterizing and differentiating landslide morphology and activity. *Geomorphology*, 73, 131–148.

Goudie, A.S. (ed.). (2004). *Encylopedia of Geomorphology*. London: Routledge.

Griffiths, J.S. (1999). Proving the occurrence and cause of a landslide in a legal context. *Bulletin of Engineering Geology and the Environment*, 58, 75–85.

(2001a). Development of a ground model for the UK Channel Tunnel portal. In *Land Surface Evaluation for Engineering Practice*, ed. J.S. Griffiths. Geological Society of London, Engineering Geology Special Publication 18, pp. 129–133.

(2001b). Mapping and engineering geology: Introduction. In *Mapping in Engineering Geology*, ed. J.S. Griffiths. Geological Society of London, Key Issues in Earth Sciences 1, pp. 1–5.

(2005). Landslides. In *Geomorphology for Engineers*, ed. P.G. Fookes, E.M. Lee and G. Milligan. Caithness, UK: Whittles, pp. 173–217.

Griffiths, J.S. and Abraham, J.K. (2008). Factors affecting the use of applied geomorphology maps to communicate with different end-users. *Journal of Maps*, 2008, 201–210.

Griffiths, J.S. and Hearn, G.J. (1990). Engineering geomorphology: A UK perspective. *Bulletin of the International Association of Engineering Geology*, 42, 39–44.

Griffiths, J.S. and Marsh, A.H. (1986). BS 5930: The role of geomorphological and geological techniques in a preliminary site investigation. In *Site Investigation Practice: Assessing BS 5930*, ed. A.B. Hawkins. Geological Society of London, Engineering Geology Special Publication 2, pp. 261–267.

Griffiths, J.S. and Stokes, M. (2008). Engineering geomorphological input to ground models: A possible approach based on earth systems. *Quarterly Journal of Engineering Geology and Hydrogeology*, 41, 1–19.

Griffiths, J.S., Brunsden, D., Lee, E.M. and Jones, D.K.C. (1995). Geomorphological investigations for the Channel Tunnel terminal and portal. *Geographical Journal*, 161, 275–284.

Griffiths, J.S., Hutchinson, J.N., Brunsden, D., Petley, D.J. and Fookes, P.G. (2004). The reactivation of a landslide during the construction of the Ok Ma tailings dam, Papua New Guinea. *Quarterly Journal of Engineering Geology and Hydrogeology*, 37, 173–186.

Griffiths, J.S., Lee, E.M., Brunsden, D. and Jones, D.K.C. (2011). The Cherry Garden landslide, Etchinghill Escarpment, south-east England. In *Geomorphological Mapping: Methods and Applications*, ed. M.J. Smith, P. Paron and J.S. Griffiths. Amsterdam: Elsevier, pp. 397–411.

Hampton, M.A., Lee, H.J. and Locat, J. (1996). Submarine landslides. *Reviews of Geophysics*, 34, 33–59.

Hart, A.B., Griffiths, J.S. and Mather, A.E. (2009). Some limitations in the interpretation of vertical stereo photographic images for a landslide investigation. *Quarterly Journal of Engineering Geology and Hydrogeology*, 42, 21–30.

Havenith, H.-B., Strom, A., Caceres, F. and Pirard, E. (2006). Analysis of landslide susceptibility in the Suusamyr region, Tien Shan: Statistical and geotechnical approach. *Landslides*, 3, 39–50.

Hearn, G.J. (2002a). Natural terrain hazard assessment: The art of applied science. In *Natural Terrain: A Constraint to Development?* Hong Kong: Institution of Mining and Metallurgy, pp. 39–60.

(2002b). Engineering geomorphology for road design in unstable mountainous areas: Lessons learnt after 25 years in Nepal. *Quarterly Journal of Engineering Geology and Hydrogeology*, 35, 143–154.

Hendron, A.J. and Patton, F.D. (1986). A geotechnical analysis of the behavior of the Vaiont slide. *Boston Society of Civil Engineering, Civil Engineering Practice*, 1, 65–130.

Highland, L.M. and Bobrowsky, P. (2008). *The Landslide Handbook: A Guide to Understanding Landslides*. US Geological Survey, Circular 1325.

Ho, H.Y., King, J.P. and Wallace, M.I. (2006). *A Basic Guide to Air Photo Interpretation in Hong Kong*. Hong Kong: Applied Geoscience Center.

Hutchinson, J.N. (1983). Methods of locating slip surfaces in landslides. *Bulletin of International Association of Engineering Geologists*, 20, 235–252.

(1988). Morphological and geotechnical parameters of landslides in relation to geology and hydrogeology. In *Landslides: Proceedings of the 5th International Symposium on Landslides, Lausanne, Switzerland*. Rotterdam: Balkema, pp. 3–35.

(1995). The significance of tectonically produced pre-existing shears. In *Proceedings of the 11th European Conference on Soil Mechanics and Foundation Engineering, Copenhagen, Vol. 4*, pp. 59–67.

(2001). Reading the ground: Morphology and geology in site appraisal. *Quarterly Journal of Engineering Geology and Hydrogeology*, 34, 5–50.

(2008). *Selected Papers on Engineering Geology and Geotechnics*. Livorno, Italy: Associazione Italiana di Geologia Applicata e Ambientale.

Jaboyedoff, M., Oppikofer, T., Abellán, A. *et al.* (2012). Use of LIDAR in landslide investigations: A review. *Natural Hazards*, 61, 5–28.

Jones, D.K.C. (1977). Man-made landforms. In *The Unquiet Landscape*, ed. D. Brunsden and J.C. Doornkamp. Newton Abbott, UK: David & Charles, pp. 149–155.

Kilburn, C.R.J. and Petley, D.N. (2003). Forecasting giant catastrophic slope collapse: Lessons from Vajont, Northern Italy. *Geomorphology*, 54, 21–32.

King, J.P. (1996). *The Tsing Shan Debris Flow*. Hong Kong Geotechnical Engineering Office, GEO Special Project Report SPR 6/96.

Korup, O. (2009). Linking landslides, hillslope erosion and landscape evolution. *Earth Surface Processes and Landforms*, 34, 1315–1317.

Lato, M., Hutchinson, J., Diederichs, M., Ball, D. and Harrap, R. (2009). Engineering monitoring of rockfall hazards along transportation corridors: Using mobile terrestrial LiDAR. *Natural Hazards and Earth System Sciences*, 9, 935–946.

Lawrance, C.J., Byard, R.J. and Bevan, P.J. (1993). *Terrain Evaluation Manual*. HMSO Transportation Research Laboratory State of Art Review 7.

Lee, E.M. (2001). Geomorphological mapping. In *Land Surface Evaluation for Engineering Practice*, ed. J.S. Griffith. Geological Society of London, Engineering Geology Special Publication 18, pp. 53–56.

(2005). Coastal cliff recession risk: A simple judgement-based model. *Quarterly Journal of Engineering Geology and Hydrogeology*, 38, 89–104.

Lee, E.M. and Jones, D.K.C. (2004). *Landslide Risk Assessment*. London: Thomas Telford.

Lee, E.M., Jones, D.K.C. and Brunsden, D. (2000). The landslide environment of Great Britain. In *Landslides in Research, Theory and Practice*, ed. E. Bromhead, N. Dixon and M.-L. Ibsen. London: Thomas Telford, pp. 911–916.

Lee, E.M., Hall, J.W. and Meadowcroft, I.C. (2001). Coastal cliff recession: The use of probabilistic prediction methods. *Geomorphology*, 40, 253–269.

Lee, E.M., Griffiths, J.S. and Fookes, P.G. (2004). Engineering geology: Geomorphology. In *Encyclopedia of Geology*, ed. R.C. Selley, L.R.M. Cocks and I.R. Plimer. Amsterdam: Elsevier, pp. 474–481.

Lefebvre, G. (1996). Soft sensitive clays. In *Landslides Investigation and Mitigation*, ed. A.K. Turner and R.L. Schuster. Washington, DC: National Academy Press, pp. 607–619.

LePrince, S., Sylvain, B., François, A. and Jean-Philippe, A. (2007). Automatic and precise orthorectification, coregistration, and subpixel correlation of satellite images, application to ground deformation measurements. *IEEE Transactions on Geoscience and Remote Sensing*, 45, 1529–1558.

Mather, A.E., Griffiths, J.S. and Stokes, M. (2003). Anatomy of a 'fossil' landslide from the Pleistocene of SE Spain. *Geomorphology*, 50, 135–149.

Matthews, J.A., Brunsden, D., Frenzel. B., Gläser, B. and Weiss, M.M. (ed.) (1997). *Rapid Mass Movement as a Source of Climatic Evidence for the Holocene*. Stuttgart: Gustav Fischer.

McInnes, R. and Jakeways, J. (ed.) (2000). *Coastal Change, Climate and Instability*. Final Technical Report on European LIFE Project, LIFE-97 ENV/UK 000510 (1997–2000).

McKean, J. and Roering, J. (2003). Objective landslide detection and surface morphology mapping using high-resolution airborne laser altimetry. *Geomorphology*, 57, 331–351.

Mitchell, C. (1991). *Terrain Evaluation: An Introductory Handbook to the History, Principles and Methods of Practical Terrain Evaluation*, 2nd edn. New York: Longman Scientific and Technical.

Nemcŏk, A., Pasek, J. and Rybar, J. (1972). Classification of landslides and other mass movements. *Rock Mechanics*, 4, 71–78.

Niethammer, U., James, M. R., Rothmund, S., Travelletti, J. and Joswig, M. (2011). UAV-based remote sensing of the Super-Sauze landslide: Evaluation and results. *Engineering Geology*, doi:10.1016/j.enggeo.2011.03.012.

Parry, S. and Ng, K.C. (2010). The assessment of landslide risk from natural slopes in Hong Kong: An engineering geological perspective. *Quarterly Journal of Engineering Geology and Hydrogeology*, 43, 307–320.

Petley, D.N. (1998). Geomorphological mapping for hazard assessment in a neotectonic terrain. *Geographical Journal*, 164, 183–201.

(2001). Hazard assessment in eastern Taiwan. In *Land Surface Evaluation for Engineering Practice*, ed. J.S. Griffith. Geological Society of London, Engineering Geology Special Publication 18, pp. 209–213.

Phipps, P.J. (2001). Terrain systems mapping. In *Land Surface Evaluation for Engineering Practice*, ed. J.S. Griffiths. Geological Society of London, Engineering Geology Special Publication 18, pp. 59–61.

(2003). Geomorphological assessment for transport infrastructure projects. *Proceedings of the Institute of Civil Engineers – Transport*, 156, 131–143.

Pickup, G., Higgins, R.J. and Warner, R.F. (1981). Erosion and sediment yield in the Fly River drainage basins, Papua New Guinea. In *Erosion and Sediment Transport in Pacific Rim Steeplands*, ed. T.R.H. Davies and A.J. Pearce. International Association of Hydrological Sciences Publication 132, pp. 438–456.

Prokesova, R., Kardos, M. and Medvedova, A. (2010). Landslide dynamics from high-resolution aerial photographs: A case study from the Western Carpathians, Slovakia. *Geomorphology*, 115, 90–101.

Rib, H.T. and Liang, T. (1978). Recognition and identification. in *Landslides: Analysis and Control*, ed. R.L. Schuster and R.J. Krizek. Washington, DC: National Academy of Sciences, pp. 33–79.

Rowlands, K.A., Jones, L.D. and Whitworth, M. (2003). Landslide laser scanning: A new look at an old problem. *Quarterly Journal of Engineering Geology and Hydrology*, 36, 155–157.

Savigear, R.A.G. (1965). A technique for morphological mapping. *Annals of the Association of American Geographers*, 55, 514–538.

Schultz, C.H. and Harper, J.A. (1996). Pittsburgh red beds cause renewed landsliding after a ca. 310 Ma pause, Allegheny County, Pennsylvania, USA. In *Landslides*, ed. J. Chacón, C. Irigaray and T. Fernández. Rotterdam: Balkema, pp. 189–196.

Schulz, W.H. (2007). Landslide susceptibility revealed by LiDAR imagery and historical records, Seattle, Washington. *Engineering Geology*, 89, 67–87.

Skempton, A.W. and Hutchinson, J.N. (1969). Stability of natural slopes and embankment foundations. In *7th International Conference on Soil Mechanics and Foundations Engineering, State-of-the-Art Volume*, pp. 291 340.

Skempton, A.W. and Vaughan, P.R. (1993). The failure of Carsington Dam. *Géotechnique*, 43, 151–173.

Skempton, A.W. and Weeks, A.G. (1976). The Quaternary history of the Lower Greensand escarpment Weald Clay vale near Sevenoaks, Kent. *Philosophical Transactions of the Royal Society, London, A*, 283, 493–526.

Smith, M.J., Paron, P. and Griffiths, J.S. (ed.) (2011). *Geomorphological Mapping: Methods and Applications*. Elsevier, Developments in Earth Surface Processes 15.

Soeters, R. and van Westen, C.J. (1996). Slope instability recognition, analysis, and zonation. In *Landslides: Investigation and Mitigation*, ed. A.K. Turner and R.L. Schuster. Washington, DC: National Academy Press, pp. 129–177.

Turner, A.K. and Schuster, R.L. (ed.) (1996). *Landslides: Investigation and Mitigation*. Washington, DC: National Academy Press.

Van Den Eeckhaut, M., Poesen, J., Verstraeten, G. *et al.* (2007). Use of LIDAR-derived images for mapping old landslides under forest. *Earth Surface Processes and Landforms*, 32, 754–769.

van Westen, C.J. and Lulie Getahun, F. (2003). Analyzing the evolution of the Tessina landslide using aerial photographs and digital elevation models. *Geomorphology*, 54, 77–89.

van Westen, C.J., Castellanos, E. and Kuriakose, S. (2008). Spatial data for landslide susceptibility, hazard, and vulnerability assessment: An overview. *Engineering Geology*, 102, 112–131.

Varnes, D.J. (1978). Slope movement types and processes. In *Landslides: Analysis and Control*, ed. R.L. Schuster and R.J. Krizek. Washington, DC: National Academy of Sciences, pp. 11–33.

Waller, A.M. and Phipps, P. (1996). Terrain systems mapping and geomorphological studies for the Channel Tunnel rail link. In *Advances in Site Investigation Practice*, ed. C. Craig. London: Thomas Telford, pp. 25–38.

Walstra, J., Dixon, N. and Chandler, J. (2007). Historical aerial photographs for landslide assessment: Two case histories. *Quarterly Journal of Engineering Geology and Hydrogeology*, 40, 315–332.

Waters, R.S. (1958). Morphological mapping. *Geography*, 43, 10–17.

Whitworth, M.C.Z., Giles, D.P. and Murphy, W. (2005). Airborne remote sensing for landslide hazard assessment: A case study on the Jurassic escarpment slopes of Worcestershire, United Kingdom. *Quarterly Journal of Engineering Geology and Hydrogeology*, 38, 285–300.

Wilder, M.G. (1985). Site and situation determinants of land use change: An empirical example. *Economic Geography*, 61, 332–344.

Wojcik, A. and Mrozek, T. (2002). Landslides in the Carpathian Flysch. In *Landslides*, ed. J. Ciesielczuk and S. Ostaficzuk. Krakow: Polish Academy of Sciences, Mineral and Energy Economy Research Institute, pp. 151–168.

WP/WLI (International Geotechnical Societies UNESCO Working Party for World Landslide Inventory) (1990). A suggested method for reporting a landslide. *Bulletin of the International Association of Engineering Geology*, 41, 5–16.

WP/WLI (International Geotechnical Societies UNESCO Working Party for World Landslide Inventory) (1995). A suggested method of a landslide summary. *Bulletin of the International Association of Engineering Geology*, 43, 101–110.

Zaruba, Q. and Mencl, V. (1969). *Landslides and Their Control*. New York: Elsevier.

(1982). *Landslides and Their Control*, 2nd edn. New York: Elsevier.

16 Developments in landslide runout prediction

SCOTT MCDOUGALL, MIKA MCKINNON, AND OLDRICH HUNGR

ABSTRACT

Runout analysis is a key component of landslide risk assessment and management, and a range of empirical and numerical methods for analyzing runout are available. Significant advances have been made in the development of numerical runout models over the past decade, particularly with respect to three-dimensional modeling capabilities. As demonstrated in a recent model benchmarking workshop, most modern numerical models are able to simulate the bulk characteristics of typical real landslides. On the other hand, progress has been slower in developing suitable methodologies for selecting input parameter values for prediction. Ideally, these methods should fit within the probabilistic framework of quantitative landslide risk assessment, allowing users to estimate the spatial probability of impact and associated hazard intensity throughout the runout zone. Recent work on advanced model calibration techniques has attempted to address this need. Simple probabilistic empirical–statistical techniques can be extended to numerical modeling applications and provide a useful reference point for these discussions.

16.1 INTRODUCTION

Flow-like landslides such as rock avalanches, debris flows, and flow slides can travel at very high velocities and impact large areas. When a potential source of a flow-like landslide is identified, a runout analysis may be required to map hazard or risk downslope or to estimate input parameters for mitigation design. Results of runout analysis may also feed into studies of possible secondary effects, such as landslide-generated waves or flooding behind landslide dams.

Several methods of analysis are available, ranging from simple empirical–statistical relationships to complex three-dimensional numerical modeling. Landslide-specific numerical models have been the focus of substantial research effort worldwide over the past decade, and model capabilities have advanced rapidly as a result. Most of the effort to date, however, has focused on developing governing equations and numerical solution methods, with the objectives of validating models and developing case-study-specific simulations. Less attention has been given to how we actually use the models in practice for runout prediction. As a result, key questions remain regarding how to select model input parameter values for prediction, and how to interpret the output and generate results that are suited to quantitative landslide risk assessment and management.

This chapter provides an overview of recent developments in empirical and numerical landslide runout prediction methods, with a focus on the use of continuum numerical models. The advantages and limitations of calibration-based methods are discussed, and some probabilistic mapping techniques are highlighted. The chapter does not provide a comprehensive review of existing numerical models; however, several references to more detailed sources of information are included.

16.2 LANDSLIDE RUNOUT MAPPING

Like other natural hazards, long-runout landslides are not deterministic processes. Their inherent variability is compounded by uncertainty due to sparse data, which is subject to interpretation and measurement error, plus our incomplete understanding of the physics of flow and limited ability to incorporate what we do understand into models. As a result, runout predictions based only on a single set of input parameter values can be misleading unless they are placed in the proper context. For example, do the results represent the researcher's "best guess" – with a subjective probability of exceedance of 50 percent, or do they

Landslides: Types, Mechanisms and Modeling, ed. John J. Clague and Douglas Stead. Published by Cambridge University Press.

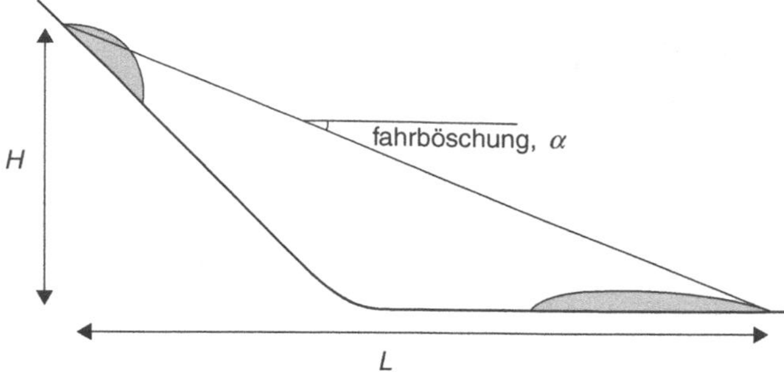

Fig. 16.1. Schematic definition of fahrböschung or angle of reach, α; H is the elevation difference between the crest of the source and the toe of the deposit and L is the length of the horizontal projection of the streamline connecting these two points.

represent a more conservative estimate – say with an exceedance probability of 10 percent or less? The difference can have significant implications for risk management and land-use decision making.

The probability of landslide impact is spatially variable. It ranges from 100 percent close to the source area to 0 percent at some distal location, which for typical flow-like landslides could be hundreds to thousands of meters downslope. The velocity and depth of the flow (i.e., hazard intensity) also vary between these locations. The challenge for landslide runout forecasters is to estimate the spatial distribution of the probability of impact and the associated hazard intensity throughout the runout zone or, where time and resources may be limited, to effectively communicate single-input results in this context.

Probabilistic prediction methods are well established for other natural hazards, for example rockfalls, snow avalanches, and hurricanes. Considerable subjective judgment is still required to select model input parameters and interpret the results, so simplicity and transparency of the methods are important considerations. Similarly, the typical practitioner does not have the resources to produce hundreds of different model runs, so computational efficiency is also a practical requirement.

Simple probabilistic empirical–statistical methods provide a useful starting point for further discussions of the probabilistic application of numerical models. Empirical and numerical methods are often used in combination to help practitioners converge on a reasonable result, but there are many instances where empirical methods can be used effectively on their own.

16.3 EMPIRICAL METHODS

16.3.1 VOLUME–FAHRBÖSCHUNG RELATIONSHIPS

The most commonly referenced empirical method of prediction is based on an apparent relationship between landslide volume and "*fahrböschung*," also known as the angle of reach. Fahrböschung was defined by Heim (1932) as the inclination of the line connecting the crest of the landslide source with the toe of the deposit, as measured along the approximate centerline of motion (Fig. 16.1). The fahrböschung is analogous to, but not strictly the same as, an average friction angle and is considered an index of the efficiency of energy dissipation. Lower fahrböschung values are associated with higher mobility.

Several researchers have used linear regression in attempts to fit a line through available volume–fahrböschung data (Scheidegger, 1973; Li, 1983; Nicoletti and Sorriso-Valvo, 1991; Corominas, 1996; Hunter and Fell, 2003). The resulting equations typically take the following power-law form:

$$\log_{10}\left(\frac{H}{L}\right) = -a\log_{10}V + b \qquad (16.1)$$

where a is the slope of the line, b is the intercept of the line at $\log_{10}V = 0$, H and L are defined in Figure 16.1, and V is the event volume. Initial and final volumes are not commonly differentiated, and uncertainties in measurements typically are not reported in these types of analyses. The negative a coefficient in Eq. 16.1 reflects an apparent inverse relationship between volume and fahrböschung. In other words, generally speaking, the larger the landslide, the higher the relative mobility.

The values of coefficients a and b and the strength and spread of the correlations differ significantly with landslide type and path morphology. Nicoletti and Sorriso-Valvo (1991) found that topography can strongly dissipate the kinetic energy of rock avalanches, depending on whether the flow is confined, unconfined, or strongly deflected. Corominas (1996) used a similar morphological classification system and also divided the events in his large database into smaller groups based on the dominant mechanism (rockfalls/avalanches, translational slides, debris flows, and earthflows), which improved the strength of the correlations. However, using Corominas' dataset, Hunter and Fell (2003) found relatively weak correlations between volume and fahrböschung for debris flows of less than 1 million m^3.

Such large data scatter can make the volume–fahrböschung approach difficult to apply in practice on a case-specific basis. As a result, Hunter and Fell (2003) recommend caution and suggest that local calibration may lessen uncertainty in a given case. Nevertheless, this explicit data scatter does have an advantage in the context of probabilistic mapping. Because it can be expressed in quantitative statistical terms, it can be used explicitly to establish limits of confidence for predictions (Hungr *et al.*, 2005; Schilling *et al.*, 2008).

An example of this approach is shown in Figure 16.2, based on the following regression equation proposed by Li (1983) for rock avalanches:

$$\log_{10}\left(\frac{H}{L}\right) = -0.153\log_{10}V + 0.664 \qquad (16.2)$$

with sample standard deviation $s = 0.118$ and correlation coefficient $r = 0.778$. Figure 16.2a shows this relation on a traditional log–log plot of volume vs. H/L. The prediction confidence limits shown in Figure 16.2a are expressed in terms of the probability of runout exceedance, where the 50 percent line represents

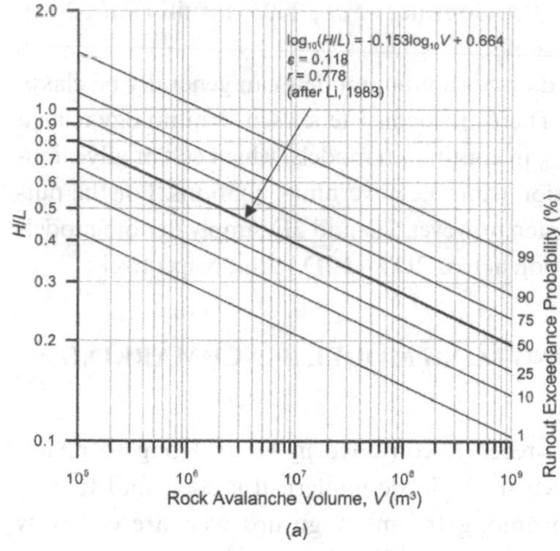

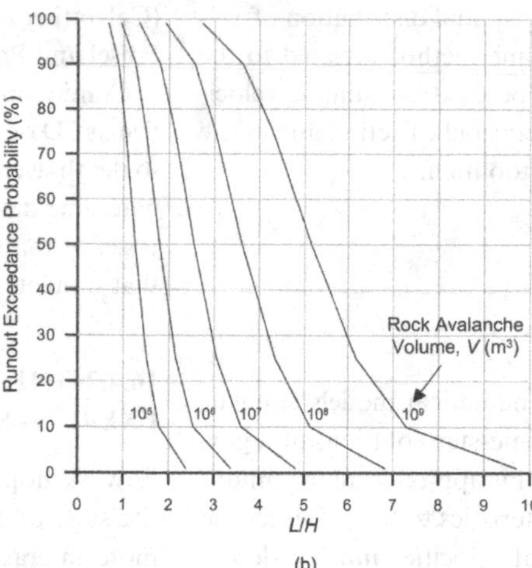

Fig. 16.2. An example of the volume–fahrböschung method for rock avalanches based on regression results presented by Li (1983). (a) Traditional log–log plot of volume vs. *H*/*L* showing prediction confidence intervals (expressed as exceedance probabilities). (b) Alternative plot showing change in predicted exceedance probability with distance from source for rock avalanches of a given order of magnitude.

the best-fit regression equation shown above. For simplicity, a normal probability distribution is assumed.

An alternative way of plotting Eq. 16.2 is shown in Figure 16.2b. This graph is arguably more intuitive for use in prediction because it relates the runout exceedance probability estimate to a readily mappable parameter, *L*/*H*. The curves in Figure 16.2b show how the probability of impact is expected to change with distance from the source area for rock avalanches of a given size.

Graphs like those shown in Figure 16.2 are useful screening-level tools, but should be used with caution. The runout exceedance probabilities shown in these graphs are based on limited data and need to be moderated by judgment.

16.3.2 VOLUME–AREA RELATIONSHIPS

Statistical correlations between landslide volume and inundation or deposit area have also been proposed (Li, 1983; Hungr, 1990; Iverson *et al.*, 1998). As in the case of the volume–fahrböschung method, the regression equations typically take the form:

$$\log_{10} A = \frac{2}{3}\log_{10} V + \log_{10} c \quad (16.3)$$

where 2/3 is the slope of the line, $\log_{10}c$ is the intercept of the line at $\log_{10} V = 0$, A is the planimetric inundation/deposit area, and V is the event volume. Equation 16.3 is also commonly expressed in the following form:

$$A = cV^{\frac{2}{3}}. \quad (16.4)$$

The slope of the line as shown in Eq. 16.3 is set to 2/3 on the basis of geometric scaling which assumes that landslides composed of similar materials will form geometrically similar deposits (Iverson *et al.*, 1998).

The relations expressed in Eqs. 16.3 and 16.4 are the basis for the US Geological Survey's GIS-based mapping program LAHARZ (Iverson *et al.*, 1998). The program also estimates the cross-sectional inundation area using the same scaling assumptions. Griswold (2004) calibrated LAHARZ using reported volume and inundation area estimates for a number of lahars, rock avalanches, and nonvolcanic debris flows. She found that lahars typically inundate areas 10 times larger than rock avalanches and debris flows of similar magnitude, with calibrated values of $c = 200$ for lahars and $c = 20$ for rock avalanches and nonvolcanic debris flows. Hungr (1990) presented a graphical interpretation of Li's (1983) data that suggests a slightly lower value of $c \approx 12$ for rock avalanches.

Like the volume–fahrböschung approach, the volume–area method is well suited to probabilistic treatment. The US Geological Survey has recently updated the LAHARZ program to allow for explicit computation of prediction confidence limits (Schilling *et al.*, 2008). A t-distribution within LAHARZ accounts for uncertainty due to the limited data sample size, as opposed to the assumed normal distribution that was used to construct the graphs shown in Figure 16.2.

The volume–area method only provides estimates of the area covered by deposits and must be used with the volume–fahrböschung method, or something similar, to estimate where the deposit will be located along the path. In LAHARZ, the starting point of deposition must be specified by the user.

16.3.3 OTHER EMPIRICAL METHODS

Other empirical–statistical relations have been proposed to estimate landslide runout distances (Hsü, 1975; Rickenmann, 1999; Legros, 2002). Methods based on the volume balance of material entrained and deposited along the path have been developed to estimate debris-flow runout distances (Cannon, 1993; Fannin and Wise, 2001). It is possible to apply all of these methods within a probabilistic prediction framework similar to the approach outlined above.

Empirical methods, however, provide only limited information on velocity, depth, and impact pressure, which are required for risk analysis or mitigation design. Rickenmann (1999) presented statistical correlations that can be used to estimate debris-flow velocity and discharge, but they only provide peak values and no information about the spatial distribution of velocity and discharge. Energy grade line methods related to the volume–fahrböschung relation can be used to estimate velocities along the path, but when constant bulk friction angles are assumed, velocity estimates may be too high.

16.4 NUMERICAL MODELS

16.4.1 MODEL DEVELOPMENT

In contrast to empirical methods, numerical models simulate the motion of a landslide through time-step solutions of a governing set of equations of motion. This approach allows unique geometry and local material characteristics to be accounted for explicitly, and it provides estimates of velocities and flow depths at different points along the path. Continuum and discontinuum numerical models have both been developed for this purpose.

Continuum models have evolved from established depth-averaged (shallow flow) hydrodynamic models by means of incremental improvements over time to account for landslide-specific behavior, including the influence of internal shear strength, spatially variable rheology, and entrainment of material from the path. Research over the past decade has focused especially on methods that are capable of simulating motion across complex 3D terrain. McDougall *et al.* (2008) presented a recent review of these developments.

Flow spreading in depth-averaged landslide continuum models is governed by internal stress gradients, and basal shear stresses provide resistance to forward motion. Different rheological models (frictional, Voellmy, Bingham) can be used to estimate the basal shear stress at a given time and location within the flow. Most modern continuum models are built on the pioneering work of Savage and Hutter (1989), who introduced the concept of strain-dependent frictional internal stresses coupled to basal shear stresses, which is based on classical earth pressure theory. Different continuum numerical solution methods are available to solve the governing equations in either their Eulerian or Lagrangian form, including finite volume, finite difference, and finite element methods. Fixed mesh, adaptive mesh, and meshless techniques (such as smoothed particle hydrodynamics) have all been used.

Discontinuum, or discrete (distinct) element, models are a multi-particle extension of the lumped-mass "bouncing ball" approach that is commonly used in rockfall analysis. In these types of models, landslides are modeled as a collection of individual particles that interact with each other and with the ground surface by falling, bouncing, rolling, and sliding, to simulate large-scale deformations. The particles may be of a variety of shapes and sizes, and a variety of inter-particle and particle-bed contact relationships can be modeled. The commercially available particle flow code (PFC) model developed by Itasca (Cundall and Strack, 1979) uses circular (in 2D) or spherical (in 3D) particles that may be bonded together or broken apart under certain conditions (Poisel and Roth, 2004). PFC has been used by several workers to simulate real landslides (Calvetti *et al.*, 2000; González *et al.*, 2003; Pirulli *et al.*, 2003; Poisel and Preh, 2008).

Continuum and discontinuum models can generally be classified as 2D or 3D. There are inconsistencies in terminology related to depth-averaging in continuum models, which essentially eliminates one dimension from the governing equations. For the purposes of this chapter, however, 2D and 3D simply denote models that simulate motion across 2D and 3D paths, respectively.

16.4.2 THE HONG KONG MODEL BENCHMARKING EXERCISES

A workshop was recently convened in Hong Kong to review the state of the art in landslide numerical models and to promote interaction among the many groups who are currently developing their own models (Hungr *et al.*, 2007). Twenty-one different teams were invited to submit modeling results for a series of selected "benchmark" cases, which included theoretical and experimental validation tests and full-scale landslide case histories. One forward-prediction case was also included in the exercises. Thirteen teams made submissions using 17 different models, including one discontinuum model. The workshop participants and key sources of more detailed information are shown in Table 16.1.

The modeling methodologies and workshop results were summarized by Hungr *et al.* (2007), who drew the following general conclusions from a subjective comparison of the submissions:

- When the same basal rheology and similar input parameter values were used, several of the models produced similar travel distances, emplacement times, and deposit shapes, in spite of the very different numerical solution methods. For example, DAN3D, which is a meshless Lagrangian model, and RASH3D, a fixed-mesh Eulerian model, produced consistently similar results.
- Despite this general good correspondence, some notable discrepancies were observed in the spatial extent and profile of the simulated deposits. Hungr *et al.* (2007) suggested that these discrepancies may be related to different assumptions about internal stress states based on the Savage–Hutter theory, which can have a strong influence on contraction and expansion of the simulated flow.
- Although 3D simulation capabilities have advanced considerably in recent years, entrainment simulation capabilities are not yet well developed. Some models require the input of user-prescribed entrainment rates, whereas other models simulate erosion based on the properties of material along the path. Simulation results appear to be sensitive to the assumed mode of entrainment.
- Depth-averaged continuum models are not suitable for simulating landslides with very steep failure surfaces or where the debris thickness is large relative to the runout distance; that is, when the assumption of shallow flow conditions may be violated.

Unfortunately, for the single forward-prediction case that was included in the benchmarking exercises, workshop

Table 16.1. *List of participants in the 2007 Hong Kong benchmarking exercises (Hungr* et al., *2007).*

Team	Model denotation	Type	Model description
University of Alberta	Wang	2D, continuum	Wang (2008)
University of Hong Kong	MADFLOW	3D, continuum	Chen and Lee (2000)
University of Milano-Bicocca and FEAT	TOCHNOG	3D [a], continuum	Roddeman (2002)
	RAMMS	3D, continuum	Christen *et al.* (2008)
Norwegian Geotechnical Institute	DAN3D (NGI)	3D, continuum	McDougall (2006)
	FLO-2D (NGI)	3D, continuum	FLO-2D Software Inc. (2007)
Technical University of Catalonia	FLATModel	3D, continuum	Medina *et al.* (2008)
Geotechnical Engineering Office, Hong Kong	3dDMM	3D, continuum	Kwan and Sun (2007)
Université Paris Diderot	SHALTOP-2D	3D, continuum	Mangeney-Castelnau *et al.* (2003)
	RASH3D (Paris)	3D, continuum	Pirulli (2005)
University of British Columbia	DAN	2D, continuum	Hungr (1995)
	DAN3D	3D, continuum	McDougall (2006)
CEDEX, Madrid	Pastor	3D, continuum	Pastor *et al.* (2009)
Vienna University of Technology	PFC	3D, discontinuum	Poisel and Preh (2008)
Kyoto University	Sassa-Wang	3D, continuum	Wang and Sassa (2002)
Politecnico di Torino	RASH3D	3D, continuum	Pirulli (2005)
University of New York at Buffalo	TITAN2D	3D, continuum	Pitman *et al.* (2003)

[a] TOCHNOG is the only continuum model listed here that does not make use of depth-averaged equations of motion. Instead, internal 3D deformation is modeled in detail.

participants were given a prescribed set of input parameter values, so different runout prediction methodologies could not be compared.

16.4.3 THE CALIBRATION-BASED APPROACH

An important distinction can be made between numerical models that require the input of measured physical parameters and models that must be calibrated through back analysis. The latter approach is rooted in the empirical methods of hydraulic engineering and appears to be emerging as the dominant approach in landslide runout modeling. All of the participants in the Hong Kong benchmarking exercises used some form of calibration to simulate the full-scale benchmark case studies (Hungr *et al.*, 2007).

In the calibration-based approach, rheological parameters are constrained by systematic adjustment through trial-and-error or statistical back analysis of full-scale prototype events. On a case-by-case basis, calibration is achieved by matching the simulated travel distance, deposit distribution, flow velocities, and emplacement times to those of the prototype landslide, and typically in that order of priority based on the general decreasing availability or reliability of data. The most common method of judging the quality of the match is simple visual comparison of a set of parametric results.

As a basic rule of thumb, the number of adjustable rheological parameters in the model should not exceed the number of calibration criteria that are used. For example, estimated travel distances and velocities are both needed to properly constrain the two-parameter Voellmy model, which contains a velocity-dependent term. Otherwise, a wide range of input parameter combinations can result in the same simulated travel distance, but with different simulated velocities. For this reason, models with a minimal number of adjustable parameters are favored.

Sample output from a case study employing the calibration approach is shown in Figure 16.3. In this case, the model DAN3D (McDougall, 2006) was used to simulate the 1855–1856 Rubble Creek landslide in southwestern British Columbia. This event involved the failure of approximately 25 million m^3 of rock from the face of a steep Quaternary volcanic feature known as The Barrier (Moore, 1976; Moore and Mathews, 1978). The debris traveled approximately 5 km down Rubble Creek Valley and came to rest on the fan at the valley mouth. Flow velocities back-calculated from superelevation measurements in three prominent channel bends ranged from 22 $m\,s^{-1}$ to 30 $m\,s^{-1}$ (Moore and Mathews, 1978).

Figure 16.3 shows the best-fit results based on a visual comparison with a surficial map presented in Moore and Mathews (1978). The two-parameter Voellmy model was used, and adjustments were made to the friction coefficient, f, which limits the slope angles on which material will come to rest, and the turbulence parameter, ξ, which limits the simulated flow velocities, to achieve a reasonable match in terms of travel distance, deposit distribution, and estimated flow velocities. The best-fit parameter combination in this case was $f = 0.07$ and $\xi = 100$ $m\,s^{-2}$, which corresponds with the calibrated values reported by Hungr and Evans (1996) in their analysis of the same event using the 2D model DAN (Hungr, 1995).

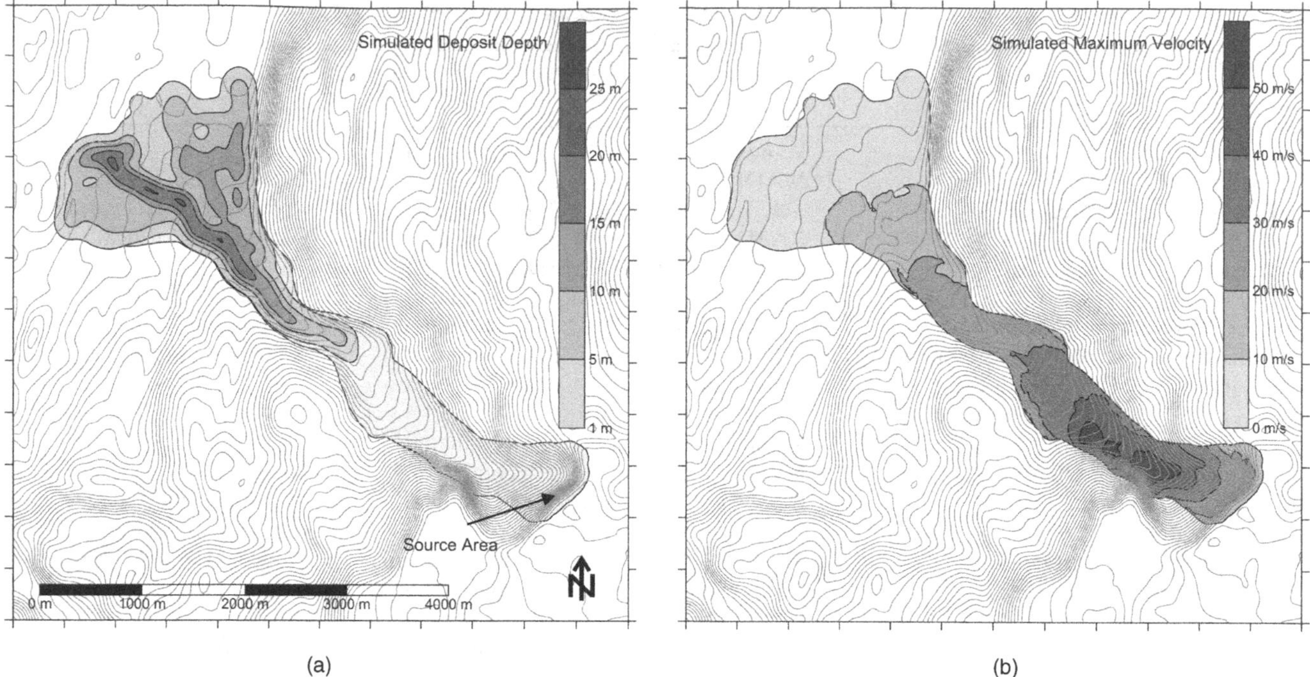

Fig. 16.3. A 3D simulation of the Rubble Creek landslide using the numerical model DAN3D. (a) Final simulated deposit distribution. (b) Maximum simulated velocities along the path. The input resistance parameter values have been adjusted by trial-and-error to achieve reasonable correspondence between the simulation results and the observed/estimated travel distance, deposit distribution, and flow velocities reported by Moore and Mathews (1978).

Model calibration is sometimes criticized as a tuning or curve-fitting exercise, in which successful simulation of almost any event can be achieved by arbitrary adjustments of variables, so that model adaptability can be mistaken for model accuracy (Iverson, 2003). This criticism can be valid on a single-case basis. Thus for prediction purposes, the key to calibration is to seek consistent patterns of rheology type and ranges of parameter values that reproduce the behavior of groups of similar events.

To date, this type of work has been fairly limited and has generally focused on qualitative assessment of the best-fit parameter values for relatively small groups of events (Hungr and Evans, 1996; Ayotte and Hungr, 2000; Revellino *et al.*, 2004; Pirulli, 2005). Typically, calibration patterns can be observed, and it is possible to make some judgment of the best-fit rheological model and parameter values for particular classes of events.

An advantage of the calibration-based approach is the ability to analyze calibration trends statistically and to apply these results to prediction in a probabilistic manner, similar to the empirical methods discussed above. One approach is to determine the best-fit parameter values for each individual case, as described above, and plot the results as a histogram, which can be fitted to a probability density function. An example based on the work of Hungr and Evans (1996) and Pirulli (2005) is shown in Figure 16.4. The probability density function can then be used to assign exceedance probabilities to parametric model runs, or to define input value probabilities for use in Monte Carlo-style predictive runs (Revellino *et al.*, 2004; Dalbey *et al.*, 2008). This approach is commonly used in rockfall mapping and has parallels in probability-of-failure slope stability analysis.

Like the fahrböschung data discussed above, frictional results such as those shown in Figure 16.4 may be volume-dependent (Pedrazzini *et al.*, 2012). Some partitioning of the calibration results according to the size of the landslide may therefore be needed.

Another approach, which is more similar to the empirical–statistical linear regression methods described above, is to seek the best-fit parameter values for each group of cases as a whole and then attempt to quantify the resulting variance. For example, in their analysis of 23 rock avalanches using the 2D model DAN, Hungr and Evans (1996) found that the total runout distance in 70 percent of the cases could be simulated within an error of approximately 10 percent using the Voellmy rheology with a single pair of input parameter values, $f = 0.1$ and $\xi = 500$ m s^{-2}. Similarly, using DAN to analyze debris avalanches and debris flows in the Campania region of Italy, Revellino *et al.* (2004) found that one pair of Voellmy parameters, $f = 0.07$ and $\xi = 200$ m s^{-2}, successfully simulated the total runout distance within an error of 4 percent in 16 out of 17 cases. Such results are useful for placing confidence bounds on predictions using best-fit calibrated values.

Unfortunately, the number of well-documented calibrated case studies is still small, and the results can be only locally or regionally applicable. Calibration results also may not be transferable between different models, considering that different

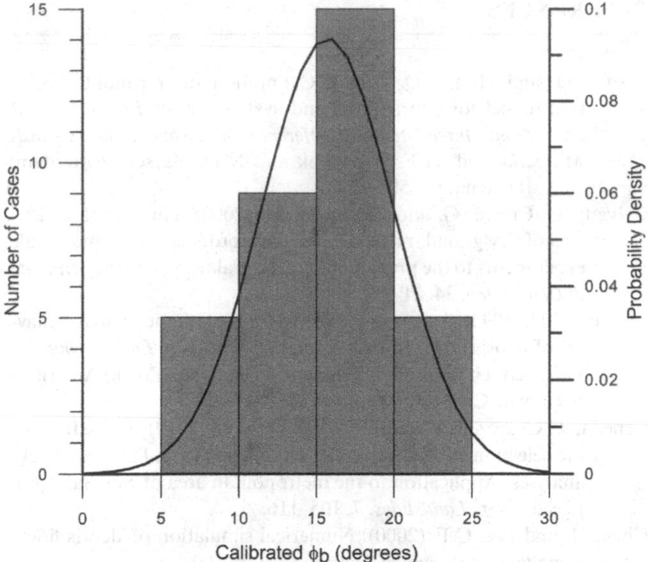

Fig. 16.4. Histogram of calibrated bulk basal friction angles based on back analyses of 34 different rock avalanches by Hungr and Evans (1996) and Pirulli (2005) using the 2D model DAN. The results are approximated using a normal distribution with a mean of 16° and a standard deviation of 4.3°.

internal stress assumptions appear to have a strong influence on model behavior (Hungr *et al.*, 2007). Another limitation of the calibration-based approach has been the lack of a broadly accepted definition of "well calibrated." The studies described above focused on runout distance as the primary calibration criterion. However, for some types of events, accurate simulation of flow depths and velocities can be just as important for predicting the consequences of a potential landslide. As described below, recent work has attempted to address some of these issues.

16.4.4 RECENT ADVANCES

Calibration work to date has relied largely on qualitative assessments of goodness of fit. This approach is fairly straightforward and repeatable in 2D analysis, but requires more subjective judgment when comparing 3D results. Often, only a visual comparison can be made, and comparison becomes challenging if more than two input parameter values are used. Galas *et al.* (2007) proposed a more objective calibration approach in which the optimal combination of input parameter values is obtained by maximizing the ratio between the intersection and union of the simulated and observed inundation areas. This method can produce similar values of the ratio for simulations that produce different results (Cepeda *et al.*, 2010), so judgment is still needed in the selection of the best-fit values.

Cepeda *et al.* (2010) proposed an alternative calibration procedure based on "receiver operating characteristic" (ROC) analysis. In this method, simulation results for a given set of input parameters are classified as "yes" or "no" depending on whether they overpredict or underpredict a given calibration criterion, such as runout distance or maximum velocity. Weights are subjectively assigned to the criteria based on their relative importance. Weighted averages of the true positive and false positive rates for each simulation are then calculated and the results are compared in ROC space (plots of true positive rate vs. false positive rate). The simulation results that plot closest to the point of perfect classification in ROC space, corresponding to a true positive rate of 1 and a false positive rate of 0, are judged to produce the best match. This method permits calibration against several criteria simultaneously and is not sensitive to the number of input parameter values that are used. It can also be extended to calibration using groups of landslides. One potential drawback of the approach, however, is the introduction of subjective judgment in the assignment of weights to the calibration criteria, which reduces transparency and may produce erroneous results that are not obvious to the user. But this problem may be overcome as users gain experience with the method and criteria weightings become more standardized.

McKinnon (2010) expanded on the work of Hungr and Evans (1996), Ayotte and Hungr (2000), Revellino *et al.* (2004), and Pirulli (2005) by systematically back-analyzing a group of 40 flow-like landslides using the 2D model DAN. Rather than focusing on the best-fit parameter combination for each individual case, a wide range of input values were used in an effort to identify and quantify systematic trends within the group as a whole. McKinnon's (2010) methodology can be summarized as follows:

1. Run every single case study with the exact same range of input parameter values, whether a given combination is expected to produce a good simulation of the event or not.
2. For each model run, normalize the simulation results with respect to each calibration criterion or index using the formula:

$$\textit{normalized index},\ \Delta = \frac{\textit{index}_{\text{modelled}} - \textit{index}_{\text{real}}}{\textit{index}_{\text{real}}} \times 100. \qquad (16.5)$$

   For example, a normalized simulated horizontal runout distance of 0 is associated with a perfect fit with respect to this criterion; a value of 10 is associated with an overprediction of 10 percent; and a value of −10 is associated with an underprediction of 10 percent.
3. For each set of input parameters, plot histograms for each calibration index showing the number of simulated case studies that fall within certain overprediction and underprediction ranges (e.g., ±10 percent, +10 to +30 percent, +30 to +50 percent, +50 to +70 percent, and +70 to +90 percent).

This approach facilitates visualization of trends and identification of best-fit parameter combinations for the group. At the same time, probability density functions can be fitted to the histogram data and can be used to assign explicit confidence limits or exceedance probability estimates to predictions. For example, if runout distance in half of the back-analyzed case studies is underpredicted using a given set of input parameter values, a forward prediction using the same input values might reasonably be associated with an exceedance probability of 50 percent, which represents the forecaster's "best guess." The corresponding sample standard deviation could then be used to contour exceedance probability estimates on either side of this value, as in the volume–fahrböschung approach shown in Figure 16.2b. A different set of input values may similarly be associated with an exceedance probability of less than 10 percent and could be used to further constrain these estimates.

McKinnon (2010) also investigated the use of the statistical analysis techniques ANOVA (analysis of variance) and cluster analysis. These techniques may help in grouping similar events and identifying common characteristics that can be used for prediction purposes.

To date, numerical prediction work has generally been limited to manual parametric model runs to simulate expected "best-guess," "lower-bound" and "upper-bound" runout behavior. Automated mapping based on Monte Carlo sampling of input parameter values can also be used, but this approach tends to be computationally expensive due to typical long model run times. Dalbey *et al.* (2008) presented several, more-efficient alternatives to traditional Monte Carlo analysis, including a recently developed method known as "polynomial chaos quadrature" (PCQ). The method was demonstrated using the 3D model TITAN2D (Pitman *et al.*, 2003), which runs on a supercomputer.

16.5 CONCLUSIONS

Several advances have been made in the field of landslide runout analysis over the past decade, especially in the development of numerical models. The recent Hong Kong benchmarking exercises produced encouraging results, suggesting that some convergence of different models is occurring. Still, new calibration and prediction methodologies have been relatively slow to develop. It is hoped that model application and prediction will receive more attention in the near future as the focus shifts away from governing equations and numerical solution methods. Prediction methodology comparisons could then be included in another round of model benchmarking exercises.

Ideally, such methods should produce results that fit within the established framework of quantitative landslide risk assessment. Probabilistic methods based on empirical–statistical correlations provide a useful reference point and can be extended to numerical modeling applications. Empirical methods can also be used in combination with numerical modeling to help practitioners draw reasonable conclusions from their results. Subjective judgment will always play a role in these types of assessments, so simple and transparent methods should be favored as much as possible. The calibration-based modeling approach provides a promising framework for such efforts, but more work is needed to refine calibration methodologies and expand the number of calibrated case studies that are available to practitioners.

REFERENCES

Ayotte, D. and Hungr, O. (2000). Calibration of a runout prediction model for debris flows and avalanches. In *Proceedings of the Second International Conference on Debris-Flow Hazards Mitigation*, ed. G.F. Wieczorek and N.D. Naeser. Rotterdam: A.A. Balkema, pp. 505–514.

Calvetti, F., Crosta, G. and Tartarella, M. (2000). Numerical simulation of dry granular flows: From the reproduction of small-scale experiments to the prediction of rock avalanches. *Rivista Italiana di Geotecnico*, 34, 21–38.

Cannon, S.H. (1993). An empirical model for the volume-change behavior of debris flows. In *Proceedings of Hydraulic Engineering '93, Vol. 2*, ed. H.W. Shen, S.T. Su and F. Wen. New York: American Society of Civil Engineers, pp. 1768–1773.

Cepeda, J., Chávez, J.A. and Cruz Martínez, C. (2010). Procedure for the selection of runout model parameters from landslide back-analyses: Application to the metropolitan area of San Salvador, El Salvador. *Landslides*, 7, 105–116.

Chen, H. and Lee, C.F. (2000). Numerical simulation of debris flows. *Canadian Geotechnical Journal*, 37, 146–160.

Christen, M., Bartelt, P., Kowalski, J. and Stoffel, L. (2008). Calculation of dense snow avalanches in three-dimensional terrain with the numerical simulation program RAMMS. In *Proceedings of the International Snow Science Workshop*, Whistler, BC, pp. 709–716.

Corominas, J. (1996). The angle of reach as a mobility index for small and large landslides. *Canadian Geotechnical Journal*, 33, 260–271.

Cundall, P.A. and Strack, O.D.L. (1979). A discrete numerical model for granular assemblies. *Geotechnique*, 29, 47–65.

Dalbey, K., Patra, A.K., Pitman, E.B., Bursik, M.I. and Sheridan, M.F. (2008). Input uncertainty propagation methods and hazard mapping of geophysical mass flows. *Journal of Geophysical Research*, 113, B05203.

Fannin, R.J. and Wise, M.P. (2001). An empirical-statistical model for debris flow travel distance. *Canadian Geotechnical Journal*, 38, 982–994.

FLO-2D Software Inc. (2007). *FLO-2D User's Manual, Version 2007.06*.

Galas, S., Dalbey, K., Kumar, D., Patra, A. and Sheridan, M. (2007). Benchmarking TITAN2D mass flow model against a sand flow experiment and the 1903 Frank Slide. In *Proceedings of the 2007 International Forum on Landslide Disaster Management*, ed. K. Ho and V. Li . Hong Kong Geotechnical Engineering Office, pp. 899–918.

González, E., Herreros, M.I., Pastor, M., Quecedo, M. and Fernández Merodo, J.A. (2003). Discrete and continuum approaches for fast landslide modelling. In *Proceedings of the First International PFC Symposium*, ed. H. Konietsky. Lisse, Netherlands: Swets and Zeitlinger, pp. 307–313.

Griswold, J.P. (2004). *Mobility statistics and hazard mapping for non-volcanic debris flows and rock avalanches*. M.Sc. thesis, Portland State University, Portland, OR.

Heim, A. (1932). *Bergsturz und Menschenleben (Landslides and Human Lives)*. Translated by N. Skermer. Vancouver: Bitech Press.

Hsü, K.J. (1975). Catastrophic debris streams (sturzstroms) generated by rockfalls. *Geological Society of America Bulletin*, 86, 129–140.

Hungr, O. (1990). Mobility of rock avalanches. In *Report of the National Research Institute for Earth Science and Disaster Prevention*, Japan, 46, pp. 11–20.

(1995). A model for the runout analysis of rapid flow slides, debris flows, and avalanches. *Canadian Geotechnical Journal*, 32, 610–623.

Hungr, O. and Evans, S.G. (1996). Rock avalanche runout prediction using a dynamic model. In *Proceedings of the 7th International Symposium on Landslides*, ed. K. Senneset. Rotterdam: A.A. Balkema, pp. 233–238.

Hungr, O., Corominas, J. and Eberhardt, E. (2005). Estimating landslide motion mechanism, travel distance and velocity. In *Landslide Risk Management*, ed. O. Hungr, R. Fell, R. Couture and E. Eberhardt. Rotterdam: Balkema, pp. 99–128.

Hungr, O., Morgenstern, N. and Wong, H.N. (2007). Review of benchmarking exercises on landslide debris runout and mobility modelling. In *Proceedings of the 2007 International Forum on Landslide Disaster Management*, ed. K. Ho and V. Li. Hong Kong Geotechnical Engineering Office, pp. 755–812.

Hunter, G. and Fell, R. (2003). Travel distance angle for "rapid" landslides in constructed and natural soil slopes. *Canadian Geotechnical Journal*, 40, 1123–1141.

Iverson, R.M. (2003). How should mathematical models of geomorphic processes be judged? In *Prediction in Geomorphology*. American Geophysical Union. Geophysical Monograph 135.

Iverson, R.M., Schilling, S.P. and Vallance, J.W. (1998). Objective delineation of lahar-hazard inundation zones. *Geological Survey of America Bulletin*, 110, 972–984.

Kwan, J. and Sun, H.W. (2007). Benchmarking exercise on landslide mobility modelling: runout analyses using 3dDMM. In *Proceedings of the 2007 International Forum on Landslide Disaster Management*, ed. K. Ho and V. Li. Hong Kong Geotechnical Engineering Office, pp. 945–966.

Legros, F. (2002). The mobility of long-runout landslides. *Engineering Geology*, 63, 301–331.

Li, T. (1983). A mathematical model for predicting the extent of a major rockfall. *Zeitschrift für Geomorphologie*, 27, 473–482.

Mangeney-Castelnau, A., Vilotte, J.P., Bristeau, O. *et al.* (2003). Numerical modelling of avalanches based on Saint Venant equations using a kinetic scheme. *Journal of Geophysical Research*, 108(B11), 2527.

McDougall, S. (2006). *A new continuum dynamic model for the analysis of extremely rapid landslide motion across complex 3D terrain*. Ph.D. thesis, University of British Columbia, Vancouver, BC.

McDougall, S., Pirulli, M., Hungr, O. and Scavia, C. (2008). Advances in landslide continuum dynamic modelling. In *Proceedings of the 10th International Symposium on Landslides and Engineered Slopes*, ed. Z. Chen, J. Zhang, Z. Li, F. Wu and K. Ho. London, Taylor and Francis, pp. 145–157.

McKinnon, M. (2010). *Landslide runout: Statistical analysis of physical characteristics and model parameters*. M.Sc. thesis, University of British Columbia, Vancouver, BC.

Medina, V., Hürlimann, M. and Bateman, A. (2008). Application of FLATModel, a 2D finite volume code, to debris flows in the northeastern part of the Iberian Peninsula. *Landslides*, 5, 127–142.

Moore, D.P. (1976). *The Rubble Creek landslide, Garibaldi, British Columbia*. M.A.Sc. thesis, University of British Columbia, Vancouver, BC.

Moore, D.P. and Mathews, W.H. (1978). The Rubble Creek landslide, southwestern British Columbia. *Canadian Journal of Earth Sciences*, 15, 1039–1052.

Nicoletti, P.G. and Sorriso-Valvo, M. (1991). Geomorphic controls of the shape and mobility of rock avalanches. *Geological Society of America Bulletin*, 103, 1365–1373.

Pastor, M., Haddad, B., Sorbino, G., Cuomo, S. and Drempetic, V. (2009). A depth-integrated, coupled SPH model for flow-like landslides and related phenomena. *International Journal for Numerical and Analytical Methods in Geomechanics*, 33, 143–172.

Pedrazzini, A., Froese, C.R., Jaboyedoff, M., Hungr, O. and Humair, F. (2012). Combining digital elevation model analysis and run-out modeling to characterize hazard posed by a potentially unstable rock slope at Turtle Mountain, Alberta, Canada. *Engineering Geology*, 128, 76–94.

Pirulli, M. (2005). *Numerical modelling of landslide runout: A continuum mechanics approach.* Ph.D. thesis, Politechnico di Torino, Turin, Italy.

Pirulli, M., Preh, A., Roth, W., Scavia, C. and Poisel, R. (2003). Rock avalanche run out prediction: Combined application of two numerical methods. In *Proceedings of the International Symposium on Rock Mechanics*, South African Institute of Mining and Metallurgy, Johannesburg, pp. 903–908.

Pitman, E.B., Nichita, C.C., Patra, A. *et al.* (2003). Computing granular avalanches and landslides. *Physics of Fluids*, 15, 3638–3646.

Poisel, R. and Preh, A. (2008). 3D landslide runout modelling using the particle flow code PFC3D. In *Proceedings of the 10th International Symposium on Landslides and Engineered Slopes*, ed. Z. Chen, J. Zhang, Z. Li, F. Wu and K. Ho. London: Taylor and Francis, pp. 873–879.

Poisel, R. and Roth, W. (2004). Run out models of rock slope failure. *Felsbau*, 22, 46–50.

Revellino, P., Hungr, O., Guadagno, F.M. and Evans, S.G. (2004). Velocity and runout simulation of destructive debris flows and debris avalanches in pyroclastic deposits, Campania region, Italy. *Environmental Geology*, 45, 295–311.

Rickenmann, D. (1999). Empirical relationships for debris flows. *Natural Hazards*, 19, 47–77.

Roddeman, D.G. (2002). *TOCHNOG User Manual: A Free Explicit/Implicit FE Program*. FEAT [available at www.feat.nl/manuals/user/user.html].

Savage, S.B. and Hutter, K. (1989). The motion of a finite mass of granular material down a rough incline. *Journal of Fluid Mechanics*, 199, 177–215.

Scheidegger, A.E. (1973). On the prediction of the reach and velocity of catastrophic landslides. *Rock Mechanics*, 5, 231–236.

Schilling, S.P., Griswold, J.P. and Iverson, R.M. (2008). Using LAHARZ to forecast inundation from lahars, debris flows and rock avalanches: Confidence limits on prediction. In *Proceedings of the American Geophysical Union 2008 Fall Meeting*, San Francisco.

Wang, F.W. and Sassa, K. (2002). A modified geotechnical simulation model for the areal prediction of landslide motion. In *Proceedings of the First European Conference on Landslides*, pp. 735–740.

Wang, X. (2008). *Geotechnical analysis of flow slides, debris flows, and related phenomena*. Ph.D. thesis, University of Alberta, Edmonton, AB.

17 Models of the triggering of landslides during earthquakes

RANDALL W. JIBSON

ABSTRACT

Several methods to assess the stability of slopes during earthquakes were developed during the twentieth century. Pseudostatic analysis was the first method; this involves simply adding a permanent body force representing the earthquake shaking to a static limit equilibrium analysis. Stress-deformation analysis, a later development, involves much more complex modeling of slopes. It uses a mesh in which the internal stresses and strains within elements are computed based on the applied external loads, including gravity and seismic loads. Stress-deformation analysis provides the most realistic model of slope behavior, but is very complex and requires a high density of high-quality soil-property data, as well as an accurate model of soil behavior. In 1965, Newmark developed a method that effectively bridges the gap between these two types of analysis. His sliding-block model is easy to apply and provides a useful index of coseismic slope performance. Subsequent modifications to sliding-block analysis have made it applicable to a wider range of landslide types. It is far easier to apply than stress-deformation analysis and yields much more useful information than pseudostatic analysis.

17.1 INTRODUCTION

Most moderate and large earthquakes trigger landslides. These triggered landslides are significant in two important contexts: (1) they commonly account for a large proportion of overall earthquake damage; and (2) sediment produced by such landslides constitutes much of the long-term sediment budget that contributes to landscape lowering. Models that predict landslide movement during earthquakes have improved steadily over the last few decades, and several models of coseismic landslide movement that allow reasonably detailed predictions of coseismic slope performance are currently available.

This chapter summarizes pertinent information about earthquake-triggered landslides and then discusses historic and current methods for assessing the coseismic performance of slopes.

17.2 CHARACTERISTICS OF EARTHQUAKE-TRIGGERED LANDSLIDES

Research on earthquake-triggered landslides has quantified many of the characteristics of these landslides and their distributions. The sections that follow briefly summarize some of this information in order to provide a general context for the subsequent discussion of modeling approaches.

17.2.1 LANDSLIDE TYPES

Earthquakes can trigger all types of landslides, but all types of landslides can also occur without seismic triggering (Jibson, 2009). Some types of landslides, however, tend to be much more abundant in earthquakes than other types. In a study of 40 major worldwide earthquakes, Keefer (1984) ranked the relative abundance of various types of landslides; he grouped these landslides into three general categories: disrupted slides and falls, coherent slides, and lateral spreads and flows. Disrupted slides and falls account for an average of 86 percent of all triggered landslides, coherent slides 8 percent, and lateral spreads and flows 6 percent. Keefer (1984, 2002) described typical characteristics of various types of earthquake-triggered landslides and their source areas. In general, slope materials that are deeply weathered, sheared, intensely fractured or jointed, or saturated are particularly susceptible to landsliding during earthquakes.

Landslides: Types, Mechanisms and Modeling, ed. John J. Clague and Douglas Stead. Published by Cambridge University Press.

Such materials are common in seismically active areas, where there is ongoing tectonic deformation.

17.2.2 LANDSLIDE-TRIGGERING CONDITIONS

In a review of intensity reports from 300 earthquakes, Keefer (1984, 2002) found that the smallest earthquake reported to have caused landslides had a magnitude of 4.0. This magnitude corresponds closely to the lower-bound triggering value of M4.3 determined by quantitative evaluation of several landslide inventories (Malamud *et al.*, 2004). Landslides of various types have different threshold-triggering magnitudes: disrupted landslides can be triggered by M4.0 earthquakes, coherent slides by M4.5–5.0 earthquakes, and lateral spreads and flows by M5.0 earthquakes. Although smaller earthquakes could conceivably trigger landslides, such triggering by very weak shaking would probably only occur on slopes where failure was imminent before the earthquake.

Harp and Wilson (1995) were the first to relate the farthest extent of landsliding with instrumental measures of earthquake shaking to establish a quantitative threshold level of ground motion needed to initiate landsliding. An analysis of data collected in southern California from the 1987 Whittier Narrows (M5.9) and Superstition Hills (M6.7) earthquakes suggests two threshold ranges for two general categories of rocks (Arias, 1970): intensities of 0.08–0.60 $m\,s^{-1}$ for Tertiary and younger rocks, and 0.01–0.07 $m\,s^{-1}$ for Mesozoic and older rocks. These ranges are controlled by the relative degree of fracturing of the two rock types and might not apply in other regions. A study relating landslide distribution to peak ground acceleration (PGA) in the 2008 Wenchuan, China earthquake indicated a lower limit of 0.05–0.07 *g* for triggering of landslides and a value of about 0.2 *g* for the most serious landsliding (Wang *et al.*, 2010).

17.2.3 GEOGRAPHIC DISTRIBUTIONS OF EARTHQUAKE-TRIGGERED LANDSLIDES

For the three categories of landslides, Keefer (1984) related earthquake magnitude to the maximum distance of triggered landslides from the earthquake source (Fig. 17.1). Upper-bound curves were well defined and were constrained to pass through the minimum threshold magnitudes indicated previously as distance approached zero. Although these upper bounds have been exceeded a few times in subsequent earthquakes (Keefer, 2002), they remain fairly reliable indicators of the maximum possible distances at which the three classes of landslides could be triggered in earthquakes of different magnitudes.

For 30 historical earthquakes, Keefer (1984) drew boundaries around all reported landslide locations and calculated the areas enclosed. His plot of area versus magnitude (Fig. 17.2) shows a well-defined upper-bound curve representing the maximum area that can be affected for a given magnitude. Rodríguez *et al.* (1999) later expanded this dataset to include 35 earthquakes, which slightly raised the upper-bound line (Fig. 17.2). Keefer

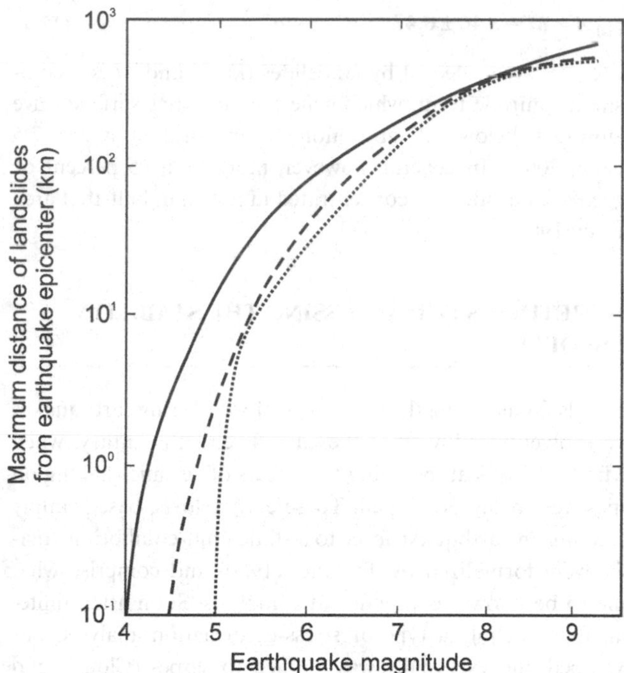

Fig. 17.1. Curves relating earthquake magnitude to the maximum epicentral distance at which landslides were triggered in 40 worldwide earthquakes (modified from Keefer, 1984). Solid line is the upper bound for disrupted landslides; dashed line is the upper bound for coherent landslides; dotted line is the upper bound for lateral spreads and flows. Magnitude is in terms of moment magnitude (**M**) for $M \geq 7.5$ and surface-wave magnitude (M_S) for $M < 7.5$.

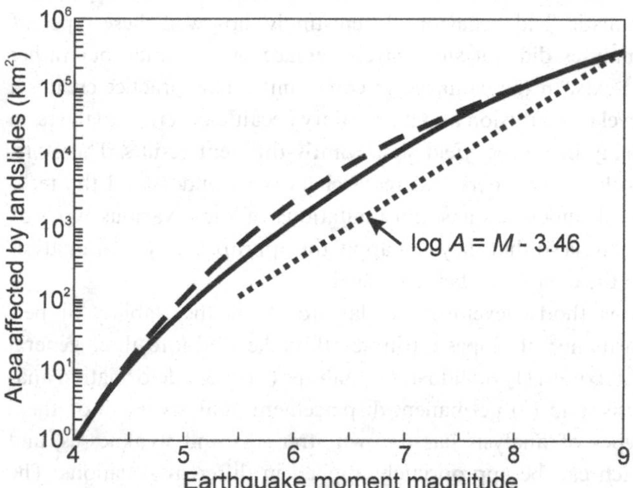

Fig. 17.2. Curves relating earthquake moment magnitude to area affected by landslides (*A*) (modified from Keefer, 2002). Solid line is the upper bound of Keefer (1984); dashed line is the upper bound of Rodríguez *et al.* (1999); dotted line is least squares regression mean from Keefer and Wilson (1989).

and Wilson (1989) fitted a regression line (Fig. 17.2) using data from 37 earthquakes, including the 30 from Keefer (1984), to predict average area affected by landslides as a function of earthquake magnitude:

$$\log_{10} A = M - 3.46 \pm 0.47 \quad (17.1)$$

where A is area affected by landslides (km^2), and M is a composite magnitude term, which generally indicates surface-wave magnitudes below 7.5 and moment magnitudes above 7.5 (Keefer, 2002). In general, however, more than 95 percent of triggered landslides are concentrated in less than half that area (Keefer, 1984).

17.3 METHODS FOR ASSESSING THE STABILITY OF SLOPES

Methods for assessing the stability of slopes during earthquakes have evolved steadily since the early twentieth century, when the first attempts at modeling the effects of seismic shaking on slopes were being developed. These early efforts, based simply on adding an earthquake force to a static limit-equilibrium analysis, were formalized by Terzhagi (1950) and comprise what came to be known as pseudostatic analysis. Soon after, finite-element modeling, a type of stress-deformation analysis, was developed and eventually was applied to slopes (Clough and Chopra, 1966). In these early years, however, this type of analysis was profoundly complex and computationally daunting. Newmark (1965) proposed a method for estimating the displacement of slopes during earthquakes that addressed some of the crude assumptions of pseudostatic analysis but was still quite simple to apply in practice, and thus overcame the difficulties of early stress-deformation analysis. Subsequent researchers refined Newmark's analysis to allow for more complex and realistic field behaviors. Interestingly, however, these types of analyses did not successively replace one another, but rather co-exist in the engineering community. This practice creates a level of confusion and uncertainty because different analyses, in many instances, yield significantly different results. Therefore, both practitioners and researchers must understand the technical underpinnings and limitations of these various types of analysis so that they can apply the appropriate type of analysis to the conditions being studied.

Methods developed to date to assess the stability or performance of slopes during earthquakes fall into three general categories: (1) pseudostatic analysis; (2) stress-deformation analysis; and (3) permanent-displacement analysis. Each of these types of analysis has its own strengths and weaknesses, and each can be appropriately applied in different situations. The sections that follow describe the historic development of each of these families of analysis and discuss their advantages and limitations.

17.3.1 PSEUDOSTATIC ANALYSIS

Stability analyses of earth slopes during earthquake shaking were initiated in the early twentieth century using what has come to be known as the pseudostatic method; the first known documentation of pseudostatic analysis in the technical literature was by Terzhagi (1950). Pseudostatic analysis models the seismic shaking as a permanent body force that is added to the force-body diagram of a conventional static limit-equilibrium analysis. Normally, only the horizontal component of earthquake shaking is modeled because the effects of vertical forces tend to average out to near zero. Consider, for example, a planar slip surface in a slope consisting of dry, cohesionless material (Fig. 17.3). The pseudostatic factor-of-safety equation is:

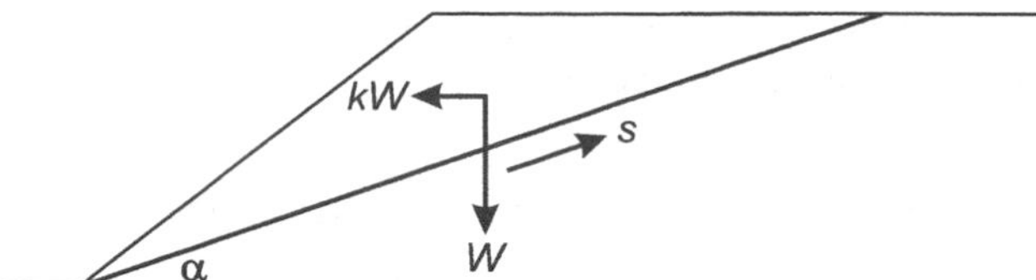

Fig. 17.3. Force diagram of a landslide in dry cohesionless soil that has a planar slip surface. W is the weight per unit length of the landslide, k is the pseudostatic coefficient, s is the shear resistance along the slip surface, and α is the angle of inclination of the slip surface.

$$FS = \left[(W\cos\alpha - kW\sin\alpha)\tan\phi\right] / (W\sin\alpha + kW\cos\alpha) \quad (17.2)$$

where FS is pseudostatic factor of safety, W is weight per unit length of slope, α is slope angle, ϕ is friction angle of the slope material, and k is the pseudostatic coefficient, defined as:

$$k = a_h / g \quad (17.3)$$

where a_h is the horizontal ground acceleration and g is the acceleration of Earth's gravity.

A common approach to pseudostatic analysis is to conduct a limit-equilibrium analysis iteratively using different values of k until $FS = 1$. The resulting pseudostatic coefficient is the yield coefficient, k_y. In the simplest sense, any ground acceleration that exceeds $k_y \times g$ is defined as causing failure. The obvious simplicity of pseudostatic analysis as an extension of static limit-equilibrium analysis makes it easy to apply, and it quickly came into common usage.

Pseudostatic analysis has some obvious drawbacks: Terzhagi acknowledged that "the conception it conveys of earthquakes on slopes is very inaccurate" (1950, p. 90). The main drawback is that including the earthquake shaking as a permanent, unidirectional body force is extremely conservative; it assumes a constant earthquake force acting only in a direction that promotes slope instability. For this reason, pseudostatic coefficients generally are selected as some fraction of the peak acceleration to account for the fact that the peak acceleration acts only briefly and does not represent a longer, more sustained, acceleration of the landslide mass. Selection of the pseudostatic coefficient is thus the most important aspect of pseudostatic analysis, but it is also the most difficult. Table 17.1 lists several recommendations for selecting a pseudostatic coefficient; significant differences in approaches and the resulting values are apparent among the studies cited. One key issue is calibration: some of these studies were calibrated for earth-dam design, in which as much as 1 m of displacement is acceptable. But these same values are commonly used in the stability assessment of natural slopes, in

Table 17.1. *Pseudostatic coefficients from various studies.*

Investigator	Recommended pseudostatic coefficient (k)	Recommended factor of safety (FS)	Calibration conditions
Terzhagi (1950)	0.1 (R-F[a] = IX) 0.2 (R-F[a] = X) 0.5 (R-F[a] > X)	>1.0	Unspecified
Seed (1979)	0.10 (M = 6.50) 0.15 (M = 8.25)	>1.15	<1 m displacement in earth dams
Marcuson (1981)	0.33–0.50 × PGA/g	>1.0	Unspecified
Hynes-Griffin and Franklin (1984)	0.50 × PGA/g	>1.0	<1 m displacement in earth dams
California Division of Mines and Geology (1997)	0.15	>1.1	Unspecified; probably based on <1 m displacement in dams

[a] Rossi–Forel earthquake intensity scale.

which the acceptable displacement might be as little as 5–30 cm (Wieczorek *et al.*, 1985; Blake *et al.*, 2002; Jibson and Michael, 2009). The most commonly used values are similar to the k = 0.15 and FS > 1.1 in general use in California (California Division of Mines and Geology, 1997) but, again, these criteria were formulated for earth dams that could accommodate about 1 m of displacement.

After summarizing a number of published approaches for determining an appropriate seismic coefficient, Kramer (1996, p. 436–437) concluded that "there are no hard and fast rules for selection of a pseudostatic coefficient for design." Justifications cited by practitioners in the engineering community of why a certain pseudostatic coefficient is used are generally some variation of "We have always done it that way" or "It seems to work." Suffice to say that in the practicing community at large, a rigorously rational basis for selecting a pseudostatic coefficient remains elusive, and engineering judgment – along with standard of practice generally – is invoked in the selection process.

Two recent studies have attempted to rationalize the selection of the pseudostatic coefficient. Stewart *et al.* (2003) developed a site screening procedure, based on the statistical relationship of Bray and Rathje (1998), wherein a pseudostatic coefficient is calculated as a function of maximum horizontal ground acceleration, earthquake magnitude, source distance, and two possible levels of allowable displacement (5 and 15 cm). Bray and Travasarou (2009) presented a straightforward approach that calculates the pseudostatic coefficient as a function of allowable displacement, earthquake magnitude, and spectral acceleration. The common basis of these rationalized approaches is calibration based on allowable displacement.

As stated previously, pseudostatic analysis tends to be overconservative in many situations, but there are some conditions in which it is not conservative at all. Slopes composed of materials that build up significant dynamic pore pressures during earthquake shaking, or that lose more than about 15 percent of their peak shear strength during shaking, are not good candidates for pseudostatic analysis (Kramer, 1996). In fact, several case studies exist of dams that passed a pseudostatic stability analysis but failed during earthquakes (Seed, 1979).

One last limitation of pseudostatic analysis is that, because it is a limit-equilibrium analysis, it tells the user nothing about what happens after equilibrium is exceeded. The analysis shows a slope to be either stable or unstable, but the consequences of instability, or even the likelihood of failure, cannot be judged.

In summary, pseudostatic stability analysis is easy to use, has a long history that provides a body of engineering judgment regarding its application, and provides a simple scalar index of stability. But this simplicity stems from a rather crude characterization of the physical processes of dynamic slope behavior that results in several drawbacks, including difficulty in rationally selecting a pseudostatic coefficient and in actually assessing the likelihood or results of failure. The use of pseudostatic analysis, while still widespread, is gradually being supplanted by the more sophisticated types of analysis discussed in the following sections.

17.3.2 STRESS-DEFORMATION ANALYSIS

In 1960, Ray Clough at the University of California developed and named the finite-element method of engineering analysis (Clough, 1960). Finite-element modeling uses a mesh to model a deformable system; the deformation at each node of the mesh is calculated in response to an applied stress. This method soon began to be applied to slopes, particularly earth dams, and provided a valuable tool for modeling the static and dynamic deformation of soil systems (Clough and Chopra, 1966). Subsequent developments in stress-deformation analysis have yielded a family of analytical methods, including finite-difference and discrete-element modeling. Several applications of finite-element modeling to earth structures have been developed and published; Kramer (1996) provided a good summary of various methods and their associated studies.

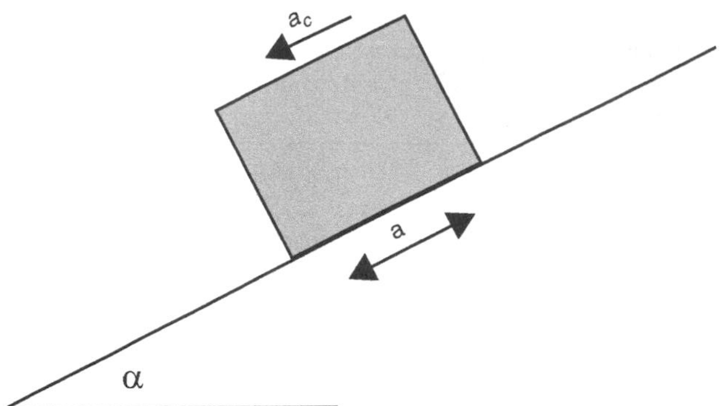

Fig. 17.4. Conceptual model of a rigid sliding-block analysis. The earthquake ground acceleration at the base of the block is denoted by a, a_c is the critical acceleration of the block, and α is the angle of inclination of the sliding surface.

Stress-deformation analysis uses highly complex models that, to be worthwhile, require a high density of high-quality data and sophisticated soil-constitutive models to predict the stress-strain behavior of the soils. The more advanced 3D models are also quite computationally intensive. For this reason, stress-deformation modeling is generally practical only for critical projects such as earth dams and slopes affecting critical lifelines or structures. Stress-deformation analysis is innately site-specific and cannot be applied to regional problems. Even as a site-specific approach, it is generally reserved for critical projects because of the difficulty in procuring the necessary data and the time and effort involved in the modeling procedure.

The advantage of stress-deformation modeling is that it gives the most accurate picture of what actually happens in the slope during an earthquake. Clearly, models that account for the complexity of spatial variability of properties and the stress-strain behavior of slope materials yield the most reliable results. But stress-deformation modeling also has drawbacks. The complex modeling is warranted only if the quantity and quality of the data merit it. These modeling procedures can be rather challenging because of: (1) the data acquisition required, commonly including undisturbed soil sampling and extensive laboratory testing of samples; (2) the need to select a suite of input ground motions; (3) the need for an accurate, nonlinear, stress-dependent, cyclic model of soil behavior; and (4) computational requirements.

17.3.3 PERMANENT-DISPLACEMENT ANALYSIS

In his 1965 Rankine lecture, Nathan Newmark (1965) introduced a method to assess the performance of slopes during earthquakes that effectively bridges the gap between overly simplistic pseudostatic analysis and highly complex stress-deformation analysis. Newmark's method models a landslide as a rigid block that slides on an inclined plane; the block has a known yield or critical acceleration, the acceleration required to overcome basal resistance and initiate sliding (Fig. 17.4). An earthquake strong-motion record of interest is selected, and those parts of the record that exceed the critical acceleration are integrated to obtain the velocity–time history of the block. The velocity–time history is then integrated to obtain the cumulative displacement of the landslide block (Fig. 17.5). The user must judge the significance of the displacement. Newmark's method and related types of analysis are commonly referred to as permanent-displacement analyses.

Several types of permanent-displacement analysis have been proposed that are designed to yield more accurate estimates of slope displacement by modeling the dynamic slope response more rigorously. More accurate modeling, of course, involves more complexity. One great advantage of Newmark's method is its theoretical and practical simplicity. This simplicity, however, is the result of many assumptions that limit the accuracy of the results. The more sophisticated methods do a better job of modeling the dynamic response of the landslide material and thus yield more accurate displacement estimates but, again, the complexity of the analysis and the difficulty in acquiring the needed input parameters involve a trade-off.

At present, analytical procedures for estimating permanent coseismic landslide displacements can be grouped into three types: (1) rigid block, (2) decoupled, and (3) coupled.

RIGID-BLOCK ANALYSIS

Newmark's (1965) original methodology is generally referred to as rigid-block analysis. To briefly reiterate, the potential landslide block is modeled as a rigid mass that slides in a perfectly plastic manner on an inclined plane (Fig. 17.4). The mass experiences no permanent displacement until the base acceleration exceeds the critical acceleration of the block, at which time the block begins to move downslope. Displacements are calculated by integrating the parts of an acceleration–time history that lie above the critical acceleration to determine a velocity–time history. The velocity–time history is then integrated to yield the cumulative displacement (Fig. 17.5). Sliding continues until the relative velocity between the block and base reaches zero.

The critical acceleration (a_c), in its simplest form, can be estimated as

$$a_c = (FS - 1)g \sin\alpha \tag{17.4}$$

where a_c is in terms of g, the acceleration of gravity, FS is the static factor of safety (the ratio of resisting to driving forces or moments in a slope), and α is the dip angle of the sliding surface (Newmark, 1965; Jibson, 1993). Equation 17.4 assumes that the seismic force is applied parallel to the slope. Critical acceleration also can be estimated iteratively in a pseudostatic analyses to determine the yield coefficient, which, when multiplied by g, is equivalent to the critical acceleration.

In rigid-block analysis, the landslide does not deform internally, experiences no permanent displacement below the critical acceleration, and deforms plastically at constant stress along a discrete basal shear surface when the critical acceleration is exceeded. Because dynamic pore pressures cannot be modeled

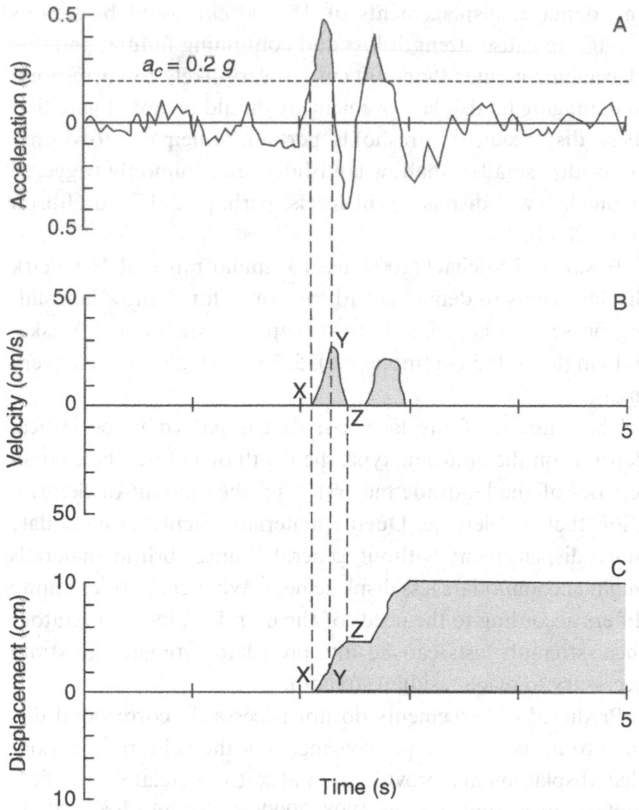

Fig. 17.5. Illustration of the Newmark integration algorithm. A, earthquake acceleration–time history with critical acceleration (dashed line) of 0.2 *g* superimposed; B, velocity of the landslide versus time; C, displacement of landslide versus time. Points X, Y, and Z are for reference between plots.

in rigid-block analysis, this analysis should be applied with great caution, or not at all, in situations where significant dynamic pore pressures can develop.

Laboratory model tests (Goodman and Seed, 1966; Wartman *et al.*, 2003, 2005) and analyses of landslides in natural slopes (Wilson and Keefer, 1983) confirm that rigid-block analysis can accurately predict slope displacements if slope geometry, soil properties, and earthquake ground motions are known. This analysis is simple to apply and provides an estimate of the inertial landslide displacement resulting from a given earthquake strong-motion record.

DECOUPLED ANALYSIS

Soon after Newmark published his rigid-block method, more sophisticated analyses were developed to account for the fact that landslides are not rigid bodies but deform internally when subjected to seismic shaking (Seed and Martin, 1966; Lin and Whitman, 1983). One such method is decoupled analysis, which estimates the effect of dynamic response on permanent sliding in a two-step procedure:

1. A dynamic-response analysis of the slope, assuming no failure surface, is performed using programs such as QUAD-4 (Idriss *et al.*, 1973) or SHAKE (Schnabel *et al.*, 1972). By estimating the acceleration–time histories at several points within the slope, an average acceleration–time history for the slope mass above the potential failure surface is developed. The average acceleration has been referred to variously as *k* (Makdisi and Seed, 1978) or HEA, the horizontal equivalent acceleration (Bray and Rathje, 1998); peak values are generally referred to as k_{max} or MHEA, the maximum horizontal equivalent acceleration. The site-response analysis requires specification of the shear-wave velocity of the material, the thickness of the potential landslide, and the damping; for an equivalent-linear analysis, modulus-reduction and damping curves are also required.
2. The resulting time history is input into a rigid-block analysis, and the permanent displacement is estimated. This approach is referred to as a decoupled analysis because the computation of the dynamic response and the plastic displacement are performed independently. Decoupled analysis thus does not account for the effects of sliding displacement on the ground motion. Studies of actual seismically triggered landslides have shown that decoupled analysis can accurately predict field behavior (Pradel *et al.*, 2005).

Makdisi and Seed (1978) published an approach to decoupled analysis of earth dams, which came into widespread use. Their design charts allow estimation of displacement as a function of slope geometry, critical acceleration, ground motion, and earthquake magnitude. This approach, however, has some limitations.

- An estimate of the ground acceleration at the ground surface or crest of the dam is required, and this calculation has significant uncertainty.
- Ranges of possible displacements can span an order of magnitude, which requires users to judge where in this range to select a design displacement.
- The analysis was designed and calibrated for earth dams, and yet it has been applied to a wide range of situations, including applications in natural slopes, where some assumptions of the analysis are not valid.

Bray and Rathje (1998) updated the Makdisi and Seed (1978) analysis for application to deeper sliding masses typical of landfills.

COUPLED ANALYSIS

In a fully coupled analysis, the dynamic response of the sliding mass and the permanent displacement are modeled together so that the effect of plastic sliding displacement on the ground motions is taken into account. Lin and Whitman (1983) pointed out that the assumptions of decoupled analysis introduce errors in the estimation of total slip and compared results for coupled and decoupled analyses. They showed that, in general, decoupled analysis yielded conservative results that were within about 20 percent of the coupled results. More recently, Rathje and

Bray (1999, 2000) compared results from rigid-block analysis with linear and nonlinear coupled and decoupled analyses.

Coupled analysis, the most sophisticated form of sliding-block analysis, uses the same inputs as does decoupled analysis, but is much more computationally intensive. Thus, in addition to the critical acceleration, the user specifies the shear-wave velocities of the materials above and below the slip surface, the damping ratio, and the thickness of the potential landslide. Either linear-elastic or equivalent-linear soil models can be used; for equivalent-linear analysis, damping and modulus-reduction curves are needed. The largest impediment to applying coupled analysis has been the lack of published software packages for its implementation, but such a package is anticipated (Jibson *et al.*, 2012).

Bray and Travasarou (2007) developed a simplified approach that used a nonlinear, fully coupled sliding-block model to produce a semi-empirical relationship for predicting displacement. The model requires specification of the yield acceleration, the fundamental period of the sliding mass (T_s), and the spectral acceleration of the ground motion at a period of $1.5T_s$.

17.3.4 INTERPRETING MODELED DISPLACEMENTS

The significance of modeled displacements must be judged by their probable effect on a potential landslide. Wieczorek *et al.* (1985) used 5 cm as the critical displacement leading to macroscopic ground cracking and failure of landslides in San Mateo County, California; Keefer and Wilson (1989) used 10 cm as the critical displacement for coherent landslides in southern California; and Jibson and Keefer (1993) used this 5–10 cm range as the critical displacement for landslides in the Mississippi Valley. In most soils, displacements in this range cause ground cracking, and previously undeformed soils can end up in a weakened or residual-strength condition. In such a case, static stability analysis in residual-strength conditions can be performed to determine stability after earthquake shaking ceases (Jibson and Keefer, 1993).

Blake *et al.* (2002) made the following recommendations for the application of rigid-block analysis in southern California:

- For slip surfaces that intersect stiff improvements (e.g., buildings, pools), median Newmark displacements should be less than 5 cm.
- For slip surfaces occurring in ductile (nonstrain-softening) soil that do not intersect with engineered improvements (e.g., landscaped areas and patios), median Newmark displacements should be less than 15 cm.
- In soils having significant strain softening (sensitivity > 2), the design should be performed using either residual strengths and allowing median displacements less than 15 cm, or using peak strengths and allowing median displacements less than 5 cm.

The California Geological Survey's (2008) guidelines for mitigating seismic hazards state that displacements of 0–15 cm are unlikely to correspond to serious landslide movement and damage; displacements of 15–100 cm could be serious enough to cause strength loss and continuing failure; and displacements greater than 100 cm are very likely to correspond to damaging landslide movement. It should be noted here that these displacement thresholds pertain principally to deeper landslides; smaller, shallow landslides are commonly triggered at much lower displacement levels, perhaps 2–15 cm (Jibson *et al.*, 2000).

Jibson and Michael (2009) used a similar range of Newmark displacements to define hazard categories for shallow landsliding on seismic landslide hazard maps of Anchorage, Alaska: 0–1 cm (low), 1–5 cm (moderate), 5–15 cm (high), >15 cm (very high).

The amount of displacement that is judged to be critical depends on the landslide type, the depth of failure, the characteristics of the landslide material, and the amount of deformation that is tolerable. Ductile materials might accommodate more displacement without general failure; brittle materials might accommodate less displacement. What constitutes failure differs according to the needs of the user. Results of laboratory shear-strength tests can be interpreted to estimate the strain necessary to reach residual strength.

Predicted displacements do not necessarily correspond directly to measurable slope movements in the field; rather, modeled displacements provide an index to correlate with field performance (Jibson *et al.*, 1998, 2000; Rathje and Bray, 2000). For the sliding-block method to be useful in a predictive sense, modeled displacements must be quantitatively correlated with field performance. Pradel *et al.* (2005) conducted such a study on a single landslide following the 1994 Northridge, California earthquake and found good agreement between the predicted and observed displacements. Jibson *et al.* (1998, 2000) compared the inventory of all landslides triggered by the Northridge earthquake to predicted Newmark displacements. The results were regressed using a Weibull model, which yielded the following equation (Jibson *et al.*, 2000):

$$P(f) = 0.335\left[1 - \exp\left(-0.048D_n^{1.565}\right)\right] \qquad (17.5)$$

where $P(f)$ is probability of failure and D_n is Newmark displacement in centimeters. This model was calibrated at regional scale using southern California data, which included primarily shallow falls and slides in brittle rock and debris, and is applicable only to these types of landslides. Regional characterization of geotechnical properties introduces significant uncertainty, which accounts for the upper bound probability of only 33.5 percent. Calibration at site-specific scale with more accurate geotechnical characterization would likely yield larger probability estimates.

17.3.5 SELECTING AN ANALYSIS

Reliable estimation of displacement depends on selecting the appropriate analysis. The best basis for selection appears to be the period ratio, T_s/T_m, where T_s is the fundamental site period and T_m is the mean period of the earthquake motion (Rathje and Bray, 1999, 2000). T_s can be estimated as:

$$T_s = 4H/V_s \tag{17.6}$$

where H is the maximum thickness of the potential landslide and V_s is the shear-wave velocity of the material above the slip surface. The mean period of the earthquake motion is defined as the average period weighted by the Fourier amplitude coefficients over a frequency range of 0.25–20 Hz (Rathje *et al.*, 1998, 2004). Mean period (T_m, in seconds) can be estimated for shallow crustal earthquakes (data from western North America), rock site conditions, and no forward directivity as a function of earthquake moment magnitude (M_w) and source distance (r, in kilometers) as follows (Rathje *et al.*, 2004):

$$\ln T_m = -1.00 + 0.18(M_w - 6) + 0.0038r \quad \text{for } M_w \le 7.25, \tag{17.7}$$

$$\ln T_m = -0.775 + 0.0038r \quad \text{for } M_w > 7.25. \tag{17.8}$$

Theoretically, coupled analysis should yield the best results because it accounts for more of the complexities of the problem than either of the other analyses. In practice, however, coupled analysis can become numerically unstable for period ratios below about 0.1, where landslide masses behave rigidly. Therefore, a general guideline for selecting a sliding-block analysis would be to use rigid-block analysis for period ratios less than 0.1 and coupled analysis for period ratios greater than 0.1.

As coupled analysis becomes more accessible, little justification will exist to continue the use of decoupled analysis. Perhaps, in light of the long historic use of decoupled analysis, one appropriate use would be for comparison with past applications. Thus, use of decoupled analysis will probably continue for some time.

17.3.6 APPLICATIONS OF PERMANENT-DISPLACEMENT ANALYSIS

Permanent-displacement analysis has been applied in a variety of ways to slope-stability problems. Most early applications dealt with the seismic performance of dams and embankments (e.g., Seed *et al.*, 1973; Makdisi and Seed, 1978; Seed, 1979; Yegian *et al.*, 1991). Permanent-displacement analysis has also been applied successfully to landslides in natural slopes (Wilson and Keefer, 1983; Pradel *et al.*, 2005), and is being used increasingly in slope design, landfills, and other engineered works (e.g., Bray and Rathje, 1998; Blake *et al.*, 2002).

Several simplified approaches have been proposed for applying permanent-displacement analysis; these involve developing empirical relationships to predict slope displacement as a function of critical acceleration and one or more measures of earthquake shaking. Many such studies plot displacement against critical acceleration ratio, the ratio of critical acceleration to peak ground acceleration (Franklin and Chang, 1977; Makdisi and Seed, 1978; Ambraseys and Menu, 1988; Jibson, 2007; Saygili and Rathje, 2008; Rathje and Saygili, 2009). Other studies have related critical acceleration ratio to some normalized form of displacement (Yegian *et al.*, 1991; Lin and Whitman, 1983). Jibson (1993, 2007) and Jibson *et al.* (1998, 2000) related Newmark displacement to critical acceleration, and Arias (1970) to intensity, which is the integral over time of the squared accelerations in a strong-motion record. As noted previously, Bray and Travasarou (2007) developed a simplified method for fully coupled analysis that predicts displacement as a function of yield acceleration, site period, and spectral acceleration. Miles and Ho (1999) and Miles and Keefer (2000) compared results from these simplified methods with their own method of rigorously integrating artificially generated strong-motion time histories throughout a study area.

The commonest application of these simplified methods is in making seismic landslide hazard maps, where regional estimation of Newmark displacement in a GIS model is required. Wieczorek *et al.* (1985) were the first to use Newmark analysis as a basis for seismic landslide microzonation, and methods for such applications have evolved steadily since that first study (Jibson *et al.*, 1998, 2000; Mankelow and Murphy, 1998; Luzi and Pergalani, 1999; Miles and Ho, 1999; Miles and Keefer, 2000, 2001; Del Gaudio *et al.*, 2003; Rathje and Saygili, 2008, 2009; Jibson and Michael, 2009). Such applications generally involve GIS modeling, in which study areas are gridded and discrete estimates of coseismic displacement are generated for each grid cell using simplified regression equations.

17.4 DISCUSSION

Pseudostatic analysis continues to be used and has a deep reservoir of engineering judgment behind it. It is conceptually simple, but the process of selecting a seismic coefficient commonly lacks a rational basis and the analysis tends to be overly conservative. Even more troubling is that this analysis, which should be conservative, sometimes is not and therefore must be calibrated to the specific conditions being modeled. Such calibration seldom happens in practice, however, because of a long-term sense of judgment regarding its application. The current availability of software that facilitates rigorous permanent-displacement analysis (Jibson and Jibson, 2003; Jibson *et al.*, 2012) renders the most powerful rationale for using pseudostatic analysis – its simplicity – invalid. Performing permanent-displacement analysis is currently little more difficult than performing pseudostatic analysis. Therefore, the most appropriate applications for pseudostatic analysis are limited to preliminary analyses and screening procedures that precede a more sophisticated analysis (Stewart *et al.*, 2003).

Stress-deformation analysis is the gold standard for single-site analysis where sufficient data exist to merit it. For critical facilities such as dams and for embankments or slopes adjacent to critical lifelines or structures, the cost and effort of stress-deformation analysis are justified.

Permanent-displacement analysis seems increasingly able to fill the gap between the two end members represented by the other two families of analysis. The development of decoupled and coupled analysis facilitates the application of permanent-displacement analysis to a variety of landslide types. Its main advantage over pseudostatic analysis is that it provides a quantitative measure of what happens after the critical acceleration is exceeded, the point at which a pseudostatic analysis simply defines failure as having occurred. Permanent-displacement analysis begins, in fact, exactly where pseudostatic analysis ends: at the moment the critical or yield acceleration is exceeded. The displacement thus modeled provides a quantitative measure – an index – of coseismic slope performance. And compared to stress-deformation analysis, permanent-displacement analysis is much easier to apply.

Newmark's (1965) method and its subsequent variations were developed to analyze the seismic behavior of earth dams and embankments (Franklin and Chang, 1977; Makdisi and Seed, 1978; Seed, 1979; Lin and Whitman, 1983). These large earth structures commonly have well-defined, homogeneous properties; they are constructed largely of relatively ductile fine-grained materials and are principally subject to deep modes of failure. These properties of large earth structures are the very reason that coupled and decoupled analysis were later developed, to better model engineered earth structures by taking into account the dynamic deformation of the soil mass and the effects of coseismic displacement on the response of the slide mass (Bray and Rathje, 1998; Rathje and Bray, 1999, 2000). These and other studies (Kramer and Smith, 1997; Wartman *et al.*, 2003, 2005; Lin and Wang, 2006) make clear that traditional Newmark analysis does not yield acceptable results for the seismic performance of large engineered earth structures. Rathje and Bray (1999, 2000) and Wartman *et al.* (2003, 2005) provide detailed treatments of specific combinations of site and shaking conditions that are, and are not, adequately modeled by rigid-block analysis.

Rigid-block analysis is best suited to a very different type of slope failure: earthquake-triggered landslides in natural slopes, an application first proposed by Wilson and Keefer (1983). As stated previously, rigid-block analysis is applicable to thinner landslides in stiffer material (T_s/T_m <0.1); in practical terms, this means fairly shallow slides and falls in rock and debris. As stated previously, Keefer's (1984, 2002) analysis of data from worldwide earthquakes indicated that more than 85 percent of earthquake-triggered landslides are of this type. Documentation of landslides from several recent earthquakes has indicated that such failures commonly make up more than 90 percent of triggered landslides (Harp *et al.*, 1981; Harp and Jibson, 1995, 1996; Keefer and Manson, 1998; Jibson *et al.*, 2004, 2006). These types of landslides are well suited to rigid-block analysis because the brittle surficial material behaves rigidly, and the relatively thin landslide masses do not experience significant site response that would modify the incident ground motions. Thus, the large majority of seismically triggered landslides in natural slopes – shallow landslides in fairly brittle surface material – are modeled well by rigid-block analysis. Larger, deeper landslides, which are much less common but potentially more destructive, should be modeled using coupled analysis.

The discontinuity between analytical methods that work for thinner landslides in stiffer materials (rigid-block analysis) and thicker landslides in softer materials (coupled analysis) presents a challenge: where exactly is the boundary between these two types of slides? A T_s/T_m ratio of 0.1 has been proposed for this boundary, but the transition is not that abrupt in nature. A robust analytical method that accommodates a full range of period ratios would be ideal, but it might be elusive for simple mathematical reasons. Rathje and Antonakos (2011) proposed a simplified empirical method that is applicable across the full range of period ratios.

17.5 CONCLUSIONS

- The three families of analyses for assessing seismic slope stability each have their appropriate application.
- Pseudostatic analysis, because of its crude characterization of the physical process, should be used only for preliminary or screening analyses.
- Stress-deformation analysis is best suited to large earth structures such as dams and embankments; it is generally too complex and expensive for more routine applications.
- Permanent-displacement analysis provides a valuable middle ground between these end members. It is simple to apply and provides far more information than a pseudostatic analysis.
- Rigid-block analysis is suitable for thinner, stiffer landslides, which typically constitute the large majority of earthquake-triggered landslides.
- Coupled analysis is appropriate for deeper landslides in softer material, which could include large earth structures and deep landslides.
- Modeled displacements provide a useful index to seismic slope performance and must be interpreted using judgment and according to the parameters of the investigation.

REFERENCES

Ambraseys, N. N. and Menu, J. M. (1988). Earthquake-induced ground displacements. *Earthquake Engineering and Structural Dynamics*, 16, 985–1006.

Arias, A. (1970). A measure of earthquake intensity. In *Seismic Design for Nuclear Power Plants*, ed. R.J. Hansen. Cambridge, MA: Massachusetts Institute of Technology Press, pp. 438–483.

Blake, T. F., Hollingsworth, R. A. and Stewart, J. P. (2002). *Recommended Procedures for Implementation of DMG Special Publication 117: Guidelines for Analyzing and Mitigating Landslide Hazards in California*. Los Angeles, CA: Southern California Earthquake Center.

Bray, J. D. and Rathje, E. M. (1998). Earthquake-induced displacements of solid-waste landfills. *Journal of Geotechnical and Geoenvironmental Engineering*, 124, 242–253.

Bray, J. D. and Travasarou, T. (2007). Simplified procedure for estimating earthquake-induced deviatoric slope displacements. *Journal of Geotechnical and Geoenvironmental Engineering*, 133, 381–392.

(2009). Pseudostatic coefficient for use in simplified seismic slope stability evaluation. *Journal of Geotechnical and Geoenvironmental Engineering*, 135, 1336–1340.

California Division of Mines and Geology (1997). *Guidelines for Evaluating and Mitigating Seismic Hazards in California*. California Division of Mines and Geology, Special Publication 117.

California Geological Survey (2008). *Guidelines for Evaluating and Mitigating Seismic Hazards in California*. California Geological Survey, Special Publication 117A.

Clough, R. W. (1960). The finite element method in plane stress analysis. In *Proceedings of the 2nd Conference on Electronic Computation*. Pittsburgh, PA: American Society of Civil Engineers, Structural Division.

Clough, R. W. and Chopra, A. K. (1966). Earthquake stress analysis in earth dams. *ASCE Journal of the Engineering Mechanics Division*, 92, 197–211.

Del Gaudio, V., Pierri, P. and Wasowski, J. (2003). An approach to time-probabilistic evaluation of seismically induced landslide hazard. *Bulletin of the Seismological Society of America*, 93, 557–569.

Franklin, A. G. and Chang, F. K. (1977). *Earthquake Resistance of Earth and Rock-fill Dams*. US Army Corps of Engineers Waterways Experiment Station, Miscellaneous Paper S-71–17.

Goodman, R. E. and Seed, H. B. (1966). Earthquake-induced displacements in sand embankments. *ASCE Journal of the Soil Mechanics and Foundations Division*, 92, 125–146.

Harp, E. L. and Jibson, R. W. (1995). *Inventory of Landslides Triggered by the Northridge, California Earthquakes*. US Geological Survey, Open File Report 95–213.

(1996). Landslides triggered by the 1994 Northridge, California earthquake. *Bulletin of the Seismological Society of America*, 86 (Suppl. 1B), S319–S332.

Harp, E. L. and Wilson, R. C. (1995). Shaking intensity thresholds for rock falls and slides: Evidence from 1987 Whittier Narrows and Superstition Hills earthquake strong-motion records. *Bulletin of the Seismological Society of America*, 85, 1739–1757.

Harp, E. L., Wilson, R. C. and Wieczorek, G. F. (1981). *Landslides from the February 4, 1976 Guatemala Earthquake*. US Geological Survey, Professional Paper 1204-A.

Hynes-Griffin, M. E. and Franklin, A. G. (1984). *Rationalizing the Seismic Coefficient Method*. US Army Corps of Engineers Waterways Experiment Station, Miscellaneous Paper GL-84–13.

Idriss, I. M., Lysmer, J., Hwang, R. and Seed, H. B. (1973). *QUAD-4: A Computer Program for Evaluating the Seismic Response of Soil Structures by Variable Damping Finite Element Procedures*. University of California, Berkeley, Earthquake Engineering Research Center Report EERC 73–16.

Jibson, R. W. (1993). Predicting earthquake-induced landslide displacements using Newmark's sliding block analysis. *Transportation Research Record*, 1411, 9–17.

(2007). Regression models for estimating coseismic landslide displacement. *Engineering Geology*, 91, 209–218.

(2009). Using landslides for paleoseismic analysis. In *Paleoseismology*, 2nd edn., ed. J. P. McCalpin. New York: Academic Press, pp. 565–601.

Jibson, R. W. and Jibson, M. W. (2003). *Java Programs for Using Newmark's Method and Simplified Decoupled Analysis to Model Slope Performance during Earthquakes*. US Geological Survey, Open File Report 03–005, version 1.1.

Jibson, R. W. and Keefer, D. K. (1993). Analysis of the seismic origin of landslides: Examples from the New Madrid seismic zone. *Geological Society of America Bulletin*, 105, 521–536.

Jibson, R. W. and Michael, J. A. (2009). *Maps Showing Seismic Landslide Hazards in Anchorage, Alaska*. US Geological Survey, Scientific Investigations Map 3077.

Jibson, R. W., Harp, E. L. and Michael, J. A. (1998). *A Method for Producing Digital Probabilistic Seismic Landslide Hazard Maps: An Example from the Los Angeles, California, Area*. US Geological Survey, Open File Report 98–113.

(2000). A method for producing digital probabilistic seismic landslide hazard maps. *Engineering Geology*, 58, 271–289.

Jibson, R. W., Harp, E. L., Schulz, W. and Keefer, D. K. (2004). Landslides triggered by the 2002 M-7.9 Denali Fault, Alaska, earthquake and the inferred nature of the strong shaking. *Earthquake Spectra*, 20, 669–691.

(2006). Large rock avalanches triggered by the M-7.9 Denali Fault, Alaska, earthquake of 3 November 2002. *Engineering Geology*, 83, 144–160.

Jibson, R. W., Rathje, E. M., Jibson, M. W. and Lee, Y. W. (2012). *SLAMMER: Seismic Landslide Movement Modeled using Earthquake Records*. US Geological Survey, Techniques and Methods.

Keefer, D. K. (1984). Landslides caused by earthquakes. *Geological Society of America Bulletin*, 95, 406–421.

(2002). Investigating landslides caused by earthquakes: A historical review. *Surveys in Geophysics*, 23, 473–510.

Keefer, D. K. and Manson, M. W. (1998). Regional distribution and characteristics of landslides generated by the earthquake. In *The Loma Prieta, California, Earthquake of October 17, 1989: Landslides*, ed. D. K. Keefer. US Geological Survey, Professional Paper 1551-C, pp. C7–C32.

Keefer, D. K. and Wilson, R. C. (1989). Predicting earthquake-induced landslides, with emphasis on arid and semi-arid environments. In *Landslides in a Semi-Arid Environment*. Riverside, CA: Inland Geological Society 2, pp. 118–149.

Kramer, S. L. (1996). *Geotechnical Earthquake Engineering*. Upper Saddle River, NJ: Prentice Hall.

Kramer, S. L. and Smith, M. W. (1997). Modified Newmark model for seismic displacements of compliant slopes. *Journal of Geotechnical and Geoenvironmental Engineering*, 123, 635–644.

Lin, J. S. and Whitman, R. V. (1983). Earthquake-induced displacements of sliding blocks. *Journal of Geotechnical Engineering*, 112, 44–59.

Lin, M. L. and Wang, K. L. (2006). Seismic slope behavior in a large-scale shaking table model test. *Engineering Geology*, 86, 118–133.

Luzi, L. and Pergalani, F. (1999). Slope instability in static and dynamic conditions for urban planning: The "Oltre Po Pavese" case history (Regione Lombardia – Italy). *Natural Hazards*, 20, 57–82.

Makdisi, F. I. and Seed, H. B. (1978). Simplified procedure for estimating dam and embankment earthquake-induced deformations. *ASCE Journal of the Geotechnical Engineering Division*, 104, 849–867.

Malamud, B. D., Turcotte, D. L., Guzzetti, F. and Reichenbach, P. (2004). Landslides, earthquakes, and erosion. *Earth and Planetary Science Letters*, 229, 45–59.

Mankelow, J. M. and Murphy, W. (1998). Using GIS in the probabilistic assessment of earthquake-triggered landslide hazards. *Journal of Earthquake Engineering*, 2, 593–623.

Marcuson, W. F. (1981). Moderator's report for session on Earth Dams and Stability of Slopes under Dynamic Loads. In *Proceedings of the International Conference on Recent Advances in Geotechnical Earthquake Engineering and Soil Dynamics, Vol. 2*. St. Louis, MO, p. 1175.

Miles, S. B. and Ho, C. L. (1999). Rigorous landslide hazard zonation using Newmark's method and stochastic ground motion simulation. *Soil Dynamics and Earthquake Engineering*, 18, 305–323.

Miles, S. B. and Keefer, D. K. (2000). Evaluation of seismic slope-performance models using a regional case study. *Environmental and Engineering Geoscience*, 6, 25–39.

(2001). *Seismic Landslide Hazard for the City of Berkeley, California*. US Geological Survey, Miscellaneous Field Studies Map MF-2378.

Newmark, N. M. (1965). Effects of earthquakes on dams and embankments. *Geotechnique*, 15, 139–159.

Pradel, D., Smith, P.M., Stewart, J.P. and Raad, G. (2005). Case history of landslide movement during the Northridge earthquake. *Journal of Geotechnical and Geoenvironmental Engineering*, 131, 1360–1369.

Rathje, E.M. and Antonakos, G. (2011). A unified model for predicting earthquake-induced sliding displacements of rigid and flexible slopes. *Engineering Geology*, 122, 51–60.

Rathje, E.M. and Bray, J.D. (1999). An examination of simplified earthquake-induced displacement procedures for earth structures. *Canadian Geotechnical Journal*, 36, 72–87.

(2000). Nonlinear coupled seismic sliding analysis of earth structures. *Journal of Geotechnical and Geoenvironmental Engineering*, 126, 1002–1014.

Rathje, E.M. and Saygili, G. (2008). Probabilistic seismic hazard analysis for the sliding displacement of slopes: Scalar and vector approaches. *Journal of Geotechnical and Geoenvironmental Engineering*, 134, 804–814.

(2009). Probabilistic assessment of earthquake-induced sliding displacements of natural slopes. *Bulletin of the New Zealand Society of Earthquake Engineering*, 41, 18–27.

Rathje, E.M., Abrahamson, N.A. and Bray, J.D. (1998). Simplified frequency content estimates of earthquake ground motions. *Journal of Geotechnical Engineering*, 124, 150–159.

Rathje, E.M., Faraj, F., Russell, S. and Bray, J.D. (2004). Empirical relationships for frequency content parameters of earthquake ground motions. *Earthquake Spectra*, 20, 119–144.

Rodríguez, C.E., Bommer, J.J. and Chandler, R.J. (1999). Earthquake-induced landslides: 1980–1997. *Soil Dynamics and Earthquake Engineering*, 18, 325–346.

Saygili, G. and Rathje, E.M. (2008). Empirical predictive models for earthquake-induced sliding displacements of slopes. *Journal of Geotechnical and Geoenvironmental Engineering*, 134, 790–803.

Schnabel, P.B., Lysmer, J. and Seed, H.B. (1972). *SHAKE: A Computer Program for Earthquake Response Analysis of Horizontally Layered Sites*. University of California, Berkeley, Earthquake Engineering Research Center, Technical Report UCB/EERC-72/12.

Seed, H.B. (1979). Considerations in the earthquake-resistant design of earth and rockfill dams. *Geotechnique*, 29, 215–263.

Seed, H.B. and Martin, G.R. (1966). The seismic coefficient in earth dam design. *ASCE Journal of the Soil Mechanics and Foundations Division*, 92, 25–58.

Seed, H.B, Lee, K.L., Idriss, I.M. and Makdisi, R. (1973). *Analysis of the Slides in the San Fernando Dams during the Earthquake of Feb. 9, 1971*. Berkeley, CA: University of California Earthquake Engineering Research Center, Report EERC 73-2.

Stewart, J.P., Blake, T.F. and Hollingsworth, R.A. (2003). A screen analysis procedure for seismic slope stability. *Earthquake Spectra*, 19, 697–712.

Terzhagi, K. (1950). Mechanism of landslides. In *Application of Geology to Engineering Practice*, ed. S. Paige. New York: Geological Society of America, pp. 83–123.

Wang, X., Nie, G. and Wang, D. (2010). Research on relationship between landslides and peak ground accelerations induced by Wenchuan earthquake. *Chinese Journal of Rock Mechanics and Engineering*, 29, 82–89.

Wartman, J., Bray, J.D. and Seed, R.B. (2003). Inclined plane studies of the Newmark sliding block procedure. *Journal of Geotechnical and Geoenvironmental Engineering*, 129, 673–684.

Wartman, J., Seed, R.B. and Bray, J.D. (2005). Shaking table modeling of seismically induced deformations in slopes. *Journal of Geotechnical and Geoenvironmental Engineering*, 131, 610–622.

Wilson, R.C. and Keefer, D.K. (1983). Dynamic analysis of a slope failure from the 6 August 1979 Coyote Lake, California, earthquake. *Bulletin of the Seismological Society of America*, 73, 863–877.

Wieczorek, G.F., Wilson, R.C. and Harp, E.L. (1985). *Map Showing Slope Stability during Earthquakes in San Mateo County, California*. US Geological Survey, Miscellaneous Investigations Map I-1257-E.

Yegian, M.K., Marciano, E.A. and Ghahraman, V.G. (1991). Earthquake-induced permanent deformations: Probabilistic approach. *Journal of Geotechnical Engineering*, 117, 35–50.

18 Slow rock-slope deformation

FEDERICO AGLIARDI, GIOVANNI B. CROSTA, AND PAOLO FRATTINI

ABSTRACT

Giant, deep-seated gravitational slope deformations (DSGSDs) affecting entire high-relief valley walls are common in alpine areas, and influence the evolution of mountain landscapes and the related hazards. In the last few years, new characterization approaches and emerging technology shed new light on the occurrence, distribution, activity, and mechanisms of these spectacular slope failures. This chapter aims to provide an overview of alpine DSGSD, as well as a discussion of outstanding issues and future research needs.

We review the definition of DSGSD and its typical features, field evidence, settings, and mechanisms. We discuss the distribution and controls on DSGSD occurrence based on the analysis of the first-ever orogen-scale inventory of these phenomena, including over 900 occurrences in the European Alps. We demonstrate that DSGSDs are widespread in active orogenic settings and chiefly occur in formerly glaciated areas under significant structural controls, and show that alpine DSGSDs are often active phenomena with engineering significance. We use the classic case study of the Cima di Mandriole sackung to illustrate the complex relations between rock structure, the morphoclimatic evolution of alpine valleys, and man-made structures, and suggest future research needs.

18.1 INTRODUCTION

The evolution of landscapes in mountain areas and related hazards are influenced by a variety of different slope instability processes, characterized by different size, location along valley flanks, styles of activity and displacement rates, sensitivity to triggering factors, and coupling with the drainage network. Some of the largest instabilities are deep-seated, slow-moving landslides in rock. The geomorphic role and impact of these large rock-slope failures have been recognized for about a century, but only in the past few decades has the phenomenon been extensively studied.

Large slope instabilities, including rockslides, rock slumps, lateral spreads, rock mass creep, and sackung (or slope sags) result from a complex interaction of different controlling features and processes. Lithology and structure (Zischinsky, 1966; Benko and Stead, 1998; Agliardi *et al.*, 2001; Ambrosi and Crosta, 2006), topographic relief and shape, state of stress (Savage *et al.*, 1986; Miller and Dunne, 1996; Martel, 2000; Molnar, 2004), weather and climate (Evans and Clague, 1994; Ballantyne, 2002), seismicity (Solonenko, 1977; Radbruch-Hall, 1978), and human activity (Heim, 1932; Macfarlane, 2008; Zangerl *et al.*, 2010) are the most important contributing factors, and constrain the geometry of failures and their evolution in space and time. A variety of data, models, and theories have been presented to explain the locations of large slope failures, their geometry, time of occurrence, triggering mechanisms, rates and types of movement, and their effects.

The response of rock slopes to changes in geometry and distribution of stress is constrained by rock mass lithology and structure at different scales. If few sets of well-spaced discontinuities are present, rock mass strength is mainly determined by the intact rock properties, and individual discontinuities constrain the location and kinematics of slope failures (e.g., planar, wedge, or toppling failures). Conversely, if several sets of closely spaced discontinuities are present, rock masses tend to behave as continua – their equivalent strength and deformability are controlled by the properties of both intact rock and the discontinuities, and can be estimated by empirically describing rock mass structure (Hoek *et al.*, 2002). In this case, the kinematics of rock-slope failure depend on rock mass strength and on slope-scale structural features, allowing for a variety of failure geometries.

Landslides: Types, Mechanisms and Modeling, ed. John J. Clague and Douglas Stead. Published by Cambridge University Press.

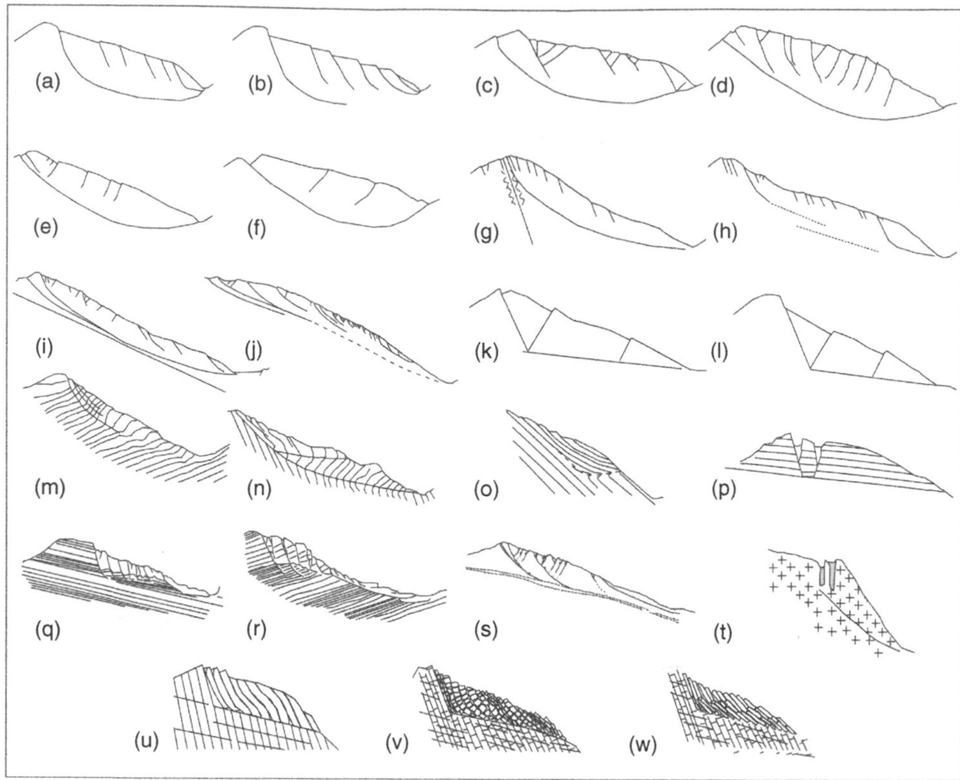

Fig. 18.1. Mechanisms of large-scale rock-slope failure proposed in the literature, ordered on the basis of failure geometry and structural controls. (a), (b), (c), (f) Bois *et al.* (2008); (d) Mahr (1977); (e), (g), (h), (i) Ambrosi and Crosta (2006); (j) Agliardi *et al.* (2001); (k), (l) Hutchinson (1988); (m), (n) Zischinsky (1966); (o), (p), (t) Chigira (1992); (q), (r) Nemcok (1972); (s) Agliardi *et al.* (2009b); (u), (v), (w) Kieffer (1998).

Master joints, faults, and folds constrain the overall stability of large rock slopes (Fig. 18.1g, i, j, s), where local relief can exceed 1000 m. In such slopes, large-scale structures or anisotropy can lead to the development of either curved (Fig. 18.1a–g), compound (Fig. 18.1h–l), or planar failure surfaces (Fig. 18.1m–r; Sjoberg, 1999; Nichol *et al.*, 2002). Large-scale toppling is common on slopes with persistent discontinuities that dip steeply into the slope. Rigid-block toppling can occur along a gently dipping basal failure surface, whereas flexural toppling dominates where the tensile strength of rock material is low enough to allow for flexure, and shear strength along discontinuities is low. Underdip toppling and slumping failures (Cruden and Hu, 1994; Kieffer, 1998) can occur in slopes with persistent discontinuities that dip steeply downslope (Fig. 18.1u–w). In this case, backward rigid or flexural block rotation can occur in association with different types of slumping (flexural, block flexural, block slumping, and kink band slumping); the type of slumping depends on rock strength and discontinuity spacing and strength (Fig. 18.1). Where the inclination of discontinuities approaches that of the slope, buckling failures can occur (Cruden and Hu, 1994). The scale and timing of landslides depend on the interplay between rock mass behavior and the types and histories of external forces. The result can be small, frequent failures (e.g., rockfalls), large-scale sudden or progressive catastrophic failures (e.g., rockslides and rock avalanches), or slow, non-catastrophic deep-seated slope deformations.

This chapter deals with slow, deep-seated, rock-slope failures, commonly referred to as deep-seated gravitational slope deformation (DSGSD), with special focus on alpine environments. DSGSD has been recognized for decades, but the phenomenon was poorly understood until about 30 years ago, when new characterization, monitoring, and modeling tools first became available and a multidisciplinary research approach began to be used (Varnes *et al.*, 1989; Bovis, 1990; Chigira, 1992; Agliardi *et al.*, 2001). The chapter provides an overview of alpine DSGSD, as well as a discussion of outstanding issues and future research needs.

18.2 DEEP-SEATED GRAVITATIONAL SLOPE DEFORMATION

18.2.1 DEFINITION AND EVIDENCE

DSGSDs are large mass movements on high-relief valley walls that extend from near the valley floor to, or beyond, the ridge crest. They have been found in most rock types and are generally characterized by discontinuous or poorly defined lateral boundaries, large volumes (commonly >0.5 km^3) and thicknesses, conspicuous surface features, and relatively low rates of movement over long periods (Varnes *et al.*, 1990; Ambrosi and Crosta, 2006). Most DSGSDs are sackungen (i.e., saggings; see Zischinsky, 1966; Hutchinson, 1988) in metamorphic, igneous, or layered sedimentary rocks in alpine areas. Lateral spreads (Fig. 18.1p–r; see Hutchinson, 1988; Cancelli and Casagli, 1995) are more common in areas with horizontal or gently dipping, thick-bedded sedimentary successions with strong competence

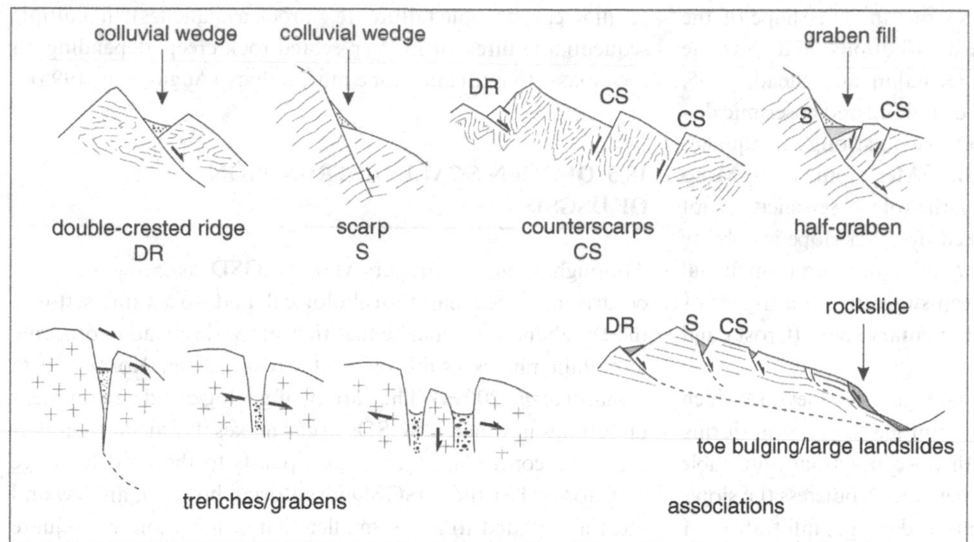

Fig. 18.2. Morpho-structural features diagnostic of deep-seated gravitational slope deformation (DSGSD) phenomena, related kinematic significance, and typical associations (modified after Agliardi *et al.*, 2001).

contrasts (e.g., sandstones overlying clays and clayey shales in the Italian Apennines).

The most common indicator of DSGSD is the presence of surface deformational features of gravitational origin (Agliardi *et al.*, 2001). Tensional features include scarps, trenches, grabens, and double or multiple ridges on the upper parts of the deforming slopes (Figs. 18.1 and 18.2). Counterscarps (antislope scarps) can be common in middle slope sectors; and compressional features, including bulging, buckling folds, and highly fractured rock masses, occur at the slope toe (Chigira, 1992). The toes of many DSGSDs lie below valley floors and are covered by alluvial or lacustrine fills. Secondary slope instabilities can be present in the lower parts of DSGSDs, in some cases leading to large catastrophic slope failures (rockslides and rock avalanches; Fig. 18.2). Surface features are often linear and resemble tectonic lineaments; in fact the lineaments, although gravitational in origin, may coincide with inherited tectonic fractures. Nevertheless, surficial DSGSD features are usually less persistent than tectonic features, have a more arcuate plan shape, and occur as swarms of scarps with associated apparent vertical offsets up to some tens of meters (Radbruch-Hall, 1978; McCalpin, 1999; Persaud and Pfiffner, 2004; Agliardi *et al.*, 2009a).

Interpreting the kinematic significance of individual surface deformational features, their spatial patterns, and their associations is crucial for understanding the overall geometry and kinematics of DSGSD, because deep exposures or site investigation data for these slope instabilities are rare. Deep DSGSD basal shear zones, however, have been observed or inferred at some dam sites (Gignoux and Barbier, 1955; Desio, 1961, Barla *et al.*, 2010) and in borehole logs and along tunnels (Ambrosi and Crosta, 2006; Bonzanigo *et al.*, 2007; Zangerl *et al.*, 2010). The internal geometry of some DSGSDs has been inferred from geophysical surveys (Brückl and Parotidis, 2005).

Although DSGSDs can be single features, two or more rock-slope failures located side-by-side and sharing part of the same boundary are common. In some cases, the boundary coincides with a tributary stream. In other cases, composite DSGSDs include rock-slope failures of different ages, with truncated or reshaped surface features providing evidence for two or more stages of instability. Such evidence indicates that some DSGSDs may span more than one glacial cycle and may be reactivated after successive glaciations. Some DSGSDs are nested features (Ambrosi and Crosta, 2006) with major lineaments bounding a series of conterminous, smaller DSGSDs. These features can extend over areas of tens of square kilometers and are commonly dissected by streams, forming sub-basins with divides contained within the larger area of instability (Ambrosi and Crosta, 2011). Conversely, persistent lineaments crossing several large DSGSDs, with no changes in direction or displacements, are commonly related to major tectonic features, such as thrusts or nappe boundaries (Ambrosi and Crosta, 2006; Agliardi *et al.*, 2009b).

18.2.2 CONTROLS AND MECHANISMS OF DSGSD

DSGSD occurrence is closely related to specific geologic and structural features, such as bedding, foliation, joints, and faults, and to certain topographic situations, such as valley bends or sites where the local slope gradient changes; they are also common in tectonically active areas (Crosta and Zanchi, 2000; Agliardi *et al.*, 2009b; Ambrosi and Crosta, 2011).

Several processes have been proposed as potential causes and triggers of DSGSD. Topographic stresses in both isotropic and anisotropic materials (Varnes *et al.*, 1989; Savage, 1994; Molnar, 2004) can interact with tectonic or locked-in stresses (Savage *et al.*, 1986; Miller and Dunne, 1996; Ambrosi and Crosta, 2011) to produce concentrations of shear stress at the toe of a slope, and extensive tensile damage along the crest. These effects scale with topographic relief and can lead to the instability of high slopes, which can be either catastrophic, progressive, or locked, depending on rock mass strength. The shape of the topography also constrains stress distributions, and thus the onset and geometry of DSGSD (Kinakin and Stead, 2005; Ambrosi and Crosta, 2011). Ground shaking and coseismic displacements along faults during moderate to strong earthquakes can accelerate or reactivate DSGSDs (McCalpin, 1999; Moro *et al.*, 2007). In other cases, however, the role of seismicity is not clear, may be negligible, or is masked by other slope instability or erosional processes (Agliardi *et al.*, 2009b). Long-term fluvial erosion at the slope toe has also been suggested as a trigger of DSGSD, particularly in layered sedimentary rocks (Crosta and Zanchi, 2000).

Alpine DSGSDs are present in many glacial valleys, in which slopes have been steepened by recurrent glacier erosion during the Pleistocene. Glaciers can predispose potentially unstable slopes to later DSGSD when they melt and debuttress the slope, and by fracture unloading and tensile damage, infiltration of meltwater into discontinuities in rock, changes in the groundwater regime, and increases in water pressure due to permafrost thaw (Crosta, 1996; Ballantyne, 2002; Holm *et al.*, 2004; Arsenault and Meigs, 2005). The efficacy of these effects is supported by numerical modeling (Brückl, 2001), cross-cutting relationships between DSGSD surface features and glacial or paraglacial features, and dating (Agliardi *et al.*, 2009b). Focused or enhanced erosion also occurs on slopes with weak lithologies, structures with favorable orientations, antecedent slide deposits, in-situ stress concentrations, or high groundwater pressures (Augustinus, 1995; Harbor, 1995).

At the scale of entire slopes, steeply dipping master fractures appear to be particularly important in constraining the localization, geometry, and surface expression of DSGSD (Agliardi *et al.*, 2001; Hippolyte *et al.*, 2006). First-order, gently dipping features such as thrust planes and nappe boundaries may control the location of shear planes and the pattern of surface features (Agliardi *et al.*, 2009a).

Geomorphological and geochronological evidence suggests a delayed or long-term evolution of DSGSD (Bovis, 1990; Ballantyne, 2002; Prager *et al.*, 2008), which may be related to the concept of progressive failure. Terzaghi (1972) emphasized that rock mass strength depends on both the frictional shear strength of discontinuity surfaces and the cohesion of intact rock bridges. Progressive failure of rock slopes requires that rock bridges fail sequentially, allowing feasible kinematic release mechanisms to develop. Eberhardt *et al.* (2004) suggested that geometric and kinematic changes experienced by natural slopes during deglaciation are less rapid than in engineered slopes and that the strength degradation process can last for thousands of years following deglaciation. Shear planes should develop initially at the toe of an unstable rock slope, where the stresses are highest, and propagate upward. However, tension cracks can appear at the top of the slope long before catastrophic failure, suggesting that the involved rock mass is dilating and deforming internally. Ambrosi and Crosta (2011) argued that instability can begin at the slope crest just after the onset of deglaciation by downwasting of glacier ice. This process may result in catastrophic failures (e.g., rock avalanches), in multiple sequential failures, or in deep-seated rock creep, depending on rock mass strength and slope morphology (Augustinus, 1995).

18.3 OROGEN-SCALE DISTRIBUTION OF DSGSD

Although some researchers view DSGSD as being rare and occurring in peculiar morphological and structural settings, there is abundant evidence that they are widespread in orogenic mountain ranges (Ambrosi and Crosta, 2006; Korup, 2006; Agliardi *et al.*, 2009a). The variety of geologic and environmental settings in which DSGSDs occur makes it difficult to understand the controlling factors and points to the need for large inventories. Existing DSGSD inventories, however, are few and most are limited to areas smaller than a few thousand square kilometers (Mortara and Sorzana, 1987; APAT, 2007; Korup and Schlunegger, 2007; Agliardi *et al.*, 2009a). These inventories have been completed by different institutions or researchers using different criteria, definitions of phenomena, mapping methods, and topographic base maps. Most DSGSD mapping is based on the recognition of distinctive surface features and criteria specific to individual areas. Consequently, many inventories are heterogeneous and incomplete, and include evidence of processes of different type and extent.

To overcome these limitations, we recently prepared an inventory of DSGSDs and large landslides covering most of the European Alps (Fig. 18.3). The inventory covers an area of about 103,000 km^2 in Italy, France, Switzerland, and Austria. It was produced using available satellite imagery and validated with data from the literature, aerial photographs, and field data. DSGSDs were recognized and mapped based on the following criteria: involvement of entire valley flanks; presence of gravitational surface features typical of DSGSD; geomorphological evidence of slope deformation and displacements along individual structures; evidence for gravitational reactivation of inherited tectonic features.

Large landslides (rockslides and rock avalanches) were also mapped and compared to the DSGSDs. Landslides were distinguished from DSGSDs based on the following criteria: smaller breadth and width-to-length ratio; limited length compared to the slope length; evidence of greater internal deformation and rock mass damage; larger runout; geomorphological evidence of past rapid displacement or present-day activity.

We mapped 904 DSGSDs, covering a total area of 5472 km^2, which is about 5 percent of the investigated area. In contrast, 791 large landslides cover 1017 km^2, or about 1 percent of the area. Individual DSGSDs range in area from 0.2 to 108 km^2. Most of the smaller ones are located in minor tributary valleys, or within larger, complex features. DSGSDs preferentially occur in moderately strong, anisotropic rock masses that are able to sustain steep high-relief slopes without failing catastrophically (Fig. 18.3). They are most common in foliated metamorphic rocks (metapelite, paragneiss; 12.7 percent areal density), flysch

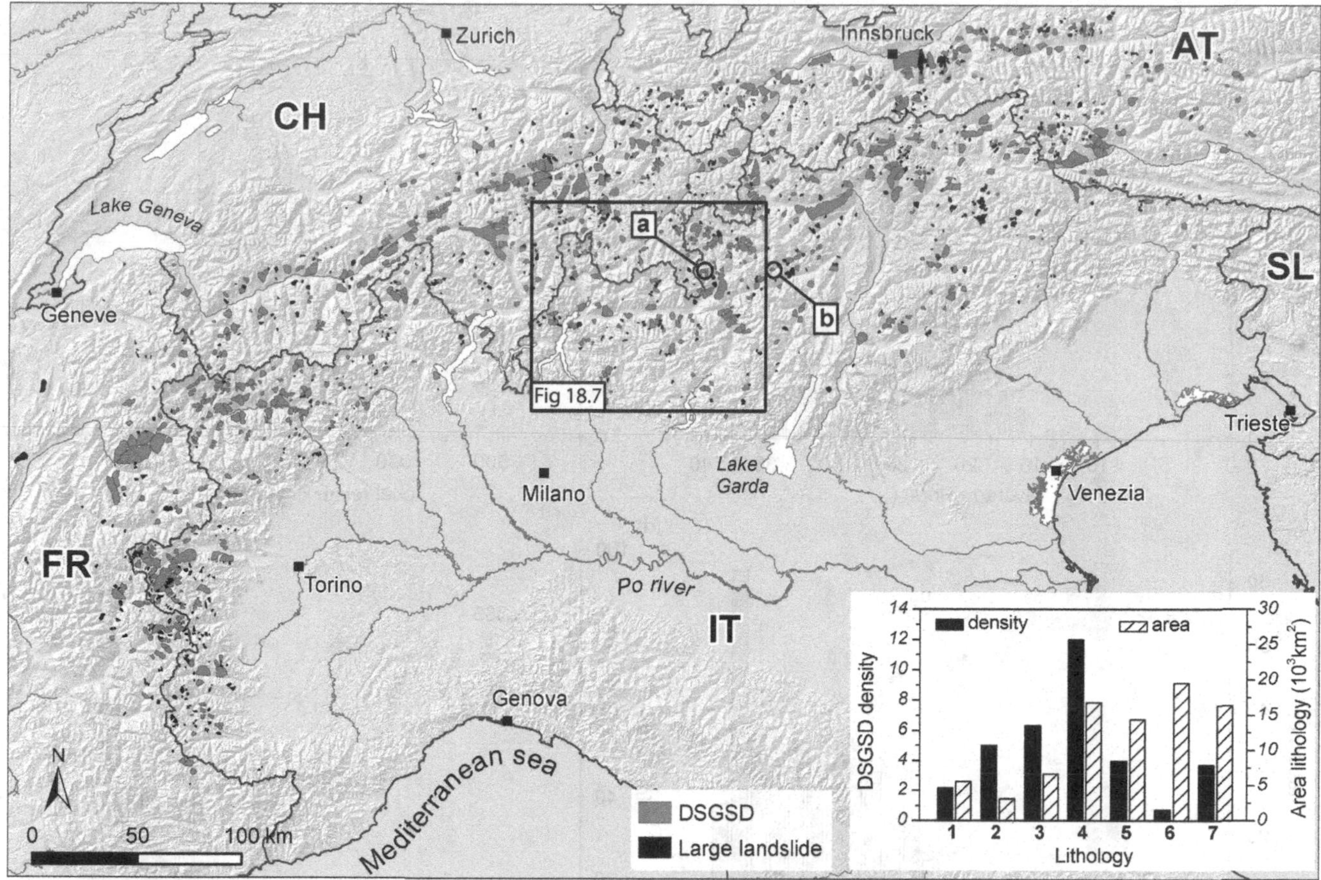

Fig. 18.3. Inventory map of DSGSD and large landslides in the European Alps. The areal density of DSGSD occurring in areas with different dominant lithologies is portrayed in the histogram (1: granitoid/metabasite; 2: volcanic rock; 3: orthogneiss; 4: metapelite; 5: flysch-type rocks; 6: carbonate rock; 7: large Quaternary accumulation; geological datasets after Swisstopo, 2007, and Bigi *et al.*, 1983: available for 725 DSGSD). The area shown in Figure 18.7 is outlined, and the locations of the (a) Mt. Farinaccio (Fig. 18.6) and (b) Cima di Mandriole DSGSDs are shown.

or layered sandstones and marls (3.9 percent), orthogneiss (6.5 percent), and volcanic rocks (about 5 percent). Few DSGSDs occur in carbonate or granitoid rocks. Morphometric analysis of the 90-m SRTM DEM revealed that slope inclination values in DSGSD areas follow a normal distribution, with mean of 24.5° (Fig. 18.4).

The DSGSDs in our inventory occur in areas with local relief (i.e., difference between maximum and minimum elevations in a circular kernel with a 5 km radius) between 1000 and 3000 m (Fig. 18.4), in agreement with results obtained in smaller study areas (Agliardi *et al.*, 2009a). These high values are typically found in tectonically active mountain belts and have been suggested as threshold values for large-scale erosion by large landslides (Montgomery and Brandon, 2002; Korup *et al.*, 2007). The role of tectonic processes is further illustrated by comparing the distribution of DSGSDs with the pattern of the present-day rock uplift (data from Kahle *et al.*, 1997). Although this comparison can be made for only part of the inventory area (20,600 km^2, 354 DSGSDs; Fig. 18.4), most DSGSDs in the Alps occur in areas with present-day uplift rates >1 mm per year; their frequency increases with uplift rate. The inventory also shows a clustering of DSGSDs (Fig. 18.3) along first-order tectonic features, mainly nappe boundaries (e.g., Isère, Susa, Rhone and Rhine, upper Ticino, upper Valtellina, and Venosta valleys). This relationship suggests that regional structures exert a passive control on the occurrence of DSGSD (Crosta and Zanchi, 2000; Ambrosi and Crosta, 2006; Hippolyte *et al.*, 2006).

During the Last Glacial Maximum, most of the areas in the Alps affected by DSGSD were covered by glaciers that reached elevations exceeding 2000 m and up to 3000 m over major valleys (Jäckli, 1970; Van Husen, 1997; Florineth, 1998; Kelly *et al.*, 2004). The upper parts of the major alpine valleys were affected by repeated re-advances of glaciers after the Last Glacial Maximum (Fig. 18.4). This observation suggests a causal relationship between glaciation and DSGSD, which is supported by field observations, absolute age dating, and numerical modeling (Augustinus, 1995; Brückl, 2001; Agliardi *et al.*, 2009b).

18.3.1 AREA–FREQUENCY STATISTICS

The mechanisms responsible for DSGSD have long been debated. These features have been interpreted to be expressions

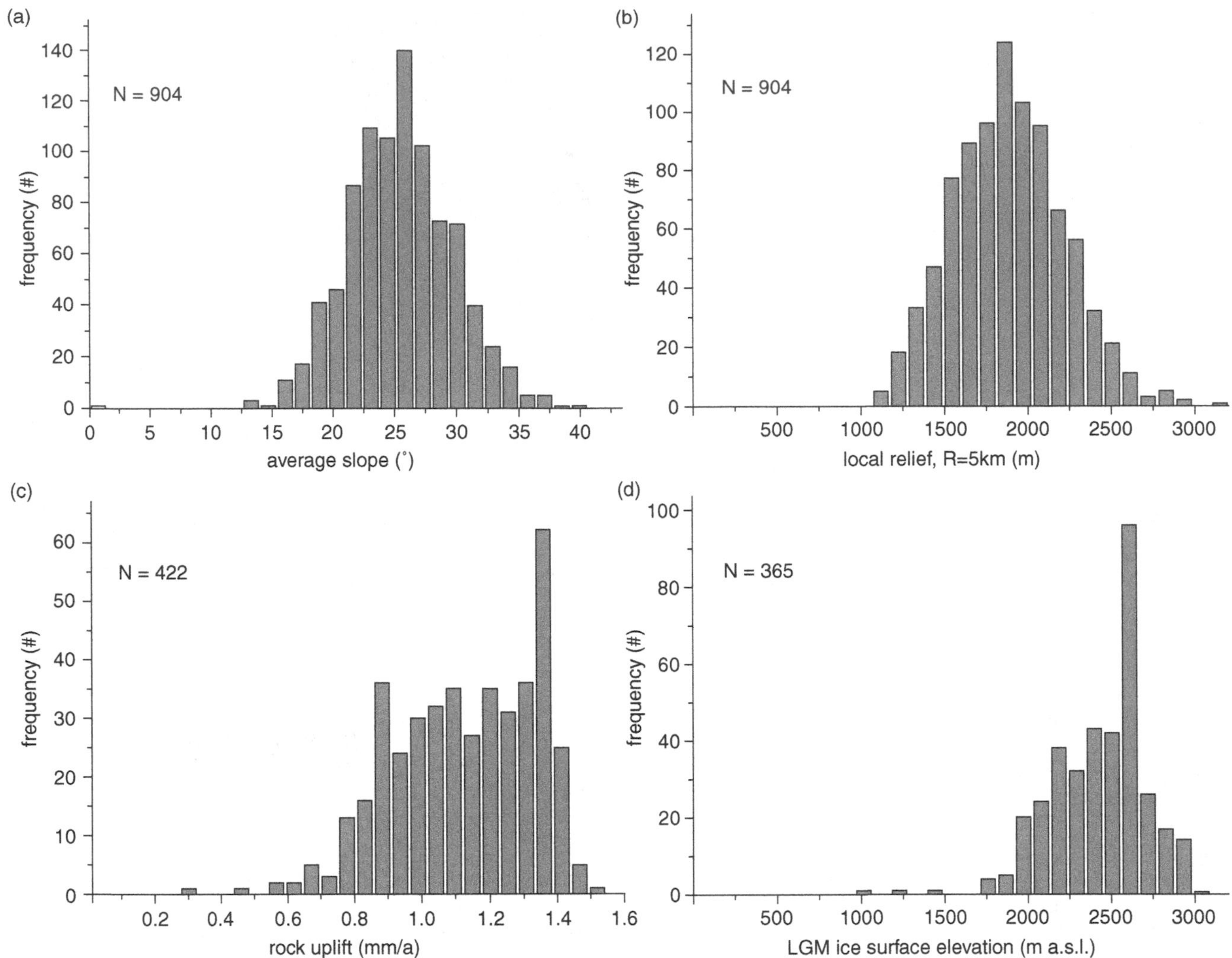

Fig. 18.4. Frequency distributions of DSGSD in the orogen-scale inventory versus: (a) average slope inclination inside affected area (topography: 90-m SRTM DEM); (b) local relief evaluated in circular moving windows with a radius of 5 km (topography: 90-m SRTM DEM); (c) modern rock uplift rate (data from Kahle *et al.*, 1997); (d) glacier surface elevations at the Last Glacial Maximum (data from Florineth, 1998, and Kelly *et al.*, 2004).

of a type of landslide (Hutchinson, 1988) or differentiated from landslides as "mass rock creep" phenomena (Radbruch-Hall, 1978). DSGSD surface lineaments have been also interpreted as active faults (Persaud and Pfiffner, 2004) or postglacial rebound faults (Ustaszewski *et al.*, 2008).

Here we compare the area–frequency distributions of DSGSD and large landslides included in the orogen-scale inventory. Several studies, carried out on landslides in different regions, have established some points of consensus on the relation between area and frequency for landslides (Hovius *et al.*, 2000; Stark and Hovius, 2001; Malamud *et al.*, 2004). First, the landslide size–frequency distribution exhibits power-law, or fractal, scaling for landslides larger than a size threshold. Second, below this threshold, the distribution deviates from a power law ("roll-over").

We examined area–frequency distributions of both DSGSD and large landslides by developing logarithmically binned, non-cumulative size–frequency distributions that report frequency density ($f = dN/dA$) as a function of landslide planar area A (dN = number of landslides with an area between A and $A + dA$). For landslides larger than a certain area, the size–frequency distribution was interpolated with a power-law function:

$$f(A) = aA^{-b}. \quad (18.1)$$

The size–frequency distributions of orogen-scale inventories follow a power-law behavior above the roll-over with exponent, $-b$, equal to 2.48 for DSGSD and 2.49 for large landslides (Fig. 18.5). These values are similar to those reported in the landslide literature (e.g., 2.11–2.48, Stark and Hovius, 2001; 2.4, Malamud *et al.*, 2004). The deflection from a power law occurs at about 1.5 km^2 for large landslides and 6 km^2 for DSGSD. This result suggests that the roll-over in the DSGSD inventory is due to a smooth transition from DSGSD to large landslides, thus representing a continuum among slope instability phenomena. For areas above the roll-over, deformation involves the entire slope, with conspicuous surface features that lead one to classify the

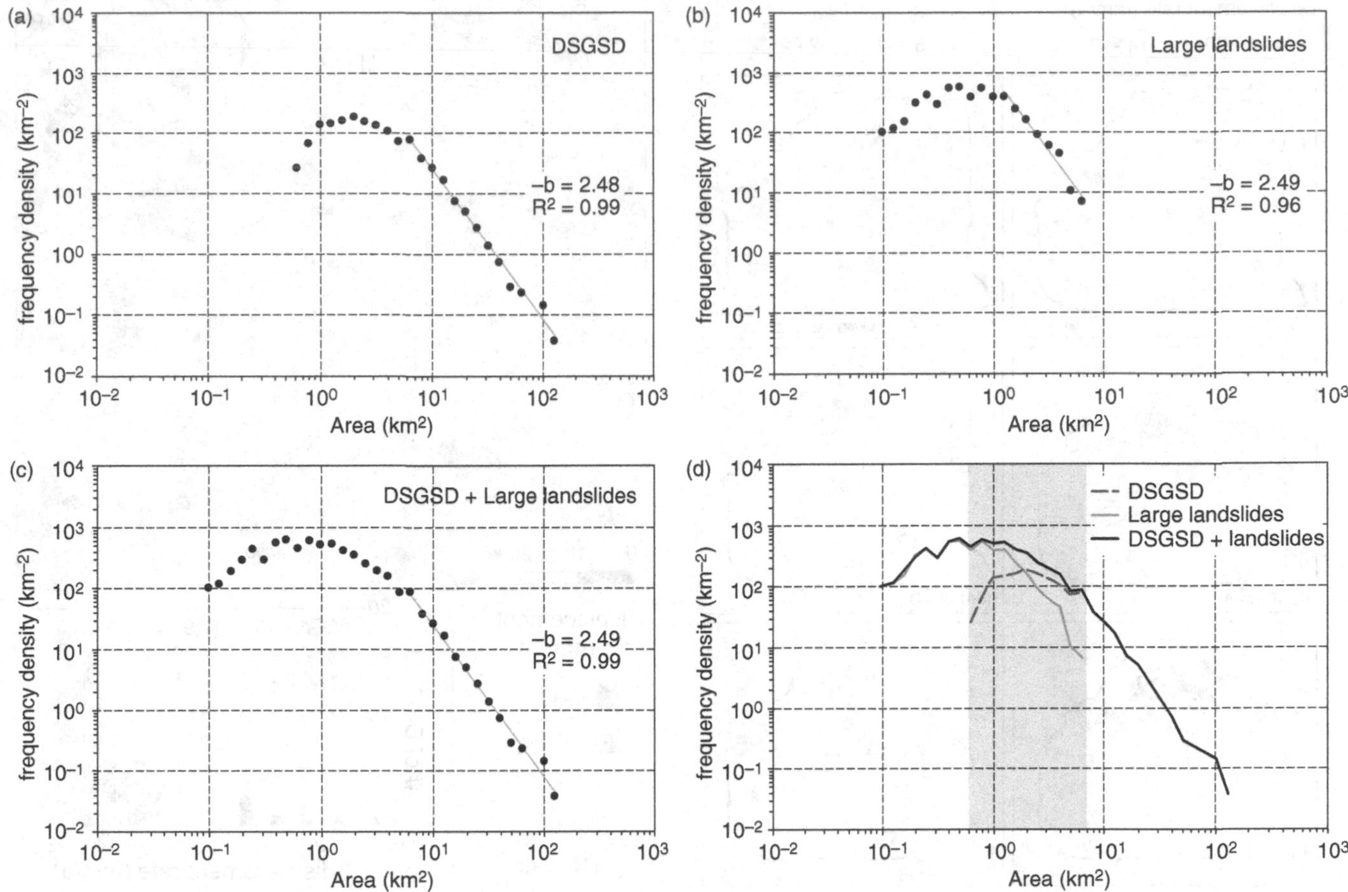

Fig. 18.5. Frequency density distribution of landslide size in the orogen-scale inventory of: (a) DSGSD (n = 904; roll-over ≈ 6 km²; modal value ≈ 2 km²); (b) large landslides (n = 791; roll-over ≈ 1.5 km²; modal value ≈ 0.5 km²); (c) joined DSGSD and large landslides (roll-over ≈ 6 km²; modal value ≈ 0.5 km²). (d) Comparison of different inventories and identification of the transition zone.

phenomenon as DSGSD. A transitional zone, marked by the coexistence of DSGSD and large landslides, occurs for instabilities between 1.5 and 6 km². These instabilities can be classified in different ways depending on their surface expression, geologic constraints, and degree of involvement of first-order geologic structures.

18.4 ACTIVITY OF DSGSDS

Until recently, DSGSDs were considered by many geologists to be relict instabilities that are not active under present climatic conditions. They were defined as old landslides or paleolandslides: scientific curiosities of little or no practical interest. Ample geomorphological and geochronological evidence, however, points to a lengthy evolution of DSGSD, with long-lasting periods of activity and reactivations, in some instances punctuated by catastrophic failures. Late-glacial triggering or continuation of DSGSD has been demonstrated with field evidence. The evidence suggests that the largest movements occurred soon after DSGSD onset, but that rates of movement varied over different parts of the instabilities and over periods of thousands of years. Beck (1968) argued that displacements along sackung scarps were episodic, whereas McCalpin and Irvine (1995) concluded that movements were continuous along individual discontinuities, with long-term displacement rates of <1 mm per year. Geodetic monitoring of DSGSD in North America over periods of up to 20 years (Varnes *et al.*, 1990; Bovis and Evans, 1995) has revealed displacements at rates up to 10 mm per year. Some infrastructure that is 50 or more years old on DSGSDs in Europe has experienced significant deformation (Ambrosi and Crosta, 2006; MacFarlane, 2008; Barla *et al.*, 2010; Zangerl *et al.*, 2010), but reliable sets of displacement data are available only for small areas of a few DSGSDs. Recently, however, new technologies, such as differential SAR interferometry (DinSAR; Gabriel *et al.*, 1989), have been successfully used to monitor DSGSD (Saroli *et al.*, 2005; Ambrosi and Crosta, 2006; Colesanti *et al.*, 2006). The "permanent scatterers" (PS) technique (Ferretti *et al.*, 2001) has provided time histories of displacements since 1991–1992 with a precision of 0.1–1.0 mm per year.

We analyzed the state of activity of a subset of our Alpine inventory – 101 DSGSDs in the Lombardy region of northern Italy (*ca.* 11,000 km² in the central Italian Alps) – using nine PS-DinSAR and SqueeSAR datasets (Ferretti *et al.*, 2011) (Table 18.1). Permanent scatterers within secondary debris accumulations, such as scree or rock glaciers, which could provide misleading information, were removed from the sample.

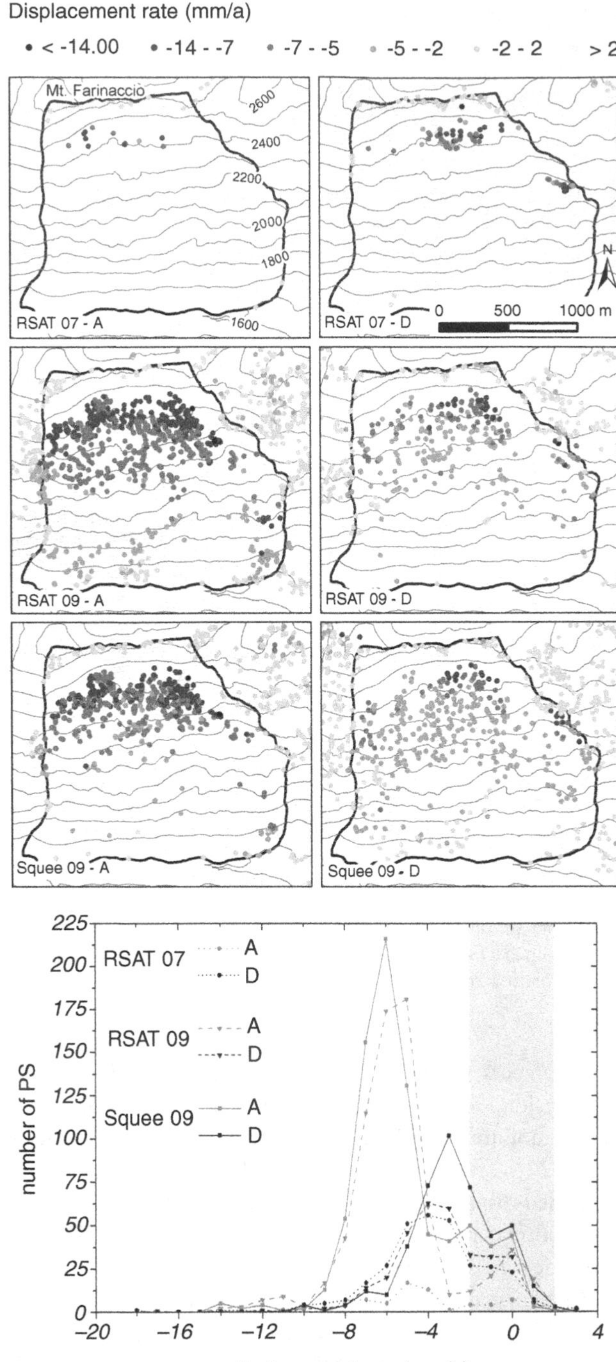

Fig. 18.6. Permanent scatterer (PS) data for the Mt. Farinaccio DSGSD (location in Fig. 18.3) and displacement rate frequency distribution for different PS datasets. RSAT 09-A was selected as the reference dataset for the statistical analysis of this DSGSD.

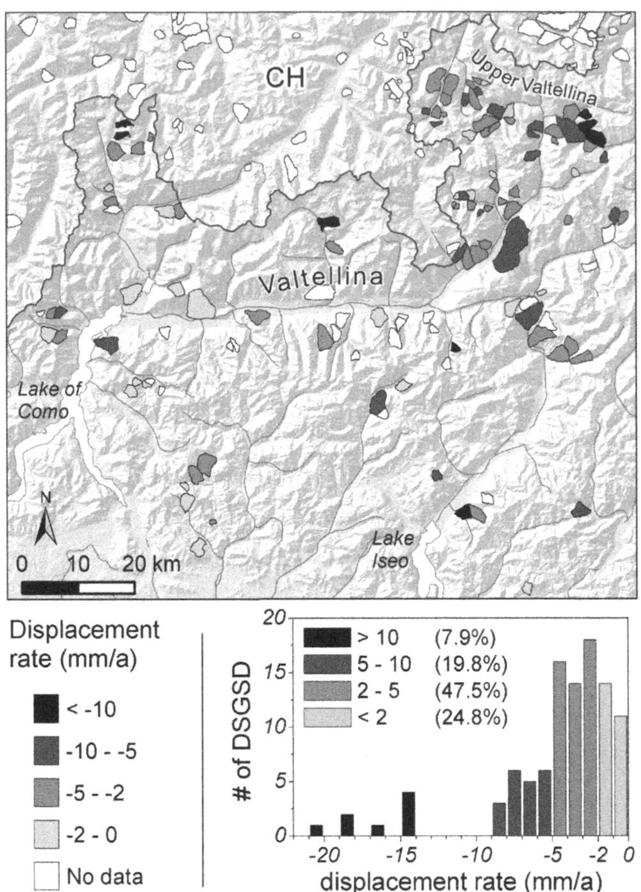

Fig. 18.7. DSGSD in the Lombardy region reclassified according to 25th percentile of PS InSAR displacement rates. Relative frequencies have been estimated excluding DSGSD without PS data (i.e., 23 percent of the sample).

For each mapped DSGSD, the frequency distribution of displacement rates was derived from the permanent scatterers (Fig. 18.6). Most frequency distributions are unimodal, with modal values differing significantly between the data acquired in ascending and descending satellite geometries. We selected the most informative datasets for statistical analysis by considering the number of permanent scatterers and the distribution of displacement rates outside the range of uncertainty (−2 to +2 mm per year) (Fig. 18.6). Minimum and maximum displacement rate values in each DSGSD area were discarded because of the possibility of outlier values in the datasets. Because most permanent scatterers are characterized by negative displacement rates (i.e., distances from the sensor increase over time along the sensor "line of sight," the displacement rate value corresponding to the lowest quartile (i.e., 25th percentile) was selected as the most representative descriptor of the degree of activity of each DSGSD. The measurements were locally validated by comparing them to long-term, ground-based monitoring data, collected over several decades at hydropower facilities (Colesanti *et al.*, 2006; Ambrosi and Crosta, 2006).

Our analysis showed that 75.2 percent of the DSGSD subset for which permanent scatterer data are available are active; 7.9 percent have 25th-percentile displacement rates in excess of 10 mm per year (Fig. 18.7), with local values of up to 30 mm per year on ridge crests. The most active DSGSDs are in the upper Valtellina Valley, where land uplift rates and seismicity are highest (Agliardi *et al.*, 2009a). Conversely, most of the DSGSDs along the main axis of Valtellina have very low displacement rates.

Table 18.1. *Basic information on PS-DinSAR datasets used for the analysis of DSGSD displacement rate (data courtesy Tele-Rilevamento Europa s.r.l. for Regione Lombardia).*

Dataset	Satellite information	Acquisition geometry	Time interval	#PS[a]
ERS1–2 – A	ERS 1, ERS 2	Ascending	1992–2000	231224
ERS1–2 – D	ERS 1, ERS 2	Descending	1992–2000	573821
RSAT 07 – A	RSAT-S3	Ascending	2003–2007	575463
RSAT 07 – D	RSAT-S3	Descending	2003–2007	736669
RSAT 09 – A	RSAT-S3	Ascending	2003–2009	221361
RSAT 09 – D	RSAT-S3	Descending	2003–2008	89440
Squee 09 – A	RSAT-S3	Ascending	2003–2009	262336
Squee 09 – D	RSAT-S3	Descending	2003–2008	135552

[a] Number of individual permanent scatterers, i.e., pixels corresponding to stable natural reflectors of the radar signal that remain coherent over a long time (Ferretti *et al.*, 2001).

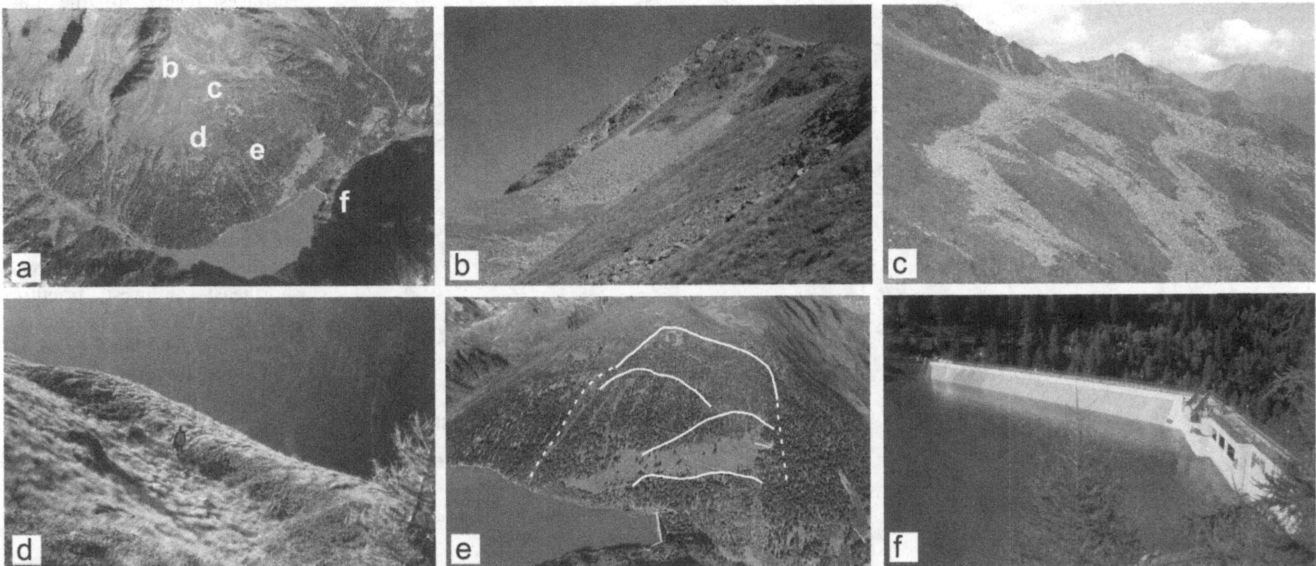

Fig. 18.8. (a) Orthophoto and close-up photographs of the Cima di Mandriole and the Pian Palù reservoir. The locations of the featured insets (b–f) are shown. (b) Western headscarp of the DSGSD trending northeast–southwest, with an estimated vertical throw about 100 m. (c) Eastern scarp swarm of the DSGSD trending northeast–southwest. (d) Gravitational counterscarp in the middle slope sector. (e) Large active rockslide occurring in the lower slope sector and impinging on the dam. (f) Pian Palù gravity dam.

18.5 THE CIMA DI MANDRIOLE CASE STUDY

Several emblematic DSGSD case studies have been described in the literature. One of them, the Cima di Mandriole sackung in Italy, illustrates the complex relations between rock structure, the morphoclimatic evolution of alpine valley slopes, and man-made structures. The site has been studied since the 1950s, and characterized in detail by Crosta *et al.* (2000), based on structural geology, geomorphology, geomechanics, monitoring, and numerical modeling. The DSGSD is located in the upper Peio Valley (Trentino-Alto Adige; Fig. 18.3) and extends over about 6 km^2. The bedrock at the site is part of the Austroalpine Campo nappe and consists of paragneiss, mylonitic orthogneiss, and quartzite. The rocks were isoclinally folded during the Variscan orogeny and then deformed by east-northeast-trending asymmetric folding during the Alpine orogeny. An antiformal structure is present at the slope scale, with foliation dipping steeply downslope in the middle and lower slope sectors. Younger brittle structures include east-northeast and northwest-trending fractures and normal and oblique dip–slip faults.

The Cima di Mandriole slope has a relief of about 1000 m, and its toe impinges against the Pian Palù reservoir and its 53-m-high dam (Figs. 18.8 and 18.9). The dam was designed in the 1940s as a dry-stone masonry gravity dam with concrete cut-off walls. During the early stages of construction, site investigations, including boreholes and 15 exploratory adits, revealed heavily fractured rock masses on the left bank of the dam site (Figs. 18.8 and 18.10b; Desio, 1961). In response, the design was modified and the dam was built using large concrete blocks with

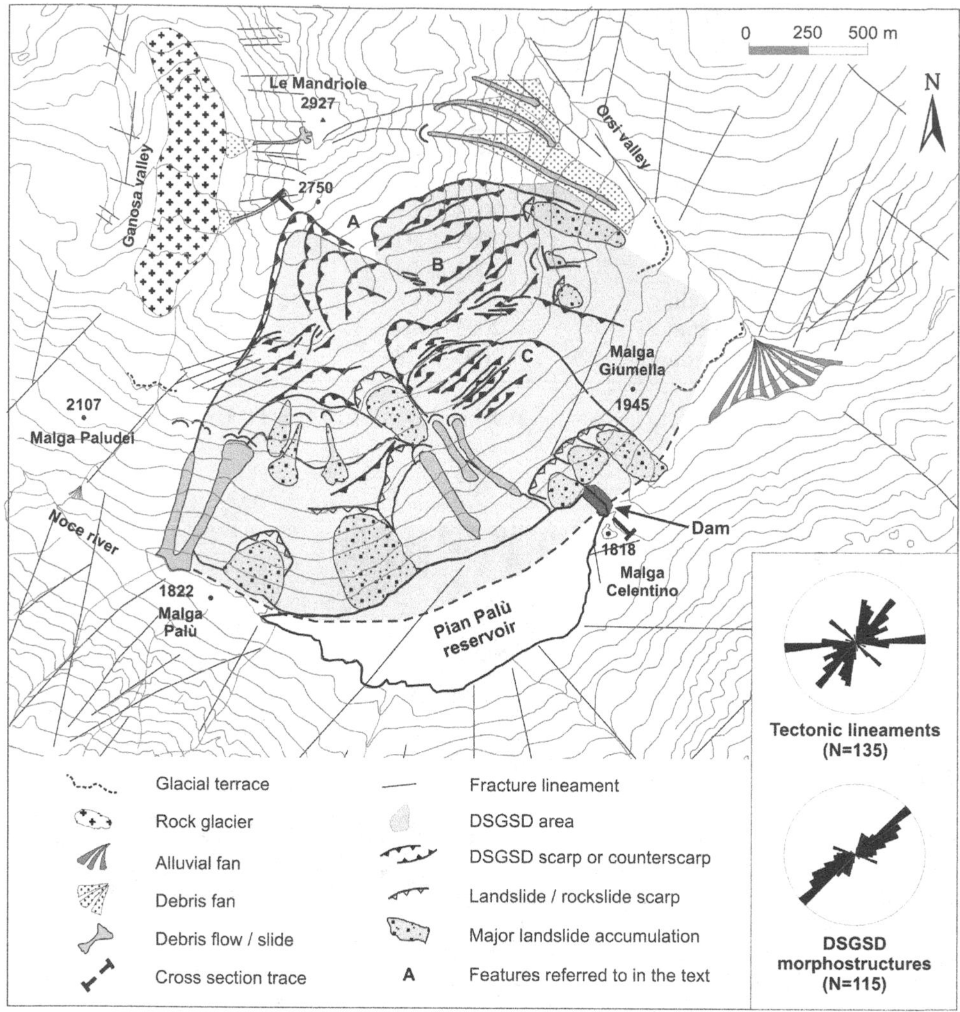

Fig. 18.9. Geomorphological and morphostructural map of the Cima di Mandriole DSGSD at Peio. (location shown in Fig. 18.3) (modified after Agliardi *et al.*, 2009a). Contour interval = 50 m. Steep head scarps with down-throws exceeding 100 m in the upper part of the slope (Fig. 18.8b, c and "A" in Fig. 18.9) follow an inherited, northeast-trending master fracture. Above 2300 m asl, the slope is crossed by swarms of rectilinear counterscarps up to 200 m long (Fig. 18.8d and "B" in Fig. 18.9) that cross-cut glacial erosional surfaces related to the Last Glacial Maximum and rock glacier deposits. Below 2300 m asl, scarps and counterscarps form gravitational half-grabens that delineate the heads of a series of large concentric landslides, with increasing morphological evidence moving downslope (Fig. 18.8e and "C" in Fig. 18.9). Most tectonic fractures trend northeast–southwest; fractures trending east–west and north–south are few but persistent. Fractures trending northeast–southwest constrain the geometry of the DSGSD, suggesting strong passive structural controls on DSGSD development.

friction joints, able to withstand large deformations. Borehole and tunnel logs revealed shear zones with clayey gouge and paragneiss blocks up to 100 m in depth at the dam cut-off trench level. The shear zones have a slightly curvilinear shape and terminate in gravelly sandy alluvial and glaciofluvial deposits at the toe of the slope. These findings, which are similar to those reported at other dam sites in Italy (Desio, 1961; Barla *et al.*, 2010), were interpreted as evidence of failure surfaces produced by large gravitational movements (Fig. 18.10).

Available data suggest that the entire slope is a DSGSD, moving over a basal shear zone up to 300 m below the ground surface. The supposed shear zone is curved upslope, but nearly planar below the lower part of the slope, consistent with the occurrence of an active wedge bounded by scarps and counterscarps in the upper sector of the slope. In the lower part of the slope, large rockslides occurred on multiple downslope-dipping scarps extending to a depth exceeding 100 m.

Cross-cutting relations between surface deformation features and glacial and periglacial deposits suggest that the DSGSD started during late-glacial time. The upper part of the slope experienced further, continuous or episodic activity in the Holocene. Precise leveling of the foundation and crest of the dam, tilting of inverse pendula, and topographic measurements of optical targets on the slope have been carried out since 1964 to monitor the movement of the slope and the behavior of the dam (Fig. 18.11). Monitoring revealed ongoing slow movement of the slope toe at an average rate of 5–6 mm per year, associated with a large landslide within the DSGSD. A sudden, stick–slip increase in displacement rates was detected in 1977, following extreme rainfall and snowmelt events (Fig. 18.11). Precise leveling carried out in the body of the dam revealed an upward tilting of the left-hand side of the dam, with a total vertical displacement of up to 10 mm, most of which occurred during 1977. The dam withstood these deformations without a loss of functionality due to its unique design (Fig. 18.11). Monitoring data demonstrated close coupling between deformation of the slope and the dam.

We performed 2D continuum stress-strain numerical modeling using the finite difference code FLAC (Itasca Consulting Group, 2005). Models were set up (Fig. 18.12a) with reference to a restored, pre-DSGSD topography, and including the relevant geomechanical and structural features. We used

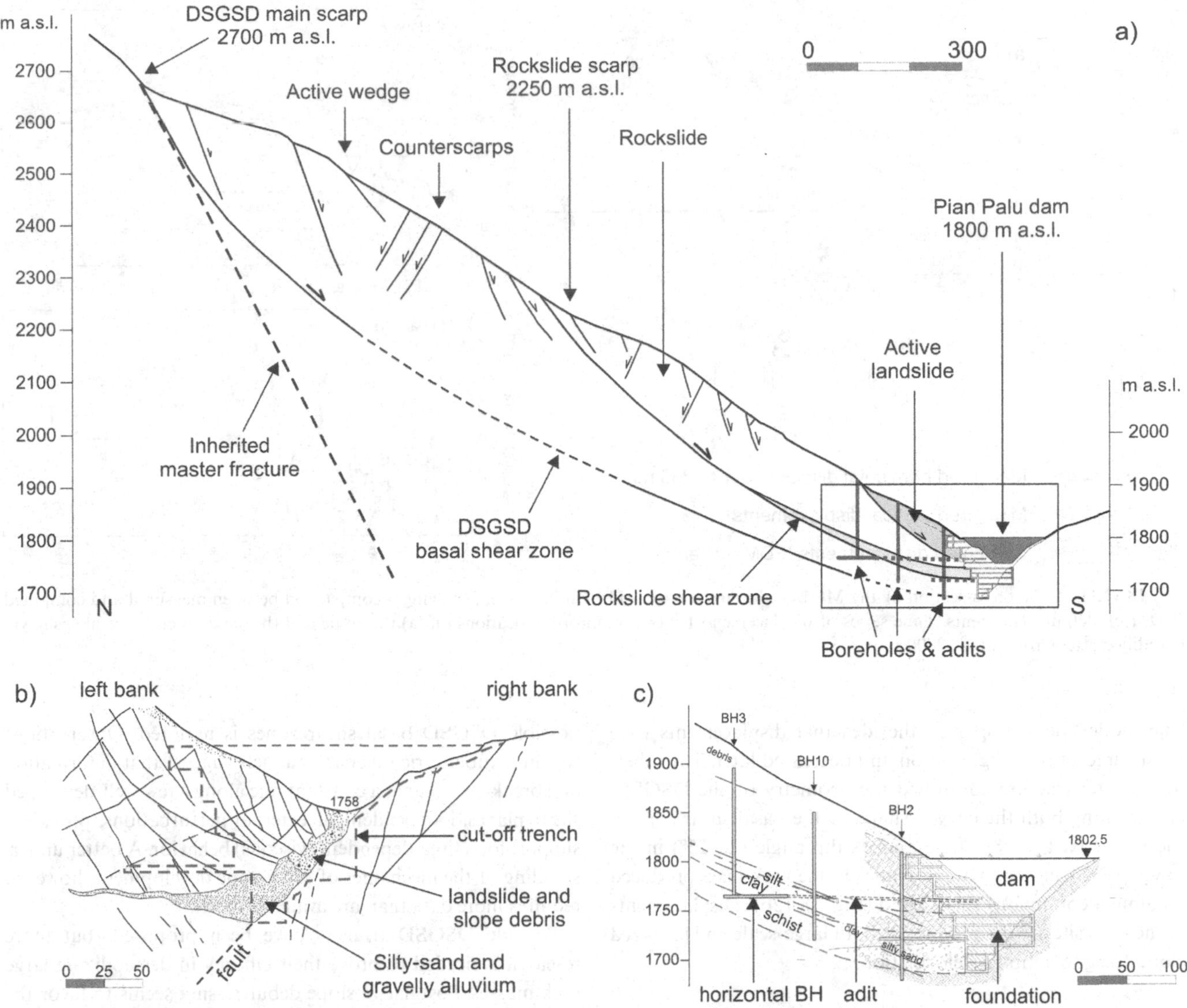

Fig. 18.10. Cross-sections showing the main features of the Cima di Mandriole sackung and its relationships with the Pian Palù dam. (a) Morphostructural cross-section and geologic model of the entire DSGSD; (b) cross-section of the dam cut-off trench section, as observed in the early construction stage, according to Desio (1961); (c) geotechnical cross-section of the dam site based on site investigations. (Modified after Crosta *et al.*, 2000.)

a Mohr–Coulomb elasto-plastic yield criterion with a ubiquitous-joint model to simulate closely spaced, northeast-trending joints parallel to the rock fabric. The role of the inherited master fracture shown in Figure 18.10 was investigated by including a discrete interface in the grid (Fig. 18.12a). Geomechanical properties of rock masses (Table 18.2) were obtained through standard rock mass characterization based on field and laboratory data (Hoek *et al.*, 2002). Ubiquitous joints were included to simulate rock mass anisotropy, and their strength was assumed to be similar to rock mass strength.

We ran a sequential model to simulate the evolution of the slope since the Last Glacial Maximum (ice surface at 2600 m asl, based on field evidence). Deglaciation was simulated by progressively removing ice layers and altering groundwater conditions accordingly. Model parameters were calibrated by the use of both morphostructural evidence and monitoring data. We computed total displacements, failure patterns, and plasticity indicators (Fig. 18.12b and 18.12c), and compared the values to field observations. Computed displacements at the slope toe were also compared with available monitoring data (Fig. 18.11).

Numerical models accounting for the effects of deglaciation, and including both slope-scale inherited structures and rock mass anisotropy (Fig. 18.12b), reproduced observed surface features and the inferred DSGSD kinematics (i.e., scarp and counterscarp locations, active wedge formation, shear strain localization along the basal failure zone and at shallower levels, and the total amount and distribution of displacements).

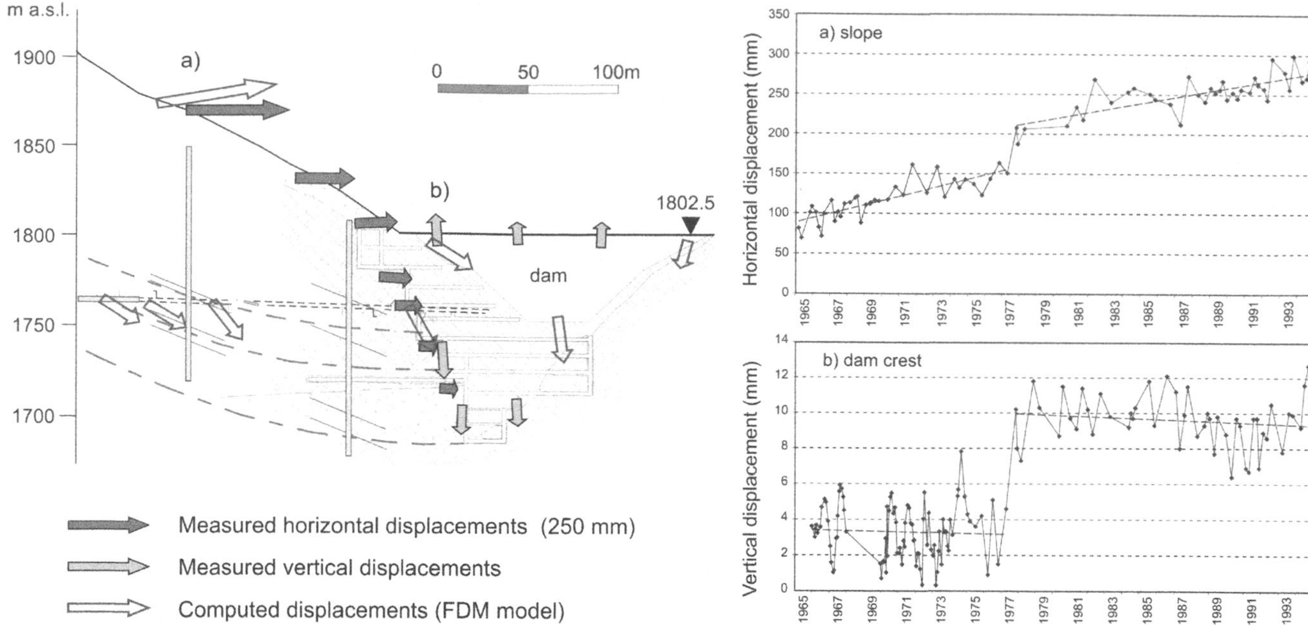

Fig. 18.11. Detailed cross-section of the Mt. Mandriole slope toe and Pian Palù dam, showing a comparison between measured and computed (FDM model) displacements. Time series of displacements for two monitoring locations on (a) the slope and (b) the dam crest are also shown. (Modified after Crosta *et al.*, 2000.)

The models also supported the idea that displacements were most rapid during deglaciation and decreased later. The inherited master fracture controlled the geometry of the DSGSD, constraining both the curved shape of the basal shear zone in the upper part of the slope and its dip angle (20–25°) in the lower part. Kinking of weak planes in the rock mass produced an almost continuous shear band (Fig. 18.12b). Displacements at the dam site are the product of both large-scale and localized slope-toe gravitational deformation.

18.6 CONCLUSIONS AND PERSPECTIVES

Recent recognition that deep-seated gravitational slope deformation is a widespread and active process in mountains has called attention to its importance both in engineering and in the evolution of the alpine landscape. Long alpine base tunnels and other large infrastructure projects highlight the need for a more complete understanding of this group of rock-slope failures.

Research is required to provide a better understanding of landforms and processes that are transitional between landslides and DSGSD. Landslides and DSGSD typically show different patterns of evolution toward failure, as well as different total displacements, displacement rates, sensitivity to triggers such as rainfall and earthquakes, and probability of catastrophic collapse. A distinction can be made on the basis of the extent of rock mass damage. Damage is greater in the case of landslides and is generally associated with rock masses of low strength, leading to more degrees of freedom, a wider range of possible failure surfaces, easier reactivation, and greater erosion by external geomorphic agents. In this context, a more detailed description of possible DSGSD basal shear zones is required. Observations within hydroelectric diversion tunnels suggest that deformation and breakage occur at a constant rate along a few well-developed shear planes, independent of external perturbations, providing support for a time-dependent (viscous) behavior. A better understanding of the mechanics of shear zone development, however, requires more data than are available today.

Possible DSGSD triggers have been proposed, but more research is needed to prove their efficacy in destabilizing large rock masses. Postglacial slope debuttressing seems to favor the triggering of deep rock-slope failures in alpine environments, but in some numerical simulations, very poor rock mass properties or significant cleft water pressures are required to destabilize slopes as they are debuttressed. These simulations highlight the need for a better understanding of groundwater in slopes during glacial periods, and how it changes during the transition to interglaciations. The incorporation of groundwater in numerical models of DSGSD may help to identify and understand differences in the evolution, distribution, and activity of DSGSD. For example, groundwater flow in tributary valleys is controlled by slope-scale rainfall and snowmelt. In contrast, groundwater flow in trunk valleys may be affected by regional factors and longer-term climate conditions. Regional groundwater flow is probably more sensitive to extreme or long-term climate change than local flow systems. Surprisingly, however, no sound data or models are available to account for the effects of glaciers, permafrost, and related groundwater circulation at different depths in slopes.

Two other important issues are the sensitivity of DSGSDs to long-term tectonic uplift and their role in the regional sediment

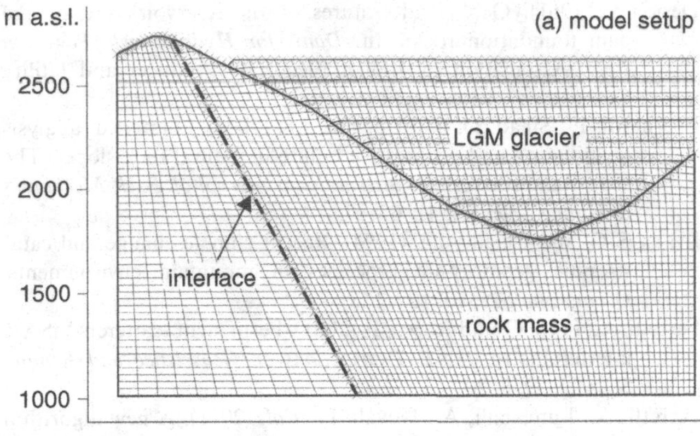

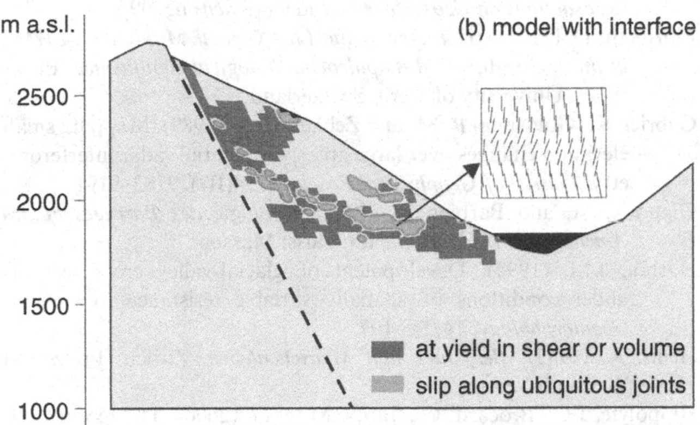

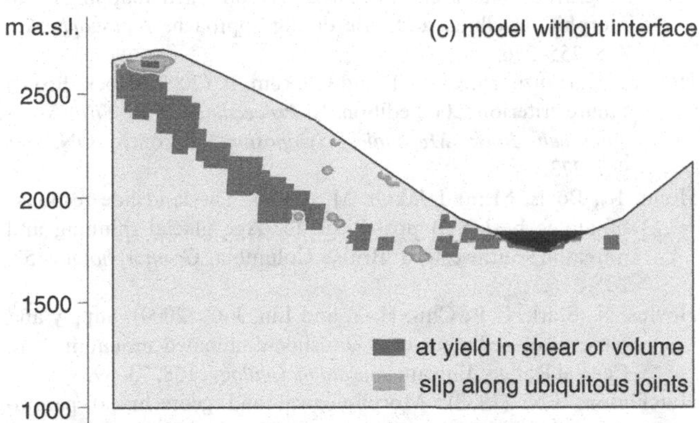

Fig. 18.12. Simplified, 2D finite difference modeling of the Cima di Mandriole sackung. (a) Model setup; (b) plasticity indicators with and (c) without slope-scale inherited structures.

budget. Slow downslope movements of large, broken masses of rocks continue for thousands of years or longer, delivering sediment to stream networks on timescales that exceed the length of the paraglacial cycle. A partial answer to this problem could come from the availability of uplift data for much larger sectors of the Alps or other mountain ranges. However, the contributions of DSGSD to the total sediment budget at both the catchment and regional scales need to be clarified. We suggest that DSGSD can play both active and passive roles at different timescales by increasing sediment availability due to enhanced

Table 18.2. *Geomechanical properties used in numerical modeling.*

Property	Value
Rock mass	
UCS of intact rock (MPa)	80
GSI (average)	53
Density (kN/m^3)	27
Bulk modulus (GPa)	10
Shear modulus (GPa)	5
Tensile strength (MPa)	0.1
Cohesion (MPa)	0.2
Friction angle (°)	32
Ubiquitous joints	
Tensile strength (MPa)	0.1
Cohesion (MPa)	0.1
Friction angle (°)	30

rock fracturing, creation of knick points along streams, triggering focused erosion in secondary valleys, modifying local drainage patterns, and damming drainage on valley floors.

ACKNOWLEDGEMENTS

We thank Regione Lombardia for providing satellite SAR datasets and the Provincia Autonoma di Trento for data related to the Cima di Mandriole case study. We are also grateful to C. Ambrosi, A. Zanchi, and R. Ignagnaro for contributing to data collection and analysis.

REFERENCES

Agliardi, F., Crosta, G. and Zanchi, A. (2001). Structural constraints on deep-seated slope deformation kinematics. *Engineering Geology*, 59, 83–102.

Agliardi, F., Zanchi, A. and Crosta, G.B. (2009a). Tectonic vs. gravitational morphostructures in the central Eastern Alps (Italy): Constraints on the recent evolution of the mountain range. *Tectonophysics*, 474, 250–270.

Agliardi, F., Crosta, G.B., Zanchi, A. and Ravazzi, C. (2009b). Onset and timing of deep-seated gravitational slope deformations in the eastern Alps, Italy. *Geomorphology*, 103, 113–129.

Ambrosi, C. and Crosta, G.B. (2006). Large sackung along major tectonic features in the central Italian Alps. *Engineering Geology*, 83, 183–200.

Ambrosi, C. and Crosta, G.B. (2011). Valley shape influence on deformation mechanisms of rock slopes. In *Slope Tectonics*, ed. M. Jaboyedoff. Geological Society of London, Special Publication 351, pp. 215–233.

APAT (2007). *Rapporto sulle Frane in Italia.* Environment Protection Agency of Italy, Report 78/2007 (in Italian).

Arsenault, A.M. and Meigs, A.J. (2005). Contribution of deep-seated bedrock landslides to erosion of a glaciated basin in southern Alaska. *Earth Surface Processes and Landforms*, 30, 1111–1125.

Augustinus, P.C. (1995). Glacial valley cross-profile development: The influence of in situ rock stress and rock mass strength, with examples from the Southern Alps, New Zealand. *Geomorphology*, 14, 87–97.

Ballantyne, C.K. (2002). Paraglacial geomorphology. *Quaternary Science Reviews*, 21, 1935–2017.

Barla, G., Antolini, F., Barla, M., Mensi, E. and Piovano, G. (2010). Monitoring of the Beauregard landslide (Aosta Valley, Italy) using advanced and conventional techniques. *Engineering Geology*, 116, 218–235.

Beck, A.C. (1968). Gravity faulting as a mechanism of topographic adjustment. *New Zealand Journal of Geology and Geophysics*, 11, 191–199.

Benko, B. and Stead, D. (1998). The Frank Slide: A re-examination. *Canadian Geotechnical Journal*, 35, 299–311.

Bigi, G., Cosentino, D., Parlotto, M., Sartori, R. and Scandone, P. (1983). Synthetic structural-kinematic map of Italy, scale 1:500,000, C. N. R. Progetto Finalizzato Geodinamica, Quaderni della Ricerca Scientifica, 114.

Bois, S., Bouissou, S. and Guglielmi, Y. (2008). Influence of major inherited faults zones on gravitational slope deformation: A two-dimensional physical modelling of the La Clapière area (southern French Alps). *Earth and Planetary Science Letters*, 272, 709–719.

Bonzanigo, L., Eberhardt, E. and Loew, S. (2007). Long-term investigation of a deep-seated creeping landslide in crystalline rock. Part 1. Geological and hydromechanical factors controlling the Campo Vallemaggia landslide. *Canadian Geotechnical Journal*, 44, 1157–1180.

Bovis, M.J. (1990). Rock-slope deformation at Affliction Creek, southern Coast Mountains, British Columbia. *Canadian Journal of Earth Sciences*, 27, 243–254.

Bovis, M.J. and Evans, S.G. (1995). Rock slope movements along the Mount Currie "fault scarp", southern Coast Mountains, British Columbia. *Canadian Journal of Earth Sciences*, 32, 2015–2020.

Brückl, E. (2001). Cause–effect models of large landslides. *Natural Hazards*, 23, 291–314.

Brückl, E. and Parotidis, M. (2005). Prediction of slope instabilities due to deep-seated gravitational creep. *Natural Hazards and Earth System Sciences*, 5, 155–172.

Cancelli, A. and Casagli, N. (1995). Classificazione e modellazione di fenomeni di instabilità in ammassi rocciosi sovrapposti ad argilliti o argille sovra consolidate. *Memorie della Società Geologica Italiana*, 50, 83–100.

Chigira, M. (1992). Long-term gravitational deformation of rock by mass rock creep. *Engineering Geology*, 32, 157–184.

Colesanti, C., Crosta, G.B., Ferretti, A. and Ambrosi, C. (2006). Monitoring and assessing the state of activity of slope instabilities by the permanent scatterers technique. In *Landslides from Massive Rock Slope Failure*, ed. S.G. Evans, G. Scarascia Mugnozza, A. Strom and R.L. Hermanns. Dordrecht, Netherlands: Springer, pp. 179–212.

Crosta, G. (1996). Landslide, spreading, deep-seated gravitational deformation: Analysis, examples, problems and proposals. *Geografia Fisica e Dinamica Quaternaria*, 19, 297–313.

Crosta, G.B. and Zanchi, A. (2000). Deep-seated slope deformations: Huge, extraordinary, enigmatic phenomena. In *Landslides in Research, Theory and Practice*, ed. E. Bromhead, N. Dixon and M.-L. Ibsen. London: Thomas Telford, pp. 351–358

Crosta, G., Zanchi, A., Agliardi, F., Frattini, P. and Ambrosi, C. (2000). *Convenzione di Studio della Frana di "Pian Palù" in sponda sinistra dell'Alta Val di Peio: Torrente Noce*. Università degli Studi di Milano-Bicocca – Provincia Autonoma di Trento, unpublished report (in Italian).

Cruden, D.M. and Hu, X.Q. (1994). Topples on underdip slopes in the Highwood Pass, Alberta, Canada. *Quarterly Journal of Engineering Geology*, 27, 57–68.

Desio, A. (1961). Geological features of the reservoirs and of the dam foundation rocks. In: *Dams for Hydroelectric Power in Italy*. Association of Italian Electricity Producer and Utility Companies.

Eberhardt, E., Stead, D. and Coggan, J.S. (2004). Numerical analysis of initiation and progressive failure in natural rock slopes: The 1991 Randa rockslide. *International Journal of Rock Mechanics & Mining Sciences*, 41, 69–87.

Evans, S.G. and Clague, J.J. (1994). Recent climate change and catastrophic geomorphic processes in mountain environments. *Geomorphology*, 10, 107–128.

Ferretti, A., Prati, C. and Rocca, F. (2001). Permanent scatterers InSAR interferometry. *IEEE Transactions on Geoscience and Remote Sensing*, 39, 8–20.

Ferretti, A., Fumagalli, A., Novali, F. *et al.* (2011). A new algorithm for processing interferometric data-stacks: SqueeSAR. *IEEE Transactions on Geoscience and Remote Sensing*, 99, 1–11.

Florineth, D. (1998). *Geometry of the Last Glacial Maximum (LGM) in the Swiss Alps and its paleoclimatological significance*. Ph.D. thesis, University of Bern, Switzerland.

Gabriel, K., Goldstein, R.M. and Zebker, H.A. (1989). Mapping small elevation changes over large areas: differential radar interferometry. *Journal of Geophysical Research*, 94(B7), 9183–9191.

Gignoux, M. and Barbier, R. (1955). *Géologie des Barrages et des Aménagements Hydrauliques*. Paris: Masson.

Harbor, J.M. (1995). Development of glacial-valley cross sections under conditions of spatially variable resistance to erosion. *Geomorphology*, 14, 99–107.

Heim, A. (1932). *Bergsturz und Menschenleben*. Zurich: Fretz and Wasmuth Verlag.

Hippolyte, J.C., Brocard, G., Tardy, M. *et al.* (2006). The recent fault scarps of the Western Alps (France): Tectonic surface ruptures or gravitational sackung scarps? A combined mapping, geomorphic, levelling, and ^{10}Be dating approach. *Tectonophysics*, 418, 255–276.

Hoek, E, Carranza-Torres, C.T. and Corkum, B. (2002). Hoek–Brown failure criterion: 2002 edition. In *Proceedings of the Fifth North American Rock Mechanics Symposium*, Toronto, ON, pp. 267–273.

Holm, K., Bovis, M. and Jakob, M. (2004). The landslide response of alpine basins to post-Little Ice Age glacial thinning and retreat in southwestern. British Columbia. *Geomorphology*, 57, 201–216.

Hovius, N., Stark, C.P., Chu, H.-Y. and Lin, J.-C. (2000). Supply and removal of sediment in a landslide-dominated mountain belt: Central Range, Taiwan. *Journal of Geology*, 108, 73–89.

Hutchinson, J.N. (1988). Morphological and geotechnical parameters of landslides in relation to geology and hydrogeology. In *Proceedings of the 5th International Symposium on Landslides, Lausanne*. Rotterdam: A.A. Balkema, pp. 3–35.

Itasca Consulting Group (2005). *FLAC Version 5.0, User Manuals*. Minneapolis: Itasca Consulting Group Inc.

Jäckli, H. (1970). *Die Schweiz zur Letzten Eiszeit, Atlas der Schweiz, Blatt 6*, Wabern-Bern: Bundesamt für Landestopographie, Map, scale 1:550,000.

Kahle, H.G., Geiger, A., Burki, B. *et al.* (1997). Recent crustal movements, geoid and density distribution: Contribution from integrated satellite and terrestrial measurements. In *Results of the NRP20 Deep Structures of the Swiss Alps*, ed. O. Pfiffner, P. Lehner, P. Heitzmann, S. Müller and A. Steck. Basel: Birkhauser Verlag, pp. 251–259.

Kelly, M.A., Buoncristiani, J.F. and Schluchter, C. (2004). A reconstruction of the Last Glacial Maximum (LGM) ice-surface geometry in the western Swiss Alps and contiguous Alpine regions in Italy and France. *Eclogae Geologicae Helvetiae*, 97, 57–75.

Kieffer, D.S. (1998). *Rock slumping: A compound failure mode of jointed hard rock slopes*. Ph.D. thesis, University of California, Berkeley, CA.

Kinakin, D. and Stead, D. (2005). Analysis of the distributions of stress in natural ridge forms: Implications for the deformation mechanisms of rock slopes and the formation of a sackung. *Geomorphology*, 65, 85–100.

Korup, O. (2006). Effects of large deep-seated landslides on hillslope morphology, western Southern Alps, New Zealand. *Journal of Geophysical Research*, 111, F01018.

Korup, O. and Schlunegger, F. (2007). Bedrock landsliding, river incision, and transience of geomorphic hillslope-channel coupling: Evidence from inner gorges in the Swiss Alps. *Journal of Geophysical Research*, 112, F03027.

Korup, O., Clague, J.J., Hermanns, R.L. *et al.* (2007). Giant landslides, topography, and erosion. *Earth and Planetary Science Letters*, 261, 578–589.

MacFarlane, D.F. (2008). Observations and predictions of the behaviour of large, slow-moving landslides in schist, Clyde Dam reservoir, New Zealand. *Engineering Geology*, 109, 5–15.

Mahr, T. (1977). Deep-reaching gravitational deformations of high mountain slopes. *International Association of Engineering Geologists Bulletin*, 16, 121–127.

Malamud, B.D., Turcotte, D.L., Guzzetti, F. and Reichenbach, P. (2004). Landslide inventories and their statistical properties. *Earth Surface Processes and Landforms*, 29, 687–711.

Martel, S.J. (2000). Modeling elastic stresses in long ridges with the displacement discontinuity method. *Pure and Applied Geophysics*, 157, 1009–1038.

McCalpin, J.P. (1999). Criteria for determining the seismic significance of sackungen and other scarp-like landforms in mountainous regions. In *Techniques for Identifying Faults and Determining their Origins*. Washington, DC: US Nuclear Regulatory Commission, pp. 2.55–2.59.

McCalpin, J.P. and Irvine, J.R. (1995). Sackungen at the Aspen Highlands ski area, Pitkin County, Colorado. *Environmental and Engineering Geoscience*, 1, 277–290.

Miller, D.J. and Dunne, T. (1996). Topographic perturbations of regional stresses and consequent bedrock fracturing. *Journal of Geophysical Research*, 101, 25523–25536.

Molnar, P. (2004). Interactions among topographically induced elastic stress, static fatigue, and valley incision. *Journal of Geophysical Research*, 109, F02010.

Montgomery, D.R. and Brandon, M.T. (2002). Topographic controls on erosion rates in tectonically active mountain ranges. *Earth and Planetary Science Letters*, 201, 481–489.

Moro, M., Saroli, M., Salvi, S., Stramondo, S. and Doumaz, F. (2007). The relationship between seismic deformation and deep-seated gravitational movements during the 1997 Umbria-Marche (central Italy) earthquakes. *Geomorphology*, 89, 297–307.

Mortara, G. and Sorzana, P.F. (1987). Fenomeni di deformazione gravitativa profonda nell'arco alpino occidentale italiano: Considerazioni lito-strutturali e morfologiche. *Bollettino della Società Geologica Italiana*, 106, 303–314 (in Italian).

Nemcok, A. (1972). Gravitational slope deformation in high mountains. In *Proceedings of the 24th International Geological Congress*, Montreal, PQ, pp. 132–141.

Nichol, S.L., Hungr, O. and Evans, S.G. (2002). Large-scale brittle and ductile toppling of rock slopes. *Canadian Geotechnical Journal*, 39, 773–788.

Persaud, M. and Pfiffner, O.A. (2004). Active deformation in the eastern Swiss Alps: Postglacial faults, seismicity and surface uplift. *Tectonophysics*, 385, 59–84.

Prager, C., Zangerl, C., Patzelt, G. and Brandner, R. (2008). Age distribution of fossil landslides in the Tyrol (Austria) and its surrounding areas. *Natural Hazards and Earth System Sciences*, 8, 377–407.

Radbruch-Hall, D. (1978). Gravitational creep of rock masses on slopes. In *Rockslides and Avalanches; Natural Phenomena*, ed. B. Voight. Amsterdam: Elsevier, pp. 607–657.

Saroli, M., Stramondo, S., Moro, M. and Doumaz, F. (2005). Movements detection of deep seated gravitational slope deformations by means of InSAR data and photogeological interpretation: Northern Sicily case study. *Terra Nova*, 17, 35–43.

Savage, W.Z. (1994). Gravity-induced stresses in finite slopes. *International Journal of Rock Mechanics and Mining Sciences*, 31, 471–483.

Savage, W.Z., Swolfs, H.S. and Powers, P.S. (1986). Gravitational stresses in long symmetric ridges and valleys. *International Journal of Rock Mechanics and Mining Sciences & Geomechanics, Abstracts*, 22, 291–302.

Sjoberg, J. (1999). *Analysis of large scale rock slopes*. Ph.D. thesis, Luleå University of Technology, Luleå, Sweden.

Solonenko, V.P. (1977). Landslides and collapses in seismic zones and their prediction. *Bulletin of the International Association of Engineering Geology*, 15, 4–8.

Stark, C.P. and Hovius, N. (2001). The characterization of landslide size distribution. *Geophysical Research Letters*, 28, 1091–1094.

Swisstopo (2007). *GeoKarten (Geologie, Tektonik, Hydrogeologie) 1:500000: Vektordaten*. Wabern-Bern: Swiss Federal Office of Topography, CD-ROM.

Terzaghi, K. (1972). Stability of steep slopes on hard unweathered rock. *Géotechnique*, 12, 251–270.

Ustaszewski, M.E., Hampel, A. and Pfiffner, O.A. (2008). Composite faults in the Swiss Alps formed by the interplay of tectonics, gravitation and postglacial rebound: An integrated field and modelling study. *Swiss Journal of Geosciences*, 101, 223–235.

Van Husen, D. (1997). *Die Ostalpen in den Eiszeiten*. Vienna: Geologische Bundesanstalt.

Varnes, D.J., Radbruch-Hall, D. and Savage, W.Z. (1989). *Topographic and Structural Conditions in Areas of Gravitational Spreading of Ridges in the Western United States*. US Geological Survey, Professional Paper 1496.

Varnes, D.J., Radbruch-Hall, D., Varnes, K.L., Smith, W.K. and Savage, W.Z. (1990). *Measurement of Ridge-spreading Movements (Sackungen) at Bald Eagle Mountain, Lake County, Colorado, 1975–1989*. US Geological Survey, Open File Report 90–543.

Zangerl, C., Eberhardt, E. and Perzlmaier, S. (2010). Kinematic behaviour and velocity characteristics of a complex deep-seated crystalline rockslide system in relation to its interaction with a dam reservoir. *Engineering Geology*, 112, 53–67.

Zischinsky, U. (1966). On the deformation of high slopes. In *Proceedings of the First Conference of the International Society for Rock Mechanics, Vol. 2*, pp. 179–185.

19 Landslide monitoring: The role of investigative monitoring to improve understanding and early warning of failure

ERIK EBERHARDT

ABSTRACT

Recent years have seen significant advances in landslide monitoring technologies. Satellite and ground-based radar systems have considerably increased the areal coverage and spatial resolution of surface displacement monitoring data. Nanotechnology has led to the development of smaller, cheaper, more reliable, and more functional borehole sensors that, together with wireless data acquisition and transmission, have significantly increased the temporal resolution of subsurface slope deformation and microseismic monitoring data. These tools provide increased capacity to detect pre-failure indicators and changes in landslide behavior. Yet the interpretation of slope monitoring data for the purpose of early warning still remains largely subjective, as geologic complexity and uncertainty continue to pose major challenges. To be effective, predictive early-warning monitoring must be preceded by investigative monitoring to provide an understanding of slope behavior over time and typical responses to external stimuli such as precipitation. This chapter reviews several recent developments in landslide monitoring technologies. It discusses the role of investigative monitoring in developing slope monitoring programs and providing early-warning alert levels. Examples are provided from several recent experimental studies involving in-situ rockslide laboratories, in which detailed instrumentation systems and numerical modeling have been used to better understand the mechanisms controlling pre-failure deformations over time and their evolution leading to catastrophic failure. These examples demonstrate that by better integrating different datasets, geologic uncertainty can be minimized and better controlled to provide improved interpretation of slope monitoring and early-warning data.

19.1 INTRODUCTION

Slope monitoring is a key component of most landslide hazard assessments, often with the objective of providing early warning of an impending failure where lives, communities, or infrastructure may be at risk. Standard practice typically involves relying on surface displacements measured over time, which are then extrapolated or analyzed for accelerations that may indicate approaching failure. Interpreting these patterns, however, is complicated by uncertainty and subjectivity deriving from complex geologic conditions and movement kinematics. Detection of accelerating behavior may either be an early warning of failure or a false alarm related to localized movements near the monitoring point. Episodic behavior is common, especially with deep-seated landslides, where changing conditions, for example groundwater fluctuations related to seasonal weather patterns, can intermittently destabilize and restabilize the slope. In many cases, slope velocities can vary by an order of magnitude without failure occurring. This chapter examines these difficulties, as well as efforts to improve early-warning monitoring by applying new technologies and research aimed at improving understanding of the geologic conditions and mechanisms responsible for deep-seated landslide movements.

19.2 APPLICATION OF EMPIRICAL APPROACHES

Techniques used for early landslide warning are largely phenomenological, relying on surface measurements of displacements over time. Most monitoring is carried out through repeat geodetic surveys of monitoring prisms, which provide

Landslides: Types, Mechanisms and Modeling, ed. John J. Clague and Douglas Stead. Published by Cambridge University Press.

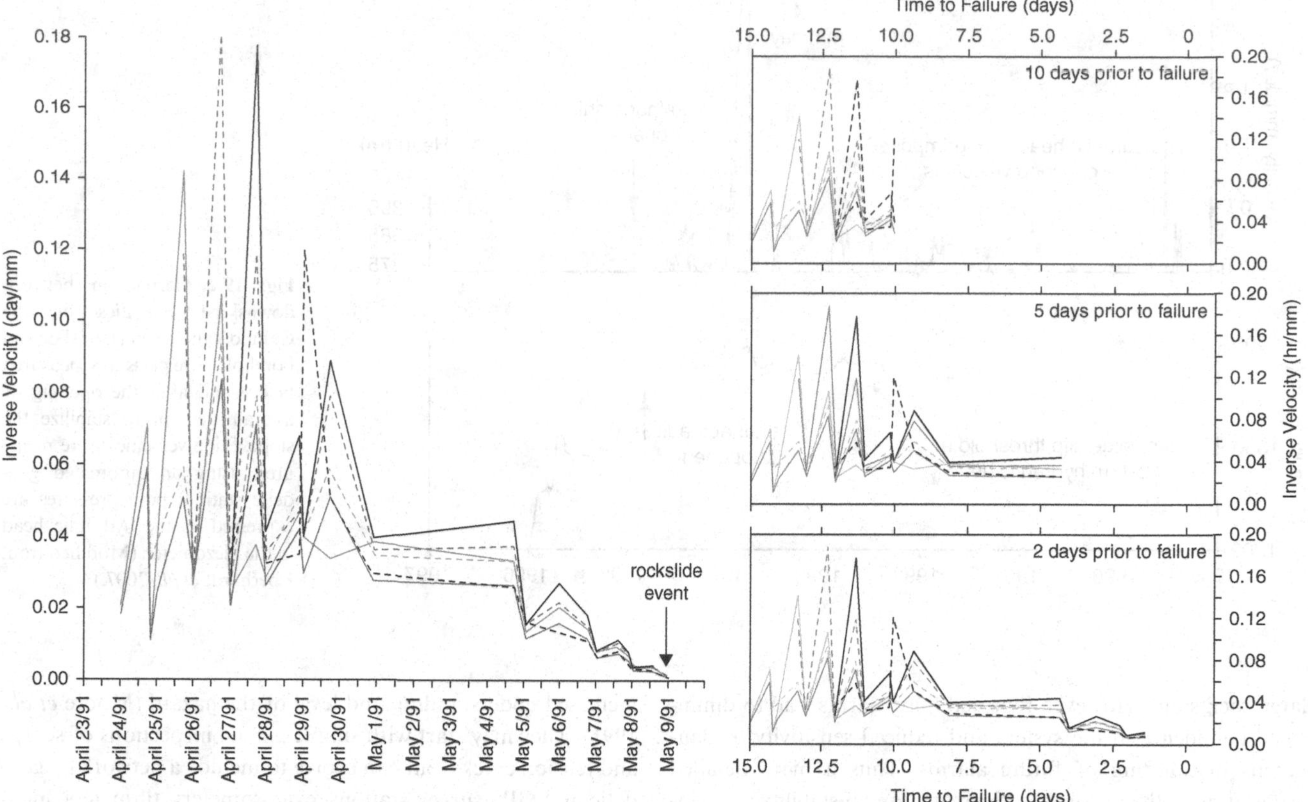

Fig. 19.1. Inverse velocity plots for the second of two rockslide events near the village of Randa based on geodetic data collected following the first event on April 18, 1991. Shown is the correlation of the predictive trend using Fukuzono's (1985) inverse velocity method relative to the date of the rockslide event. Also shown are the data trends at 10, 5, and 2 days before the failure. (Data source: Ischi *et al.*, 1991.)

precise data on magnitudes and rates of horizontal and vertical ground movements. Extensometers and crackmeters are also used to measure movements across open tension cracks and scarps delimiting the boundary of the potential slide mass. Displacement rates are then analyzed for accelerations that exceed set thresholds. Thresholds are commonly defined over time and rely on the experience of the analyst.

Much work has been carried out on displacement rates of large rockslides, in part because of their destructive potential (Bhandari, 1988; Crosta and Agliardi, 2003; Rose and Hungr, 2007; Eberhardt, 2008). The inverse velocity method of Fukuzono (1985) is a commonly applied technique. Based on Saito's (1965) attempts to compare slope displacement records with laboratory-derived creep rupture curves, Fukuzono suggested that the time of failure could be predicted by extrapolating the inverse velocity of slope displacement to its intersection with the time axis. The method is illustrated in Figure 19.1, which shows geodetic data collected following the first of two large rockslides near the village of Randa in the southern Swiss Alps. Except for some precursory activity in the form of rockfalls and water flow from daylighting fractures, the initial rockslide, which involved 20 million m^3 of massive crystalline rock, was unexpected (Schindler *et al.*, 1993). During a detailed investigation and monitoring of the area behind the newly formed scarp, a second rockslide of 10 million m^3 occurred. The inverse velocity plot in Figure 19.1 shows that an accurate prediction of the time of the second rockslide could have been made almost to the hour. However, as shown by the snapshots of the data 10, 5, and 2 days before failure (Fig. 19.1), the predictive trend is less evident. A correct prediction may have been possible several hours prior to failure, but any one of a number of accelerations recorded in the 2-week period beforehand would have produced a false prediction.

Empirical approaches such as that shown in Figure 19.1 are inherently holistic and disregard details pertaining to the underlying slope failure mechanism. The analysis is carried out in the same way whether the displacement measurements are derived from geodetic monitoring of survey prisms distributed across the slope or an extensometer positioned across a tension crack. Generally, the kinematics and causes of failure are not well understood, and instead the surface manifestations of the instability (i.e., surface displacements) are relied upon for predictive analysis. Not surprisingly, only a few cases have been reported where these techniques have been successfully applied in prediction (see Rose and Hungr, 2007); most cases involve back analysis.

Despite these shortcomings, displacement monitoring is an important component of most early-warning strategies. At least it can call attention to changes in velocity that may precede sudden failure (e.g., Salt, 1988; Crosta and Agliardi, 2003). Displacement monitoring also addresses economic realities in terms of what is typically feasible for on-site monitoring of a

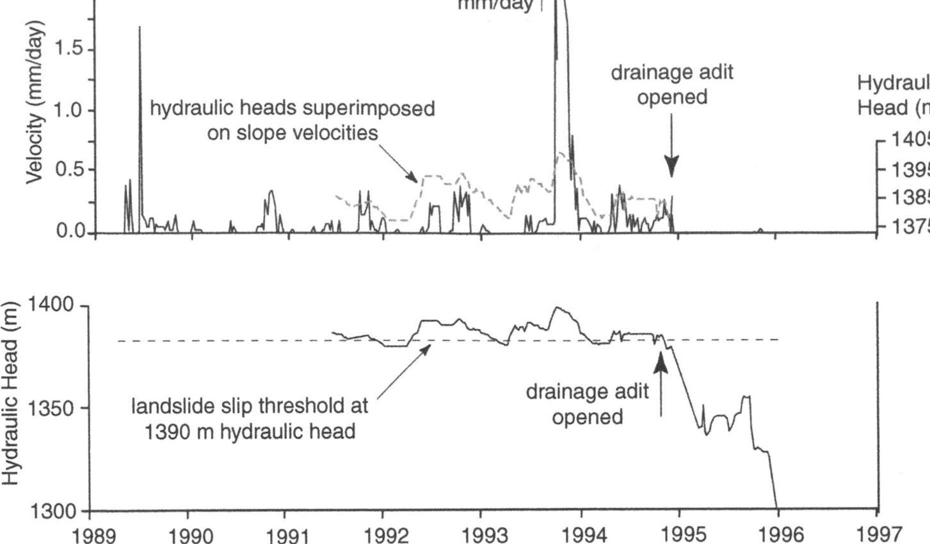

Fig. 19.2. Correlation between downslope velocities of the Campo Vallemaggia landslide and borehole pore pressures measured before and after the opening of a drainage adit to stabilize the slope. Slide velocities were measured using an automated geodetic station; pore pressures are expressed as the hydraulic head in the piezometer. (Modified from Eberhardt *et al.*, 2007.)

large rock slope. However, repeated false alarms lead to diminished confidence in the system and reduced sensitivity or dangerous discounting of future alarms. Thus a more detailed accounting of the subsurface kinematics and instability mechanism is required to better interpret the surface displacement signal. For most rock-slope instabilities, the presence of rock mass fabric and discontinuities, multiple moving blocks, and internal shear surfaces impose complexity on the slope monitoring signal, making the interpretation of early-warning data difficult.

19.3 INVESTIGATIVE MONITORING

For slope monitoring to be effective, the data should first be used to gain an understanding of three-dimensional behavior of the failing rock mass. The monitoring network thus serves two purposes (Moore *et al.*, 1991):

1. *investigative monitoring*, to provide an understanding of the behavior of the slope over time and typical responses to external stimuli such as precipitation and temperature
2. *predictive monitoring*, to provide a warning of a change in behavior that might facilitate evacuation or intervention to prevent hazardous sliding.

Investigative monitoring provides information on the boundaries of the landslide, both at surface and at depth, and may confirm or refute geologic interpretations of the slide kinematics and mode of instability (e.g., toppling, rotation, translation, or a complex combination of two or more of these modes). The rate of movement and its relationship and sensitivity to external destabilizing forces (e.g., precipitation, groundwater fluctuations, and earthquakes) can be determined. Investigative monitoring programs can be modified over time to reflect the increased understanding and level of the hazard (Moore *et al.*, 1991). They may start with simple annual inspections of scarps and tension cracks, but can evolve to include a network of geodetic and GPS survey stations, extensometers, tiltmeters, inclinometers, and piezometers.

Two examples are provided here. Campo Vallemaggia is an 800 million m^3 deep-seated, slow-moving landslide in strongly fractured and weathered crystalline rocks in southern Switzerland. Slope displacements have continuously damaged roads and buildings in two small villages situated on the slide mass. Intervention and mitigation to stabilize the landslide required an understanding of the controlling factors, and an emergency plan was prepared in the event that evacuation was deemed necessary. Yet implementing the emergency plan was not straightforward, as decision makers had to contend with the unpredictable nature of the landslide; when velocities reached alarming levels and an evacuation of the villages seemed warranted, the movements would return to normal. This situation led to a detailed investigative monitoring program involving geological and geophysical surveys, borehole inclinometer and pore-pressure measurements, and surface geodetic monitoring (Bonzanigo *et al.*, 2007). Inspection of the displacement record over time showed an episodic stick–slip behavior; accelerated movements were associated with periods of intense precipitation, leading to alarm. Detailed inclinometer and piezometer measurements collected over a 5-year period indicated a relation between movement rates and pore pressures that exceeded a threshold level (Fig. 19.2). These measurements were subsequently used to justify proceeding with the construction of a deep drainage adit that successfully stabilized the landslide (Eberhardt *et al.*, 2007).

The Hochmais-Atemkopf rockslide in northern Tyrol, Austria, is another prominent deep-seated, slow-moving landslide in crystalline rocks. Detailed investigative monitoring was

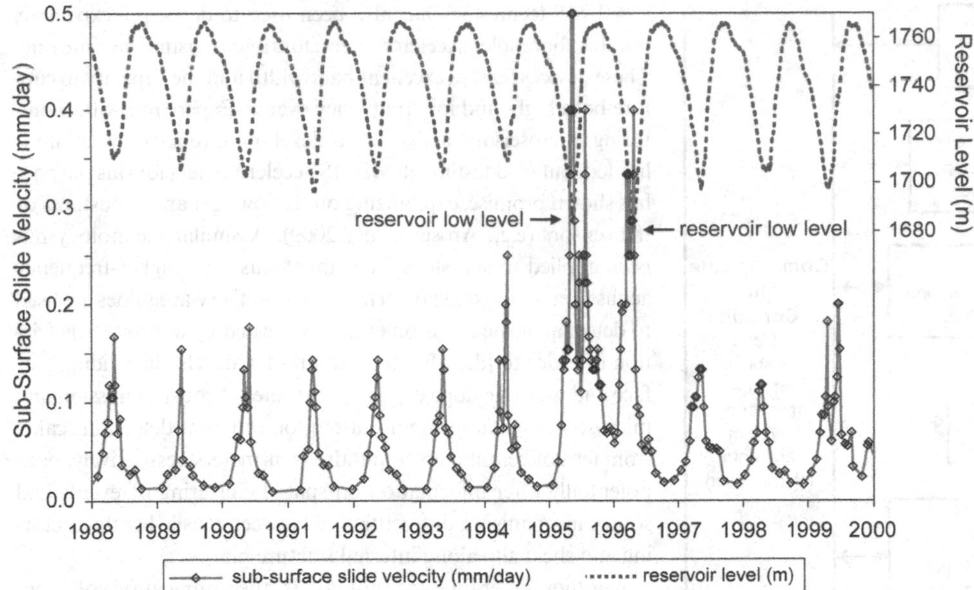

Fig. 19.3. Correlation between reservoir level and subsurface extensional velocities of the Hochmais-Atemkopf rockslide system. Slide velocities were measured within an investigation adit using a wire-extensometer spanning the active sliding zone below the most active slide body. Note the association of the highest velocities in this data record with the two lowest reservoir levels. (Modified from Zangerl *et al.*, 2010.)

initiated because of its location above a reservoir. Geological and geomorphological mapping combined with surface and subsurface investigative monitoring indicated that the rockslide comprises several imbricated sliding masses characterized by different sliding patterns and deformation rates (Zangerl *et al.*, 2010). The most active sliding mass episodically accelerated and decelerated. A key finding of the monitoring program was that the acceleration phases only partly correlated with the period of spring snowmelt and precipitation (Zangerl *et al.*, 2010). Rather, slope velocities began to increase much earlier in late winter when the reservoir level was at its lowest point (Fig. 19.3). Reservoir drawdown was thus an important controlling factor.

In both cases, mitigation and early-warning decisions were made on the basis of results from investigative monitoring that provided a more thorough understanding of the deformation mechanisms (Eberhardt *et al.*, 2007; Zangerl *et al.*, 2010). Without this investigation phase, it is unlikely that predictions or early-warning thresholds could have been set with confidence.

19.4 NEW TECHNOLOGIES FOR EARLY-WARNING MONITORING

Landslides can change their behavior within a few months, weeks, or days. Thus, the effectiveness of a predictive monitoring system depends on the duration and frequency of monitoring. In order for decision makers to be able to react correctly to predictive monitoring data, the probability of faulty alarms or misleading instrument readings must be minimized by the use of appropriate and reliable instruments (Eberhardt *et al.*, 2008). Significant advances have been made in landslide monitoring technologies over the past several years, especially with the use of radar. Radar offers high-resolution, full-area spatial coverage of a landslide, in contrast to geodetic or GPS point measurements. Satellite-based interferometric synthetic aperture radar (InSAR) has become an effective, regional-scale monitoring tool capable of covering areas of thousands of square kilometers (e.g., Meisina *et al.*, 2005). However, it has limited application as an early-warning monitoring tool because the data are not continuously acquired; instead, image acquisitions are separated by weeks.

Ground-based radar, on the other hand, offers continuous scanning and real-time monitoring over distances of up to 4 km with sub-millimeter accuracy. Commercially available synthetic aperture radar (SAR) systems such as Ingegneria dei Sistemi's IBIS-L (Rödelsperger *et al.*, 2010) are capable of performing real-time, near-continuous, line-of-sight monitoring of a landslide, day or night, and in all weather conditions (−50°C to 50°C). Microwave interferometry is used to measure displacements of the land surface by comparing phase information of the backscattered electromagnetic waves collected at different times. Similar technologies, including Groundprobe's slope stability radar (SSR), use real-aperture radar from a stationary platform rather than synthetic-aperture radar from a moving platform. With SSR, the system is typically 30–3500 m from the foot of the slope and the region of interest is continuously scanned. The phase differences for the same pixels on two successive scans provide a highly accurate measure of the amount of movement between the scans (Harries and Roberts, 2007). The combination of near real-time measurement, sub-millimeter precision, and broad area coverage has established SSR as a key tool for managing unstable rock slopes, especially in open-pit mines (e.g., Cahill and Lee, 2006; Little, 2006; Day and Seery, 2007). Figure 19.4 presents the hazard management framework developed by Harries (2009) for ground-based radar. It emphasizes the sequential nature of the approach, in which the context of the problem is first established and the hazard is then identified, analyzed, evaluated, and treated.

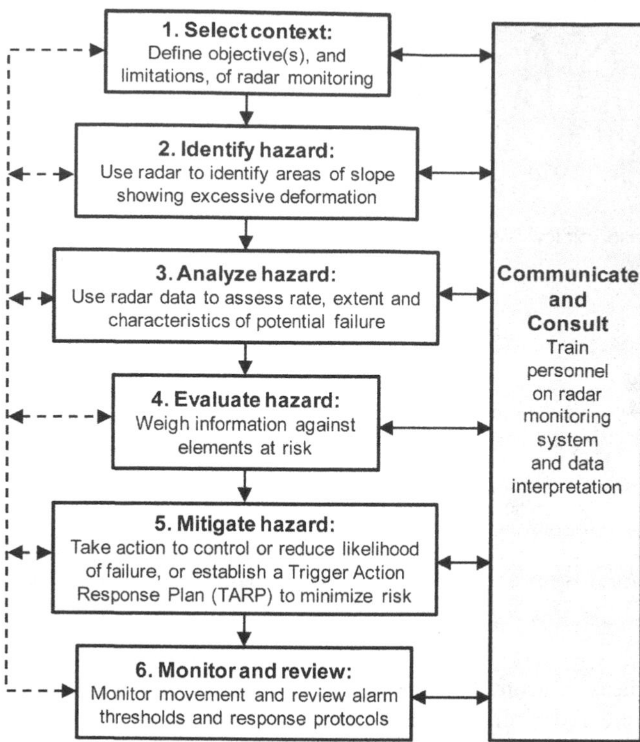

Fig. 19.4. Rock slope hazard management framework for ground-based radar monitoring. (Modified from Harries, 2009.)

New developments in the use of radar include efforts to extend its use beyond that of a line-of-sight monitoring instrument to one that can resolve important 3D kinematic information (Severin *et al.*, 2011). A high-resolution 3D displacement map can now be constructed from continuous differential movements that are locally influenced by heterogeneity and major geologic structures such as faults (Fig. 19.5). Severin *et al.* (2011) have shown that this novel application of radar allows better definition of areas of high displacement (Fig. 19.5c) and an improved understanding of the slope kinematics (Fig. 19.5d). Movements between survey prisms no longer need to be extrapolated, and small areas of accelerating movement that may pose potential safety concerns can be identified and monitored.

Technological advances have also been made in subsurface monitoring (see Eberhardt and Stead, 2011). Borehole data are generally not relied upon for early warning, but they are extremely valuable for investigative monitoring. Developments in MEMS (micro-electrical mechanical systems) technology, which uses sub-micrometer electrical and mechanical components, have led to the use of smaller, lighter, more functional, and more reliable sensors that can be produced for a fraction of the cost of conventional transducers. Abdoun and Bennett (2008) describe a wireless MEMS-based borehole deformation system (ShapeAccelArray), consisting of a rope-like array of rigid segments containing triaxial MEMS accelerometers connected by flexible joints. Like an inclinometer, the accelerometers measure millimeter-scale deformations, but over larger deformations (up to tens of centimeters) and in real time (high-frequency sampling).

MEMS technology has also been used to develop inexpensive, durable borehole accelerometers for microseismic monitoring. These devices feature excellent bandwidth and the capacity to capture both high- and low-frequency events. Experience with monitoring microseismic activity in a developing rock-slope failure is limited, but field testing of MEMS accelerometers for this purpose has shown promise, capitalizing on the low cost and robustness of the sensors (e.g., Arosio *et al.*, 2009). A similar technology has been applied to soil slopes, but this focuses on higher-frequency acoustic emissions (Dixon *et al.*, 2003). Active waveguides are used to detect acoustic emission events generated by inter-particle friction in order to identify the location of a developing sliding surface and monitor slope displacement rates. Acoustic emission and microseismic monitoring are useful on different detection scales, from tens of centimeters to hundreds of meters, respectively; both potentially offer information on spatial clustering of events and source mechanisms, differentiating between tensile brittle fracturing and shear slip along internal fracture planes.

Another emerging technology is the application of fiber optics in high-resolution displacement transducers, piezometers, strain gauges, and temperature gauges. Of particular interest for landslide monitoring is the use of fiber Bragg grating (FBG) strain sensors as point measurement devices (Chen *et al.*, 2008; Moore *et al.*, 2010) and distributed Brillouin scattering sensors for monitoring distributed strains along the sensing fiber (Inaudi and Glisic, 2008). Moore *et al.* (2010) describe detailed field testing of a borehole FBG fiber optic strain monitoring system to resolve micrometer-scale deformations resulting from episodic rockslide movements over periods of seconds to hours. They report performance advantages, including high resolution, rapid sampling rates, multiplexing potential, and insensitivity to electrical disturbances such as lightning strikes. Cost, power consumption, conformance between the sensor and deforming borehole, durability, and a restricted range before breaking, limit the use of fiber optics in long-term monitoring (Moore *et al.*, 2010). In contrast, a low-cost, more durable early-warning system based on time-domain reflectometry (TDR) has been proposed by Thuro *et al.* (2011). The use of a metal-based coaxial cable instead of brittle glass fiber in TDR allows more strain and a greater range before breaking, but does so at the expense of measurement resolution.

19.5 INVESTIGATIVE MONITORING AND IN-SITU ROCKSLIDE LABORATORIES

The variable behavior of large landslides in response to seasonal precipitation patterns or changing reservoir levels poses challenges for both landslide early warning and understanding landslide failure mechanisms. Quantitative methods used to assess the stability of a landslide typically focus on the acting forces, but ignore factors related to the temporal evolution of strength degradation, internal shearing, and progressive failure, all of which contribute to the final collapse of the slope. Reliance on displacement monitoring for early warning and the difficulty in interpreting the data without a clear understanding

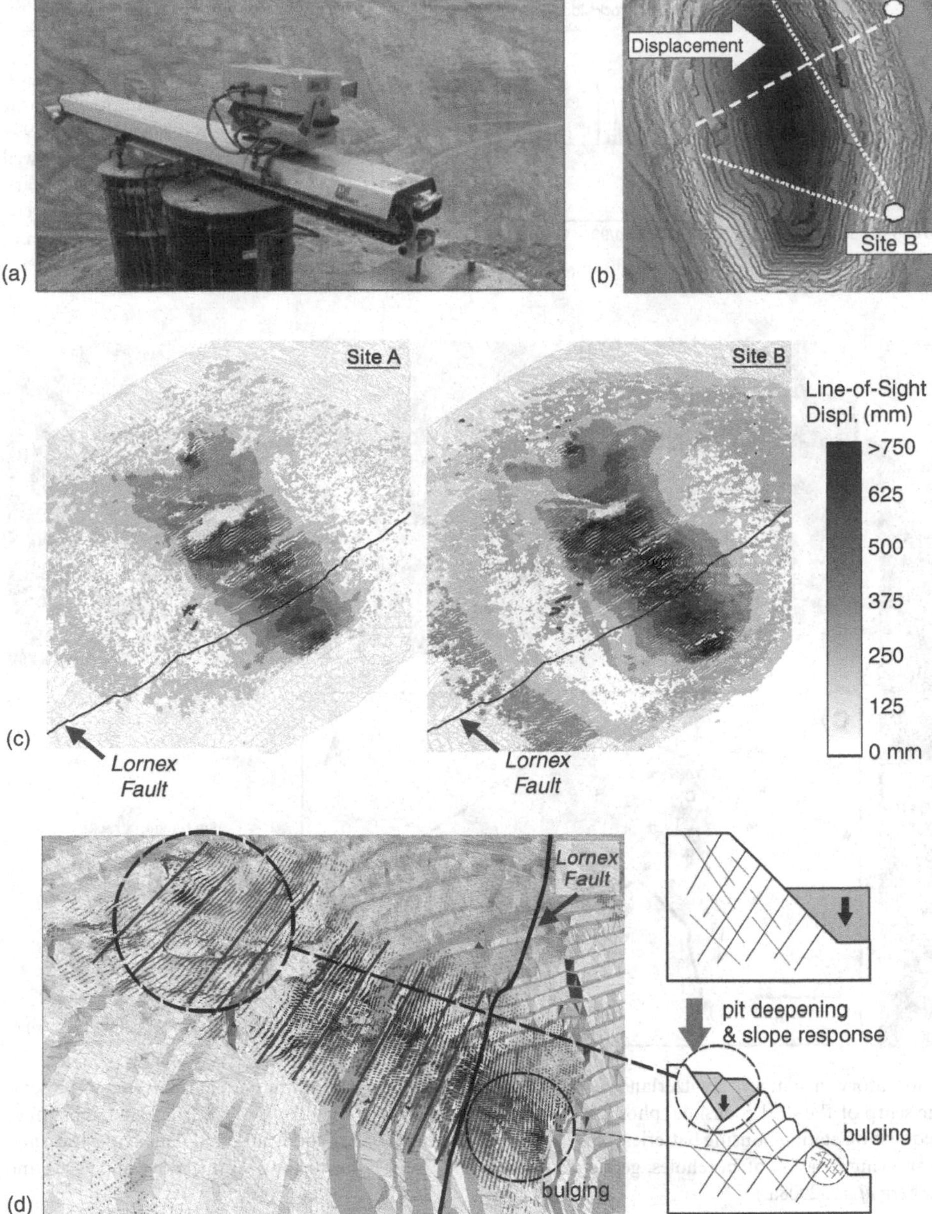

Fig. 19.5. Simultaneous deployment of two IBIS-M radar systems for "stereo" experiment: (a, b) setup at two different vantage points; (c) respective line-of-sight displacements recorded; and (d) combined 3D displacement map showing high-resolution vectors for displacements greater than 250 mm, and interpreted slope kinematics. (Data provided by J. Severin.)

of the underlying mechanisms contribute to variable, and often unreliable, early-warning predictions. The 1991 Randa rockslide occurred during a period of heavy snowmelt and precipitation; however, analysis of climatic data from previous years showed that the amount of water infiltrating the slope was not exceptional (Fig. 19.6; Eberhardt *et al.*, 2001). In many similar cases in which landslides have suddenly accelerated to failure, precipitation is not exceptional. These observations suggest that time-dependent mechanisms relating to rock mass strength degradation and progressive failure bring the slope to the point of failure, with snowmelt and precipitation providing the final impetus for catastrophic failure (Eberhardt *et al.*, 2004b).

The need to better understand complex landslide behavior in order to improve predictive monitoring has catalyzed several recent large multidisciplinary studies, including the Randa In-Situ Rockslide Laboratory in Switzerland (Loew *et al.*, 2012, Chapter 24, this volume), the Turtle Mountain Field Laboratory in western Canada (Froese *et al.*, 2012, Chapter 25, this volume), and the Åknes/Tafjord Project in Norway (Blikra, 2012, Chapter 26, this volume). The first of these, the Randa In-Situ Rockslide Laboratory, was developed, in part, to improve early-warning capabilities in the presence of non-persistent discontinuities, multiple moving blocks, and internal shearing in massive, steep crystalline rock slopes. A

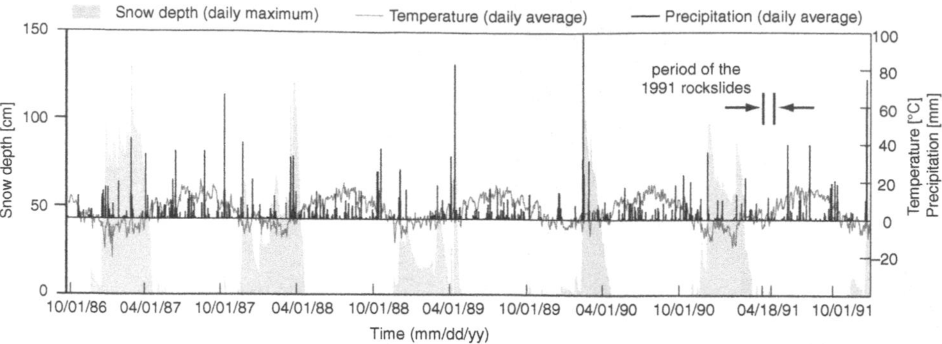

Fig. 19.6. Snow depth, temperature, and precipitation records for the period preceding the 1991 Randa rockslide events, as well as those for the previous years. Data provided by MeteoSchweiz. (Modified from Eberhardt *et al.*, 2001.)

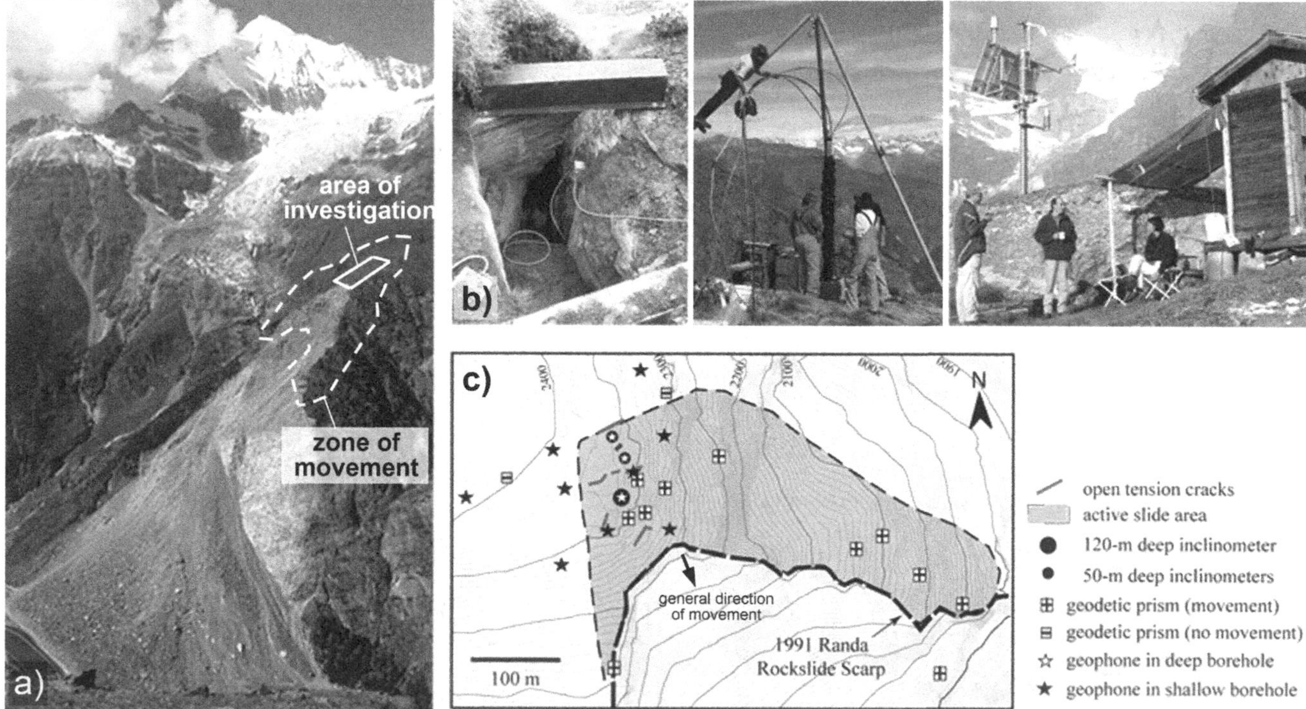

Fig. 19.7. The Randa In-Situ Rockslide Laboratory in southern Switzerland. (a) Area of investigation (solid white line) and outline of present-day instability (white dashed line) above the scarp of the 1991 rockslide (photograph by H. Willenberg). (b) Installation of surface and subsurface monitoring instruments and central data acquisition station housing batteries, power generation sources (solar and wind), and data acquisition and transmission hardware. (c) Plan view map showing location of boreholes, geodetic reflectors, and geophones relative to the active slide area and open tension cracks. (Modified from Willenberg *et al.*, 2008a.)

high alpine facility was constructed above the scarp of the 1991 rockslide (Fig. 19.7a), where ongoing movements of 1–2 cm per year are being recorded in a gneissic rock mass with a volume of up to 10 million m^3. This facility includes a variety of instrumentation systems designed to measure temporal and 3D spatial relationships between fracture systems, displacements, pore pressures, and microseismicity (Fig. 19.7b, c). The monitoring has been complemented by detailed geological mapping, acquisition of borehole televiewer data, and surface and borehole seismic and radar surveys to determine the distribution of fractures, identify key geologic features, and develop a 3D geological model of the unstable rock mass (Willenberg *et al.*, 2008b).

The Åknes/Tafjord Project involved a similar, multidisciplinary field campaign to investigate a 30–40 million m^3 deep-seated rockslide moving at 2–4 cm per year high above a fjord. The rockslide, if sudden failure occurred, could generate a tsunami that would impact people and infrastructure along the fjord (Blikra *et al.*, 2005). GPS, resistivity, georadar, reflection and refraction seismic, and airborne laser scanning surveys were carried out to define the subdomains within the sliding body and characteristics of the internal and basal sliding surfaces (Ganerød *et al.*, 2008). Based on these findings, subsequent work was focused on the shear strength characteristics of a composite sliding surface involving shear fractures filled with gouge, non-persistent joints, and intact rock bridges, as well as the potential

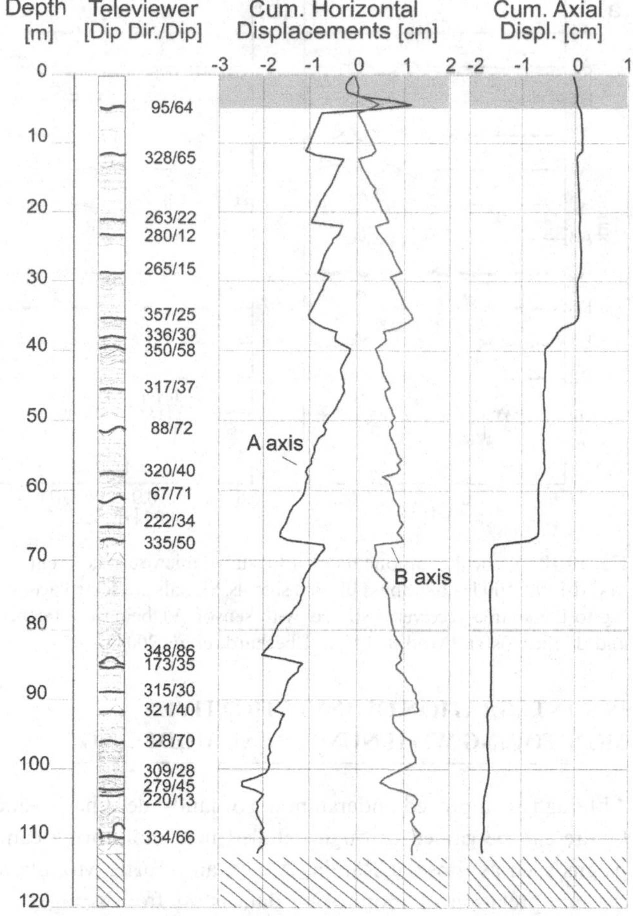

Fig. 19.8. Integrated borehole dataset from Randa, correlating optical televiewer log and traces of active fractures with cumulative inclinometer and axial extensometer displacements over a 5-year period. The inclinometer profiles are top-down integrated with the positive A-axis approximately aligned in the direction of overall slope movement. (Modified from Willenberg *et al.*, 2008a.)

for strength degradation over time and corresponding failure scenarios (Grøneng *et al.*, 2010).

19.5.1 DISPLACEMENTS AND 3D KINEMATICS

Resolving the complex displacement fields generated by multiple moving blocks and relating these to the geology and fracture network is a key objective in the design of an instrumentation network. At Turtle Mountain, a combination of surface-based crackmeters, wire-line extensometers, and tiltmeters were used, together with remote sensing data, to differentiate moving subdomains and to better understand the real-time monitoring data relative to each subdomain (Froese *et al.*, 2009). At Randa, surface displacement monitoring was supplemented with deep inclinometer measurements. The deepest inclinometer (120 m) was also fitted with a borehole extensometer to resolve the profile of 3D displacement vectors at depth (Fig. 19.8). Biaxial in-place inclinometers were installed and positioned at depths coinciding with key fractures identified through borehole televiewer surveys.

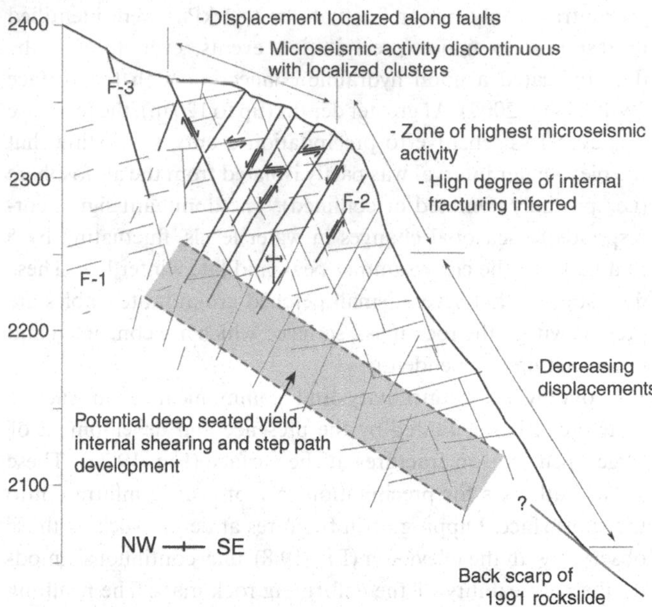

Fig. 19.9. Kinematic model of ongoing block movements at Randa relative to 3D network of discontinuities and faults mapped using geological and geophysical methods. (Modified from Willenberg *et al.*, 2008a.)

Integration of the different datasets into a 3D geological model showed that the displacements recorded on surface and at depth were localized across active discontinuities and that the kinematic behavior of the slope is dominated by complex internal block movements rather than those of a coherently sliding single mass (Willenberg *et al.*, 2008a). Near-surface block movements are dominated by normal fault slip, with blocks between these fractures rotating downslope (Fig. 19.9) and producing a number of graben structures visible at the surface.

19.5.2 PORE-PRESSURE RESPONSES AT DEPTH

Fracture networks affect the distribution of pore pressures at depth and thus have a coupled relationship with unstable rock-slope movements. The design of the monitoring networks at Randa, Åknes, and Turtle Mountain included borehole monitoring of pore pressures. At Randa, piezometers were positioned and packed off along zones of potentially higher fracture permeability identified from borehole televiewer data. The data showed several water tables distributed within the rock mass and different types of pore-pressure interaction with infiltrating surface water and atmospheric pressure variations (Willenberg, 2004). This situation is common in deep-seated landslides in crystalline rock, where preferential fracture permeability and hydraulic barriers (e.g., from fault gouge) result in isolated compartments of groundwater flow and reaction delays between surface precipitation and pressure responses at depth. There may be a poor correlation between slope movements and precipitation events in such situations (Bonzanigo *et al.*, 2007).

Major fluctuations in pore pressures (>10 kPa) were detected at shallow depths (<50 m) at Randa during spring snowmelt.

In contrast, only minor fluctuations (>2 kPa) were identified in response to major precipitation events, even though the data indicated a good hydraulic connection with the surface (Willenberg, 2004). At greater depths (up to 120 m), the response was even less sensitive to precipitation events, suggesting that the piezometer interval was partly isolated from the atmosphere (i.e., partially confined or confined). The dominant signal corresponds to seasonal changes in water levels, fluctuating by 5 kPa between the early summer peak and late winter low. These data suggest that several small perched groundwater tables are present within the rock mass, some of which are connected and others acting independently.

Groundwater connectivity and communication in such a system can be facilitated by the presence and development of large open tension fractures at the surface (Fig. 19.7c). These act as pathways for precipitation to more easily infiltrate into the subsurface. Dipping open fractures at depth, such as those observed with the televiewer (Fig. 19.8), may continuously modify the permeability of the deforming rock mass. The resulting complex, heterogeneous, hydrogeologic situation introduces a variety of reaction delays between surface precipitation and movement at depth along sliding surfaces. Precipitation quickly infiltrates into upper portions of the slide mass, resulting in sudden movements, some of which are localized; whereas seasonal fluctuations in the groundwater drive more extensive and continuous deep-seated movements over a longer period of time (Bonzanigo *et al.*, 2007).

19.5.3 MICROSEISMICITY

An additional key monitoring strategy at Randa, Åknes, and Turtle Mountain is the use of microseismic instruments. At Randa, detection and location of spatially clustered microseismic events provide information on zones of activity related to slip along existing fractures and newly generated fractures. The microseismic network includes three 28 Hz triaxial geophones mounted in three deep boreholes (50–120 m deep), nine 8 Hz triaxial geophones mounted in shallow boreholes (up to 5 m deep), and two 24-channel seismographs (Fig. 19.7c). The 11 geophones were distributed to ensure that the seismic sources could be reliably constrained within the area of the landslide (Spillmann *et al.*, 2007). A key finding of the microseismic monitoring program at Randa was that the higher frequency content of the recorded events was strongly attenuated (Fig. 19.10), pointing to the presence of large open fractures at depth. Open fractures were inferred to depths of up to 100 m from the borehole televiewer and pore-pressure data (Willenberg *et al.*, 2008b).

The data indicate that seismic events occur mainly in the area where the slope is moving, and are most numerous near the retrogressive backscarp of the 1991 failure and where the mapped fault density is highest, indicating a high degree of fracturing of the rock mass (Spillmann *et al.*, 2007). Many of the active fractures are aseismic. Seismicity tends to be localized in patches along fractures, suggesting that movement is impeded by asperities such as rock bridges (Willenberg *et al.*, 2008a).

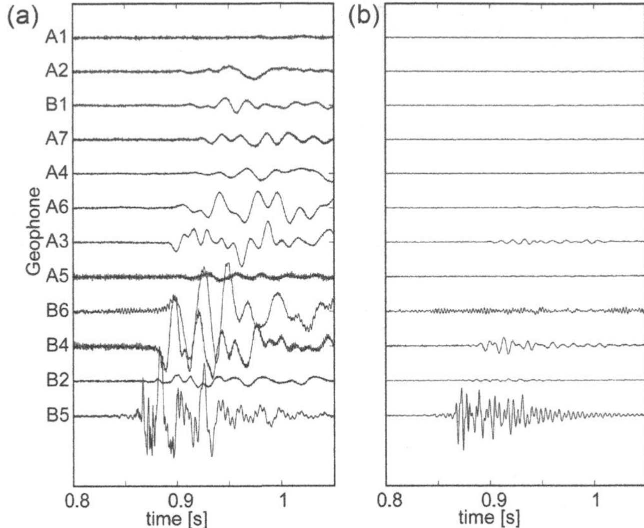

Fig. 19.10. Vertical components of a locatable microseismic event: (a) raw; (b) 100–500 Hz bandpass filtered signals. Signals are sorted according to the source–receiver distance, with sensor A1 being the farthest and B5 the closest. (Modified from Eberhardt *et al.*, 2004a.)

19.6 INTEGRATION OF INVESTIGATIVE MONITORING WITH NUMERICAL MODELING

Although an improved understanding of landslide behavior and failure can be gained through detailed field monitoring campaigns such as those at Randa, Åknes, and Turtle Mountain, the interpretation of monitoring data is far from straightforward. Numerical modeling serves as an important additional investigative tool; its role is to provide insights into the complex cause-and-effect relations that govern the behavior of a landslide. Investigative monitoring provides a means to calibrate and constrain sophisticated 2D and 3D numerical models (Kalenchuk *et al.*, 2012, Chapter 28, this volume), whereas investigative numerical models are used to interpret and understand complex field measurements (Eberhardt and Willenberg, 2005; Watson *et al.*, 2006). Thus emphasis is placed on integrated studies that include field measurements and numerical stability analyses to provide a more complete understanding of the hazard and lead to the establishment of more informed early-warning thresholds and response protocols.

Numerical modeling studies have been carried out at Randa, Åknes, and Turtle Mountain, in order to understand the block kinematics and deformation mechanisms contributing to the displacements determined by monitoring (Eberhardt *et al.*, 2004b; Willenberg, 2004; Eberhardt, 2008; Grøneng *et al.*, 2010; Brideau *et al.*, 2011). Figure 19.11 is an example of output from a series of distinct element models generated for the Randa study, incorporating the active geologic structures identified through mapping and investigative monitoring. The blocks between these structures are modeled using a Mohr–Coulomb elasto-plastic, strain-softening, constitutive model to simulate a brittle rock response. Properties are scaled to those for an equivalent continuum to account for smaller-scale discontinuities not explicitly

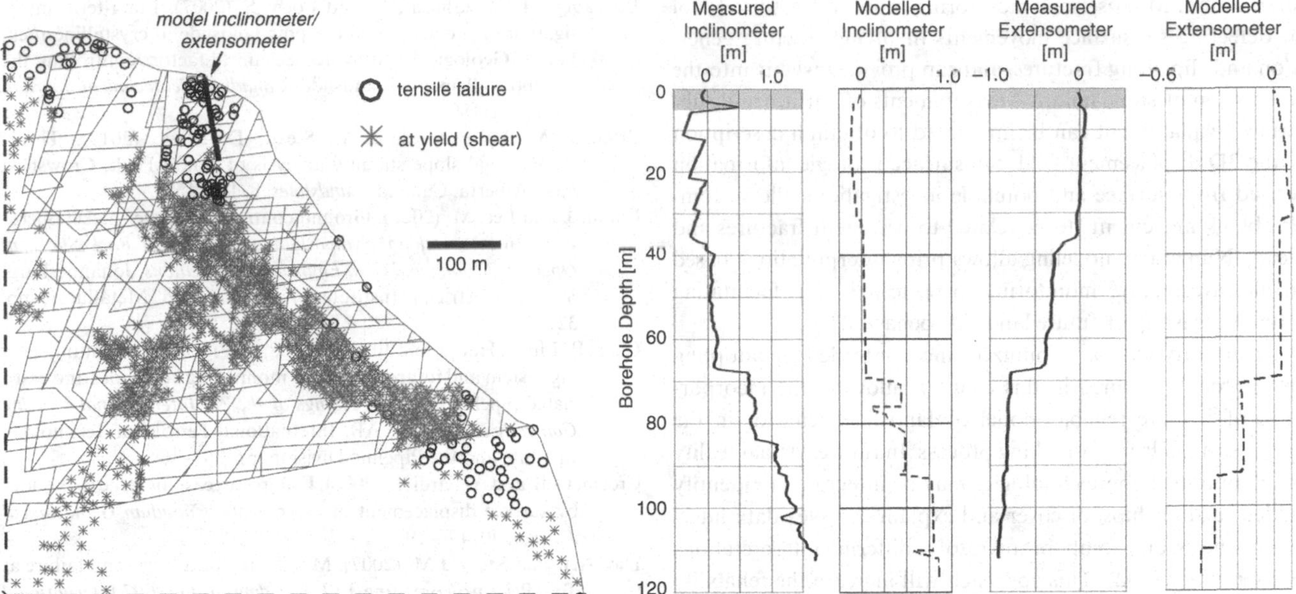

Fig. 19.11. Distinct element modeling of complex rock-slope displacements at Randa and comparison between measured and modeled cumulative displacement profiles, assuming slip along discontinuities and elasto-plastic block deformation. Note that model boundaries extend beyond those shown, as indicated by the dashed boundary line. (Modified from Eberhardt, 2008.)

included in the distinct element model. The results indicate the potential for a deeper-seated rupture surface extending below the investigation boreholes (Fig. 19.11), which agrees with the explicit brittle fracture modeling results of Eberhardt *et al.* (2004b). Deformation patterns in the model are in agreement with the measured toppling and translation movements of blocks in the upper part of the slope. Below this, increasing stresses with depth and extensional downslope strains promote a more complex type of deformation, involving rock mass yield, brittle fracture, slip, and opening along existing structures, which combine to produce a small outward rotation and translation of the deforming rock mass. This pattern of deformation agrees with the kinematic model developed by Willenberg *et al.* (2008a) on the basis of mapping and monitoring data (Fig. 19.9). Although these processes cannot be confirmed, due to the limited depth of the investigation boreholes, the results illustrate the benefits of incorporating a mechanistic understanding of rock-slope failure when interpreting monitoring data.

Ongoing work at Randa (Loew *et al.*, 2012, Chapter 24, this volume) includes the use of remote sensing tools to better constrain and interpret the extent and kinematics of the deformations (Gischig *et al.*, 2011a). Ground-based InSAR is being used to monitor the inaccessible steep slope face formed by the 1991 rockslide and is providing important constraints on the location of the basal rupture and lateral release surfaces. LIDAR and photogrammetry are likewise being used to collect data from inaccessible portions of the slope face. A recent kinematic analysis of these data by Gischig *et al.* (2011a) suggests that a rupture surface transitions from stepped-planar sliding at the base, to failed rock bridges in the center, to the opening of tension cracks in the vertical backscarp. Toppling is the dominant mode on the upper part of the slope, with translational sliding below, in agreement with the kinematic model of Willenberg *et al.* (2008a).

Researchers are also examining correlations between seasonal temperature cycles and surface deformation trends at Randa (Gischig *et al.*, 2011b). This work suggests that thermomechanical effects are especially important at Randa because of the limited amount of groundwater in the unstable rock mass. Similarly, preliminary findings at Turtle Mountain suggest that thermal cycling contributes more to measured slope deformation than heavy precipitation (Moreno and Froese, 2007).

19.7 CONCLUSIONS AND WAY FORWARD

Geologic complexity and uncertainty are significant obstacles to landslide hazard assessment, monitoring, and early warning. Predictive slope monitoring must be preceded by investigative monitoring to provide an understanding of slope behavior over time, as well as the typical responses to external events such as heavy precipitation and seasonal fluctuations in groundwater. Emerging monitoring technologies, including ground-based radar, are helping to provide real-time, high-resolution deformation data. More confidence in interpreting the kinematics of movement and approaching failure can be gained if several instrumentation systems are used to provide different spatial and temporal coverage, as well as redundancy (Froese and Moreno, 2011).

Experimental schemes integrating state-of-the-art site characterization, monitoring, and numerical modeling are being used to better understand the mechanisms controlling slope deformation. Microseismic monitoring is an underutilized tool for detecting subsurface movements involving fracture generation and slip along fractures, and can provide insights into the progression of slope failure. Measurements of surface and subsurface displacement can be integrated to obtain a description of the 3D displacement field. Subsurface geologic information derived from surface and borehole investigations allows complex block movements to be related to persistent fractures and shears. Numerical modeling allows prior interpretations based on the investigative monitoring to be tested, thus facilitating better forecasting of future landslide behavior.

Uncertainty can be minimized and controlled by adopting the approach outlined in this chapter, allowing the incorporation of a more reliable model of landslide behavior in the early-warning decision-making process. In future, virtual reality and data immersion technologies may be integrated to identify hidden relationships, discover and explain complex data interdependencies, and compare and resolve differing interpretations (Kaiser *et al.*, 2002). This approach will increase the reliability of data and practitioners' confidence in them, facilitating the development of monitoring procedures, thresholds, alert levels, and notification and emergency response protocols (e.g., Froese and Moreno, 2011).

ACKNOWLEDGEMENTS

I acknowledge and thank the researchers whose results I drew upon for the material presented here, including Luca Bonzanigo, Christian Zangerl, and Jordan Severin, and those connected to the Randa In-Situ Rockslide Laboratory team, including Simon Loew, Heike Willenberg, Keith Evans, Hansruedi Mauer, Tom Spillmann, Björn Heincke, Alan Green, and Doug Stead. Special thanks are also extended to Doug Stead and John Clague for their valuable review comments.

REFERENCES

Abdoun, T. and Bennett, V. (2008). A new wireless MEMS-based system for real time-deformation monitoring. *Geotechnical News*, 26(1), 36–40.

Arosio, D., Longoni, L., Papini, M. *et al.* (2009). Towards rockfall forecasting through observing deformations and listening to microseismic emissions. *Natural Hazards and Earth System Sciences*, 9, 1119–1131.

Bhandari, R.K. (1988). Special lecture: Some practical lessons in the investigation and field monitoring of landslides. In *Proceedings of the 5th International Symposium on Landslides*, ed. C. Bonnard. Rotterdam: A.A. Balkema, pp. 1435–1457.

Blikra, L.H. (2012). The Åknes rockslide, Norway. In *Landslides: Types, Mechanisms and Modeling*, ed. J.J. Clague and D. Stead. Cambridge, UK: Cambridge University Press, pp. 323–335.

Blikra, L.H., Longva, O., Harbitz, C. and Løvholt, F. (2005). Quantification of rock-avalanche and tsunami hazard in Storfjorden, western Norway. In *Landslides and Avalanches*, ed. K. Senneset, K. Flaate and J.O. Larsen. London: Taylor & Francis Group, pp. 57–64.

Bonzanigo, L., Eberhardt, E. and Loew, S. (2007). Long-term investigation of a deep-seated creeping landslide in crystalline rock. Part 1. Geological and hydromechanical factors controlling the Campo Vallemaggia landslide. *Canadian Geotechnical Journal*, 44, 1157–1180.

Brideau, M.-A., Pedrazzini, A., Stead, D. *et al.* (2011). Three-dimensional slope stability analysis of South Peak, Crowsnest Pass, Alberta, Canada. *Landslides*, 8, 139–158.

Cahill, J. and Lee, M. (2006). Ground control at Leinster nickel operations. In *International Symposium on Stability of Rock Slopes in Open Pit Mining and Civil Engineering Situations*. Johannesburg: Southern African Institute of Mining and Metallurgy, pp. 321–334.

Chen, P., Liu, J., Hao, J. and Jing, H. (2008). A Fiber Bragg Grating sensing system and its application to monitoring landslides and associated pipelines. In *Proceedings of the 7th International Pipeline Conference*. Calgary, AB: International Petroleum Technology Institute and the Pipeline Division, pp. 363–368.

Crosta, G.B. and Agliardi, F. (2003). Failure forecast for large rock slides by surface displacement measurements. *Canadian Geotechnical Journal*, 40, 176–191.

Day, A.P. and Seery, J.M. (2007). Monitoring of a large wall failure at Tom Price iron ore mine. In *Proceedings of the 2007 International Symposium on Rock Slope Stability in Open Pit Mining and Civil Engineering*, ed. Y. Potvin. Perth: Australian Centre for Geomechanics, pp. 333–340.

Dixon, N., Hill, R. and Kavanagh, J. (2003). Acoustic emission monitoring of slope instability: Development of an active waveguide system. *Proceedings of the Institution of Civil Engineers, Geotechnical Engineering*, 156, 83–95.

Eberhardt, E. (2008). Twenty-Ninth Canadian Geotechnical Colloquium: The role of advanced numerical methods and geotechnical field measurements in understanding complex deep-seated rock slope failure mechanisms. *Canadian Geotechnical Journal*, 45, 484–510.

Eberhardt, E. and Stead, D. (2011). Geotechnical instrumentation. In *SME Mining Engineering Handbook* (3rd edn.), ed. P. Darling. Engelwood, CO: Society for Mining, Metallurgy & Exploration, pp. 551–572.

Eberhardt, E. and Willenberg, H. (2005). Using rock slope deformation measurements to constrain numerical analyses, and numerical analyses to constrain rock slope deformation measurements. In *Proceedings of the 11th International Conference on Computer Methods and Advances in Geomechanics*, ed. G. Barla and M. Barla. Bologna: Pàtron Editore, pp. 683–692.

Eberhardt, E., Willenberg, H., Loew, S. and Maurer, H. (2001). Active rockslides in Switzerland: Understanding mechanisms and processes. In *International Conference on Landslides: Causes, Impacts and Countermeasures*, ed. H.H. Einstein, E. Krauter, H. Klapperich and R. Pöttler. Essen: Verlag Glückauf Essen, pp. 25–34.

Eberhardt, E., Spillmann, T., Maurer, H. *et al.* (2004a). The Randa Rockslide Laboratory: Establishing brittle and ductile instability mechanisms using numerical modelling and microseismicity. In *Landslides: Evaluation and Stabilization*, ed. W.A. Lacerda, M. Ehrlich, S.A.B. Fontoura and A.S.F. Sayão. Leiden: A.A. Balkema, pp. 481–487.

Eberhardt, E., Stead, D. and Coggan, J.S. (2004b). Numerical analysis of initiation and progressive failure in natural rock slopes: The 1991 Randa rockslide. *International Journal of Rock Mechanics and Mining Sciences*, 41, 69–87.

Eberhardt, E., Bonzanigo, L. and Loew, S. (2007). Long-term investigation of a deep-seated creeping landslide in crystalline rock. Part 2. Mitigation measures and numerical modelling of deep drainage at Campo Vallemaggia. *Canadian Geotechnical Journal*, 44, 1181–1199.

Eberhardt, E., Watson, A.D. and Loew, S. (2008). Improving the interpretation of slope monitoring and early warning data through better understanding of complex deep-seated landslide failure mechanisms. In *Landslides and Engineered Slopes: From the Past to the Future*, ed. Z. Chen, J. Zhang, Z. Li, F. Wu and K. Ho. London: Taylor & Francis, pp. 39–51.

Froese, C.R. and Moreno, F. (2011). Structure and components for the emergency response and warning system on Turtle Mountain, Alberta, Canada. *Natural Hazards*, doi:10.1007/s11069-011-9714-y.

Froese, C.R., Moreno, F., Jaboyedoff, M. and Cruden, D.M. (2009). 25 years of movement monitoring on South Peak, Turtle Mountain: Understanding the hazard. *Canadian Geotechnical Journal*, 46, 256–269.

Froese, C.R., Charrière, M., Humair, F., Jaboyedoff, M. and Pedrazzini, A. (2012). Characterization and management of rockslide hazard at Turtle Mountain, Alberta, Canada. In *Landslides: Types, Mechanisms and Modeling*, ed. J.J. Clague and D. Stead. Cambridge, UK: Cambridge University Press, pp. 310–322.

Fukuzono, T. (1985). A new method for predicting the failure time of a slope. In *Proceedings of the 4th International Conference and Field Workshop on Landslides*. Tokyo: National Research Center for Disaster Prevention, pp. 145–150.

Ganerød, G.V., Grøneng, G., Rønning, J.S. *et al.* (2008). Geological model of the Åknes rockslide, western Norway. *Engineering Geology*, 102, 1–18.

Gischig, V., Amann, F., Moore, J.R. *et al.* (2011a). Composite rock slope kinematics at the current Randa instability, Switzerland, based on remote sensing and numerical modeling. *Engineering Geology*, 118, 37–53.

Gischig, V.S., Moore, J.R., Evans, K.F., Amann, F. and Loew, S. (2011b). Thermomechanical forcing of deep rock slope deformation. 2. The Randa rock slope instability. *Journal of Geophysical Research*, doi:10.1029/2011JF002007.

Grøneng, G., Lu, M., Nilsen, B. and Jenssen, A.K. (2010). Modelling of time-dependent behavior of the basal sliding surface of the Åknes rockslide area in western Norway. *Engineering Geology*, 114, 414–422.

Harries, N. (2009). Rock slope hazard management associated with rapid brittle failure in open pit mines. In *Slope Stability 2009: Proceedings of the International Symposium on Rock Slope Stability in Open Pit Mining and Civil Engineering*. Santiago: Universidad de Los Andes, e-paper.

Harries, N.J. and Roberts, H. (2007). The use of slope stability radar (SSR) in managing slope instability hazards. In *Rock Mechanics: Meeting Society's Challenges and Demands*, ed. E. Eberhardt, D. Stead and T. Morrison. London: Taylor & Francis, pp. 53–59.

Inaudi, D. and Glisic, B. (2008). Distributed fibre-optic sensing for long-range monitoring of pipelines. *Monitor*, 2008 (March), 35–41.

Ischi, H., Keusen, H.R. and Scheller, E. (1991). *Randa, Kt. Wallis: Bergsturz Grossgufer vom April/Mai 1991: Zusammenfassender Bericht über die Aktivitäten der Geotest AG*. Martigny: Geotest AG.

Kaiser, P.K., Henning, J.G., Cotesta, L. and Dasys, A. (2002). Innovations in mine planning and design utilizing collaborative virtual reality (CIRV). In *Proceedings of the 104th CIM Annual General Meeting*. Vancouver: Canadian Institute of Mining, Metallurgy and Petroleum, e-paper.

Kalenchuk, K.S., Hutchinson, D.J., Diederichs, M. and Moore, D. (2012). Downie Slide, British Columbia, Canada. In *Landslides: Types, Mechanisms and Modeling*, ed. J.J. Clague and D. Stead. Cambridge, UK: Cambridge University Press, pp. 345–358.

Little, M.J. (2006). Slope monitoring strategy at PPRust open pit operation. In *International Symposium on Stability of Rock Slopes in Open Pit Mining and Civil Engineering Situations*. Johannesburg: Southern African Institute of Mining and Metallurgy, pp. 211–230.

Loew, S., Gischig, V., Willenberg, H., Alpiger, A. and Moore, J.R. (2012). Randa: Kinematics and driving mechanisms of a large complex rockslide. In *Landslides: Types, Mechanisms and Modeling*, ed. J.J. Clague and D. Stead. Cambridge, UK: Cambridge University Press, pp. 297–309.

Meisina, C., Zucca, F., Fossati, D., Ceriani, M. and Allievi, J. (2005). PS InSAR integrated with geotechnical GIS: Some examples from southern Lombardia. In *Geodetic Deformation Monitoring: From Geophysical to Engineering Roles*, ed. F Sansò and A.J. Gil,. Berlin: Springer, pp. 65–72.

Moore, D.P., Imrie, A.S. and Baker, D.G. (1991). Rockslide risk reduction using monitoring. In *Proceedings of the Canadian Dam Association Meeting*. Whistler, BC: Canadian Dam Safety Association, pp. 245–258.

Moore, J.R., Gischig, V., Button, E. and Loew, S. (2010). Rockslide deformation monitoring with fiber optic strain sensors. *Natural Hazards and Earth System Sciences*, 10, 191–201.

Moreno, F. and Froese, C.R. (2007). Turtle Mountain Field Laboratory (TMFL). Part 2. Review of trends 2005 to 2006. In *Proceedings of the First North American Landslide Conference*, ed. V.R. Schaefer, R.L. Schuster and A.K. Turner. Association of Environmental and Engineering Geologists, Publication 23.

Rödelsperger, S., Läufer, G., Gerstenecker, C. and Becker, M. (2010). Monitoring of displacements with ground-based microwave interferometry: IBIS-S and IBIS-L. *Journal of Applied Geodesy*, 4, 41–54.

Rose, N.D. and Hungr, O. (2007). Forecasting potential rock slope failure in open pit mines using the inverse-velocity method. *International Journal of Rock Mechanics and Mining Sciences*, 44, 308–320.

Saito, M. (1965). Forecasting the time of occurrence of a slope failure. In *Proceedings of the 6th International Conference on Soil Mechanics and Foundation Engineering*. Toronto: University of Toronto Press, pp. 537–541.

Salt, G. (1988). Landslide mobility and remedial measures. In *Proceedings of the 5th International Symposium on Landslides*, ed. C. Bonnard. Rotterdam: A.A. Balkema, pp. 757–762.

Schindler, C., Cuénod, Y., Eisenlohr, T. and Joris, C.-L. (1993). Die Ereignisse vom 18 April und 9 Mai 1991 bei Randa (VS): Ein atypischer Bergsturz in Raten. *Eclogae Geologicae Helvetiae*, 86, 643–665.

Severin, J., Eberhardt, E., Leoni, L. and Fortin, S. (2011). Use of ground-based synthetic aperture radar to investigate the complex 3-D kinematics of a large open pit slope. In *Proceedings of the 12th International Congress on Rock Mechanics*, Beijing.

Spillmann, T., Maurer, H., Green, A.G. *et al.* (2007). Microseismic investigation of an unstable mountain slope in the Swiss Alps. *Journal of Geophysical Research*, 112(B07301), doi:10.1029/2006JB004723.

Thuro, K., Singer, J. and Festl, J. (2011). Towards a low cost 3D early warning system for unstable alpine slopes: The Aggenalm landslide monitoring system. In *Slope Stability 2011, International Symposium on Rock Slope Stability in Open Pit Mining and Civil Engineering*, ed. E. Eberhardt and D. Stead. Vancouver: Canadian Rock Mechanics Association, e-paper.

Watson, A.D., Martin, C.D., Moore, D.P., Stewart, T.W.G. and Lorig, L.L. (2006). Integration of geology, monitoring and modeling to assess rockslide risk. *Felsbau*, 24(3), 50–58.

Willenberg, H. (2004). *Geologic and kinematic model of a complex landslide in crystalline rock (Randa, Switzerland)*. D.Sc. thesis, Swiss Federal Institute of Technology, Zurich.

Willenberg, H., Evans, K.F., Eberhardt, E., Spillmann, T. and Loew, S. (2008a). Internal structure and deformation of an unstable crystalline rock mass above Randa (Switzerland). Part II. Three-dimensional deformation patterns. *Engineering Geology*, 101, 15–32.

Willenberg, H., Loew, S., Eberhardt, E. *et al.* (2008b). Internal structure and deformation of an unstable crystalline rock mass above Randa (Switzerland). Part I. Internal structure from integrated geological and geophysical investigations. *Engineering Geology*, 101, 1–14.

Zangerl, C., Eberhardt, E. and Perzlmaier, S. (2010). Kinematic behaviour and velocity characteristics of a complex deep-seated crystalline rockslide system in relation to its interaction with a dam reservoir. *Engineering Geology*, 112, 53–67.

20 Groundwater in slopes

LUCIANO PICARELLI, SERGE LEROUEIL, LUCIO OLIVARES, LUCA PAGANO, PAOLO TOMMASI, AND GIANFRANCO URCIUOLI

ABSTRACT

Groundwater flows in the pores and fractures of soils and rocks, driven by changing mechanical and environmental boundary conditions, for example air temperature, humidity, and precipitation. Because the groundwater environment continuously changes, the steady-state condition is only a theoretical concept: the movement of water in soils and rocks is transient, and any change is delayed with respect to the trigger. This chapter focuses on the role of time and the effects of anthropogenic, natural, and climatic processes on groundwater flow and slope behavior. The unsaturated condition is considered a normal soil state and the saturated condition a special case.

20.1 THE ROLE OF GROUNDWATER IN SLOPE STABILITY

Geoscientists have long been aware of the role that groundwater plays in the behavior of slopes. The influence of weather is, of course, pre-eminent, and the ground surface can behave either as a source of water or as a hot and dry boundary capable of extracting water from the subsoil, depending on season. The changing nature of this boundary implies continuous modifications in both the physical state of soil (e.g., water content, degree of saturation) and the state of stress.

Precipitation is the main cause of slope failure (Sidle and Ochiai, 2006). Through infiltration, it causes a generalized increase in pore water pressures, leading to an increase in the driving forces on a slope and a decrease in the resisting forces. The consequence is a general reduction in stability conditions. Conversely, lack of precipitation and high evapotranspiration cause a decrease in soil water content and reduced pore water pressures, and therefore lower driving forces and higher resisting forces. The effects on groundwater of both precipitation and evapotranspiration are normally delayed. A significant consequence is that the groundwater flow is transient, and thus the steady-state flow represents an unusual state, which can be established only under special circumstances. In cold environments, snow melting and freezing play a similar role with respect to groundwater flow.

Changes in the groundwater flow regime can also be caused by anthropogenic activities and geologic processes. Cuttings and surcharges, erosion and sedimentation, impounding of artificial reservoirs, land flooding, rapid drawdown, and other processes cause delayed changes in the groundwater flow system and have mechanical effects on slopes.

Today, mathematical models and the availability of powerful computational tools help account for these processes and allow a better appraisal of the role of time. On the other hand, the presence of saturated and unsaturated zones is no longer a computational problem due to improved knowledge of the properties and behavior of unsaturated soils. The steady-state condition, however, is still commonly assumed in numerical analyses that are performed to solve engineering problems. The important role of time thus vanishes from the computation and sometimes even from the mind of the investigator. The effects of anthropogenic and natural processes, and the role of time, on groundwater and slope behavior are the focus of this chapter.

20.2 BASIC GOVERNING EQUATIONS

The behavior of partially saturated soils can be related to two independent stress variables, the *net stress* (Jennings and Burland, 1962; Fredlund and Morgenstern, 1977):

$$\sigma^*_{ij} = \sigma_{ij} - \delta_{ij} u_a \tag{20.1}$$

Landslides: Types, Mechanisms and Modeling, ed. John J. Clague and Douglas Stead. Published by Cambridge University Press.

and the *matric suction* (hereafter called suction):

$$s = u_a - u_w. \tag{20.2}$$

In these expressions, σ_{ij} is the total stress tensor, u_a is pore air pressure, u_w is pore water pressure in unsaturated soil, σ^*_{ij} is the net stress tensor, and δ_{ij} is the Kronecker operator.

In applications in which the soil does not collapse upon wetting, and often for practical reasons, it is convenient to adopt the following expression, which combines Eqs. 20.1 and 20.2 into the equivalent effective stress tensor:

$$\sigma'_{ij} = \sigma_{ij} - \delta_{ij}[\chi u_w + (1-\chi)u_a] = [\sigma_{ij} - \delta_{ij}u_a] + [\delta_{ij}\chi(u_a - u_w)] = \sigma^*_{ij} + \delta_{ij}\chi s \tag{20.3}$$

where χ is a coefficient ranging from 0 for dry conditions to 1 for saturated conditions (Bishop, 1959). For saturated soils (null suction or $\chi = 1$), both approaches lead to Terzaghi's effective stress principle.

Fredlund *et al.* (1978) proposed the following extension of the Mohr–Coulomb shear strength expression for unsaturated soils:

$$\tau_{\text{lim}} = c' + (\sigma - u_a)\tan\phi' + (u_a - u_w)\tan\phi^b \tag{20.4}$$

where τ_{lim} is shear strength, σ is total stress normal to the shear plane, c' is soil skeleton cohesion, ϕ' is soil skeleton friction, and ϕ^b is the friction angle related to suction ($\tan\phi^b = \chi \tan\phi'$; Eq. 20.3). The suction-dependent component of the shear strength $(u_a - u_w)\tan\phi^b$ is known as the apparent cohesion (Taylor, 1948). For $(u_a - u_w) = 0$, Eq. 20.4 becomes the well-known Mohr–Coulomb shear strength criterion for saturated soils.

Although more representative, but more complex, equations exist, Eq. 20.4 is generally adopted for stability analysis. It requires an evaluation of total stresses and pore water and pore air pressures. These values can be obtained by coupling the equations that govern the flow of the two fluids through the porous continuum and those that govern the deformation of the soil skeleton.

The flow equations include mass balance of materials or phases, constitutive laws, thermodynamic balance, and initial and boundary conditions. The problem is often simplified by assuming null air pressure and isothermal conditions, which allows the set of governing equations of the air mass balance and thermodynamic balance to be omitted. In this case, the net stresses become equal to the total stresses.

Following the isothermal approach proposed by Alonso *et al.* (1988) and assuming water incompressibility, water flow is governed by the mass balance equation:

$$-\frac{\partial v_i}{\partial x_i} = n\frac{\partial(S_r)}{\partial t} + S_r\frac{\partial(n)}{\partial t} \tag{20.5}$$

where x_i is the geometric coordinate, t is time, v_i is specific discharge, n is the soil porosity, and S_r is the degree of saturation of soil. Equation 20.5 requires the adoption of two constitutive laws. The first one relates the degree of saturation to suction through the water retention curve:

$$S_r = f(u_a - u_w) = f(-u_w). \tag{20.6}$$

The second one relates specific discharge, v_i, to hydraulic conductivity of the soil, K_{ij}, and hydraulic head, h, through a generalization of Darcy's law:

$$v_i = -K_{ij}(n, S_r)\frac{\partial h}{\partial x_j}. \tag{20.7}$$

Equations 20.6 and 20.7 take into account the hydraulic properties of the soil through the function, f, and the hydraulic conductivity coefficients, K_{ij}.

Any change in boundary conditions causes changes in hydraulic heads and pore pressures, which drive the system toward a new virtual steady-state condition. The duration of this transient stage depends on: (1) the amount of water required to bring the system into a new equilibrium condition; (2) the length of flow paths; and (3) the hydraulic conductivity of the soil. Equation 20.5 shows that the first component depends on the expected variation in the volumetric water content, $n{\cdot}S_r$. The variation in the degree of saturation is controlled by the water retention curve, and the variation in the porosity is controlled by the induced state of stress and by the deformability of the soil. The change in porosity can be calculated using the following equations:

$$\textit{equilibrium}: \frac{\partial \sigma_{ij}}{\partial x_j} + b_i = 0, \tag{20.8}$$

$$\textit{compatibility}: d\varepsilon_{ij} = \frac{1}{2}\left(\frac{\partial u_i}{\partial x_j} + \frac{\partial u_j}{\partial x_i}\right), \tag{20.9}$$

constitutive law, which can be expressed through the one-tensorial approach:

$$d\sigma'_{ij} = C_{ijkl}\,d\varepsilon_{kl} \tag{20.10a}$$

or through the two-tensorial approach:

$$d\sigma^*_{ij} = C^*_{ijkl}\,(d\varepsilon_{kl} - d\varepsilon_{kl0}). \tag{20.10b}$$

In these expressions: b_i is the body force; u_i and u_j are components of soil displacement; $d\varepsilon_{kl}$ and $d\varepsilon_{kl0}$ are change in, respectively, soil strain and suction-dependent component of strain; and C_{ijkl} and C^*_{ijkl} are fourth-order stiffness tensors, in the one-tensorial and in the two-tensorial approach.

The hydrologic response (Eqs. 20.5–20.7) and the mechanical soil response (Eqs. 20.8–20.10) are coupled: the solution of the hydraulic equations requires the solution of mechanical equations and vice versa. The manipulation of such expressions under some simplified assumptions leads to some well-known equations, such as the Terzaghi and Richards equations, which are widely used in soil mechanics.

Any analysis requires the definition of initial and boundary conditions. Initial conditions are expressed in terms of physical properties of the soil (n and S_r), hydraulic heads, state of stress, and state variables defined by the constitutive laws (Eqs. 20.6,

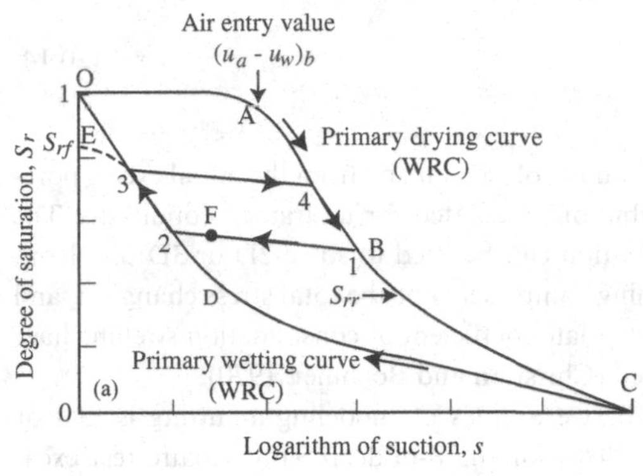

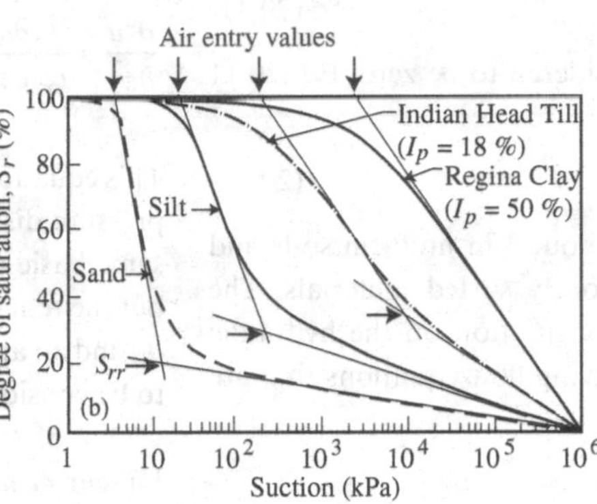

Fig. 20.1. (a) The characteristic water retention curve. (b) Typical drying water retention curves of soils (from Vanapalli *et al.*, 1999).

20.7, and 20.10). Mechanical boundary conditions are typically related to changes in slope geometry or in loading conditions, and thus are expressed in terms of changes in total stresses. Hydraulic boundary conditions are expressed in terms of fluid pressure or flow and may change with time. In some cases, hydraulic heads are imposed, for example as by a reservoir, or may be determined from site measurements. Generally, a null specific discharge is imposed in the direction normal to the boundaries of zones characterized by much smaller hydraulic conductivity. In other cases, the discharge through a surface can be separately calculated and assigned as a boundary condition, as in the case where the ground surface is subjected to outward flow due to evapotranspiration or inward flow induced by infiltration. This can be imposed only while pore water pressure at the boundary is negative or nil; if the calculated value is positive, the analysis must be repeated to force null pore water pressure (Rulon and Freeze, 1985).

In some practical problems, the analysis is carried out with the simplified assumption of a rigid soil skeleton. In such a case, the governing equations reduce to Eqs. 20.5–20.7, and the duration of the transient stage is nil in saturated soils and is controlled by the change in the volumetric water content and the hydraulic properties in unsaturated soils. In several cases, the domain is fully saturated and the duration of the transient process is controlled solely by changes in porosity. If changes in boundary conditions do not significantly affect the total stresses, the equilibrium condition is satisfied *a priori* and governing equations do not include Eq. 20.8.

20.3 FEATURES OF THE POROUS MEDIUM RELEVANT FOR ANALYSIS

The most important soil characteristics for groundwater modeling of unsaturated soils are: the water retention curve, which relates the degree of saturation, S_r, to suction, $u_a - u_w$, in both drying and wetting processes (Fredlund and Rahardjo, 1993); the saturated hydraulic conductivity, K_{sat}; the unsaturated hydraulic conductivity, K_{unsat}, which can be defined from K_{sat} and the water retention curve; and the compressibility/swelling characteristics of the soil.

20.3.1 WATER RETENTION CURVE

When a saturated soil is progressively dried, it remains saturated until the matric suction $(u_a - u_w)_b$ is reached, at which point (the air entry value) air starts penetrating the soil (point A, Fig. 20.1a). As the drying process proceeds, the degree of saturation decreases, while suction increases along the drying water retention curve (from O to A to B to C, at which point it is completely dry). Generally, the curve has a major change in slope at S_{rr}, the so-called "residual degree of saturation." If, after drying at C, the soil is wetted, it follows the wetting water retention curve (from C to D to O or E), which generally shows some hysteresis. If wetting is very slow, the water retention curve may attain full saturation at zero suction (O); but if it is relatively rapid, as in most cases, some air is trapped in the soil and the water retention curve extends toward a degree of saturation, S_{rf}, called submergence or satiated degree of saturation. If the soil is dried to point B (or 1), then wetted, and then dried again, it can follow a path such as 1–2–3–4–1. Therefore, in-situ, it can be at a point such as F in Figure 20.1a.

The water retention curve depends on the type of soil. Figure 20.1b shows drying water retention curves for different soils. Generally, the finer the soil, the larger the air entry value.

20.3.2 SATURATED HYDRAULIC CONDUCTIVITY

The saturated hydraulic conductivity, K_{sat}, varies approximately with the square of the size of the pores and thus the size of the particles. It ranges through several orders of magnitude, from 10^{-12} to 1 $m\,s^{-1}$. It also generally decreases with decreasing void ratio and depends on the soil fabric, temperature, fluid chemistry, and other factors; in some cases, it is anisotropic (Leroueil and Hight, 2002; see Freeze and Cherry, 1979, for details).

20.3.3 UNSATURATED HYDRAULIC CONDUCTIVITY

The hydraulic conductivity of unsaturated soils, K_{unsat}, can be expressed as a function of the saturated hydraulic conductivity, K_{sat}, and of the degree of saturation, S_r. According to Brooks and Corey (1964):

$$K_{unsat} = K_{sat}[(S_r - S_{rr})/(1 - S_{rr})]a^k. \quad (20.11)$$

For soils for which S_{rr} can be considered to be zero, Eq. 20.11 becomes:

$$K_{unsat} = K_{sat}\, S_r a^k. \quad (20.12)$$

The parameter a^k has a value of about 3 in uniform soils and larger values in well-graded or poorly sorted materials. The strong influence of the degree of saturation on the hydraulic conductivity is important when solving flow equations that are highly nonlinear.

20.3.4 COMPRESSIBILITY/SWELLING CHARACTERISTICS

Except in swelling soils, strains induced by pore water pressure changes are commonly small. Therefore, compressibility/swelling characteristics must be associated with small strains. This is particularly true for slopes in clays that are generally overconsolidated (Leroueil, 2001).

20.4 EFFECTS OF CUTTINGS AND DRAWDOWN ON GROUNDWATER IN SLOPES

Analyses of the consequences of cuttings or drawdown are part of engineering practice. In both cases, a decrease in rapid total stress can induce a change in pore pressures, followed by consolidation or swelling. If the soil has a low coefficient of consolidation/swelling, it takes time for pore pressures to reach a new equilibrium.

In principle, the study of these processes requires a coupled analysis of water flow and effective stress and strain. The analysis is often simplified by separating the calculation of excess pore pressures induced by a virtual undrained (i.e., instantaneous) removal of total stress and the solution of the following consolidation/swelling problem. For saturated soils, excess pore water pressures can be calculated from the *a* and *b* parameters (Henkel, 1959). However, more sophisticated coupled analyses can account for the actual rate of stress change.

20.4.1 CUTTINGS

A classic reference case is for 1D conditions, saturated and isotropic soil, and an instantaneous vertical stress decrease. In this case, the swelling process induced by soil removal can be determined from the consolidation equation (Terzaghi, 1943), which comes directly from the relevant hydraulic and mechanical equations (combining Eqs. 20.5 and 20.7):

$$\nabla^2 u = \frac{1}{c_{vs}}\left(\frac{\partial u}{\partial t} - \frac{\partial \sigma_z}{\partial t}\right). \quad (20.13)$$

In this expression, u is the pore pressure in saturated soil, σ_z the vertical total stress and c_{vs} the coefficient of swelling. For constant σ_z, Eq. 20.13 takes the form:

$$\frac{\partial^2 u}{\partial z^2} = \frac{1}{c_v}\frac{\partial u}{\partial \tau}. \quad (20.14)$$

This equation can be solved starting from the initial excess pore-pressure distribution calculated for undrained conditions. The same basic equation can be used to solve 2D or 3D problems, but the remaining components of the total stress change, σ_x and σ_y, and an appropriate coefficient of consolidation/swelling have to be considered (Christian and Boehmer, 1970).

One of the best examples of modeling a cutting is that of Lafleur *et al.* (1988) for the 8-m-deep Saint-Hilaire test excavation in saturated homogeneous firm clays. The excavation, bounded by four slopes of 18°, 27°, 34°, and 45°, was carried out over a period of about 1 month. The pore water pressure variations in the 18° and 27° slopes, which were stable, were reproduced by a coupled analysis carried out with the Abaqus code for linear-elastic soil (Laflamme and Leroueil, 1999). Unloading was simulated through different undrained steps, assuming constant hydraulic boundary conditions ($u = 0$ at the faces of the excavation). The coefficient of consolidation was established on the basis of laboratory and in-situ investigations, including cross-hole tests. The best fit with observed hydraulic heads, however, was obtained using a coefficient of consolidation/swelling 2.3 times higher than the measured one, which is a relatively small value for this kind of problem.

Results of the analysis are shown in Figure 20.2. The figure shows a good agreement between measured and computed pore water pressures, both during excavation and swelling. The results support the validity of the assumptions that the process was controlled by the small strain stiffness of the soil, which remained in the elastic domain during the entire swelling process, and of the condition $u = 0$ on the faces of the excavation.

In general, the factor of safety (FS) decreases as pore water pressures increase toward steady-state conditions (Bishop and Bjerrum, 1960). In cases where stress levels are high, the analysis should be carried out through coupled hydromechanical elastic-plastic models.

Potts *et al.* (1997) carried out a number of numerical analyses using the Imperial College Finite Element Program (ICFEP) to simulate the delayed effects of cutting in the London Clay. They used typical hydraulic and mechanical soil parameters and considered different slope geometries. They assumed different values of the coefficient of earth pressure at rest, K_0, and constant suction at the ground surface; in particular, the soil was assumed to be either softening or non-softening. The analyses demonstrated that a slope collapse could occur during the swelling phase, as suggested by experience. In particular, they provided evidence for the development of the failure, which starts from the toe and spreads into the slope as swelling continues. With dilative soils, however, excess pore pressures can temporarily decrease in the shear zone, even as swelling continues, acting as a partial brake on the post-failure movements. Another important point arising from the analysis is that suction on the ground surface can significantly delay the time to failure.

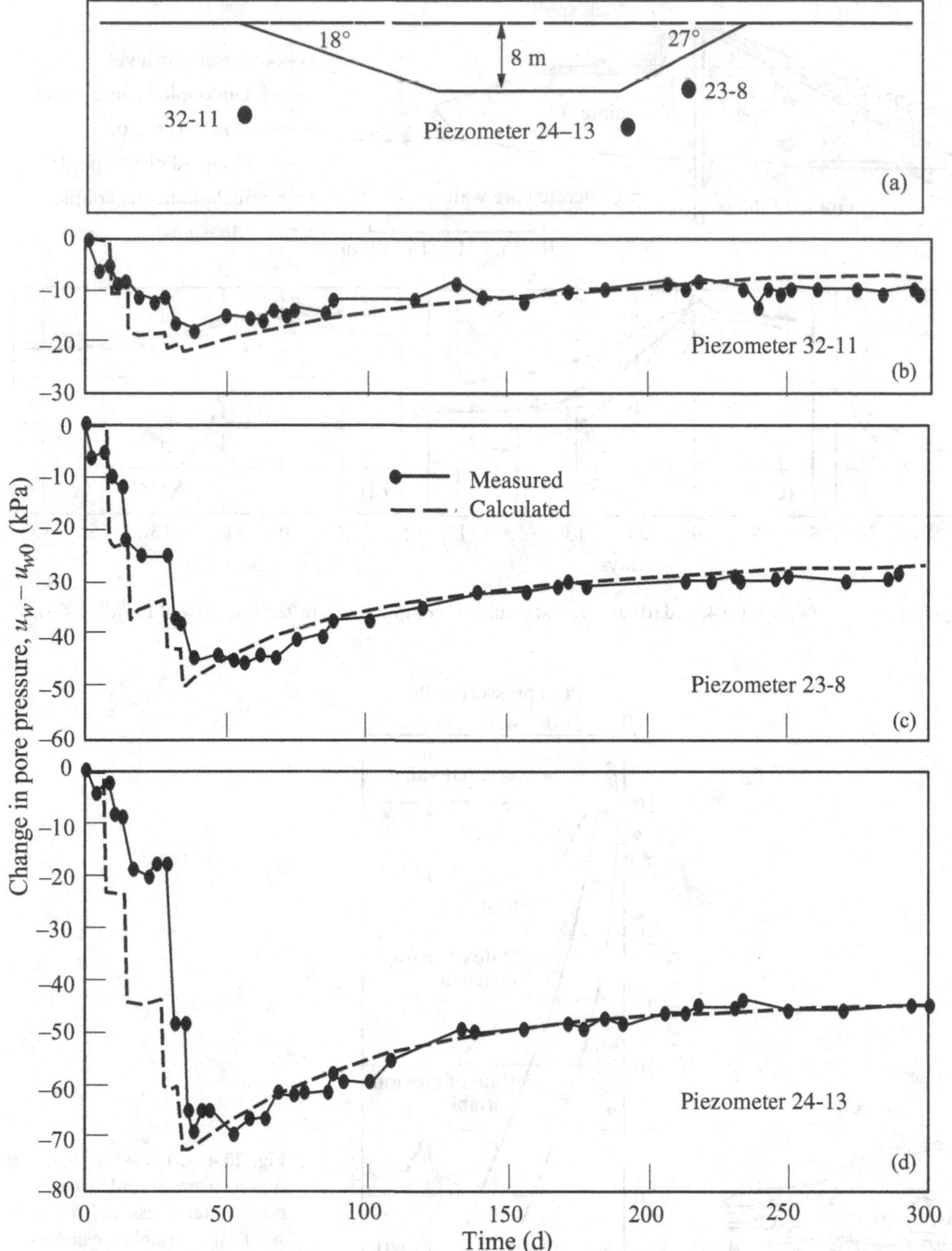

Fig. 20.2. (a) Saint-Hilaire test excavation. Pore water pressures measured (solid lines) and calculated (dashed lines) from three piezometers: (b) 32–11, (c) 23–8, and (d) 24–13 (Laflamme and Leroueil, 1999).

20.4.2 RAPID DRAWDOWN

Rapid drawdown of reservoirs can lead to unfavorable conditions in the dam or the reservoir slopes. The problem can be investigated by: (1) *a simple groundwater flow analysis*, in which changes in pore pressure in the dam are calculated with the hydraulic equations; or (2) *a coupled analysis*, using both hydraulic and mechanical equations in which the deformability of the soil is taken into account (Pinyol, 2010). In the former case, drawdown causes a progressive decrease in pore water pressures due to drainage. In principle, this analysis can be performed assuming saturated/unsaturated soil behavior and adopting the actual water retention curve of the soil in order to account for the decrease in the degree of saturation in some parts of the embankment. The latter approach implies calculation of excess pore pressures generated by the decrease in total stress and their subsequent dissipation.

Alonso and Pinyol (2009) studied the effects of rapid drawdown at Shira Dam in northern Scotland. They used three approaches in their study:

1. pure seepage and saturated/unsaturated soil behavior
2. instantaneous drawdown and dissipation of induced excess pore pressures

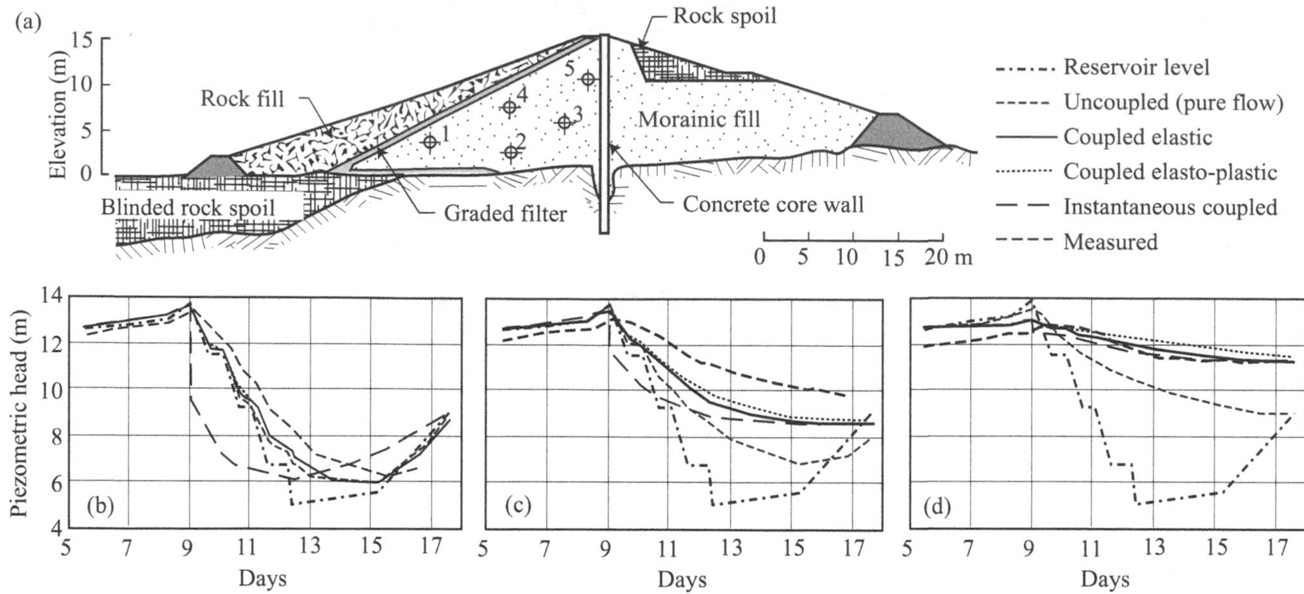

Fig. 20.3. (a) The Shira Dam. Readings at piezometers (b) 1, (c) 4, and (d) and 5, and results of computations (after Alonso and Pinyol, 2009).

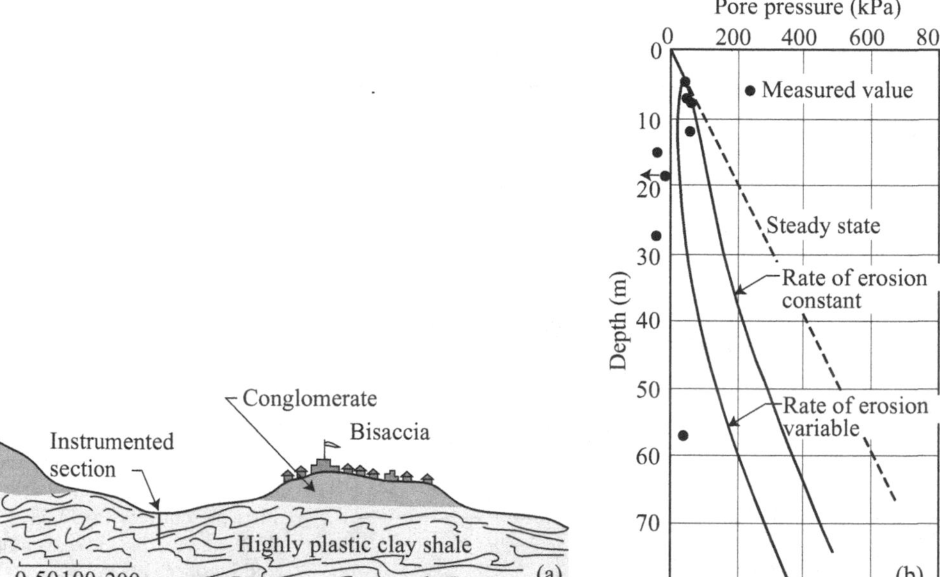

Fig. 20.4. (a) Bisaccia Hill and (b) measured and calculated pore water pressures for constant and variable erosion rates (Picarelli and Urciuoli, 1993).

3. a coupled analysis reproducing the actual change in total stresses, induced pore pressures, and their subsequent dissipation.

Pore pressures recorded by piezometers 1, 4, and 5 during drawdown and the subsequent rise of the reservoir are shown in Figure 20.3b, c, d, along with the results of calculations carried out using the code CODE-BRIGHT developed at the Universitat Politècnica de Catalunya in Barcelona. The computed pore pressures differed considerably depending on the modeling approach used: the best fits to the measured data were obtained using a coupled saturated/unsaturated analysis with an elasto-plastic constitutive law. In terms of the stability conditions of the embankment slopes, the classical undrained approach, with instantaneous drawdown, provided the most conservative results over the short term (day 9 in Fig. 20.3), as the calculated pore pressures were much higher than those measured or calculated using the other approaches at the end of the drawdown phase (days 12–13).

20.5 EFFECTS OF EROSION ON GROUNDWATER IN SLOPES

The effects of erosion are similar to those of cuttings, but the unloading process is much slower and therefore should not trigger excess pore pressures. However, there are cases of relatively

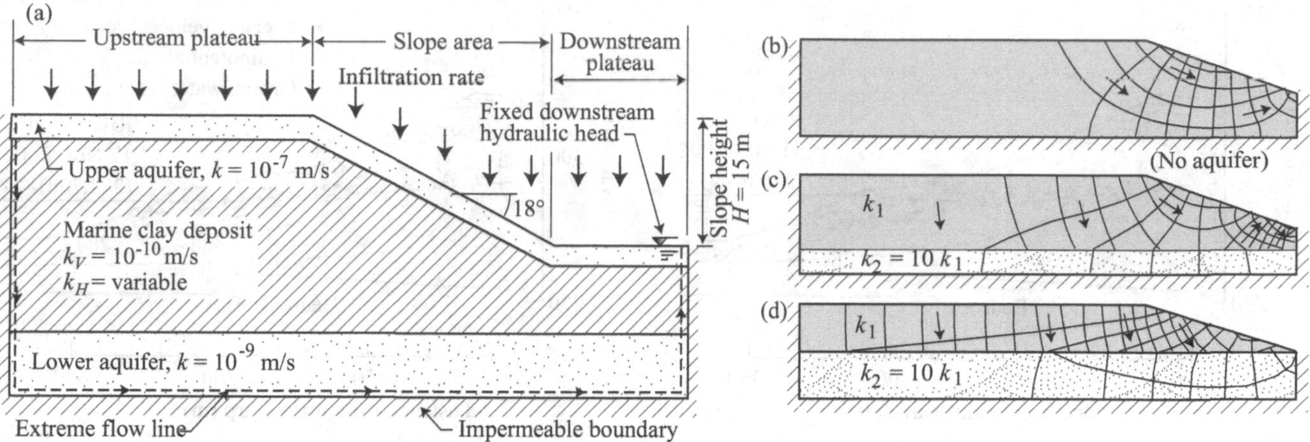

Fig. 20.5. (a) Geometrical scheme and assumptions in the simulation of seepage conditions in the banks of the St. Lawrence River valley. (b), (c), (d) Seepage conditions as a function of the position of the river valley within the stratigraphic sequence (Lafleur and Lefebvre, 1980).

rapid erosion of thick deposits of highly plastic clay that have led to the build-up of negative excess pore pressures.

Di Nocera *et al.* (1995) describe a sequence of conglomerates overlying a thick formation of highly plastic clay shales in southern Italy. These rocks have been eroded along two parallel faults over the past 300,000 years, and the Bisaccia Hill now stands between the two faults. Erosion has now extended into the shales at the base of the hill (Fig. 20.4a).

A 2D numerical analysis was carried out to investigate the effects of erosion. The simulation was started 300,000 years ago with a horizontal ground surface located at the same elevation as the present-day top of Bisaccia Hill. The water table was located at the top of the shales, 100 m below the ground surface. Two different rates of soil removal were used: (1) a constant rate, and (2) a rate three times faster in the shales than in the conglomerates. Analyses were carried out using a coupled approach and both the CRISP (Britto and Gunn, 1987) and Abaqus (Hibbit *et al.*, 2010) codes. The soil behavior was simulated using the elasto-plastic Cam Clay model, based on the results of an extensive, careful laboratory testing program. Figure 20.4b shows the measured and calculated pore water pressure profiles for a vertical section in the center of the valley. The comparison shows the very low values of pore water pressures, with even local negative values, depending on both ongoing erosion and the very low hydraulic conductivity of the rock ($K_{sat} = 10^{-12}$ m s^{-1}). The inferred pore water regime provides the only explanation for the present stability of the hillslopes, which in the case of steady-state pore water pressures (Fig. 20.4b) would be highly unstable. This special situation is not unique; comparable data have been obtained for the St. Leo area in Italy and an area in South Dakota (Neuzil, 1993) where similar erosion processes are active. In most cases, however, the change in the stress state occurs too slowly to trigger negative excess pore pressures; rather, the main consequence is to modify water flow patterns as the ground surface changes.

Another classic example is the postglacial erosion of valleys by the St. Lawrence River and its tributaries in Canada (Lafleur and Lefebvre, 1980). The river and its tributaries have incised thick deposits of Champlain clay and the permeable units that overlie and underlie it: alluvial sands above and a till sheet or fissured bedrock below. The riverbed is located at different depths in the stratigraphic sequence, depending on its location. Lafleur and Lefebvre (1980) performed numerical 2D analyses for unconfined steady-state flow in saturated rigid soil to focus on the effect of the position of the riverbed on local groundwater flow (Fig. 20.5a). They found that the location of the riverbed strongly affects the groundwater flow system. As the riverbed extends deeper into the clays (Fig. 20.5b, c), an upward flow develops under the toe of the slope; the gradient increases significantly as the riverbed approaches the base of the clays (case c). Where the riverbed cuts into the basal layer, vertical water flow takes place toward the till (case d). These results, which have been confirmed by local monitoring at different sites, highlight the dynamic evolution of the groundwater regime.

Other researchers have studied the effects of river erosion on the groundwater regime. Eshraghian *et al.* (2008), for example, describe transient groundwater flow in the banks of the Thompson River, British Columbia. The groundwater flow system is controlled by the level of the river; when the river is lowest, seepage occurs from the river banks, whereas when the river is high, water flows from the river toward the banks, leading to a rise in the groundwater table. As drawdown begins, adjacent unstable slopes are subjected to significant seepage forces (Fig. 20.6). These forces, along with changes in slope geometry due to erosion and changing water support on the toe of the slope, have important effects on the stability of the slopes. Misfeldt *et al.* (1991) describe a similar situation along the South Saskatchewan River at the site of the Hepburn landslide.

20.6 EFFECTS OF CLIMATE ON GROUNDWATER IN NATURAL SOIL SLOPES

Changes in hydrologic boundary conditions during alternating periods of rainy and dry weather cause continuous fluctuations in pore water pressures and consequent changes in porosity and

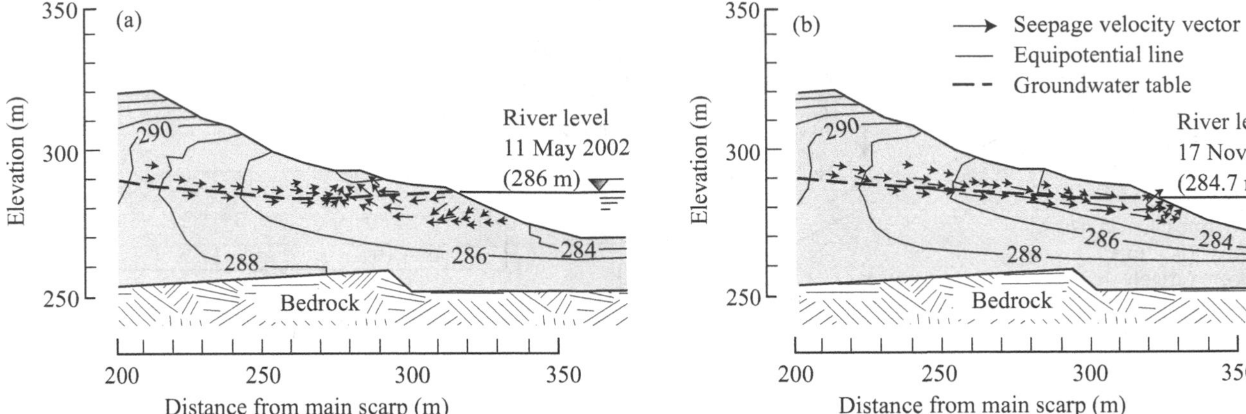

Fig. 20.6. Influence of the level of the Thompson River, British Columbia, on the groundwater regime in the river banks: (a) rising river level; (b) end of river drawdown (Eshraghian *et al.*, 2008).

in degree of saturation of soils. Here, we consider the simple case of the infinite slope, followed by some examples of natural slopes in both unsaturated and saturated soils.

20.6.1 THE CASE OF THE INFINITE SLOPE

A useful approach for analyzing a transient groundwater regime is to equalize the component of pore water pressure that exceeds the steady-state value (Lambe and Whitman, 1969). In an infinite slope, pore water pressures and degree of saturation are uniform along planes parallel to the ground surface. The component of water flow parallel to the slope is therefore a function of the difference in elevation between points located along the same line and is not time-dependent. Therefore, the transient character of flow due to changing hydrologic conditions depends only on variations in the hydraulic head in the direction, n, normal to the slope. The isobars of excess pore water pressures are parallel to the slope; thus the analysis of the transient flow can be treated as a 1D problem.

The effects of any hydrologic perturbation in saturated soils can be investigated using the classical consolidation theory with proper boundary conditions. Cavalera (1977) investigated the propagation in the subsoil of a sinusoidal variation, $V(t)$, of the pore water pressure at the top of a horizontal deposit lying on an impermeable horizontal boundary. The analysis was carried out by applying the theory of heat conduction, with a governing equation that is identical to the 1D consolidation equation for saturated soil (Eq. 20.14). It showed that pore water pressure changes affect a part of the deposit, the thickness of which depends on both the duration of the perturbation and the soil properties. Specifically, the shorter the period, the thinner the layer involved in the process; the same result is obtained for lower values of the coefficient of consolidation/swelling. Such results again highlight that the rainfall history can modify the groundwater flow system to a depth that depends both on soil properties and on the frequency and duration of the precipitation events. This effect explains why precipitation-induced landslides are shallow, and deep failures are generally the result of more complex hydraulic and mechanical processes.

The propagation of a hydrologic perturbation in partially saturated soils is also delayed, depending on the time required to cause a change in water content. Two simple cases of 1D flow are reported below to emphasize the effect of the hydraulic properties of the soil on the depth subjected to changing water contents within a given time span and the suction time history at different depths. The problem can be solved using the Richards equation (1931):

$$\frac{\partial}{\partial x}\left(k_x(h)\cdot\frac{\partial h}{\partial x}\right)+\frac{\partial}{\partial y}\left(k_y(h)\cdot\frac{\partial h}{\partial y}\right) + \frac{\partial}{\partial z}\left(k_z(h)\cdot\frac{\partial h}{\partial z}\right)=m_w(h)\cdot\gamma_w\cdot\frac{\partial h}{\partial t} \tag{20.15}$$

where m_w is the coefficient of volumetric water content change corresponding to the slope of the water retention curve, which globally accounts for both water content and porosity changes induced by changes in pore water pressure. The solution has been obtained using the SEEPW code (GEO-SLOPE, 2004).

The first example (Collins and Znidarcic, 2004) is a case of vertical infiltration, here taken as representative of the infiltration component normal to the slope, from the top of a 4-m-thick soil column having a permeable base subject to null suction. The soil has a hydraulic conductivity that depends on the degree of saturation. The initial suction profile is linear, representing a hydrostatic condition. Continuous infiltration is simulated by imposing a water film at the ground surface (u = 0). Due to water ingress and consequent progressive saturation of the soil column, the suction profile changes continuously. Figure 20.7 shows the pressure heads at different times calculated for two typical materials: a fine-grained soil (Fig. 20.7a: $K_{sat} = 1.5 \times 10^{-8}$ m s^{-1}) and a coarse-grained soil (Fig. 20.7b: $K_{sat} = 1.5 \times 10^{-6}$ m s^{-1}). Infiltration takes place more rapidly in the coarse soil, due to its higher hydraulic conductivity. However, in both cases, wetting of the deepest part of the column requires considerable

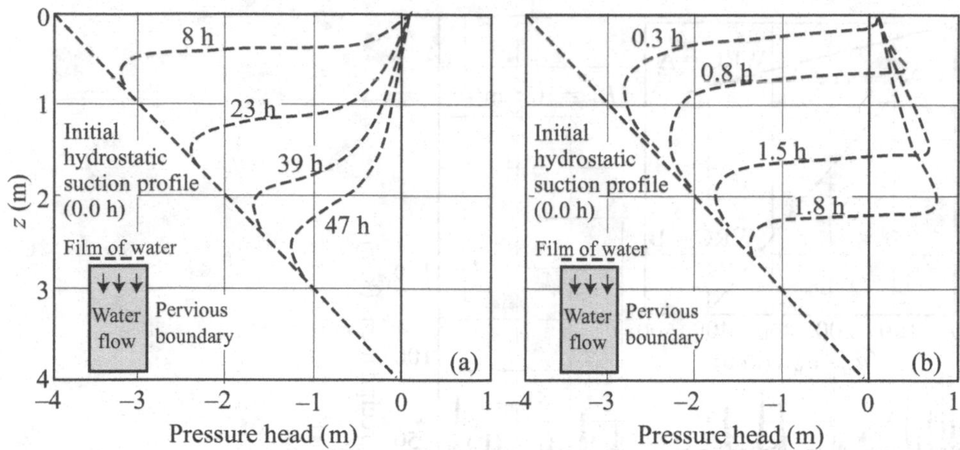

Fig. 20.7. Effects of vertical water infiltration in a column of (a) fine-grained and (b) coarse-grained soil (modified from Collins and Znidarcic, 2004).

time. This example highlights the role of hydraulic conductivity, which governs the duration of the rainfall that can significantly modify the pore water pressure regime. The shape of the water retention curve, however, also has an important influence, as shown in the following case.

The second example shows the effects of a given rainfall history on an infinite column of soil with different hydraulic properties (Fig. 20.8). Soil properties have been idealized to allow easier appraisal of their effects on the hydrologic response of the system. The adopted water retention curves are represented by straight lines: the first line is for a soil (A in Fig. 20.8a) in which the degree of saturation changes only slightly even under strong suction changes; the second line represents a completely different material (B in Fig. 20.8a) in which the degree of saturation changes significantly with small suction increases. Three different values of hydraulic conductivity ($K_1 = 10^{-5}$ m s^{-1}; $K_2 = 10^{-6}$ m s^{-1}; $K_3 = 10^{-7}$ m s^{-1}) that do not vary during infiltration, regardless of suction, have been used to isolate the influence of the water retention curve on the soil response. In all computations, the water table is located 20 m below ground level and the initial condition is represented by a hydrostatic suction profile.

Figure 20.8 shows the calculated suction history at a depth of 1 m, where the initial suction is 186 kPa. Each plot compares the values obtained for the same hydraulic conductivity and for the two water retention curves. For any rainfall, soil A displays a sharper suction decrease than soil B, due to the smaller amount of water required for a pore water pressure change. In the case of the highest value of K (Fig. 20.8a), the suction change rapidly disappears after rainfall, no matter which water retention curve is used. As a result, suction continuously fluctuates around the steady-state value. The effect of the water retention curve is revealed by different peaks in suction, which are higher for soil A due its stiffer hydrologic response. On the other hand, in the case of the lowest conductivity (Fig. 20.8c), suction progressively changes with rainfall, departing from the steady-state value. Again, the peaks are higher for soil A; for soil B, the pore water pressure trends toward a more stable value, with lower changes between successive rainfall events.

In conclusion, a high hydraulic conductivity and a gently sloping water retention curve lead to shorter transient groundwater responses. The opposite occurs for low hydraulic conductivities and steeper water retention curves, for which the steady state is only a virtual condition. The selection of the initial conditions is consequently crucial for the analysis.

20.6.2 NATURAL SOIL SLOPES

The groundwater flow system in natural slopes is complex and strongly dependent on the lithological and geomorphological contexts in which they are located (Hodge and Freeze, 1977). Additional complexities come from local changes in permeability due to inhomogeneity and anisotropy of earth materials and on the hydrologic conditions at the soil–atmosphere interface.

SATURATED SOILS

A classic example of the dependence of the groundwater regime in saturated soils on hydrologic boundary conditions is described by Kenney and Lau (1984). The banks of Wabi Creek in Ontario are partly covered by water, the level of which varies with the season. At any time, part of the bank is wetted and part of it is exposed to air. In addition, during winter, the ground surface is frozen. As a consequence, the hydraulic boundary conditions differ through time; in particular, the ground surface is impervious in winter, but is a source of water in the other seasons, due to snowmelt and rainfall. The hydraulic heads in the subsoil over a 10-year monitoring period were provided by 65 piezometers. Figure 20.9 summarizes the amplitude of fluctuations of the piezometric level at each instrumented point, showing the influence of depth. Shallow layers experienced large changes in pore water pressures, whereas deeper levels showed negligible changes.

UNSATURATED SOILS

Site monitoring with classical geotechnical instruments and weather stations is a useful approach for investigating the relationship between weather and groundwater conditions and for

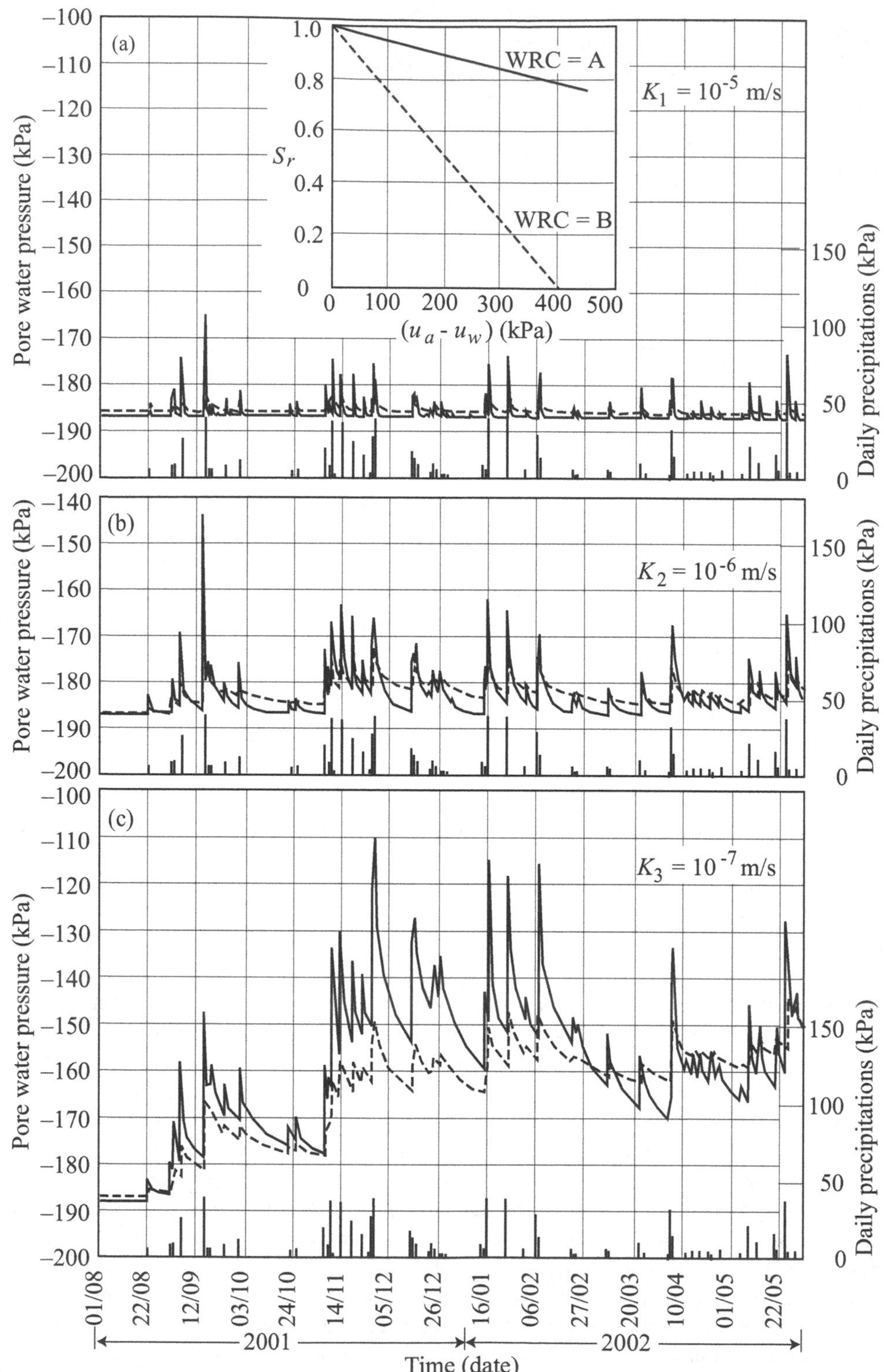

Fig. 20.8. Calculated suction history at a depth of 1 m in an unsaturated soil layer subject to rainfall, for two soils (A and B in inset) with different water retention curves and different hydraulic conductivities.

setting up advanced numerical models and efficient early-warning systems. Excellent data on unsaturated soils have become available over the past several decades (Brand *et al.*, 1984; Rahardjo *et al.*, 2005), complemented recently by experimental results using lysimeters and small-scale slope models (Olivares and Damiano, 2007).

Large amounts of data are being collected at instrumented sites located around Naples, where a thin cover of unsaturated

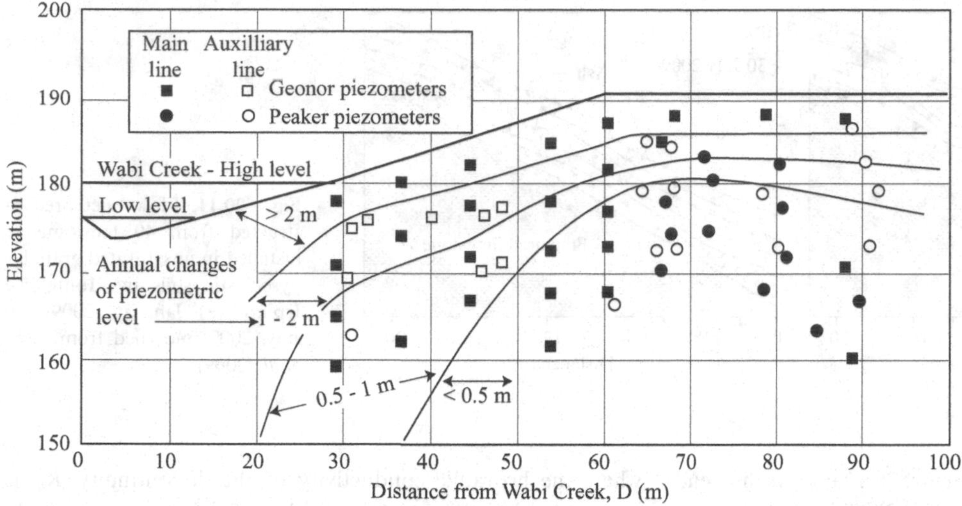

Fig. 20.9. Fluctuations in piezometric heads in the banks of Wabi Creek (modified from Kenney and Lau, 1984).

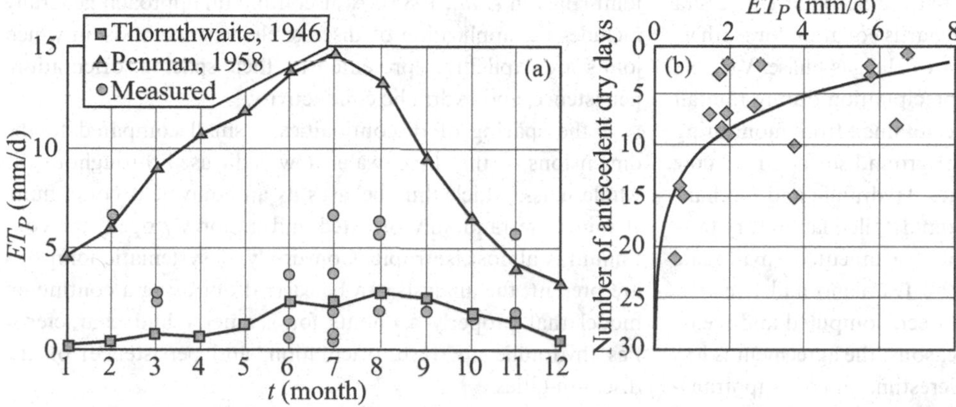

Fig. 20.10. (a) Calculated upward flux rate at the Cervinara site between 2002 and 2006 at depths between 60 cm and 90 cm; and (b) influence of the number of antecedent dry days (after Damiano *et al.*, 2012).

granular soils of pyroclastic origin (air-fall tephra) rests on fractured limestones (Olivares *et al.*, 2003; Papa *et al.*, 2009).

Based on data provided by tensiometers installed at depths of 60–90 cm at the Cervinara site in Italy, Damiano *et al.* (2012) determined the hydraulic flux from the soil toward the ground surface during dry periods. The calculations are based on Darcy's law and suction readings, and account for the hydraulic conductivity of the soil (Fig 20.10a). Hydraulic conductivity was estimated from the Brooks and Corey (1964) expression, as a function of the degree of saturation of the soil. This, in turn, was determined for the soil through the water retention curve, based on measured values of suction. Damiano *et al.* (2012) compared the hydraulic flux to the monthly potential evapotranspiration using the Thornthwaite (1946) and Penman (1958) expressions. Such expressions yield different values of evapotranspiration, but the influence of the temperature and humidity, which were measured at a weather station at the study site, is evident in both cases. In contrast, the measured evapotranspiration values, between 1 and 8 mm per day, do not display a clear trend but fall between the two theoretical curves, being closer to values derived using the Thornthwaite expression. In fact, the hydraulic flux seems to mostly depend on the number of antecedent dry days, regardless of the season; the higher the number of dry days, the lower the flux rate (Fig. 20.10b). This result suggests that the influence of the soil conductivity is greater than that of the weather conditions. This type of data provides critical information about conditions at the uppermost boundary during dry phases and is therefore useful when designing and running numerical models.

Figure 20.11 shows the very different groundwater flow systems that can arise under different weather conditions. The figure shows velocity vectors obtained at another instrumented site in the same region (Monteforte Irpino) as the Cervinara site, but during different seasons. The velocity vectors were determined based on the equipotential lines obtained from 40 tensiometers. The figure shows that the groundwater regime is characterized by downward infiltration in wet periods (January, Fig. 20.11a) and by evapotranspiration in dry periods (July, Fig. 20.11b). It is informative to compare Figure 20.11 with Figure 20.6, which also shows the role of seasonal changes in hydrologic boundary conditions, but governed by river level.

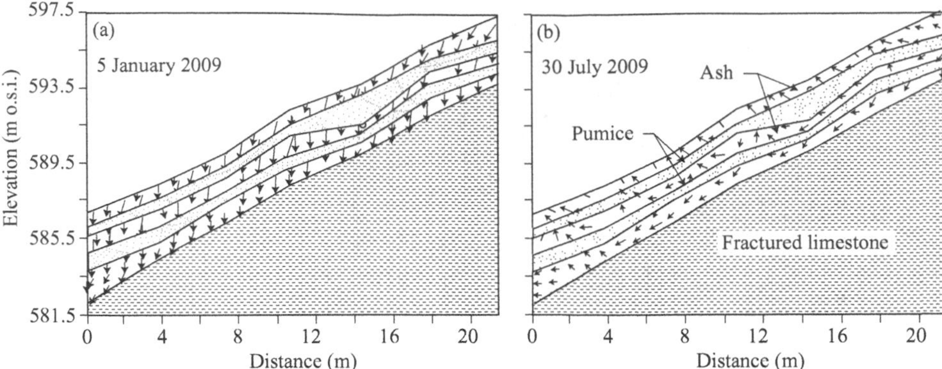

Fig. 20.11. Flow net reconstructed from 40 tensiometers installed in unsaturated granular pyroclastic soils in Monteforte Irpino: (a) January, 2009; (b) July, 2009 (modified from Papa *et al.*, 2009).

Returning to the Cervinara site, Figure 20.12a shows the trend of suction from January 2006 to August 2007. These data were compared with the results of numerical analyses carried out with the IMOD-3D code developed at the Seconda Università di Napoli, which implements the Richards equation for isothermal conditions, but excludes the flux of the gas phase. Variable boundary conditions (intensity of precipitation during rainfall events and upward mean water flux obtained from monitoring during dry days) were imposed at the ground surface. Free flow was imposed at the other boundaries. Hydraulic and mechanical soil properties were obtained from detailed laboratory tests and infiltration experiments on an instrumented small-scale physical model (Picarelli *et al.*, 2006; Damiano and Olivares, 2010). There is good agreement between computed and measured values of suction for the wet seasons; the agreement is less for dry periods, probably due to underestimation of evapotranspiration. Nevertheless, the data permit values of the factor of safety to be calculated; the values significantly decrease at shallow depths during wet seasons (Fig. 20.12b).

Such data show that correct modeling of the hydrologic response of slopes in unsaturated soil is only possible if evapotranspiration is taken into account. When developing conceptual methods, a major effort must be made to account for the influence of all atmospheric variables on the fluxes of moisture between the soil and the atmosphere. Such methods can help with early warnings of slope failure and also land-use planning based on reliable scenarios of future climate change.

20.6.3 ROCK SLOPES

ROLE OF ROCK MASS STRUCTURE

The groundwater flow system in rock slopes strongly depends on the spatial distribution of discontinuities. When discontinuity spacing is large compared to the dimensions of the slope, flow is channelized along single discontinuities. In this case, a realistic approach to the analysis of the groundwater flow system is the application of a discontinuous model, in which flow, q, through a discontinuity under a hydraulic gradient i, is calculated as:

$$q = K_d e i \qquad (20.16)$$

where the hydraulic conductivity of the discontinuity, K_d, is proportional to the square of the hydraulic aperture, e, which, in turn, depends on the real aperture and roughness of the joint (Barton *et al.*, 1985). A discontinuum approach generally includes the application of discrete element methods, in which joints are explicitly represented by their spacing, orientation, persistence, and hydraulic conductivity.

If the spacing of discontinuities is small compared to the dimensions of the slope, water flow is diffused throughout the whole mass, which thus behaves as an equivalent continuum. If joints are randomly oriented and uniformly open, the continuum is almost isotropic. Conversely, if systematic joint sets are present, the analysis can be carried out using a continuum model that properly accounts for geometry and characteristics (hydraulic aperture, orientation, and persistence) of the discontinuities.

Different approaches have been used to introduce rock mass structure into continuum models, but not all are suited to seepage in slopes. When joints are interconnected and open, the simplest approach is to introduce K_d and joint spacing and orientation into an equivalent permeability tensor (Priest, 1993), assuming a rigid saturated rock matrix. K_d can be estimated from structural data or back-calculated from borehole packer tests if joints in the pressurized segment of the borehole have been adequately investigated (Golder Associates, 2010). In order to reproduce irregular and poorly connected networks of discontinuities in continuum models, some authors (e.g., Dershowitz *et al.*, 2004) have discretized the continuous finite element models (FEM) or finite difference models (FDM) using a synthetic network of lines and planes of finite extent based on properties derived from structural mapping (joint dispersion, intensity, continuity, and termination). The hydraulic properties of the joints are then introduced into grid elements, thus reproducing discontinuities. This type of discretization and the classical discontinuum approach require that structural data be available with continuity through the whole rock mass. Therefore, they are seldom used in natural slopes where only local structural data are available.

Rock mass structure also affects pore water pressure measurements. Sullivan (2007) noted, with reference to a surface mine, that piezometer measurements depict the actual hydraulic behavior

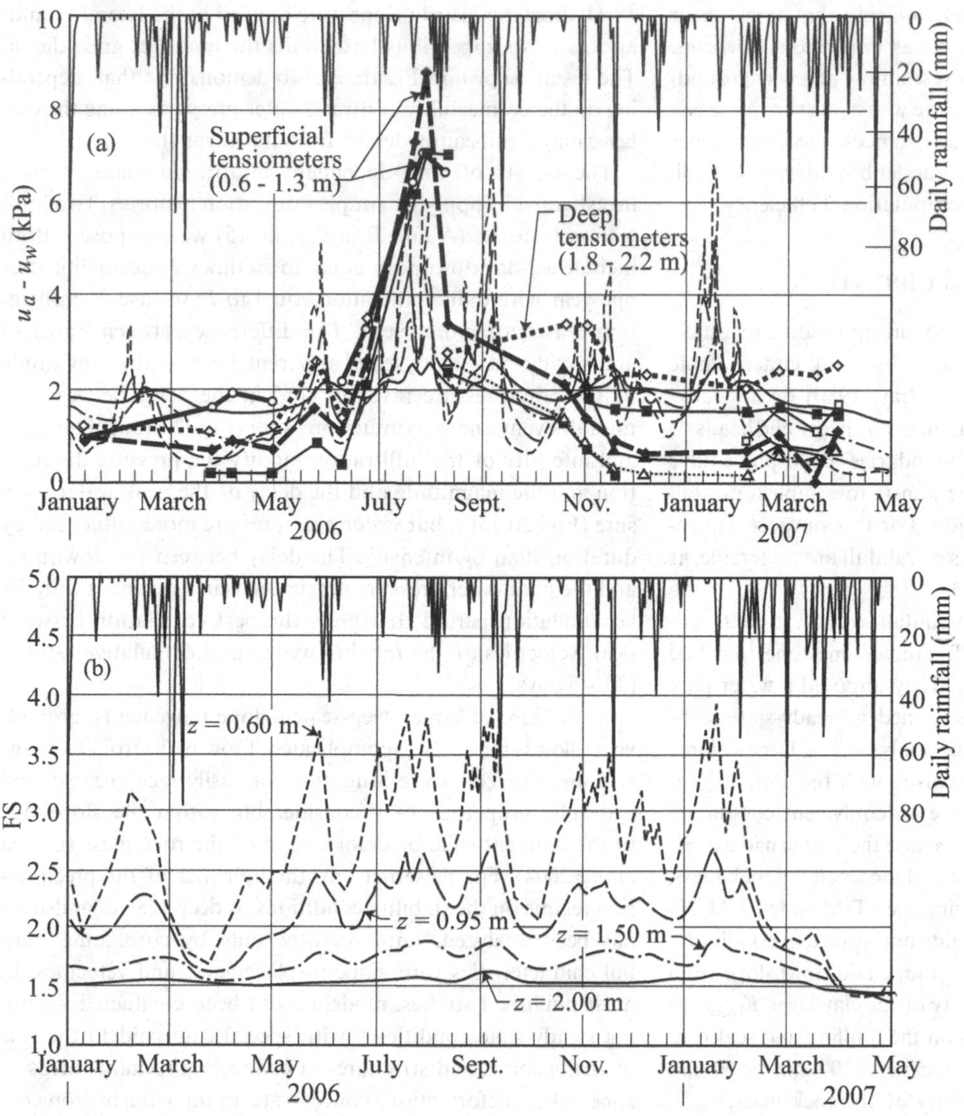

Fig. 20.12. (a) Measured (dots) and calculated (solid lines) values of suction at the Cervinara site and (b) factor of safety of the slope based on measured values of suction (after Damiano *et al.*, 2012).

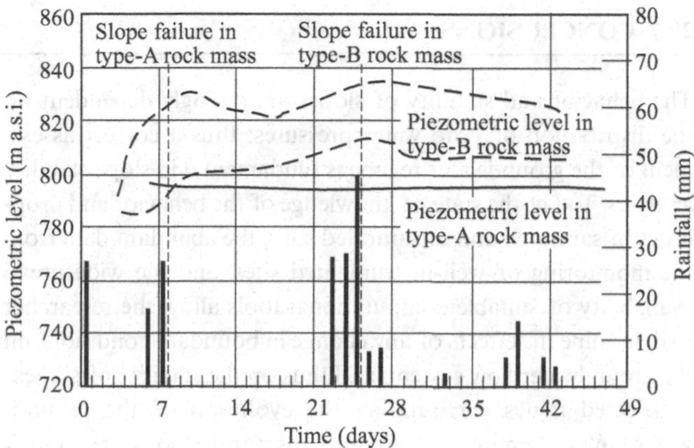

Fig. 20.13. Response of piezometers in a rock mass with closely spaced (type B, dashed line) and very widely spaced (type A, solid line) joints (modified from Sullivan, 2007).

of the rock mass in zones with closely spaced joints (type B in Fig. 20.13), where flow involves large volumes of the rock mass. Conversely, in zones with very widely spaced joints (for joint spacing classification, see ISRM, 1978), flow occurs within discrete joints (type A in Fig. 20.13), and water pressure changes can easily be missed, even though the behavior of the slope is undoubtedly correlated with rainfall.

Another source of complexity is the strong contrast in hydraulic conductivity caused by major structures (Rulon and Freeze, 1985). These conditions exist in rock masses with alternating layers of different permeability or with isolated clay layers. In this respect, shear zones produced by tectonic or gravitational movements have a hydraulic conductivity that may be either lower or higher than that of the surrounding rock mass, whether alteration has occurred in the shear zone or not. The distribution of water pressures and of rock mass quality at Campovallemaggia, Switzerland (Bonzanigo *et al.*, 2007) and

Rosone, Italy (Castelli *et al.*, 2009) indicate that thick shear zones with high contents of fines act as impervious horizons, limiting infiltration at depth and sustaining perched groundwater. The slopes experience rapid pore water pressure increases after intense rainfall. In these circumstances, the shear zones can be characterized as impervious model boundaries (Castelli *et al.*, 2009), providing improved computational efficiency.

TRANSIENT FLOW AFTER PRECIPITATION

Pore water pressure changes due to precipitation are generally introduced into numerical models by using either simple groundwater conditions (Hoek and Bray, 1980) or a steady-state seepage analysis in which increments of hydraulic heads or inflows are applied at the model boundaries. Both procedures can be ineffective, especially if the aim is to establish rainfall thresholds for failure or reactivation. For this purpose, transient analyses applying time histories of rainfall are preferable, as shown in the following case study.

Graziani *et al.* (2010) introduced rainfall to analyze the reactivation of a large rock wedge in fractured limestone that had slid on a persistent clay-filled joint. They imposed a water pore pressure distribution and then performed a steady-state seepage analysis and a transient seepage analysis. The large volume of the wedge (140,000 m^3) in comparison with the average joint spacing (0.3 m) allowed them to use an equivalent continuum modeling approach. Furthermore, because the wedge had experienced significant displacements that had loosened the rock mass, they adopted an isotropic model using the FDM code FLAC3D (Itasca, 2006), both under steady-state and transient conditions. The model extended about 150 m upslope and downslope from the wedge. The hydraulic conductivity of the clay layer, K_{layer}, was measured through laboratory tests on the infilling and scaled to account for the minimum element thickness (200 mm vs. 50 mm in reality). The hydraulic conductivity of the rock mass, K_{mass}, was evaluated from joint spacing and aperture. The code allowed simulation of flow through the unsaturated upper part of the rock mass, assuming that hydraulic conductivity varies with saturation degree, S_r, according to the equation:

$$K(S_r) = K_{sat} K^{\wedge}(S_r) \qquad (20.17)$$

where $K^{\wedge}(S_r) = [S_r^2(3 - 2S_r)]$. The choice of the $K(S_r)$ function should be based on the equivalent porosity of the rock mass, and its saturated hydraulic conductivity should be measured through in-situ tests or back analyses of the response of piezometers to rainfall under transient conditions (Castelli *et al.*, 2009). Because piezometers could not be installed within the sliding mass, the $K(S_r)$ function was chosen on the basis of considerations of rock mass structure and the relevant literature.

For case 1, as shown in Figure 20.14a, a hydrostatic distribution was imposed within the tension crack delimiting the back of the wedge; pressure linearly decreases along the clay layer, from the hydrostatic value at the crack base to zero at the layer outcrop. Steady-state analyses (case 2, see Fig. 20.14b) were performed, assuming a uniform inflow over the slope surface equal to that calculated from 20 years of average daily rainfall, $\bar{I}$. Hydrostatic distributions were applied at the lateral boundaries, and an impervious base limits the model at great depth. The results shown in Figure 20.14b demonstrate that, depending on the permeability contrast, water pressures along the clay layer may significantly depart from the linear state.

The effects of time-dependent infiltration (case 3) were investigated by applying simple infiltration histories. Two infiltration histories (A and B in Fig. 20.15) were imposed. Both include a 2-day downpour equal to 20 times $\bar{I}$, occurring after one year with daily infiltration equal to $\bar{I}$; in case A, infiltration ceases after the event. The difference between transient and steady-state analyses is apparent both in the magnitude of the pore-pressure response and in the delay between the rainfall event and maximum pressure (Fig. 20.15a). Duration and intensity of the infiltration modify the pressure distribution and the magnitude and the delay of the peak water pressure (Fig. 20.15b), but water pressures are more influenced by duration than by intensity. The delay between the downpour and the pore water pressure maximum is in agreement with the accumulation period that gives the best correlation between peak velocities of the moving wedge and cumulative rainfall (30–45 days).

In the case of large, deep-seated slope movements, groundwater flow is much more complicated. Flow paths from recharge and unstable zones are long and not easily recognizable, and hydraulic properties vary considerably within the slope due to the different state of deformation of the rock mass (Cappa *et al.*, 2004). For these reasons, the influence of the precipitation regime on the stability conditions of deep-seated landslides has been analyzed, until recently, only by correlating rainfall characteristics with slope displacements and velocities. In most published studies, modeling has been conducted assuming steady-state conditions, with special care paid to the role of major joints and structures in connecting recharge areas to zones where deformations concentrate. In this case, hydromechanical coupling can play an important role in accounting for variations of hydraulic conductivity related to zones of compression and unloading (Cappa *et al.*, 2004)

20.7 CONCLUSIONS

The behavior and stability of slopes are strongly dependent on the distribution of pore water pressures; thus a correct assessment of the groundwater regime is fundamental in slope stability analyses. Today, the state of knowledge of the behavior and properties of saturated and unsaturated soils, the abundant data from the monitoring of well-instrumented sites, and the widespread availability of suitable computational tools allow the researcher to determine the effects of any change in boundary conditions on the groundwater flow system, and thus on the stability of slopes.

Selected issues surrounding the evolution of the groundwater flow system in saturated and unsaturated soils and in rock masses have been reviewed, stressing several unexpected aspects of the complex flow pattern of water and its consequences for slope behavior. These lessons should enable geotechnical

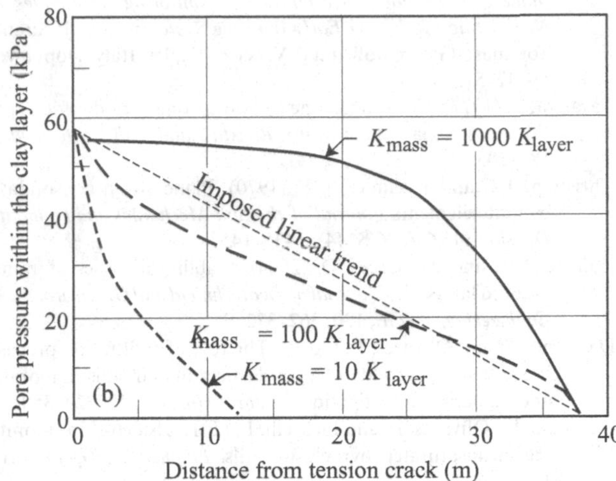

Fig. 20.14. (a) Imposed pore water pressure distribution at the base of a wedge, assuming hydrostatic conditions in the tension crack and linear head loss along the basal layer. (b) Variation in pore water pressure along the dip direction of the basal clay layer obtained through steady-state analyses (modified from Graziani *et al.*, 2010).

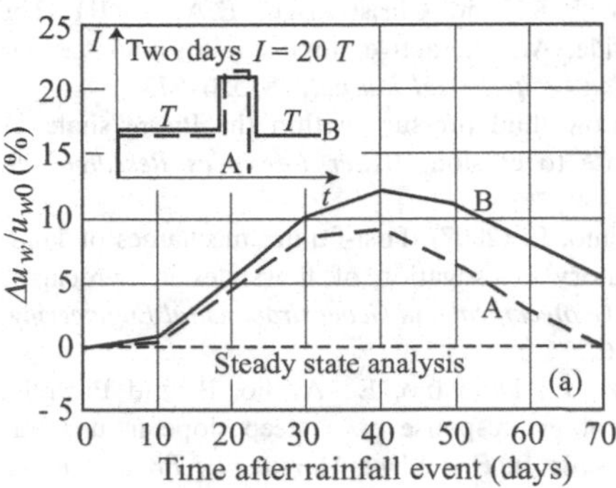

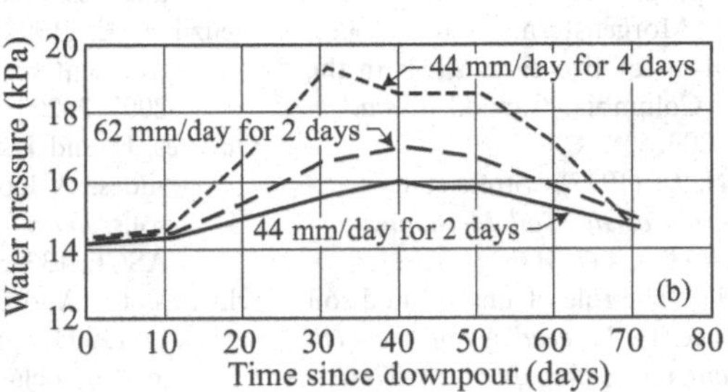

Fig. 20.15. (a) Water pressure along the dip direction of the basal clay layer vs. time, assuming two infiltration histories A and B. (b) Water pressures along the dip direction of the basal clay layer vs. time, obtained from transient analyses with different duration and intensities of infiltration (modified from Graziani *et al.*, 2010).

engineers and engineering geologists to solve complex technical problems and to understand the effects of geologic phenomena and climate on slope behavior. Advanced models can help in short-term predictions of pore-pressure changes and subsequent implementation of early-warning systems, as well as in long-term forecasting of groundwater trends and establishing appropriate criteria for land management.

REFERENCES

Alonso, E. and Pinyol, N. (2009). Slope stability under rapid drawdown conditions. In *Rainfall-Induced Landslides: Mechanisms, Monitoring Techniques and Nowcasting Models for Early Warning Systems*, ed. L. Picarelli, P. Tommasi, G. Urciuoli and P. Versace. Naples, Italy: Doppiavoce, pp. 11–27.

Alonso, E.E., Battle, F., Gens, A. and Lloret, A. (1988). Consolidation analysis of partially saturated soils. Application to earth dam construction. In *Proceedings of the 6th International Conference on Numerical and Analytical Methods in Geomechanics, Innsbruck, Vol. 1*, pp. 1303–1308.

Barton, N.R., Bandis, S.C. and Bakhtar, K. (1985). Strength, deformation, and conductivity coupling of rock joints. *International Journal of Rock Mechanics, Mineral Science & Geomechanics Abstracts*, 22(3), 121–140.

Bishop, A.W. (1959). The principle of effective stress. *Teknisk Ukeblad*, 106, 859–863.

Bishop, A.W. and Bjerrum, L. (1960). The relevance of the triaxial test to the solution of stability problems. In *Proceedings of the ASCE Research Conference on Shear Strength of Cohesive Soils, Boulder, CO*. New York: ASCE, pp. 473–501.

Bonzanigo, F., Eberhardt, E. and Loew, S. (2007). Long term investigation of a deep-seated creeping landslide in crystalline rock. Part I. Geological and hydromechanical factors controlling the Campovallemaggia landslide. *Canadian Geotechnical Journal*, 44, 1157–1180.

Brand, E.W., Premchitt, J. and Phillipson, H.B. (1984). Relationship between rainfall and landslides in Hong Kong. In *Proceedings of the 4th International Symposium on Landslides. Toronto, ON, Vol. 1*, pp. 377–384.

Britto, B.M. and Gunn, M.J. (1987). *Critical State Soil Mechanics via Finite Elements*. Chichester, UK: Ellis Horwood Ltd.

Brooks, R.H. and Corey, A.T. (1964). *Hydraulic Properties of Porous Media*. Fort Collins, CO: Colorado State University, Hydrology Paper 3, pp. 1–27.

Cappa, F., Guglielmi, Y., Soukatchoff, V.M. *et al.* (2004). Hydromechanical modeling of a large moving rock slope inferred from slope levelling coupled to spring long-term hydrochemical monitoring: example of the La Clapière landslide (Southern Alps, France). *Journal of Hydrology*, 291, 67–90.

Castelli, M., Duca, S., Pisani, G. and Scavia, C. (2009). Groundwater flow in a large landslide: The Rosone case study. In *Rainfall-*

Induced Landslides: Mechanisms, Monitoring Techniques and Nowcasting Models for Early Warning Systems, ed. L. Picarelli, P. Tommasi, G. Urciuoli and P. Versace. Naples, Italy: Doppiavoce, pp. 42–51.

Cavalera, L. (1977). Consolidazione per variazione periodica delle pressioni interstiziali al contorno. *Rivista Italiana di Geotecnica*, 11, 187–205.

Christian, J.T. and Boehmer, J.W. (1970). Plane strain consolidation by finite elements. *Journal of the Soil Mechanics and Foundation Division, ASCE*, 96(SM4), 1435–1457.

Collins, B.D. and Znidarcic, D. (2004). Stability analyses of rainfall induced landslides. *Journal of Geotechnical and Geoenvironmental Engineering, ASCE*, 130, 362–372.

Damiano, E. and Olivares, L. (2010). The role of infiltration processes in steep slope stability of pyroclastic granular soils: Laboratory and numerical investigation. *Natural Hazards*, 52, 329–350.

Damiano, E., Olivares, L. and Picarelli, L. (2012). Steep-slope monitoring in unsaturated pyroclastic soils. *Engineering Geology*, 137, 1–12.

Dershowitz, W.`S., La Pointe, P.`R. and Doe, T.`W. (2004). Advances in discrete fracture network modeling. In *Proceedings of the U.S. EPA/NGWA Fractured Rock Conference: State of the Science and Measuring Success in Remediation, Portland, ME*.

Di Nocera, S., Fenelli, G.B., Iaccarino *et al.* (1995). An example of the geotechnical implications of geological history. In *Proceedings of the 11th European Conference on Soil Mechanics and Foundation Engineering, Copenhagen, Vol. 8*, pp. 39–48.

Eshraghian, A., Martin, C.D. and Morgenstern, N.R. (2008). Movement triggers and mechanisms of two earth slides in the Thompson River valley, British Columbia. Canada. *Canadian Geotechnical Journal*, 45, 1189–1209.

Fredlund, D.G. and Morgenstern, N.R. (1977). Stress-state variables and unsaturated soils. *Journal of the Soil Mechanics and Foundation Division, ASCE*, 103(GT5), 447–466.

Fredlund, D.G. and Rahardjo, H. (1993). The role of unsaturated soil behaviour in geotechnical practice. In *Proceedings of the 11th Southeast Asian Geotechnical Conference, Singapore*, pp. 37–49.

Fredlund, D.G., Morgenstern, N.R. and Widger, R.A. (1978). The shear strength of unsaturated soils. *Canadian Geotechnical Journal*, 15, 313–321.

Freeze, R.A. and Cherry, J.A. (1979). *Groundwater*. Englewood Cliffs, NJ: Prentice Hall.

GEO-SLOPE (2004). *Seepage Modelling with SEEP/W: User's Guide Version 6.16*. Calgary, AB: GEO-SLOPE International.

Golder Associates (2010). *Report on Fractured Bedrock Field Methods and Analytical Tools*. Burnaby, BC: Golder Associates, Volume II, Appendices

Graziani, A., Rotonda, T. and Tommasi, P. (2010). Fenomeni di scivolamento planare in ammassi stratificati: Situazioni tipiche e metodi di analisi. In *Problemi di Stabilità nelle Opere Geotecniche*, ed. G. Barla and M. Barla. Bologna: Pàtron, pp. 93–124.

Henkel, D.J. (1959). The relationships between the strength, pore water pressure and volume-change characteristics of saturated clays. *Géotechnique*, 9, 119–135.

Hibbitt, D., Karlsson, B. and Sorensen, P. (2010). *Abaqus Theory Manual* [available at www.abaqus.com].

Hodge, R.A.L. and Freeze, R.A. (1977). Groundwater flow systems and slope stability. *Canadian Geotechnical Journal*, 14, 466–476.

Hoek, E. and Bray, J. (1980). *Rock Slope Engineering*. London: Institute of Mining and Metallurgy.

ISRM (1978). Suggested methods for quantitative description of discontinuities in rock masses. *International Journal of Rock Mechanics, Mineral Science & Geomechanics Abstracts*, 6, 349–368.

Itasca (2006). *FLAC3D: User's Manual Version 3.1*. Minneapolis, MN: Itasca.

Jennings, J.E.B. and Burland, J.B. (1962). Limitations to the use of effective stresses in partly saturated soils. *Géotechnique*, 12, 125–144.

Kenney, T.C. and Lau, K.C. (1984). Temporal changes of groundwater pressure in a natural slope of nonfissured clay. *Canadian Geotechnical Journal*, 21, 138–146.

Laflamme, J.F. and Leroueil, S. (1999). *Analyse des pressions interstitielles mesurées aux sites d'excavation de saint-hilaire et de rivière Vachon, Québec*. Ministère des Transports du Québec, Report GTC-99-10.

Lafleur, J. and Lefebvre, G. (1980). Groundwater regime associated with slope stability in Champlain clay deposits. *Canadian Geotechnical Journal*, 17, 44–54.

Lafleur, J., Silvestri, V., Asselin, R. and Soulié, M. (1988). Behaviour of a test excavation in soft Champlain Sea clay. *Canadian Geotechnical Journal*, 25, 705–715.

Lambe, T.W. and Whitman, R.V. (1969). *Soil Mechanics*. New York: John Wiley & Sons.

Leroueil, S. (2001). Natural slopes and cuts: movement and failure mechanisms. *Géotechnique*, 51, 197–243.

Leroueil, S. and Hight, D.W. (2002). Behaviour and properties of natural soils and soft rocks. In *Characterisation and Engineering Properties of Natural Soils, Vol. 1* ed. T.S. Tan, K.K. Phoon, D.W. Hight and S. Leroueil. Rotterdam: Balkema, pp. 229–254.

Misfeldt, G.A., Sauer, E.K. and Christiansen, E.A. (1991). The Hepburn landslide: An interactive slope-stability and seepage analysis. *Canadian Geotechnical Journal*, 28, 556–573.

Neuzil, C.E. (1993). Low fluid pressure within the Pierre shale: A transient response to erosion. *Water Resources Research*, 29, 2007–2020.

Olivares, L. and Damiano, E. (2007). Post-failure mechanics of landslides: A laboratory investigation of flowslides in pyroclastic soils. *Journal of Geotechnical and Geoenvironmental Engineering*, ASCE, 133, 51–62.

Olivares, L., Andreozzi, L., Damiano, E., Avolio, B. and Picarelli, L. (2003). Hydrologic response of a steep slope in unsaturated pyroclastic soils. In *Fast Slope Movements: Prediction and Prevention for Risk Mitigation*, ed. L. Picarelli. Bologna: Patron, pp. 391–397.

Papa, R., Pirone, M., Nicotera, M. and Urciuoli, G. (2009). Meccanismi di innesco di colate di fango in piroclastiti parzialmente sature. In *Rainfall-Induced Landslides: Mechanisms, Monitoring Techniques and Nowcasting Models for Early Warning Systems*, ed. L. Picarelli, P. Tommasi, G. Urciuoli and P. Versace. Naples, Italy: Doppiavoce, pp. 91–106.

Penman, H.L. (1958). Natural evaporation from open water, bare soil and grass. *Proceedings of the Royal Society of London A*, 193, 120–140.

Picarelli, L. and Urciuoli, G. (1993). Effetti dell'erosione in argilliti di alta plasticità. *Rivista Italiana di Geotecnica*, 17(1), 29–47.

Picarelli, L., Evangelista, A., Rolandi, G. *et al.* (2006). Mechanical properties of pyroclastic soils in Campania Region. In *Characterisation and Engineering Properties of Natural Soils, Vol. 4*, ed. T.S. Tan, K.K. Phoon, D.W. Hight and S. Leroueil. Rotterdam: Balkema, pp. 2331–2384.

Pinyol, N. (2010). *Landslides in reservoirs: A coupled thermo-hydro-mechanical approach*. Ph.D. thesis, Universitat Politècnica de Catalunya, Barcelona.

Potts, D.M., Kovacevic, N. and Vaughan, P.R. (1997). Delayed collapse of cut slopes in stiff clay. *Géotechnique*, 47, 953–982.

Priest, D. (1993). *Discontinuity Analysis for Rock Engineering*. London: Chapman & Hall.

Rahardjo, H., Lee, T.T., Leong, E.C. and Reazaur, R.B. (2005). Response of a residual soil slope to rainfall. *Canadian Geotechnical Journal*, 42, 340–351.

Richards, L. A. (1931). Capillary conduction of liquids in porous mediums. *Physics*, 1, 318–333.

Rulon, J. and Freeze, A. (1985). Multiple seepage faces on layered slopes and their implications for slope-stability analysis. *Canadian Geotechnical Journal*, 22, 347–356.

Sidle, R. C. and Ochiai, H. (2006). *Landslides: Processes, Prediction and Land Use*. American Geographical Union, Water Resources Publication 18.

Sullivan, T. D. (2007). Hydromechanical coupling and pit slope movements. In *Slope Stability: Proceedings of the International Symposium on Rock Slope Stability in Open Pit Mining and Civil Engineering*, ed. Y Potvin. Perth: Australian Centre for Geomechanics, pp. 3–43.

Taylor, D. W. (1948). *Fundamentals of Soil Mechanics*. New York: Wiley.

Terzaghi, K. (1943). *Theoretical Soil Mechanics*. New York: Wiley.

Thornthwaite, C. W. (1946). An approach toward a rational classification of climate. *Geographical Review*, 38, 55–95.

Vanapalli, S. K., Fredlund, D. G. and Pufahl, D. E. (1999). The influence of soil structure and stress history on the soil-water characteristics of a compacted till. *Géotechnique*, 49, 143–159.

21 Soil slope stabilization

EDWARD N. BROMHEAD, SEYYEDMAHDI HOSSEYNI, AND NOBUYUKI TORII

ABSTRACT

This chapter contains a discussion of methods for stabilizing potentially unstable slopes and slowly moving landslides. The methods described include the use of earthworks; surface water and deeper drainage; walls, piles and shear keys; pre-tensioned ground anchors; and a variety of geotechnical processes.

Some of the reasons for underperformance in stabilization works are described. Where possible, simple rules are given, and suggestions are made as to how to analyze the effectiveness of the stabilization measures.

21.1 INTRODUCTION

Unstable, or potentially unstable, slopes in soils and weak rocks can be treated in many ways to reduce risk. Engineered stabilization works are often expensive, and can be difficult to construct; therefore, before contemplating this approach, all other possibilities should be exhausted. Common alternatives are to do nothing, or alternatively to abandon the site and relocate whatever is at risk (Bromhead, 1997). While it is possible to stabilize a *potentially* unstable slope by the means described below, the following discussion is written from the perspective of stabilizing slopes that have failed or are in the process of failing. The stabilization process aims to prevent further movements of a mass of soil that is moving at a manageably slow rate along definite surfaces of rupture. Accordingly, deflection structures that resist or divert fast-moving landslides such as debris flows are excluded. It is assumed that, *a priori*, the location of surfaces of rupture (Hutchinson, 1983), together with groundwater levels and their future behavior, are known, so that appropriate stability calculations can be made with simple limit equilibrium methods.

Stabilization works fall into a number of categories, listed in decreasing order of importance, and treated at commensurate length.

- Earthworks solutions, where stability is obtained by altering the ground profile – moving soil from where it reduces stability to where it improves it, or by importing or exporting soil to improve the overall balance. Small landslides can also be stabilized by removing slip debris entirely, and filling the resulting space with better-quality fill that may be reinforced and provided with an internal drainage system.
- Improvement of soil strength in-situ by reducing the groundwater pressures through drainage.
- Construction of structures through any basal slip surfaces in order to provide additional shear resistance. Many of these structures are *passive* – they provide a reaction to thrusts exerted by the landslide – but they might also be *preloaded* or *active* (e.g., ground anchors), where the preloading force directly counters the thrust of the landslide. Preloading reduces ground movements more rapidly than passive systems, as the latter only build up reactions as they are deformed.
- Geotechnical procedures and processes such as grouting, electro-osmosis, and heat treatment of the soils improve the strength of the ground locally or globally. These procedures have been used as treatment measures with varying degrees of success, but are not commonplace.

The failure of natural slopes may not be blamed on the engineer, but the failure of engineered slopes will be. Failed stabilization attempts are worse than failures of engineered slopes, and therefore some common defects in stabilization are listed and described below in the hope that they will be avoided in future.

Stabilization schemes must be feasible, practical, effective, and robust. *Feasible* means that the method can be employed on that particular site, and *practical* means that the necessary plant,

Landslides: Types, Mechanisms and Modeling, ed. John J. Clague and Douglas Stead. Published by Cambridge University Press.

equipment, materials, and personnel can be deployed safely and economically. Works must be *effective*, in that they prevent further movement, or reduce movement to tolerable levels, for the duration of the design life of the scheme. The *robustness* (or *resilience*) of a scheme is its ability to continue to operate should some part of it malfunction: the whole should not suddenly collapse. The idea is encapsulated in the term "fail-safe," as used in other branches of engineering.

If, on the other hand, a stabilization scheme is critically dependent on one element, that element must be designed with an appropriate reserve of safety, and receive regular monitoring, together with planned maintenance and performance reviews. Further intervention may then be required at a future point if parts of a scheme show evidence of degradation, and this will require access to the site to be maintained.

21.2 STABILIZATION BY EARTHWORKS

A large part of the effect of stabilization by earthworks is immediate, arising simply from the weight of the fill placed or soil removed. This effect operates throughout the service life of the slope, although it can be affected by later erosion or earth-moving. Steps can be taken to protect against such incidents.

Earthworks solutions often need space outside of the immediate footprint of the slide: for example to fill over the toe, or grade back the rear scarp. The stability of the adjacent ground cannot always be relied upon, and earthworks may cause failures in adjacent previously stable land.

Ideally there is a balance between the quantities of cut and fill; but, if not, a nearby site must be found from where suitable fill can be sourced or where unwanted material may be deposited. Disposal of earthfill containing chemical, biological, or radioactive contaminants is particularly difficult. Some plant species may also be defined as contaminants (e.g., Japanese knotweed – *Fallopia japonica*, *Polygonum cuspidatum*, or *Reynoutria japonica* in the USA and Europe).

21.2.1 EXCAVATION AT THE HEAD OF SLIDE/SLOPE

Removal of the head of a slide mass is an obvious action. However, unloading the head of a landslide is not always possible: in a moderate to large slide there is the question of physical access and safety; there is always the prospect of increasing the effective height of the rear scarp, and correspondingly decreasing its stability; and, at the end of the process, a large flat area is created in which water may collect (known as ponding), requiring surface water drainage.

Drainage channels cut through the toe of a landslide or caused by gully erosion constitute unloading of the toe and must therefore be taken into account in the overall equilibrium calculation.

In compound slides that exhibit a graben, the optimum area for excavation is self-defining: it is the graben. If the original ground level needs to be retained, then various types of lightweight replacement fill can be used. Polystyrene foam blocks are lightweight and can be placed with little effort, but they are costly. Other possible lightweight fills include compressed bales of rubber tires ("tire bales") and proprietary sintered aggregates made from glass or clay products. It is important that any fill put onto the head of a slide is low-density, so that there is no nett loading. Most lightweight fills are flammable, and some are not even chemically inert.

21.2.2 COMPLETE OR PART REPLACEMENT

It is only practical to dig out landslides in their entirety when they are small. Many slips in infrastructure earthworks are small enough to use this approach. In traditional UK railway practice, the landslide debris was usually taken off-site for disposal, and the cavity packed with "spent ballast" or recycled track-bed materials. The free-draining characteristics of this mainly gravel-sized, angular, crushed rock material, with its high native angle of shearing resistance, and its ready availability, made it ideal for this purpose. However, such materials are now considered to be contaminated, and fresh rockfill is used instead of waste material. Crusher-run materials from quarries may be used. Rockfill is permeable and, with mainly angular particles, exhibits a high angle of shearing resistance. Geofabrics may be used as separators to prevent fines washing in, or traditional filter media can be employed.

It is essential that any slip surfaces are removed and broken up so that residual strength no longer applies. This means excavating a larger hole than the slide mass. Often the undrained strength of the parent material may hold the cavity open satisfactorily, but the side slopes need to be checked, so that there is a safe working environment on-site. Excavating into competent ground beneath and around the slip may also enable geogrid reinforcement to be anchored outside the slip area. Drainage layers can be incorporated into the backfill if it is not free-draining.

Sometimes it is necessary to reinstate the original earthwork profile, but elsewhere it might be useful to adopt a flatter overall angle or to provide a toe berm. Toe berms may be created and extended laterally to each side of the reinstatement area to improve stability there, as a precautionary measure.

Re-use of excavated material requires some storage on-site, and room in which to select or reject material. Clay soils are often completely unsuitable for re-use unless dried or made suitable by the incorporation of lime or cement. The action of recompacting incorporates air, and the undrained strength of the resulting fill may rise relative to the original clay. However, long-term changes in soil moisture may ultimately remove any strength gain from recompaction. If reconstructed to the original profile, stability can only then be maintained if the repaired section contains reinforcement and drainage.

21.2.3 FILL AT TOE

Earthfills may be placed to form a debris trap area and not bear directly on the slide, or may be provided to deflect slides

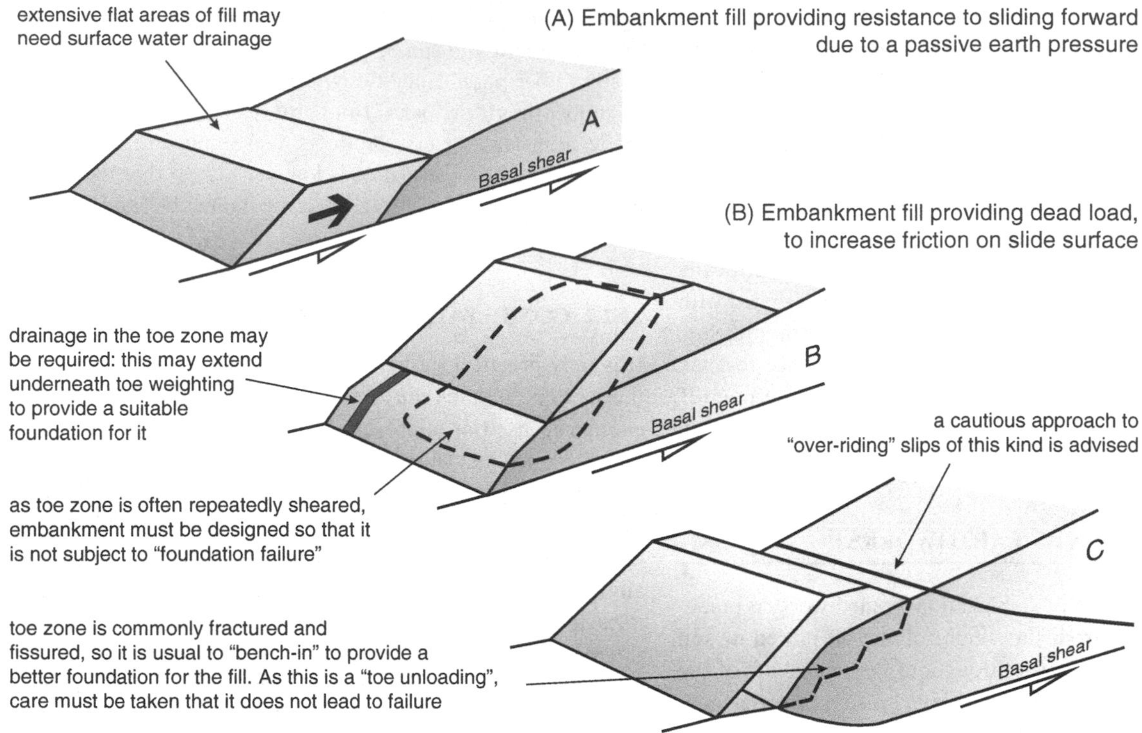

Fig. 21.1. Toe loads of different shapes and positions on the landslide toe give rise to different potential problems in practice (after Bromhead, 2005).

around and away from buildings and infrastructure. This useful arrangement is, however, not stabilization.

Figure 21.1 shows three different types of toe fill. In the first of these (indicated by the letter "A"), the fill is mainly constructed on flat ground beyond the toe of the slide. Some lateral pre-stress is provided by compaction. However, the main resistance is only obtained when the fill is compressed by movements of the slide, as it is a reaction, akin to generating a passive earth pressure, not an action. The resistance mobilized may be within the fill, or generated by sliding at the fill–foundation contact, or both. As strain dependency is never modeled in limit equilibrium methods, this effect must be considered separately. If a fill of this type is provided to give additional stability before undertaking an activity such as driving a tunnel into the toe zone of a landslide (which constitutes toe unloading), then the system may be subject to some movement before the full restraint is obtained.

The second sketch (marked "B") in Figure 21.1 is of a toe load that increases the normal stress on the slip surface. If the toe of the landslide rises underneath the fill, the weight of the fill decreases the nett destabilizing force for the landslide as a whole. If the slip surface is horizontal (as shown), the weight does not reduce the nett destabilizing force, and if the slip surface is inclined downward toward its breakout position, the weight of the fill may initially contribute a nett destabilization. In all cases, these fills contribute an increase to the shear resistance. This effect is contingent on the ground consolidating under the weight of the fill. The toe zone of a landslide is unlikely to be dilatant under the load of a fill, so the case where pore pressures initially decrease under the loading is not considered here.

Section B of Figure 21.1 also highlights the need to provide drainage in the toe of a landslide before a fill can be built on it. Embankment fill slopes must not fail in a foundation failure mode through the landslide toe. Some landslides have a toe zone networked with shear surfaces as the landslide was repeatedly thrust and overthrust. Any one of these secondary shear surfaces may be remobilized if they are in a critical position or orientation. The disturbance in a landslide toe zone will cause it to be weak, even if slickensided shears are absent.

Stabilization scheme designers need to be aware of the risk of secondary failures (see Fig. 21.1, sketch B). Reinforcements in the face of a toe loading embankment are helpful to prevent local instability. Section C of Figure 21.1 illustrates a fill that straddles the toe breakout position of the landslide. In order to fail, the shear displacement must pass through the toe fill. As illustrated, some of the weight of the fill is directly over a rising segment of the slip surface.

The dashed line in Figure 21.1C indicates the extent of benching-in that the construction team might do, not least to remove fissured ground, topsoil, and vegetation, and therefore to provide a good foundation contact. What passes for good construction practice in preparation of the contact by removal of topsoil or compressible materials may constitute unloading of the toe of the slope, which is potentially disastrous to overall stability. Site preparation for filling therefore needs to be approached with caution. Also shown in section C of Figure 21.1, but equally applicable in the other two sections, is the possibility that a new toe appears in the form of an overriding slip surface. The effect of the toe-weighting fill is localized, and its effect will only be felt in the upper parts of the landslide if the landslide body is rather rigid (e.g., a block-slide) and not if it is a landslide predominantly formed of debris. The corresponding underriding slip surface that passes beneath both the fill and the original basal shear surface is increasingly less likely to occur as consolidation takes hold, and the gain in effective stresses underneath the fill improves shear strength.

For landslides at risk of erosion, protection may be required in the form of a concrete, masonry, or gabion wall. The toe of a stabilizing earthfill may be supported by a crib or gabion wall to decrease land take. Space may be found, even in a narrow cutting, for a toe fill if part of the floor of the cutting can be taken out of service; this may require single-track working of a railway over the affected length, or taking over a lane or hard-shoulder of a highway (Biczysko, 1981).

21.2.4 SLOPE ANGLE REDUCTION

Slope angle reduction or flattening works best where the predominant mode of failure is shallow, but it does not work so well where the mode of failure is deep, for example where the slide has a rotational or compound form. Flattening a slope can unload the toe of an undetected major slide system, which is particularly problematic if the pore water pressures in the slope are independent of where the ground surface is located. Flatter slopes also reduce runoff, and present an increased area for infiltration. Extensive regrading is also not possible without removing vegetation cover, which ordinarily worsens stability by changing infiltration and evapotranspiration characteristics. Regrading breaks up both artificial and natural land drainage networks, and bare earth slopes are very susceptible to erosion.

It may not prove practical to reduce the slope angle of large natural slopes subject to shallow landslides.

21.2.5 OVERALL DESIGN VIABILITY CHECK FOR EARTHWORKS

Excavations can be analyzed by changing the ground profile and re-analyzing. The changes can remove sections of slip surface, but they are usually steep sections. The strength of the replacement lightweight fill must be modeled conservatively. If the excavation, with or without replacement using lightweight fills, is over only part of the footprint of a landslide, a problem arises with calculating the effect on the landslide as a whole, and a 3D analysis may be required.

Fills are modeled by adding extra soil zones. They may not have the same effect on every section through a landslide. It is helpful to establish early on what the likely maximum improvement in factor of safety is by analyzing the largest conceivable toe fill. Whatever improvement can be obtained may not be fully realized until consolidation has taken place.

In very large slides, the effectiveness of a toe weighting may differ considerably from section to section. Approximate 3D analysis is done by amalgamating the effects on a number of sections, usually by averaging. However, real 3D analysis is a practical proposition and may be fully merited for a large landslide. For small slides, investigations and analyses are likely to have been concentrated along a single cross-section. Where this *principal cross-section* is through the axis of the landslide, the effectiveness of the toe fill may diminish toward the sides of the slide, and the overall improvement in stability will be less than predicted.

21.2.6 DETAILED DESIGN ASPECTS FOR EARTHWORKS

Earthfills are unlikely to be denser than the natural soils on a site. Very heavy fills, which may be useful in congested sites, require the use of exotic materials – for example, stacked concrete blocks or scrap iron – but expense is a factor. Lightweight materials with densities of about half that of the natural soil are also expensive and, again, these densities are achieved by the use of generally exotic materials. Tire bales and dense polystyrene foam may be light, but they need protection from fire at the surface. In some jurisdictions, the use of waste materials in construction attracts landfill taxes. Gedney and Weber (1978) describe a fill constructed from wood fiber. Although no doubt this fill is lightweight, its long-term durability is questionable.

21.2.7 USEFUL RULES OF THUMB

The neutral line concept (Hutchinson, 1977, 1984b) is useful in planning both toe-weighting fills and excavations at the head of a slide. Clearly, loading at the head of a slide is bad for its stability, and loading at the toe is good. Excavations at these places have the opposite effects. There must therefore be a location somewhere in the slide where a smallish load or excavation would have no nett effect on stability. The trace of such points on a plan of the landslide is termed the *neutral line*, and the *neutral point* is the corresponding position on a cross-section of the landslide. The position of this point occurs directly above the point(s) at which the basal slip surface of a landslide has a certain inclination. This inclination is revealed from theory by adding a small load to the landslide, and equating the change in factor of safety to zero.

Figure 21.2 shows a stylized cross-section of a rotational landslide with two neutral points marked on it. Fills are always good and excavations or cuts always bad to the left of where the slip surface inclination $\alpha = 0$, and fills always reduce stability and cuts improve it to the right of the point where the slip surface inclination $\alpha = \frac{\tan\phi'}{F}$ in the figure. These points should

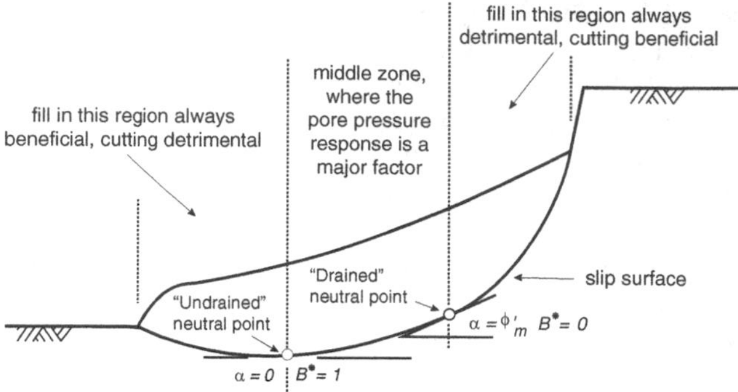

Fig. 21.2. Cut and fill – the neutral line approach (after Hutchinson, 1977).

be identified during stability analysis and earthworks located accordingly. In between the two limits, the effect depends on the pore water pressure response to loading, and will vary as pore-pressure equalization takes place.

The neutral point occurs where the slip surface is horizontal (Fig. 21.2), when the loading is undrained (i.e., where the increase in *vertical total stress* is offset by an equivalent change in *pore water pressure*, so that the loading causes no change in *effective stress*). Correspondingly, where the loading is fully drained, which is when there is no change in *pore water pressure* so that the increase in *total stress* is mirrored by the increase in *effective stress*, the neutral point occurs where the slip surface is inclined at an angle a equal to the mobilized angle of shearing resistance, ϕ'_m, such that:

$$\tan\alpha = \tan\phi'_m = \frac{\tan\phi'}{F} \tag{21.1}$$

As noted by Bromhead (2004) the neutral point method causes no difficulties if applied to a slope where $F = 1$, and is done for a single load increment, but it does pose a problem if load is applied in increments, causing F to vary, and thus the position of the neutral point to change. The ambiguity is resolved if one considers the neutral point for the desired, or *target*, F.

If it is necessary to load upslope of the appropriate neutral point, then the decrease in stability must be overcome by appropriate increases in loading downslope of the neutral point. If the trace of the neutral point in plan is made (thus defining the neutral line) this – or indeed multiple neutral lines – can define corridors within which earthwork cuts and fills can most sensibly be contemplated.

The application of the neutral point to three-dimensional analysis is complicated by the fact that some landslides may change their vector direction of movement in response to a localized loading. In such cases the criteria for finding the location of the neutral point must be applied in the direction of ultimate movement (Bromhead, 2004).

21.2.8 ERRORS AND PROBLEMS WITH EARTHWORKS

It is well known that limit equilibrium slope stability analyses may not function correctly when the analyst places fill material with a high angle of shearing resistance straddling the toe breakout zone of a landslide, so that the toe of the landslide is projected up through the fill. Aspects of this effect were discussed as early as 1967 by Whitman and Bailey.

Embankment fills may consist of material that has a significantly higher angle of shearing resistance ϕ' than the landslide debris. The difference in ϕ' is particularly large if the face of the fill is armored against erosion with large stone or concrete blocks. When analyzing the stability of the landslide and toe fill system using a limit equilibrium approach, the investigator may be tempted simply to project the slip surface upward through the high ϕ' fill. The effect of this projection is sometimes to cause convergence problems in the analysis but, more commonly, to indicate very high improvements in safety factor with only small segments of projected slip surface through the high ϕ' fill. This result is a consequence of the intrinsic assumptions in the limit equilibrium method. The issue is avoided if large ϕ' values are not used in this setting, and modified geometry or pre-computed strength properties are used for the slip surface extension.

The undrained loading effect, whereby the toe fill load increases pore pressures in its foundation, may mean that the toe fill effectiveness is lower initially than after consolidation has taken place. The time required for consolidation may require a larger fill in the short term than would be required in the long term. In contrast, excavations experience an undrained unloading effect whereby pore water pressures decrease initially, thus increasing the initial effectiveness of excavation-based schemes.

Examples of major errors in the use of earthworks include:

- Loading at the head of a small landslide in order to construct an access track or a piling platform. Access tracks are required so that soil nailing equipment can be deployed to stabilize the rear scarp of the slide; piling platforms are required to provide a working platform for the safe operation of top-heavy piling rigs (Fig. 21.3).
- Loading at the head of a landslide to buttress the rear scarp, thus causing movements in the slide itself.
- Excavation at the toe of a landslide in the form of benching, so that a toe fill could be properly placed and the topsoil removed (Fig. 21.1).
- Excavation into the toe of a slope that contained an unrecognized earlier failure (Chandler, 1979). Figure 21.3 also shows an excavation for a tunnel portal provoking slide movements.
- Construction of a toe-weighting fill on a landslide by hydraulic fill (from a dredger) which liquefied.
- Earthworks interrupting a natural drainage system, causing the migration of water to somewhere where it can cause instability.

As some soils exhibit an aggressive chemistry, for example pyritic shales that oxidize and release iron sulfates, they may prove to be unsuitable as fills. Moreover, excavating cuttings or even borrow pits may cause groundwater pollution from runoff.

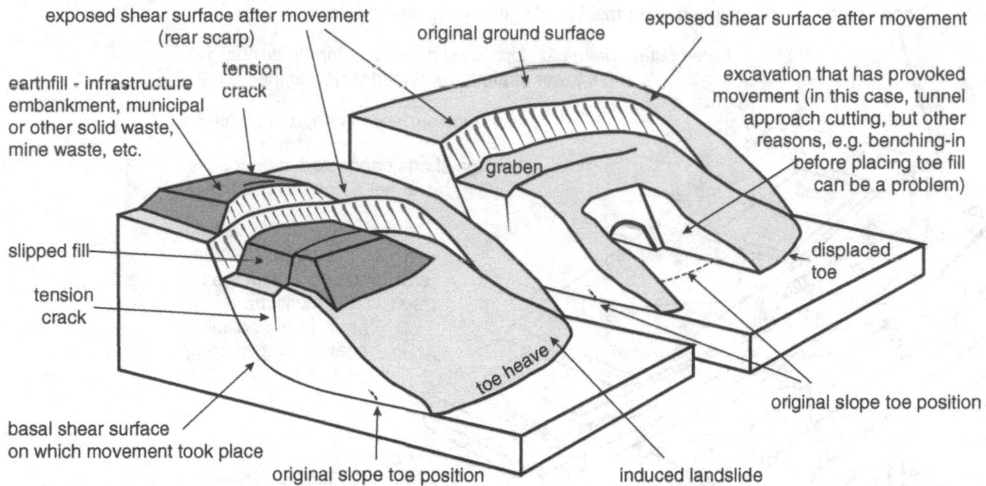

Fig. 21.3. Left: Load from fill on a pre-existing slide provoking movement. Right: Excavation into slide toe. Sometimes it is not obvious before the event that the excavations are into an old landslide. Tunneling on its own may unload the toe, as may gulley erosion, infrastructure cuttings, trenches to install services, etc.

21.3 STABILIZATION BY DRAINAGE

Drainage makes use of the principle of effective stress; that is, a decrease in the water pressure in a soil increases its resistance to shear. This effect is true both when the soil is unsheared and is therefore operating with peak strength, and also when it has been sheared and is operating with residual strength along pre-existing shear surfaces.

Drainage cannot be employed if there is little or no easily removable water in the slope. It may, in any case, require a considerable time for the effect to take hold. Less obvious is the inevitable result that the removal of water from a soil slope will cause consolidation settlements that, in large, slow-moving, landslides, may be similar to – or even exceed – the slide deformations and deformation rates for a number of years. Moreover, there are the problems of transporting the collected water across the edges of any unstable area, and of disposing of it safely. In some cases, disposal of drainage water poses stability risks; in others it may be prohibited by local planning ordinances concerned with watercourses and water quality, or by the need to avoid pollution.

Drainage measures may be subdivided into those that prevent the infiltration of water in the first place and those that remove that which is already there. Barriers to infiltration perform a related function. Surface barriers include some types of vegetation, and underground barriers include clay cut-offs, dam cores, and grouting. Barriers reduce pore pressures downstream, but invariably increase them upstream. Some drains may be used to lower high groundwater pressures during construction, such as those induced by undrained loading, whereas some drains have to function continuously throughout the design life of the scheme and therefore need to be provided with filters.

21.3.1 SURFACE WATER DRAINAGE

Surface water may be controlled by trimming slopes to appropriate gradients, blanketing them with grass or other vegetation, and providing a system of ditches to collect and transport surface water before it infiltrates. If the purpose of a ditch is to transport water, it is better if it is lined. Lining with slabs is sometimes used, but even small movements can disrupt the pointing between the slabs and allow water leakage. Timber-lined channels are also prone to failures. High discharges have a considerable scour potential at discharge points from channeled drainage, and stilling arrangements may be required. Sediments and vegetation collect where there are changes of gradient or direction in surface water drainage systems, so that regular maintenance is necessary. Galvanized corrugated steel half-moon channel sections are useful, because the corrugations provide roughness. Other proprietary lining systems include concrete channel sections that overlap and also provide a rough cascade profile that has the additional benefits of inducing turbulence to slow the flow, and oxidation that can help reduce the oxygen demand of some types of contaminated flow. Rubberized conveyor belt can be used as a channel lining; it is flexible and seamless, although of questionable long-term durability.

Porous clay pipes and clay tiles with an inverted V-shape (tile drains) have traditionally been used in land drainage: their modern equivalents are perforated plastic pipes laid in machine-dug trenches and backfilled. Figure 21.4 shows a compendium of shallow and deep drainage systems, including surface water drainage. Secondary drains in herringbone or chevron patterns shorten the path of downslope overland flow and thus reduce the potential for gully or rill erosion. Gullying is also inhibited by a good cover of well-rooted vegetation or the provision of mats on the slope surface. Matting systems are supplied by a number of competing commercial providers, and may be either biodegradable or permanent. Biodegradable mats are used to retain soil and seeds until the vegetation develops adequate roots and can then retain soil particles.

21.3.2 DEEP DRAINAGE – WITHOUT ACCESS

Deep drainage may be accomplished with drains drilled in from the surface. These drains are generally either sub-horizontal,

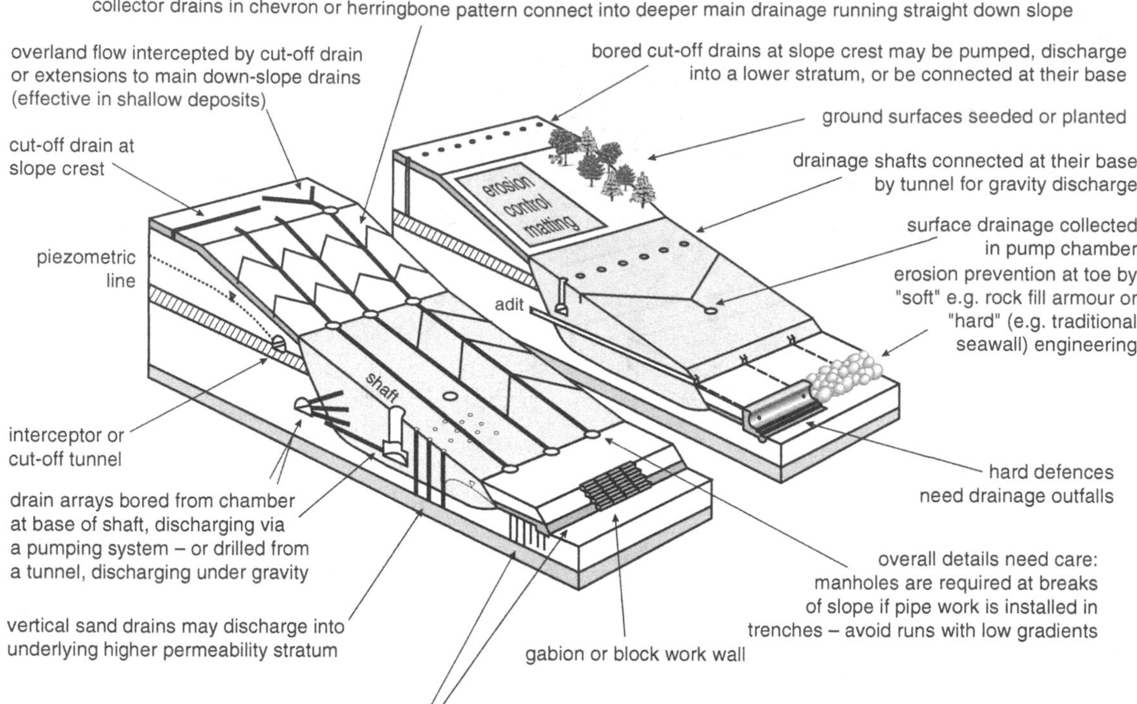

Fig. 21.4. A medley of shallow and deep drainage solutions for a slope. The upper slopes have been regraded, and need protection against erosion; the lower slopes have also been flattened and definitely need surface water drainage. The toe of the slope may be provided with erosion protection if the toe load could be scoured away.

with only a slight gradient, or largely vertical. Vertical drains penetrating an underlying permeable stratum or adit may be able to empty under gravity; otherwise they need to be pumped (Bianco and Bruce, 1991).

Many permanent drains require a liner to keep the hole open. In the case of sub-horizontal bored drains, the liner takes the form of a plastic tube with push-fit or threaded connectors. Some tubes have holes or slots in the top half, and a solid bottom. Holes are formed by drilling, punching, or sawing. These pipes have to be correctly oriented when they are installed. Pipes with a regular coverage of water entry holes all over rely on the water finding it easier to exit along the pipe than to leak out.

The invention of geofabric filters has simplified the protection of perforated drainage pipes from siltation. Geofabric filters can be pre-formed into tubing, which is threaded onto the outside of a pipe and secured by adhesive tape. Although external geofabric tubing is easily torn on installation, damage to the filter fabric can be minimized where it is closely fitted to the tube, and the tube is joined with threaded couplers that have a larger outside diameter than the pipe itself. Alternatively, a geofabric tube inside the plastic hole former is kept in shape by a spine made of plastic spiders. This system can be removed for filter replacement, and is less vulnerable to damage during installation, although many engineers will consider the filter to be in the wrong place.

Drilled drains are subject to blockage by precipitation, with calcite and iron compounds being the main culprits. Although flushing with weak acids may alleviate the problem, it is probably simpler to install additional drains rather than to attempt to maintain those that are dysfunctional. Inspection of these drains is restricted to monitoring discharges, although piezometric monitoring in the adjacent soils, or even in particular drain runs, is possible. Piezometers, however, may also progressively degrade due to precipitates.

Vertical drains may be narrow and provided with permeable fillers or wicks, or larger in diameter and left open. Open drainage shafts need to be securely capped for the safety of animals and people. Syphonic systems can empty comparatively shallow drains, but deeper drains need to be equipped with a power supply and electric pumps.

21.3.3 DEEP DRAINAGE – INSPECTABLE

Very large-scale drainage systems usually have tunnels, adits, and shafts that are entered during construction or operation; and members of the public may gain illicit access later. Trespassers are unlikely to have good lighting, or to follow proper ventilation procedures, and so are at more risk than inspectors, who have appropriate safety training and equipment. Securely locked doors and manhole covers are therefore essential.

Working in underground and confined spaces requires design to minimize trip and fall hazards, with linings to provide ground support, provision of barriers, lighting (including emergency systems), ventilation, and staff training in safe working practices, including escape and rescue. Drainage systems not only collect water, but may also collect harmful gases, for example, radon, carbon dioxide and monoxide, methane, and other poisonous, combustible, or explosive gases. Exhaust gases from construction plant may collect at the base of excavations. Gases may come out of solution in the groundwater, or may emanate from the soil or rock in which the drainage tunnels are constructed. Where hazardous gases are present, workers need to carry and be trained in the use of gas detectors and breathing apparatus, and the tunnels need to be flushed with fresh air for an appropriate length of time before entry. Combustible gases demand that workers be equipped appropriately with nonferrous tools and safety lamps, and that they do not carry smoking materials underground.

Tunnels, adits and shafts are effective deep drainage measures, as they have a large carrying capacity and are unlikely to become blocked by siltation or precipitation of minerals. Their effectiveness is extended if arrays of bored drains are drilled from them. Shafts may have one or more rooms from which arrays or fans of sub-horizontal drains are bored. Systems of shafts may be interconnected, with pumping only from the shaft with the lowest collection point, or pumps may be provided for each.

Adits can be constructed with a fall, so that water drains under gravity. Tunnels may have falls to both entrances or just to one. No tunnels or shafts can drain water to levels deeper than their lowest outfall point. It is therefore preferable, if at all possible, to position the tunnels underneath the slip surface, if necessary extending the influence of the drainage into the slip mass with inclined bored drains drilled from the tunnel.

21.3.4 DRAINAGE TO REMOVE EPHEMERAL WATER PRESSURES

Some drainage systems only remove water pressures set up under applied loads, and once the process is complete, the drains are of no further use. Examples of such drains include *sand drains*, the sand-filled boreholes that were constructed in the past to serve this purpose, or *wick drains*, with geofabric wicks punched in the ground by a mandrel and not placed in bored holes. Although wick drains accelerate the dissipation of pore water pressure, the disturbance accompanying installation can increase the settlement, and disruption of permeable bedding (smearing) may result in poor connection of the wick to the soil.

21.3.5 TREES AND VEGETATION

Trees and other plants covering a slope help in reducing the amount of water that reaches the ground surface, accelerating the return of water to the atmosphere by transpiration, and also by binding the surface of the soil with their roots. However, as plants depend on soil moisture, they also break up the soil mass with their roots. Similar disturbance is caused by burrowing animals. This assists in infiltration of whatever moisture does reach the ground. Removal of vegetation cover leads to decreased stability, but vegetation can only become established if a slope is reasonably stable. The use of vegetation cover alone is unlikely to appeal to engineers seeking to stabilize a given landslide.

21.3.6 OVERALL DESIGN VIABILITY CHECK

The effect of drainage is modeled by lowering the piezometric line relative to its original position. There are no adverse effects likely for other surfaces or parts of the slope unless there is leakage.

It is impossible to remove all pore water pressures from a slope, but a very useful analysis to carry out is one without water pressures. This yields a theoretical upper bound to the effectiveness of drainage. If the resulting factor of improvement is not significantly greater than the target for the slope stabilization scheme as a whole, then drainage alone is not the answer.

21.3.7 DETAILED DESIGN ASPECTS

Drains can silt up or become blocked. For many types of drain, geofabric filters are used to prevent the ingress of small particles. Trench drains, which are little more than a trench filled with gravel, rockfill, or even demolition rubble, may be lined with a sheet of geofabric obtained from a roll, and overlapped at junctions. The geofabric is often overlapped at the top (Fig. 21.5) and buried under soil or gravel to prevent it from being disturbed by people or animals.

Drainage systems may collect contaminated water that needs treatment; for example, wastewater discharges from septic tanks. Mine waste leachate may contain minerals from processing, for example cyanides, or from the oxidation of the source rock when exposed to air (acid mine drainage). Leachate from municipal waste contains harmful chemicals from the waste stream. The appropriate treatment differs from site to site. Ground deformations disrupt containment systems and almost always open up the ground to allow infiltration of surface water and release of contaminants.

21.3.8 USEFUL RULES OF THUMB

As the unit weight of soil is about twice the unit weight of water, except in peat, the ratio of factors of safety (FoS) in frictional soils between the no-water case and the groundwater at ground level case is about 2:1. Commonly encountered piezometric conditions fall somewhere between these two extremes.

Drains oriented normal to the plane of a typical analysis cross-section have the same effect at all cross-sections, whereas drains oriented parallel to the plane of the cross-section show a variation in effect – most where there is a drain and least between the drains (Fig. 21.5). A gravel- or rubble-filled trench drain that penetrates a shallow landslide can bring the local piezometric surface down to its basal shear surface, and also replaces part of it with material that has a higher ϕ'. Drains that do not penetrate so far are much less effective. Shallow land

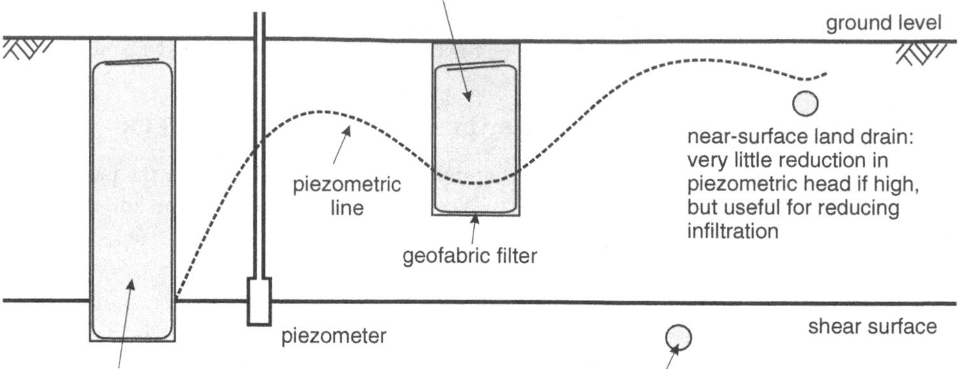

Fig. 21.5. Piezometric variation between drains and effectiveness of different types of shallow drainage. The 3D effect of drains can be limited.

drains are the least effective of all, but do have a role in preventing waterlogging of the surface which can feed deeper infiltration after the rain or snowmelt has finished. Drains drilled beneath a basal shear surface may prove surprisingly ineffective if the in-situ soil is much less permeable than the landslide debris above, as the lower-permeability solid material acts as a cut-off. However, drainage that promotes the downward flow of groundwater is more effective in reducing pore pressures than drains that cause a horizontal flow.

21.3.9 ERRORS AND PROBLEMS WITH DRAINAGE

If the outfall of a drainage system is blocked accidentally or maliciously, the build-up of water pressure may be worse than having no drainage at all. Leakage from drains, or *ponding*, if drain runs are disturbed creates adverse conditions.

Some major errors in construction or design include:

- Failure to connect surface water or other drains to a proper outfall before heavy rain.
- Erosion resulting from peak drain discharge.
- Use of high water-flush pressures when drilling sub-horizontal tube drains, raising the groundwater pressures.
- Generation of consolidation settlements from drainage or dewatering.
- Defects arising from modeling with the wrong pore water pressures. If the base case pore pressures (i.e., prior to drainage) are assumed to be high, then the effect of drainage is overestimated.

21.4 STABILIZATION WITH EMBEDDED WALLS, PILES, AND SHEAR KEYS

Piles, piers, buttresses, or walls constructed through a landslide and into the underlying bedrock create a *passive system.* Initially, and when undeflected, passive systems have little effect on stability, but slow movements generate reaction forces that ultimately lead to stabilization. As loads are transferred into the ground underneath the basal shear of the landslide, these systems work best when that ground is strong, stiff, and unlikely to develop its own instability. Also, the stiffer the structure, the less movement is typically required to mobilize its resistance.

A compendium of passive structures is shown in Figure 21.6, with an active system of stressed anchors in the foreground. Limited space in the diagram precludes showing the range of types of embedded walls, and the set of contiguous bored piles in the diagram is an *aide memoire* for the range of embedded wall systems. Embedded walls may be stiffened with buttresses, counterforts, and some types of reinforced concrete capping beams.

A major advantage of structural restraints is that they are installed within the landslide footprint, and also they can be designed to work despite high piezometric conditions, so that drainage is not essential. A difficulty with many passive systems lies in the complex calculations of resistance and soil-structure interaction that are required in designing them.

The structural behavior is usually tension, shear, and bending, and the geometry of a typical landslide militates against behavior in compression, although some piled systems may have a few piles which act in compression, especially where they are installed through a steeply dipping basal shear surface toward the head of a landslide, or where they are installed at an angle to the vertical.

21.4.1 SHEAR KEYS OR BUTTRESSES

Shear keys or buttresses are structures that are installed towards the toe of a slide (Figs. 21.6 and 21.7) and are unreinforced or lightly reinforced concrete, although masonry or even rockfill have been used, the latter performing a dual function as a drain.

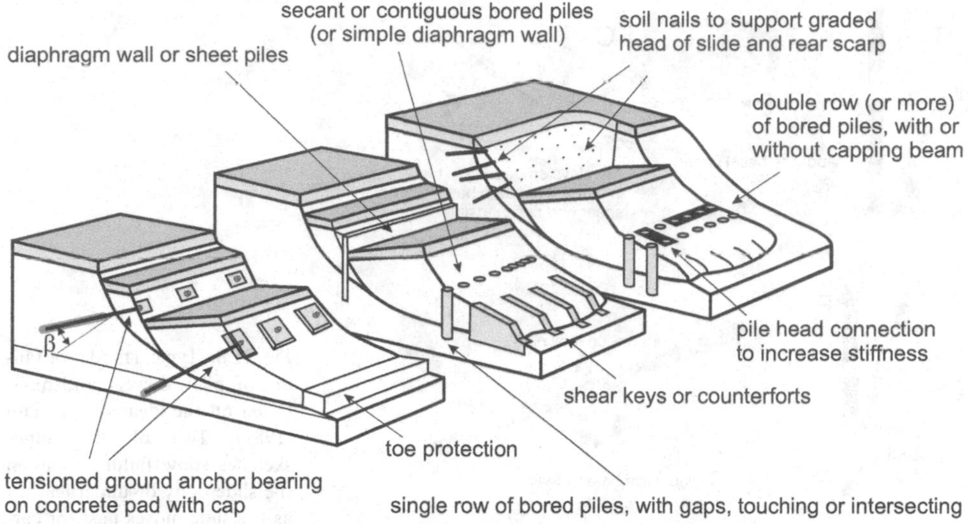

Fig. 21.6. Foreground left: anchored solution to an unstable slope. Middle and rear right: buttress and pile solutions for slope stabilization. (After Bromhead, 2005.)

Fig. 21.7. Unreinforced concrete buttresses on the Danube Ship Canal, Romania. These concrete buttresses are finished and awaiting infilling with rockfill.

The effect on a landslide is increased shear resistance in the appropriate part of the basal shear surface. The buttress resists movement both by basal shear where it is founded in undisturbed soil beneath the sliding surface of the landslide, and by passive earth pressure. Typically, basal shear is the main resistance element because the toe of the buttress may be located where it is difficult to obtain much passive earth pressure. Ordinarily, a buttress would be sized to make bearing capacity at its toe a noncritical case.

21.4.2 PILES

Piling systems are increasingly being used for slope stabilization (Viggiani, 1981; Poulos, 1995). Specialist piling companies have developed cost-effective systems for installing piles, and this makes the approach attractive.

Small-diameter piles may be arranged in a three-dimensional fan, like the roots of a tree (root piles), joined at the head to increase stiffness. Shorter piles are drilled with a continuous flight auger system using a long mast and auger, as this is quick relative to other bored-pile construction techniques. Percussive bored piling systems use a much lighter and smaller tripod rig that can be deployed in situations where the heavy rigs with tall masts cannot. Percussively bored piles are restricted in length only by the difficulties of drilling at depth and the length of the wire rope used.

21.4.3 DOWELS

Dowels are small-diameter piles that operate in the shear mode (see "A" in Fig. 21.8). They are used in rock slopes as unstressed rock bolts.

21.4.4 SOIL NAILS

Soil nails are grouted small-diameter boreholes containing a single reinforcing bar. They are inclined at an angle to the slip surface so that landslide movements induce a tension in the nail. As the nail crosses the basal shear, some shear deformation is inevitably also induced. A load-spreading plate is employed at the head of the nail, and this is used in conjunction with a geogrid on the ground surface to further assist in distributing the nail tension. Some tensioning is possible if the head of the soil nail is secured with a nut acting on the spreader plate. Soil nails have become a popular method of preventing shallow movements in the face of cut slopes, where the calculations are done with notional slip surfaces.

Soil nails are difficult to install long enough or strong enough to act on their own to stabilize a moderate or large landslide, but they are highly effective in pinning back small landslides on steep slopes. In Figure 21.6 they are shown retaining a small

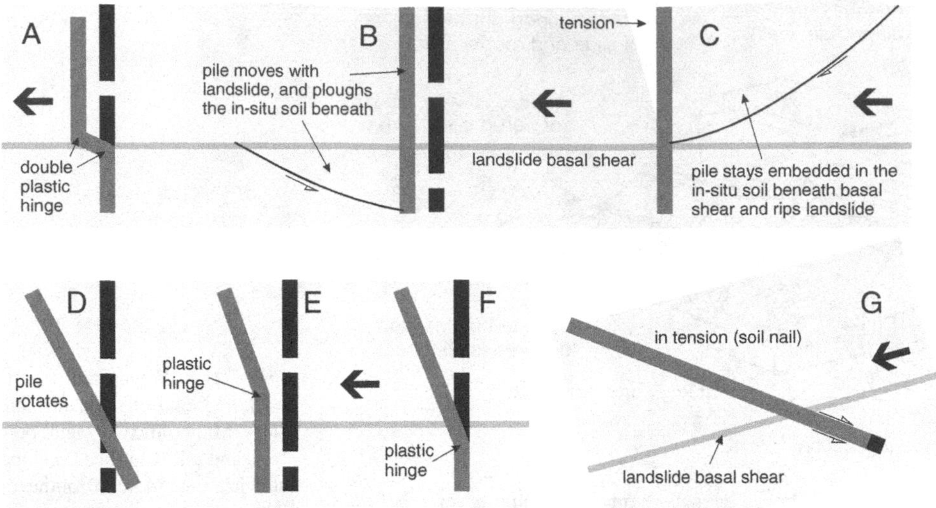

Fig. 21.8. Typical modes of failure of piles, dowels, and nails, based on the ideas of Viggiani (1981). Two of the upper sketches show failure zones in the slide body or underneath it as the slide moves past, or carries, a rigid pier or pile.

secondary slip, and also providing additional stability to a regraded face.

21.4.5 WALLS

Piles installed with space between them rely on arching to prevent the soil flowing through the gaps. They must be fully embedded because a difference in level for the ground surface upslope and downslope of the row of piles would merely facilitate the passage of soil between them.

Some walls, such as soldier-pile walls or plank-and-post walls are basically a set of piles that do have this difference in level, and the gaps are bridged with planks to prevent the loss of retained soil. Walls can also be constructed from contiguous or secant bored piles, diaphragm walling systems, and driven sheet piles; they can be constructed from reinforced concrete, masonry, gabions, and cribs. Steel sheet piles are flexible, subject to corrosion, and may declutch (i.e., tear apart at the interlocks) when driven or following ground movement. A series of failures in a coastal slope involving steel sheet piles installed to stabilize it is described by Barton and Garvey (2011).

Care should be taken to differentiate between embedded walls for landslide stabilization and retaining walls. Retaining walls are usually designed to resist earth pressures generated from soil loading relatively close to the wall and not those generated by landslide movement. Accordingly, the main use of retaining walls in a landslide should be to retain a face for a toe fill. In that case, a crib or gabion wall may be all that is needed. Examples of the use of reinforced earth and gabions to support the edge of a stabilizing fill are given by Gedney and Weber (1978).

In most cases, anchoring a wall or providing it with tie-backs improves its stability against overturning, so that the wall can more easily resist the slide movement. Tie-backs should be secured in the in-situ soil or rock underneath the landslide.

21.4.6 ISOLATED PIERS

If piles are too widely spaced, they may not stabilize a landslide, but instead they may simply act as stable islands as the slide moves past them.

It may be possible to construct a bridge that will let the landslide pass beneath it, much as sheds allow slides to pass overhead. Ideally, bridge structures have no piers in the landslide, but such piers can be designed (Hutchinson, 1984a) to resist soil pressures as the landslide moves past (Fig. 21.8). In effect, the pier plows through the soil of the landslide, pushing a bow wave ahead of it. As the soil then fails in a passive earth pressure mode, the pier must be designed to resist a substantial set of passive pressures from this three-dimensional failure.

21.4.7 OVERALL DESIGN VIABILITY CHECK

There is no theoretical upper bound to the effect achievable with structural solutions: only economics and the site practicalities limit what may be achieved. Buttresses need sizeable open excavations in the toe of a landslide that are only safe if supported and drained. The temporary works case always needs to be considered with any large excavation into a landslide. Alternatively, shear keys may be cast in-situ in excavations formed by a diaphragm walling technique or with bored piles.

Piling rigs are both heavy and readily toppled, so that it is usual to construct a piling platform to bear and spread their weight. The piling platform should not itself reduce the stability of a slope. Ideally, piling should be deployed where there is already relatively flat ground on the landslide and where the piling platform is situated over a low-angle part of the basal shear of the landslide. It is difficult to construct piles on a steep slope, although if the slip surface is not particularly deep and relatively small-diameter piles are required, it may be possible to use a lightweight rig or one that operates from a crawler tractor.

Small-diameter piles, dowels, and soil nails have many generic similarities in terms of how they are constructed.

Driven piles have been used in landslide stabilization, but engineers are rightly nervous about vibration or the possibility of dynamically induced pore-pressure effects and disturbance of the soil mass. For very shallow slope failures that are little more than movement of the topsoil, plate piles have been employed successfully. These are little more than galvanized steel plates welded to angle sections, driven into the ground in a staggered grid to provide more resistance to sliding.

21.4.8 DETAILED DESIGN ASPECTS

Viggiani (1981) summarized the modes in which a pile installed through the basal shear surface of a landslide might deflect and fail, and therefore how it might deliver a reaction force into the landslide. These modes are illustrated in Figure 21.8 for an initially vertical isolated pile through a horizontal basal shear, with movement of the landslide from right to left, as shown by the arrows. Diagram "A" shows a pile that is gripped so that it cannot rotate either in the landslide or in the in-situ soil beneath. This pile fails with two plastic hinges immediately above and below the basal shear. The other two diagrams on the top row show a pile that is rigid and does not rotate but is also too strong to fail with plastic hinges. In diagram "B" it is dragged by the landslide so that it plows the underlying in-situ soil. In diagram "C," the underlying soil is strong enough to resist failure, and instead the pile remains stationary while the landslide moves past. In this latter case, the soil of the landslide must fail, and the pile rips a passage through it. In the lower row, diagram "D" shows a stiff pile rotating, and "E " and "F" show it failing with a plastic hinge respectively above or below the basal shear surface. It is possible, in principle, to compute the loads from the landslide that cause these different types of failure, and to find the mechanism with the lowest load. The reliable reaction into the landslide for a group may be less than the sum of the resistances calculated for individual piles. These calculations are often easiest to do using the undrained strength of the soil, but the analyst must take care that resistances are not lower in the drained state.

Diagram "G" shows a soil nail acting in tension; the basal shear surface is here intentionally shown inclined to differentiate this case from the piles. The bending and shear resistances may be comparatively small. A pile may also act in tension or compression, especially if installed as one of a set of inclined or root piles.

Bored piles used in slope stabilization may need to be heavily reinforced with steel tubes or H sections instead of bar reinforcement in order to resist bending moments. Large-diameter shafts can be constructed with much larger diameters and depths than can be obtained with piling rigs by shaft-sinking methods.

The stiffness of a pile system can be substantially increased by connecting the pile heads together in an upslope direction to form portal frames or more complicated, and therefore stiffer, structures. Connecting the heads of a row of piles along the contours of a slope, and thus normal to the slide direction, does not have this beneficial effect. Stiffening the top of a pile system with a structural connection to upslope or downslope piles increases the resistance to failure modes "D," "E," and "F" in Figure 21.8.

21.4.9 USEFUL RULES OF THUMB

The neutral point can be used to find optimum positions to locate piling platforms. The resisting and driving force terms are equal in a landslide with $F = 1$, that is: $D = R = \Sigma W \sin\alpha$. The total additional shear resistance to be provided by the structural system can be computed approximately as a proportion of this, so that additional resistance to get $F = 1.25$ is therefore 25 percent of D. Some methods of stability analysis, e.g., Morgenstern and Price's (1965) method, do not calculate the driving force explicitly, but it or its equivalent can be recovered from the average shear stress and length of the slip surface on a cross-section.

21.4.10 ERRORS AND PROBLEMS

It is usually a mistake to use a gabion, crib, or other wall to stabilize a landslide, because the design of these features is based on local earth pressure and not the accumulated thrust from a landslide.

It is also a major mistake to undertake limit equilibrium-based stability analysis with odd-shaped blocks of soil in the analysis that are given concrete properties. The resulting factors of safety can be extremely high, and unrelated to the stability of slip surfaces that go under, over, or around the block in question. Published examples of this error are understandably few and far between, although the last few diagrams in Wolfe *et al.* (1996) make this mistake.

21.5 STABILIZATION WITH GROUND ANCHORS

The fundamental point of a stressed ground anchor (Fig. 21.6) is that it forms an *active* system, where the load is put into the system at the outset, and it does not require landslide movement to induce a reaction. Ground anchors are installed in boreholes drilled through the landslide mass into the solid ground below the basal shear surface. The anchor is usually a wire strand tendon that is sheathed to protect it from corrosion, although alloy bars are sometimes used for low loads. This anchor is grouted into the bottom section of the borehole with a cementitious grout. A bearing pad of reinforced concrete is provided to spread the anchor load over part of the ground surface so that the soil does not suffer a local bearing capacity failure when the anchor load is applied (Fig. 21.6). The tendon is stressed by jacking against the concrete pad. Systems used for jacking and monitoring the load are the same as those used in post-tensioned concrete structures. When the load is fully applied, the tendon is secured, and the jacking system is removed. The anchor tendon is usually secured at the anchor head with a wedge locking arrangement; tapered wedges grip the individual anchor strands. Anchor details are given by Barley (1991) and Cornforth (2005).

Tendons that lose load need to be re-tensioned. The loss of load may be due to creep in the anchorage, take-up of load at the anchor head wedge system, creep in the tendon, or consolidation deformations in the ground under the load-spreading concrete pad. In the worst cases, re-tensioning exercises must be carried out periodically.

Ground anchors are most useful with moderate-sized landslides (Fig. 21.6). With large landslides it can be difficult to provide the requisite number of anchors, their lengths, and the total force that needs to be applied. Small landslides may not offer suitable locations for the bearing pads, and are usually more easily stabilized by another means. The long-term durability of anchor tendons is suspect. Some engineers do not favor leaving highly stressed elements in the ground, where maintenance may be neglected.

Anchors can be used to supplement the existing factor of safety, in which their brittle failure characteristics are of little concern. Should one anchor in a system lose its tension due to corrosion or ground consolidation, the reserve of stability is diminished, but the system as a whole keeps the slope stable. However, if ground anchors are used to replace some other resistance (for example to permit excavation) the stability of the resulting scheme is critically dependent on the performance of the anchor system. This is because the loss of a number of anchors may allow the slope to move so that the remaining anchors are overstressed and the whole system of landslide and anchors fails together. All anchor systems need to be designed with some redundancy, so that individual anchors can be taken out of service for corrosion checks, general performance testing, re-stressing, and also to cope with a number of anchors underperforming.

The designer has many options available: numerous small anchors or fewer larger anchors; their inclination; how to secure them at depth; and how to arrange the load-spreading pads at the surface. The first of the choices is most easily made on the grounds of the equipment and facilities available, and the technical difficulties associated with drilling the holes and finding suitable strata at depth in which to secure the anchor ends. The anchor length at depth is determined from the pull-out resistance of the grouted capsule. Some anchors are therefore installed with under-reamed or enlarged cavities.

Load-spreading pads can be made from pre-cast concrete units. However, pre-cast units do not make perfect contact with the soil beneath, and therefore bed-in or deflect significantly when the load is first applied. Settlements vary from pad to pad, so that a row of anchor pads installed in this way commonly presents an irregular appearance after stressing. In that respect, cast in-situ pads can perform better. In addition, cast in-situ pads can be much larger than pre-cast units, as they do not need to be craned into place.

21.5.1 OVERALL DESIGN VIABILITY CHECK

The orientation of the anchors determines whether they mainly increase the normal stresses on a potential slip surface or counteract the shear stresses on that surface. Any method of stabilization that increases normal stresses could generate pore water pressures, such that the normal effective stresses are not increased initially and the anchor only becomes effective as consolidation takes place. However, the force component that counteracts destabilizing forces is likely to take effect as the anchor is stressed.

In order to incorporate anchor loads into a design, it is necessary to use a limit equilibrium slope stability analysis code that supports arbitrarily inclined point loads. Only anchor forces above the slip surface need to be considered when analyzing the effect of anchors on a particular landslide. However, the analyst must not forget to remove the anchor loads, or to offset them with an equal and opposite load at the embedded end of the anchor, when considering the prospect of failures at greater depth than the ends of the anchors.

21.5.2 DETAILED DESIGN ASPECTS

The optimum angle, β_{opt}, (for minimum force to generate a particular improvement in stability) for a ground anchor system relative to the slip surface it passes through, is determined from:

$$\tan \beta_{opt} = \frac{\tan \phi'}{F}. \tag{21.2}$$

This optimum angle inevitably makes the anchors become sub-horizontal, and therefore longer than if a steeper and nonoptimum angle were chosen. Figure 21.6 shows the β angle for one of a pattern of anchors. Several practical problems to do with installation are exacerbated if the anchors become longer, and so nonoptimum inclinations are often chosen.

The embedded end of the anchor tendon is usually stripped of its protective coat, the individual strands are splayed and degreased, and the whole set of exposed wires is encapsulated in a casing that simplifies feeding the anchor into its borehole. This casing is filled with material, commonly epoxy-based, that secures the individual strands. The capsule is grouted at the end of the hole. Encapsulation in this manner improves the corrosion resistance at the down-hole end of the anchor. The length of the grouted section is greater in soils than in rocks.

At the other end of the tendon, the corrosion protection must also be stripped back to allow the steel strand to be gripped. Corrosion protection at this end of the anchor will always prove to be a problem. Thick layers of grease and other protective systems are used to inhibit corrosion. This end of the anchor must be protected by a weatherproof cap.

21.5.3 USEFUL RULES OF THUMB

Computer software is now so fast that many analysts will simply try different load configurations and magnitudes until a working system is found. The approximations inherent in limit equilibrium methods, however, make the results indicated for very small anchors unreliable.

The resisting and driving forces (R and D, respectively) are broadly equivalent in a slowly moving slide. In a very simple analysis, $D = R = \Sigma W \sin\alpha$. Hence, with an assumed angle β (the average angle that the anchors make with the slip surface, see Fig. 21.6) and the anchor load P, the desired factor of safety is approximately given by:

$$F = \frac{D + P\cos\beta\tan\phi'}{D - P\sin\beta} \qquad (21.3)$$

from which the necessary anchor force P can be recovered. In a very large landslide, the magnitude of D may make it impractical to provide enough anchor force. However, on the basis of this first estimate, it may be possible to see whether anchors alone will provide a suitable solution. If so, the actual pattern of anchors together with their inclinations and loads will need to be determined through more formal analyses. While there is an optimum inclination, β, for a ground anchor that will maximize its effectiveness, this angle typically requires long anchors at low inclination, whereas most design favors short anchors at orientations that are easier and cheaper to construct.

21.5.4 ERRORS AND PROBLEMS

If each anchor bears on its own pad, then any excessive pad deformation under load relates to that pad and anchor only. However, it is commonly found economical to have more than one anchor per pad, especially with lightly loaded anchors. Uneven deformations of multi-anchor pads cause differential relaxation in anchor loads. Initially, the stressing of one anchor will relax the loads in other anchors that have been pre-tensioned. Later, when all anchors bearing on a multi-anchor pad are carrying their full load, an anchor that goes out of service will affect the others adversely, either by allowing the anchor pad to rotate under imbalanced loading or by load transfer to other anchors. Reinforced concrete structural grillages are sometimes employed to spread the load between anchors attached to the nodes of the grillage. However, such grillages can propagate failures within anchor groups.

21.6 STABILIZATION USING GEOTECHNICAL PROCESSES

Geotechnical processes are methods of increasing the shear resistance of the soil, commonly by altering the physics or chemistry of the soil within the footprint of the landslide. Many of the techniques are only practical in certain highly specific circumstances, and some claims for them are of dubious validity. The likelihood of getting a satisfactory result is increased where these techniques have been used successfully in similar geologic settings; correspondingly, a novice user is unlikely to be able to apply an unfamiliar technique successfully.

A variety of materials can be used to form unreinforced columns in an unstable slope. Beles and Stanculescu (1958) describe piles formed by burning clay in-situ, using gas that would otherwise have been flared off from an oilfield. Railway practice in Britain in the nineteenth century included baking clay in-situ using coal fires in pits, although the hazard from fires in combustible embankment fills was a considerable disincentive to the use of this practice, and with the increasing cost of fuels, the technique is unlikely to become popular again. Ground freezing uses expensive-to-run construction plant and is a temporary expedient.

A method used in UK railway practice in the past, but not currently in favor, is to grout a slipped area with cementitous grout. In poorly compacted fill, or fill that is networked with hollows due to dissolution, decomposition, animal burrows, or combustion, the grout flows under pressure into and fills the cavities, making the entire mass stronger. It is difficult to see the mechanism involved when grouting clay fills, but there are numerous discontinuities that become filled with grout, including the basal slip surface. Cementitious grouts are usually calcium-rich and can cause ion exchange in some clays, improving their shear strength.

Ion exchange is also a mechanism involved both when lime-filled piles are constructed and also in the re-use of excavated material combined with lime or cement (cement-stabilized soil). Lime and cement also take up free water from the soil. Electro-osmosis has been used to drive water from a soil mass, but the method is not in widespread use.

The fundamental issue with this family of solutions is that they need to be applied by appropriately experienced practitioners. There is no clearly defined theoretical or mechanical basis for their operation, and there is an understandable concern about their long-term performance.

21.7 SUMMARY

Engineers concerned with slope stabilization can usually only proceed from an understanding of the failure. This understanding normally requires a good topographic survey from which cross-sections can be drawn, because stability analyses continue to be done in two dimensions. Sections need to be in the direction of movement. Three-dimensional analysis is practical, but is only merited when there is sufficient understanding of the landslide process. Sometimes a landslide is part of a larger system, and care must be taken not to improve local stability at the cost of inducing instability on a more regional scale.

REFERENCES

Barley, A.D. (1991). Slope stabilisation by new ground anchorage systems in rocks and soils. In *Slope Stability Engineering: Applications and Developments*, ed. R.J. Chandler. London: Thomas Telford, pp. 335–340.

Barton, M.E. and Garvey, P.M. (2011). Reactivation of landsliding following partial cliff stabilisation at Barton-on-Sea, Hampshire. *Quarterly Journal of Engineering Geology and Hydrogeology*, 44, 233–248.

Beles, A. A. and Stanculescu, I. I. (1958). Thermal treatment as a means of improving the stability of earth masses. *Géotechnique*, 8, 158–165.

Bianco, B. and Bruce, D. A. (1991). Large landslide stabilisation by deep drainage wells. In *Slope Stability Engineering: Applications and Developments*, ed. R. J. Chandler. London: Thomas Telford, pp. 319–326.

Biczysko, S. (1981). Relic landslip in West Northamptonshire. *Quarterly Journal of Engineering Geology*, 14, 169–174.

Bromhead, E. N. (1997). The treatment of landslides. *Geotechnical Engineering*, 125, 85–96.

Bromhead, E. N. (2004). Landslide slip surfaces: Their origins, behaviour and geometry. In *Proceedings of the 9th International Symposium on Landslides, Rio de Janeiro*. Amsterdam: Balkema, pp. 3–22.

Bromhead, E. N. (2005). Geotechnical structures for landslide risk reduction. In *Landslide Hazard and Risk*, ed. T. Glade, M. Anderson and M. Crosier. New York: John Wiley & Sons, pp. 549–594.

Chandler, R. J. (1979), Stability of a structure constructed in a landslide. In *Proceedings of the 7th European Conference on Soil Mechanics & Foundation Engineering*, **3**, 175–182.

Cornforth, D. H. (2005), *Landslides in Practice: Investigation, Analysis and Remedial Options in Soils.* New York: Wiley.

Gedney, D. S. and Weber, W. G. (1978). Design and construction of soil slopes. In *Landslides: Analysis and Control*, ed. R. L. Schuster and R. J. Krizek. National Academy of Sciences, Transportation Research Board, Special Report 176, pp. 172–192.

Hutchinson, J. N. (1977). Assessment of the effectiveness of corrective measures in relation to geological conditions and types of slope movement. *Bulletin of the International Association of Engineering Geology*, 16, 131–155.

(1983). Methods of locating slip surfaces in landslides. *Bulletin of the Association of Engineering Geology*, 20, 235–252.

(1984a). Landslides in Britain and their countermeasures. *Journal of the Japan Landslide Society*, 21, 1–25.

(1984b). An influence line approach to the stabilisation of slopes by cuts and fills. *Canadian Geotechnical Journal*, 21, 363–370.

Morgenstern, N. R. and Price, V. E. (1965). The analysis of the stability of general slip surfaces. *Géotechnique*, 15, 79–93.

Poulos, H. G. (1995). Design of reinforcing piles to increase slope stability. *Canadian Geotechnical Journal*, 32, 808–818.

Viggiani, C. (1981). Ultimate lateral load on piles used to stabilize landslides. In *Proceedings of the 10th International Conference on Soil Mechanics and Foundation Engineering, Stockholm*, pp. 555–560.

Whitman, R. V. and Bailey, W. A. (1967). Use of computers for slope stability analysis. *ASCE Journal of the Soil Mechanics Division*, 93, 475–498.

Wolfe, W. E., Busby, H. R. and Kim, S. H. (1996). Highway embankment repair and stabilisation using power plant waste products. In *Landslides: Proceedings of the 7th International Symposium on Landslides, Vol. 3*, ed. K. Senneset. Amsterdam: Balkema, pp. 1823–1828.

22 Rockfall characterization and modeling

PAOLO FRATTINI, GIOVANNI B. CROSTA, AND FEDERICO AGLIARDI

ABSTRACT

Rockfalls pose a significant threat to life and property, although significant advances in rockfall protection have been made in the past decade. Determining rockfall processes and related hazard, however, remains a difficult task because of the complexity and intrinsic stochastic nature of the physics involved. The appropriate application of rockfall modeling tools requires a thorough understanding of their logic, assumptions, advantages, and limitations, as well as careful assessment of rockfall sources, block and slope characteristics, and model calibration data. This chapter provides a discussion of major issues in rockfall definition, characterization, and modeling, with special emphasis on rockfall runout. Our discussion is supported by modeling examples carried out using the 3D simulator Hy-STONE. Different modeling approaches are critically evaluated, including the empirical shadow angle method, and 2D and 3D mathematical models. Application of the shadow angle concept requires the user to be aware of several issues related to definition of the shadow angle and the effects of morphological constraints. Most limitations of empirical approaches can be overcome with mathematical models that account for slope morphology and roughness, energy dissipation at impact or by rolling, and the effects of vegetation, block fragmentation, and block–structure interaction. We discuss different modeling approaches and calibration problems and the important dependency of model parameters and results on correct characterization of the topography.

22.1 INTRODUCTION

Rockfalls are widespread and can threaten people, structures, and lifelines. They are particularly important hazards in mountainous areas. Rockfalls typically have small volumes, but are frequent on some slopes, and can have long runout. Protection from rockfalls requires the characterization of rockfall hazard for different scenarios, spatial distribution, frequency and intensity of impacts, vulnerability of exposed elements, expected costs, and a cost–efficiency analysis of mitigation options (Agliardi *et al.*, 2009). In practice, the problem is mostly dealt with through susceptibility–hazard assessments (Crosta and Agliardi, 2003; Jaboyedoff *et al.*, 2005) or engineered countermeasures (Ritchie, 1963; Nichol and Watters, 1983; Calvetti and di Prisco, 2009; Volkwein *et al.*, 2009).

Rockfall hazard is the probability that a specific location will be reached by rock fragments with a particular intensity (Crosta and Agliardi, 2003; Jaboyedoff *et al.*, 2005). This probability is a combination of onset probability (i.e., the probability of rockfall occurrence) and runout probability (i.e., the probability that falling blocks will reach specific locations; Fig. 22.1). Rockfall hazard assessment thus requires the initial identification of potential release locations and the probability that rockfalls of different sizes will occur. This probability can be estimated by using magnitude–frequency relationships derived from event catalogs (Hungr *et al.*, 1999; Dussauge *et al.*, 2003). In practice, however, catalogs are generally incomplete due to undersampling and censoring effects. In these cases, only a spatial onset probability, which depends on local geomechanical and morphological conditions, can be defined. Once rockfall sources have been characterized, the runout of falling blocks needs to be determined to assess the extent of the threatened area, the probability that blocks will reach given locations, and the velocity and kinetic energy of those blocks.

Rockfall kinematics and dynamics depend on block geometry, slope topography, surficial geology, and vegetation. Most rockfall analyses require the use of models to simulate the relevant physical processes. The models rely on several assumptions

Landslides: Types, Mechanisms and Modeling, ed. John J. Clague and Douglas Stead. Published by Cambridge University Press.

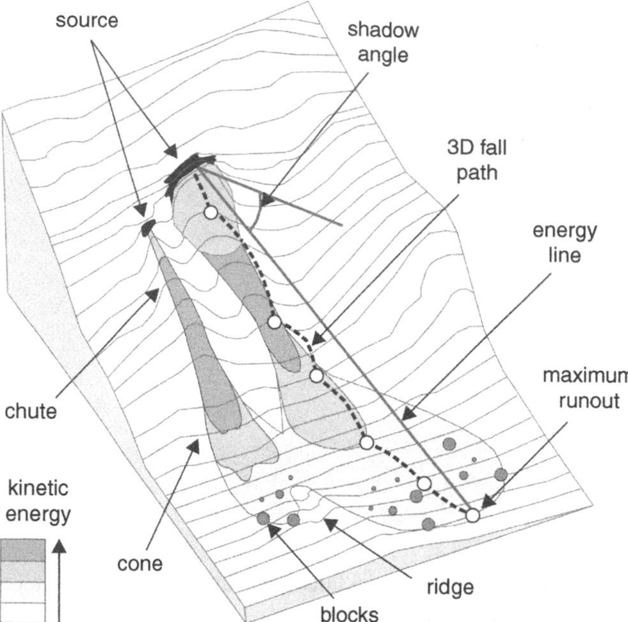

Fig. 22.1. Schematic illustration of key rockfall concepts and terms (modified after Jaboyedoff *et al.*, 2001).

about the rockfall process and the assigned values of controlling variables. The accuracy of the modeling results depends on the ability of the analyst to account for rockfall physics, topography, and related uncertainties (Agliardi and Crosta, 2003).

This chapter provides an up-to-date overview of rockfall processes and modeling approaches and tools, with special focus on rockfall runout. Major modeling issues and research needs are addressed, and some questions are asked, the answers to which represent the frontiers for rockfall research in the future.

22.2 DEFINITION AND PROCESSES

Rockfall can be defined as the downward movement of detached rock fragments (a single block, several blocks, or a fragmented mass of rock) by freefalling, bouncing, rolling, and sliding (Cruden and Varnes, 1996). Rochet (1987) classified rockfalls by their volume and degree of particle interaction: block falls (volume 10^{-2} to 10^{2} m^3), mass falls (10^2–10^5 m^3), very large mass falls (10^5–10^7 m^3), and mass displacement (more than 10^7 m^3). Here we deal with rockfalls that have volumes less than 10^5 m^3, also known as "fragmental rockfalls" due to the negligible amount of interaction among falling blocks (Evans and Hungr, 1993).

Rock can detach due to seismic ground motion, rainfall, freeze–thaw cycles, root growth or leverage in high winds, or progressive weathering of rock masses. Rockfalls originate from cliffs of different size, lithology, and structure. Their occurrence is controlled by the geometrical and hydromechanical properties of discontinuities, such as orientation, density, persistence, and roughness. Hence, the geomechanical characterization of discontinuity sets and block size and shape is extremely important.

Once failure occurs, blocks move downslope following irregular paths (Fig. 22.1) and gain velocity as potential energy is converted into kinetic energy. Total energy is progressively dissipated due to impact and rolling on the slope, until the blocks stop. To describe the position and velocity of a block at any time, the analyst must model trajectories and energy loss, both of which depend on slope and block characteristics. Three-dimensional slope geometry and the lateral and downslope variations in roughness are the main controls on rockfall paths. Chutes, channels, convexities, and longitudinal ridges affect trajectories and the partitioning of kinetic energy into translational and rotational components. Rockfall runout on a rough topography results in lateral dispersion – the deviation of the trajectories from the steepest gradient. Empirical observations (Bozzolo and Pamini, 1986; Evans and Hungr, 1993; Azzoni *et al.*, 1995) and parametric modeling studies (Crosta and Agliardi, 2004) show that lateral dispersion values are typically about 20 percent of the slope length, but can be smaller for short and steep slopes and larger for irregular or channeled slopes. The geotechnical properties of the slope material, including grain size, void ratio, and water content, affect energy loss during rockfall descent. For example, rock surfaces do not allow significant energy damping at impact or by rolling, whereas talus material or gravelly layers dissipate a considerable amount of energy. Soft ground allows significant block penetration at impact, affecting the angular velocity of blocks as they rebound. The shape, mass, and mechanical properties of blocks (e.g., strength, elastic moduli, microcracking) affect their inertia, energy partitioning, and possible fragmentation.

In principle, it should be possible to describe rockfall dynamics if one knows the initial conditions, 3D slope geometry, and energy loss at impact or by rolling. In practice, however, most of the required parameters are uncertain in both space and time. The size, shape, and geomechanical properties of the blocks and the exact location of source areas are usually poorly known. Furthermore, the geometrical and mechanical characteristics of the surface material may differ markedly along a slope, requiring a stochastic analytical approach (Bourrier *et al.*, 2008).

22.3 SOURCE ZONE CHARACTERIZATION

Rockfall sources can be identified using several approaches. Geomorphological mapping of potential or active source areas and past rockfall evidence are based on air photo interpretation and field surveys (Agliardi and Crosta, 2003; Frattini *et al.*, 2008). Although subjective, this approach can be effective when employed by experts. Moreover, it accounts implicitly for all the relevant controls, for example lithology and structure, and enables the analyst to link mapped rockfall sources to evidence of runout, such as block accumulations, isolated boulders, and impact marks. A pure morphometric approach based on slope angle thresholds is straightforward, but suffers from several limitations. Threshold slope angles are defined subjectively and need site-specific calibration to account for different rock mass strength. Moreover, the accuracy of slope angles derived from digital elevation models (DEM) depends on spatial resolution; low-resolution DEMs typically yield lower slope gradients (Chang and Tsai, 1991). When high-resolution data such as those provided by LIDAR or terrestrial laser scanning (TLS) are used, slope angles exceeding specified thresholds may be associated with artifacts resulting from buildings or vegetation filtering. Approaches based on the decomposition of slope frequency distributions have been used to identify specific morphologies, such as cliff areas (Loye *et al.*, 2009). Rockfall sources can be also identified and classified by heuristic or statistical ranking of selected descriptors (Pierson *et al.*, 1990; Cancelli and Crosta, 1993; Rouiller *et al.*, 1998; Frattini *et al.*, 2008). When suitable data can be collected, rockfall sources can be identified and evaluated through rock-slope stability analysis. Limit equilibrium or numerical stability calculations for plane, wedge, or toppling failure of blocks are performed locally because they require site-specific information, mainly joint orientation, roughness, persistence, and cliff size. However, simplified methods of kinematic stability analysis have been developed for regional-scale assessments (Rouiller *et al.*, 1998; Gokceoglu *et al.*, 2000; Günther *et al.*, 2004). These methods are based on the recognition of regional joint patterns and regional rock mass strength characteristics, and they may be supported by structural analysis based on remotely sensed data (Tonon and Kottenstette, 2006; Jaboyedoff *et al.*, 2007; Sturzenegger *et al.*, 2007; Abellán *et al.*, 2011).

The way that rockfall sources are mapped has a significant effect on runout modeling results. Sources can be points (e.g., isolated outcrops or localized failures), lines (e.g., cliff-top envelopes), or areas. Point sources may be appropriate when source locations are well known or when back analysis of past events is carried out. Linear sources are generally the tops of rocky cliffs, which implicitly assigns the maximum possible potential energy to the blocks. As a consequence, the highest fall velocities and impact energies will be simulated, and these sources favor rock fragmentation and decreasing energy restitution when velocity-dependent damping relationships are used (Pfeiffer and Bowen, 1989). Polygon sources allow simulation of more complete distributions of block sizes and trajectories, thus resulting in a more realistic distribution of stopping points. On the other hand, a distribution of release points that is too wide may mask the spatial pattern of rockfall trajectories, which is crucial to understanding rockfall behavior.

22.4 MODELING OF ROCKFALL RUNOUT

Different approaches for simulating rockfall runout have been developed during the past few decades. Today, there is a wide range of commercially available software and advanced research modeling tools. Models are characterized by different physical assumptions, computational efficiency, and data analysis capabilities. It is crucial that potential users be aware of such differences and of the suitability of models for different applications or modeling conditions. Models are generally classified according to the underlying physics and their spatial dimension, and can be used at different scales depending on the nature and scope of the project.

22.4.1 EMPIRICAL MODELS

Simple empirical–statistical and semi-empirical runout models based on the "energy line" concept (*Fahrböschung*; see Heim, 1932) have been widely used. Included in this group is the shadow angle method (Scheidegger, 1973; Lied, 1977; Onofri and Candian, 1979; Evans and Hungr, 1993). Empirical models are based on the assumption that the maximum distance that individual rocks can travel is related to the intersection with the topography of an "energy line" characterized by a specific inclination from the horizontal (i.e., the shadow angle; see Fig. 22.1). The shadow angle integrates the loss in frictional energy due to different motion types over the entire length of a rockfall path. Statistical analysis of the runout of a large number of blocks establishes shadow angles corresponding to different runout probabilities (Onofri and Candian, 1979). This approach has been widely used in practical and research applications due to its easy implementation in geographic information systems (GIS) for regional-scale analysis (Jaboyedoff and Labiouse, 2011). Values of the shadow angle have been derived from rockfall inventories collected in different geologic and geomorphological settings (Fig. 22.2a, b, and references therein). The empirical nature of this approach requires the settings in which specific shadow angles apply to be well known. However, it is often difficult to determine how these values were obtained. Few authors describe the morphoclimatic and mechanical conditions in which rockfalls occur, the type of events (i.e., individual block falls, large falls with fragmentation), the characteristics of vegetation along the path, or the grain size of slope scree. Furthermore, it is not always stated that the sources were actually identified or, alternatively, were described by a proxy, such as the cliff-top envelope.

The practical value of reported shadow angles in different settings may be limited or compromised by specific morpho-topographic features, such as cliff height or valley confinement. Site-specific features could result in a large overestimation of rockfall runout (Fig. 22.2c) or unrealistic run-up (Fig. 22.2d). Moreover, it is often unclear whether the researcher used the actual plan length of the fall path or a straight-line distance to estimate the shadow angle (Fig. 22.2f).

22.4.2 MATHEMATICAL MODELS

Mathematical models provide a physically sound description of rockfall trajectories and of the velocity, energy, and height of blocks at each location in the model domain. Such descriptions are obtained by solving different sets of kinematic or dynamic equations of motion, depending on the adopted physical framework (i.e., lumped mass, rigid body, hybrid/mixed) and spatial

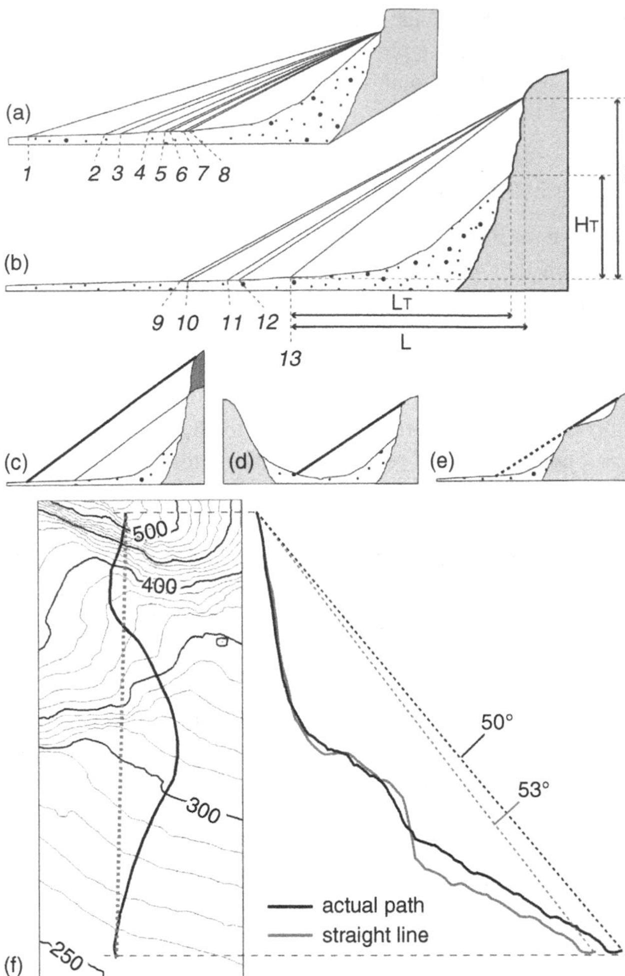

Fig. 22.2. Published values of minimum shadow angles (i.e., maximum runout), defined with reference to (a) talus apex, (b) cliff top or actual source area. 1 – Domaas, 1994 (17°); 2 – Holm and Jakob, 2009 (21°); 3 – Wieczorek *et al.*, 1998 (22°); 4 – Evans and Hungr, 1993 (rockfall fragments on glaciers, 24°); 5 – Copons and Villaplana, 2008 (25.5°); 6 – Meissl, 1998 (26°); 7 – Evans and Hungr, 1993 (27.5°); 8 – Lied, 1977 (28°); 9 – Onofri and Candian, 1979 (28.5°); 10 – Meissl, 1998 (29°); 11 – Toppe, 1987 (32°); 12 – Heinimann *et al.*, 1998 (33°); 13 – Copons *et al.*, 2009 (36.9°). The applicability of the shadow angle approach is limited by problems related to (c) the height of rocky cliffs, (d) valley morphology constraints, (e) the effect of complex topography, and (f) whether a straight line or the actual runout path is considered to define the angle.

framework (2D or 3D). Different motion types and processes (freefall, impact, rolling, sliding) are associated in space and switch through time as the rockfall proceeds. Transitions between different motion types are controlled by user-defined rules or threshold values. The same computed runout distance may thus result from different combinations of motions.

Two-dimensional simulation tools have been developed based on lumped-mass (Piteau and Clayton, 1976; Rochet, 1987; Stevens, 1998), hybrid (Pfeiffer and Bowen, 1989; Jones *et al.*, 2000) and rigid-body approaches (Bozzolo and Pamini, 1986; Azzoni *et al.*, 1995). Two-dimensional models simulate rockfall on slope profiles sampled along the rockfall path. This approach is suitable when the geometrical and dynamic effects of the 3D topography can be ignored, for example on planar cliff-talus slopes, on slopes characterized by low roughness (Crosta and Agliardi, 2004), or on slopes with little or no vegetation. Two-dimensional modeling may be suitable when back-analyzing events for which the fall path is known. Conversely, when significant lateral dispersion or a sinuous fall path is expected, subjective *a priori* choice of the "most probable" fall path may introduce considerable uncertainty or error.

Three-dimensional models are increasingly being used to simulate rockfalls (Guzzetti *et al.*, 2002; Agliardi and Crosta, 2003; Crosta *et al.*, 2004; Dorren *et al.*, 2006; Lan *et al.*, 2007). They overcome the limitations of 2D models by exploiting the three-dimensional form of the topography provided by DEMs. This is not the case, however, when a hydrologic approach (i.e., flow direction) is used to derive fall paths, when several unconstrained empirical rules are combined to decide possible block behaviors, or when results are interpolated between different trajectories. Three-dimensional models are powerful, but delicate, tools and require a complete understanding of process and model parameters.

22.4.3 A CASE STUDY

Here we discuss major rockfall modeling issues, starting with simulations performed using a 3D model (Hy-STONE; Crosta *et al.*, 2004; Frattini *et al.*, 2008; Agliardi *et al.*, 2009). The model core is a hybrid algorithm based on a three-dimensional extension of Pfeiffer and Bowen (1989) and Azzoni *et al.* (1995). Freefall, impact and rolling can be simulated with different damping relations. Topography is described by a DEM, and all relevant parameters are spatially distributed. The stochastic nature of rockfall processes and parameters is accommodated by slope morphology and roughness, and by the random sampling of most parameters from different probability density distributions (uniform, normal, exponential). Specific model components explicitly account for the interactions between blocks and countermeasures or structures. An elasto-viscoplastic damping model (di Prisco and Vecchiotti, 2006) has also been incorporated into the model to account for impact on soft ground. In addition, the geometric and dynamic effects of vegetation and fragmentation processes can be simulated (Crosta *et al.*, 2006).

These issues are illustrated with 3D rockfall modeling performed for the Mt. San Martino–Corno di Medale cliffs in Italy (Lecco, Como Lake). The cliffs are sub-vertical and up to 400 m high. They are developed in gently north- to northwest-dipping Middle Triassic massive dolostones and underlying bedded limestones. The lower part of the cliff is mantled with Quaternary deposits that slope 33–38° and include cemented talus, glacial, glaciofluvial, and alluvial deposits, and scree. The cliff has historically produced rockfalls; one on February 22, 1969 caused eight fatalities. A full-scale rockfall field test was carried out here by Broili (1973). Agliardi and Crosta (2003) used aerial photographs and a field survey to characterize the cliffs, identify rockfall sources, and map rockfall deposits. A DEM with a 2 × 2 m cell size, derived from airborne LIDAR data (courtesy Regione Lombardia), was used in the study. Simulations, calibrated with geomorphological evidence (i.e., mapped fall paths and fallen blocks), back analysis of earlier (1954, 1969) rockfalls, and available experimental data (Broili, 1973; Agliardi and Crosta, 2003) provide the basis for the following discussion of general rockfall modeling issues.

22.5 KEY ISSUES IN ROCKFALL RUNOUT MODELING

22.5.1 APPLICABILITY OF THE SHADOW ANGLE CONCEPT

Three-dimensional numerical modeling results can be used to evaluate the meaning and applicability of simple empirical approaches. Figure 22.3a, b shows the rockfall trajectories computed for the Mt. San Martino–Corno di Medale case study by calibrating Hy-STONE to account for velocity-dependent energy losses on impact. Trajectories have been reclassified according to the angular difference between shadow angle values defined by considering either straight-line or the actual computed fall paths (Fig. 22.3b). In this case study, "actual path" shadow angles are up to 3.5° lower than the corresponding straight-line ones, excluding those trajectories that are constrained by protection works (Fig. 22.3c). Therefore, the predictive use of "actual-path" shadow angle values in settings where simple, straight-line fall paths are favored may overestimate the maximum expected rockfall runout. Conversely, the use of straight-line shadow angle values in areas with complex, irregular, or channeled topography may result in underestimates of runout (Fig. 22.2e). Zoning decisions in the latter instance could be unsafe. In any case, the use of straight-line shadow angles in complex settings could result in an incorrect estimate of the magnitude and spatial distribution of block velocity and energy along the paths.

Another limitation of the shadow angle concept is grounded in its oversimplification of rockfall physics. The approach assumes sliding friction, with energy loss along the slope described by a single equivalent friction coefficient. With this assumption, velocity would be expected to progressively decrease as the slope becomes less steep; that is, in the direction of the toe. Experimental evidence (Pfeiffer and Bowen, 1989), however, suggests that the energy lost at impact is proportional to the velocity of impact (i.e., the higher the velocity, the more energy is lost). In typical cliff-talus slope systems, high-velocity vertical impacts dominate at the talus apex, thus most block velocity is lost in the proximal part of the fall path (Evans and Hungr, 1993). If the shadow angle is measured from the cliff top (Fig. 22.3d), a shorter runout and higher shadow angle might be expected with a higher cliff height. Empirical observations (Domaas, 1994) and simulation results obtained by including velocity-dependent restitution coefficients support this supposition (Fig. 22.3d). As a consequence, the use of shadow angles for high rock walls will result in unrealistically long expected runouts, or even run-up on opposite valley sides (Fig. 22.2c, d). A possible option for overcoming this problem is to use shadow angles keyed to the talus apex rather than the cliff top (Fig. 22.2a). Unfortunately, in areas with complex morphology, clear identification of the talus apex corresponding to specific blocks at the slope toe can be difficult.

22.5.2 ENERGY LOSS MODELING

Among the different processes involved in a rockfall, block impact and rebound are the most complex. In most cases, plastic deformation of substrate and the geometry of impact can be cumulatively accounted for by contact functions that relate block kinematics (velocity) or dynamics (energy) before and after the impact (Falcetta, 1985; Bozzolo and Pamini, 1986; Pfeiffer and Bowen, 1989; di Prisco and Vecchiotti, 2006; Bourrier *et al.*, 2008). Such functions are expressed as restitution and friction coefficients, and are commonly modeled as material constants.

Several investigators have made significant efforts to derive accurate values for these coefficients (Pfeiffer and Bowen, 1989; Azzoni and de Freitas, 1995; Wong *et al.*, 2000; Chau *et al.*, 2002; Labiouse and Heidenreich, 2009). They obtained the values using numerical models or experimental observations from laboratory or in-situ rockfall tests, while accounting for different controlling factors. The latter include: the type and thickness of the impact layer and surface material; the presence of pore water in liquid or solid state; block mass, shape, lithology, weathering and discontinuities; local roughness at different scales; the geometry of the block-to-surface contact; impact mode, angle, and velocity; depth of penetration; and the presence and type of vegetation. As a consequence, the coefficients of restitution cannot be deterministically defined. This inability to precisely quantify coefficients of restitution has a major influence on model calibration and final results, which also depend on subjective attribution made by the modeler using expert knowledge or available data. Literature values are frequently used without a complete evaluation of site conditions and of the consequences of these choices on the final results.

An alternative to restitution coefficients for simulating impacts is the use of stochastic impact laws (Bourrier *et al.*, 2009) or more sophisticated deterministic approaches based on either finite element or discrete element numerical methods (Cundall, 1971). Even with these approaches, however, the feasibility of the analysis and the reliability of results carry significant uncertainties.

A constitutive model recently developed by di Prisco and Vecchiotti (2006) has been incorporated into Hy-STONE to simulate the response of homogeneous soil layers to the impact of spherical rigid boulders. The model has proven to be reliable against laboratory and in-situ tests, simulating correctly the penetration depth, exit velocity, and rebound height. Input data for simulation include local slope gradient, boulder size, impact velocity, and values of the soil parameters commonly used for

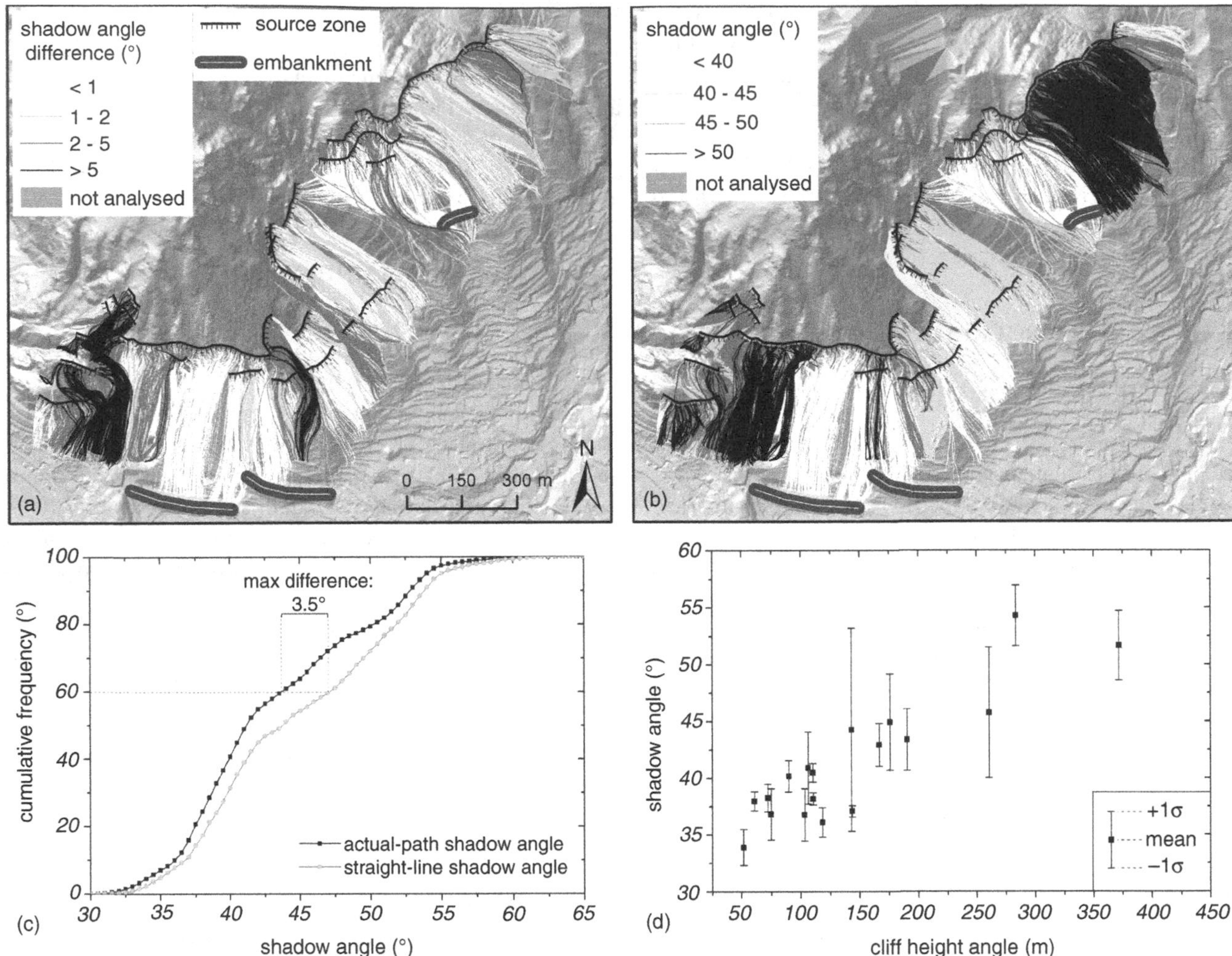

Fig. 22.3. (a) Rockfall trajectories classified according to the difference between shadow angles calculated according to either a straight-line or actual plan runout path. (b) Rockfall trajectories classified according to the value of the shadow angle of the actual path. (c) Cumulative frequency distribution of straight-line and actual-path shadow angle values. (d) Dependence of the actual-path shadow angle on the height of the source cliffs: 10 spherical blocks from each source cell (total = 21,880), with a uniform block radius of 0.5 m, were simulated.

geotechnical characterization of granular materials, which can be obtained by field observations and in-situ testing. Time-dependence of the mechanical response of soil layers results from the significant soil fabric rearrangement induced by block penetration. Two failure mechanisms are thus considered: a viscoplastic mechanism and a coupled Coulomb friction mechanism, the latter for trajectories inclined by a certain amount normal to the impact surface. A numerical, iterative solution of the nonlinear differential equation system has been developed within Hy-STONE and provides the following outputs: block displacement and the exit velocity vector, the dynamic load on the soil, and depth of penetration. These outputs permit an accurate description of impact energy loss on soft ground (Fig. 22.4) and provide the basic information for the design of rock shelters or embankments.

In experimental situations, the size of the falling block with respect to the typical grain size of slope surface material can strongly affect rockfall trajectory, impact geometry, and energy dissipation. This effect has been observed for spheres falling along a granular slope (Rioual *et al.*, 2000; Bourrier *et al.*, 2008; Ammi *et al.*, 2009), and it is even more important for blocks with irregular shapes and those smaller than the average surface blocks. For example, if a large block hits small blocks, part of the energy is spent rearranging the latter (Bourrier *et al.*, 2008). Conversely, an impact against larger blocks will have more of an elastic nature and cause a restitution of most of the energy. The orientation of the impacting surface on these blocks can be different from the local slope, with unpredictable effects on the fall path.

In view of the issues discussed in this section, impact restitution and rolling friction coefficients are not valid material properties. They mostly depend on the experimental conditions under which they are derived or on mathematical description of both topography and physical processes. In particular, when mathematical simulations are performed, all material parameters should be considered to be both model- and scale-dependent.

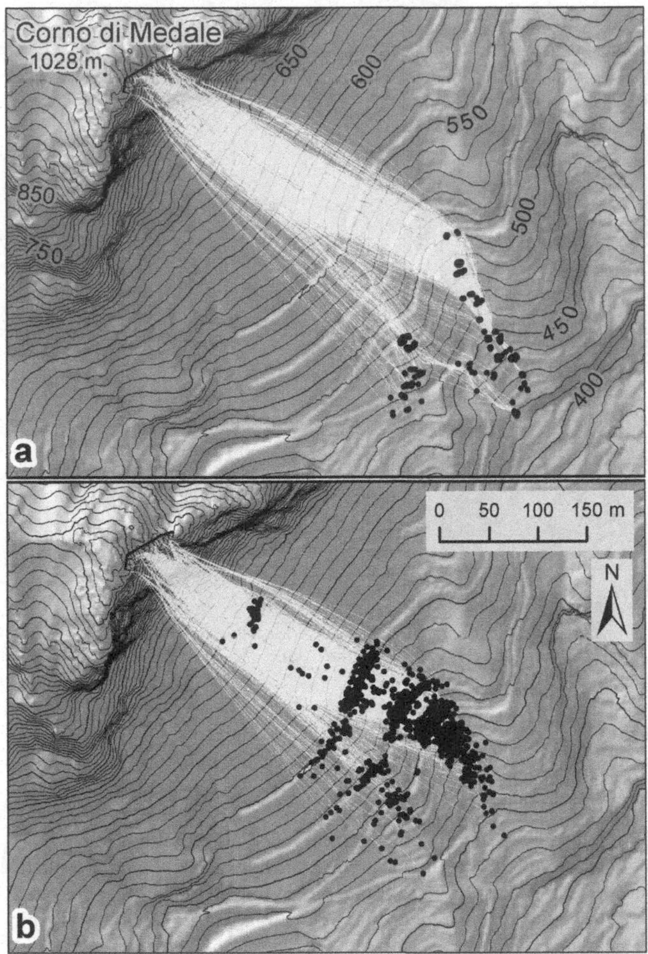

Fig. 22.4. Model trajectories (white lines) and stopped blocks (black dots) determined using (a) a hybrid model with velocity-dependent restitution and friction coefficients, and (b) a hybrid viscoplastic impact model. The viscoplastic model better predicts the distribution of stopping points along slopes in granular material. The simulation involved 2150 spherical blocks, with an exponential radius distribution (modal value = 0.5 m).

22.5.3 EFFECTS OF TOPOGRAPHIC MODEL RESOLUTION AND ROUGHNESS

High-resolution DEMs derived from LIDAR and terrestrial laser scanning are increasingly being used in rockfall modeling. These techniques allow a very accurate description of topographic surfaces and typically provide rougher surface models than other techniques for generating DEMs. Small steps in the slope geometry can be reproduced, and topographic roughness is explicitly introduced into the model. Therefore, the introduction of additional parameters to account for topographic roughness may be redundant or erroneous when high-resolution LIDAR or laser scanning data are used. This redundancy may cause unrealistic block projections or trajectories, or changes in the type of motion, or too large a reduction in the simulated runout distance. Instead, impact restitution and rolling friction coefficients should be suitably scaled to account for the DEM roughness at different resolutions. To illustrate this concept, we reproduced the same runout pattern portrayed in Figure 22.3 with different DEM resolution, by resampling the original DEM (cell: 2 m) with 6-m, 10-m, and 20-m cell sizes (Fig. 22.5). The dependence of calibrated coefficients of restitution and friction on the DEM cell is mostly due to the smoothing effect associated with larger cell sizes. This smoothing reduces the energy loss component due to DEM roughness. As a consequence, different sets of input values would be required to obtain the same pattern of impact marks and the same runout, 3D trajectory in plan, and motion (bouncing, rolling, and sliding).

DEM resolution also controls the spatial distribution of rockfall trajectories in 3D simulations. In particular, lateral dispersion of trajectories can result from topographic roughness. DEM roughness effects on the simulated dispersion of trajectories and distribution of kinematic quantities (e.g., velocity and height) have previously been investigated (Crosta and Agliardi, 2004), but the increased availability of very high-resolution DEMs provides a more explicit description of microtopography, potentially providing an improvement in understanding its effects on rockfall trajectories. Back analysis of three events shows a sharp reduction in lateral dispersion as DEM resolution decreases (Fig. 22.6). This effect could be corrected by using a stochastic approach that allows for small random variations of trajectories in order to simulate sub-grid effects. The stochastic part of the simulation would depend on the increasing disorder and chaotic effects introduced by roughness and on the ability to account for block size and block shapes different from those generally used in models (spherical, cylindrical). The inclusion of complex block shapes would require a much larger number of simulations to obtain a robust set of results, as well as a much greater computational effort. A related problem is changes in shapes of blocks during rockfall. Accounting for these issues will become increasingly important as mathematical modeling shifts from 2D to 3D simulation, where detailed descriptions of topography become routine.

22.5.4 MODEL CALIBRATION

Parameter calibration is an important task in rockfall modeling. Most model parameters are empirical or semi-empirical, and provide a phenomenological account of complex physical processes involved in rockfall motions (freefall, bouncing, rolling). The exact sequence of motions along a single fall path is rarely known, and several combinations of them may result in the same observed runout (Agliardi and Crosta, 2003). Therefore, it is difficult to know in advance which parameters will be most suitable in a given slope sector. This issue is critical because different sequences of motion would produce different spatial patterns of velocity and bounce height, even where the maximum runout is the same. Therefore, as many calibration constraints as possible are required in order to obtain the best approximation of rockfall dynamics.

Calibration approaches proposed in the literature use back analysis to achieve the best fit between simulation results and

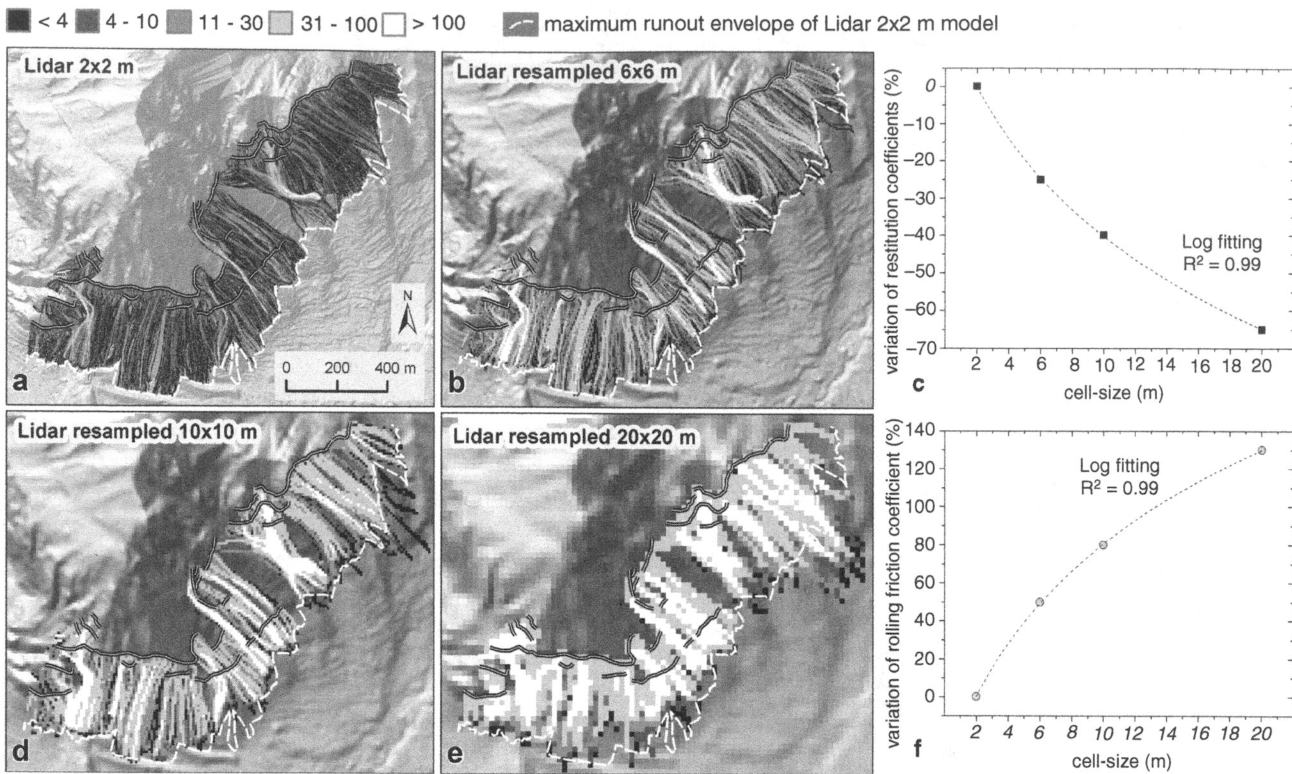

Fig. 22.5. Rockfall models with (a) 2 × 2 m LIDAR DEM and with DEMs obtained by resampling the original DEM at resolutions of (b) 6 × 6 m, (d) 10 × 10 m, and (e) 20 × 20 m. Models for resampled DEMs were calibrated to obtain the same runout as the first model. Differences in (c) normal and tangential restitution coefficients and (f) rolling friction coefficients that are necessary to calibrate the resampled models (in percentage with respect to coefficients of model (a) vs. cell size. Points in (c) and (f) are interpolated using a logarithmic function: ten spherical blocks from each source cell (total = 21,880), with a uniform block radius of 0.5 m, were simulated.

geomorphological evidence (Agliardi and Crosta, 2003), event data (Agliardi *et al.*, 2009), or experimental data derived from half-scale (Chau *et al.*, 2002; Labiouse and Heidenreich, 2009) or full-scale (Broili, 1973; Pfeiffer and Bowen, 1989; Azzoni *et al.*, 1995; Jones *et al.*, 2000) tests. Geomorphological evidence useful for calibration includes the runout of blocks of different size, the extent of active scree slopes, and older rockfall accumulations. Event data include impact hollows, rolling and sliding scours, damage to vegetation and structures, and source location and extent. These data can provide information about the type of motion and the length or height of bounces. Experimental data, commonly provided by high-speed camera recordings, provide measurements of block shape, velocity, impact and rebound angles, and fragmentation. These data, however, are rarely available, due to budget constraints, and are difficult to extrapolate to different morphological settings.

When event data are lacking, model calibration can be based on runout criteria, but the constraints on block kinematics along the fall paths may be broad or unreliable. If event or experimental data are available, additional constraints on the type of motion and jump lengths can be provided by the spatial distribution of impact or stopping points.

Even when high-quality data are available, model calibration is difficult and may take most of the time available for a sound rockfall simulation. Simpler models are generally easier to calibrate, but may not yield reliable results. In the case of 2D models, a comparison can be made between the observed and simulated runouts within a specific slope sector. For example, a good indication of model performance might be that 90 percent of the simulated blocks come to rest within a slope segment where the actual blocks stopped. Nevertheless, additional information would be required to ensure that the true dominant motion type has been came to rest. The same approach could be used in 3D by counting cells or windows to measure the distribution of simulated and observed impacts and stopping points. In this case, meaningful results could be obtained if actual rockfall sources contributing to model cells have been identified. When multiple sources contribute to model cells, which is commonly the case in 3D modeling, the number of blocks impacting or arrested in each cell depends on the effect of topographic convergence or divergence, the spatial distribution of surface characteristics, and model resolution. In these situations, a case-by-case calibration approach is required to exploit the available data.

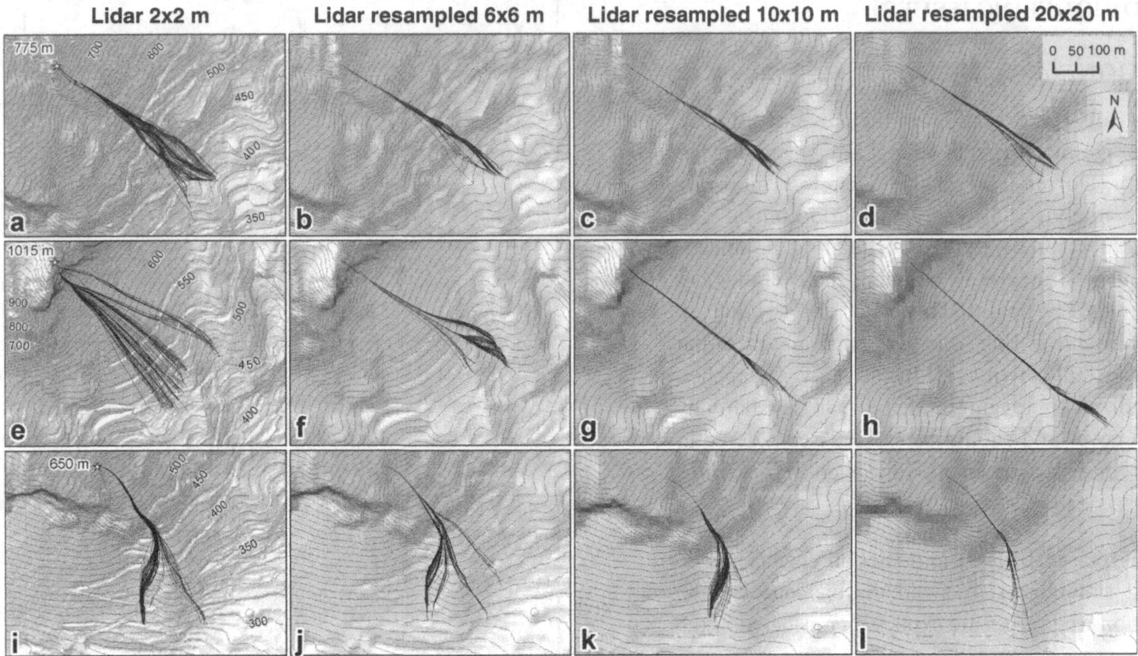

Fig. 22.6. Example of a 3D simulation performed using topography of different detail: (a), (e), (i) 2 × 2 m; (b), (f), (j) 6 × 6 m; (c), (g), (k) 10 × 10 m; (d), (h), (l) 20 × 20 m cell size. Lower-resolution DEMs were produced by resampling the original 2 × 2 m DEM. Each simulation was performed with 100 spherical blocks from each point source cell, with an exponential radius distribution (modal value = 0.5 m).

Fig. 22.7. (a) Rockfall model with no forest cover. The envelope of the associated runout area is portrayed in the diagrams as a dashed black line. (b), (c) Model results obtained by accounting for the forest cover by modifying the restitution and friction coefficients. Changes in coefficient values are shown in terms of the percentage change with respect to reference model (a) values. (d) Map of forest type and density. (e), (f) Model results obtained by explicitly accounting for forest cover effects. Forest density in model (f) is twice that considered in model (e). Ten spherical blocks from each source cell (total = 21,880), with a uniform block radius of 0.5 m, were simulated.

22.6 ADVANCED MODELING ISSUES

22.6.1 VEGETATION

Rockfall modeling rarely takes account of the presence of forest. Yet vegetation has significant effects on rockfall energy dissipation, and trajectory (Dorren *et al.*, 2005). Therefore, modeling tools capable of predicting the role of vegetation on rockfall dynamics and runout would be useful for evaluating the efficacy of forest management strategies in reducing the overall rockfall risk in alpine areas. The potential beneficial effects, however, can be limited or variable in time, due to changes in vegetation cover. Episodic rockfall activity, wildfires, snow avalanches, landslides, windstorms, or logging activities can substantially change vegetation cover and should be carefully considered in rockfall scenarios for risk assessment. In this context, various questions arise.

- Should bare soil/rock restitution and friction coefficients be used in rockfall simulations to avoid double-counting the effect of vegetation when trees are explicitly considered in the analysis?
- What is the effect of roots?
- How can the temporal variability of vegetation be incorporated into rockfall models?

The availability of powerful computers now allows researchers to integrate vegetation within 2D and 3D models (Zinggeler *et al.*, 1991; Dorren *et al.*, 2004; Crosta *et al.*, 2006; Lundström *et al.*, 2009).

Modeling vegetation requires information about the density and position of trees on the slope, their height and width, and the types of impact. Because it is difficult to accurately define the exact position of individual trees, they are commonly placed randomly over a slope according to a specified density value. LIDAR and multispectral or hyperspectral data, however, now afford the opportunity to map the exact position and the height and the width of trees (Evans *et al.*, 2001; Koukoulas and Blackburn, 2005). Different types of impact can be defined depending on the relative positions of the blocks and the trees. Deterministic or stochastic approaches can be used to model impacts, considering possible combinations of tree positions, impact block directions, and tree characteristics, including stem width and height and maximum absorbable energy (Dorren *et al.*, 2005, Crosta *et al.*, 2006).

Hy-STONE simulations accounting for the effects of vegetation either through calibrated restitution coefficients or through explicit vegetation modeling are shown in Figure 22.7. In these simulations, the energy lost by impact on tree stems is greatest for central impacts, and decreases according to a Gaussian distribution away from the stem axis. This approach is realistic when the block is relatively small or similar in size to the width of the tree. Otherwise, the block will be minimally affected by its impact on the stem. The energy that can be absorbed by a tree is assumed to decrease linearly moving up the stem. And the angular deflection of the block on impact

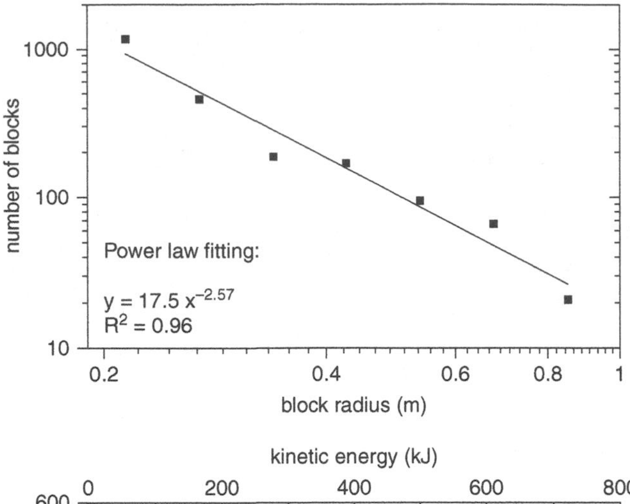

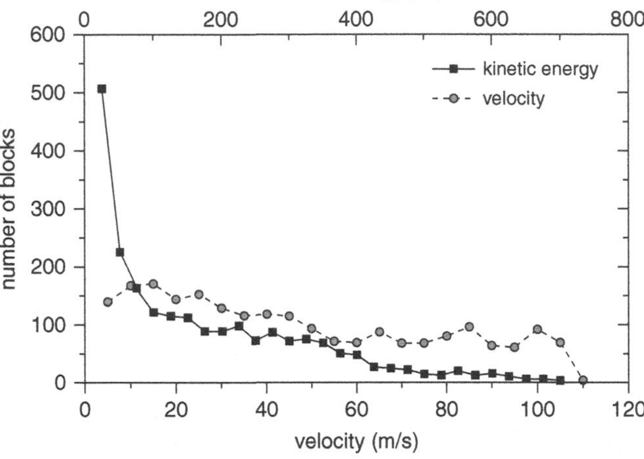

Fig. 22.8. (a) Non-cumulative frequency distribution of radii of fragments produced by impacts of 32 blocks (logarithmic bins). The distribution follows a power law. (b) Non-cumulative frequency distribution of kinetic energy and velocity of fragments produced by impacts.

is assumed to vary according to the type of impact (central, lateral, scour).

Figure 22.7 shows the sensitivity of the model to both restitution/friction coefficients and tree simulations. The distribution of blocks on the slope could be used to estimate the most active rockfall paths and a possible reduction in forest cover due to rockfall activity.

22.6.2 FRAGMENTATION

Blocks commonly fragment during rockfall events, but fragmentation is rarely accounted for in rockfall hazard assessments or the design of countermeasures (Nocilla *et al.*, 2008; Giacomini *et al.*, 2009; Wang and Tonon, 2009). Blocks can fragment on impact due to the extreme impulse stresses; the probability of fragmentation depends on block size and velocity, and rock strength and elastic properties (Chau *et al.*, 2003). The spacing, persistence, and strength of block discontinuities influence the fragmentation process. Fragmentation can also produce flying rocks, which are generally characterized by fan-like trajectories,

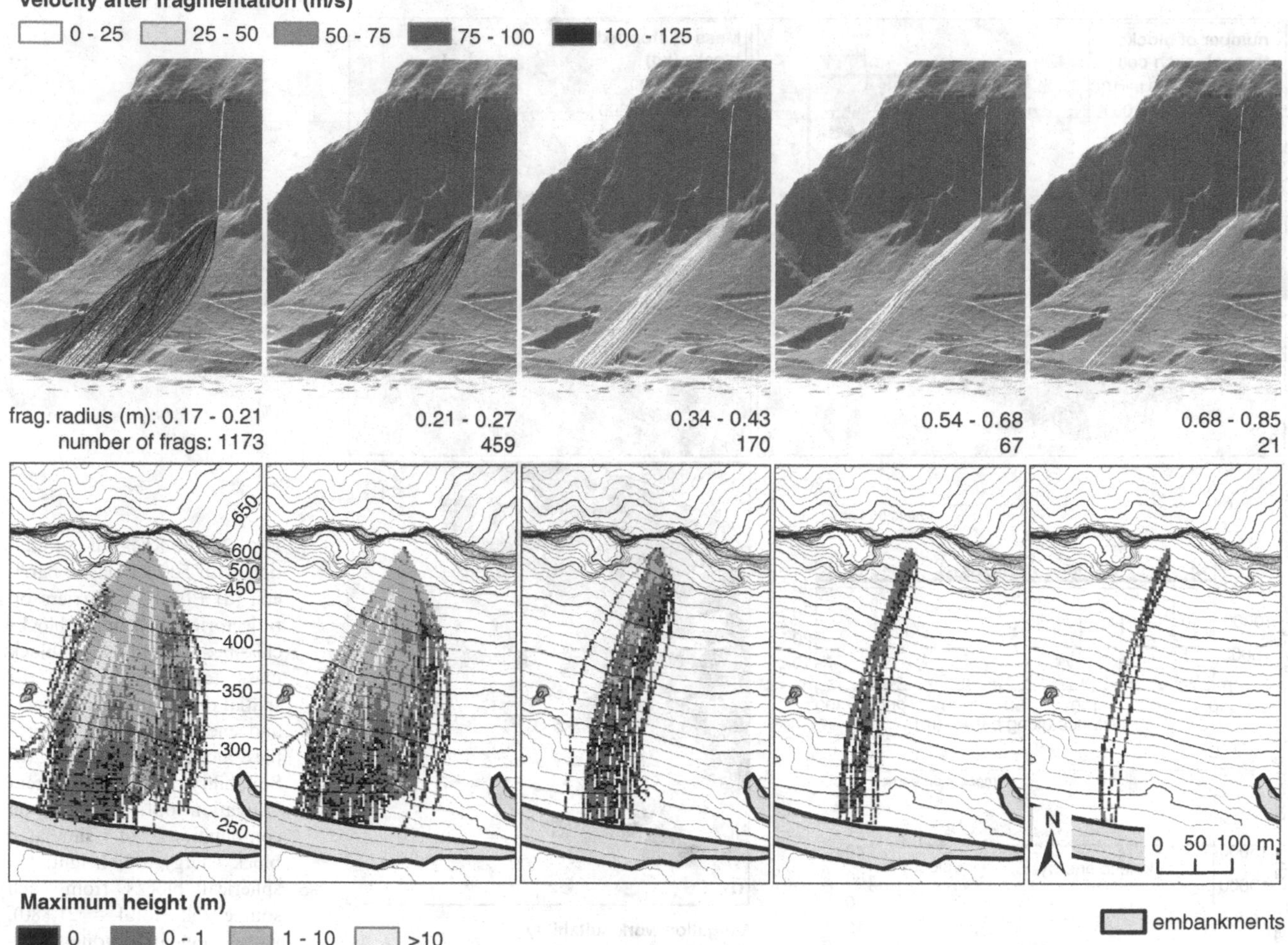

Fig. 22.9. Example of a simulation that takes into account block fragmentation. The trajectories of fragments with different radii (five classes) are shown (see also Fig. 22.8). Velocity and height of fragment trajectories have been calibrated based on in-situ test observations (Broili, 1973). Fifty spherical blocks with an exponential radius distribution (modal value = 1 m) were launched from a point source.

low trajectories, and high velocities (Broili, 1977; Wieczorek and Snyder, 1999; Wieczoreck *et al.*, 2000). Flying rocks can extend laterally over a large angle from the fragmentation point, making a correct prediction of the subsequent rockfall path difficult. Block fragmentation is generally the cause of extreme behavior, major damage, and injury and loss of life, and can compromise protection structures. Consequently, the prediction of the possible size, number of fragments, and fragment trajectories generated upon impact can help in hazard zonation and the design and placement of protection systems. Nevertheless, a complete understanding of the fragmentation process during rockfalls has not yet been achieved, full-scale experimental data are sparse, and few simulation models that consider fragmentation have been proposed (Crosta *et al.*, 2006; Wang *et al.*, 2010). Major issues in developing such models include: the definition of threshold conditions for the onset of fragmentation, controlling block properties (e.g., Young modulus, compressive strength, Poisson ratio), the expected distribution of fragment sizes, and the transfer of kinetic energy from the parent block to the fragments.

We tested the effect of fragmentation on the results of our 3D simulations (Figs. 22.8 and 22.9) using Hy-STONE. The model detects fragmentation conditions based on a fracture energy criterion, modified from Yashima *et al.* (1987). At each fragmentation point, we determined the distribution of fragment sizes and the kinetic energy of each generated fragment according to user-defined distribution constraints. Stochastic behavior is accounted for by repeating the fragmentation simulation at each relevant point. The frequency of fragment mass, as a percentage of the total mass, is distributed according to the relation:

$$R(d)\% = \left[\frac{d_i}{d^*}\right]^n \tag{22.1}$$

where d^* is the diameter of the initial block and d_i is the diameter of the ith fragment. The number of fragments can then be computed according to a mass conservation criterion, and the energy of each generated fragment is calculated assuming that the available kinetic energy after impact is equally distributed among all fragments (Fig. 22.8a). We then computed the exit

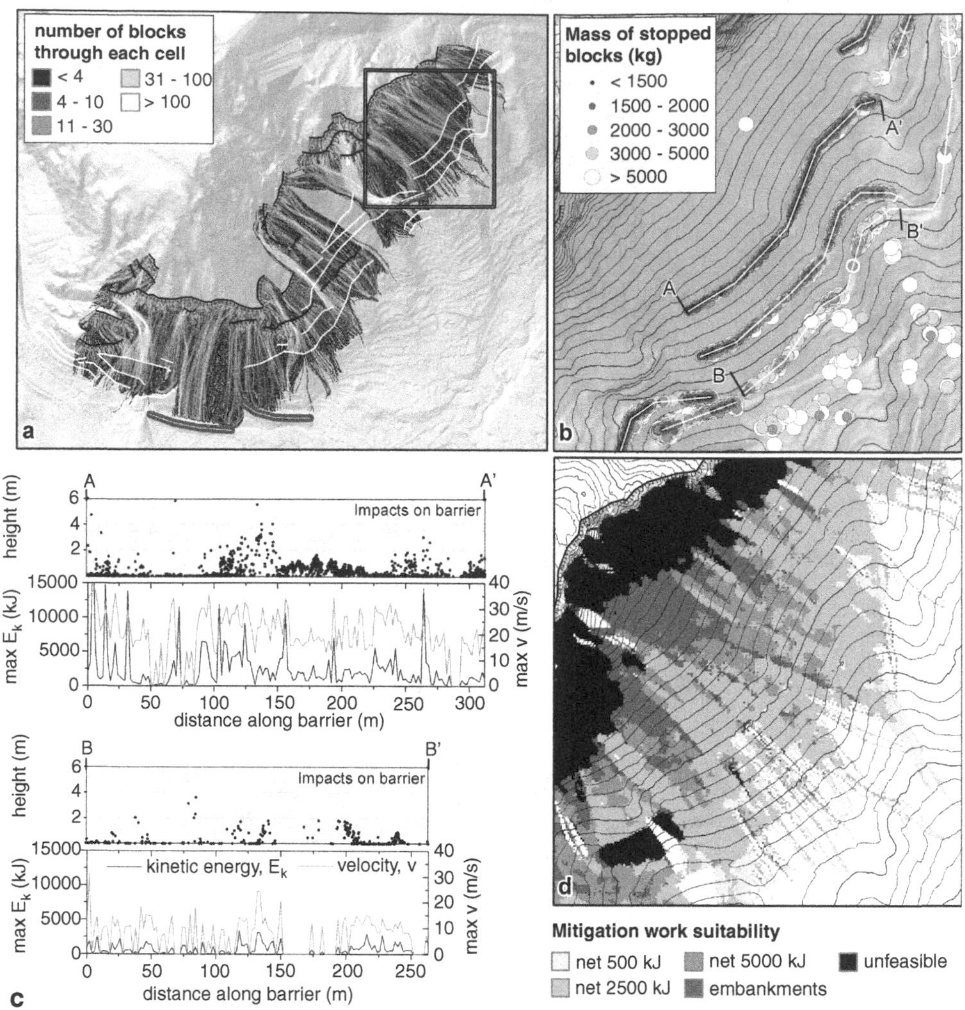

Fig. 22.10. Rockfall simulation with 500 kJ elastic barriers: (a) Effect of barriers on trajectories. (b) Stopped blocks classified according to block mass. (c) Plots showing the spatial distribution of impacts and energy and velocity profiles along two barriers. (d) Suitability map for different mitigation options based on combinations of simulated block energy and height: ten spherical blocks from each source cell (total = 21,880), with exponential radius distribution (modal value: 0.5 m), were simulated.

velocity and direction of each fragment according to momentum conservation criteria and to some imposed constraints on the deviation on the horizontal plane and projection angle on the vertical plane (Fig. 22.8b). Figure 22.9 shows the results of the simulation of 50 spherical blocks released from a single location, of which 32 fragmented. The results show that block fragmentation has an effect on the runout extent and on the spatial distribution of velocities and heights of the flying rocks. The largest fragments, however, display behavior that is more similar to that of the parent blocks.

22.7 CONCLUSIONS

The availability of improved data collection techniques and powerful simulation tools is providing a better understanding of rockfall processes. Researchers and practitioners now have the tools to assess rockfall hazard and improve risk-based countermeasures. Three-dimensional modeling can be performed to test the suitability, optimal placement, and required performance of specific types of structural countermeasures, including flexible barriers, embankments, and shelters (Fig. 22.10). Simulated fall paths and the distribution of kinetic energy and fly height, both in plan and along profiles, can be evaluated, allowing the sectors most prone to rockfall to be identified and lateral dispersion effects to be determined. The availability of new tools, however, does not reduce the importance of careful field data collection and evaluation by qualified professionals. Complex models require more and better field observations and a sound stochastic approach.

Users must be fully aware of the logic, advantages, and deficiencies of the available mathematical models. Different models are applicable to different spatial scales. Each model requires a different assessment of rockfall sources, block and slope characteristics, and specific types of input data for calibration. Complex models require an additional calibration effort and specific strategies for analyzing the results. The costs and benefits of different modeling approaches should be evaluated against the scale and objective of the modeling (i.e., regional assessment, local hazard or risk assessment, countermeasure design). In some cases, advanced modeling approaches, such as the calculation of fragmentation, have moved beyond the experimental knowledge required to validate and calibrate the model. These issues highlight the need for new research aimed at providing, on one hand, a deeper knowledge of complex rockfall processes and, on the other, specific guidelines both to assist rockfall practitioners in applying increasingly powerful analytical tools and to avoid their misuse.

REFERENCES

Abellán, A., Vilaplana, J.M., Calvet, J., García-Sellés, D. and Asensio, E. (2011). Rockfall monitoring by Terrestrial Laser Scanning: Case study of the basaltic rock face at Castellfollit de la Roca (Catalonia, Spain). *Natural Hazards and Earth System Sciences*, 11, 829–841.

Agliardi, F. and Crosta, G. (2003). High resolution three-dimensional numerical modelling of rockfalls. *International Journal of Rock Mechanics and Mining Sciences*, 40(4), 455–471.

Agliardi, F., Crosta, G.B. and Frattini, P. (2009). Integrating rockfall risk assessment and countermeasure design by 3D modelling techniques. *Natural Hazards and Earth System Sciences*, 9, 1059–1073.

Ammi, M., Oger, L., Beladjine, D. and Valance, A. (2009). Three-dimensional analysis of the collision process of a bead on a granular packing. *Physical Review E: Statistical, Nonlinear, and Soft Matter Physics*, 79, 021305.

Azzoni, A. and de Freitas, M.H. (1995). Experimentally gained parameters, decisive for rock fall analysis. *Rock Mechanics and Rock Engineering*, 28, 111–124.

Azzoni, A., La Barbera, G. and Zaninetti, A. (1995). Analysis and prediction of rock falls using a mathematical model. *International Journal of Rock Mechanics and Mining Sciences & Geomechanical Abstracts*, 32, 709–724.

Bourrier, F., Nicot, F. and Darve, F. (2008). Physical processes within a 2D granular layer during an impact. *Granular Matter*, 10, 415–437.

Bourrier, F., Eckert, N., Nicot, F. and Darve, F. (2009). Bayesian stochastic modeling of a spherical rock bouncing on a coarse soil. *Natural Hazards and Earth System Sciences*, 9, 831–846.

Bozzolo, D. and Pamini, R. (1986). Simulation of rock falls down a valley side. *Acta Mechanica*, 63, 113–130.

Broili, L. (1973). In situ tests for the study of rock fall. *Geologia Applicata e Idrogeologia*, 8, 105–111 (in Italian).

Broili, L. (1977). Relations between scree slope morphometry and dynamics of accumulation processes. In *Proceedings of the Meeting on Rock Fall Dynamics and Protective Work Effectiveness*, Bergamo, Italy: ISMES, pp. 11–23.

Calvetti, F. and di Prisco, C. (2009). An uncoupled approach for the design of rockfall protection tunnels. *Structural Engineering International*, 19, 342–347.

Cancelli, A. and Crosta, G. (1993). Hazard and risk assessment in rockfall prone areas. In *Risk Reliability in Ground Engineering*, ed. B.O. Skip. London: Thomas Telford, pp. 177–190.

Chang, K.T. and Tsai, B.W. (1991). The effect of DEM resolution on slope and aspect mapping. *Cartography and Geographic Information Science*, 18, 69–77.

Chau, K.T., Wong, R.H.C. and Wu, J.J. (2002). Coefficient of restitution and rotational motions of rockfall impacts. *International Journal of Rock Mechanics and Mining Sciences*, 39, 69–77.

Chau, K.T., Wu, S.Z., Zhu, W.C., Tang, C.A. and Yu, T.X. (2003). Dynamic fracture and fragmentation of spheres. In *16th ASCE Engineering Mechanics Conference*. Seattle: University of Washington.

Copons, R. and Vilaplana, J.M. (2008). Rockfall susceptibility zoning at a large scale: From geomorphological inventory to preliminary land use planning. *Engineering Geology*, 102, 142–151.

Copons, R., Vilaplana, J.M. and Linares, R. (2009). Rockfall travel distance analysis by using empirical models (Solà d'Andorra la Vella, Central Pyrenees). *Natural Hazards and Earth System Sciences*, 9, 2107–2118.

Crosta, G.B. and Agliardi, F. (2003). A new methodology for physically based rockfall hazard assessment. *Natural Hazards and Earth System Sciences*, 3, 407–422.

(2004). Parametric evaluation of 3D dispersion of rockfall trajectories. *Natural Hazards and Earth System Sciences*, 4, 583–598.

Crosta, G.B., Agliardi, F., Frattini, P. and Imposimato, S. (2004). A three-dimensional hybrid numerical model for rockfall simulation. *Geophysical Research Abstracts*, 6, 04502.

Crosta, G.B., Frattini, P., Imposimato, S. and Agliardi, F. (2006). Modeling vegetation and fragmentation effects on rockfalls. *Geophysical Research Abstracts*, 8, 07694.

Cruden, D.M. and Varnes, D.J. (1996). Landslides types and processes. In *Landslides: Investigation and Mitigation*, ed. A.K. Turner and R.L. Schuster. Transportation Research Board, Report 247, pp. 36–71.

Cundall, P.A. (1971). A computer model for simulating progressive large scale movements in blocky rock systems. In *Proceedings of the Symposium of the International Society of Rock Mechanics*, Nancy, France, pp. 129–136.

di Prisco, C. and Vecchiotti, M. (2006). A rheological model for the description of boulder impacts on granular strata, *Geotechnique*, 56, 469–482.

Domaas, U. (1994). *Geometrical Methods of Calculating Rockfall Range*. Norwegian Geotechnical Institute, Report 585910–1.

Dorren, L.K.A., Maier, B., Putters, U.S. and Seijmonsbergen, A.C. (2004). Combining field and modeling techniques to assess rockfall dynamics on a protection forest hillslope in the European Alps. *Geomorphology*, 57, 151–167.

Dorren, L.K.A., Berger, F., le Hir, C., Mermin, E. and Tardif, P. (2005). Mechanisms, effects and management implications of rockfall in forests. *Forest Ecology and Management*, 215, 183–195.

Dorren, L.K.A., Berger, F. and Putters, U.S. (2006). Real size experiments and 3D simulation of rockfall on forested and non-forested slopes. *Natural Hazards and Earth System Sciences*, 6, 145–153.

Dussauge, C., Grasso, J.-R. and Helmstetter, A. (2003). Statistical analysis of rockfall volume distributions: Implications for rockfall dynamics. *Journal of Geophysical Research*, 108(B6), 2286.

Evans, D.L., Roberts, S.D., McCombs, J.W. and Harrington, R.L. (2001). Detection of regularly spaced targets in small-footprint LiDAR data: Research issues for consideration. *Photogrammetric Engineering & Remote Sensing*, 67, 1133–1136.

Evans, S.G. and Hungr, O. (1993). The assessment of rock fall hazard at the base of talus slopes. *Canadian Geotechnical Journal*, 30, 620–636.

Falcetta, J.L. (1985). Un nouveau modèle de calcul de trajectoires de blocs rocheux. *Revue Française de Géotechnique*, 30, 11–17 (in French).

Frattini, P., Crosta, G.B., Carrara, A. and Agliardi, F. (2008). Assessment of rockfall susceptibility by integrating statistical and physically based approaches. *Geomorphology*, 94, 419–437.

Giacomini, A., Buzzi, O., Renard, B. and Giani, G.P. (2009). Experimental studies on fragmentation of rock falls on impact with rock surfaces. *International Journal of Rock Mechanics and Mining Sciences*, 46, 708–715.

Gokceoglu, C., Sonmez, H. and Ercanoglu, M. (2000). Discontinuity controlled probabilistic slope failure risk maps of the Altindag (settlement) region in Turkey. *Engineering Geology*, 55, 277–296.

Günther, A., Carstensen, A. and Pohl, W. (2004). Automated sliding susceptibility mapping of rock slopes. *Natural Hazards Earth System Sciences*, 4, 95–102.

Guzzetti, F., Crosta, G., Detti, R. and Agliardi, F. (2002). STONE: A computer program for the three-dimensional simulation of rockfalls. *Computers and Geosciences*, 28, 1081–1095.

Heim, A. (1932). *Bergsturz und Menschenleben*. Zurich, Switzerland: Fretz and Wasmuth Verlag.

Heinimann, H.R., Holtenstein, K., Kienholz, H., Krummenhacher, B. and Mani, P. (1998). *Methoden zur Analyse und Bewertung von Naturgefahren*. Umwelt-Materialien 85, Naturgefahren. Bern, Switzerland: BUWAL (in German).

Holm, K. and Jakob, M. (2009). Long rockfall runout, Pascua Lama, Chile. *Canadian Geotechnical Journal*, 46, 225–230.

Hungr, O., Evans, S.G. and Hazzard, J. (1999). Magnitude and frequency of rockfalls and rock slides along the main transportation corridors of south-western British Columbia. *Canadian Geotechnical Journal*, 36, 224–238.

Jaboyedoff, M. and Labiouse, V. (2011). Technical Note: Preliminary estimation of rockfall runout zones. *Natural Hazards and Earth System Sciences*, 11, 819–828.

Jaboyedoff, M., Baillifard, F., Hantz, D., Heidenreich, B. and Mazzoccola, D. (2001). Terminologie. In *Prevention des Mouvements de Versants et des Instabilités de Falaises*. Programme Interreg IIc: "Falaises", Report 48–57.

Jaboyedoff, M., Dudt, J.P. and Labiouse, V. (2005). An attempt to refine rockfall zoning based on kinetic energy, frequency and fragmentation degree. *Natural Hazards and Earth System Sciences*, 5, 621–632.

Jaboyedoff, M., Metzger, R., Oppikofer, T. *et al.* (2007). New insight techniques to analyze rock-slope relief using DEM and 3D-imaging cloud points: COLTOP-3D. In *Rock Mechanics: Meeting Society's Challenges and Demands*, ed. E. Eberhardt, D. Stead and T. Morrison. London: Taylor and Francis, pp. 61–68.

Jones, C.L., Higgins, J.D. and Andrew, R.D. (2000). *Colorado Rock Fall Simulation Program Version 4.0*. Denver, CO: Colorado Department of Transportation, Colorado Geological Survey.

Koukoulas, S. and Blackburn, G.A. (2005). Mapping individual tree location, height and species in broadleaved deciduous forest using airborne LiDAR and multi-spectral remotely sensed data. *International Journal of Remote Sensing*, 26, 431–455.

Labiouse, V. and Heidenreich, B. (2009). Half-scale experimental study of rockfall impacts on sandy slopes *Natural Hazards and Earth System Sciences*, 9, 1981–1993.

Lan, H., Martin, C.D. and Lim, C.H. (2007). RockFall analyst: A GIS extension for three-dimensional and spatially distributed rockfall hazard modeling. *Computers & Geosciences*, 33, 262–279.

Lied, K. (1977). Rockfall problems in Norway. In *Proceedings of the Meeting on Rock Fall Dynamics and Protective Work Effectiveness*. Bergamo, Italy: ISMES, pp. 51–53.

Loye, A., Jaboyedoff, M. and Pedrazzini, A. (2009). Identification of potential rockfall source areas at a regional scale using a DEM-based geomorphometric analysis. *Natural Hazards and Earth System Sciences*, 9, 1643–1653.

Lundström, T., Jonsson, M.J., Volkwein, A. and Stoffel, M. (2009). Reactions and energy absorption of trees subject to rockfall: A detailed assessment using a new experimental method. *Tree Physiology*, 29, 345–359.

Meissl, G. (1998). *Modellierung der Reichweite von Felsstürzen. Fallbeispiele zur GIS-gestützten Gefahrenbeurteilung aus dem Bayerischen und Tiroler Alpenraum*. Innsbrucker Geographische Studien, Band 28 (in German).

Nichol, M.R. and Watters, R.J. (1983). Comparison and effectiveness of rock fall mitigation techniques applied by states in the USA and Canada. In *Proceedings of the 20th Annual Engineering Geology and Soil Engineering Symposium*. Boise, ID: Idaho Transportation Department, Division of Highways, pp. 123–142.

Nocilla, N., Evangelista, A. and Scotto di Santolo, A. (2008). Fragmentation during rock falls: Two Italian case studies of hard and soft rocks. *Rock Mechanics and Rock Engineering*, 42, 815–833.

Onofri, R. and Candian, C. (1979). *Indagine sui Limiti di Massima Invasione dei Blocchi Rocciosi Franati durante il Sisma del Friuli del 1976*. Udine, Italy: CLUET (in Italian).

Pfeiffer, T.J. and Bowen, T.D. (1989). Computer simulations of rockfalls. *Bulletin of the Association of Engineering Geologists*, 26, 135–146.

Pierson, L.A., Davis, S.A. and Van Vickle, R. (1990). *Rockfall Hazard Rating System Implementation Manual*. US Department of Transportation, Federal Highway Administration Report FHWA-OREG-90–01.

Piteau, D.R. and Clayton, R. (1976). Computer rockfall model. In *Proceedings of the Meeting on Rock Fall Dynamics and Protective Work Effectiveness*. Bergamo, Italy: ISMES, pp. 123–125.

Rioual, F., Valance, A. and Bideau, D. (2000). Experimental study of the collision process of a grain on a two-dimensional granular bed. *Physical Review E*, 62, 2450–2459.

Ritchie, A.M. (1963). Evaluation of rockfall and its control. *Transportation Research Record*, 17, 13–28.

Rochet, L. (1987). Application des modèles numériques de propagation à l'étude des éboulements rocheux. *Bulletin Liaison Pont Chaussée*, 150/151, 84–95 (in French).

Rouiller, J.D., Jaboyedoff, M., Marro, C., Philippossian, F. and Mamin, M. (1998). *Pentes Instables dans le Pennique Valaisan. Matterock: Une Méthodologie d'Auscultation des Falaises et de Détection des Éboulements Majeurs Potentiels.* Zurich: VDF Hochschulverlag AG and ETH.

Scheidegger, A.E. (1973). On the prediction of the reach and velocity of catastrophic landslides. *Rock Mechanics*, 5, 231–236.

Stevens, W. (1998). *RocFall: A tool for probabilistic analysis, design of remedial measures and prediction of rockfalls*. M.A.Sc. thesis, University of Toronto, Toronto, ON.

Sturzenegger, M., Yan, M., Stead, D. and Elmo, D. (2007). Application and limitations of ground-based laser scanning in rock slope characterization. In *Rock Mechanics: Meeting Society's Challenges and Demands*, ed. E. Eberhardt, D. Stead and T. Morrison. London: Taylor and Francis, 29–36.

Tonon, F. and Kottenstette, J.T. (2006). Laser and photogrammetric methods for rock face characterization: A workshop. In *Laser and Photogrammetric Methods for Rock Face Characterization*, ed. F. Tonon and J.T. Kottenstette. Alexandria, VA: American Rock Mechanics Association, pp. 5–10.

Toppe, R. (1987). Terrain models as a tool for natural hazard mapping. In *Avalanche Formation, Movement and Effects*, ed. B. Salm and H. Gubler. International Association of Hydrological Sciences, Publication 162, pp. 629–638.

Volkwein, A., Roth, A., Gerber, W. and Vogel, A. (2009). Flexible rockfall barriers subjected to extreme loads. *Structural Engineering International*, 19, 327–331.

Wang, Y. and Tonon, F. (2009). Discrete element modeling of impact fragmentation in rock fall analysis. In *Proceedings of the 43rd US Rock Mechanics Symposium and 5th U.S.–Canada Rock Mechanics Symposium*, Asheville, NC, Paper 09–153.

Wang, Y., Tonon, F., Crosta, G.B., Agliardi, F. and Zavodni, Z.M. (2010). 3-D modeling of rockfall fragmentation at impact using a discrete element model. In *Proceedings of the 44th US Rock Mechanics Symposium and 5th U.S.–Canada Rock Mechanics Symposium*, Salt Lake City, UT, Paper 10–204.

Wieczorek, G.F. and Snyder, J.B. (1999). *Rock Falls from Glacier Point above Camp Curry, Yosemite National Park, California*. US Geological Survey, Open File Report 99–385.

Wieczorek, G.F., Morrissey, M.M., Iovine, G. and Godt, J. (1998). *Rock-Fall Hazards in the Yosemite Valley*. US Geological Survey, Open File Report 98–467, scale 1:12,000.

Wieczorek, G.F., Snyder, J.B., Waitt, R.B. *et al.* (2000). The unusual air blast and dense sandy cloud triggered by the 10 July 1996, rock fall at Happy Isles, Yosemite National Park, California. *Geological Society of America Bulletin*, 112, 75–85.

Wong, R.H., Ho, K.W. and Chau, K.T. (2000). Shape and mechanical properties of slope material effects on the coefficient of restitution on rockfall study. In *Proceedings of the 4th North American Rock Mechanics Symposium NARMS 2000*, Seattle, pp. 507–514.

Yashima, S., Kanda, Y. and Sano, S. (1987). Relationships between particle size and fracture energy or impact velocity required to fracture as estimated from single particle crushing. *Powder Technology*, 51, 277–282.

Zinggeler, A., Krummenacher, B. and Kienholz, H. (1991). Steinschlagsimulation in Gebirgswäldern. *Berichte und Forschungen*, 3, 61–70 (in German).

23 The 2006 Eiger rockslide, European Alps

MICHEL JABOYEDOFF, MARC-HENRI DERRON, JULIEN JAKUBOWSKI, THIERRY OPPIKOFER, AND ANDREA PEDRAZZINI

ABSTRACT

In July 2006, approximately 2 million m^3 of massive limestone began to move on the east flank of the Eiger in central Switzerland. For more than two years after the initial failure, the rock mass moved at rates of up to 70 cm per day. A detailed analysis of the structures and velocities of the different moving blocks was conducted with the aid of terrestrial laser scanning. The moving rock mass included a rear block that subsided, pushing a frontal block forward. Movement directions were controlled by discontinuity sets that formed wedges bounded on one side by sub-vertical bedding planes. The instability was, until recently, buttressed by a glacier. Slope observations and results of continuum and discontinuum modeling indicate that the structure of the rock mass and topography were the main causes of the instability. Progressive weathering and mechanical fatigue of the rock mass appear to have led to the failure. A dynamic analytical model further indicates that the rockslide was primarily controlled by a reduction in the strength of discontinuities, the effects of ice deformation, and – to a limited extent – groundwater flow. This study shows that realistic and simple instability models can be constructed for rock-slope failures if high-resolution data are available.

23.1 INTRODUCTION

In June 2006, a rock mass with a volume of approximately 2 million m^3 began to move on the east flank of the Eiger in central Switzerland (Oppikofer *et al.*, 2008). This event was well publicized because of the exceptionally fast movement rates and frequent rockfalls, and because a catastrophic rockfall had killed two people traveling the Gotthard Highway one month earlier, closing the route for five weeks (Liniger, 2006).

Oppikofer *et al.* (2008) employed a terrestrial laser scanner to produce 3D images of the rock slope, probably the first time that the movement of a rockslide had been periodically monitored using this tool. The main structures controlling the vectors and rates of displacement of the different blocks were mapped using terrestrial laser scanning (TLS), leading to the creation of a bilinear wedge failure model with sliding of a frontal block activated by a rear subsiding block.

The 2006 Eiger rockslide exhibits several features that are common to many rockslides. First, it affected a prominent spur in a glacial valley and, second, it occurred shortly (*ca.* 150 years) after a period of major glacier retreat, during which most of the slope became ice-free. Deglaciation on a more regional scale occurred in alpine valleys at the end of the Pleistocene (Portmann, 1977; Florineth and Schlüchter, 1998) and probably induced similar instabilities in steep rock slopes. Deglaciation processes and progressive rock mass degradation are considered important preparatory factors in the development of both large and small rock-slope instabilities (Agliardi *et al.*, 2001; Ballantyne, 2002; Eberhardt *et al.*, 2004, Ambrosi and Crosta, 2006; Fischer *et al.*, 2010). In Swiss mountain valleys, glacier erosion and retreat lead to important changes in slope stability in two main ways: by unloading the slope, causing stress relief in the rock mass (Ballantyne, 2002; Cossart *et al.*, 2008; Ambrosi and Crosta, 2011; Ghirotti *et al.*, 2011), and by erosional oversteepening of valley walls, which results in stress concentrations and rock mass damage at the toe of the slope (Eberhardt *et al.*, 2004).

The 2006 Eiger rockslide provides a unique opportunity to analyze the destabilization of a rock slope in an alpine environment. The evolution of the displacement rates of the slowly moving rock mass prior to collapse is complex and difficult to model. We modeled the slope movements using Phase2 (Rocscience, 2010) and UDEC (Itasca, 2004), and provided

Landslides: Types, Mechanisms and Modeling, ed. John J. Clague and Douglas Stead. Published by Cambridge University Press.

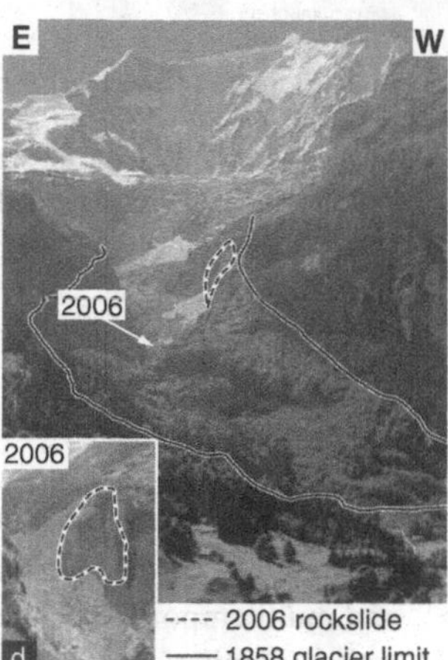

Fig. 23.1. (a) Photograph of the Eiger rockslide (September 22, 2006), showing its main features. Inset shows geographical location. (b) Photograph of the blocks remaining on October 21, 2008. (c) Historic photograph of Lower Grindelwald Glacier in 1858 (© Frédéric Martens, Alpine Club Library London, modified from OcCC, 2007). (d) Photograph of Lower Grindelwald Glacier in 1974 and in 2006 (inset), showing the location of the Eiger rockslide (© Heinz J. Zumbühl (1974) and Samuel U. Nussbaumer (2006), modified from OcCC, 2007). (e) Photograph of the July 13, 2006 partial collapse of the northern block (B. Petroni).

an analytical solution to the implied stresses based on a two-dimensional bilinear wedge failure model (Sarma, 1979; Norrish and Wyllie, 1996) using an optimization procedure (Bardet and Kapuskar, 1989). The results are consistent with initial failure along a steep back crack and a progressive weakening of the basal sliding zone over time.

23.2 GEOLOGY AND MORPHOLOGY

Mount Eiger is located in the central Swiss Alps near the village of Grindelwald. It is part of the autochthonous Mesozoic sedimentary cover of the crystalline Aar Massif, which dips steeply to the northwest (Günzler-Seiffert and Wyss, 1938). At the rockslide site, the Mesozoic cover consists mainly of banded, tabular, Upper Jurassic Limestone with sub-vertical bedding (discontinuity set S0). Regionally, S0 has been deformed into an antiformal syncline.

The Eiger rockslide is located on a prominent spur (Fig. 23.1) where the adjacent U-shaped valley becomes a narrow deep valley occupied by Lower Grindelwald Glacier and its torrent (Fig. 23.1a, d). This morphologic change is likely due to differences in rock mass strength between the crystalline basement and Triassic to Middle Jurassic sedimentary rocks to the south and massive Upper Jurassic Limestone to the north (Günzler-Seiffert and Wyss, 1938). The Lower Grindelwald Glacier may have highly stressed this rock spur. In 1860, during the Little Ice Age, the level of the Lower Grindelwald Glacier was approximately 200 m higher than at present and reached the top of the back scarp of the rockslide (Messerli *et al.*, 1978; OcCC, 2007) (Fig. 23.1c). The rate of thinning and retreat of the glacier accelerated after 1935 (Messerli *et al.*, 1978; Werder *et al.*, 2010) (Fig. 23.1d). The 2006 Eiger rockslide was probably caused by the removal of this glacier buttress and the subsequent stress relief within the rock mass (Oppikofer *et al.*, 2008).

23.3 METHODS

23.3.1 LASER SCANNING TECHNOLOGY

Laser scanning is a widely used remote sensing tool for acquiring point clouds of the ground surface (Baltsavias, 1999; Lichti *et al.*, 2002). A laser beam is emitted in a precisely known direction, backscattered by the ground surface, and the return pulse is recorded by the scanner. The distance between the instrument and the ground surface is calculated using the time of flight of the pulse. Mirrors inside the instrument deflect the outgoing laser beam in different directions, allowing the area of interest to be swept (Lichti *et al.*, 2002). The Optech ILRIS-3D laser scanner used for this study has a wavelength of 1500 nm, an acquisition rate of about 2500 points per second, and a practical range on rock slopes of about 600–800 m. A terrestrial laser scanner has a fixed position during data acquisition, providing a better point accuracy (1σ = *ca.* 1.5 cm at a distance of 300–400 m; Oppikofer *et al.*, 2009) than aerial laser scanners (1σ = 5–20 cm; Baltsavias, 1999).

23.3.2 TLS DATA TREATMENT AND ANALYSIS

Typical TLS data acquisition and analysis procedures include the following (Conforti *et al.*, 2005):

1. Acquisitions from different positions and different view directions.
2. Manual cleaning of scans and removal of vegetation.
3. Co-registration of the scans using a three-step procedure: (1) a rough manual matching; (2) manual identification of common points in different scans; and (3) minimization of co-registration errors using a point-to-surface iterative closest point (ICP) algorithm (Besl and McKay, 1992; Teza *et al.*, 2007). This procedure is also used for multitemporal scans by limiting the iterative alignment procedure to the assumed stable part around the rockslide (Teza *et al.*, 2007; Oppikofer *et al.*, 2008, 2009).
4. Georeferencing of the TLS point cloud by co-registering it on an airborne laser scanning point cloud or by using ground control points.
5. Detailed structural analysis of the georeferenced TLS point cloud (Sturzenegger *et al.*, 2007; Sturzenegger and Stead, 2009), either by fitting planes on points forming a discontinuity (Sturzenegger and Stead, 2009) or by using special software, such as Coltop3D (Jaboyedoff *et al.*, 2007, 2009). Coltop3D computes the spatial orientation of each point relative to its neighborhood. The principal discontinuity sets and their mean orientations are derived from an orientation-specific coloring of the TLS point cloud (Jaboyedoff *et al.*, 2007; Oppikofer *et al.*, 2009).
6. Displacements between different TLS datasets can be determined using the ICP algorithm (Oppikofer *et al.*, 2008, 2009). Positive differences mean that the data points are situated above or in front of the reference surface, for example due to the forward movement of a slope. Negative differences indicate that the data points are behind or below the reference, for example due to vertical settlement or rockfalls.
7. Displacement vectors are created by selecting identical points in different point clouds. Errors on the measured displacement vectors are relatively high (length: 1σ = 10 cm; orientation: 1σ = 2–3°; Oppikofer *et al.*, 2008).
8. The roto-translation matrix technique provides a detailed displacement analysis, including both translation and rotation of individual rockslide blocks (Monserrat and Crosetto, 2008; Oppikofer *et al.*, 2009).
9. The rockfall volume is estimated by comparing the pre-rockfall and post-rockfall surfaces (Rabatel *et al.*, 2008; Abellán *et al.*, 2010).

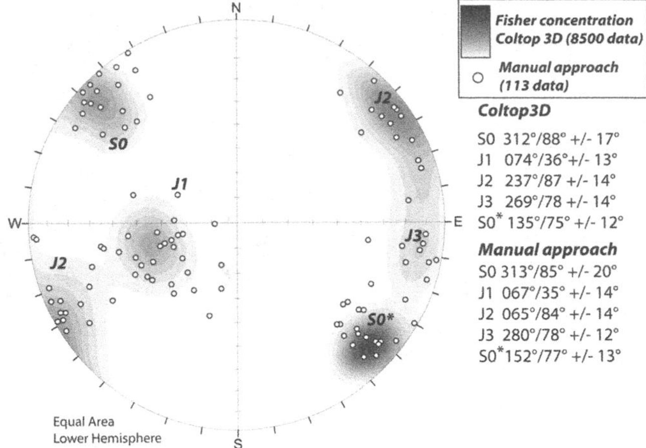

Fig. 23.2. Stereonet (lower hemisphere) showing results of the structural analyses. The Fisher concentration corresponds to poles of orientations extracted by Coltop 3D analysis. The points are poles of orientation obtained by manual fitting in PolyWorks (InnovMetric, 2011). Numerical values and two-standard-deviation variability are shown.

10. A geometric model of instability is created by combining the geometry of the main discontinuities, the displacements, and multitemporal cross-sections parallel to the sliding direction (Oppikofer *et al.*, 2008, 2009).

For this study, data were treated and analyzed using PolyWorks® software (InnovMetric, 2011).

23.4 STRUCTURAL INTERPRETATION

Four main discontinuity sets, including bedding, were detected on the TLS point cloud using Coltop3D and manual plane fitting (Fig. 23.2). The presence of meter-scale, open folds results in significant variation in the mean orientation of the bedding planes. In the area of the rockslide, folding is responsible for two distinct bedding orientations (S0 and S0* in Fig. 23.2). The bedding is characterized as an undulating systematic discontinuity set with wide to very wide spacing (International Society of Rock Mechanics, 1978).

The east flank of the valley has a stepped morphology linked to the presence of sub-vertical joint sets or faults (J2 and J3), with less-steep discontinuities (J1) forming the flatter areas. These discontinuity sets are also apparent at the outcrop scale based on the TLS analysis. Joint sets J2 and J3 have a wide spacing with a medium to high persistence, and are dominantly planar and rough. The geometrical characterization of discontinuity set J1 is difficult because its orientation is sensitive to modification by glacial erosion. Based on surface morphology, however, J1 can be described as moderate to widely spaced, with medium to high persistence.

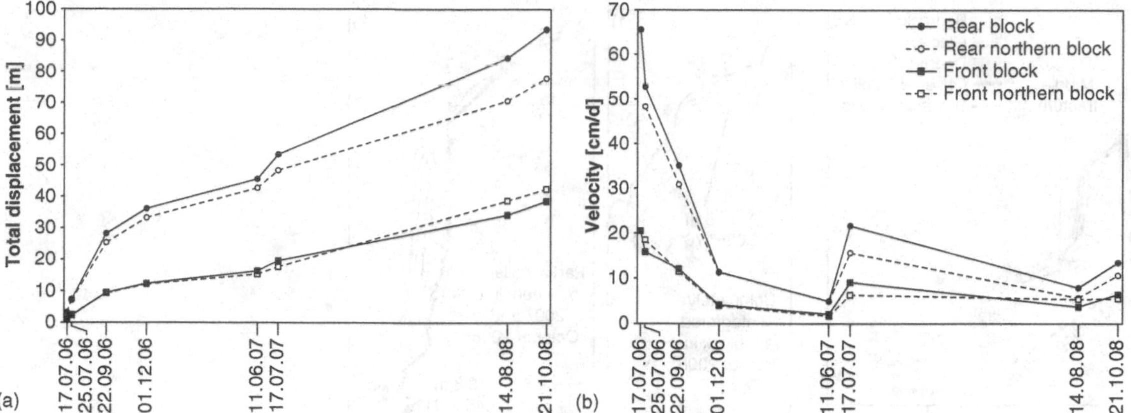

Fig. 23.3. Graphs of (a) total displacement since the beginning of the TLS monitoring and (b) velocity versus the date of the TLS acquisition. (Modified from Oppikofer *et al.* (2008), with permission from Macmillan Publishers Ltd.)

The shape of the Eiger rockslide is clearly controlled by the orientations of these regional discontinuity sets. The back scarp has a saw-tooth shape controlled by the intersection of J2 and S0. Bedding planes also control the lateral extent of the instability and the collapse of individual rock columns.

23.5 EVOLUTION OF THE SLOPE INSTABILITY

After heavy rainfall in August 2005, eyewitnesses reported the first rockfalls on the south flank of the rock spur. Slope movements became apparent on June 10, 2006, with large rockfalls of several hundred cubic meters. Two steep valley-parallel cracks opened at that time. One of these cracks formed the 250-m-long back scarp that separates the spur from the stable rock mass. The other crack separates the spur into a front and a rear block (Fig. 23.1a). The two blocks are further separated by a bedding-parallel fracture into a southern and a northern part.

From June to August 2006, the slope movements were rapid, averaging several decimeters per day, and they caused frequent rockfalls and partial collapses. Opening of the back crack was initially measured by hand, but total station surveys were later made. The installed targets, however, were soon destroyed by rockfalls and collapses, and thus terrestrial laser scanning was adopted to monitor the slope movements.

Sequential TLS point clouds allowed the progress of the slope movements to be recorded in 3D, with displacements of the rear block of up to 80 cm per day and of the front block of up to 20 cm per day (Oppikofer *et al.*, 2008). A large (169,000 m^3) partial collapse of the northern block occurred on the evening of July 13, 2006, preceded by sequential rockfalls and displacement of a fast-moving column at rates of up to 125 cm per day. Photographs of this collapse reveal that it started with failure of this fast-moving column (Fig. 23.1e); this finding is important information for rockslide monitoring and especially for spatial and temporal failure prediction (Oppikofer *et al.*, 2008; Abellán *et al.*, 2009, 2010).

The shortest distance comparisons provided a first assessment of slope movements and highlighted the different displacement directions and velocities of the rear block relative to the front one, with mainly downward displacements in the rear and out-of-slope displacements on the basal sliding surface (Oppikofer *et al.*, 2008). Periodic monitoring of slope movements continued for several months and showed a decrease in the velocities of both the front and the rear blocks. During the first year of TLS monitoring, the total displacements were 15 m for the front block and 50 m for the rear block (Fig. 23.3a). These high displacements were accompanied by the progressive lateral disintegration and break-up of the rear block. Slope movements continued between the summer of 2007 and fall 2008, and rockfall activity again increased in 2008, leading to a partial collapse of the front block (*ca.* 60,000 m^3). The front block continued to break up, with several collapses, until October 21, 2008, at which time only a blade-shaped, 90-m-high needle remained (Fig. 23.1b).

Displacement vectors were measured for the different blocks in order to provide a more precise analysis of the slope movements. Due to disintegration and break-up of the blocks, it became increasingly difficult to locate suitable point pairs to measure displacement vectors. Therefore, after June 2007 we applied the roto-translation matrix technique to determine the displacement vectors.

The sliding direction of the front block differed little throughout the measurement period (trend/plunge: 042°/34° ± 9°), but its average velocity decreased from 20 cm per day in the summer of 2006 to 2 cm per day in the winter of 2006/7, before rising again to 4–6 cm per day in July 2007 (Fig. 23.3b). The front of the northern block displayed a similar velocity pattern (Fig. 23.3b) and a constant displacement direction between July 2006 and July 2007 (038°/39° ± 11°) before moving more toward the north in the summer of 2008.

The sliding direction of the rear block varied only slightly between July and September 2006 (mean orientation: 007°/68°),

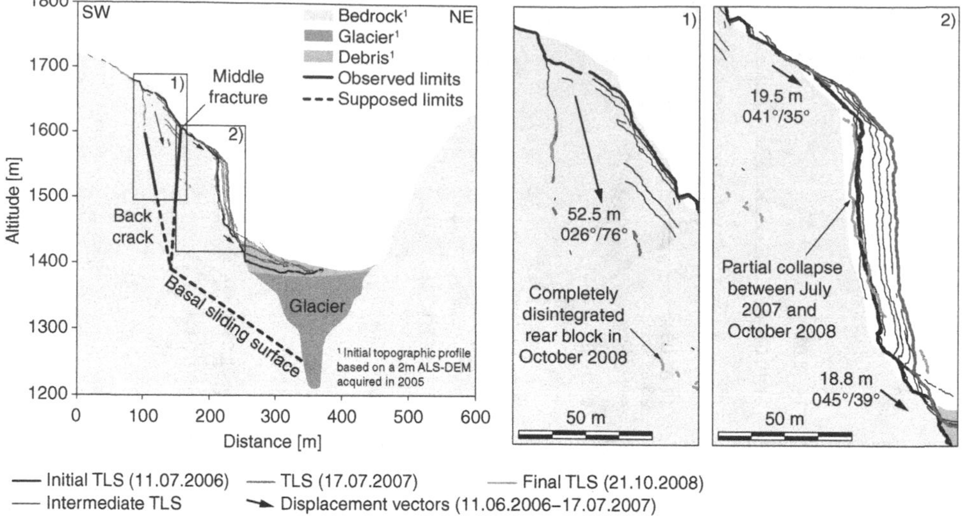

Fig. 23.4. Profile across the Eiger rockslide parallel to the displacement direction of the front block. The cross-section series is based on TLS data. The July 11, 2006, July 17, 2007, and October 21, 2008 profiles are displayed with thicker lines than the other TLS acquisitions. Note the complete disintegration of the rear block in 2008 and the partial collapse on the front block between July 2007 and October 2008 (apparent retreat of the rock wall). (Modified from Oppikofer *et al.* (2008), with permission from Macmillan Publishers Ltd.)

before moving more toward the east. This change is likely due to the block disintegration and toppling toward the east. In the final monitoring period (August–October 2008), the sliding direction was similar to that in the summer of 2006 (007°/77°). The rear of the north block exhibited a more variable displacement pattern (on average 043°/75°). However, the pattern of velocity vs. time of both back blocks was similar to that of the front blocks (Fig. 23.3b). The velocity of the rear block decreased from 65 to 5 cm per day between July 2006 and June 2007, then increased up to 22 cm per day in the summer of 2007, before reaching a steady displacement rate of 8–13 cm per day between July 2007 and October 2008 (Fig. 23.3b).

Structural analysis of discontinuity sets (Fig. 23.2) and the displacement vectors show that the movements were strongly controlled by intersections of discontinuities. The rear parts of the rockslide moved along the intersection line of S0 and J2, while the front block moved forward parallel to the intersection line of S0 and J1 (Oppikofer *et al.*, 2008).

On the basis of the measured 3D displacements and the observed structures, Oppikofer *et al.* (2008) proposed a sliding mechanism for the Eiger rockslide, involving a bilinear wedge failure (Fig. 23.4). They concluded that the front blocks slid along J1 and S0 discontinuity sets, forming a wedge, while the rear blocks moved downward parallel to discontinuity set J2. The rear blocks, which are the active ones, provided a driving force, pushing the front blocks along their sliding surface (Sultan and Seed, 1967).

23.6 DESCRIPTION OF GEOMECHANICAL MODELS

Three different series of models were created based on continuum, discontinuum, and analytical dynamic approaches.

1. First we used continuum models to simulate the evolution and stability conditions of the valley during deglaciation.
2. Second, we used a distinct element code to investigate, in more detail, the potential failure mechanism and the influence of pre-existing discontinuity sets. Pore water pressure was not considered in either model.
3. Third, based on previous slide initiation modeling and TLS analysis, we used a 2D bilinear wedge failure model (Norrish and Wyllie, 1996) to perform a back analysis of the rockslide that accounted for groundwater conditions, shear strength degradation, and progressive volume loss.

23.6.1 CONTINUUM MODELING

We carried out a multi-stage, stress-strain analysis using the finite element code Phase² 7.0 (Rocscience, 2010). Progressive deglaciation since the mid-nineteenth century was simulated in six stages, starting with maximum ice cover in 1860 (Fig. 23.1c). A Mohr–Coulomb elasto-plastic yield criterion was used to simulate the response of the rock mass to the removal of supporting ice (Cai *et al.*, 2007). Due to the difficulty in physically accessing the failure, we characterized the rock mass using the geological strength index (GSI) (Hoek *et al.*, 2002). The characterization is based on photographs, TLS data, and inspection of rock outcrops near the TLS acquisition stations. Joint surface conditions, in particular joint alteration, are poorly known, yet need to be critically considered in the modeling.

Data from 15 uniaxial compression tests of rocks of similar lithology to those at Mt. Eiger (J.-F. Mathier, personal communication, 2010) have been used to determine uniaxial compressive strength and the Young modulus. These tests showed a range of intact uniaxial compressive strength from 65 to 180

Table 23.1. *Rock mass properties used in the continuum model.*

	Stage 0 GSI = 70 D = 0.7 mi = 12	**Stage 1** GSI = 65 D = 0.7 mi = 12	**Stage 2** GSI = 60 D = 0.7 mi = 12	**Stage 3** GSI = 55 D = 0.8 mi = 12	**Stage 4** GSI = 50 D = 0.8 mi = 12	**Stage 5** GSI = 50 D = 1.0 mi = 12
Equivalent friction angle (°)	45	43	40	36	33	29
Equivalent cohesion (MPa)	2.9	2.3	1.9	1.5	1.3	1.0
Tensile strength (MPa)	1	0.9	0.85	0.8	0.75	0.75
Poisson ratio	0.3	0.3	0.3	0.3	0.3	0.3
Young modulus (GPa)	25	25	25	25	25	25

Note: Disturbance factor was changed to a depth of 150 m only.

Table 23.2. *Rock mass properties used in the discontinuum models.*

Parameters	Malm Limestone	Ice
Uniaxial compressive strength	90–110	
Young modulus (GPa)	25	10
Poisson ratio	0.33	0.3
Density (kg/m^3)	2700	900
Intact rock friction angle (°)	42	–
Intact rock cohesion (MPa)	2	–
Joint normal stiffness (GPa m^{-1})	10	–
Joint shear stiffness (GPa m^{-1})	1	–
Joint friction angle (°)	30–35	–
Joint cohesion (MPa)	0.2–0.8	–
Joint dilatation angle (°)	5	–
Joint tensile strength (MPa)	0.5	–
Joint residual cohesion (MPa)	0.1	–
Joint residual friction angle (°)	15	–

MPa, with mean and median values close to 100 MPa. The estimated mean Young modulus is 25,000 MPa.

We progressively lowered the rock mass strength from a GSI value of 70, the maximum estimated value in the area, to 50, corresponding to the observed present-day value. To account for glacially induced stress relaxation, we altered the disturbance factor, *D* (Hoek *et al.*, 2002). This factor represents the degree of disturbance of the rock mass to a depth of 150 m, which is based on observations by Bucher and Loew (2009) of the maximum depth of formation of postglacial sheeting joints (Table 23.1).

We estimated the Mohr–Coulomb parameters and rock mass strength using the Hoek–Brown criterion (Table 23.2). However, the Hoek–Brown criterion for hard rock masses at lower stresses tends to overestimate strength (Hajiabdolmajid *et al.*, 2002; Diederichs, 2003; Carter *et al.*, 2008). For this reason, the Mohr–Coulomb parameters must be carefully considered in modeling and should ideally be checked with reliable laboratory tests.

Mechanical properties of ice were obtained from the literature (Schulson, 1999; Gagnon and Gammon, 1995). To allow for ductile deformation, the ice was modeled using a simple Mohr–Coulomb elasto-plastic constitutive law.

We assumed an in-situ stress ratio (horizontal to vertical stress) of 0.5, which is consistent with values derived from previous geomechanical analyses in the Swiss Alps (Eberhardt *et al.*, 2004; Fischer *et al.*, 2010) and with fault planc solutions of earthquakes that suggest a strike–slip to normal mechanism for faults in the area (Kastrup *et al.*, 2004). We investigated the influence of joint set J1 on strength (ratio of rock strength to induced stress at every point in the mesh) and tensile stress distributions using a ubiquitous-joint model (anisotropic continuous material).

23.6.2 DISCONTINUUM MODEL

Discontinuum modeling was performed using the 2D distinct element code UDEC (Itasca, 2004), which provides a realistic analysis of a discontinuous medium such as jointed rock (Eberhardt *et al.*, 2004; Brideau *et al.*, 2006). The goal of this analysis was to better define the failure mechanism and the geometry of the failure surfaces. The models assume that the rock mass and remaining ice behave as elasto-plastic Mohr–Coulomb materials (Table 23.2). We modeled the behavior of the discontinuity sets using Mohr–Coulomb slip with residual strength and assigned strength properties of the rock mass based on average GSI values (stage 2 in Table 23.1). Intact rock bridges along large-scale discontinuities were modeled indirectly by assigning higher initial strength properties to discontinuities, as described by Jennings (1970) and recently applied by Fischer *et al.* (2010) and Gischig *et al.* (2011). Using this approach, we calculated joint set properties based on a combination of discontinuity and intact rock properties. The location and the mean orientation of the rear and the middle fracture are known and were introduced into the model. On the other hand, the location of the basal sliding surface and its geometrical relation to the middle and the rear surface are unknown. In the first series of model runs, we introduced the basal failure surface at different depths. We then introduced discontinuity sets J1 and J2 into the lower part of the model to test whether a step-like basal surface would develop. The UDEC models were tested by comparing the orientations of the calculated displacement

Table 23.3. *Parameters of the model in Figure 23.5.*

Parameter	Value	Unit
Back crack length l_1	255	m
Middle failure length l_2	220	m
Basal line length l_3	235	m
Glacier length l_4	165	m
Glacier height h_4	105	m
Back crack angle a_1	77	deg
Dip angle of the middle failure surface a_2	86	deg
Dip angle of the basal failure surface a_3	34	deg
Water density ρ_w	1000	kg·m^{-3}
Rear and front block densities ρ_{RB}, ρ_{FB}	2700	kg·m^{-3}
Initial effective friction angle	42	deg

vectors with the corresponding vectors obtained through TLS. Similarly, the velocity ratio between the rear and the front blocks obtained by TLS was compared with that obtained by modeling. This ratio provides important information about the interaction of the two blocks and the geometrical intersection of the middle and the basal surfaces. It should be noted, however, that the absolute velocity values provided by the two methods are not directly comparable, because the modeling yields computer time-step-calculated velocities, whereas TLS provides real velocities.

23.6.3 ANALYTIC MODEL

An analytical model was employed to study the development of progressive failure and deformation in a dynamic way by testing different friction laws (see Fig. 23.5 and Table 23.3 for the model geometry), including ice deformation. Rainfall infiltration and strength degradation were also considered. The moving wedges are assumed to slide against each other without losing contact (Coulthard, 1979; Sarma, 1979; Stamatopoulos and Petridis, 2006). Normal contact stresses are constant along the sliding surfaces and are assumed to be compressive only. Shear stresses (τ) depend on sliding velocities and normal contact stresses (σ_n) according to the following equation:

$$\tau(u,\sigma_n) = \mu(u)\sigma_n \qquad (23.1)$$

where μ is a friction coefficient function that depends only on the sliding velocity (u). The function μ, however, strongly depends on materials on both sides of the sliding surface. We assume here that the friction coefficient decreases with sliding velocity, as argued by Spinnler (2001) in a mechanical design context. A simple model is proposed (Fig. 23.6a), whereby friction coefficient decreases with velocity at a rate that depends on a characteristic velocity u_0:

$$\mu(u) = \mu_1 + e^{-\frac{u}{u_0}}(\mu_0 - \mu_1) \qquad (23.2)$$

where μ_0 is the static friction coefficient and μ_1 is the asymptotical friction coefficient. Evolution over time of the static friction angle ϕ (related to μ_0 as $\mu_0 = \tan(\phi)$), which simulates the progressive degradation of discontinuity strength (Alonso and Pinyol, 2010), is assumed to be the same for all sliding surfaces. Because strength degradation may cause the velocity to increase, the peak friction angle $\phi(t)$ is decreased during the simulations to a given final value (Fig. 23.6b):

$$\phi(t) = \phi_0 + \left(\frac{t}{t_d}\right)^a (\phi_1 - \phi_0) \qquad (23.3)$$

where ϕ_0 and ϕ_1 are initial and final friction angles, t_d the simulation duration, and a is an exponent that allows a faster decrease of friction angle at the beginning of the scenario ($a < 1$) or at the end ($a > 1$). This time-dependency was chosen because we made no *a priori* hypothesis that the friction angle is linearly dependent on deformation. Displacements have a significant influence on strength, but it is unclear when, and at what rate, strength degradation occurs in order to explain the late acceleration of the blocks without an assumption about the underlying mechanism. We also tested a constant friction angle (i.e., no strength degradation).

Another possible explanation for the increase in sliding velocity in the spring and summer of 2007 is an increase in pore water pressure. We propose to use a simple groundwater model because there is little information about the permeability and porosity of the rock slope. The model stipulates, first, that the water table during a period of rain is zero at $t = t_0$, because at this time the water has not yet infiltrated the rock slope, and is also zero at $t = \infty$, because water has completely drained from the slope. Second, the model stipulates that water table must be positive and continuous after the start of the rain at $t = t_0$ to $t = \infty$. In addition, full infiltration is assumed; that is, all the rainfall infiltrates the slope. A groundwater model that meets these requirements is given by Eq. 23.4. Pore water pressure within the rock mass is defined by an impulse response function $\rho(t)$ that represents the degree of filling of the reservoir (volume of water contained in the rock mass), resulting from a rainfall of intensity I_0, occurring at $t = t_0$. Such behavior is expressed by the difference between a filling term of time constant τ_{t1} (relaxation time) and a draining term of time constant $\tau_{t2} \geq \tau_{t1}$:

$$\rho(t) = I_0\left(e^{-\frac{t-t_0}{\tau_{t2}}} - e^{-\frac{t-t_0}{\tau_{t1}}}\right) \qquad (23.4)$$

If $\tau_{t2} = \tau_{t1}$, the water never fully fills the rock mass; if $\tau_{t2} \gg \tau_{t1}$, water almost never drains. Figure 23.6c shows an example of the dynamic behavior of $\rho(t)$, in which groundwater remains in the rockslide mass after one specific rainfall (day 21). τ_{t1} is kept constant (1 day), whereas τ_{t2} varies from 1 day to 30 days. The water table and pore pressure also depend on the secondary porosity of the rock (η) created by fracturing. This parameter

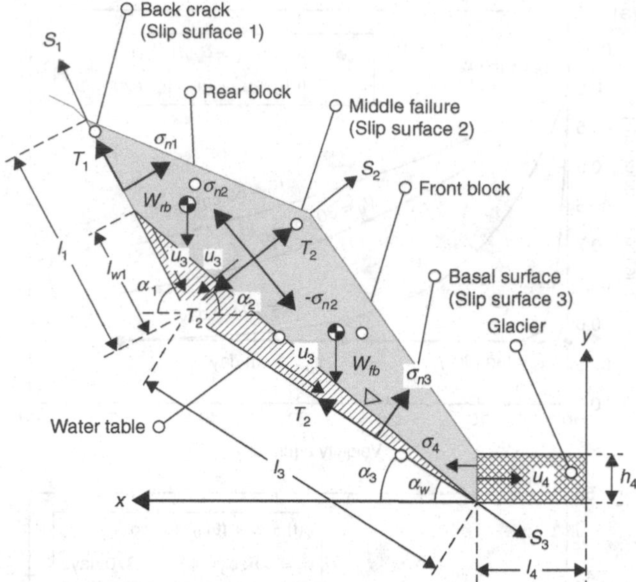

Fig. 23.5. Simplified geometric representation of the bilinear-wedge model of the Eiger rockslide. The sliding part of the system is divided into a rear block (upper wedge), a front block (lower wedge), and the glacier. The level of the water table decreases linearly toward the glacier. The rear and front blocks are subject to normal contact stress (σ_n), shear stress (τ), weight (W), and pore water pressure (not shown).

is used to create the relationship between the quantity of water stored in the rock mass and the groundwater level. Input values of I_0 are proportional to the measured precipitation (data provided by Meteoswiss).

The resisting stress exerted by a glacier depends on the horizontal velocity of the ice through Glen's flow law (Glen, 1955):

$$\dot{\varepsilon} = \frac{u_x}{l_4} = Be^{-\frac{Q}{RT}} \sigma_4^n \tag{23.5}$$

where $\dot{\varepsilon}$ is the glacier strain, l_4 is the length of the ice that is deforming, σ_4 is its normal stress, T is its temperature, and B, Q, R, and n are constants, the values of which are considered unknowns in the back analysis.

Parameters were determined using optimization methods (Bardet and Kapuskar, 1989; Zhang *et al.*, 2010). The error between calculated and measured horizontal velocities of the moving part of the system was minimized by calibrating a given set of parameters using the Matlab-2009 *fmincon* function (Mathworks, 2009).

Two sets of calibration parameters, which vary during the optimization, were considered successively. The first calibration focused on time-dependent strength degradation without the influence of pore pressure by using (1) two parameters from the strength degradation law in Eq. 23.3 (ϕ_1 and a), (2) two parameters from the friction law in Eq. 23.2 (the ratio of static-to-dynamic friction coefficients μ_1/μ_0, and the limit velocity u_0), and (3) parameter n of Glen's flow law. The last parameter is important because the front of the moving mass is constrained by the scree deposit, which changes the stress condition in the deforming ice. The infiltration law need not be parameterized in this first calibration because the pore pressure is neglected. The second calibration investigated whether pore pressure triggered the increase in sliding velocity at the end of the modeling period (400 days; July 17, 2006 to August 21, 2007), with no consideration of strength degradation. The calibrated parameters comprise (1) rock porosity (η), (2) the ratio of the static-to-dynamic friction coefficients μ_1/μ_0, and (3) parameter n of Glen's flow law. All rainwater is assumed to enter the slope. Filling and draining time constants, τ_{t1} and τ_{t2}, are set to 1 day and 10 days, respectively. In both back analyses, collapses of the rear and front blocks are taken into account by linearly decreasing their mass to, respectively, 40 percent and 70 percent of the initial mass.

We then set up equations describing the equilibrium of forces in the X and Y directions for both the rear and front blocks (see Fig. 23.5). Velocity is influenced by the contribution of normal stresses over the sliding surfaces, but not by their distribution. Moment equilibrium does not have to be achieved, as solving for it only influences the distribution of the normal contact stress. The resulting wedge model at time t contains four unknowns: the normal contact stresses at sliding surfaces σ_{n1}, σ_{n2}, σ_{n3}, and the horizontal velocity of the moving part of the system, u_x.

23.7 MODELING RESULTS

23.7.1 INITIATION PROCESSES

CONTINUUM MODELS

Finite element analysis demonstrates the influence of progressive deglaciation on the stress and strain distribution in the Lower Grindelwald Glacier valley. The coupled influence of glacial unloading and reduction of rock mass strength leads to a shear zone near the base of the slope, which transforms into a tensile failure zone sub-parallel to the valley (central and upper portion of Fig. 23.7A). The location of the maximum shear strain indicators agrees well with the geometry of the instability, although in the rear portion of the slope it seems to lie 20–25 m (10 percent) behind the observed back scarp. Even if the present-day rock mass strength is assumed to be lower (GSI 50), the tensile and shear yield indicators are not continuous along the central portion of the slope. This result suggests that large slope failure cannot develop without an external driver or the presence of persistent discontinuities.

Numerical simulations are sensitive to the anisotropy of rock mass strength. Modeling results obtained by considering anisotropy in joint set J1 are presented in Figure 23.7B. The introduction of ubiquitous joints with the same orientation as J1, with an assumed residual friction angle of 30° and residual cohesion 0.1 MPa along the fractures, amplifies the results obtained using isotropic continuum modeling. Areas with low strength factors, where the stress in the material exceeds the material strength, coincide with the maximum extent of the expected rockslide.

Continuum modeling can be used to examine the evolution of stresses and strains and also to delimit the potential extent of an instability (Eberhardt *et al.*, 2004). However, the results in the Eiger case provide a simplified representation of the failure mechanism. In particular, continuum modeling fails to show the development of the middle fracture and the successive differentiation of the deforming rock mass in two distinct blocks.

DISCONTINUUM MODELS

We used UDEC to test different geometrical configurations in an attempt to reproduce the movements induced by progressive deglaciation along discontinuity sets and the orientation and magnitude of displacement vectors documented by TLS. A series of sensitivity models were developed to determine the equivalent cohesion and friction angle. During deglaciation, the rock slope becomes unstable when the peak friction angle has a value of 30–35° and an apparent cohesion of 0.1–0.5 MPa. Early displacements along pre-existing fractures take place simultaneously in the upper and lower portion of the slope, which implies that the discontinuity sets at that time are not fully persistent and that progressive fracturing of rock bridges occurred during deglaciation.

We developed a second set of UDEC models to better define the geometry of the landslide and the failure mechanism. In particular, we investigated the orientation and the termination of the middle fracture, and the shape and location of the basal failure surface. In the first model, the middle fracture terminated on the basal failure surface (Fig. 23.8a). Results indicate that this configuration causes the two blocks to rotate and translate simultaneously, with similar velocities. The velocity ratio between the rear and the frontal block is close to 0.8. These results are not in agreement with TLS data, indicating different displacement behavior of the two blocks and a velocity ratio of about 0.36. The second model (Fig. 23.8b) is characterized by a middle fracture dipping 85° into the slope, with the orientation of joint set J2 and terminating on the rear fracture. The frontal-to-rear velocity ratio and the orientation of the displacement vectors in this model are in agreement with TLS results.

The exact location of the basal failure surface is difficult to delineate because no borehole data are available and its location is independent of the observed displacement vectors. Nevertheless, TLS data indicate a cumulative total settlement of the rear block of more than 50 m. Models including a joint set that dips in the same direction as the failure surface indicate that most of the displacements occur 40–60 m below the intersection of the rear and middle fractures (Fig. 23.8c, d)

Larger displacements are obtained when the basal failure surface is closer to the surface. Based on these observations, we suggest that the basal failure surface may be located 50–60 m below the intersection of the rear and the middle fractures.

We also analyzed the failure surface geometry using UDEC and assuming both planar and stepped failure surfaces. A stepped failure surface was modeled by introducing discontinuity sets J1 and J2 (mean spacing of 10 m for both sets) in the lower portion of the slope. Results indicate that movements take place along different planes and at different heights, creating a thick and dislocated, stepped-like failure surface (Fig. 23.8c, d). The directions of the displacement vectors in the area

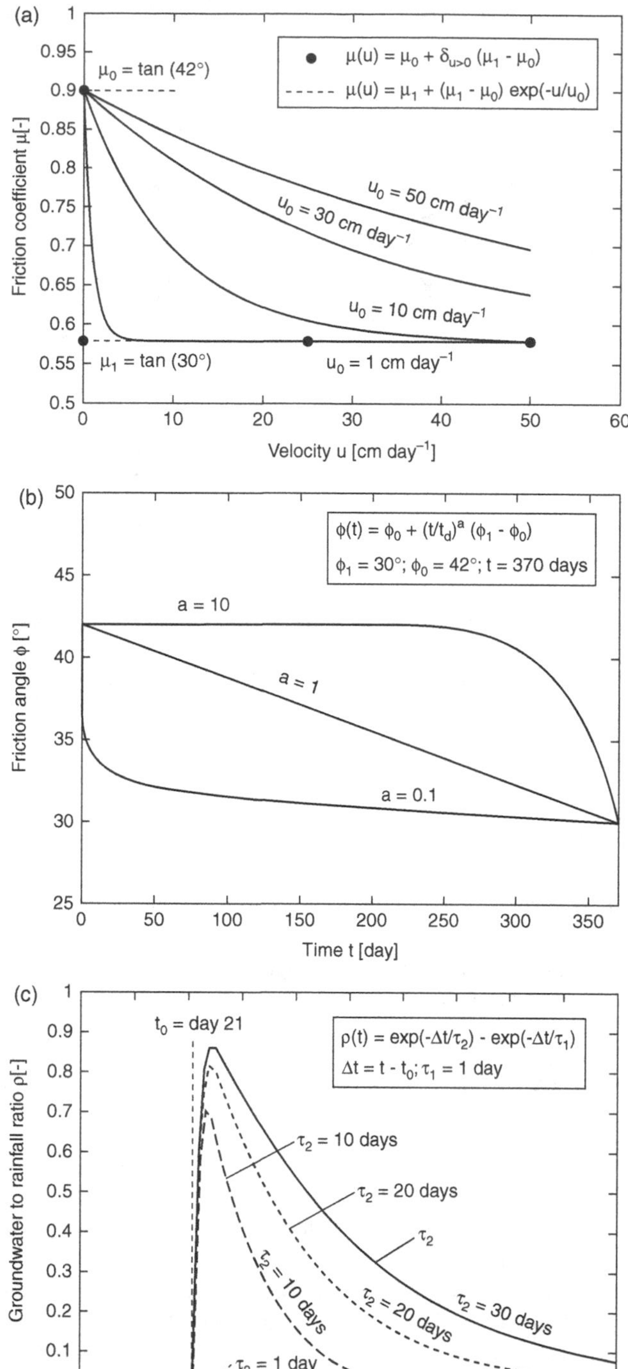

Fig. 23.6. Models used in the back analysis of the Eiger rockslide: (a) friction laws (Eq. 23.1); (b) strength degradation law (Eq. 23.2); and (c) infiltration law (Eq. 23.4).

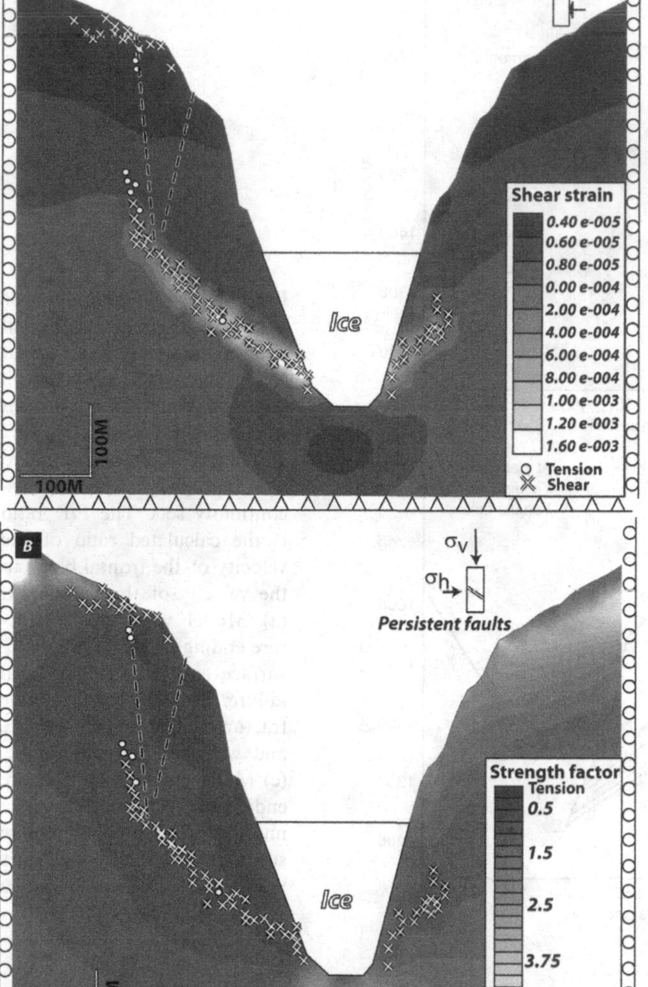

Fig. 23.7. Finite element modeling of the initiation processes of the Eiger rockslide. (A) Shear strain calculation for the last phase of deglaciation. A shearing zone formed at the bottom of the slope, and a tension zone developed higher on the slope, parallel to the valley. (B) Strength factor calculation (shear strength divided by the shear stress) obtained by introducing weakness planes corresponding to J1. The modeled unstable area matches the documented Eiger rockslide area.

of the rear block are in good agreement with the TLS results. However, the directions of the displacement vectors for the front block, particularly those close to the middle fracture, have a steeper plunge than the vectors indicated by TLS observations. The modeled front/rear velocity ratio is also higher (0.65) than the ratio derived from TLS observations. These results indicate that the basal failure surface may follow a number of planes of weakness parallel to joint set J1, with no or little influence exerted by the steep discontinuity set J2.

23.7.2 DYNAMIC EVOLUTION

We performed two types of back analysis as described in Section 23.6.3. The model of time-dependent strength degradation without pore pressure reproduces the early decrease in velocity with no effect from strength degradation or pore water pressure (Fig. 23.9a, b; Table 23.4). The velocity decrease is mainly due to the loss of mass from the rear block. If the volume of the rear block is kept constant, no deceleration is observed in the first 150 days of the model run. However, if neither strength degradation nor pore water is invoked, the acceleration at the end of the modeling period cannot occur.

For ϕ_0 values of 42°, 38°, and 35°, strength degradation results in a late increase in velocity (Table 23.5 and Fig. 23.9). If only pore water pressure is introduced (Table 23.5), the calculated velocities fit the early measured velocities, but not the late-stage ones, and the late acceleration, which should occur at around day 320, is not simulated by the model for any value of ϕ_0 (Fig. 23.9c, d). These results suggest that strength degradation can better explain the late acceleration of the blocks than an increase in pore water pressure. Opening of fractures, loss of contact, and shearing of first-order roughness elements likely explain strength degradation along the failure surface.

23.8 DISCUSSION

The geomorphic situation at the Eiger is comparable to that of the 1991 Randa rockslide. Both rockslides occurred at spurs on high steep cliffs (Sartori *et al.*, 2003; Pedrazzini *et al.*, 2010). The spurs consist of rocks that are resistant to glacial erosion, but, as suggested by Sartori *et al.* (2003), they are subject to higher stresses induced by glacier flow than other locations. The stress field created by these conditions appears to promote failure along pre-existing structures.

The Eiger rockslide illustrates the role of rock fatigue in causing rock-slope failures (Jaboyedoff *et al.*, 2004; Stead *et al.*, 2006). Although this valley became deglaciated *ca.* 6000 years ago (Portmann, 1977), the spur did not fail until 2006. It remains unclear, however, why the spur failed in 2006 and what triggered the initiation of the movement that led to the final collapse. It is clear that debuttressing of the slope in the late nineteenth and twentieth centuries (Fig. 23.1c, d) brought the slope very close to the threshold of instability, because no exceptional precipitation or earthquakes occurred prior to the onset of movements in 2006.

The stress and strain distributions within the rock slope were favorable for instability (Fig. 23.7A), but the failure process would have occurred over a significantly longer period of time without the pre-existing weaknesses in the rock mass. Medium to highly persistent and moderate to widely spaced discontinuity sets define the size of the instability. The strain within the rock slope was accommodated by pre-existing discontinuities, notably set J2, the sliding surface discontinuity set J1, and set S0, which played a primary role in the lateral release and

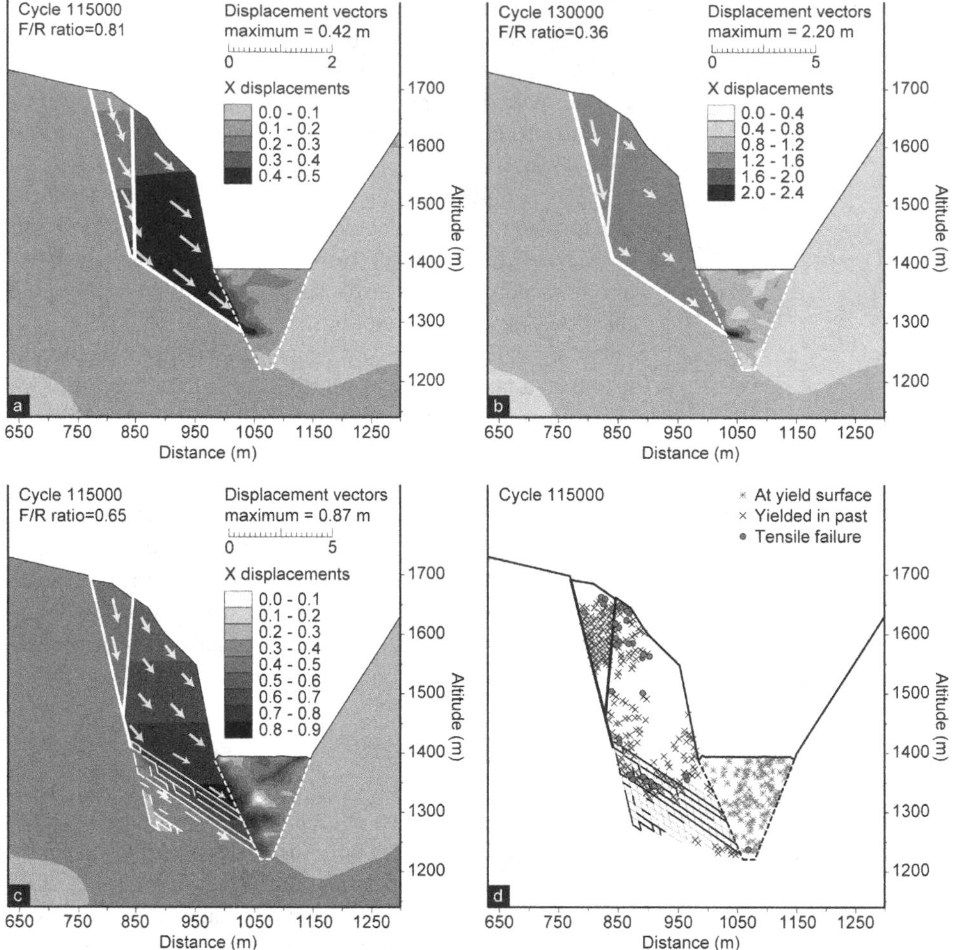

Fig. 23.8. Discontinuum models of progressive failure associated with glacier thinning and retreat (pre-failure conditions). For all models, white arrows represent calculated displacement vectors and white lines correspond to activated, fully persistent discontinuity sets. The F/R ratio is the calculated ratio of the velocity of the frontal block to the velocity of the rear block. (a) Model with middle fracture ending on the basal failure surface and with planar basal failure. (b) Model with middle fracture ending on rear fracture and with planar basal failure. (c) Model with middle fracture ending on rear fracture and a multi-step-path basal failure surface. (d) Plasticity indicators within the UDEC model for a stepped-like basal surface.

break-up and participated in the formation of the wedge S0/J1 (Oppikofer *et al.*, 2008), which controlled the sliding direction of the front block. The displacement direction of the subsiding block was controlled by the wedge J2/S0 (Oppikofer *et al.*, 2008) and is close to the dip direction of J2. The sliding directions of both blocks are close to the dip directions of sets J1 and J2, respectively; thus we felt justified in modeling the basal sliding surface and the back scarp as 2D planar surfaces.

The Mt. Eiger rockslide did not move significantly before July 2006, which implies that the discontinuity properties changed slowly, likely due to progressive breakage of rock bridges by localization of stress, which eventually created through-going continuous structures (Fig. 23.10A–C). When through-going structures were established, the back section of the rock mass behaved as a subsiding wedge that pushed the front block forward. The base of the subsiding wedge may have been located well above the main basal sliding surface. The latter was controlled by the wedge J1/S0, without a stepped sliding surface, as shown in Figures 23.4 and 23.8b.

Velocities were very high at the beginning, which is uncommon for a rockslide. The velocities then decreased, primarily due to the loss of mass of the rock blocks. The analytical model shows that the other main change is the reduction in the friction angle of the basal sliding surface. A gouge may have formed along the slip surface, making a strong decrease in ϕ possible (Fig. 23.9b). Alternatively, local rock disintegration changed the stress regime within the moving rock mass. The latter is supported by observations that the rock mass was highly fragmented and progressively collapsed (Fig. 23.1e).

23.9 CONCLUSIONS

Subsiding wedges provide the driving force for many rock-slope failures (e.g., Turtle Mountain; Agliardi *et al.*, 2001; Froese *et al.*, 2009, 2012, Chapter 25, this volume). The well-constrained geometry of the 2006 Mt. Eiger rockslide allowed us to test a dynamic model for this type of failure. We built a geologically realistic model using an optimization procedure and parameters based on simple physical laws. This approach may prove helpful in forecasting rock-slope movements and improving understanding of failure mechanisms (Zhang *et al.*, 2010).

The reduction in the mass of the rear block at the Eiger seems to have been sufficient to cause an initial decrease in displacement

Table 23.4. *Results of the analytical model based on strength degradation, using a strength degradation law, an exponential friction law, no hydrological law, and the Glen's glacier flow law.*

Decision variables	Peak friction angle = 42°	Peak friction angle = 38°	Peak friction angle = 35°	Peak friction angle = 30°
ϕ_1	26.23	19.53	10.19	15.01
a	8.77	4.23	5.56	10 (upper bound)
μ_1/μ_0	0.71	0.58	0.50 (lower bound)	1 (upper bound)
u_0	1.87	4.04	5.5080	0.93
n (initial Glen's law)	2.11	2.02	1.99	2.07
Function of error	0.176764 after 59 iterations and 492 function evaluations.	0.176764 after 59 iterations and 492 function evaluations.	0.188382 after 48 iterations and 369 function evaluations.	0.425383 after 18 iterations and 138 function evaluations.

Notes: Calibration parameters: final friction ratio – ϕ_1; strength degradation parameter – a; static to limit friction ratio – μ_1/μ_0; limit velocity for exponential friction law – u_0; and Glen's flow law exponent – n. Collapse ratio (rear/front) = 0.4/0.7. Glen's flow law parameters: B = 22.2, Q = 28,000 cal/mol/K, T = 273°K.

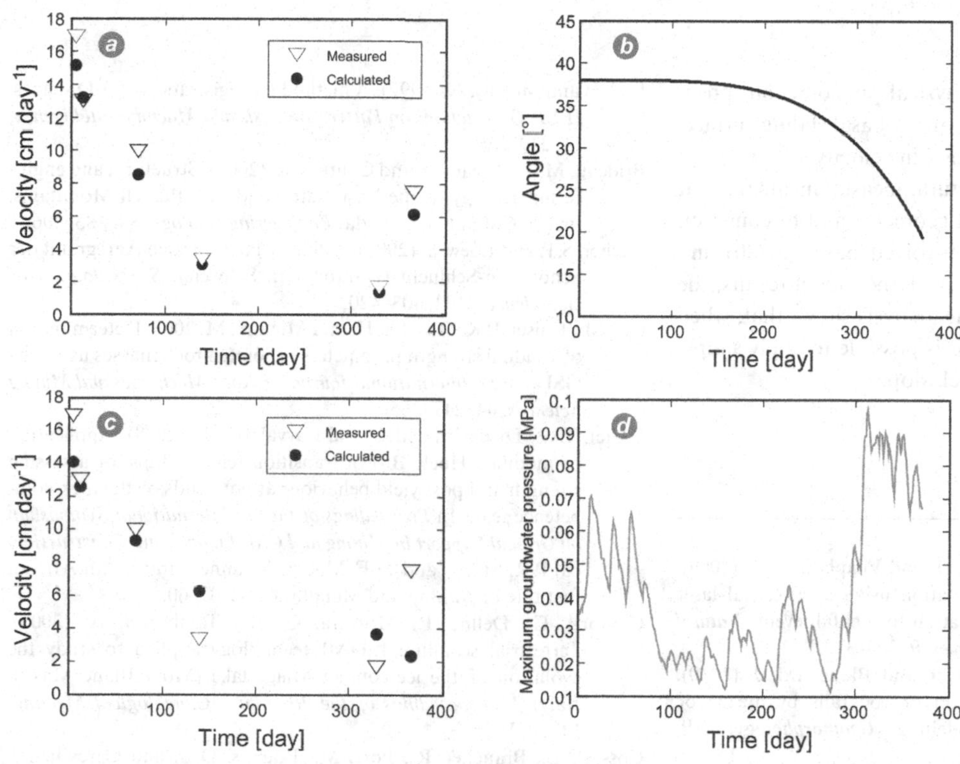

Fig. 23.9. Numerical results obtained using an initial peak friction of 38°. Velocity histories with (a) strength degradation, (b) friction angle, (c) non-zero pore pressure and no strength degradation, and (d) pore pressure time-dependence.

Table 23.5. *Results of the analytical model based on pore water pressure, no strength degradation, the exponential friction law, and hydrological law (with porosity).*

Decision variable	Peak friction angle = 42°	Peak friction angle = 38°	Peak friction angle = 35°
Porosity	0.1	0.02	0.02
μ_1/μ_0	0.75	0.26	0.76
u_0	1.00	0.5949	0.1 (lower bound)
τ_{t1}	2.76	1.46	0.90
$\Delta\tau_t$	10.42	16.23	1.03

Notes: Calibration parameters: rock porosity; static to limit friction ratio (μ_1/μ_0); limit velocity for exponential friction law (u_0); filling time constant (τ_{t1}); and dewatering and filling time constants difference ($\Delta\tau_t$). Collapse ratio (rear/front) = 0.4/0.7 and the Glen's flow law parameters: B = 22.2, Q = 28,000 cal/mol/K, T = 273°K.

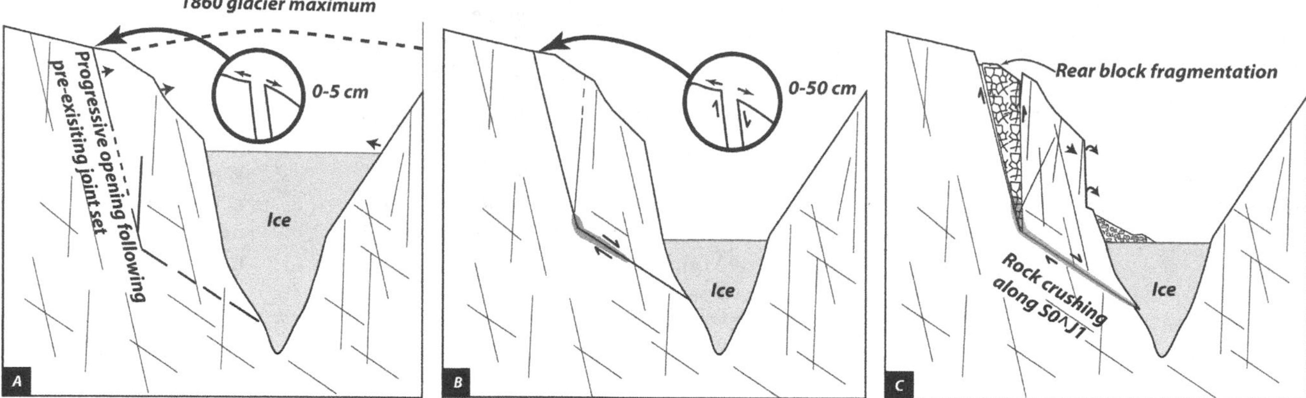

Fig. 23.10. Progressive evolution of the Eiger rockslide. (A) Opening of the rear fracture before July 2006. (B) Progressive increase in the rate of displacement associated with the establishment of a basal failure surface. (C) Fragmentation of the rear block, accompanied by a decrease in velocity.

velocity. However, an additional physical phenomenon, probably a decrease in the friction angle of the basal sliding surface, is required to explain the later increase in velocity.

The displacement data and structural measurements made it possible to obtain a reliable model geometry and to constrain the mechanism of sliding, which involved basal (J1/S0) and rear (J2/S0) surfaces. The agreement of the model results, the structural analysis, and the dynamic analysis shows that, when high-resolution data are available, it is possible to apply simple geomechanical models to failing rock slopes.

REFERENCES

Abellán, A., Jaboyedoff, M., Oppikofer, T. and Vilaplana, J. M. (2009). Detection of millimetric deformation using a terrestrial laser scanner: Experiment and application to rockfall event. *Natural Hazards and Earth System Sciences*, 9, 365–372.

Abellán, A., Vilaplana, J. M., Calvet, J. and Blanchard, J. (2010). Detection and spatial prediction of rockfalls by means of terrestrial laser scanning modelling. *Geomorphology*, 119, 162–171.

Agliardi, F., Crosta, G. and Zanchi, A. (2001). Structural constraints on deep-seated slope deformation kinematics. *Engineering Geology*, 59, 83–102.

Alonso, E. E. and Pinyol, N. M. (2010). Criteria for rapid sliding. I. A review of Vaiont case. *Engineering Geology*, 114, 198–210.

Ambrosi, C. and Crosta, G. B. (2006). Large sackung along major tectonic features in the central Alps. *Engineering Geology*, 83, 183–200.

(2011). Valley shape influence on deformation mechanisms of rock slopes. In *Slope Tectonics*, ed. M. Jaboyedoff. Geological Society of London, Special Publication 351, pp. 215–233.

Ballantyne, C. K. (2002). Paraglacial geomorphology. *Quaternary Science Reviews*, 21, 1935–2017.

Baltsavias, E. P. (1999). Airbone laser scanning: Basic relations and formulas. *ISPRS Journal of Photogrammetry & Remote Sensing*, 54, 199–214.

Bardet, J.-P. and Kapuskar, M. M. (1989). A simplex analysis of slope stability. *Computers and Geotechnics*, 8, 329–348.

Besl, P. and McKay, N. (1992). A method for registration of 3-D shapes. *IEEE Transactions on Pattern Analysis and Machine Intelligence*, 14, 239–256.

Brideau, M.-A., Stead, D. and Couture, R. (2006). Structural and engineering geology of the East Gate Landslide, Purcell Mountains, British Columbia, Canada. *Engineering Geology*, 84, 183–206.

Bucher, S. P. and Loew, S. (2009). Talklüfte im Zentralen Aaregranit der Schöllenen-Schlucht (Kanton Uri, Schweiz), *Swiss Journal of Geosciences*, 102, 403–420.

Cai, M., Kaiser, P. K., Tasaka, H. and Minami, M. 2007. Determination of residual strength parameters of jointed rock masses using the GSI system. *International Journal of Rock Mechanics and Mining Sciences*, 44, 247–265.

Carter, T. G., Diederichs, M. S. and Carvalho, J. L. (2008). Application of modified Hoek–Brown transition relationships for assessing strength and post yield behaviour at both ends of the rock competence scale. In *Proceedings of the 6th International Symposium on Ground Support in Mining and Civil Engineering Construction*, ed. P. R. Stacey and D. F. Malan. Johannesburg: South African Institute of Mining and Metallurgy, pp. 37–60.

Conforti, C., Deline, P., Mortara, G. and Tamburini, A. (2005). Terrestrial scanning LiDAR technology applied to study the evolution of the ice-contact Miage lake (Mont Blanc Massif, Italy). In *Proceedings of the 9th Alpine Glaciological Meeting*, Milan, Italy.

Cossart, E., Braucher, R., Fort, M., Bourlès, D. L. and Carcaillet, J. (2008). Slope instability in relation to glacial debuttressing in alpine areas (Upper Durance catchment, southeastern France): Evidence from field data and ^{10}Be cosmic ray exposure ages. *Geomorphology*, 95, 3–26.

Coulthard, M. A. (1979). *Back Analysis of Observed Spoil Failures Using a Two-Wedge Method*. Technical Report 83, CSIRO Division of Applied Geomechanics, Melbourne, Australia.

Diederichs, M. S. (2003). Rock fracture and collapse under low confinement conditions. *Rock Mechanics and Rock Engineering*, 36, 339–381.

Eberhardt, E., Stead, D. and Coggan, J. S. (2004). Numerical analysis of initiation and progressive failure in natural rock slopes: The 1991 Randa rockslide. *International Journal of Rock Mechanics and Mining Sciences*, 41, 69–87.

Fischer, L., Amann, F., Moore, J. and Huggel, C. (2010). Assessment of periglacial slope stability for the 1988 Tschierva rock avalanche (Piz Morteratsch, Switzerland). *Engineering Geology*, 116, 32–43.

Florineth, D. and Schlüchter, C. (1998). Reconstructing the Last Glacial Maximum (LGM) ice surface geometry and flowlines in the central Swiss Alps. *Eclogae Geologicae Helvetiae* 91, 391–407.

Froese, C., Moreno, F., Jaboyedoff, M. and Cruden, D.M. (2009). 25 years of movement monitoring on the South Peak of Turtle Mountain: Understanding the hazard. *Canadian Geotechnical Journal*, 46, 256–269.

Froese, C.R., Charrière, M., Humair, F., Jaboyedoff, M. and Pedrazzini, A. (2012). Characterization and management of rockslide hazard at Turtle Mountain, Alberta, Canada. In *Landslides: Types, Mechanisms and Modeling*, ed. J.J. Clague and D. Stead. Cambridge, UK: Cambridge University Press, pp. 310–322.

Gagnon, R.E. and Gammon, P.H. (1995). Triaxial experiments on iceberg and glacier ice. *Journal of Glaciology*, 41, 528–540.

Ghirotti, M., Martin, S. and Genovois, R. (2011). The Celentino deep-seated gravitational slope deformation (DSGSD): Structural and geomechanical analyses (Peio Valley, NE Italy). In *Slope Tectonics*, ed. M. Jaboyedoff. Geological Society of London, Special Publications 351, pp. 235–252.

Gischig, V., Amann, F., Moore, J.R. *et al.* (2011). Composite rock slope kinematics at the current Randa instability, Switzerland, based on remote sensing and numerical modelling. *Engineering Geology*, 118, 37–53.

Glen, J.W. (1955). The creep of polycrystalline ice. *Proceedings of the Royal Society of London. Series A, Mathematical and Physical Sciences*, 228, 519–538.

Günzler-Seiffert, H. and Wyss, R. (1938). *Swiss Geological Atlas, Sheet 396 Grindelwald (Atlas Sheet 13)*. Swisstopo, scale 1:25,000.

Hajiabdolmajid, V., Kaiser, P.K. and Martin, C.D. (2002). Modelling brittle failure of rock. *International Journal of Rock Mechanics and Mining Sciences*, 39, 731–741.

Hoek, E., Carranza-Torres, C.T. and Corkum, B. (2002). Hoek–Brown failure criterion: 2002 edition. In *Proceedings of the 5th North American Rock Mechanics Society*, Toronto, ON, pp. 267–273.

InnovMetric (2011). *PolyWorks: 3D Scanner Software – 3D Scanning Software – 3D Digitizers* [available at www.innovmetric.com/polyworks/3D-scanners/home.aspx?lang=en, accessed July 6, 2011].

International Society of Rock Mechanics (ISRM) (1978). Suggested methods for quantitative description of discontinuities in rock masses. *International Journal of Rock Mechanics and Mining Sciences & Geomechanics Abstracts*, 15, 319–358.

Itasca (2004). *UDEC: Universal Distinct Element Code (Version 4.0)*. Minneapolis, MN: Itasca Consulting Group.

Jaboyedoff, M., Baillifard, F., Bardou, E. and Girod, F. (2004). Weathering, cycles of saturation–unsaturation, and strain effects as principal processes for rock mass destabilization. *Quarterly Journal of Engineering Geology and Hydrogeology*, 37, 95–103.

Jaboyedoff, M., Metzger, R., Oppikofer, T. *et al.* (2007). New insight techniques to analyze rock-slope relief using DEM and 3D-imaging cloud points: COLTOP-3D software. In *Rock Mechanics: Meeting Society's Challenges and Demands*, ed. E. Eberhardt, D. Stead and T. Morrison. London: Taylor & Francis, pp. 61–68.

Jaboyedoff, M., Couture, R. and Locat, P. (2009). Structural analysis of Turtle Mountain (Alberta) using digital elevation model: Toward a progressive failure. *Geomorphology*, 103, 5–16.

Jennings, J.E. (1970). A mathematical theory for the calculation of the stability of open cast mines. In *Planning Open Pit Mines*, ed. P.W.J. van Rensburg. Johannesburg: South African Institute of Mining and Metallurgy, pp. 87–102.

Kastrup, U., Zoback, M.L., Deichmann, N. *et al.* (2004). Stress field variations in the Swiss Alps and the northern Alpine foreland derived from inversion of fault plane solutions. *Journal of Geophysical Research*, 109, B01402, doi:10.1029/2003JB002550.

Lichti, D.D., Gordon, S.J. and Stewart, M.P. (2002). Ground-based laser scanners: Operations, systems and applications. *Geomatica*, 56, 21–33.

Liniger, M. (2006). Die Herausforderung der Gefahrenprognose bei Massenbewegungen: Rutsch-und Sturzprozesse. *Bulletin for Applied Geology*, 11(2), 75–88.

Mathworks (2009). *Matlab 9*. Natick, MA: The MathWorks Inc.

Messerli, B., Messerli, P., Pfister, C. and Zumbühl, H.J. (1978). Fluctuations of climate and glaciers in the Bernese Oberland, Switzerland and their geological significance, 1600 to 1975. *Arctic and Alpine Research*, 10, 247–260.

Monserrat, O. and Crosetto, M. (2008). Deformation measurement using terrestrial laser scanning data and least squares 3D surface matching. *ISPRS Journal of Photogrammetry and Remote Sensing*, 63, 142–154.

Norrish, N.I. and Wyllie, D.C. (1996). Rock slope stability analysis. In *Landslides: Investigation and Mitigation*, ed. A.K. Turner and R.L. Schuster. Washington, DC: Transportation Research Board, Special Report 247, pp. 391–425.

OcCC (2007). *Climate Change and Switzerland in 2050: Expected Impacts on Environment, Society and Economy*. Bern, Switzerland: OcCC/ProClim.

Oppikofer, T., Jaboyedoff, M. and Keusen, H.R. (2008). Collapse at the eastern Eiger flank in the Swiss Alps. *Nature Geoscience*, 1, 531–535.

Oppikofer, T., Jaboyedoff, M., Blikra, L., Derron, M.-H. and Metzger, R. (2009). Characterization and monitoring of the Åknes rockslide using terrestrial laser scanning. *Natural Hazards and Earth System Sciences*, 9, 1003–1019.

Pedrazzini, A., Jaboyedoff, M., Derron, M.-H., Abellán, A. and Vega Orozco, C. (2010). Reinterpretation of displacements and failure mechanisms of the upper portion of Randa rock slide. In *Proceedings of the 63rd Canadian Geotechnical Conference & 6th Canadian Permafrost Conference*, Calgary, AB, pp. 913–921.

Portmann, J.-P. (1977). Variations glaciaires historiques et préhistoriques dans les Alpes Suisses. *Revue du Club Alpin Suisse*, 4, 145–172.

Rabatel, A., Deline, P., Jaillet, S. and Ravanel, L. (2008). Rock falls in high-alpine rock walls quantified by terrestrial LIDAR measurements: A case study in the Mont Blanc area. *Geophysical Research Letters*, 35, L10502, doi:10.1029/2008GL033424.

Rocscience (2010). *Phase² Version 7.0*. Toronto, ON: Rocscience Inc.

Sarma, S.K. (1979). Stability analysis of embankments and slopes. *Journal of the Geotechnical Engineering Division*, 105, 1511–1524.

Sartori, M., Baillifard, F., Jaboyedoff, M. and Rouiller, J.-D. (2003). Kinematics of the 1991 Randa rockslides (Valais, Switzerland). *Natural Hazards and Earth System Sciences*, 3, 423–433.

Schulson, E.M. (1999). The structure and mechanical behaviour of ice. *Journal of the Minerals, Metals and Materials Society*, 51(2), 21–27.

Spinnler, G. (2001). *Conception des Machines: Principes et Applications*. Lausanne, Switzerland: Presses Polytechniques et Universitaires Romandes.

Stamatopoulos, C. and Petridis, P. (2006). Back analysis of the Lower San Fernando Dam slide using a multi-block model. In *Geohazards*, ed. F. Nadim, R. Pöttler, H. Einstein, H. Klapperich and S. Kramer. Lillehammer, Norway: Engineering Conferences International Symposium Series P0.

Stead, D., Eberhardt, E. and Coggan, J.S. (2006). Developments in the characterization of complex rock slope deformation and failure using numerical modelling techniques. *Engineering Geology*, 83, 217–235.

Sturzenegger, M. and Stead, D. (2009). Quantifying discontinuity orientation and persistence on high mountain rock slopes and large landslides using terrestrial remote sensing techniques. *Natural Hazards and Earth System Sciences*, 9, 267–287.

Sturzenegger, M., Yan, M., Stead, D. and Elmo, D. (2007). Application and limitations of ground-based laser scanning in rock slope characterization. In *Rock Mechanics: Meeting Society's Challenges and Demands*, ed. E. Eberhardt, D. Stead and T. Morrison. London: Taylor & Francis, pp. 29–36.

Sultan, H. A. and Seed, H. B. (1967). Stability of sloping core earth dams. *Journal of the American Society of Civil Engineers*, 93(4), 45–47.

Teza, G., Galgaro, A., Zaltron, N. and Genevois, R. (2007). Terrestrial laser scanner to detect landslide displacement fields: A new approach. *International Journal of Remote Sensing*, 28, 3425–3446.

Werder, M. A., Bauder, A., Funk, M. and Keusen, H.-R. (2010). Hazard assessment investigations in connection with the formation of a lake on the tongue of Unterer Grindelwaldgletscher, Bernese Alps, Switzerland. *Natural Hazards and Earth System Sciences*, 10, 227–237.

Zhang, J., Tang, W. H. and Zhang, L. M. (2010). Efficient probabilistic back-analysis of slope stability model parameters. *Journal of Geotechnical and Geoenvironmental Engineering*, 136, 99–109.

24 Randa: Kinematics and driving mechanisms of a large complex rockslide

SIMON LOEW, VALENTIN GISCHIG, HEIKE WILLENBERG, ANDREA ALPIGER, AND JEFFREY R. MOORE

ABSTRACT

In this chapter we summarize integrated investigations carried out at the Randa rock-slope instability in Matter Valley, Switzerland, between 2000 and 2010. We present a 3D geometric and structural model of the current instability, which consists of 5–6 million m^3 of crystalline rock. We also document the complex kinematic behavior and discuss the driving factors for observed slope movements. We show that both the May 1991 failure and the current instability are bounded laterally by the same large-scale fault, and at the base by a planar or stepped rupture surface daylighting at the contact with the Randa orthogneiss. The spatial distribution of current displacements indicates toppling in the upper section of the instability between 2200 and 2400 m asl and sliding in the lower area between 1900 and 2200 m asl. Continuous displacement time series from the surface and deep boreholes show increasing deformation rates when ground surface temperatures decrease in fall and a decrease after snowmelt in spring as the rock warms. We have not detected displacement signatures related to heavy rainstorms or snowmelt. Mapping of the locations of springs and recordings of borehole pore water pressure demonstrate locally perched groundwater in open fractures, and a low regional groundwater table located at or below the basal rupture surface. Numerical modeling results support the hypothesis that thermo-mechanical coupled deformation resulting from annual temperature changes and critically stressed fractures in a complex topography is the primary mechanism driving deep-seated displacements at Randa.

24.1 OVERVIEW OF INVESTIGATIONS AT THE RANDA IN-SITU LABORATORY

Catastrophic release of fractured rock masses generally follows accelerated displacements and is typically not associated with an obvious trigger. In addition, the time of failure is generally challenging to predict. Without an external trigger, an unstable rock slope must fail progressively due to cumulative long-term damage that reduces the strength of the rock mass; this damage is however difficult to observe (Eberhardt *et al.*, 2004). Only in a few cases have preceding displacements and environmental factors been measured in sufficient detail to capture the main causes and mechanisms of large rock-slope failures (Schindler *et al.*, 1993; Krähenbühl, 2004, 2006).

To improve our understanding of the kinematics and mechanisms that drive failure of large, brittle, jointed rock-slope instabilities without a fully developed pre-existing failure surface, the Chair of Engineering Geology at ETH Zurich has created a long-term research facility in the Swiss Alps, called the Randa In-Situ Rockslide Laboratory. The test site was established at an elevation of 2300–2400 m asl (Figs. 24.1 and 24.2) within a slowly moving, crystalline rock mass located behind the head scarp of a 30 million m^3 failure that occurred in 1991 close to the village of Randa. The currently unstable rock mass under investigation has a volume of about 5–6 million m^3 and is moving toward the 1991 scarp (displacement trend 125–147°, plunge 25–35°) at rates up to 30 mm per year (Gischig *et al.*, 2011a). The first elements of this laboratory were installed in 2001 and included three deep boreholes: SB120, which is 120 m in length; and SB50N and SB50S, each of which are 50 m in length.

Understanding the kinematics of unstable rock masses requires a detailed and reliable description of the boundary of the instability and the internal 3D geologic structure. Therefore, the first phase of investigations carried out at Randa between 2001 and 2005 focused on surface geological mapping (Willenberg *et al.*, 2008a), geophysical borehole logging (Willenberg *et al.* 2008a), a 1D borehole radar reflection survey (Spillmann *et al.*, 2007a), and 3D surface georadar (Heincke *et al.*, 2005, 2006a)

Landslides: Types, Mechanisms and Modeling, ed. John J. Clague and Douglas Stead. Published by Cambridge University Press.

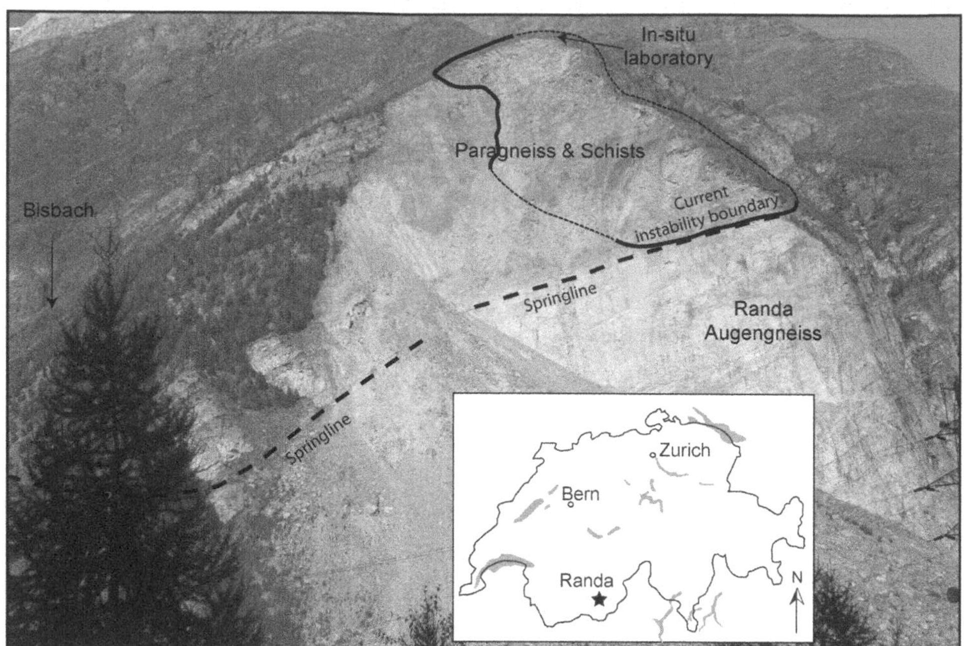

Fig. 24.1. Location of the study area (inset), view of the 1991 scarp, geologic and hydrogeologic framework, current Randa instability, and relative location of the Randa in-situ laboratory.

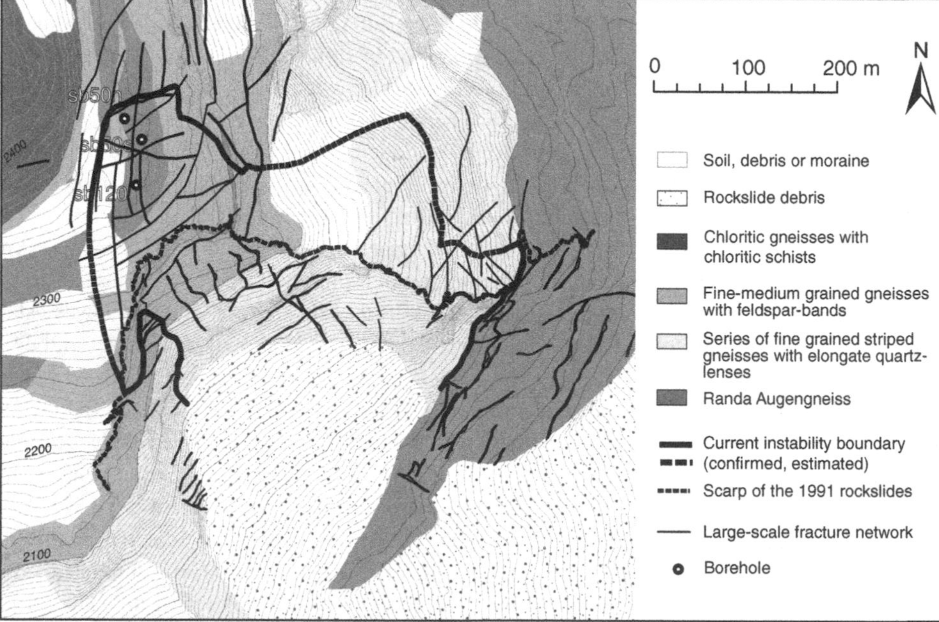

Fig. 24.2. Geological map of the Randa slope instability, including traces of faults and fracture zones and boundary of the unstable rock mass. Also shown are the locations of the three deep boreholes. (Modified from Willenberg *et al.*, 2008a.)

and seismic refraction surveys (Heincke *et al.*, 2006b). These structural investigations were later complemented with a systematic study of fractures visible on the steep, 800-m-high cliff forming the scarp of the 1991 failures (Gischig *et al.*, 2011a). These investigations were facilitated by analysis of high-resolution digital surface models (DSM) generated from helicopter-based LIDAR (point density of 1–3 points per m^2) and oblique undistorted optical images (Eisenbeiss, 2009).

The kinematics of the current instability were assessed through systematic analysis of 3D displacements measured at all accessible active fractures, both at the surface and in deep boreholes (Willenberg *et al.*, 2008b). In addition, geodetic networks at two different scales, as well as ground-based radar interferometry measurements (Gischig *et al.*, 2009), allowed analysis of the displacement field over the entire exposed surface of the instability. The final kinematic interpretation was supported by stereographic analysis and numerical modeling.

The driving mechanisms responsible for the displacements were evaluated based on hydrogeologic (Willenberg *et al.*, 2008b; Alpiger, 2010), microseismic (Spillmann *et al.*,

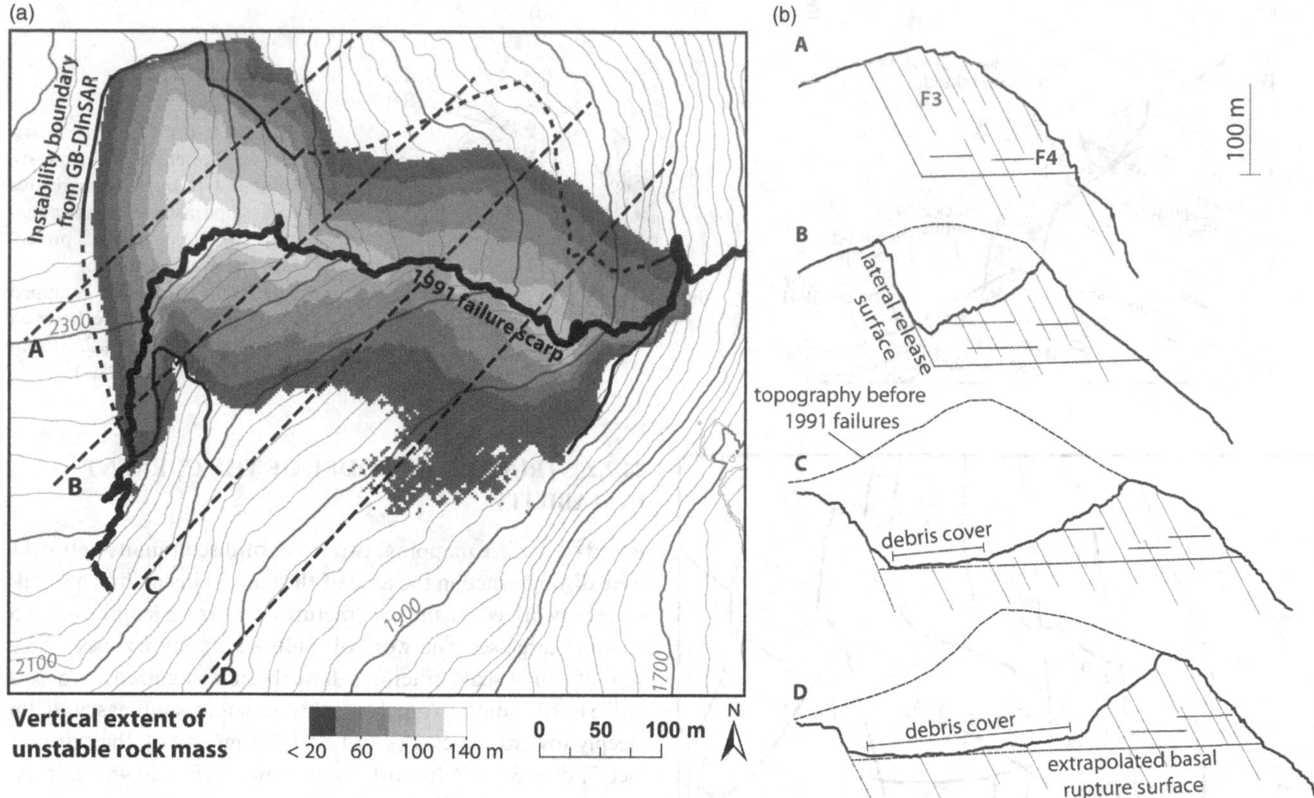

Fig. 24.3. (a) Map of vertical thickness of the current Randa instability, and (b) vertical cross-sections showing original topography, fracture sets forming basal and lateral release surfaces, geometry of the May 1991 slope failure, and basal rupture surface of the current instability. (From Gischig *et al.*, 2011a.)

2007b), continuous deformation (Moore *et al.*, 2010), climatic, and thermal measurements, and were supported by coupled thermo-mechanical numerical modeling (Gischig *et al.*, 2011b, 2011c).

In this chapter, we present selected results from previous and ongoing investigations critical for understanding the displacement mechanisms of the large, complex rock-slope instability at Randa. The results presented here are unique in terms of the comprehensive nature of the structural and kinematic investigations, and lead to conclusive results implicating thermo-mechanical forcing as the primary mechanism driving the current displacements at Randa.

24.2 GEOLOGICAL AND STRUCTURAL MODEL

24.2.1 GEOLOGIC SETTING AND 1991 ROCKSLIDES

The north–south-trending Matter Valley is one of the deepest glacially eroded valleys in the Alps, with an elevation difference at Randa of up to 3000 m over a horizontal distance of only 5000 m. Above the village of Randa, on the western wall of the Matter Valley, two large catastrophic rock-slope failures of approximately 23 and 7 million m³ took place in April and May of 1991. The Randa in-situ laboratory was subsequently constructed behind the head scarp of the May 1991 failure within a newly observed retrogressive instability. As shown in Figures 24.2 and 24.3, the 1991 rockslides, as well as the current instability, are located on a ridge trending perpendicular to the main valley axis. The ridge is dissected by several faults dipping steeply toward the southeast, east, and northwest, and cross-cutting relatively massive orthogneiss in the lower section (the so-called Randa augengneiss) and heterogeneous paragneisses with minor schists in the upper regions above 1900 m asl (Figs. 24.2 and 24.5). The foliation, related brittle-ductile shear zones, and lithological boundaries dip 20–30° to the west-southwest, obliquely into the slope (Fig. 24.4).

The 1991 rockslides were both multiphase events lasting several hours. The first release in April was preceded by rock bursts at the toe of the slope, followed by translational sliding on a persistent northeast-dipping basal fault and non-persistent southeast-dipping fractures (Wagner, 1991; Sartori *et al.*, 2003). The second release in May began with dynamic rock bursts and water jets at the foot of the failing rock mass, and rapid opening of fractures propagating in northeast–southwest directions in the head scarp area (Schindler *et al.*, 1993). The upper part of the May 1991 release area is a large asymmetric wedge located at the center of the original ridge, with a steep release

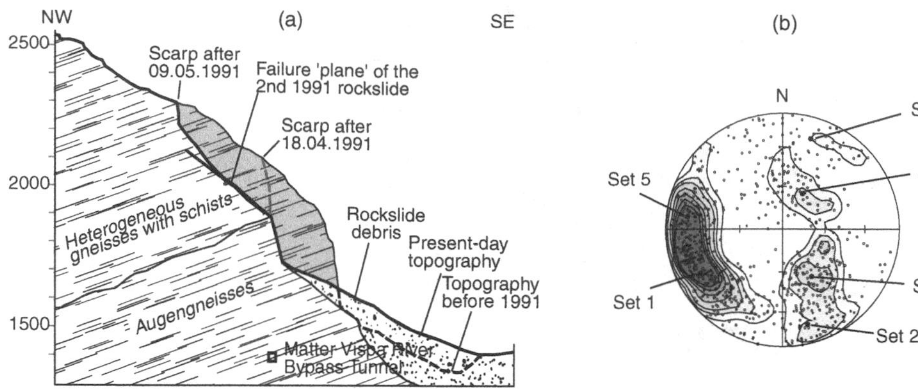

Fig. 24.4. (a) Geologic cross-section oriented northwest–southeast through retrogressive rock-slope instabilities at Randa, and (b) stereographic projection (lower hemisphere) of small-scale fractures mapped at the ground surface between 1900 and 2380 m asl. (From Willenberg *et al.*, 2008a.)

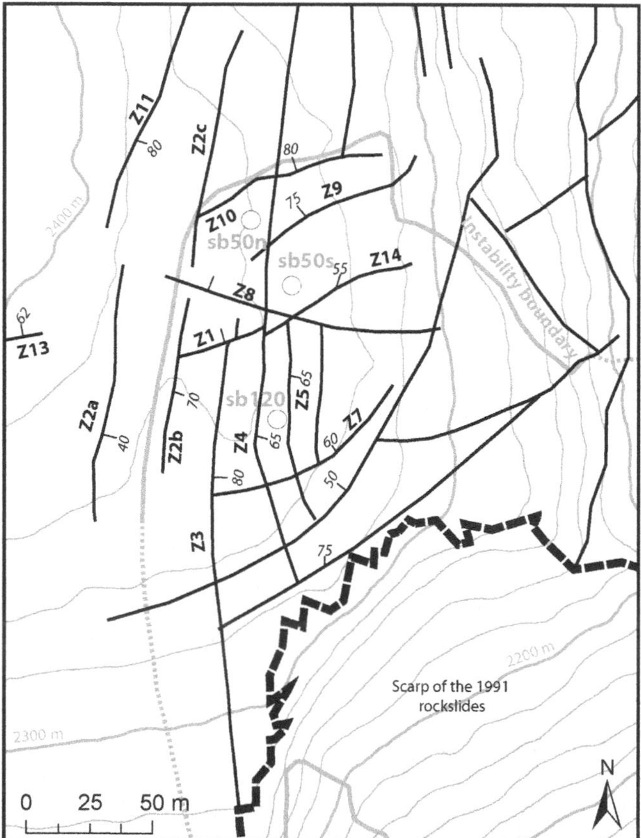

Fig. 24.5. Map of faults and fracture zones of the F2 and F3 sets in and around the unstable rock mass above the 1991 rockslide scarp.

surface in the west formed by a large planar fault dipping 70° E and a stepped rupture plane in the north (Fig. 24.3). A basal sliding surface dipping *ca.* 40° toward 135° developed in the lower part of the May 1991 release area. This surface becomes less constrained by the fault to the west as elevation decreases (Gischig *et al.*, 2011a). Along a northwest–southeast-striking section, the May 1991 release can be characterized as (1) planar sliding at the toe, (2) stepped-planar sliding and failed rock bridges in the central portion, and (3) tensile failure near the sub-vertical head scarp.

24.2.2 STRUCTURAL MODEL OF THE CURRENT INSTABILITY

Based on surface mapping, two scales of discontinuity networks were distinguished in the area of the current instability: a small-scale network consisting of fractures with trace lengths up to 5 m, and a large-scale network of faults and fracture zones. Three sets of small-scale fractures have been recognized: two sets strike north–south to north-northwest–south-southeast and dip steeply toward the east (sets 1 and 5, Fig. 24.4b); the third set (set 7) dips 30–60° toward the northwest (Fig 24.4b), normal to the measured direction of current surface displacements. As seen in Figure 24.4b, no fracture set in the accessible, mapped part of the instability (mainly above 2300 m asl) dips in the direction of surface displacements, that is to the southeast. The calculated mean trace length of 1–2 m suggests limited persistence of these small-scale fractures (Willenberg *et al.*, 2008a).

The large-scale fracture network (trace lengths greater than 10 m) includes fracture zones, brittle faults, and brittle-ductile shear zones with specific architecture and spatial orientation. The F1 set comprises mostly brittle-ductile shear zones oriented parallel to the foliation and dipping at 20–30° to west-southwest. The F2 set includes brittle faults, brittle-ductile shear zones, and fracture zones dipping steeply toward the north or northwest (Z1, Z6–10, Z12–13; see Fig. 24.5) and coinciding with open cracks at the ground surface. F2 faults show clear indications of shear (striations), but no fine-grained or clayey gouge. Anastomosing fractures dissect the shear zone into small lenticular bodies. F3 discontinuities strike north–south and dip steeply or moderately toward the east (Z2–Z5, Z11; see Fig. 24.5). Important additional structural information for the lower, inaccessible parts of the instability was gained from helicopter-based LIDAR and photogrammetry data acquired an oblique view angle. This orientation provided optimal observations of fracture patterns and the properties of the May 1991 release surface (Gischig *et al.*, 2011a). In contrast to the structures mapped at higher elevations, where no faults or fractures dip out of the slope at angles favorable for sliding, such a set (called F4) can be identified in the 1991 head scarp. F4 discontinuities in the paragneisses dip 45° to south-southeast and form stepped shear surfaces with a trace length of more than 200 m.

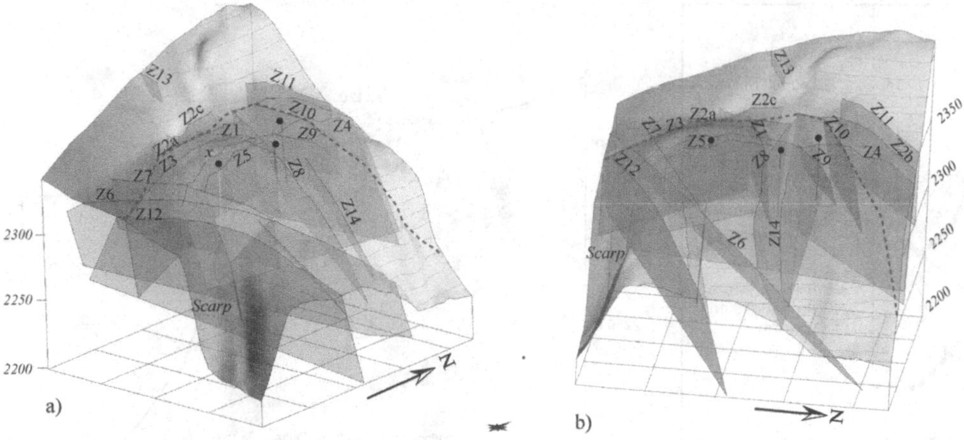

Fig. 24.6. Oblique views of a 3D model of the fault pattern in the upper part of the instability, derived from integrated geological mapping and borehole and surface radar investigations. (From Willenberg *et al.*, 2008a.)

Useful subsurface information on the large-scale fracture network was obtained from borehole optical televiewer logs and single-hole georadar reflection measurements conducted in the three boreholes (Spillmann *et al.*, 2007a). To compensate for the non-uniqueness associated with the rotational symmetry of single-hole radar data, Spillmann *et al.* (2007a) attempted to match the key parameters of all reflections with those of the faults and fracture zones observed at the surface or within boreholes. Using this approach, the sources of most radar reflections could be identified. These reflections were likely generated at faults and fracture zones Z2, Z3, Z4, Z6, Z7, Z8, Z9, and Z10, and could be correlated with larger structures seen in optical televiewer logs. The minimum feature lengths, interpreted from borehole radar data, ranged from 13 to 85 m.

Steeply dipping fracture zones are generally difficult to detect using traditional surface-based, ground-penetrating radar. To address this issue, Heincke *et al.* (2006a) developed a novel 3D migration scheme based on computation of semblance values. Together with topographic migration schemes, detailed 3D models of large-scale discontinuities could be derived for two subdomains of the mapping area. Figure 24.6 shows a view of a 3D model representing the fault pattern in the upper part of the instability, derived from integrated geological mapping and borehole and surface radar investigations (Willenberg *et al.*, 2008a).

The *P*-wave velocity structure at the study site was investigated using a new 3D tomographic seismic refraction technique (Heincke *et al.*, 2006b). Inversion of first-arrival travel-time picks revealed a broad zone of remarkably low seismic velocities (<1500 m s^{-1}) compared to the surrounding areas (>3800 m s^{-1}) (Fig. 24.7). This ultra-low-velocity zone extends to more than 25 m depth over an area of 200 × 100 m and encompasses the currently active part of the instability, as well as a significant portion of the adjacent slope to the west that is not currently moving. An explanation of such low velocities requires that 17 percent of the investigated volume be air-filled voids (fractures). The depth of very low *P*-wave velocities (1500–2700 m s^{-1}) in the moving rock mass is on average about 35 m. Lineaments in the

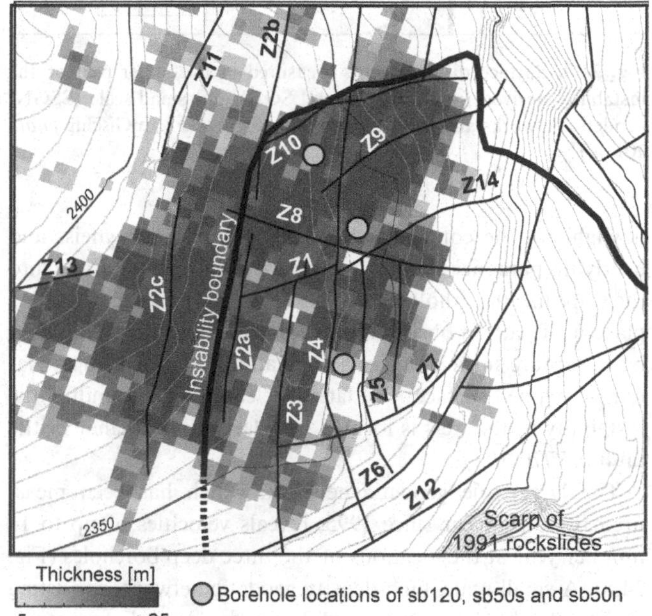

Fig. 24.7. Thickness of ultra-low-velocity zones (<1500 m s^{-1}) determined from seismic refraction tomography, with overlay of surface faults and fracture zones and boundary of the current instability. (Modified from Heincke *et al.*, 1996a and Willenberg *et al.*, 2008a.)

low-velocity zone strike parallel to the F2 fault set and can be traced to more than 50 m depth.

24.3 DISPLACEMENTS AND SLOPE KINEMATICS

24.3.1 SPATIAL DISPLACEMENT PATTERNS

Ground-based radar interferometry measurements conducted between 2005 and 2007 provided clear identification of a lateral release plane and basal rupture surface, both of which coincide with the locations of those exploited by the May 1991 failure (Figs. 24.2 and 24.3). The basal shear surface (135°/40°)

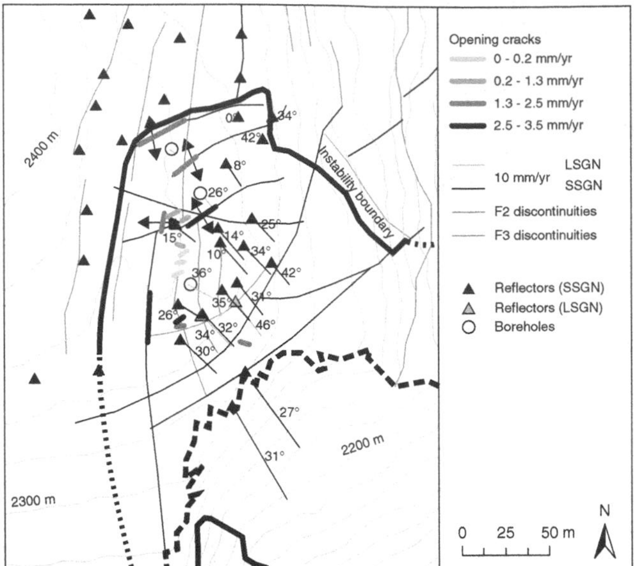

Fig. 24.8. Surface displacements measured in the upper part of the instability derived from large-scale (LSGN) and small-scale (SSGN) geodetic networks and surface crack monitoring. (From Gischig *et al.*, 2011a.)

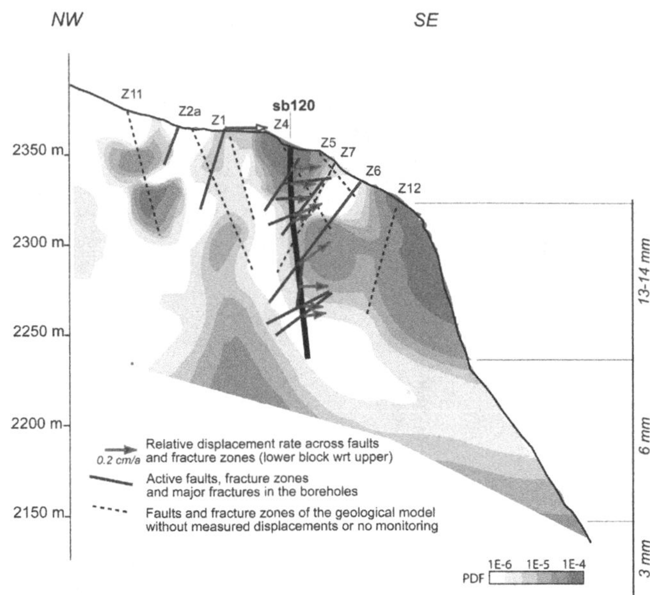

Fig. 24.9. Cross-section through borehole sb120 (black) showing displacement vectors as measured along active faults and fracture zones (movements of the lower block with respect to the upper block), and cumulative probability density function (PDF) of micro-earthquake hypocenter locations as background. Active faults that intersect the surface or borehole are shown by solid lines; dashed lines denote faults where no ongoing dislocation was detected. (From Willenberg *et al.*, 2008b.)

daylights at the contact between the Randa augengneiss and overlying paragneiss, but dips toward the valley and not into the slope like the lithologic contact (Fig. 24.4). The uniform displacement pattern observed immediately above the daylighting basal rupture surface suggests translational sliding along a planar or stepped surface. The lateral release plane bounding the instability to the west is formed by a large-scale, planar brittle fault (95°/70°).

The large-scale geodetic network, which has been measured twice per year since 1995, reveals velocities of up to 14 mm per year at the locations of the three deep boreholes (Fig. 24.8). A small-scale, high-density geodetic network, consisting of 20 reflectors inside the instability and 14 on adjacent stable ground, was surveyed monthly between September 2008 and August 2009. This network covered a larger area of the accessible instability and showed maximum velocities of 30 mm per year at the May 1991 head scarp. Rates less than 14 mm per year were measured toward the northern boundary of the instability (Fig. 24.8) and along the ridge extending down to the Randa augengneiss boundary (Figs. 24.2 and 24.3).

Surface deformation measured by the geodetic networks, fracture benchmark quadrilaterals, and crack extensometers demonstrates that displacements within the unstable rock mass are controlled by the large-scale fracture network (Figs. 24.2, 24.5, and 24.6). This relation is supported by borehole measurements; in most cases active displacements occur within one inclinometer measurement interval (0.61 m), and axial deformations also appear to be highly localized. In all cases, displacements can be correlated with distinct features seen in borehole televiewer images. In contrast, areas of high-density, small-scale fractures generally do not correlate with active displacement zones.

Displacement vectors along individual active fractures have been determined from combined borehole inclinometer and axial strain data (Fig. 24.9). Identified dislocation zones correlate in most cases with distinct faults of the F2 set, as observed in borehole televiewer images. Willenberg *et al.* (2008b) separated displacements measured at 0–80 m depth in SB120 into shear and normal components. In this interval, they mainly observed shear displacements in a normal-faulting sense. Below 80 m, only a few steeply dipping fractures showed significant openings. Absolute displacements measured at the ground surface, combined with relative displacements in the upper 80 m of borehole SB120, suggest that block toppling along F2 fractures is the main kinematic failure mode in the upper part of the unstable rock mass, with a transition to stepped-planar sliding at greater depth.

Indirect evidence for the spatial distribution of displacements at depth can be derived from systematic analysis of microseismic events recorded by geophones distributed throughout the unstable rock mass (Spillmann *et al.*, 2007b). A nonlinear probabilistic technique was used to locate 223 predominantly rockslide-related events, based in part on the heterogeneous 3D velocity model of Heincke *et al.* (2006b). This analysis showed that microseismicity is concentrated within two main zones: one following the 1991 rockslide scarp and the other coinciding with the highest densities of fracture zones and faults. Most microseismic events occurred within 50–100 m of the ground surface.

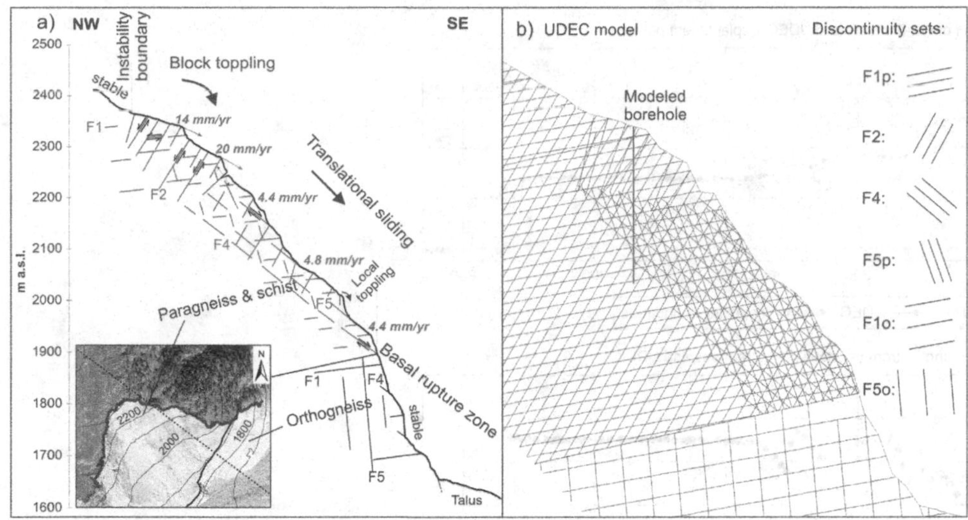

Fig. 24.10. (a) Conceptual kinematic model of the current Randa instability, and (b) geometry of the central portion of the 2D UDEC model. (From Gischig et al., 2011a.)

24.3.2 KINEMATIC ANALYSIS

Stereographic kinematic analysis (after Hocking, 1974; Goodman and Bray, 1976; Wyllie and Mah, 2004) was performed using all discontinuity orientations and two slope orientations of 140°/75° (upper portion of the scarp above 2200 m asl) and 155°/55° (lower portion of the scarp between 1900 and 2200 m asl). Lateral constraints on individual zones between active fractures were ignored; consequently the approach does not reproduce true conditions, but simply indicates which kinematic modes are possible. The analysis showed that, for friction angles of 30–35°, the suggested F2 toppling mode in the upper part of the instability and sliding along F4 fractures in the lower part of the instability are the dominant kinematic failure modes (Gischig *et al.*, 2011a).

Kinematic failure modes were further investigated with a 2D discontinuum model (UDEC), which represents the location and geometry of active fracture sets, including individual faults with known location and dip from 3D structural analysis (Fig. 24.10). All discontinuity sets were modeled as fully persistent, and intact rock bridges were simulated using variable discontinuity strength properties, as suggested by Jennings (1970). Slip along discontinuities was allowed, constrained by a Mohr–Coulomb failure criterion that includes residual strength. Elastic properties of rock mass blocks were derived from laboratory tests on intact core pieces and geological strength index (Hoek *et al.*, 2002) assessment (Gischig *et al.*, 2011a). In-situ stresses were initialized through successive excavation of topography, starting from a rectangular block with initial upper elevation at 2800 m asl and a horizontal/vertical stress ratio (k_0) of 0.6.

Figure 24.11 shows normalized accumulated displacements after the final excavation stage, representing movements following the 1991 failures. Modeled displacements correspond well with observed patterns, both at the ground surface and in the 120-m-deep borehole. The modeled cumulative maximum displacement of 0.75 m is comparable to the accumulated opening of monitored cracks (*ca.* 1.5 m, including displacements that occurred before the 1991 failure). The model also reproduces the observed boundaries of the instability, sliding-mode failure between 1850 and 2100 m along F4 fractures, and toppling along F2 fractures above 2200 m (Gischig *et al.*, 2011a).

24.4 ENVIRONMENTAL FACTORS DRIVING ROCK-SLOPE DISPLACEMENTS

24.4.1 CONTINUOUS DISPLACEMENT MONITORING

Continuous strain measurements from surface crack extensometers and borehole inclinometers reveal cyclical annual displacement trends superimposed on relatively constant long-term values (typically *ca.* 2 mm per year; Willenberg *et al.*, 2008b; Gischig *et al.*, 2011c). Surface extensometers indicate that cracks open in winter, when ground temperatures drop (Fig. 24.12b), and begin to close shortly after the onset of snowmelt (black bars in Fig. 24.12d, e). Conditions are relatively stable between June and October (Fig. 24.12a). Inclinometer data from 68 m depth in borehole SB120 show a similar, but weaker, seasonal pattern of displacements; this pattern is more clearly shown in the plot presented in Figure 24.12d, which has been processed to remove the linear displacement trend.

24.4.2 HYDROLOGIC OBSERVATIONS AND EFFECTS

Records of snowmelt and rainfall were systematically analyzed for the years 2002–2010. As shown by Alpiger (2010), neither heavy rainstorms (>25 mm per day) in summer nor snowmelt in spring show significant correlation with observed temporal displacement patterns across active surface or borehole discontinuities.

Pore pressures were measured with vibrating wire piezometers in slotted inclinometer casings at the bottom of each borehole. The annulus was grouted with cemented geotextile packers above these 3-m-long slotted sections. At the bottom of borehole SB50S, the pressure head is equivalent to a 3–4 m high column

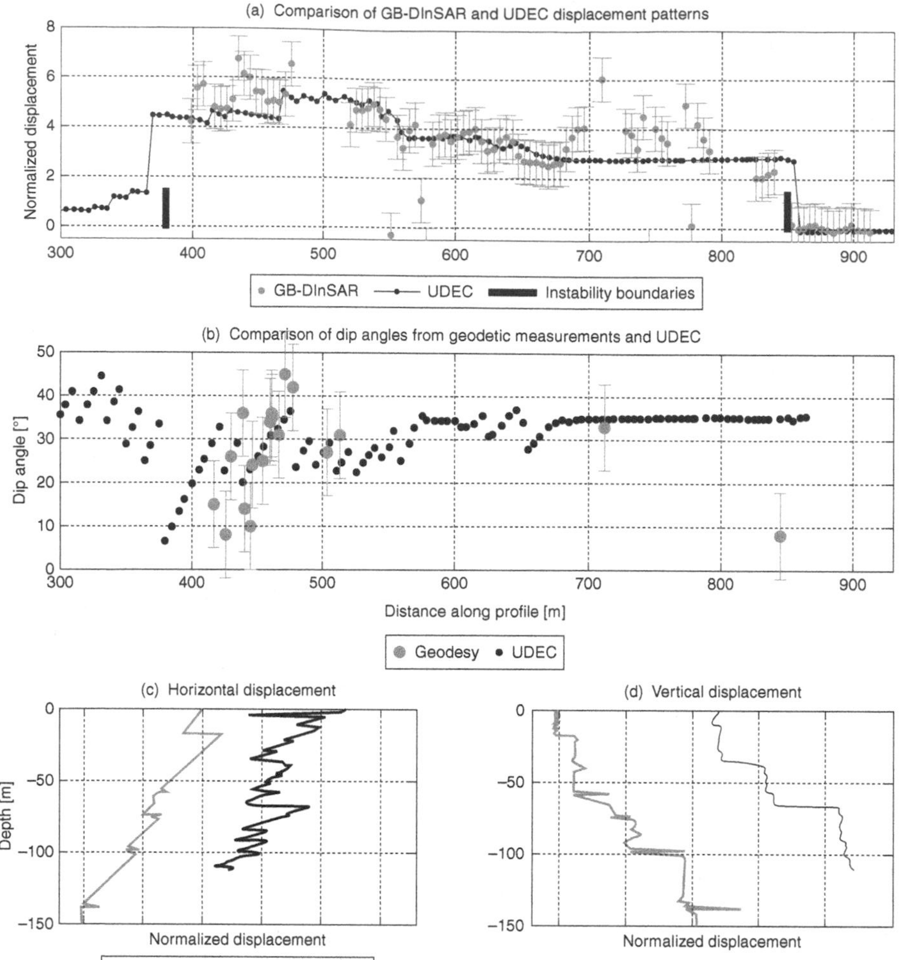

Fig. 24.11. (a) Comparison of modeled normalized surface displacements and radar interferometry measurements. (b) Plunge angles derived from 3D geodetic measurements and UDEC model. (c), (d) Comparison of measured and modeled inclinometer and extensometer surveys in borehole SB120. (From Gischig *et al.*, 2011a.)

of water (Fig. 24.12e). Seasonal fluctuations in this borehole include rapid pressure increases of 1–2 m over a period of 1 week to 1 month at the beginning of snowmelt, and decimeter-scale responses to major precipitation events (Willenberg, 2004; Alpiger, 2010). Because snow thickness and snowmelt are measured at higher elevations (2750–2910 m asl), the entire duration of pore-pressure increase may correspond to the local snowmelt period at the Randa in-situ laboratory (Fig. 24.12e). Borehole SB120, in contrast to SB50S, does not show any water pressures, but only small pressure variations that correlate with barometric pressure changes (Fig. 24.12e). No significant attenuation or time lag has been observed with respect to air pressure. Similarly, air pressure changes coincide with measured pore pressures in borehole SB50S, without attenuation or phase shift. These observations suggest that perched groundwater fills fractures during periods of high surface infiltration. A deep (>120 m) groundwater table is supported by the local ridge topography and many open fractures down to at least 85 m depth, as observed during drilling, borehole logging, and geophysical investigations.

Changes in groundwater seepage rate and spring line elevation Fig. 24.1) were mapped in the field and, with images of the entire slope, captured daily using a time-lapse camera. These data show no groundwater discharge during winter or summer from the May 1991 failure scarp. Only in springtime, after the onset of snowmelt, do several springs develop along foliation-parallel (F1, 240°/20°) discontinuities within the orthogneiss or along its upper boundary (Fig. 24.13). All of the main springs are located at the contact between the Randa orthogneiss and the overlying paragneisses; no groundwater seepage was identified above this line (Fig. 24.13). A seepage or spring line can be followed from the basal rupture surface in the steep rock face about 1 km to the south, everywhere following the approximate upper boundary of the orthogneiss (Fig. 24.1). In the south, close to the Bisbach tributary, the springs are accessible and located in debris or stable orthogneiss.

Water sampled from these springs near the Bisbach tributary has relatively high electrical conductivity (200–450 μS cm^{-1}), indicating longer travel distances in stable fractured bedrock. This supposition is supported by chemical measurements carried out on groundwater from inflows to a 3-km-long bypass tunnel of the Matter Vispa River constructed in intact rock below the Randa rockslides (Fig. 24.4; Girod, 1999). Water sampled in the northern paragneiss section of this tunnel at a depth of up to 450 m below ground surface has electrical conductivity ranging

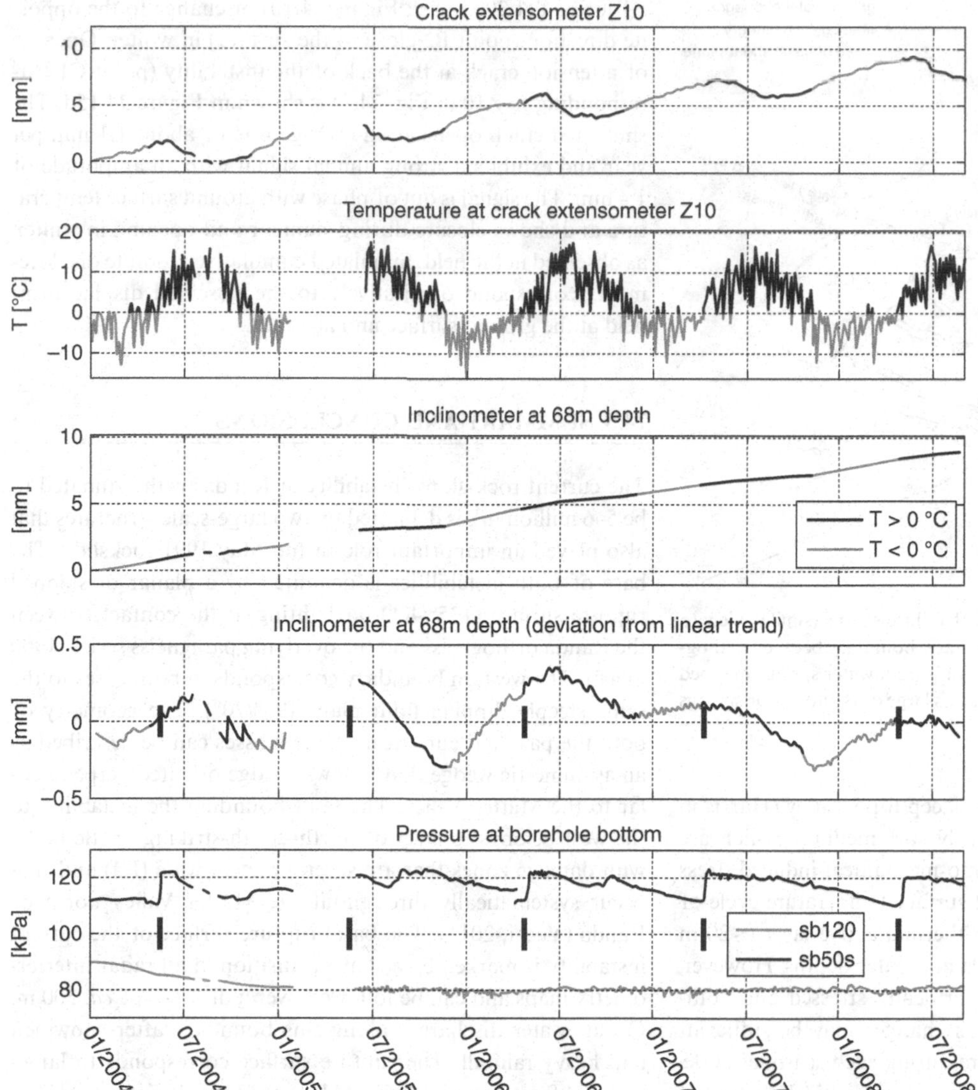

Fig. 24.12. (a) Surface crack extensometer, (b) surface temperature, (c), (d) SB120 borehole inclinometer, and (e) SB120 and SB50S borehole pressure data measured between 2004 and 2010. Black bars indicate the onset of snowmelt as indicated by piezometer data. (From Gischig *et al.*, 2011c.)

from 180 to 280 μS cm^{-1}, whereas groundwater from the Randa augengneiss section, which is about 900 m deep, has higher conductivity, ranging from 250 to 400 μS cm^{-1}.

Figure 24.14 shows a conceptual hydrogeological model of the current instability at Randa. The low regional groundwater table explains the negligible impact of surface hydrologic perturbations on slope and crack displacements. Only close to the basal rupture surface or along individual cracks containing perched water can snowmelt-induced increases in the water table cause fracture slip due to local reductions in effective normal stress and shear strength.

24.4.3 THERMAL OBSERVATIONS AND THERMO-MECHANICAL FORCING

Figure 24.12 suggests that temperature, rather than groundwater pressure, may play the dominant role in driving ongoing slope movements at Randa. Displacement data measured with different independent monitoring techniques, both at ground surface and at depth, reveal a consistent seasonal pattern, which is contradictory to most other reported slope instabilities where groundwater pressure drives movements (e.g., Bonzanigo *et al.*, 2007). Displacement rates increase in winter and decrease in spring after snowmelt. This pattern is similar to that observed at the Checkerboard Creek landslide in Canada, where Watson *et al.* (2004) showed that displacement rates down to 26 m depth, well below the thermal active layer, are maximum in winter and minimum in summer. They interpreted this behavior as arising from the contraction of blocks during winter, causing a decrease in normal stresses along steeply dipping discontinuities and favoring slip. Gischig *et al.* (2011b) used conceptual, coupled 2D thermo-mechanical simulations to show that such effects can drive slope deformations to >100 m depth. The simulations indicated that thermo-elastic expansion and contraction in the shallow near-surface can induce stress changes at depths well below the thermal active layer. Factors these allow these stress

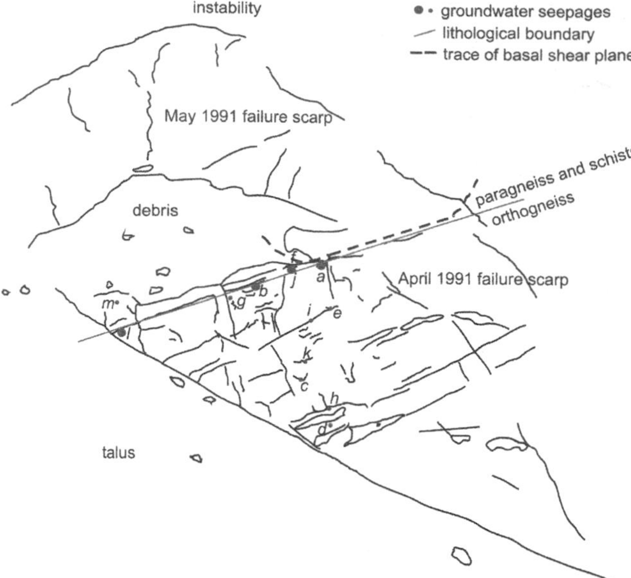

Fig. 24.13. Perspective view of the 1991 failure scarps (similar view to Fig. 24.1), trace of the basal rupture surface, boundary between orthogneiss and paragneiss, and the locations of groundwater springs mapped in spring 2010. Main springs are marked as large dots and minor springs as small dots.

changes to propagate to depth are steep topography (Harrison and Herbst, 1977) and a sufficiently stiff medium, which are both met at Randa. Modeled thermo-mechanical-induced stress changes resulting from an annual surface temperature cycle of ±10°C reach up to 1 MPa in the thermal active layer (<20 m depth), but only about 10–100 kPa at greater depths. However, provided that discontinuities are critically stressed and sufficiently close to failure, such stress changes may be sufficient to induce small incremental failure during each seasonal cycle. If a sufficient number of discontinuities are critically stressed, thermo-mechanical effects can drive progressive movement of the entire slope.

The 2D numerical model developed for brittle-plastic kinematic analysis (Fig. 24.10) was used by Gischig *et al.* (2011c) to study thermo-mechanical reactions of the rock mass at the surface and at depth at the current Randa instability. Thermal properties were derived from analysis of rock temperature data and from published literature. A sinusoidal temperature time history with a 1-year period and 10°C amplitude was applied to the ground surface (Fig. 24.15a). Figure 24.15b shows the modeled horizontal displacements across three toppling discontinuities in borehole SB120 at 14, 48, and 68 m depth (left-lateral shear sense). A clear annual signal is visible at all depths, but the different intervals show different phase shifts with respect to the temperature signal at the ground surface. Although the modeled local rates are lower than those measured across active discontinuities (1–2 mm per year), they approach the same order of magnitude. Shear displacements along the basal rupture surface are shown in Figure 24.15c (right-lateral shear sense). Displacement rates increase in summer and either cease (point B1 located at 200 m asl, Figure 24.10) or change to the opposite direction (point B2 close to the surface) in winter. Opening of a tension crack at the back of the instability (point C1 corresponding to Z10 in Fig. 24.5) is shown in Figure 24.15d. The simulated crack opens at an average rate of about 1.1 mm per year and exhibits a strong annual signal with an amplitude of 1.4 mm. The signal is out of phase with ground surface temperature and shows closure during summer and opening in winter, as observed in the field. Simulated cumulative absolute displacements correspond qualitatively to the observed displacement field at the ground surface and at depth.

24.5 SUMMARY AND CONCLUSIONS

The current rock-slope instability at Randa, with estimated to be 5–6 million m^3, is delimited by two large-scale structures that also played an important role in the May 1991 rockslide. The base of both instabilities is delimited by a planar or stepped rupture surface (135°/40°) daylighting at the contact between the Randa orthogneiss and the overlying paragneiss series (1900 m asl). The western boundary corresponds in both cases to the same steeply dipping fault plane (95°/70°). The geometry of both the past and current unstable masses can be described as an asymmetric wedge that follows a ridge oriented perpendicular to the Matter Valley. The fault bounding the instability to the west belongs to a set of north–south-striking brittle faults with damage zones that are several meters wide (F3) and that occur systematically throughout the Matter Valley north of Randa (Yugsi, 2011). The basal rupture surface of the current instability is marked by a sharp transition in all radar interferometry maps and can be followed over a distance of *ca.* 200 m. Groundwater discharges along this boundary after snowmelt and heavy rainfall. The rupture surface corresponds to large-scale F4 fractures, which formed stepped basal rupture surfaces in the April and May 1991 failures. At the regional scale, F4 fractures are significantly less common than F3 fractures, and have an irregular spatial distribution (Yugsi, 2011). These findings support the more general conclusion that critically oriented large-scale faults exert first-order control on slope stability.

The spatial distribution of displacements within the current instability at Randa shows coherent zones moving at rates up to 30 mm per year that have side lengths of several decameters and are separated by persistent active fractures. Toppling is the dominant kinematic failure mode in the upper portion of the instability, above 2200 m asl, with translational sliding below. A reliable 3D model of all active fractures has allowed us to reproduce the observed spatial displacement pattern derived from geodetic and radar interferometry measurements in a 2D discontinuum numerical model.

Continuous displacement data from inclinometers, extensometers, and fiber optic strain sensors, measured at the surface and at depths of 12–68 m in borehole SB120, show increasing deformation rates when the ground surface temperature drops in fall, and a decrease after snowmelt in spring as the rock warms.

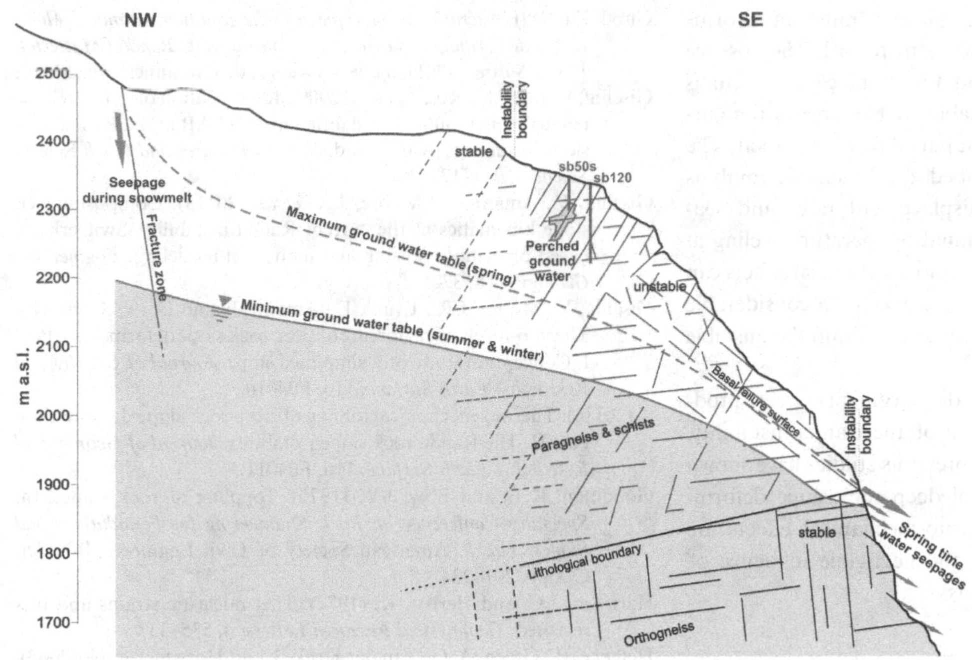

Fig. 24.14. Conceptual hydrogeological model of the Randa rock slope and current instability.

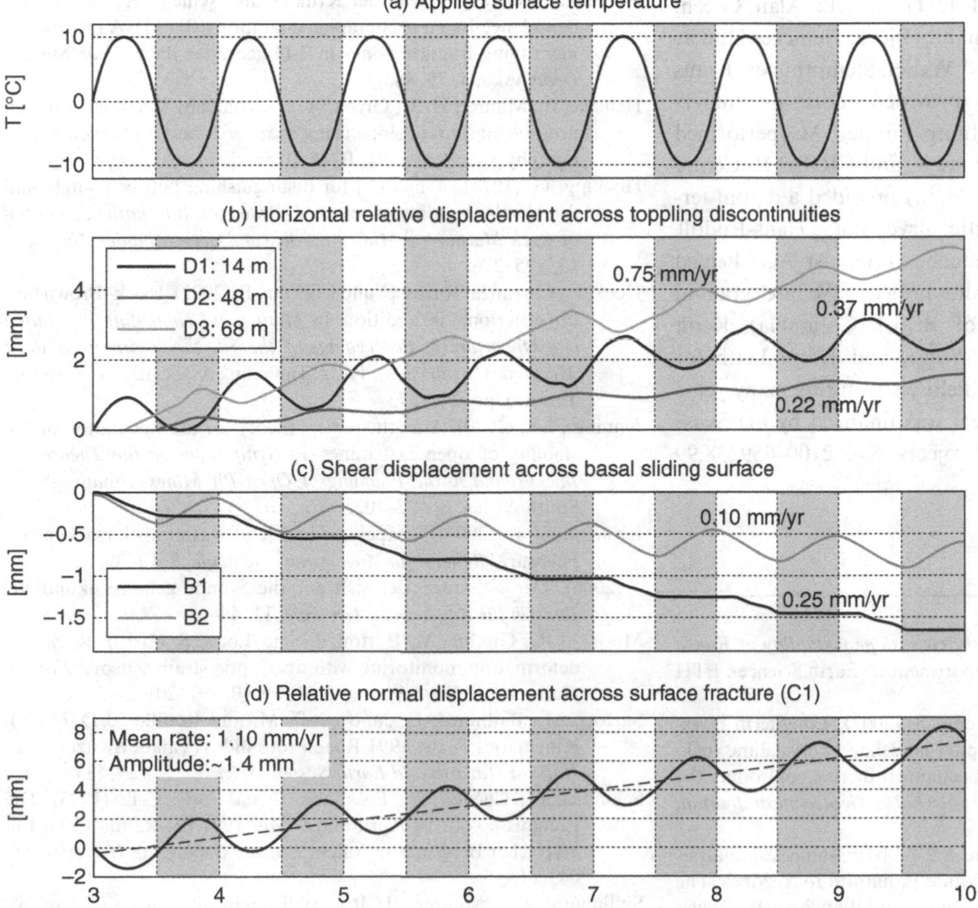

Fig. 24.15. (a) Imposed temperature cycles at the ground surface. (b) Time series of simulated horizontal displacements across discontinuities at depth in borehole SB120 (left lateral shear is positive). Gray shading indicates times when temperatures at the surface are below 0°C. (c) Time series of shear displacement at two points along the basal rupture surface (right-lateral shear is positive). Times of increased deformation occur in summer as opposed to the points in (a). (d) Modeled crack opening at the surface corresponding approximately to the location of crack Z10. (From Gischig *et al.*, 2011c.)

No correlations were found between heavy summer rainstorms or spring snowmelt and the displacement record. The absence of significant hydraulic control on the displacement field is explained by a low groundwater table that intersects the current instability only in its lowermost part during snowmelt. The same discontinuum model developed for kinematic analysis reproduced order-of-magnitude displacement rates and seasonal values solely by applying annual temperature cycling at the ground surface. The fact that thermo-mechanical effects can be detected at Randa points to the presence of a considerable amount of critically stressed discontinuities within the unstable rock mass.

The largest thermo-mechanical displacements are a product of toppling in the upper portion of the Randa instability, between 2250 and 2350 m asl. Few previous studies have shown that similar thermal effects control deep rock-slope deformation. In the case of Randa, thermo-mechanical effects are clearly observable, due in part to the negligible influence of groundwater within the dry rock mass.

ACKNOWLEDGEMENTS

This multidisciplinary project benefited from the advice and support of other researchers at ETH Zurich: Alan Green, Hansrudolf Maurer, Tom Spillmann, Björn Heincke, Hilmar Ingensand, Henry Eisenbeiss, and Walter Stempfhuber. Klaus Brauch (Terratec, Heitersheim) provided tools for analyzing optical televiewer images, Stump Foratec AG performed inclinometer and extensometer surveys, and CREALP (Centre de Recherche sur l'Environment Alpin) provided a digital terrain model and long-term geodetic survey data. Hans-Rudolf Keusen and Bernhard Krummenacher (Geotest AG) helped with planning and supported the project. We are grateful to Erik Eberhardt (University of British Columbia), Keith Evans, Florian Amann, Freddy Yugsi, and Kerry Leith for long and fruitful discussions and field visits during many years of research at Randa. This project was financed by the Swiss National Science Foundation (Projects No. 2100–059238.99 and 200020–112073).

REFERENCES

Alpiger, A. (2010). *Hydrogeology of the rock slope instability at Randa (Switzerland)*. B.Sc. thesis, Department of Earth Sciences, ETH Zurich, Switzerland.

Bonzanigo, L., Eberhardt, E. and Loew, S. (2007). Long-term investigation of a deep-seated creeping landslide in crystalline rock. Part I. Geological and hydromechanical factors controlling the Campo Vallemaggia landslide. *Canadian Geotechnical Journal*, 44, 1157–1180.

Eberhardt, E., Stead, D. and Coggan, J.S. (2004). Numerical analysis of initiation and progressive failure in natural rock slopes: The 1991 Randa rockslide. *International Journal of Rock Mechanics and Mining Sciences*, 41, 68–87.

Eisenbeiss, H. (2009). *UAV photogrammetry*. Ph.D. thesis, ETH Zurich, Switzerland.

Girod, F. (1999). *Altération météorique de roche granitique en milieu Alpin: Le cas de l'orthogneiss associé à l'éboulement de Randa (Mattertal, Valais, Suisse)*. Ph.D. thesis, Université de Lausanne, Switzerland.

Gischig, V., Loew, S., Kos, A. *et al.* (2009). Identification of active release planes using ground-based differential InSAR at the Randa rock slope instability, Switzerland. *Natural Hazards and Earth System Sciences*, 9, 1–12.

Gischig, V., Amann, F., Moore, J.R. *et al.* (2011a). Composite rock slope kinematics at the current Randa instability, Switzerland, based on remote sensing and numerical modeling, *Engineering Geology*, 118, 37–53.

Gischig, V., Moore, J.R., Evans, E., Amann, F. and Loew, S. (2011b). Thermo-mechanical forcing of deep rock slope deformation. Part I. Conceptual study of a simplified slope. *Journal of Geophysical Research – Earth Surface*, 116, F04010.

(2011c). Thermo-mechanical forcing of deep rock slope deformation. Part II. The Randa rock slope instability. *Journal of Geophysical Research – Earth Surface*, 116, F04011.

Goodman, R.E. and Bray, J.W. (1976). Toppling of rock slopes. In: *Specialty Conference on Rock Engineering for Foundations and Slopes, Vol. 2*. American Society of Civil Engineers. Boulder, CO, pp. 201–234.

Harrison, J.C. and Herbst, K. (1977). Thermoelastic strains and tilts revisited. *Geophysical Research Letters*, 4, 535–537.

Heincke, B., Green, A.G., van der Kruk, J. and Horstmeyer, H. (2005). Acquisition and processing strategies for 3-D georadar surveying a region characterized by rugged topography. *Geophysics*, 70, K53–K61.

Heincke, B., Green, A., van der Kruk, J. and Willenberg, H. (2006a). Semblance-based topographic migration (SBTM): A method for identifying fracture zones in 3-D georadar data. *Near Surface Geophysics*, 4, 79–88.

Heincke, B., Maurer, H.R., Green, A. *et al.* (2006b). Characterizing an unstable mountain slope using shallow 2- and 3-D seismic tomography. *Geophysics*, 71, B241–B256.

Hocking, G. (1974). A method for distinguishing between single and double plane sliding of tetrahedral wedges. *International Journal of Rock Mechanics, Mining Science and Geomechanics Abstracts*, 13, 225–226.

Hoek, E., Carranza-Torres, C. and Corkum, B. (2002). Hoek–Brown failure criterion: 2002 edition. In *Mining and Tunnelling Innovation and Opportunity: Proceedings of the 5th North American Rock Mechanics Symposium, Vol. 1*. Tunneling Association of Canada, Toronto, pp. 267–273.

Jennings, J.E. (1970). A mathematical theory for the calculation of the stability of open cast mines. In *Symposium on the Theoretical Background to the Planning of Open Pit Mines*, Johannesburg, South Africa, pp. 87–102.

Krähenbühl, R. (2004). Temperatur und Kluftwasser als Ursachen von Felssturz. *Bulletin für Angewandte Geologie*, 9, 19–35.

(2006). Der Felssturz, der sich auf die Stunde genau ankündigte. *Bulletin für Angewandte Geologie*, 11, 49–63.

Moore, J.R., Gischig, V., Button, E. and Loew, S. (2010). Rockslide deformation monitoring with fiber optic strain sensors. *Natural Hazards and Earth System Sciences*, 10, 191–201.

Sartori, M., Baillifard, F., Jaboyedoff, M. and Rouiller, J.D. (2003). Kinematics of the 1991 Randa rockslides (Valais, Switzerland). *Natural Hazards and Earth System Sciences*, 3, 423–433.

Schindler, C., Cuénod, Y., Eisenlohr, T. and Joris, C.L. (1993). Die Ereignisse vom 18 April und 9 Mai 1991 bei Randa (VS): Ein atypischer Bergsturz in Raten. *Eclogae Geologicae Helvetiae*, 86, 643–665.

Spillmann, T., Maurer, H.R., Willenberg, H. *et al.* (2007a). Characterization of an unstable rock mass based on borehole logs and diverse borehole radar data. *Journal of Applied Geophysics*, 61, 16–38.

Spillmann, T., Maurer, H.R., Heincke, B., Willenberg, H. and Green, A. (2007b). Microseismic monitoring of an unstable rock mass. *Journal of Geophysical Research – Solid Earth*, 112, B07301.

Wagner, A. (1991). *Bergsturz Grossgufer Randa: Etude Structurale et Géomécanique*. Sion, Switzerland: Centre de Recherches Scientifiques Fondamentales et Appliquées de Sion.

Watson, A.D., Moore, D.P. and Stewart, T.W. (2004). Temperature influence on rock slope movements at Checkerboard Creek. *In Proceedings of the 9th International Symposium on Landslides*, Rio de Janeiro, Brazil, pp. 1293–1304.

Willenberg, H. (2004). *Geologic and kinematic model of a complex landslide in crystalline rock (Randa, Switzerland)*. Ph.D. thesis, Swiss Federal Institute of Technology, Zurich, Switzerland.

Willenberg, H., Loew, S., Spillmann, T. *et al.* (2008a). Internal structure and deformation of active slope instability in crystalline rock at Randa, Switzerland. I. 3D rock mass structure from integrated geological and geophysical investigations. *Engineering Geology*, 101, 1–14.

Willenberg, H., Evans, K., Loew, S. and Eberhardt, E. (2008b). Internal structure and deformation of an active slope instability in crystalline rock at Randa, Switzerland. II. 3D deformation pattern and kinematic model. *Engineering Geology*, 101, 15–32.

Wyllie, D.C. and Mah, C.W. (2004). *Rock Slope Engineering: Civil and Mining* (4th edn.). New York: Spon Press.

Yugsi, M.F.X. (2011). *Structural control of multi-scale discontinuities on slope instabilities in crystalline rock (Matter Valley, Switzerland)*. Ph.D. thesis, ETH Zurich, Switzerland.

25 Characterization and management of rockslide hazard at Turtle Mountain, Alberta, Canada

COREY R. FROESE, MARIE CHARRIÈRE, FLORIAN HUMAIR, MICHEL JABOYEDOFF AND ANDREA PEDRAZZINI

ABSTRACT

In 1903, more than 30 million m^3 of rock fell from the east slopes of Turtle Mountain in Alberta, Canada, causing a rock avalanche that killed about 70 people in the town of Frank. The Alberta Government, in response to continuing instabilities at the crest of the mountain, established a sophisticated field laboratory where state-of-the-art monitoring techniques have been installed and tested as part of an early-warning system. In this chapter, we provide an overview of the causes, trigger, and extreme mobility of the landslide. We then present new data relevant to the characterization and detection of the present-day instabilities on Turtle Mountain. Fourteen potential instabilities have been identified through field mapping and remote sensing. Lastly, we provide a detailed review of the different in-situ and remote monitoring systems that have been installed on the mountain. The implications of the new data for the future stability of Turtle Mountain and related landslide runout, and for monitoring strategies and risk management, are discussed.

25.1 INTRODUCTION

Turtle Mountain in southwest Alberta, Canada (Fig. 25.1) became famous in 1903 when more than 30 million m^3 of rock fell from the east flank of the mountain and overran part of the town of Frank, killing about 70 people (McConnell and Brock, 1904; Charrière, 2011). Continuing instability at the crest of Turtle Mountain led the Alberta Government, in 2003, to establish a state-of-the-art field laboratory to monitor the hazard (Moreno and Froese, 2006; Froese *et al.*, 2009). The South Peak area (Fig. 25.1) is now monitored to better characterize deformation along deep tension cracks at the top of the mountain and to provide early warning of an impending landslide.

This chapter summarizes research carried out at Turtle Mountain since the 1903 landslide. In the first part of the chapter, we summarize changes in our understanding of the geology of Turtle Mountain, in particular elements of the geology that are relevant to the 1903 rock avalanche. Different ideas about the causes and trigger of the landslide and its extreme mobility are also briefly discussed. The second part of the chapter focuses on new data that have delineated and characterized ongoing instabilities on Turtle Mountain. It summarizes the methods that have been adopted to detect and characterize the instabilities, estimate their volumes, and assess their potential runout. The results of combined field observations, analysis of a high-resolution digital elevation model (HRDEM), and numerical modeling are presented. A detailed review of different in-situ and remote monitoring systems currently installed on the mountain, and the implications of the new data for improvement of monitoring strategies and risk management, are discussed in the final part of the chapter.

25.1.1 GEOLOGIC SETTING

Turtle Mountain is developed in an anticline that lies in the hanging wall of the Livingstone thrust fault at the east front of the Canadian Rocky Mountains (Langenberg *et al.*, 2007). Paleozoic carbonate rocks overlie the thrust, and Mesozoic sedimentary rocks underlie it (Figs. 25.2 and 25.3). The Turtle Mountain anticline is a noncylindrical, asymmetric inclined anticline with east vergence (Humair *et al.*, 2010). The fold axis is difficult to delineate on the flank of Turtle Mountain due to the significant fracturing in the hinge area. Below North Peak, the anticline is cut by the Turtle Mountain thrust fault, and only the west limb of the structure remains (Figs. 25.2 and 25.3).

Landslides: Types, Mechanisms and Modeling, ed. John J. Clague and Douglas Stead. Published by Cambridge University Press.

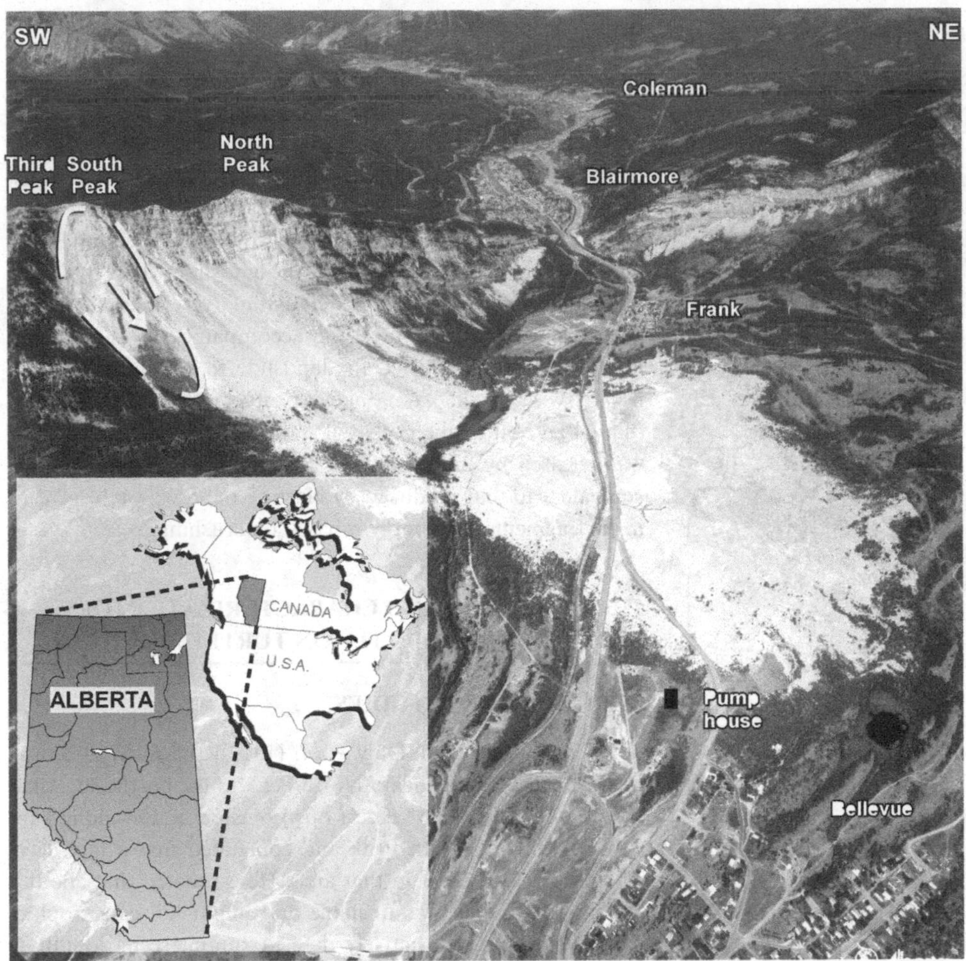

Fig. 25.1. Aerial view of the study area (map in inset).

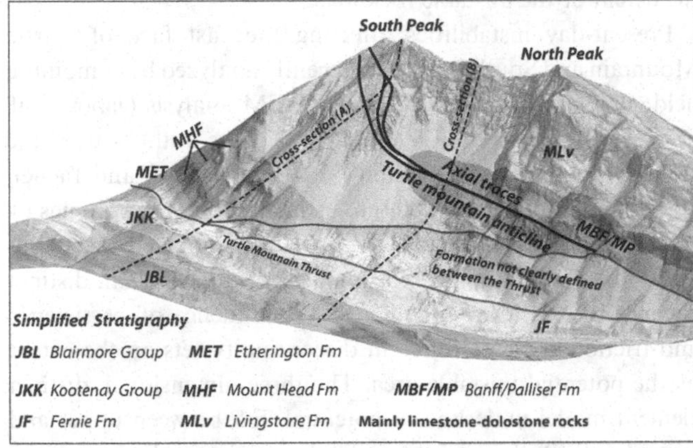

Fig. 25.2. Simplified three-dimensional geology of Turtle Mountain (modified from Langenberg *et al.*, 2007) Cross-sections A and B are shown in Figure 25.3.

25.2 PREVIOUS STUDIES

25.2.1 EVOLUTION OF UNDERSTANDING OF THE GEOLOGY OF TURTLE MOUNTAIN

McConnell and Brock (1904) first investigated the geology of Turtle Mountain and the cause of the 1903 landslide. They interpreted the source area of the landslide to be part of a monocline of Paleozoic limestone dipping 50° to the west (Cruden and Krahn, 1973). Daly *et al.* (1912) concluded that the failure occurred due to planar sliding along steep joints orthogonal to the bedding planes. They also highlighted the presence of the coal mine at the base of the mountain as contributing to the failure. Allan (1933) was the first to propose that the landslide occurred on the limb of an anticline. Based on Daly's cross-section, Terzaghi (1950) proposed that progressive weakening of the rock mass led to a decrease in the factor of safety at Turtle Mountain.

Cruden and Krahn (1973, 1978) argued that slickensides along bedding planes favored a flexural-slip folding as the mechanism for the genesis of the Turtle Mountain anticline. They also postulated that the 1903 failure followed bedding planes on the east limb of the structure, with a rear release controlled by two steep joint sets. Based on this geological model, several geomechanical models were developed over the next four decades to describe the failure and to define the main important triggering factors. First, Cruden and Krahn (1973, 1978) used limit equilibrium analysis to show that the slope was in a marginal state of stability prior to the failure. Benko and Stead (1998), and subsequently Cruden and Martin (2007), used distinct element/

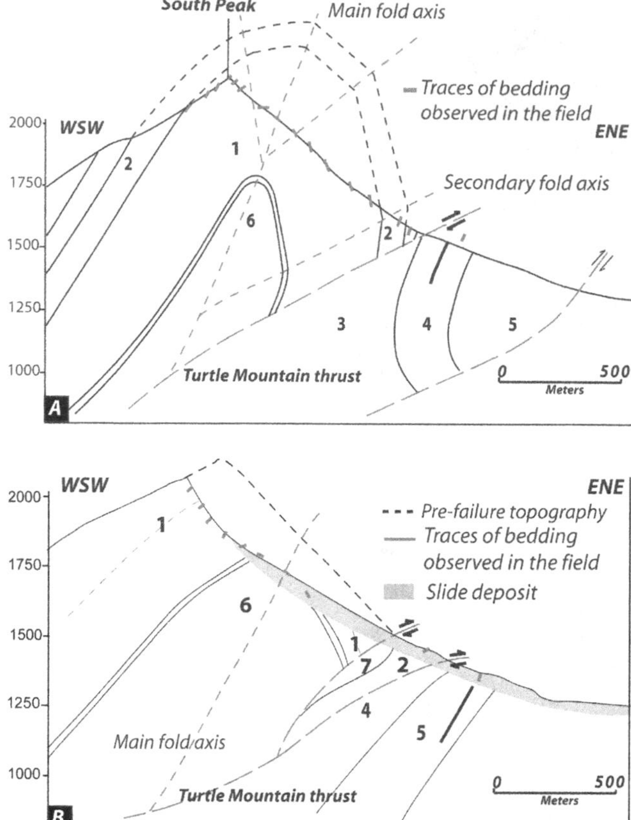

Fig. 25.3. Schematic cross-sections through the Frank Slide. (A) Profile across South Peak. (B) Profile across the Frank Slide source area. Legend for cross-sections: 1 – Livingstone Formation; 2 – Mt. Head Formation; 3 – Fernie Formation; 4 – Kootenay Group; 5 – Blairmore Group; 6 – Banff and Palliser formations; 7 – Etherington Formation. (Modified from Pedrazzini *et al.*, 2012.)

finite difference and finite element models to analyze factors that predispose the slope to instability, such as the joint orientations, the fold shape, and mining near the base of the mountain. Jaboyedoff *et al.* (2009) proposed a composite mechanism involving progressive toppling of wedges along gently dipping surfaces formed by persistent discontinuities in the upper part of the scarp, enhanced by water pressure fluctuations. Other recent studies have focused more on analysis of the joint sets at South Peak (Fossey, 1986; Couture *et al.*, 1998; Langenberg *et al.*, 2007).

25.2.2 EVOLUTION OF UNDERSTANDING OF FRANK SLIDE RUNOUT

Several theories have been suggested to explain the long runout of the Frank Slide. Based on observations of the deposit and eyewitness testimony, Kent (1966) proposed that the rock avalanche traveled due to fluidization by an air mechanism. Cruden and Hungr (1986) rejected this hypothesis based on observations of inverse grading in the deposit, which did not support an upward flow of air. They observed that part of the deposit margin consists of finer sediment than that present elsewhere and termed this zone the "splash area." They suggested that Melosh's (1987) acoustic fluidization theory best explained the mobility of the rock avalanche. Drapeau (1997) and Couture *et al.* (1998) investigated the distribution of block sizes in the debris sheet using aerial photographs. They found that block diameter ranges from 0.25 to 11 m and decreases from the proximal part of the deposit toward the periphery. Hungr and Evans (2004) argued that the mobility of the landslide is best explained by liquefaction of the saturated alluvium over which it traveled. The rise in pore water pressure, accompanied by a reduction in effective stress, allowed the landslide to travel farther than it otherwise would have done. Locat *et al.* (2006) estimated that 24 percent of the potential energy of the failed rock mass was released by fragmentation, assuming that the energy was equivalent to that produced by blasting the rock, or 16 percent if the fragmentation energy is based on crushing.

25.3 DETECTION AND CHARACTERIZATION OF POTENTIAL INSTABILITY ON TURTLE MOUNTAIN

25.3.1 PREVIOUS STUDIES

Based on the work of Daly *et al.* (1912), the town of Frank was relocated from beneath North Peak to the northeast. Allan (1931, 1933) identified several open tension cracks along subvertical joints at South Peak and concluded that a high level of hazard was present in that area. He estimated that about 5 million m^3 of rock was unstable and might fail catastrophically, with several subjectively defined runout scenarios. BGC Engineering (2000) confirmed Allan's interpretation, but revised the extent of the unstable rock mass.

Present-day instabilities affecting the east face of Turtle Mountain and South Peak were recently analyzed by combining field survey measurements and HRDEM analysis (Jaboyedoff *et al.*, 2009; Pedrazzini *et al.*, 2012). These authors used the sloping local base-level method (SLBL; Jaboyedoff and Tacher, 2006; Jaboyedoff *et al.*, 2009) to estimate potential volumes of future landslides. Brideau *et al.* (2010) investigated the stability of the South Peak area using a three-dimensional, distinct element model. They highlighted the influence of persistence and friction angle of different discontinuity sets on the extent of the potential unstable area. The three-dimensional, distinct element model underlines a potential link between upper and lower South Peak instabilities if fully persistent discontinuities are assumed. Recently, potential unstable zones were identified in the lower Third Peak area (Pedrazzini *et al.*, 2012) and North Peak area (Humair, 2011).

25.3.2 PRESENT-DAY ANALYSIS AND ADOPTED METHODOLOGIES

Detection and mapping of the discontinuity sets at Turtle Mountain has involved the use of classical field survey methods and large-scale HRDEM analysis using COLTOP 3D

Table 25.1. *Mean strike and dip of the joint sets in the seven structural domains (SD; see Fig. 25.4).*

SD	BP	J1	J2	J3	J3′	J4	J5	J6	J8	P1	P2
1	261/62	020/38	056/36	107/37	137/46	167/60	244/53	–	351/80	–	–
2	290/48	021/68	056/66	106/42	132/48	187/61	275/72	318/60	191/89	–	–
3	288/44	015/51	069/58	107/36	137/59	183/60	287/46	317/57	181/82	–	–
4	282/53	008/63	052/49	100/38	141/42	200/58	278/69	324/69	356/85	–	–
5	–	011/46	–	–	113/36	–	–	–	013/89	105/68	289/69
6	109/61	003/43	043/48	270/40	120/57	223/53	–	331/50	190/87	–	–
7	108/51	004/65	025/55	274/38	105/74	222/67	–	322/46	022/84	–	–

Note: Possible errors about the mean values = ±15°.

(Jaboyedoff *et al.*, 2007, 2009). Kinematic analyses (Brideau *et al.*, 2010; Pedrazzini *et al.*, 2011) have been done following the methods proposed by Richards *et al.* (1978). A friction angle of 30–35° has been used on the basis of previous studies (Benko and Stead, 1998; Cruden and Martin, 2007; Brideau *et al.*, 2010). In addition to the kinematic analyses, two-dimensional distinct element modeling has been performed using UDEC (Itasca, 2004) to assess more quantitatively the role of fractures on rock-slope failure in the North Peak area. UDEC simulates the response to static or dynamic loading of a discontinuous material, in this case a jointed rock mass that is considered to be an assemblage of discrete blocks. The mechanism of failure is assumed to be controlled only by the discontinuities; thus rigid-block conditions are assumed.

The extent of the unstable zones and their volumes have been calculated based on (1) the SLBL method and (2) a rigorous geometrical reconstruction. The SLBL method was applied to the Frank Slide using the base-level concept initially defined in geomorphology (Mills, 2003). The method produced a curved surface from the digital elevation model through an iterative procedure (Jaboyedoff *et al.*, 2009).

Rock volumes were calculated by fitting planes with orientations that correspond to the main discontinuity sets delimiting the unstable zones. TINs (triangular irregular networks) were created from the constructed surfaces delineated by the discontinuity sets. Volumes were calculated in a GIS environment by subtracting each TIN from the present-day topographic surface represented by the HRDEM (Pedrazzini *et al.*, 2012).

25.3.3 FRACTURE PATTERN

Structural analysis of both limbs of the Turtle Mountain anticline yielded 2000 measurements, from which 10 discontinuity sets were identified. Bedding plane (BP) and joint sets J1, J2, J3, J3′, J4, and J8 are present over almost the entire investigated area, whereas joint sets J5, J6, and P1–P2 are more limited spatially (Table 25.1). Joint sets J5 and J6, however, are present in both fold limbs, whereas set P1–P2 is only present in the hinge area. Results obtained by HRDEM analysis are in good agreement with the results of field surveys, although joint sets J5, J8, and P1–P2 were not detected. The investigated area has been divided into seven homogeneous fracture zones or structural domains (SD; Fig. 25.4). Each structural domain is characterized by different fracture orientations, persistence, spacing, and rock mass conditions.

25.3.4 INSTABILITY IN SOUTH PEAK AREA

The South Peak area can be divided into two zones: upper South Peak and lower South Peak. Upper South Peak is located on the west limb of the Turtle Mountain anticline, within SD 2 and SD 3; lower South Peak contains the fold hinge and the east limb of the anticline and corresponds to SD 5 and SD 7 (Fig. 25.4).

Displacement vectors of moving rock masses bounded by open cracks are based on measurements derived from photogrammetric targets and extensometers (Froese *et al.*, 2009), as well as field measurements (Pedrazzini *et al.*, 2012). Most of the cracks in the saddle follow discontinuity sets J2, J3, and J8 (Humair, 2011). Based on these observations, upper South Peak can be subdivided into three unstable sectors, characterized by specific deformation modes (Fig 25.5A):

- sector 1 – superficial toppling
- sector 2 – frontal wedge sliding
- sector 3 – subsidence.

Constrained by the displacement vectors (Fig. 25.5B), the kinematic analysis revealed that the discontinuity sets involved in the instability are J1, J2, J3, J6, J8, and BP (Froese *et al.*, 2009; Humair, 2011; Pedrazzini *et al.*, 2011). Sector 1 is controlled mainly by joint set J2, but locally by joint sets J1 and J3. Movements in sector 2 follow joint set intersections J2–J6 or J3–J6 and, to a lesser extent, J2–BP and J3–BP. Sector 3 moves along rear release surfaces controlled by joint set J2. Volume estimates based on the projection of the involved discontinuity sets range from 0.2 to 0.57 million m^3 (Table 25.2; Fig. 25.6). The proposed geometrical model was confirmed by Brideau *et al.* (2010) based on three-dimensional distinct element modeling.

The lower South Peak area is close to the hinge of the anticline and is heavily fractured and covered by rockfall debris originating from the cliffs above (Pedrazzini *et al.*, 2012). Many open fractures are present in this area and seem to be related to gravitational movement to the northeast. Kinematic analyses

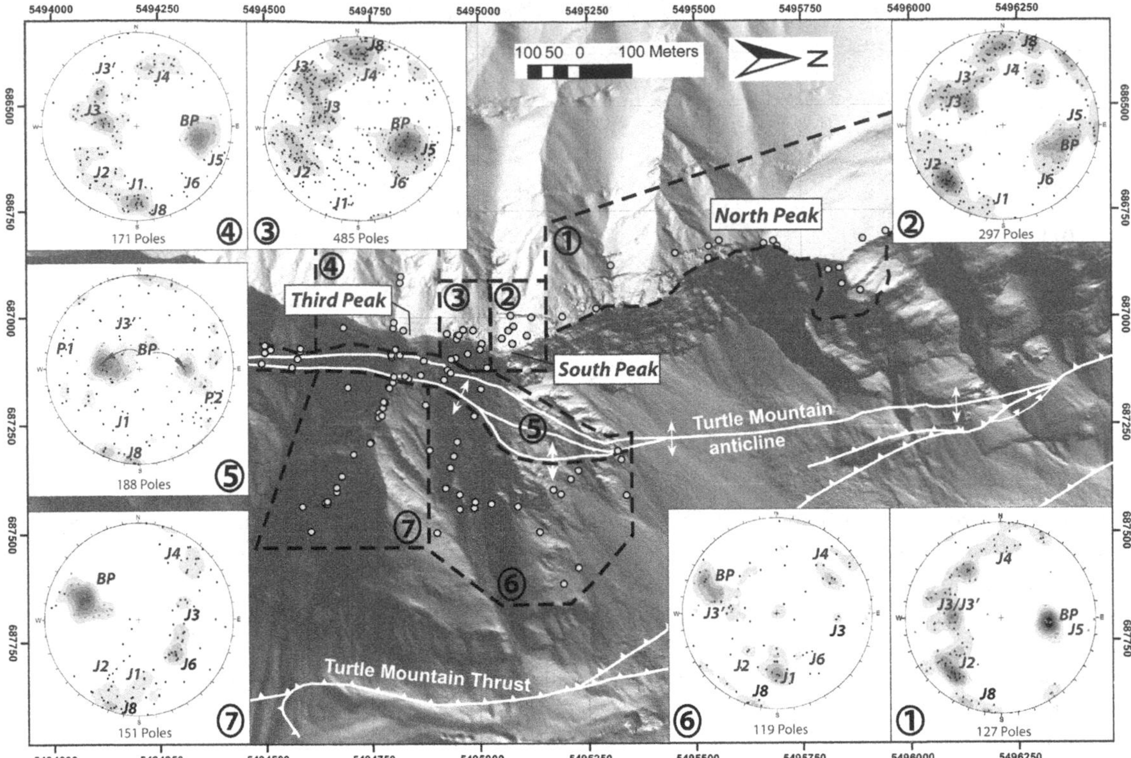

Fig. 25.4. Structural domains identified between Third Peak and North Peak, and their related characteristics. The dashed lines represent zones containing the instabilities discussed in the text.

and field measurements of crack orientations suggest that the lower South Peak area contains at least six unstable zones, all of which appear to be moving toward the northeast, mainly through planar sliding on joint set J1. The rear release for the moving blocks is controlled by joint set J2 and, to a lesser extent, by J3 or J3′, whereas set BP provides the lateral release (Fig. 25.7). Large-scale planar sliding on bedding seems unlikely due to the steep dip of the bedding relative to the slope.

The volume of the six unstable zones in the lower South Peak area ranges from 0.12 to 1.89 million m^3 (Table 25.2, Fig. 25.7). The total volume of unstable rock in this area is about 5.5 million m^3 (Pedrazzini *et al.*, 2012).

25.3.5 INSTABILITIES IN THIRD PEAK AREA

Investigations were also carried out in two areas on Third Peak: (1) upper Third Peak, which contains the hinge of Turtle Mountain anticline and parts of the west and east fold limbs (SD 4, 5, and 7; Fig. 25.4); and (2) lower Third Peak in the east limb of the fold (SD 7; Fig. 25.4). Instabilities on upper Third Peak are controlled by planar sliding on BP or J1 in the fold hinge area or by toppling on BP in the west fold limb. Only small instabilities (<10,000 m^3) were identified in the hinge zone (Pedrazzini *et al.*, 2012). The rapid change in the orientation of bedding in the fold hinge indicates that large gravitational movements are unlikely in that area. Instability in the lower Third Peak area is evidenced by open tension cracks with a total displacement of 20 cm (Pedrazzini *et al.*, 2012). This area is marked by rugged slope morphology, shallow rock-slope instabilities, and deep-seated gravitational slope deformation. Based on the tension crack displacement vectors and kinematic analysis, movement is most likely to be occurring along a stepped-like planar failure controlled by discontinuity set J1, with BP and J2 or J3′ acting, respectively, as lateral and rear release surfaces. We estimate the total volume of unstable rock in the Third Peak area to be 2.5–3.0 million m^3.

25.3.6 INSTABILITIES IN NORTH PEAK AREA

An investigation of the rocky spur northeast of North Peak (Fig. 25.8) was undertaken when open fractures were discovered behind the peak during field mapping. There is currently no monitoring in this area, which is located on the west limb of the anticline in zone SD 1 (Fig. 25.4). Kinematic analysis revealed potential planar sliding along J2, J3, J3′, and J4, as well as potential wedge sliding along joint set intersections J2–J3, J2–J4, and J3–J4. Joint sets J2, J3, and J3′ are visible at the top of the spur.

We assessed the influence of the joint sets on slope stability in the North Peak area using the Universal Distinct Element Code model UDEC (Itasca, 2004). We used the same mechanical proprieties as Brideau *et al.* (2010). All models were run for 25,000

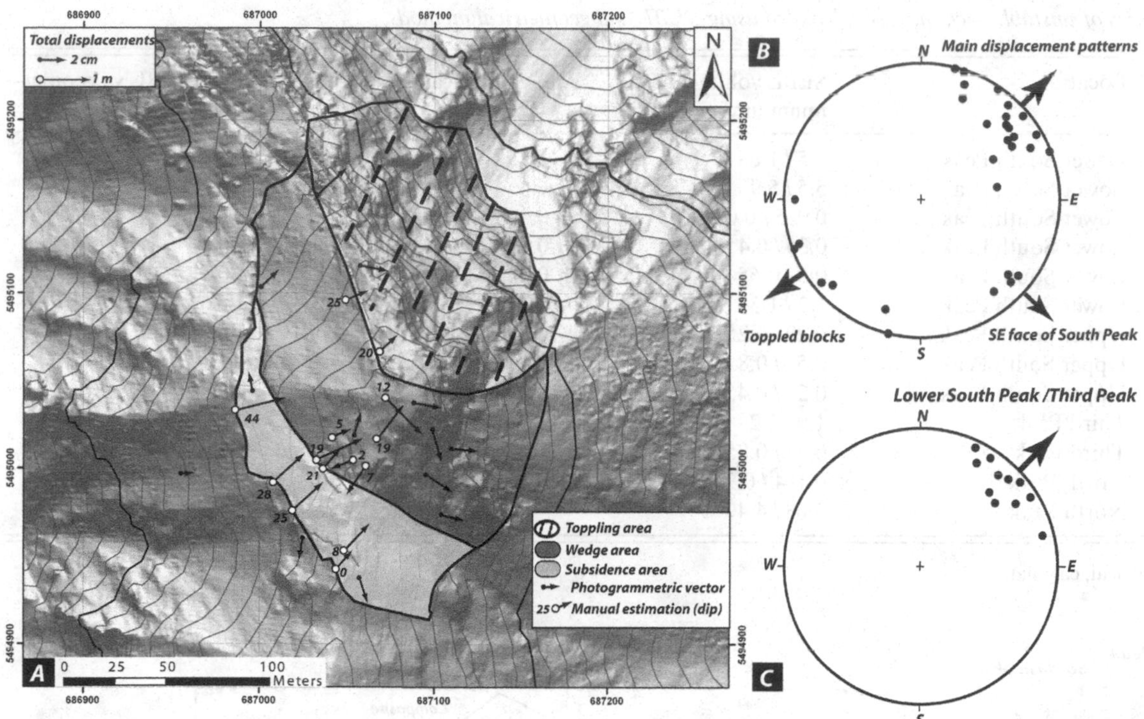

Fig. 25.5. (A) Map of deformation mode sectors and comparison of displacement vector rates measured using photogrammetric techniques between 1984 and 2005 (Froese *et al.*, 2009) and total amount of opening of open cracks, measured manually. (B) Stereographic projection of manually measured displacement vectors for the upper South Peak area (equal area/lower hemisphere projection). (C) Stereographic projection of the manually measured displacement vectors for the lower South Peak and Third Peak areas (equal area/lower hemisphere projection). (Modified from Pedrazzini *et al.*, 2012.)

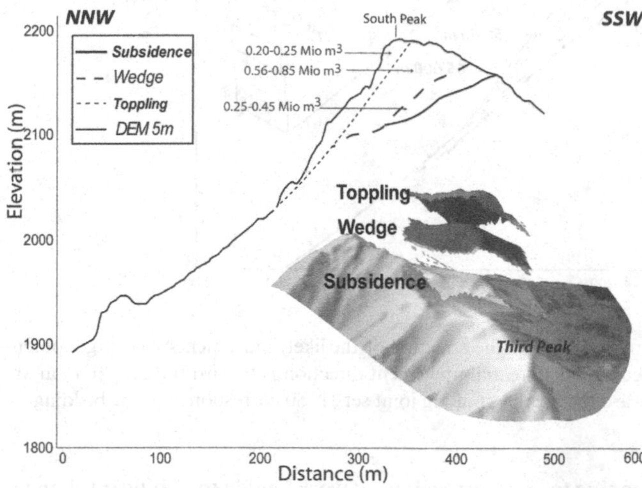

Fig. 25.6. Representation of the SLBL calculation for the upper South Peak instability. The cross-section is oriented along the line of intersection of the planes forming the moving wedge. (Modified from Pedrazzini *et al.*, 2011.)

calculation steps, assuming an elastic constitutive criterion for the rock blocks.

UDEC modeling supports the field observations that joint sets J2, J3, and J3′ play an active role in instability in the upper part of the spur, whereas joint set J4 is more important in the central and basal portions of the cliff (Fig. 25.8D). The failure mechanism can be described as a biplanar sliding: the spur subsides along joint sets J2, J3, or J3′ at the top, while J4 forms the basal sliding surface with J8 acting as a lateral release surface and BP as a rear release surface. Numerical modeling also indicated that only the surficial portion of the cliff is likely to collapse, in agreement with field observations. Following Brideau *et al.* (2010), we defined the volume of the unstable mass as the cumulative volume of blocks displaying displacements greater than 1 m after 25,000 calculation steps. Simulated unstable volumes range from 0.1 to 0.25 million m^3 (Table 25.2). The differences between the simulations stem from differences in the values assigned to the persistence of discontinuity set J4 and the friction angle of the joints, which are the two most important factors in the numerical modeling. Figure 25.8D shows the extent of the unstable area for one of six models.

The SLBL method was also used to evaluate the volume of the unstable rock mass in the North Peak area (Table. 25.2). Morphological features, including open cracks and the zone of surficial instability, provide constraint on the simulated unstable area. The volume estimates range from 0.18 to 0.25 million m^3, which is in agreement with the estimates based on UDEC analyses (Fig. 25.8A). A second series of SLBL models provided estimates of the volume of rock that would be involved in the collapse of the entire spur (Fig. 25.8B, C). Such an event

Table 25.2. *Volumes of unstable rock masses calculated using SLBL and geometrical methods.*

Symbol	Location	SLBL volume (Mm^3) minimum/maximum	Geometrical calculation	Mean SLBL vs geometrical (%)
LSP-1	Lower South Peak	1.5 / 1.65	1.85	−18
LSP-2	Lower South Peak	5.5 / 5.9	6.4	−12
LSP-3	Lower South Peak	0.065 / 0.08	0.065	10
LSP-4	Lower South Peak	0.35 / 0.45	0.48	−20
LSP-5	Lower South Peak	0.5 / 0.65	0.74	−30
LSP-6	Lower South Peak	1.2 / 1.35	1.42	11
USP-1	Upper South Peak	0.20 / 0.25	0.2	11
USP-2	Upper South Peak	0.57 / 0.85	0.8	−13
USP-3	Upper South Peak	0.25 / 0.45	0.5	−43
3dP-1	Third Peak	1.9 / 2.2	2.35	−15
3dP-2	Third Peak	0.57 / 0.85	1.0	−41
NP-1	North Peak	0.184 / 0.249	0.102 / 0.247[1]	−19
NP-2	North Peak	2.68 / 4.4	–	–

[1]UDEC, not geometrical, calculation.

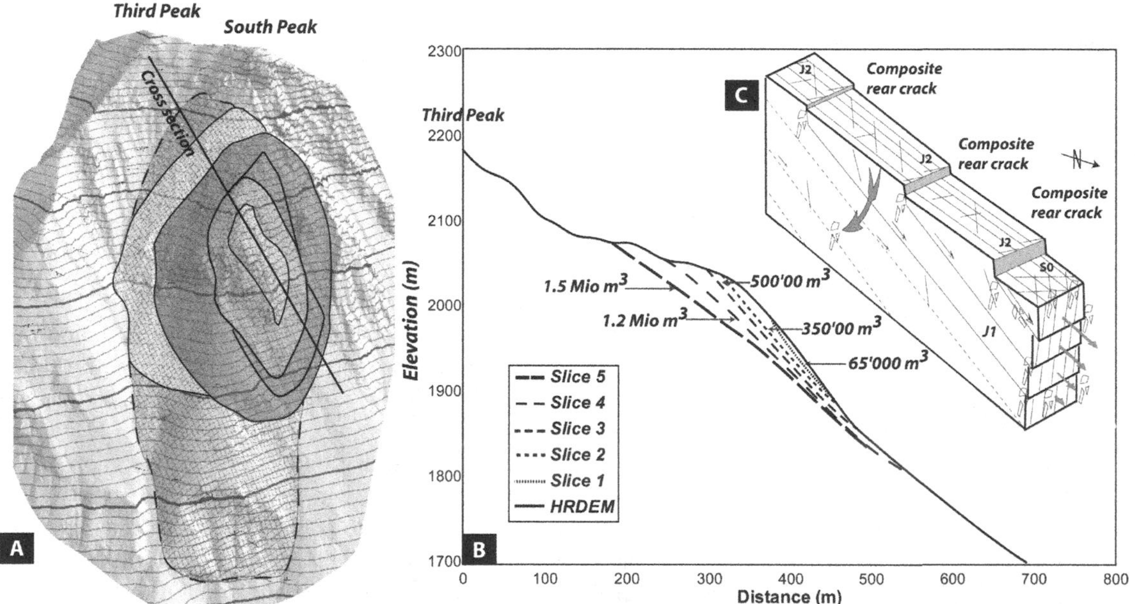

Fig. 25.7. (A) Instability areas detected in the lower South Peak area. (B) Cross-sections in the direction of the likely movements. The larger potential unstable volume (dotted lines in A) is not shown in the cross-section because its potential movement direction is toward the east. (C) Failure model for the lower South Peak area; the rock mass seems to be moving toward the northeast along joint set J1. S0 corresponds to the bedding.

would be catastrophic, but would require a major reduction in the joint surface shear strength (i.e., a lower friction angle) and an increase in the joint persistence. In this scenario, the failure could involve 2.7–4.4 million m^3 of rock.

25.4 HAZARD RISK MANAGEMENT

Different approaches to managing the risk of a future landslide have been tried in the years since the 1903 disaster. The approaches have ranged from land-use planning and hazard zoning to monitoring. Recent work has led to significant changes in hazard management. The following sections describe the evolution of the historic and modern work undertaken to characterize and monitor slope movements at Turtle Mountain, and the use of runout models in land-use planning.

25.4.1 MONITORING

Three levels of monitoring have been applied to the peaks and east face of Turtle Mountain since the 1903 landslide and can be grouped as follows (Larocque, 1977).

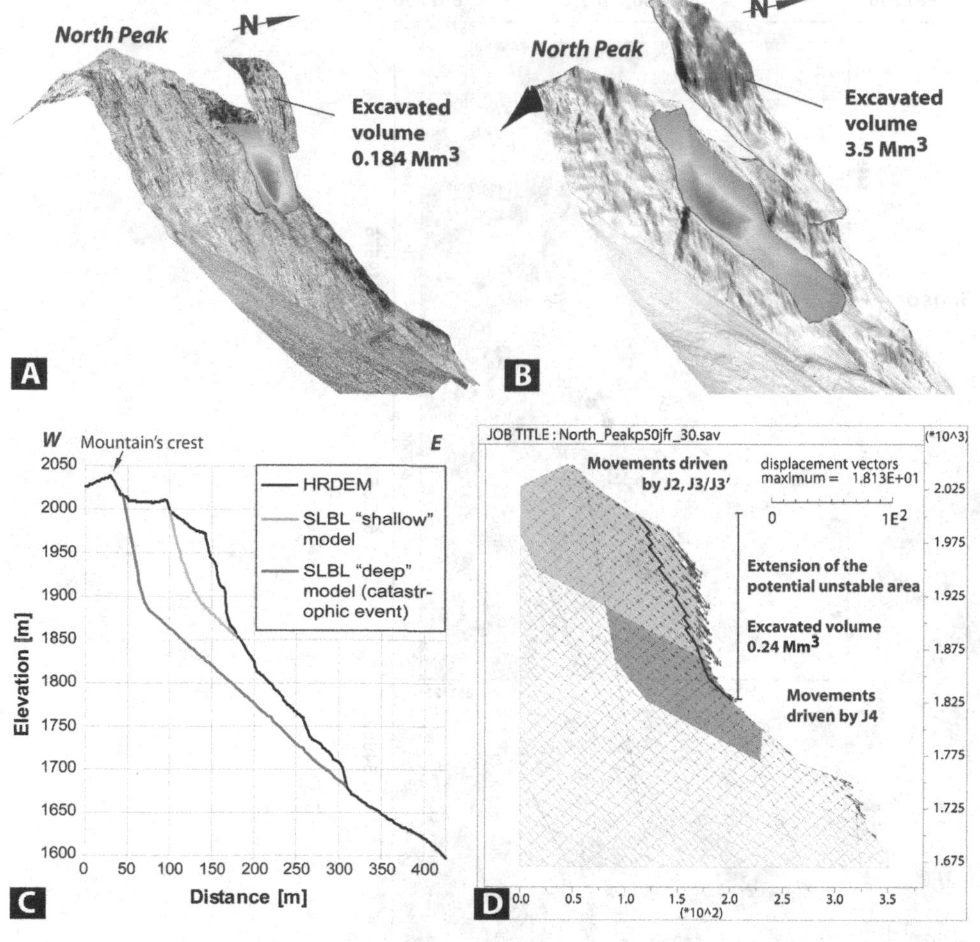

Fig. 25.8. (A) View toward the North Peak area; the extent of the shallow unstable zone is shown. (B) Result of the SLBL calculation for a deep failure model, showing the location of the failure surface. (C) Result of the SLBL calculation for deep and shallow models. (D) UDEC results for shallow models showing the control on movement exerted by joint sets J2 and J3/J3′ at the top of the cliff and set J4 at the base of the cliff.

- **Level 1**: A surveillance system is established to detect the initial stages of instability and to measure geotechnical parameters that are specifically required for design.
- **Level 2**: This is initiated when movements are observed in the Level 1 system; a more detailed system is deployed to better understand the characteristics of the movements.
- **Level 3**: Deployment of a system to provide warnings when movements documented by the Level 2 system are too large to control.

ON-MOUNTAIN MONITORING

Since the 1930s, much of the deformation monitoring on the mountain has involved repeat surveys of specific points, principally around deep fissures on South Peak. Methods that have been used include manual tape measurements at 18 locations (Allan, 1933), surveys of a set of trilateral signs (Cruden, 1986), and a series of 23 photogrammetric targets (Fraser and Gruendig, 1985), and installation of two moire crack gauges (Cruden, 1986; Kostak and Cruden, 1990). All of these methods can be classified as Level 1, based on the criteria outlined by Larocque (1977). Froese *et al.* (2009) completed a detailed review and interpretation of the data from these point surveys.

In the early 2000s, a series of telemetered point targets was installed on South Peak. This network consists of tiltmeters, crack meters, surface wire extensometers, differential GPS, and EDM targets (Fig. 25.9; Moreno and Froese, 2006; Read *et al.*, 2005). The intent was that this nework would function as a Level 3 system and provide warning, but in fact the movements were so low and poorly understood that it serves as a Level 1 system.

A lesson learned from monitoring between 2003 and the present is that harsh weather conditions at the top of Turtle Mountain greatly impact the reliability of the data stream and the lifespan of the sensors. The Alberta Geological Survey has documented the data performance and reliability issues associated with the on-mountain sensors on an annual basis over this period (Moreno and Froese, 2006, 2007, 2009a, 2009b; Froese and Moreno, 2011). In addition, with an improved understanding of the complexity of the deformation zones and mechanics, the use of directionally dependent sensors also created issues. For example, some sensors were only able to detect displacements in one or two directions. Based on these findings and the small movements that are presently occurring (millimeter-level), the focus of the current monitoring program is a Level 1 characterization and not a warning system.

The data from monitoring points installed prior to 2007 and an updated understanding of potentially unstable rock volumes

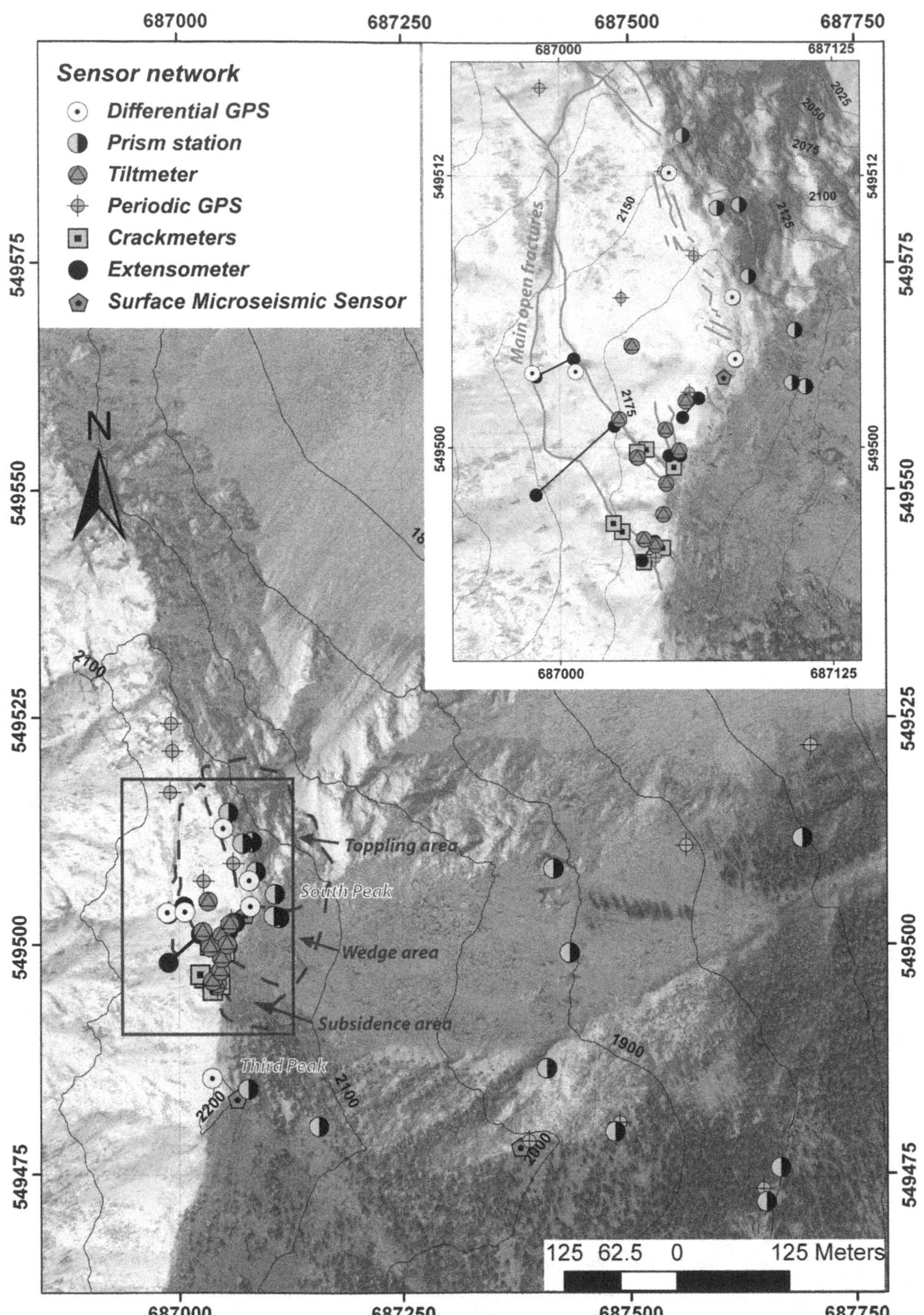

Fig. 25.9. Overview of the sensor network on Turtle Mountain.

and movement directions (Pedrazzini *et al.*, 2012) led to an expansion of the monitoring network, with a focus on directionally independent sensors and remote monitoring technologies. The network now includes 20 mirror prisms on the east face of the mountain (Bjorgan and Froese, 2008), 23 metal pedestals on which dGPS measurements are taken on a semi-annual basis (Teskey and Eberling, 2010; Moreno and Froese, 2009c), and 12 continuous dGPS stations (Bjorgan and Froese, 2008). The dGPS stations are attached to concrete columns placed in zones where the structural mapping and surface morphology indicate that movement is occurring. We have found dGPS to be particularly useful on Turtle Mountain because it provides measurements of deformations in three dimensions, and provides a year-round reliable data stream. The limited resolution of the dGPS data, however, does not allow observation of the onset of very small movements. A summary of upgrades and commentary on data quality are provided by Moreno and Froese (2009c).

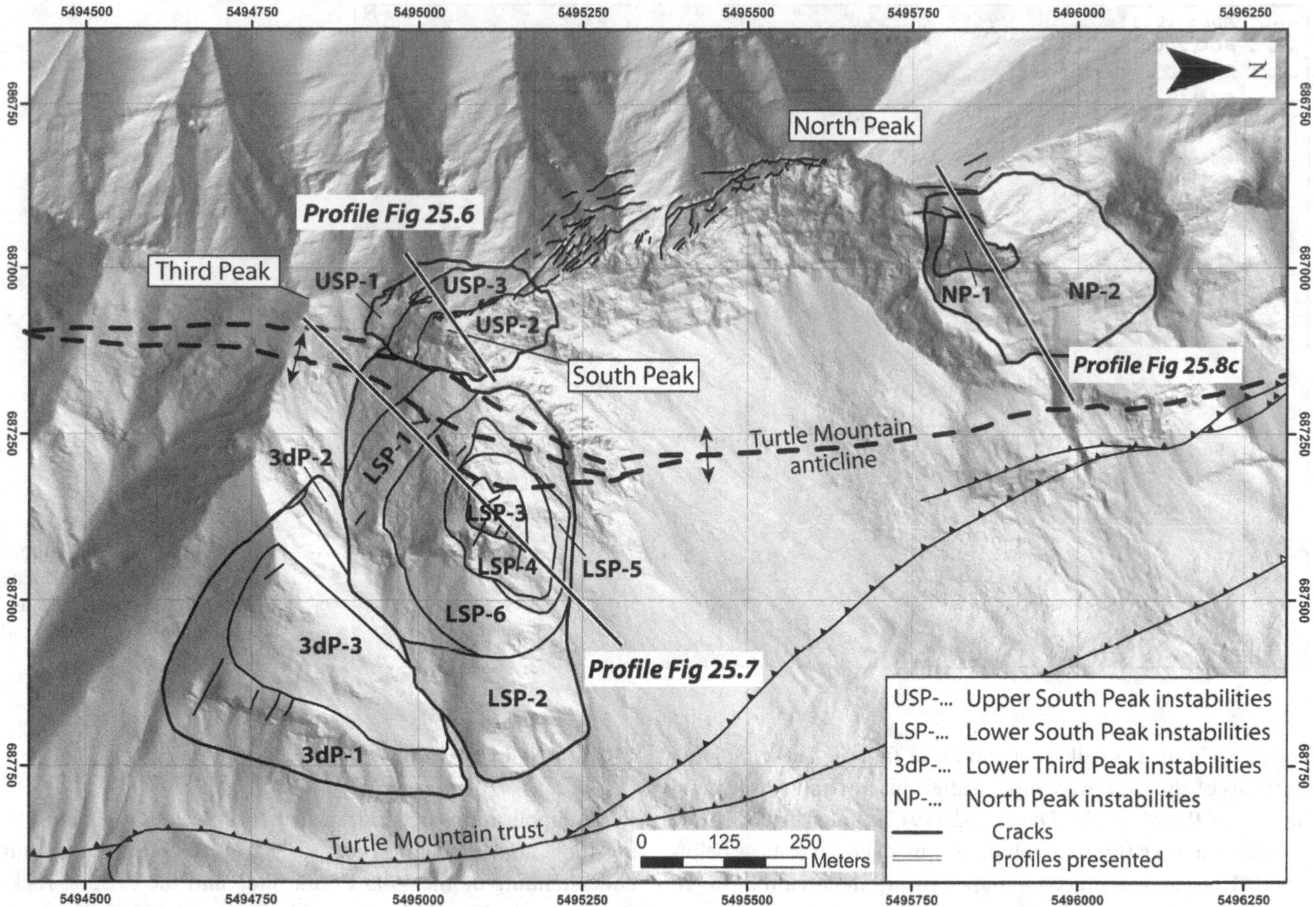

Fig. 25.10. The 14 unstable zones identified in the Third, South and North Peak areas.

REMOTE MONITORING

The peak and east face of Turtle Mountain pose significant issues for monitoring because of the seasonally harsh climate and steep slopes. Snow loading and lightning strikes damage sensors, requiring that personnel access potentially dangerous areas. Unfortunately, the use of "off-mountain" monitoring technologies, such as electronic distance measurement (EDM), is limited because prisms are often covered with snow or ice in winter and can only be employed where there is a clear line of sight between the prisms and the survey instrument. Readings cannot be made when it is dark or in periods of inclement weather. Therefore, more effort was made after 2007 to use technologies that could remotely monitor the east face of the mountain.

In 2009, the Alberta Geological Survey and the University of Alberta installed a ground-based InSAR system (Dehls *et al.*, 2010) in the valley, approximately 2.7 km east of Turtle Mountain. A significant advantage of ground-based InSAR over other laser-type monitoring technologies (EDM, TLS) is that it can acquire data at night and in fog or rain, thus providing a more reliable data stream (Rodelsperger *et al.*, 2010).

The InSAR system at Turtle Mountain was installed at the same location as the EDM total station, a dGPS base station, a meteorological station, and a web camera (webcam) from which photographs of the face of the mountain are collected and archived every 10 minutes. All data can thus be reviewed to assess the source of potential anomalies in the ground-based InSAR data. To date, the InSAR monitoring campaign at Turtle Mountain has consisted of a 3-month calibration campaign in fall 2009 and a 4-month monitoring campaign in spring and summer 2010. In 2009, the InSAR data were compared with the webcam images and the meteorological station data acquired at the base of the valley and the mountain peak to calibrate the system and processing software for atmospheric effects. Atmospheric corrections were then applied to the data collected in 2010. The 2010 data provided a broad overview of movements on the east face of the mountain; the main areas of movement were the talus slopes on the lower two-thirds of the face.

25.4.2 HAZARD ZONING

As described in previous sections, 14 new areas of potential instability have been discovered on Turtle Mountain (Fig. 25.10), in addition to the area initially recognized by Allan (1931). After the 1903 landslide, North Peak was the focus of

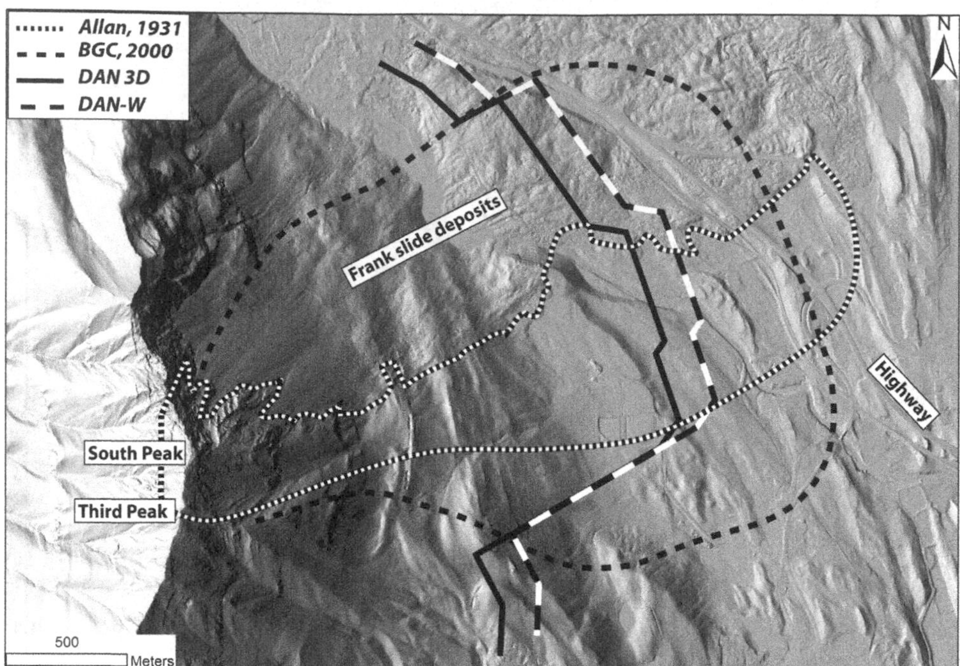

Fig. 25.11. Updated runout scenario and previous estimate of runout for a single large landslide (modified from Pedrazzini *et al.*, 2012).

concern about a possible second large failure. The remaining portions of the town were moved after this possibility was highlighted in the report by Daly *et al.* (1912). Later, Allan (1931) identified a large (up to 5 million m^3) unstable rock mass below South Peak and produced a map showing the area that might be impacted in the event of a sudden failure. He provided no explanation of the method he used to produce the map, but he likely based it on the extent of the 1903 debris sheet and the topography below South Peak. In 1999, the Government of Alberta contracted BGC Engineering (Vancouver, BC) to reassess Allan's estimated failure volume and runout using techniques available at that time.

Most recently, dynamic modeling has been completed to determine the consequences of potential failures, and thus assist land-use planning, in the valley below Turtle Mountain (Hungr, 2008). Both a three-dimensional model (DAN3D) and a two-dimensional model (DANW) were used for the analyses (see Hungr and McDougall, 2009, for a description of the two models). The models were calibrated by back-analyzing the 1903 landslide and other smaller rock avalanches. Maximum runout distances for 12 of the 14 source zones (Fig. 25.10) identified during recent surveys are listed in Table 25.3 and shown in Figure 25.11. Maximum 2D runout distances are typically greater than the 3D runout distances, due to the concentration of energy along a narrow path in the 2D analysis. The direction of movement of the 3D landslide and the assumed 2D profile commonly did not agree, showing the sensitivity of the 3D models to topographic details. As shown in Figure 25.11, runout limits for the 12 failures on South Peak and Third Peak are less extensive than the limits defined by Allan (1931) and BGC Engineering (2000), and therefore present a more optimistic case for considering land-use restrictions in the valley.

25.5 CONCLUSIONS

Recent applications of modern structural techniques, monitoring tools, and modeling methods have greatly increased our understanding of the 1903 Frank Slide and the existing rock-slope hazard at Turtle Mountain. New geospatial modeling tools, coupled with a high-resolution LIDAR digital elevation model (HRDEM) and field mapping, have provided a detailed structural evaluation of the entire mountain and new interpretations of the potentially unstable zones and their volume. Detailed structural analysis has highlighted the influence of folding and post-folding deformation on the development of brittle rock tectonic features that played a primary role in the 1903 landslide and continue to control ongoing slow movements.

The runout scenarios of Allan (1931, 1933) and BGC Engineering (2000), which until recently have been the basis for discussions and emergency planning with municipal and provincial emergency management officials, are based on the assumption of one large (*ca.* 5 million m^3) rock-slope failure (Froese *et al.*, 2005; Froese and Moreno, 2011). These scenarios were made without the advantage of the mapping technologies and runout models that are available today. More recent work indicates that present instabilities on Turtle Mountain involve rock masses of much smaller volume, implying less runout than had previously been envisioned. This information has been presented formally to the municipal and provincial emergency management officials. Although documented instabilities on Turtle Mountain indicate a series of small failures, the potential for a single, large failure cannot be completely ruled out. As a result, the current plan for emergency response and evacuation continues to use the larger-volume runout estimate provided by BGC Engineering (2000).

Table 25.3. *Potential source volumes identified by HRDEM analysis and calculated using SLBL (from Pedrazzini* et al., *2012).*

Symbol	Location	Expanded volume ($m^3 \times 10^6$)	Bulk friction angle (°)
LSP-1	South Peak	1.89	25.0
LSP-2	Lower South Peak	6.59	20.0
LSP-3	South Peak	0.12	36.0
LSP-4	South Peak	0.46	30.4
LSP-5	South Peak	0.63	29.7
LSP-6	South Peak	1.44	26.2
USP-1	Upper South Peak	1.99	24.5
USP-2	Upper South Peak	0.30	31.0
USP-3	Upper South Peak	1.37	26.3
3dP-1	Third peak	2.59	23.2
3dP-2	Third Peak	0.05	37.0
3dP-3	Third Peak	1.37	26.3

The Turtle Mountain monitoring system has been extended to the lower Third Peak, east face, and lower South Peak areas in order to characterize potential movements there. The current system covers the above areas and consists of an array of overlapping GPS monitoring points and a series of 20 mirror prisms that are monitored from a robotic total station located on the valley floor, approximately 3 km from the east face of the mountain. Movement rates of up to 5 mm per year have been documented at the crest of Turtle Mountain with the EDM system, photogrammetric targets, dGPS, and the ground-based InSAR. The different monitoring results are in agreement, indicating that these technologies are able to accurately measure very slow movements. The results increase our confidence that movements elsewhere on the east face of the mountain can be detected. The fact that, to date, no displacements have been observed elsewhere on the mountain indicates that any movements, if they are occurring, are less than a few millimeters per year.

REFERENCES

Allan, J.A. (1931). *Report on the Stability of Turtle Mountain, Crowsnest District, Alberta.* Edmonton, AB: Department of Public Works.

(1933). *Report on Stability of Turtle Mountain, Alberta and Survey of Fissures between North Peak and South Peak*. Edmonton, AB: Alberta Provincial Archives, Alberta Department of Public Works.

Benko, B. and Stead, D. (1998). The Frank Slide: A reexamination of the failure mechanism. *Canadian Geotechnical Journal*, 35, 299–311.

BGC Engineering (2000). *Geotechnical Hazard Assessment of the South Flank of Frank Slide, Hillcrest, Alberta*. Report to Alberta Environment, 00–0153.

Bjorgan, G. and Froese, C.R. (2008). Combined GPS and EDM monitoring on Turtle Mountain, Alberta. In *Proceeding of the 61st Canadian Geotechnical Conference and 9th Joint CGS/IAH-CNC Groundwater Conference*, Edmonton, AB, pp. 1236–1242.

Brideau, M.-A., Pedrazzini, A., Stead, D. *et al.* (2010). Three dimensional slope stability analysis of South Peak, Crowsnest Pass, Alberta, Canada. *Landslides*, 8, 139–158.

Charrière, M. (2011). *Granulometrical, geological and morphological description of the Frank Slide deposit (Alberta, Canada)*. M.Sc. dissertation, University of Lausanne, Switzerland.

Couture, R., Locat, J., Drapeau, J.P., Evans, S.G. and Hadjigeorgiou, J. (1998). Evaluation de la granulométrie à la surface de débris d'avalanche rocheuse par l'analyse d'images. In *Proceedings of the 8th International Congress of the International Association for Engineering Geology and the Environment, Vol. 2*, ed. D.P. Moore and O. Hungr. Rotterdam: A.A. Balkema, pp. 1383–1390.

Cruden, D.M. (1986). *Monitoring the South Peak of Turtle Mountain, 1980 to 1985*. Alberta Environment, Research Management Division and Earth Sciences Division, RMD Report 86/37.

Cruden, D.M. and Hungr, O. (1986). The debris of the Frank Slide and theories of rockslide-avalanche mobility. *Canadian Journal of Earth Sciences*, 23, 425–432.

Cruden, D.M. and Krahn, J. (1973). A re-examination of the geology of Frank slide. *Canadian Geotechnical Journal*, 10, 581–591.

(1978). Frank Rockslide, Alberta, Canada. In *Rockslides and Avalanches: Vol.1 – Natural Phenomena*, ed. B. Voight. Amsterdam: Elsevier, pp. 97–112.

Cruden, D.M. and Martin, C.D. (2007). Before the Frank Slide. *Canadian Geotechnical Journal*, 44, 765–780.

Daly, R.A., Miller, W.G. and Rice, G.S. (1912). *Report of the Commission Appointed to Investigate Turtle Mountain, Frank, Alberta*. Canadian Geological Survey, Memoir 27.

Dehls, J., Giudici, D., Mariotti, F. *et al.* (2010). Monitoring Turtle Mountain using ground-based synthetic aperture radar (GB-InSAR). In *Proceedings of the 63rd International Canadian Geotechnical Conference and 6th Canadian Permafrost Conference*, Calgary, AB, pp. 1635–1640.

Drapeau, J.P. (1997). *Évaluation de la Distribution Granulométrique des Débris d'un Ecroulement Rocheux à l'Aide d'un Système d'Analyse d'Images Digitalisées*. Rapport de Projet de Fin d'Études, Université Laval, Département de Géologie et de Génie Géologique, Montréal, PQ.

Fossey, K.W. (1986). *Structural geology and slope stability of the southeast slope of Turtle Mountain, Alberta*. M.Sc. thesis, University of Alberta, Edmonton, AB.

Fraser, C.S. and Gruendig, L. (1985). The analysis of photogrammetric deformation measurements on Turtle Mountain. *Photogrammetric Engineering and Remote Sensing*, 51, 207–216.

Froese, C.R. and Moreno, F. (2011). Structure and components for the emergency response and warning system on Turtle Mountain, Alberta, Canada. *Natural Hazards*, doi:10.1007/s11069–011–9714-y.

Froese, C.R., Murray, C., Cavers, D.S. *et al.* (2005). Development of a warning system for the south peak of Turtle Mountain. In *Landslide Risk Management*, ed. O. Hungr, R. Fell, R.R. Couture and E. Eberhardt. Leiden: A.A. Balkema, pp. 705–712.

Froese, C.R., Moreno, F., Jaboyedoff, M. and Cruden, D.M. (2009). 25 years of movement monitoring on South Peak, Turtle Mountain: Understanding the hazard. *Canadian Geotechnical Journal*, 46, 256–269.

Humair, F. (2011). *Turtle Mountain anticline (Alberta, Canada): Rock slope stability related fracturing–folding, fracturing, rock mass condition, slope stability investigations and geological modeling*. M.Sc. thesis, University of Lausanne, Switzerland.

Humair, F., Charrière, M., Pedrazzini, A. *et al.* (2010). The Frank Slide (Alberta, Canada): From the contributing factors to the processes of propagation. In *Proceedings of the 63rd International Canadian Geotechnical Conference and 6th Canadian Permafrost Conference*. Calgary, AB, pp. 1613–1621.

Hungr, O. (2008). *Geotechnical Research (2008): Turtle Mountain, Frank, Alberta: Runout Analyses of Potential Landslides on South and Third Peaks*. Unpublished report prepared for Alberta Geological Survey.

Hungr, O. and Evans, S. G. (2004). Entrainment of debris in rock avalanches: An analysis of long run-out mechanism. *Geological Society of America Bulletin*, 116, 1240–1252.

Hungr, O. and McDougall, S. (2009). Two numerical models for landslide dynamic analysis. *Computers and Geosciences*, 35, 978–992.

Itasca (2004). *UDEC Version 4.0 Universal Distinct Element Code*. Minneapolis, MN: Itasca Consulting Group Inc.

Jaboyedoff, M. and Tacher, L. (2006). Computations of landslide slip surface using DEM. Geological Society of London, William Smith Meeting, 2004.

Jaboyedoff, M., Metzger, R., Oppikofer, T. *et al.* (2007). New insight techniques to analyze rock slope relief using DEM and 3D-imaging cloud points: COLTOP-3D software. In *Rock Mechanics: Meeting Society's Challenges and Demands, Vol 1*, ed. E. Eberhardt, D. Stead and T. Morrison. London: Taylor & Francis, pp. 61–68.

Jaboyedoff, M., Couture, R. and Locat, P. (2009). Structural analysis of Turtle Mountain (Alberta) using digital elevation model: Towards a progressive failure. *Geomorphology*, 103, 5–16.

Kent, P. E. (1966). The transport mechanism in catastrophic rock falls. *Journal of Geology*, 74, 79–83.

Kostak, B. and Cruden, D. M. (1990). The moire crack gauges on the crown of the Frank Slide. *Canadian Geotechnical Journal*, 27, 835–840.

Langenberg, C. W., Pana, D., Richards, B. C., Spratt, D. A. and Lamb, M. A. (2007). *Structural Geology of the Turtle Mountain Area near Frank, Alberta*. Energy Resources Conservation Board, EUB/AGS Science Report 2007–01.

Larocque, G. (1977). Monitoring. In *Pit Slope Manual*, ed. Mining Research Laboratories (Canada). Canada Centre for Mineral and Energy Technology, CANMET Report 77-1-5.

Locat, P., Couture, R., Leroueil, S., Locat, J. and Jaboyedoff, M. (2006). Fragmentation energy in rock avalanches. *Canadian Geotechnical Journal*, 43, 830–851.

McConnell, R. G. and Brock, R. W. (1904). *Report on the Great Landslide at Frank, Alberta*. Canada Department of the Interior, Annual Report for 1903.

Melosh, H. J. (1987). The mechanics of large rock avalanches. *Geological Society of America, Reviews in Engineering Geology*, 7, 41–49.

Mills, H. H. (2003). Inferring erosional resistance of bedrock units in the east Tennessee mountains from digital elevation data. *Geomorphology*, 55, 263–281.

Moreno, F. and Froese, C. R. (2006). *Turtle Mountain Field Laboratory, Monitoring and Research Summary Report, January to December 2005*. Alberta Energy and Utilities Board, EUB/AGS Earth Sciences Report 2006–7.

(2007). *Turtle Mountain Field Laboratory Monitoring and Research Summary Report, 2005*. Alberta Energy and Utilities Board, EUB/AGS Earth Sciences Report 2006–07.

(2009a). *ERCB/AGS Roles and Responsibilities: Manual for the Turtle Mountain Monitoring Project, Alberta*. Energy Resources Conservation Board, ERCB/AGS Open File Report 2009–06.

(2009b). *ERCB/AGS System Manual for the Turtle Mountain Monitoring Project, Alberta*. Energy Resources Conservation Board, ERCB/AGS Open File Report 2009–14.

(2009c). *Turtle Mountain Field Laboratory, Alberta (NTS 82G): 2008 Data and Activity Summary*. Energy Resources Conservation Board, ERCB/AGS Open File Report 2009–15.

Pedrazzini, A., Jaboyedoff, M., Froese, C. R, Langenberg, C. W. and Moreno, F. (2011). Structural analysis of Turtle Mountain; Origin and influence of fractures in the development of rock failure. In *Slope Tectonics*, ed. M. Jaboyedoff. Geological Society of London, Special Publication 351, pp. 163–183.

Pedrazzini, A., Froese, C. R., Jaboyedoff, M., Hungr, O. and Humair, F. (2012). Combining digital elevation model analysis and run-out modeling to characterize hazard posed by a potentially unstable rock slope at Turtle Mountain, Alberta, Canada. *Engineering Geology* 128, 76–94.

Read, R. S., Langenberg, W., Cruden, D. *et al.* (2005). Frank Slide a century later: The Turtle Mountain monitoring project. In *Landslide Risk Management*, ed. O. Hungr, R. Fell, R. R. Couture and E. Eberhardt. Leiden: A. A. Balkema, pp. 713–723.

Richards, L. R., Leg, G. M. M. and Whittle, R. A. (1978). Appraisal of stability conditions in rock slopes. In *Foundation Engineering in Difficult Ground*, ed. F. G. Bell. London: Newnes-Butterworths, pp. 449–512.

Rodelsperger, S., Laufer, G., Gerstenecker, C. and Becker, M. (2010). Monitoring of displacements with ground-based microwave interferometry: IBIS-S and IBIS-L. *Journal of Applied Geodesy*, 4, 41–54.

Terzaghi, K. (1950). Mechanism of landslides. In *Application of Geology and Engineering Practice*, ed. S. Paige. Boulder, CO: Geological Society of America, pp. 83–123.

Teskey, W. and Eberling, A. (2010). *Turtle Mountain Deformation Monitoring, 2007–2010, Final Report*. Unpublished report prepared by the Department of Geomatics Engineering, University of Calgary for Alberta Geological Survey.

26 The Åknes rockslide, Norway

LARS HARALD BLIKRA

ABSTRACT

The Åknes rockslide is a large, slow-moving landslide in Precambrian gneiss in Sunnylvsfjorden in western Norway. It has a volume of more than 50 million m^3 and parts are moving at velocities of 2–8 cm per year. If the sliding mass were to fall into the fjord, it would generate large tsunami waves. Given the hazard, a major site investigation was conducted and a monitoring program was established in cooperation with a number of national and international groups. The monitoring program integrated a variety of surface and subsurface instruments, including extensometers, crackmeters, tiltmeters, single lasers, GPS, total station, ground-based radar, geophones, climate station, and borehole inclinometers and piezometers. Reliable power and communications systems operate the instruments and transmit data. Movement data collected to date demonstrate continuous movement throughout the year, but with significant seasonal differences. During spring snowmelt and heavy precipitation events, the rate of movement can increase to 1 mm per day, which is 10 times the annual mean. Preliminary early-warning levels and associated actions have been implemented based on data from the Åknes rockslide and information on historic rockslides elsewhere in coastal Norway.

26.1 INTRODUCTION

The Åknes rockslide is located on the northwest flank of Sunnylvsfjorden in western Norway (Fig. 26.1). It has an estimated volume of up to 54 million m^3 and is moving at a velocity of up to 8 cm per year. Catastrophic failure of the rock mass would trigger a devastating tsunami in the fjord (Blikra, 2008; Glimsdal and Harbitz, 2011). Because of the size of the moving rock mass, remedial engineering measures are not feasible and the risk must be managed by implementing an effective warning system. As pointed out by Crosta and Agliardi (2003), remedial countermeasures are generally not useful for large rockslides, due to the extremely high kinetic energy involved.

A major investigation, monitoring, and early-warning program was begun at Åknes in 2004. The operational monitoring and early-warning system is now administered by the Åknes/Tafjord Early Warning Center on a permanent basis. At the moment, emergency planning entailing evacuation, road closure, and other active measures is the only effective tool to reduce the consequences of a rockslide-generated tsunami. One of the most challenging tasks of the program was to develop realistic monitoring thresholds that are not too conservative, resulting in excessive false alarms, but that would provide adequate warning should a catastrophic failure occur (Blikra, 2008; Froese and Moreno, 2011). This chapter presents the concept and principles of the investigation program and the implemented monitoring and early-warning systems. It also provides an overview of the rockslide and selected surface and subsurface deformation data.

26.2 OVERVIEW OF GEOLOGY AND GEOMORPHOLOGY

The Åknes rockslide is located in the Western Gneiss Region (WGR) of Norway and is seated in medium-grained granitic and granodiorite gneisses of Proterozoic age (Braathen *et al.*, 2004; Ganerød *et al.*, 2008). The gneisses contain bands and lenses of mafic material. They have well-developed foliation and mineral banding (Braathen *et al.*, 2004) and numerous centimeter- to decameter-scale tight folds. At Åknes, biotite-rich layers up to 20 cm thick coincide with zones of high fracture frequency (Ganerød *et al.*, 2008). The sliding surfaces are

Landslides: Types, Mechanisms and Modeling, ed. John J. Clague and Douglas Stead. Published by Cambridge University Press.

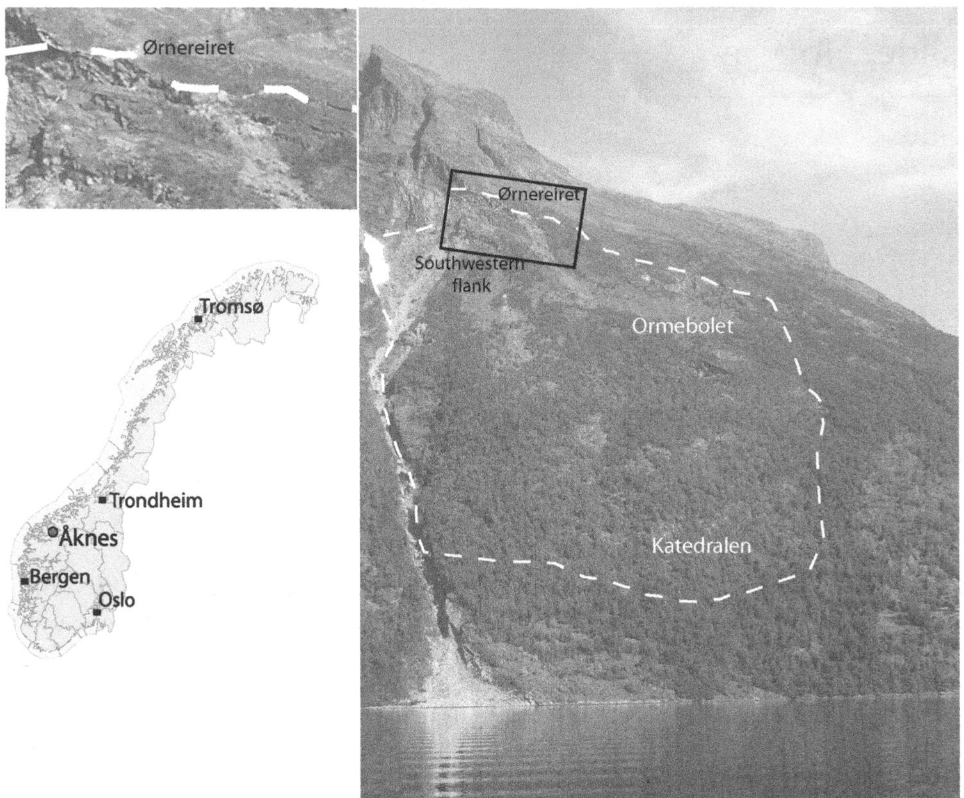

Fig. 26.1. Location (inset map) and view of the Åknes rockslide showing the extent of the unstable area. The inset image is from the upper area, with well-defined open tension fractures and the graben.

likely located within these mica-rich layers. The foliation generally slopes parallel to the surface. Well-defined, very steep, sharp folds are related to the tension cracks at the top of the landslide (Ganerød *et al.*, 2008; Jaboyedoff *et al.*, 2011).

Morphological investigations have revealed several characteristic features of the landslide (Fig. 26.2), including a prominent upper fracture system that can be followed for more than 500 m. The slope-parallel foliation and weak biotite-rich layers control the large-scale displacement dynamics. A large depression or graben has developed in the upper west corner of the rockslide (Fig. 26.1, detail). The total vertical displacement measured here is 20–30 m. Tension fractures are also present in the upper and the middle parts of the slope. Prominent slide scarps characterize the east side of the deep canyon that defines the west boundary of the landslide. Historic data indicate that slides occurred on the upper part of the slope in the late 1800s, 1940, and 1960. Small slide scars also characterize the lower part of the rockslide. Large blocks have been pushed or squeezed out of the slope in two areas, one in the middle and the other on the lowermost part of the rockslide (Fig. 26.2). Springs discharge water on the lowermost part of the slope at about 100 m asl, and there are also smaller springs in the middle part of the landslide (Frei, 2008).

26.3 INVESTIGATIONS

The principles of the investigation program and selected monitoring data are presented here. The methods described have proven to be important in evaluating the geometry and volume of the moving rock mass and in designing the monitoring system.

26.3.1 PRINCIPLES OF THE INVESTIGATION PROGRAM

The investigation program was important for several reasons.

- First, we need to define rockslide scenarios in order to model possible tsunamis. Realistic volume estimates and failure scenarios are essential for risk management planning for numerous municipalities along the fjord.
- Second, we need to know the location and extent of the unstable area and the displacement pattern in order to design and implement an effective monitoring program.
- Third, we need an extensive knowledge base to conduct reliable and real-time monitoring and to provide early warning if required.

A thorough understanding of deformation dynamics is especially important during critical events, when decisions regarding alarm levels and evacuation have to be made at short notice. There are no guidelines or handbooks defining the specific requirements for the type and level of investigations needed to perform reliable monitoring and to provide early warning of large landslides. Because of the high risk related to large rockslides and tsunamis, our requirements are demanding. It must

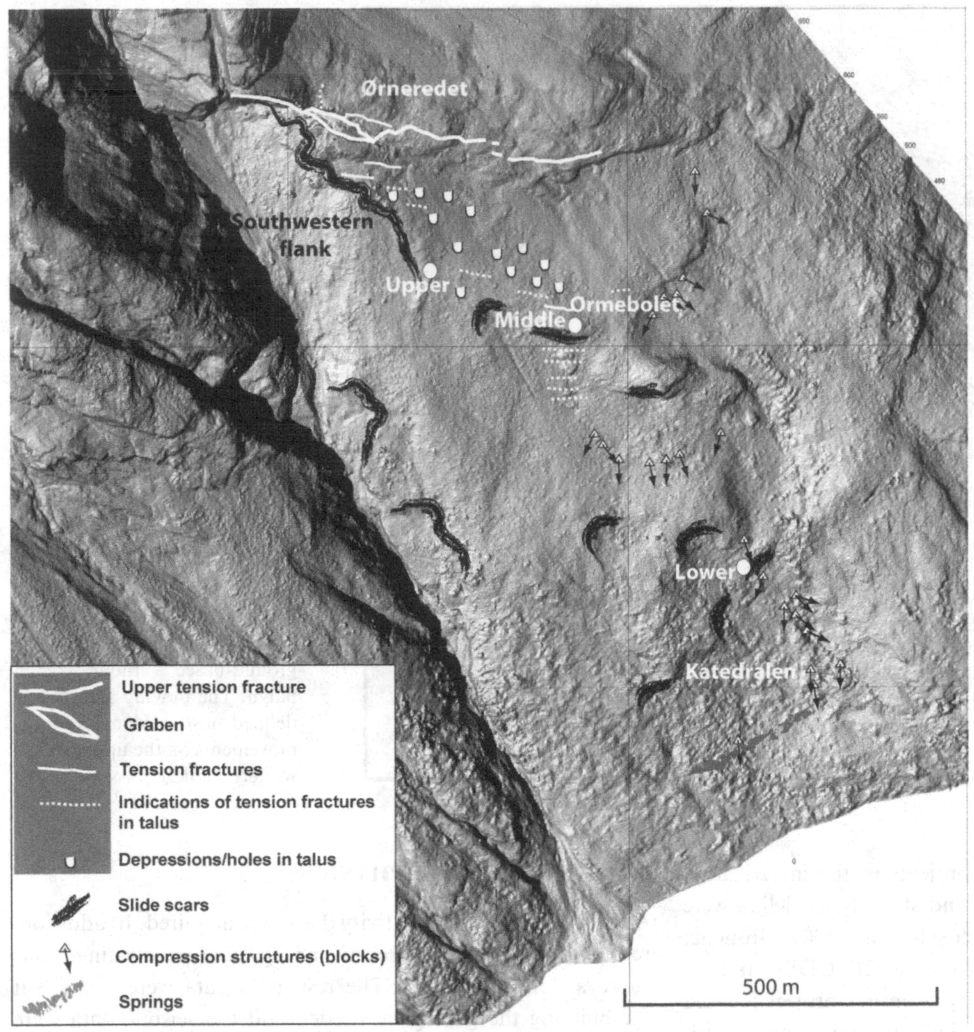

Fig. 26.2. Shaded relief map showing morphological features of the Åknes rockslide (modified from Blikra, 2008). White circles are borehole locations (upper, middle and lower location).

be stressed however, that the investigations need to be based on the local conditions.

SURFACE INVESTIGATIONS

We required high-resolution topographic data and thus carried out an airborne LIDAR campaign by helicopter. Based on this survey, we produced topographic maps with 1 m contour intervals. Detailed geological data were also required, including structural geology, geomorphology, and hydrology. Surface displacements were investigated in great detail using satellite-based and ground-based radar, ground-based LIDAR, and periodic measurements using GPS and total station. The surface investigations allowed us to define the distribution of the unstable zones and provided an overview of the deformation in different sectors of the landslide.

SUBSURFACE INVESTIGATIONS

We also required subsurface geological data, including the depth of the sliding zones and the related deformation, in order to obtain a realistic 3D geometric model of the instability. These data were acquired by drilling and geophysical surveys. Extensive 2D resistivity measurements were made at Åknes and supplemented with refraction and reflection seismic surveys (Rønning *et al.*, 2006, 2007). Boreholes were drilled, and near-continuous core acquired, at three sites on the landslide. In-hole geophysical measurements were made in the three boreholes (Ganerød *et al.*, 2007). Televiewer logging of boreholes provides additional information on subsurface rock structure, including fractures and foliation, but was only performed in the upper part of one of the boreholes.

LABORATORY ANALYSIS AND MODELING

To perform stability modeling, we collected rock samples from the surface and from drill cores for analysis of shear strength and other input parameters. Although a stability model cannot conclusively demonstrate that a slope will ultimately fail, it does provide an understanding of the critical parameters for failure (sensitivity analysis). Modeling of landslide runout must be performed as part of the investigation program. At Åknes,

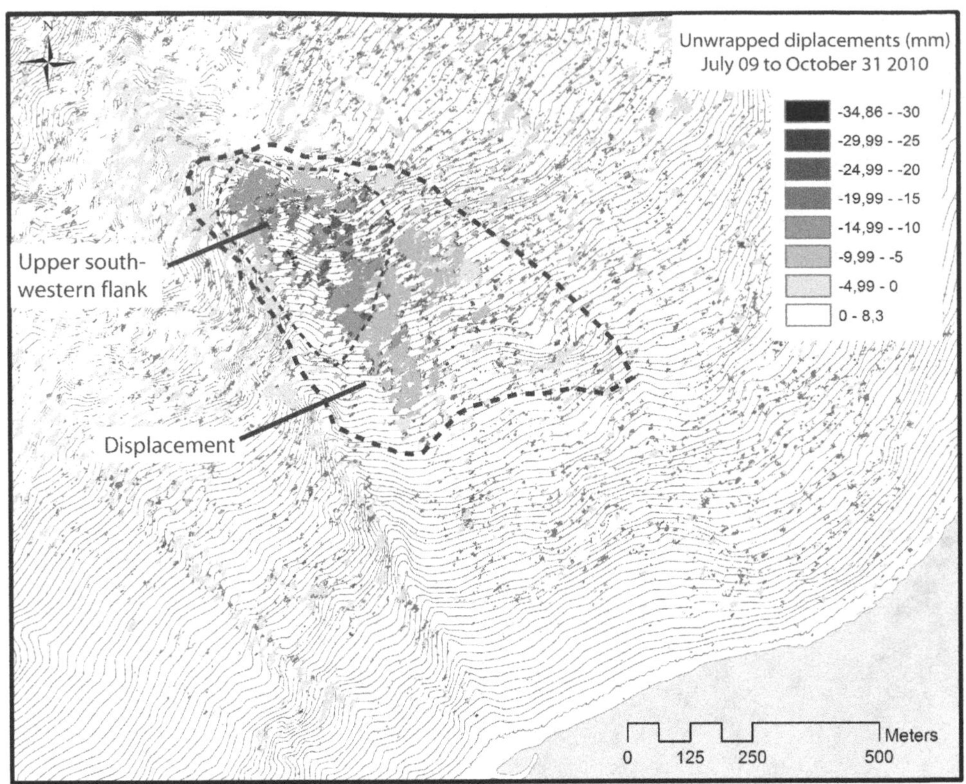

Fig. 26.3. Velocities measured for the period July 9 to October 31, 2010 during the LiSALab ground-based radar campaign. The moving area is well defined; also evident are large movements on the upper southwestern flank.

we linked several Ph.D. research projects to the investigation program, and laboratory analysis and stability modeling were done as part of these projects (Kveldsvik *et al.*, 2008; Grøneng, 2010; Grøneng *et al.*, 2010). Codes such as 2D UDEC and 3D FLAC were used to model stability. Another project focused on slide and tsunami modeling and included laboratory experiments (Glimsdal and Harbitz, 2011).

26.3.2 SURFACE DISPLACEMENTS

Movement of the Åknes slope has been measured using GPS, total station, ground-based radar, ground-based LIDAR, InSAR, extensometers, and single lasers. The first deformation measurements were made using classical methods, including GPS, total station, and ground-based radar. These data showed displacements in large parts of the unstable area, with exceptionally high velocities of 6–8 cm per year on the southwestern flank (detail in Fig. 26.1, Fig. 26.2). We also measured displacements of up to 15 cm per year in the center of the upper graben. Movement rates in areas below and east of the graben were 2–4 cm per year. These early data have been confirmed using InSAR and ground-based LIDAR (Oppikofer, 2009). A recent ground-based radar campaign from Oaldsbygda, on the other side of the fjord, has also confirmed the high velocities in the upper part of the landslide (Fig. 26.3). Displacements in the lower part of the unstable area have not yet been recorded. However, geological mapping, boreholes, and geophysical surveys all show that this area has deformed gravitationally. This area may have been active during an earlier phase of the landslide.

26.3.3 SURFACE GEOPHYSICS

Almost 12 km of 2D resistivity data were acquired. In addition, refraction and reflection seismic and ground-penetrating radar surveys were performed. The resistivity data were important for building the geological model, and the seismic data were required to link the resistivity values to seismic velocities, in part because both dry highly fractured rocks and more competent nonfractured rocks have high resistivity values. The following model was developed to explain the subsurface geophysical characteristics (Figure 26.4; Rønning *et al.*, 2006):

- an *upper zone* of debris and dry and fractured bedrock with high resistivity and low seismic velocity (0 to 40–50 m depth)
- a *middle zone*, from 40–50 to 120 m depth, with low resistivity and higher seismic velocities, which is probably fractured, water-saturated rock
- a *lower zone* with high resistivity and high seismic velocities, indicative of less-fractured bedrock.

26.3.4 CORE DRILLING

An important part of the investigations at Åknes has been the drilling program. Seven boreholes were drilled to 150–200 m depth at three locations (Fig. 26.2). Core was logged during drilling, and measurements were made of water pressure, water flow, the system pressure, and penetration velocity. These data provided important information on hydrologic conditions and on the occurrence of large fractures. There was a marked reduction in water pressure and system pressure at 35–50 m depth

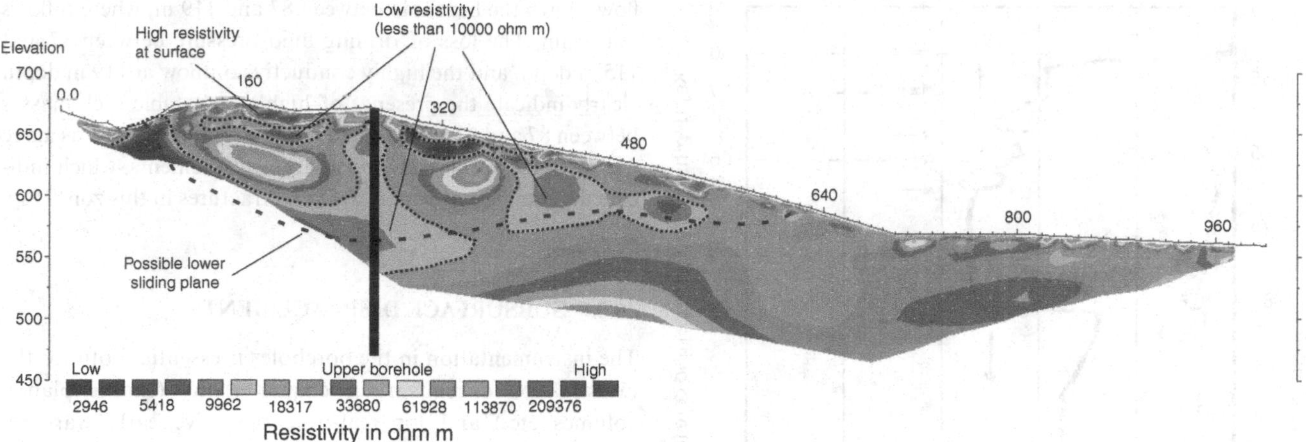

Fig. 26.4. 2D resistivity cross-section at the upper borehole (Rønning *et al.*, 2007). The stippled line delineates the possible depth of the unstable area.

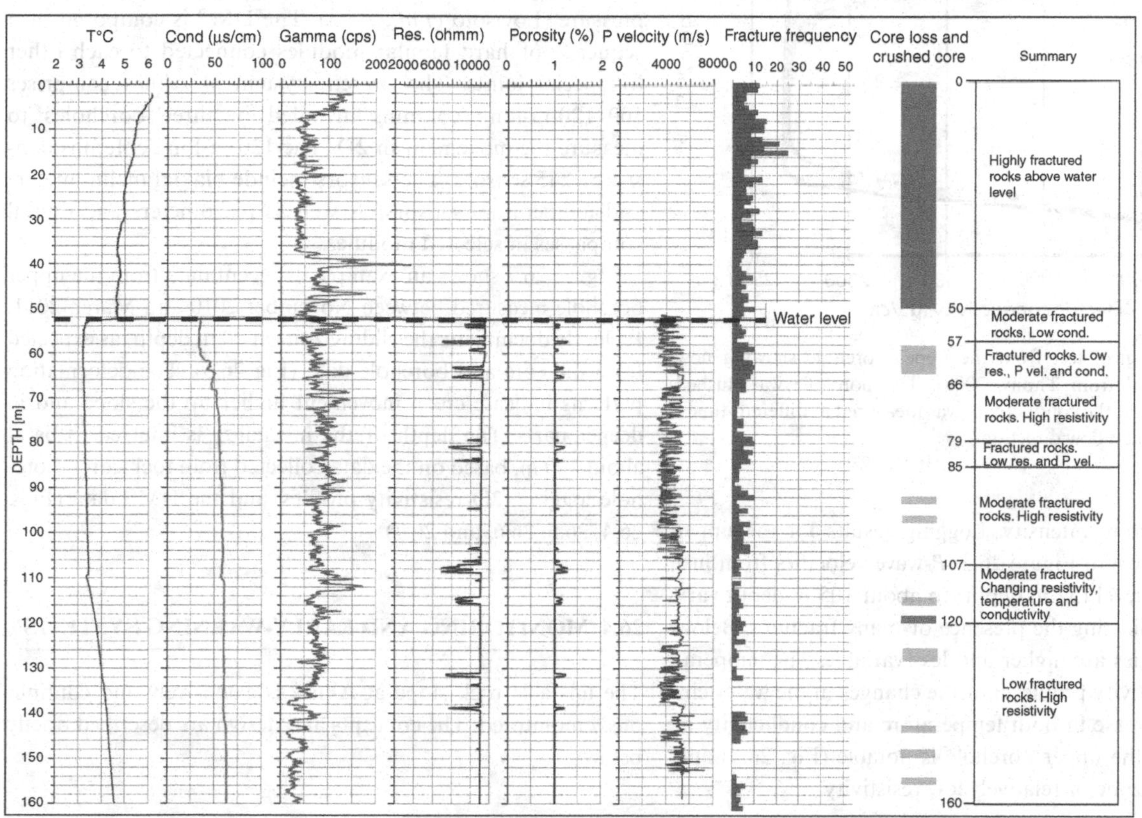

Fig. 26.5. Geophysical log and summary of data from the cores (modified from Rønning *et al.*, 2006).

during drilling of the upper borehole, when large open fractures were encountered. Similar reduced water pressure could be observed at 120 m depth.

The cores were of excellent quality and were carefully logged to provide information about rock quality and the characteristics of fractures and weak zones (Ganerød *et al.*, 2007). Some zones were totally crushed, and others had well-defined breccias with silt and clay; we experienced some core loss in zones of crushed rock. The fracture frequency in cores provided a good indication of down-hole variation in rock quality (Fig. 26.5).

26.3.5 BOREHOLE LOGGING

All boreholes were logged using a number of different methods before in-situ instrumentation was installed. Resistivity profiles showed the locations of zones with water or water-filled fractures, and *P*-wave velocity profiles provided an overview of rock

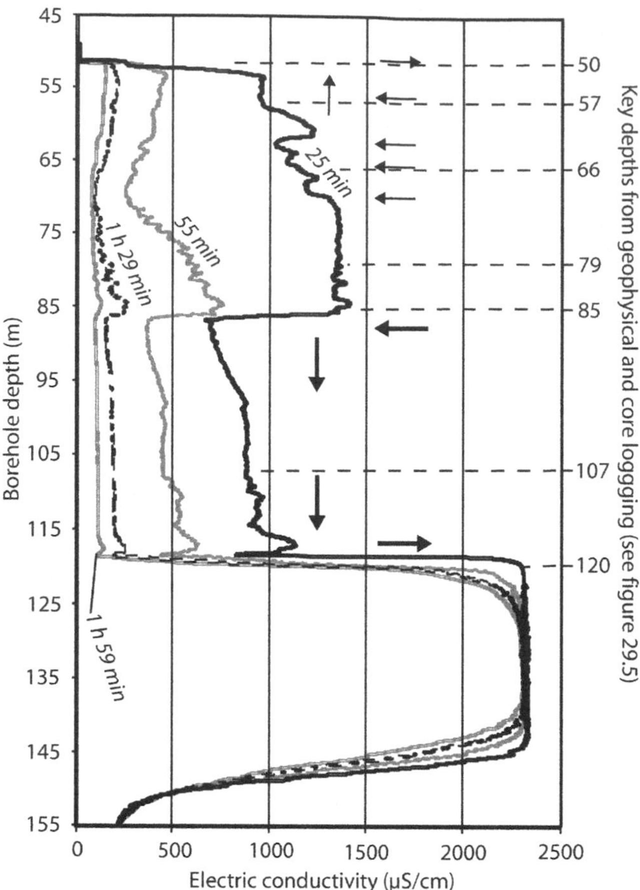

Fig. 26.6. Hydrological test from the upper borehole showing flow directions (modified from Thöny, 2008). The borehole was flushed with saline water, and the conductivity was measured at different times. Arrows show the main flow directions.

quality and fracture intensity. Logging results for the upper borehole are shown in Figure 26.5. *P*-wave velocities from just below the water level (52 m depth) to about 105 m depth vary considerably, indicating the presence of many fractures. Below this zone, velocities are higher and less variable. The temperature and conductivity profiles indicate changes in the water circulation; the increase in both temperature and conductivity at 110 m depth in the upper borehole is notable (Fig. 26.5) and coincides with a zone of relatively low resistivity.

Groundwater flow was characterized using a combination of heat pulse flow-meter logging and a new protocol known as dynamic fluid electric conductivity (DFEC) (Thöny, 2008). Large-scale tracer experiments were also performed. The dynamic fluid electric conductivity logging was conducted using the following procedure: (1) the borehole was first flushed with water of a higher salinity than the water in the system; and (2) the conductivity was then logged several times after the flushing to observe time-dependent changes in salinity. The tests located zones of stagnant and circulating water (Fig. 26.6). The water flows out from the borehole at the water level (52 m depth), flows into the borehole between 55 and 87 m depth, and flows down the borehole between 87 and 119 m, where it flows out again. The loss of drilling fluid pressure between 87 and 115 m depth and the highly conductive outflow at 119 m depth clearly indicate the presence of highly permeable rock masses between 87 and 120 m depth (Thöny, 2008). These results agree well with the surface 2D resistivity measurements, which indicate the presence of water-saturated fractures in this zone (Fig. 26.4).

26.3.6 SUBSURFACE DISPLACEMENTS

The instrumentation in the boreholes is essential both for the investigation of subsurface characteristics (sliding planes, volumes etc.) and for real-time operative early warning. Three boreholes have been instrumented using the DMS (differential monitoring of stability) system. The DMS is a multi-parametric column for investigation and permanent monitoring of subsurface movements, temperature, and water pressure (Lovisolo *et al.*, 2003). The DMS is composed of a sequence of hard tabular modules connected to each other by special joints. The present system at Åknes comprises 100–120-m-long columns installed in three boreholes to measure the movement in 2D. The 120-m-long columns consist of 245 sensors. The sensors include biaxial inclinometers and temperature sensors, as well as piezometers and digital compasses in selected modules.

Figure 26.7 shows an example of recent data from the upper borehole measured between November 2010 and March 2011. It clearly documents the sliding zone at 50 m depth, as expected from the core and borehole data (Fig 26.5). The deformation patterns indicate creep movement higher up the slope and in deeper parts. The depth of the instability is inferred to be at about 120 m, based on the data collected from rock cores, borehole logging, 2D resistivity profiles, and the DMS data (Figs. 26.4, 26.5, 26.6, and 26.7)

26.4 MONITORING AND EARLY-WARNING SYSTEMS

The unstable rock slope at Åknes is extensively and continuously monitored. The concepts and design are described briefly below.

26.4.1 ÅKNES MONITORING

The monitoring system was designed to be robust, while providing the range of data required for stability analysis and decision-making during periods of accelerated movement. The system includes surface and subsurface sensors designed to provide a full understanding of deformation and the controlling mechanisms. The sensors monitor all important sectors or parts of the unstable area on the surface and in boreholes, and measure both local deformation across individual fractures and total displacements. They include pore-pressure sensors and a full climate station. The system includes redundant, or back-up,

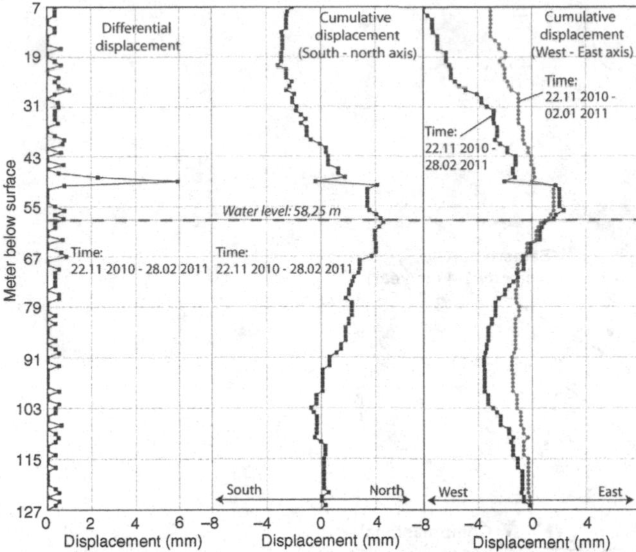

Fig. 26.7. Data from the DMS (differential monitoring of stability) column in the upper borehole, acquired from November 22, 2010 to February 28, 2011. The 120-m-long column extends from 7 m to 127 m depth. The left curve shows the differential displacements of each sensor, and the middle and the right curves show cumulative displacements along two different axes. The right curve also shows the displacement over a shorter time span. The three-month measuring period is sufficient to define the main sliding surface at 50 m depth. There also appears to be movement deeper in the slope, but the displacements of individual sensors are too small to confirm this over the short period of measurement.

instruments for all critical areas. The infrastructure includes a back-up power supply and data communication capabilities. An overview of the monitoring system is provided in Figures 26.8 and 26.9).

SURFACE MONITORING

Monitoring to assess total displacements includes a permanent GPS network with eight antennas (Fig. 26.9A), a total station measuring distances to 30 prisms (Fig. 26.9B), and a ground-based radar located on the opposite side of the fjord to measure distances to eight reflectors. Due to large atmospheric changes and relatively long distances, it has been difficult to obtain reliable measurements from the total station, especially during the winter (Nordvik and Nyrnes, 2009). The data from the ground-based radar suffer similar problems, and reliable results depend on reference reflectors being located nearby and at the same elevations as the monitoring points.

A series of small crackmeters and a microseismic network of eight three-component geophones were installed in the upper landslide area (Fig. 26.8). The microseismic sensors are stable and give reliable data. A series of surface 2D tiltmeters were placed throughout the landslide area.

Local measurements across the fractures were made using three large extensometers and two single lasers (Fig. 26.9C). The extensometers provide excellent data with limited noise. The single lasers are relatively stable, but data acquired during bad weather, especially during periods of snow and fog, are unreliable.

SUBSURFACE MONITORING

The three boreholes are instrumented with DMS columns (Fig. 26.9D), which are linked to a real-time operational early-warning system (Fig. 26.8). Biaxial inclinometers and temperature sensors are spaced at 1-m intervals on the columns. Additional piezometer sensors and digital compasses are placed in selected meter modules. The system is stable and gives high-resolution continuous data, as can be seen in Figure 26.7.

26.4.2 HAZARD AREAS, ALARM LEVELS, AND EARLY WARNING

The definition of hazard zones at Åknes is based on extensive work on landslide and tsunami modeling (Glimsdal and Harbitz, 2011). Probabilities for different scenarios have been evaluated on the basis of the actual deformation on the slope and historical and geological documentation of rockslide events in the fjord. The largest modeled landslide at Åknes is 54 million m^3 and is estimated to have an annual probability of >1/5000. A scenario involving a failure of 18 million m^3 is estimated to have an annual probability of >1/1000. The largest rockslide would have dramatic consequences: the tsunami triggered by such a landslide would affect a large number of communities along the fjord (Fig. 26.10), with an estimated run-up of 85 m at Hellesylt and 70 m at Geiranger.

The monitoring and early-warning system is operated by the Åknes/Tafjord Early Warning Center. All monitoring data are processed and analyzed at the Center to assess movement behavior and velocity trends, from which threshold values for warning levels have been established. We use velocity and acceleration as the criteria for the different warning levels (Fig. 26.11).

The velocity thresholds are different from instrument to instrument, and also differ according to the time period over which the data are evaluated (Table 26.1). Extensometer data over a period as short as one hour can serve as the basis for a warning, whereas data from the total station and GPS are assessed over a longer period, due to larger fluctuations and noise. The normal situation at Åknes is the "green" level, with movements typically about 0.1–0.2 mm per day (Fig. 26.11). The next highest level ("blue") comes into effect when the velocity increases substantially due to seasonally wet weather, and the "yellow" level is reached when the velocity increases beyond the seasonal fluctuations. The "orange" level is reached when movements accelerate, and the "red" level is reached when a catastrophic failure is imminent – at which time an evacuation is ordered. Specific actions are implemented for the different warning levels and involve the Early Warning Center, police, county governor, municipalities, road authorities, coastguards, and power companies.

The operational system is based on the following routines:

- daily check of all sensors and data by the geologist on duty
- daily technical checks by the technical person on duty

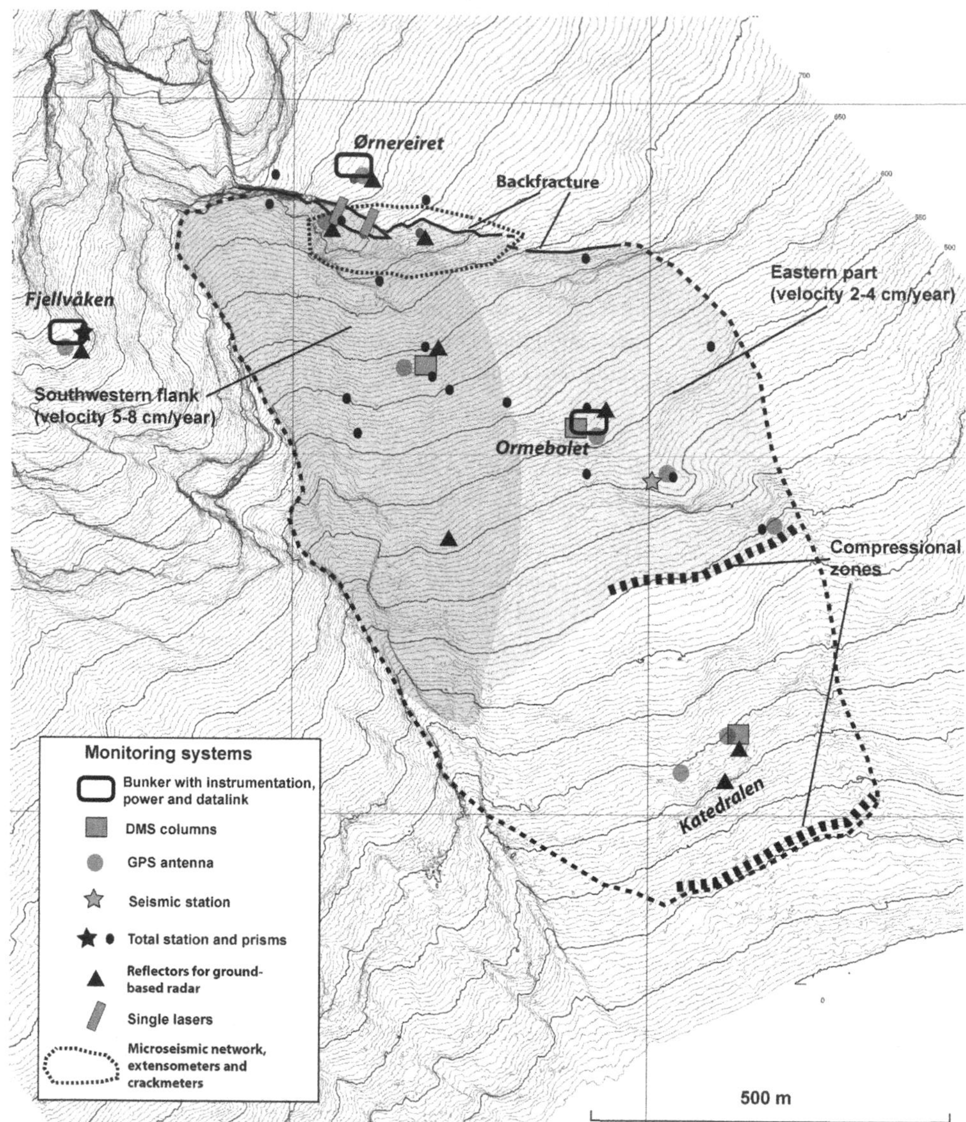

Fig. 26.8. Overview of the monitoring systems at Åknes. The main deformation sectors of the rockslide are also shown.

- mobile phone text messages (SMS) on data acquired from selected sensors and technical system failures
- long-term contracts with monitoring companies that have different operational responsibilities.

The use of SMS and e-mail messages is difficult due to the different types of noise in the data. It is important to use thresholds that do not result in too many false alarms but instead provide warning of real events. SMS messages are never used alone to change warning levels, but provide important support for the people on duty. One of the challenges is that sensors with large noise levels, for example the single lasers, create a large number of messages and cannot be used alone to raise the warning level.

Two different methods are used to issue warnings. First, a system of phone messages is generated based on a continuously updated database, in addition to automatic SMS messages based on regional coverage from the mobile network companies. Second, an electronic warning siren in all the villages situated in the tsunami hazard zone can be activated. In addition, a large effort has been made to plan and establish evacuation routes for inhabitants of the tsunami hazard zones. The hazard zones have been established on the basis of landslide scenarios and the results of the tsunami modeling (Fig. 26.10; Glimsdal and Harbitz, 2011).

26.4.3 DISPLACEMENT MODEL AND DEFORMATION CHARACTERISTICS

Numerous geological, geophysical, and geotechnical studies have been completed on the Åknes rockslide to better understand the movement mechanism and to locate sliding surfaces (Blikra, 2008; Ganerød *et al.*, 2008; Kveldsvik, 2008; Oppikofer, 2009; Grøneng, 2010; Kristensen *et al.*, 2010). The mode of deformation and displacement is not yet fully understood, but

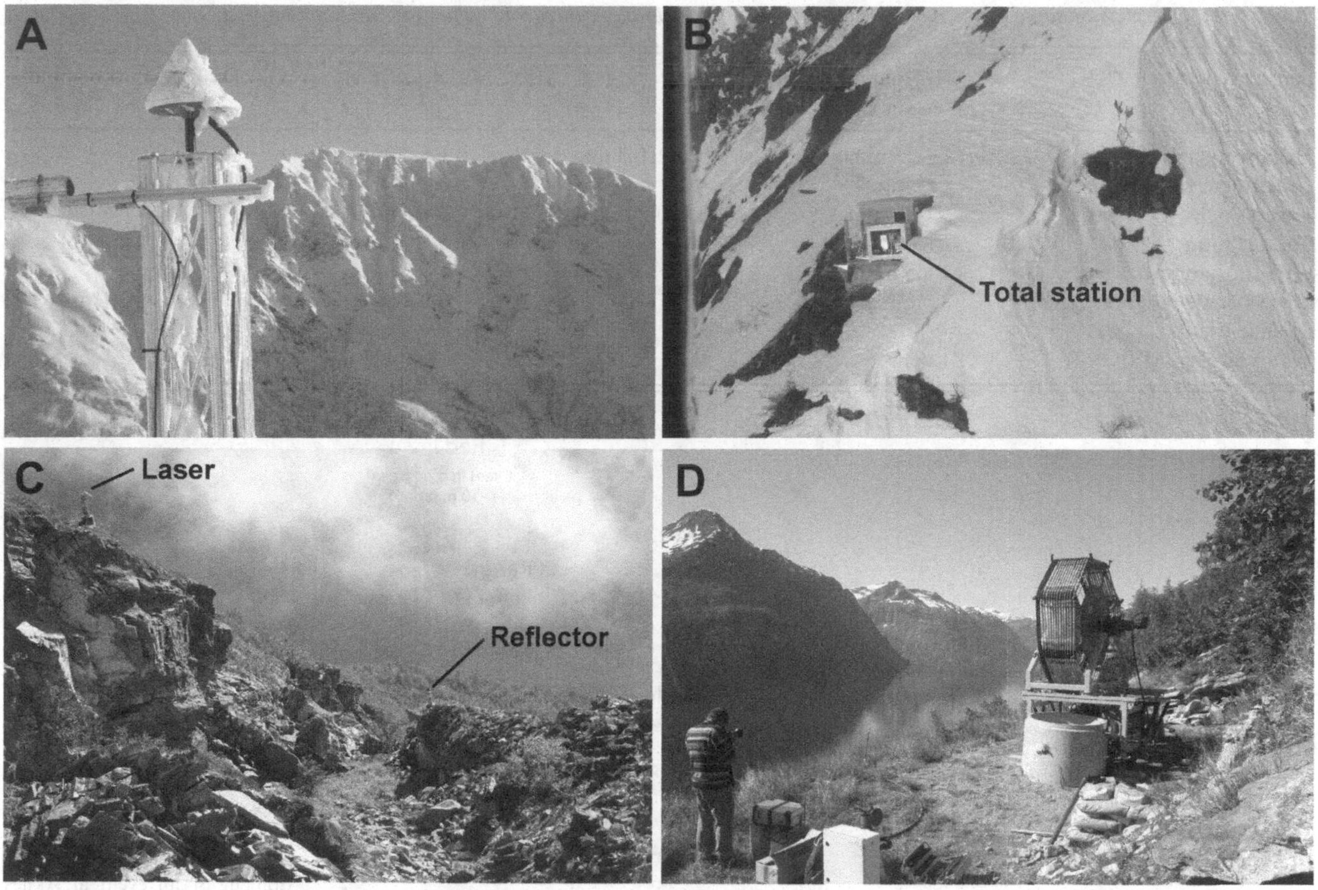

Fig. 26.9. Selected monitoring systems at Åknes: (A) GPS antenna, (B) total station, (C) single laser at the deep upper fracture, (D) borehole DMS instrumentation.

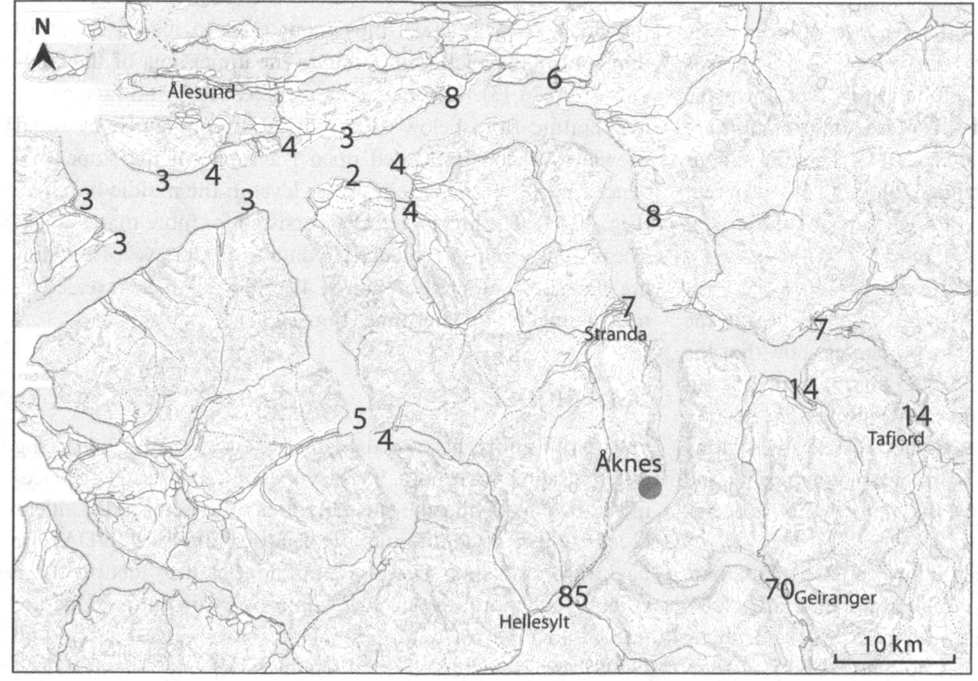

Fig. 26.10. Estimated tsunami run-up heights (m) for the largest landslide scenario at Åknes (54 million m^3).

Table 26.1. *Overview of some of the monitoring systems with preliminary threshold values for SMS messages.*

Sensor type	Normal movement / seismic events / water level		Noise	Daily SMS threshold
Extensometers	16–25 mm a^{-1}	< 0.2–0.3 mm day^{-1}	1.5–3 mm	1.5–3 mm
Crackmeters	1–3 mm a^{-1}	< 0.1 mm day^{-1}	0.2–0.4 mm	0.3–0.6 mm
Lasers	50–70 mm a^{-1}	0.1–0.2 mm day^{-1}	3–30 mm	3–7 mm
Total station	10–60 mm a^{-1}		5–20 mm	20 mm
GPS (horizontal)	1–65 mm a^{-1}	< 0.1– 0.25 mm day^{-1}	4–10 mm	5–10 mm
DMS (single inclinometers)	30 mm a^{-1}	–	0.2–0.5 mm	1.3 mm or 0.1° m^{-1}
Seismic events	–	2–30 events	–	50 events
Water level (DMS)	1–6 m	–	–	2 m

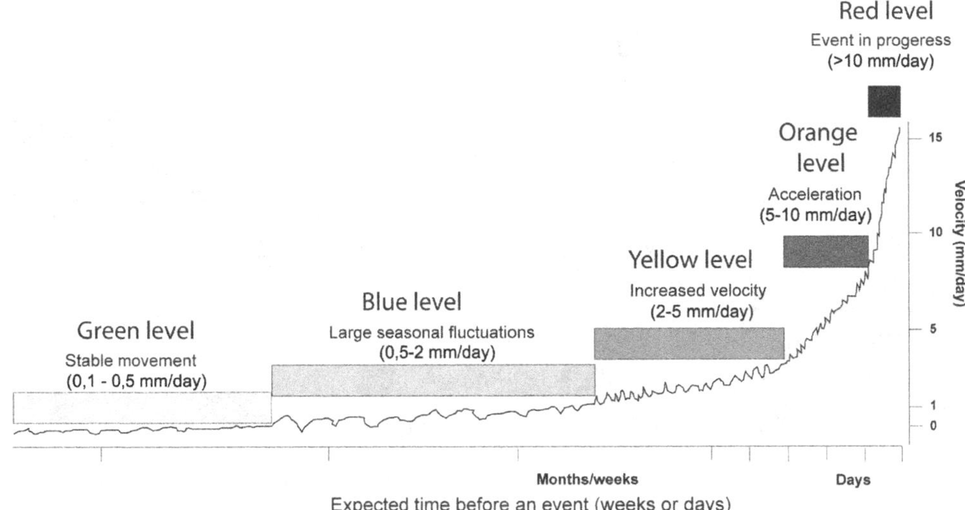

Fig. 26.11. Schematic diagram showing the possible development of an event at Åknes (Blikra, 2008). Alarm levels used for early warning are indicated.

the instability is clearly controlled by the structure of the gneissic rocks: the steep back scarp follows near-vertical folds, whereas the sliding zones are parallel to the foliation farther down the slope. However, the geometry and subsurface deformation are much more complex (Jaboyedoff *et al.*, 2011). The rockslide is composed of several semi-independent blocks with different surface movement directions and different rates of movement at depth.

The average annual displacement rate ranges from 6–8 cm per year on the upper southwest flank to about 2 cm per year in the middle and east sector (Fig. 26.8). The boreholes show that the upper sliding plane is situated between 25 and 55 m depth, but sliding also occurs deeper than 100 m (Rønning *et al.*, 2006). The drill cores show several brecciated and crushed zones along some of the foliation planes. The borehole logging results and borehole deformation data also show that deformation is occurring at considerable depths (Fig. 26.12).

Long-term movement rates are roughly constant, but there are shorter-term seasonal fluctuations (Braathen *et al.*, 2004; Blikra, 2008; Kveldsvik, 2008; Kristensen *et al.*, 2010; Grøneng *et al.*, 2011). Velocities are lowest in mid-winter and mid-summer. Rates increase during periods of heavy precipitation and snowmelt. In May 2006, movement rates increased about 10 times above the background rate in the upper area of the landslide (Fig. 26.13). The increase was preceded by an increase in temperature from below 0°C to 10°C, accompanied by rapid snowmelt. Water infiltrated open fractures on the slope, evidenced by a 4 m increase in water level in the middle borehole (Fig. 26.13). The higher velocity persisted for more than a week. The change was also evidenced in data from the level of the sliding plane in the upper borehole at 49–50 m depth, as revealed by measurements at a later time (Fig. 26.14).

26.4.4 MODELING

Substantial efforts have been made to model the landslide, using different numerical codes. Kveldsvik *et al.* (2008) used displacement data to divide the unstable area into semi-independent blocks and a discontinuous deformation analysis (DDA) for backward modeling. The block boundaries they identified are reliable, but more displacement data over the entire slope are needed to perform a proper DDA analysis. Kveldsvik *et al.* (2008) also used the Universal Distinct Element Code (UDEC) to study 2D slope stability. They compared a number of possible

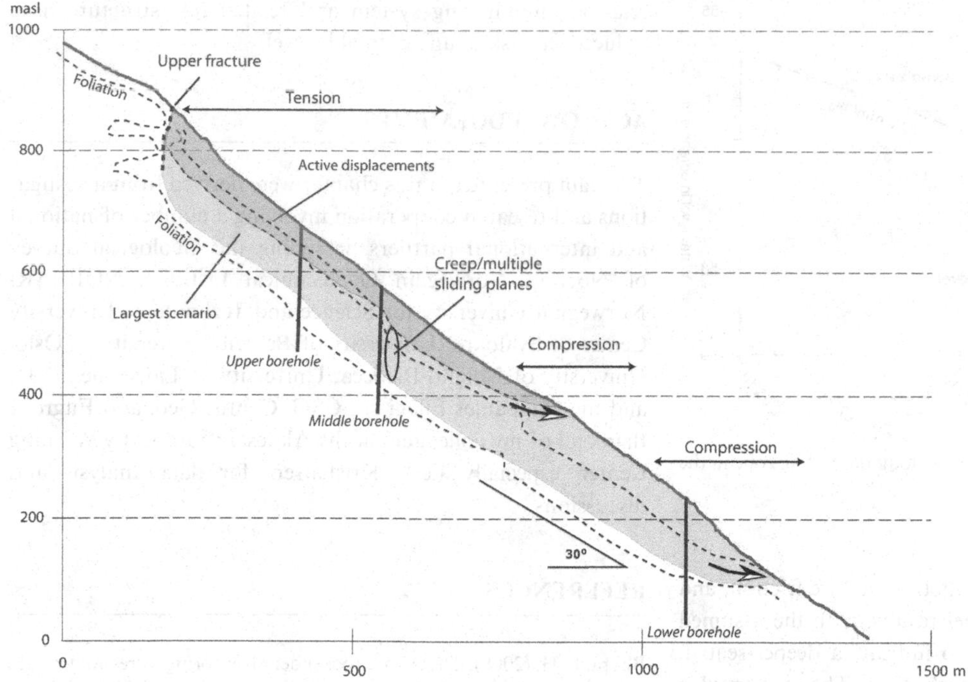

Fig. 26.12. Schematic 2D deformation model shown on a 2D resistivity profile.

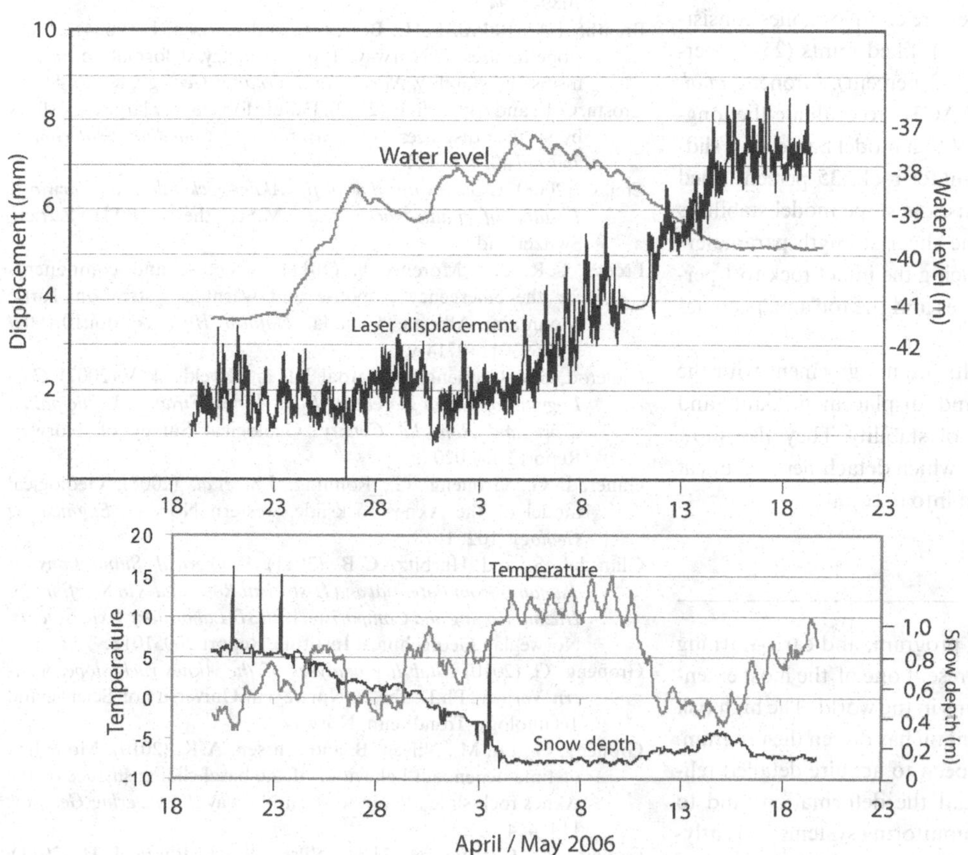

Fig. 26.13. Increased movement of the Åknes rockslide in April and May 2006, shown by data from a single laser. Also shown are water level measurements in the middle borehole and temperature and snow-depth data from the meteorological station near the top of the slope.

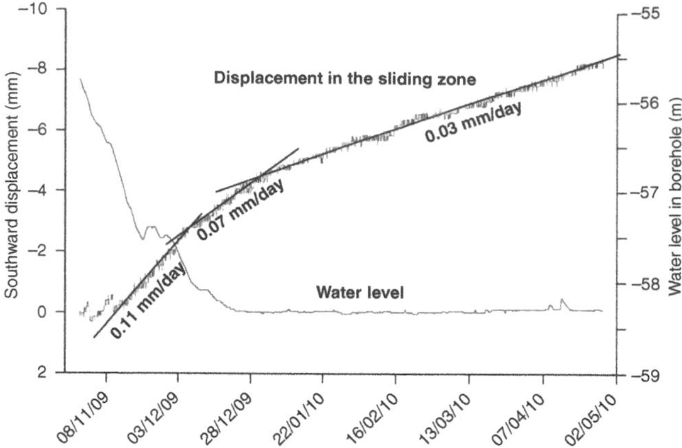

Fig. 26.14. Water levels and displacements along the sliding zone in the upper borehole (49–50 m depth).

scenarios by changing fracture geometry, fracture friction, and groundwater conditions. The model results, with the assumed shear strength properties, appear to indicate a deeper-seated, rather than a shallow instability mechanism. The authors also concluded that a shallow slide could lead to a more deep-seated failure. Grøneng (2010) studied the shear strength of sliding planes based on field observations, drill-core data, and back analysis. The sliding planes at Åknes are complex zones consisting of unfilled joints (62–74 percent), filled joints (25–35 percent), and bridges of intact rock (1–3 percent). Grøneng *et al.* (2010) used the numerical code FLAC3D to evaluate the long-term stability of the Åknes slope. With a model based on a sliding zone composed of 3 percent intact rock, 35 percent filled joints, and 62 percent unfilled joints, the slope model stabilizes in about 100 years. However, if the shear strength parameters are lowered over time (e.g., by reducing the intact rock to 1 percent), the rate of movement increases and the toe area becomes unstable.

In summary, the modeling results are in agreement with the documented field observations and displacement data, and show that the slope is at the limit of stability. They also show that the large landslide scenario, in which detachment occurs at depths below 100 m, must be taken into account.

26.5 CONCLUSIONS

The site investigation, monitoring program, and early-warning protocols carried out at Åknes represent one of the most extensive rockslide management programs in the world. The high risk posed by a landslide-generated tsunami has driven the program. The main aims of the work have been to acquire detailed reliable knowledge of the geology and the deformation, and to implement robust and redundant monitoring systems and early-warning protocols and related infrastructure. The work has shown the importance of both surface and subsurface investigations and monitoring. There are still some unanswered questions about the geologic deformation, but the now-established real-time monitoring system and related infrastructure have reduced the risk to an acceptable level.

ACKNOWLEDGEMENTS

The data presented in this chapter were derived from investigations and research cooperation involving a number of national and international partners, including the Geological Survey of Norway, Norwegian Geotechnical Institute, NORSAR, Norwegian University for Science and Technology, University Centre of Svalbard, University of Bergen, University of Oslo, University of Milano-Bicocca, University of Lausanne, ETH, and the companies Ellegi srl, CSG, Cautus Geo, and Fugro. I thank all of my colleagues at the Åknes/Tafjord Early Warning Center, especially Lene Kristensen, for data analysis and discussions.

REFERENCES

Blikra, L. H. (2008). The Åknes rockslide: Monitoring, threshold values and early-warning. In *Landslides and Engineered Slopes – From the Past to the Future: Proceedings of the 10th International Symposium on Landslides and Engineered Slopes*, ed. Z. Chen, J.-M. Zhang, K. Ho, F.-Q. Wu and Z.-K. Li. CRC Press, pp. 1089–1094

Braathen, A., Blikra, L. H., Berg, S. S. and Karlsen, F. (2004). Rock-slope failures of Norway: Type, geometry, deformation mechanisms and stability. *Norwegian Journal of Geology*, 84, 67–88.

Crosta, G. C. and Agliardi, F. (2003). Failure forecast for large rock slides by surface displacement *measurements. Canadian Geotechnical Journal* 40, 176–191.

Frei, C. (2008). *Groundwater flow at the Åknes rockslide site (Norway): Results of multi-tracer test.* M.Sc. thesis, ETH Zurich, Switzerland.

Froese, C. R. and Moreno, F. (2011). Structure and components for the emergency response and warning system on Turtle Mountain, Alberta, Canada. *Natural Hazards*, doi:10.1007/s11069–011–9714-y.

Ganerød, G. V., Grøneng, G., Aardal, I. B. and Kveldsvik, V. (2007). *Core Logging of Seven Boreholes from Åknes, Stranda Municipality, Møre and Romsdal County*. Geological Survey of Norway, Report 2007.020.

Ganerød, G., Grøneng, G., Rønning, J. S. *et al.* (2008). Geological model of the Åknes rockslide, western Norway. *Engineering Geology*, 102, 1–18.

Glimsdal, S. and Harbitz, C. B. (2011). *Numerical Simulations of Tsunamis from Potential and Historical Rock Slides in Storfjorden: Hazard Zoning and Comparison with 3D Laboratory Experiments*. Norwegian Geotechnical Institute, Report 20051018.

Grøneng, G. (2010). *Stability analyses of the Åknes rock slope, western Norway*. Ph.D. thesis, Norwegian University of Science and Technology, Trondheim, Norway.

Grøneng, G., Lu, M., Nilsen, B. and Jenssen, A. K. (2010). Modelling of time-dependent behaviour of the basal sliding surface of the Åknes rock slide area in western Norway. *Engineering Geology*, 114, 414–422.

Grøneng, G., Christiansen, H. H., Nilsen, B. and Blikra, L. H. (2011). Meteorological effects on seasonal displacements of the Åknes rockslide, western Norway. *Landslides*, 8, 1–15.

Jaboyedoff, M., Oppikofer, T., Derron, M.-H. *et al.* (2011). Complex landslide behavior and structural control: A 3D conceptual model of Åknes rockslide, Norway. In *Slope Tectonics*, ed. M. Jaboyedoff. Geological Society of London, Special Publication 351, pp. 147–161.

Kristensen, L., Blikra, L.H. and Hole, J. (2010). *Åknes: State of Instrumentation and Data Analysis.* Åknes Report 02.2010.

Kveldsvik, V. (2008). *Static and dynamic stability analyses of the 800 m high Åknes rock slope, western Norway*. Ph.D. thesis, Norwegian University of Science and Technology, Trondheim, Norway.

Kveldsvik, V., Einstein, H.H., Nilsen, B. and Blikra, L.H. (2008). Numerical analysis of the 650,000 m² Åknes rock slope based on measured displacements and geotechnical data. *Rock Mechanics and Rock Engineering*, 22, 689–728.

Lovisolo, M., Ghirotto, S., Scardia, G. and Battaglio, M. (2003). The use of differential monitoring stability (DMS) for remote monitoring of excavation and landslide movements. In *Proceedings of the Sixth International Symposium on Field Measurements in Geomechanics*, ed. I. Myrvoll. Oslo: Balkema, pp. 519–524.

Nordvik, T. and Nyrnes, E. (2009). Statistical analysis of surface displacements: An example from the Åknes rockslide, western Norway. *Natural Hazards and Earth System Sciences*, 9, 713–724.

Oppikofer, T. (2009). *Detection, analysis and monitoring of slope movements by high-resolution digital elevation models*. Ph.D. thesis, University of Lausanne, Switzerland.

Rønning, J.S., Dalsegg, E., Elvebakk, H., Ganerød, G. and Tønnesen, J.F. (2006). *Geofysiske Målinger Åknes og Tafjord, Stranda og Nordal Kommuner, Møre og Romsdal.* Geological Survey of Norway, Report 2006.002 (in Norwegian).

Rønning, J.S., Dalsegg, E., Heincke, B.H. and Tønnesen, J.F. (2007). *Geofysiske Målinger på Bakken og ved Hegguraksla, Stranda og Norddal Kommuner, Møre og Romsdal.* Geological Survey of Norway, Report 2007.023 (in Norwegian).

Thöny, R. (2008). *Dynamic fluid electric conductivity logging for identification and characterization of preferential groundwater flow in the Åknes rockslide (Norway)*. M.Sc. thesis, ETH Zurich, Switzerland.

27 A seismometric approach for back-analyzing an unusual rockfall in the Apennines of Italy

GIANLUCA BIANCHI FASANI, CARLO ESPOSITO, LUCA LENTI, SALVATORE MARTINO, MASSIMO PECCI, AND GABRIELE SCARASCIA MUGNOZZA

ABSTRACT

This chapter describes a rockfall that occurred at Corno Grande, the highest peak in the Apennines, in Italy on August 22, 2006. A limestone block with an estimated volume of about 30,000 m^3 fell from the sub-vertical northeast slope of Corno Grande. Although relatively small, the rockfall covered a large area (*ca.* 35,000 m^2) with debris and generated a giant abrasive dust cloud that affected an area of *ca.* 110,000 m^2 at the base of the slope. The dust cloud extended 3 km from the source slope and reached the village of Casale San Nicola. The A24 motorway was temporarily closed due to limited visibility. The rockfall was recorded by a seismometer array located in the nearby Gran Sasso underground laboratories (LNGS – Laboratori Nazionali del Gran Sasso). We processed the seismic data to assess the evolution of the rockfall in terms of mean velocity, impacts, and energy dissipation. The field surveys and data derived from the seismometer array allowed us to constrain the kinematic evolution of the failure. At least three debris impacts can be seen in the available data; they triggered an air blast and the dust cloud.

27.1 INTRODUCTION

On August 22, 2006, a limestone block with an estimated volume of about 30,000 m^3 fell from the sub-vertical northeast wall of Corno Grande ("Paretone") in the Gran Sasso d'Italia Massif, Italy. Corno Grande is the highest peak in the Italian Apennines. Rockfalls are common on the steep, jointed calcareous slopes of the central Apennines and have volumes ranging from hundreds to thousands of cubic meters. The 2006 event, although relatively small in size, was exceptional because of the air blast and dust cloud that it produced (Fig. 27.1).

Fig. 27.1. Panoramic view of Corno Grande and the Gran Sasso d'Italia during the 2006 rockfall.

Other similar rockfall events have been studied and reported elsewhere. Of particular note is the rockfall in Yosemite National Park, California documented by Morrissey *et al.* (1999) and Wieczorek *et al.* (2000). They proposed a model to explain the generation of the air blast and the abrasive power of the dust cloud. Morrissey *et al.* (1999) highlight the importance of at least two separate high-energy impacts at the base of

Landslides: Types, Mechanisms and Modeling, ed. John J. Clague and Douglas Stead. Published by Cambridge University Press.

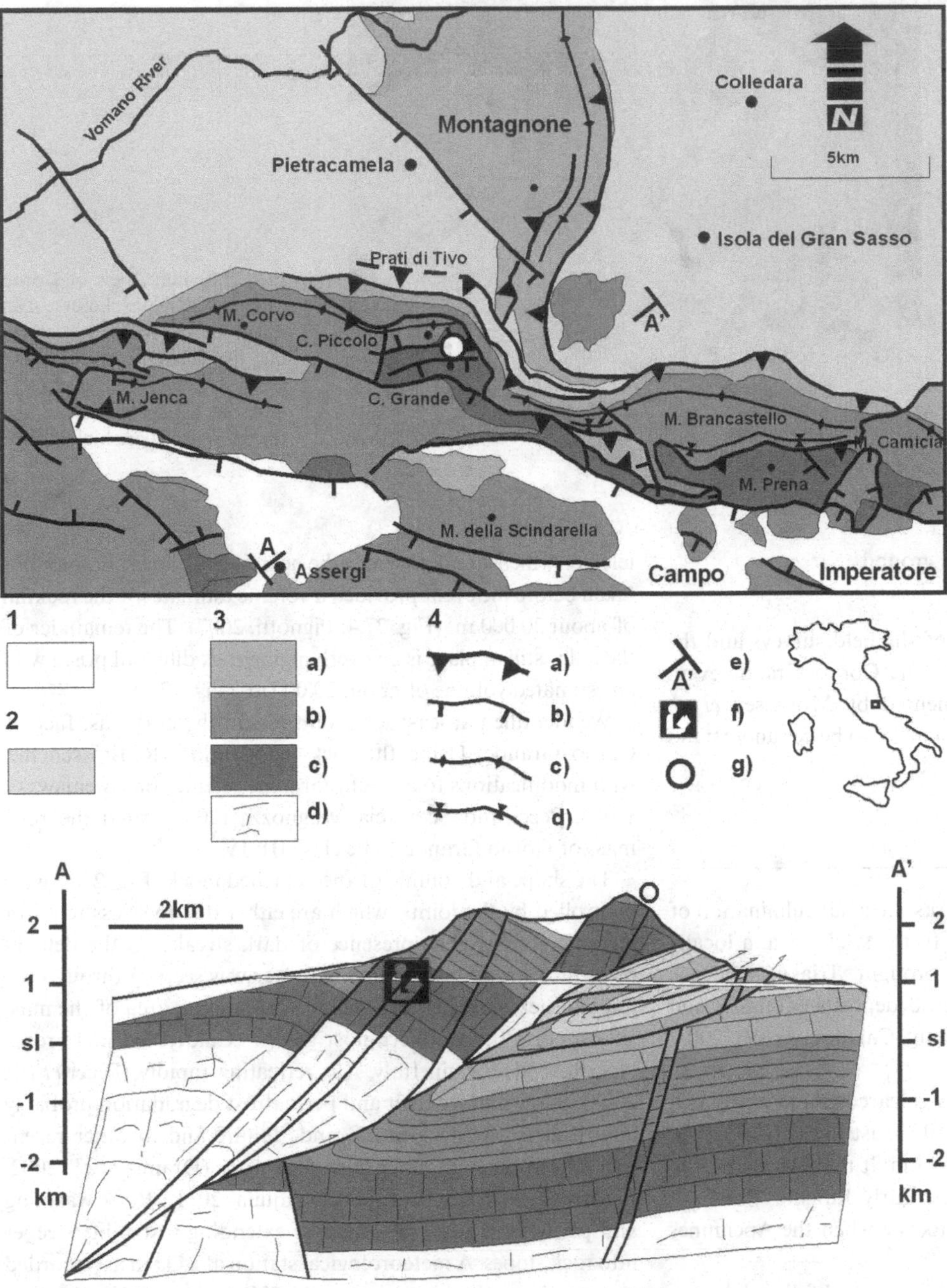

Fig. 27.2. Structural geology of the Gran Sasso d'Italia area and cross-section along the Gran Sasso Tunnel. 1 – Post-orogenic deposits: continental deposits (Quaternary). 2 – Syn-orogenic deposits (siliciclastic deposits): pre-evaporitic, evaporitic. and post-evaporitic members (Upper Miocene–Lower Pliocene). 3 – Pre-orogenic carbonate succession: a) carbonate and shale (Miocene); b) slope-basin carbonate succession (Middle Jurassic–Oligocene); c) carbonate platform succession (Triassic–Lower Jurassic); d) dolomite (Triassic); 4 – a) thrust fault; b) normal fault; c) anticline; d) syncline; e) cross-section along the Gran Sasso Tunnel; f) LNGS laboratory; g) detachment area. (Modified after Calamita *et al.* 2002.)

the slope; the dust cloud generated by the first impact allowed the second one to encounter a denser atmosphere, reducing the speed of sound and facilitating the generation of an air blast. Wieczorek *et al.* (2000) note the importance of the geomorphic setting, specifically the steepness of the slope and the possibility of significant freefall, which favor high impact velocities and the generation of an air blast.

We documented the Corno Grande rockfall by integrating the results of field surveys with available seismic records of the event. Field surveys provide information on:

- the volume of the detached rock mass
- the main discontinuity sets and related hydraulic conditions
- paths of the falling blocks
- the deposition area
- the extent of the air blast-damaged area.

The surveys were supported with photographs taken during the event and interviews with local inhabitants. The event was recorded at an underground seismic station of the Laboratori Nazionali del Gran Sasso (LNGS-INFN). We acquired and processed the seismogram to constrain:

- the duration of the rockfall and its impacts at the base of the slope
- the evolution of the rockfall

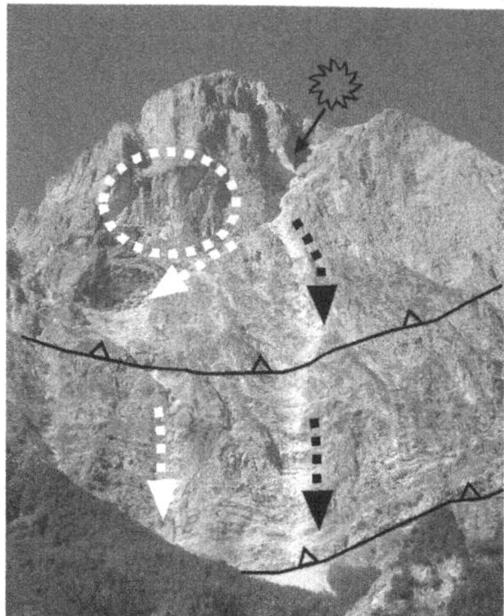
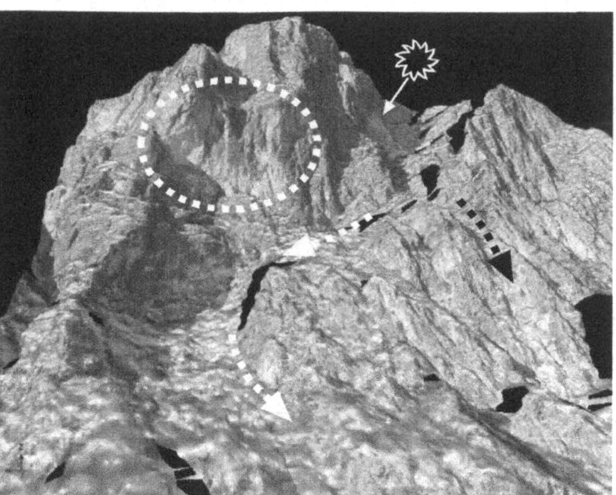

Fig. 27.3. Left: View of Corno Grande. Right: Laser scan image of the rock wall, showing the Jannetta route (white arrows). The "Farfalla" scar and the detachment area and the main thrust are highlighted.

- the total energy transferred to the ground
- the average velocity.

Comparison of the main findings of the field surveys and the seismogram allowed us to compare the Corno Grande event with the Yosemite rockfall documented by Morrissey *et al.* (1999) and Wieczorek *et al.* (2000) in order to better understand the cause of the destructive dust cloud.

27.2 GEOLOGIC SETTING

The Gran Sasso d'Italia Massif marks the axial culmination of the Apennine fold-and-thrust belt. In the study area, a locally north-vergent thrust system has brought Triassic–Miocene carbonate units over Miocene clastic deposits (Ghisetti and Vezzani, 1990, and references therein; Calamita *et al.*, 2002) (Fig. 27.2).

The steep source slope comprises calcareous and marly calcareous rocks that are cut by several thrust faults (Fig. 27.3). The east face coincides with a normal fault that elevated Gran Sasso and Corno Grande during the Early Jurassic. The fault was reactivated during the Late Miocene when the Apennines were compressed.

A thick apron of historic and prehistoric rockfall and debris-flow deposits covers the base of the slope (D'Alessandro *et al.*, 2003). Some of the debris was left by a large rockfall in 1897 that left a scar known as "Farfalla" due to its butterfly shape (circle in Fig. 27.3). Similar debris is common some distance from Gran Sasso (Bianchi Fasani *et al.*, 2008) and may have been deposited by rock avalanches from the same source slope.

27.3 FIELD INVESTIGATIONS AND DESCRIPTION OF THE ROCKFALL

The rockfall occurred at 07:30 GMT, when a block detached from the "Guglia Bambù," a fault-bounded rock pillar. Laser-GPS telemetry measurements from the nearest peak, and photographs taken before the event provided a volume estimate for the rockfall of about 30,000 m^3 (Fig. 27.4; Pignotti, 2007). The remainder of the pillar still in place is an overhanging rock dihedral prism with an estimated volume of about 20,000 m^3 (Fig. 27.4).

We identified at least four joint sets in the northeast face of Corno Grande. Using the rock mass rating (RMR) scheme, with modifications to account for joint orientation (Bieniawski, 1989), Pecci and Scarascia Mugnozza (2007) rated the rock mass of Corno Grande to be class III–IV.

The shape and volume of the detached block (Fig. 27.4) were controlled by the joints, which are either due to stress relief or tectonic activity. The presence of dark streaks on the detachment face strongly suggests that water was seeping through the jointed rock mass prior to failure. The summit area of the massif supports the southernmost glacier (Calderone) in Europe; like other glaciers in Italy, it is retreating rapidly (Pecci *et al.*, 2008). Snow and ice melt and permafrost degradation probably contributed to the Corno Grande failure and to other recent rockfalls in the high mountains of Europe (Dramis *et al.*, 2002; Fischer *et al.*, 2006; Pecci and D'Aquila, 2011). Rock warming and permafrost thaw may also be extending instability deeper into rock slopes. A meteorological station at 2433 m asl recorded an annual average temperature of 1.7°C during 2002/3. Extreme temperatures during that period were −13.2°C (March 25, 2002) and +15.1°C (July 10, 2002) (Fig. 27.5). The station shows a bimodal distribution of precipitation, with maximum values during the spring and in August.

After detaching, the rock mass fragmented in its 1100 m fall to the base of the slope below. Field evidence suggests a first phase of 120-m freefall from the detachment area to a rock ledge. Thereafter, the debris split into two parts: the first part descended along an almost straight trajectory, mainly by freefall and bouncing; the second part streamed down a chute (Jannetta route; Fig. 27.3), entraining additional rock debris. In both cases, the debris reached the base of the slope, where it spread over an area of about 35,000 m^2. Part of the debris

Fig. 27.4. (A) Rock face showing the detached block; (B) unstable pillar.

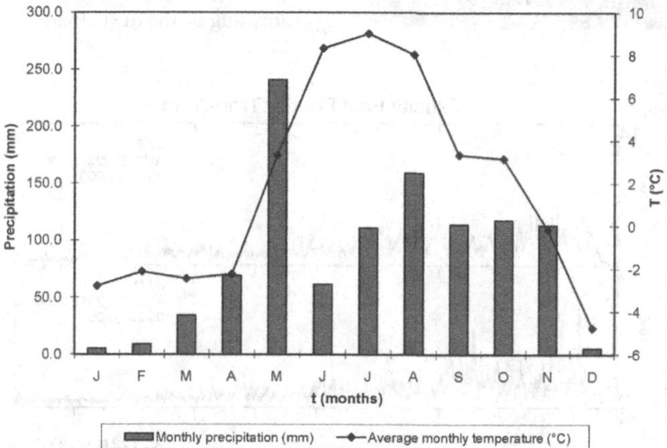

Fig. 27.5. Record of monthly precipitation and average monthly temperature at the meteorological station near the "Carlo Fanchetti" Hut of the Italian Alpine Club between January 8, 2002 and January 6, 2003 (after Pecci and Scarascia Mugnozza, 2007; courtesy of Ente Parco del Gran Sasso e Monti della Laga).

Fig. 27.6. The rockfall deposit and the area affected by the air blast (white dashed line).

flowed into the upper section of a deeply incised valley (Fig. 27.6). The debris ranged in size from fine sand to blocks. The larger blocks came to rest 250 m from the impact point; the finer material was deposited up to 3 km from the source.

The photographic sequence (Fig. 27.7) and eyewitness reports indicate that, shortly after failure, the rock mass hit the first rock ledge and generated a large, dense dust cloud. The cloud reached as far as the village of Casale San Nicola and the A24 motorway (Fig. 27.7); the latter was temporarily closed due to limited visibility.

Hundreds of trees and bushes were uprooted or snapped off over an area of about 110,000 m^2 around the impact zone (Fig. 27.6). The cause of this destruction was an air blast generated by the main impact of the debris along and at the base of the steep slope. The air blast carried the mixture of abrasive dust and rock fragments.

27.4 ANALYSIS OF THE SEISMIC RECORD

The rockfall was recorded by the seismic array UNDERSEIS, installed within the Gran Sasso laboratories (Fig. 27.2). The record is a three-component velocigram in $\mu m\,s^{-1}$ with a nominal duration of about 300 s.

Fig. 27.7. Photographs showing the evolution of the rockfall and spreading of the dust cloud.

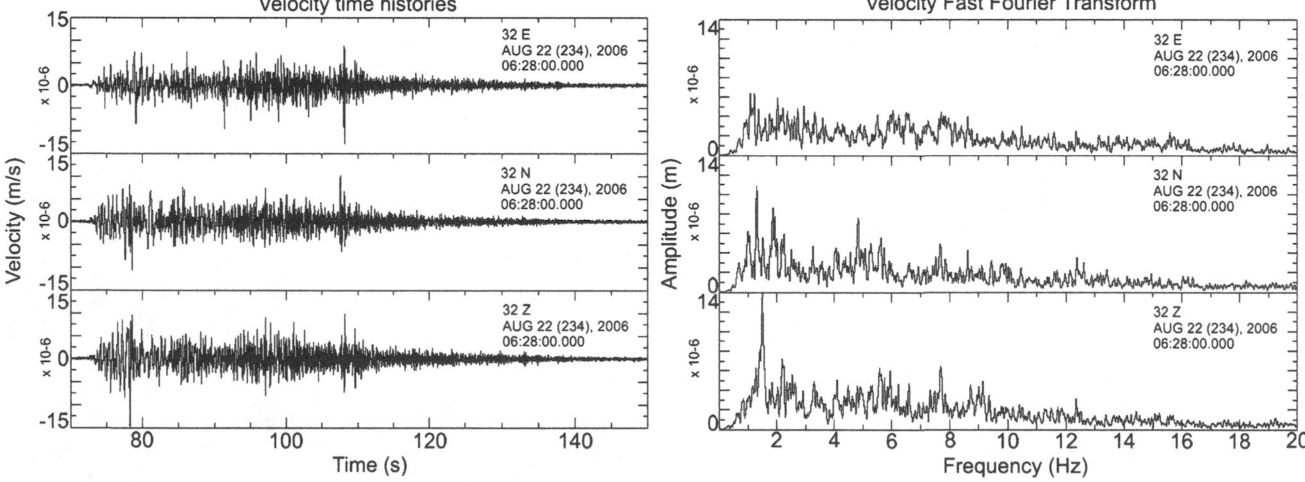

Fig. 27.8. Time history and Fast Fourier Transform (FFT) of the horizontal, NS, EW, and up (Z) components of the vibrational signal recorded during the rockfall on August 22, 2006 at the LNGS velocimetric station.

The rockfall occurred on the mountain slope, whereas the seismogram was recorded within the mountain. Consequently, neither the event source nor the receiver location corresponds to the usual configuration of an earthquake source and a free-field seismometer. Nevertheless, the receiver was so close to the event source (about 2 km) that we can ignore attenuation when identifying the delay among the more intense vibrational events.

Seismic analysis code (SAC; see www.iris.edu/manuals/sac/SAC_Home_Main.html) was used to process the velocimetric record in both time and frequency domains. The four components of the record determined with SAC are:

1. the variation of velocity with time
2. the instantaneous position of the velocity vector in the horizontal plane
3. the fast Fourier transform (FFT) of the velocimetric signal computed within adjacent and narrow time windows
4. the fast Fourier transform of the entire record.

An energy analysis was also performed by computing the instantaneous kinetic energy of each component and integrating it through the duration of the recorded event. The duration of the event was about 35 s (Fig. 27.8). The maximum velocity

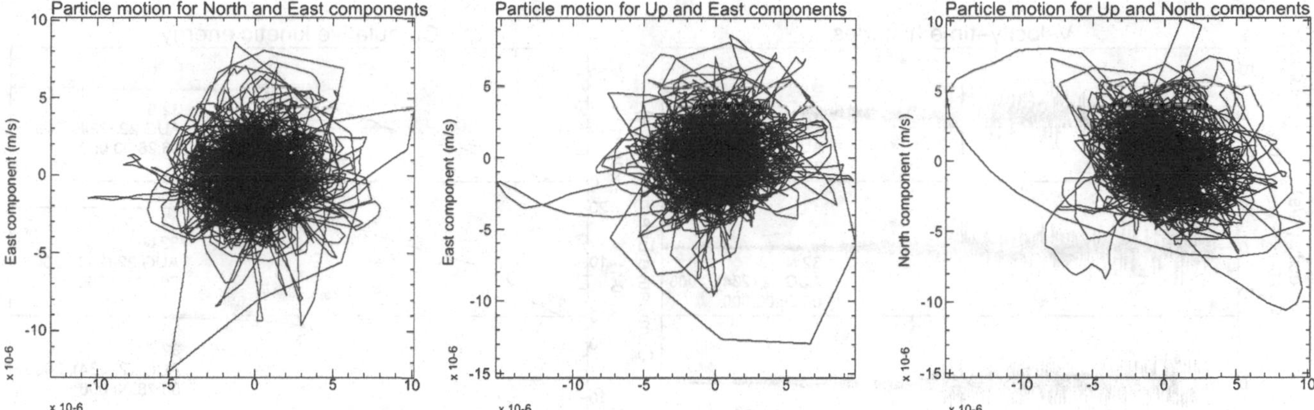

Fig. 27.9. Particle motions in the horizontal (NS–EW) and vertical (Z–EW, Z–NS) planes for each component of the ground motion recorded on August 22, 2006 at the LNGS velocimetric station.

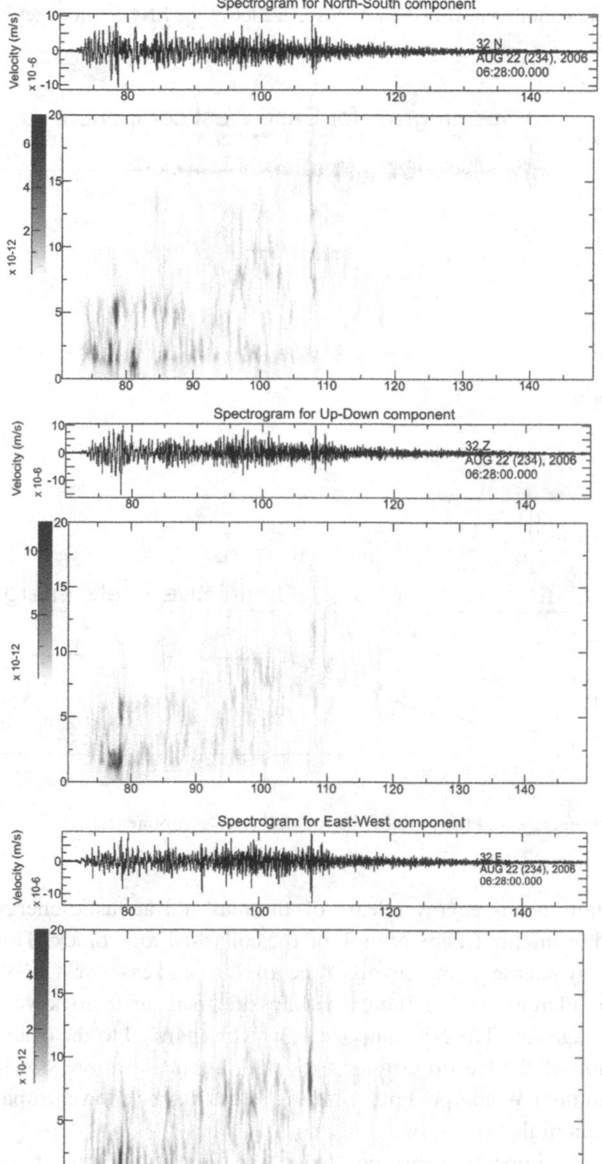

Fig. 27.10. Spectrograms for the components of the ground motion recorded on August 22, 2006 at the LNGS velocimetric station.

peak (*ca.* 12 μm s^{-1} with a computed peak ground acceleration (PGA) of about 6×10^{-5} *g*) was reached within 5 s. During this 5 s interval, the cumulative kinetic energy increased by one order of magnitude.

The particle motions (Fig. 27.9) obtained for each component did not define a preferential direction of ground motion. Similar shapes and intensities are indicated by both the FFTs and the spectrograms obtained for the recorded ground motion components (Figs. 27.10 and 27.11).

The rockfall record clearly differs from that of an earthquake in three respects.

1. No P-waves and S-waves can be distinguished.
2. The spectral contribution of low frequencies (i.e., down to 1 Hz) is higher for all components during the first 5 seconds of the event.
3. Higher frequencies are particularly intense after 20 seconds.

Of particular interest are the differences between spectrograms obtained for each component of the ground motion and the related cumulative energy over time (Figs. 27.10 and 27.12). The first significant energy increase (up to one order of magnitude) is associated with a more intense, low-frequency event (lower than 1 Hz) within the first 5 s in all spectrograms except that for the north–south (NS) horizontal direction of ground motion. The NS spectrogram shows a significant spectral content within the first 10 s. A second, minor, but still sharp, energy increase occurs about 107 s from the beginning of the record and corresponds to an intense high-frequency event (5–10 Hz) in the spectrograms.

27.5 DISCUSSION

Here we attempt to quantify the rockfall process, in terms of time, velocity, and energy dissipation, by coupling field evidence with our interpretation of the seismic records. Taking $t_0 = 74$ s in the seismogram as the starting point (i.e., the time the first signal appears in the velocimetric record), an intense low-frequency event is recorded at the time $t_1 = 79$ s (Fig. 27.12). The duration of the event, about 2.5 s, is consistent with an abrupt

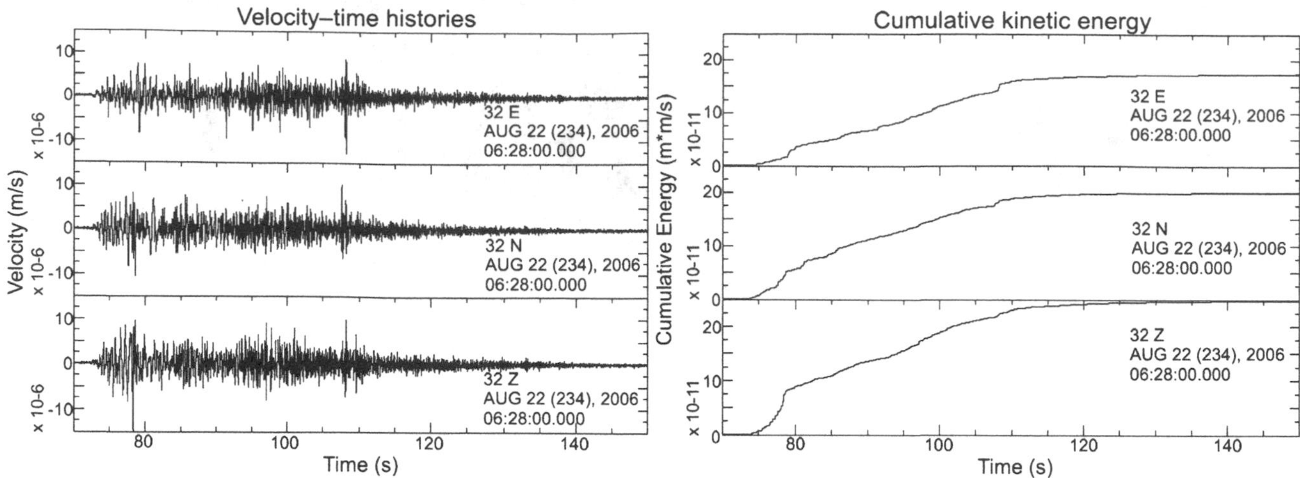

Fig. 27.11. Cumulative kinetic energy computed for each component of the ground motion recorded on August 22, 2006 at the LNGS velocimetric station.

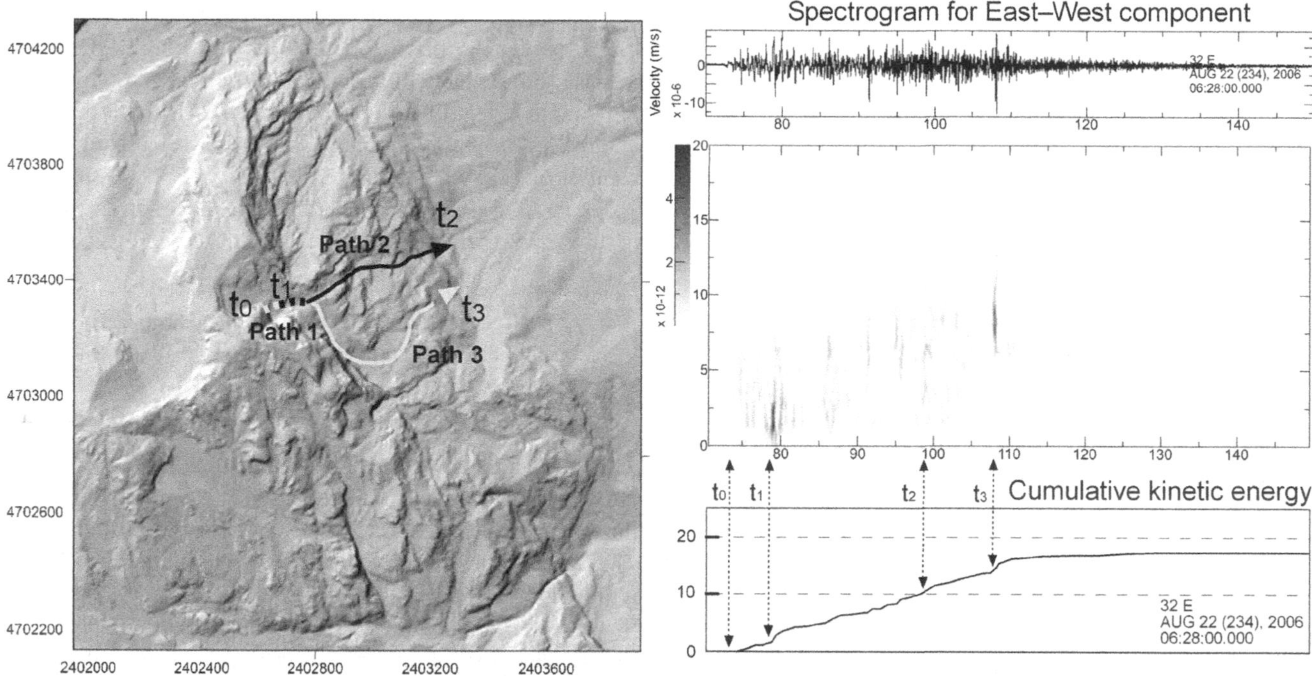

Fig. 27.12. Correlation between time histories, spectrograms, cumulative kinetic energy, and the path of the rockfall (EW components).

impact with high energy dissipation. The interval of 5 s between t_0 and t_1 is consistent with an initial phase of freefall of the failed rock mass, followed by impact on the rock ledge at about 2500 m asl (Fig. 27.13). The rock face in the section between the detachment area and the rock ledge is nearly vertical, thus favoring a freefall motion. A block falling approximately 130 m along this part of the path would require about 5 s to reach the ledge. The velocity at impact is calculated to be 50 $\mathrm{m\,s^{-1}}$. A key characteristic of this portion of the record is the sharp decrease, by about half an order of magnitude, of the spectral amplitude in the frequency range 0–6 Hz (Figs. 27.10 and 27.12). We attribute the sharp decrease in the spectral amplitude to a quasi-instantaneous energy release of thermal and acoustic energy and to intense fragmentation of the collapsed rock block. This energy release mainly involved the up (UP) and east–west (EW) ground motion components, perhaps due to an air-to-rock wave propagation. The NS component, which is parallel to the orientation of the Corno Grande rock wall, attenuates more slowly than the EW component, reflecting a rock-to-rock wave propagation in that direction.

After impacting the rock ledge, the fragmented rock mass split into two parts. Some of the debris traveled down a chute, entraining additional rock debris. The high-frequency phase in the seismogram (Figs. 27.10 and 27.13), which appears after the

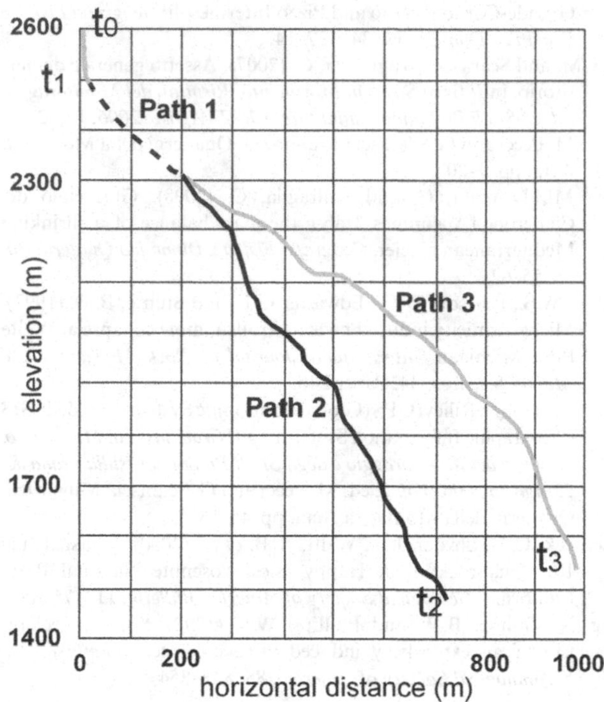

Path	Length (m)	Slope (°)	Δt (s)	vi (m/s)	vm (m/s)
1	112	83	$t_{0\text{-}1}$ 5	50	-
2	1031	58	$t_{1\text{-}2}$ 15	-	51
3	1182	44	$t_{1\text{-}3}$ 25	-	43

Fig. 27.13. Rockfall path profiles; vi = instantaneous velocity; vm = mean velocity. See Figure 27.12 for location.

initial impact on the ledge, is consistent with an avalanche of blocks within a finer matrix in a confined channel.

Two high-frequency events are recognized within the spectrograms, most clearly in the EW component of the ground motion, at about 100 s (t_2 = 21 s) and 107 s (t_3 = 28 s) (Fig. 27.12). These events are likely the impacts of the two streams of rockfall debris at the base of the slope. Reconstruction of the two paths highlights significant differences in their slopes and the time it took debris to reach the base of the slope (Fig. 27.13); the later arrival involved greater energy dissipation (Fig. 27.12). Mean velocities of about 50 and 45 m s^{-1} were derived for debris traveling along the two paths (Fig. 27.13), by dividing the length of the paths by the time lag between, respectively, t_1 and t_2 and t_1 and t_3. These values are consistent with field observations, strengthening the interpretation that the recorded high-frequency events were the result of two main impacts.

27.6 CONCLUDING REMARKS

We combined evidence from field observations and a seismogram to highlight the sequence of events related to a large rockfall at Corno Grande in the Gran Sasso d'Italia Massif. The seismogram was interpreted in terms of the time of the failure, the velocity of the sliding mass, and energy dissipation. The event sequence is highlighted by three impacts, which contributed to the generation of an abrasive dust cloud that spread over a large area around the source slope.

In the future, an effort should be made to analyze the seismic signature of unstable rock slopes before they fail. Documenting the state of a slope that is approaching catastrophic failure is a complex and challenging task; stress conditions and the behavior of joints can constrain pre-failure processes such as the generation of new cracks, opening or closing of joints, and the readjustment of the stress field within a rock mass. Studies have been undertaken in mines and on natural slopes to monitor collapses using acoustic and seismic instruments (Miller *et al.*, 1989; Lei *et al.*, 2004; Lai *et al.*, 2006; Paskaleva *et al.*, 2006; Deparis *et al.*, 2008). Controlled explosions have also been used to document slope failure (Phillips *et al.*, 1997; Yang *et al.*, 1998), and laboratory tests have been performed on brittle materials to analyze possible correlations between acoustic emissions and observed microfracturing (Ganne *et al.*, 2007).

REFERENCES

Bianchi Fasani, G., Esposito, C., Scarascia Mugnozza, G., Stedile, L. and Pecci, M. (2008). The 22 August, 2006, anomalous rockfall along the Gran Sasso NE wall (Central Apennines, Italy). In *Landslides and Engineered Slopes – From the Past to the Future: Proceedings of the 10th International Symposium on Landslides and Engineered Slopes*, ed. Z. Chen, J.-M. Zhang, K. Ho, F.-Q. Wu and Z.-K. Li. CRC Press, pp. 355–360.

Bieniawski, Z.T. (1989). *Engineering Rock Mass Classifications: A Complete Manual for Engineers and Geologists in Mining, Civil and Petroleum Engineering*. John Wiley & Sons, Canada.

Calamita, F., Scisciani, L., Adamoli, M., Ben M'Barek, M. and Pelorosso, M. (2002). Il sistema a thrust del Gran Sasso d'Italia (Appennino Centrale). *Studi Geologici Camerti*, 1, 19–32.

D'Alessandro, L., De Sisti, G., D'Orefice, M., Pecci, M. and Ventura, R. (2003). Geomorphology of the summit area of the Gran Sasso d'Italia (Abruzzo Region, Italy). *Geografia Fisica e Dinamica Quaternaria*, 26, 125–141.

Deparis, J., Jongmans, J., Cotton, F. *et al.* (2008). Analysis of rock-fall and rock-fall avalanche seismograms in the French Alps. *Bulletin of the Seismological Society of America*, 98, 1781–1796.

Dramis, F., Fazzini, M., Pecci, M. and Smiraglia, C. (2002). The geomorphological effects of global warming in the Calderone Glacier area (Central Apennine, Italy). In *Climate Changes, Active Tectonics and Related Geomorphic Effects in High Mountain Belts and Plateaux*. Addis Adaba, Ethiopia.

Fischer, L., Kääb, A., Huggel, C. and Noetzli, J. (2006). Geology, glacier retreat and permafrost degradation as controlling factors of slope instabilities in a high-mountain rock wall: The Monte Rosa east face. *Natural Hazards and Earth System Sciences*, 6, 761–772.

Ganne, P., Vervoort, A. and Wevers, M. (2007). Quantification of pre-peak brittle damage: Correlation between acoustic emission and observed micro-fracturing. *International Journal of Rock Mechanics and Mining Sciences*, 44, 720–729.

Ghisetti, F. and Vezzani, L. (1990). Stili strutturali nei sistemi di sovrascorrimento della Catena del Gran Sasso (Appennino Centrale). *Studi Geologici Camerti*, Special Volume 1990, 37–50.

Lai, X.P., Cai, M.F. and Xie, M.W. (2006). In situ monitoring and analysis of rock mass behavior prior to collapse of the main transport roadway in Linglong Gold Mine, China. *International Journal of Rock Mechanics and Mining Sciences*, 43, 640–646.

Lei, X., Masuda, K., Nishizawa, O. *et al.* (2004). Detailed analysis of acoustic emission activity during catastrophic fracture of faults in rock. *Journal of Structural Geology*, 26, 247–258.

Miller, A., Richards, J.A., McCann, D.M., Browitt, C.W.A. and Jackson, P.D. (1989). Microseismic techniques for monitoring incipient hazardous collapse conditions above abandoned limestone mines. *Quarterly Journal of Engineering Geology*, 22, 1–18.

Morrissey, M.M., Savage, W.Z. and Wieczorek, G.F. (1999). Air blast generated by rockfall impacts: Analysis of the 1996 Happy Isles event in Yosemite National Park. *Journal of Geophysical Research*, 104, 23189–23198.

Paskaleva, I., Aronov, A.G., Seroglazov, R.R. and Aronova, T.I. (2006). Characteristic features of induced seismic processes in mining regions exemplified by the potassium salt deposits in Belarus and Bulgaria. *Acta Geodaetica Geophysica Hungarica*, 41, 293–303.

Pecci, M. and D'Aquila, P. (2011). Geomorphological features and cartography of the Gran Sasso d'Italia Massif between Corno Grande-Corno Piccolo and Pizzo Intermesoli. *Geografia Fisica e Dinamica Quaternaria* 34, 127–143.

Pecci, M. and Scarascia Mugnozza, G. (2007). Assetto generale del territorio. In *Il Gran Sasso in Movimento. Risultati del Monitoraggio e degli Studi Preliminari sulla Frana del 22 Agosto 2006, Vol. 2*, ed. M. Pecci and G. Scarascia Mugnozza. Quaderni della Montagna Acta, pp. 9–20.

Pecci, M., D'Agata, C. and Smiraglia, C. (2008). Ghiacciaio del Calderone (Apennines, Italy): the mass balance of a shrinking Mediterranean glacier. *Geografia Fisica e Dinamica Quaternaria*, 31, 55–62.

Phillips, W.S., Pearson, D.C., Edwards, C.L. and Stump, B.W. (1997). Microseismicity induced by a controlled, mine collapse at White Pine, Michigan. *International Journal of Rock Mechanics and Mining Sciences*, 34, Paper 246.

Pignotti, S. (2007). Rilievi GPS (Global Positioning System) e modelli GIS (Geographic Information System). In *Il Gran Sasso in Movimento. Risultati del Monitoraggio e degli Studi Preliminari sulla Frana del 22 Agosto 2006, Vol. 2*, ed. M. Pecci and G. Scarascia Mugnozza. Quaderni della Montagna Acta, pp. 45–58.

Wieczorek, G.F., Snyder, J.B., Waitt, R.B. *et al.* (2000). Unusual July 10, 1996, rockfall at Happy Isles. Yosemite National Park, California. *Geological Society of America Bulletin*, 112, 75–85.

Yang, X., Stump, B.W. and Phillips, W.S. (1998). Source mechanism of an explosively induced mine collapse. *Bulletin of the Seismological Society of America*, 88, 843–854.

28 Downie Slide, British Columbia, Canada

KATHERINE S. KALENCHUK, D. JEAN HUTCHINSON, MARK DIEDERICHS, AND DENNIS MOORE

This chapter is dedicated to John Francis Psutka (1953–2011), whose work contributed greatly to it. He was an accomplished engineering geologist who loved to work on Downie and many other mountain slopes bordering hydroelectric reservoirs in British Columbia.

ABSTRACT

The Downie Slide case history demonstrates the importance of detailed site investigation and long-term monitoring, as well as the need for reinterpretation of the local geologic, structural, and groundwater conditions, and reassessment of slope stability as new information becomes available. Forty-five years of investigation, monitoring, and assessment have been carried out at Downie Slide. The long history of studies, coupled with recent technological advances in analytical tools, has enabled us to better understand the behavior of large complex landslides.

28.1 INTRODUCTION

Downie Slide, the largest known active slope instability in the world, is located 64 km north of the Revelstoke Dam on the west bank of the Revelstoke Reservoir, in the Columbia River valley, British Columbia, Canada (Fig. 28.1). This massive, active, composite, extremely slow-moving rockslide has a volume of 1.5 billion m^3. It measures 3300 m from toe to head scarp and 2400 m along the toe, with a maximum thickness of approximately 250 m. Downie Slide is thought to have been initiated by glacial retreat, approximately 10,000–12,000 years ago (Piteau *et al.*, 1978; Brown and Psutka, 1980). Total slope displacements are estimated between 250 and 300 m (Patton and Hodge, 1975), but there is no evidence that the slide failed rapidly or blocked the valley (BC Hydro, 2010).

Downie Slide was first identified in the 1950s during reconnaissance mapping for potential dam sites on the Columbia River (Blown, 1966). Interest in this massive slope instability grew through the 1960s, when the potential hazard that a landslide of this magnitude poses to hydroelectric operations was recognized. This chapter focuses on the site investigations, landslide stability analyses, and hazard mitigation that were initiated prior to the reservoir filling (1984/5) and that are still ongoing today.

28.2 SITE INVESTIGATION

Several holes were drilled near river level during early mapping in the 1950s (Blown, 1966). CASECO Consultants Ltd., designers of the Mica Dam, which is located 70 km upstream from Downie Slide, initiated an extensive exploration program in 1965, when the potential for Downie Slide material to block the valley and therefore flood the toe of the Mica Dam was being investigated. This program involved drilling five exploration boreholes totaling about 1000 m in length. Three holes were drilled on the landslide to obtain groundwater profiles and to install inclinometers, and two non-instrumented exploration holes were drilled on the opposite side of the valley.

Site investigation of the landslide continued through the 1970s and 1980s by BC Hydro, under the guidance of the Downie Slide Review Panel. The panel of experts also oversaw design and construction of an extensive drainage system. By 1985, reservoir filling was completed and the involvement of the review panel was reduced to a review of annual reports.

With each drilling program, complementary investigations and studies were also carried out, including:

- *surface surveys* (geological mapping, aerial photography studies, surface displacement surveys, meteorological observations, and stream monitoring)
- *subsurface investigations* (seismic refraction surveys, inclinometer, extensometer, and piezometer installations, downhole geophysics, groundwater monitoring, and permeability testing)

Landslides: Types, Mechanisms and Modeling, ed. John J. Clague and Douglas Stead. Published by Cambridge University Press.

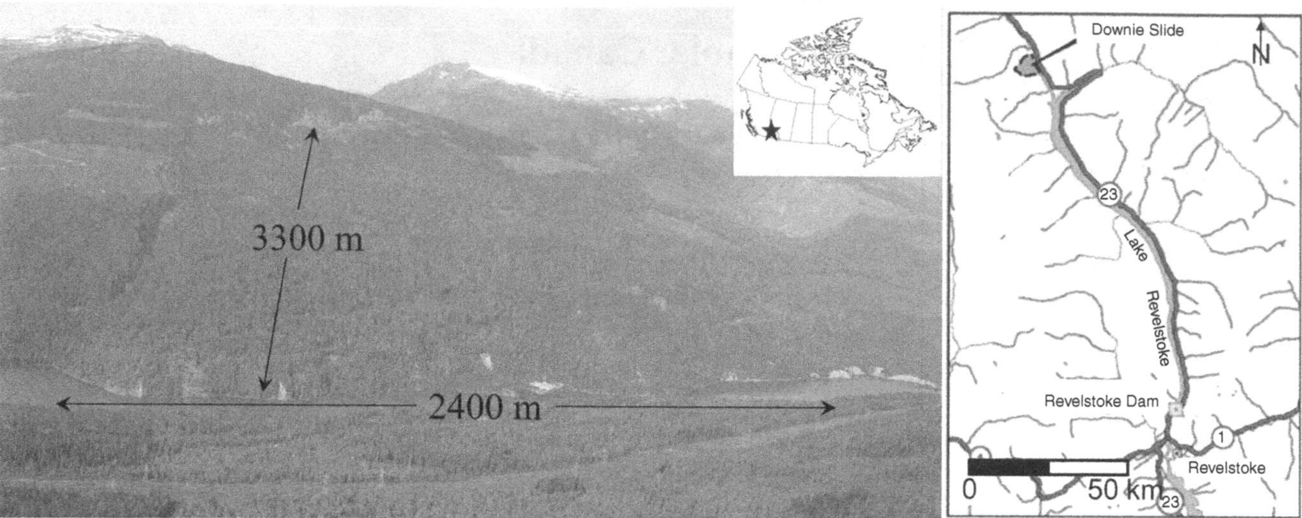

Fig. 28.1. Left: Overview photo of Downie Slide viewed from the east side of the valley, August, 2008. Right: Location of Downie Slide, north of the town of Revelstoke, British Columbia, Canada.

- *material property evaluation* (mineralogy and material strength laboratory testing of gouge and rock samples, and petrographic studies)
- *modeling*, including scaled physical model and mathematical studies of slide-generated waves, groundwater modeling, and slope stability studies.

Seismic monitoring was attempted, but was unsuccessful because ground movements could not be differentiated from surface disturbances such as tree movements in the wind.

Over the operating life of the Revelstoke Reservoir, site investigation has continued during annual site inspections and periodic reviews of monitoring data. Most recently, in 2009, a new geological and geomorphological assessment was conducted, involving surficial mapping, an aerial LIDAR (light, distance, and ranging) survey, and high-resolution aerial photography.

28.3 HAZARD MANAGEMENT AND MITIGATION

Two hydroelectric projects, Mica and Revelstoke dams, operate upstream and downstream, respectively, from Downie Slide. Although the likelihood of rapid failure of Downie Slide is now considered to be low, the possibility still remains. One consequence of a rapid slide failure is the potential formation of a slide-generated wave that could impact Revelstoke Dam. Another consequence of rapid slide failure is blockage of the valley with a landslide dam and subsequent uncontrolled rise of the reservoir level upstream that could impact the toe of Mica Dam. Furthermore, the eventual failure of a landslide dam could cause a downstream surge of water and debris toward Revelstoke Dam.

The magnitude of kinetic energy associated with slow-moving, massive landslides makes it impossible to completely stop their movement. Therefore, approaches to hazard mitigation of such landslides should aim to improve stability by reducing and controlling movement rates. Monitoring should be implemented to track changes in the slope behavior. Emergency warning and response plans should be developed, for situations when the acceptable displacement rate is exceeded, which are commensurate with the potential consequences of failure.

Groundwater pressures, geomechanical properties, geologic setting, and topography are all significant contributors to the ongoing instability of Downie Slide. For such a large landslide, the only factor that can be controlled is the groundwater pressure, and therefore drainage was the only means available to reduce slope movement rates. BC Hydro committed to develop a drainage system that more than offset the decrease in stability caused by submergence of the toe, and to install a comprehensive monitoring system to evaluate the effects of remedial measures. An extensive underground drainage system was constructed between 1974 and 1982. Two adits totaling more than 2.43 km and in excess of 13.6 km of drainholes were installed. The adits are primarily located within the landslide mass itself, and the drainholes extend into the rock mass above and below the basal shear zone. As required by the BC Comptroller of Water Rights, the reservoir was raised in stages as the performance of the slide and drainage works was monitored. In 1984, the expert review panel to the Comptroller concluded that "the effects of the reservoir at the slide were of small scale" and that "the drainage monitoring and surveillance system had functioned as planned" (BC Hydro, 1987). Groundwater pressures slowly rose in the slide in the 1990s, thus additional drainholes were installed in 2007/8 to ensure compliance with the recommendation that "annual peak water levels do not exceed the values assumed for stability calculations" (BC Hydro, 1987).

28.4 SITE CHARACTERIZATION

Regional and local geological and morphological information has been collected from borehole geology logs, geological mapping of the drainage adits, geological and geomorphological mapping of surficial conditions, aerial photograph interpretation, and analysis of LIDAR data. This information has been documented in reports by numerous authors (Wheeler, 1965; Jory, 1974; BC Hydro, 1974, 1976, 1978, 1983; Brown and Psutka, 1980; Bourne and Imrie, 1981; Kalenchuk *et al.*, 2012a).

28.4.1 GEOLOGIC SETTING

In the Columbia River valley area, a metasedimentary sequence consisting primarily of thinly bedded quartzites, semipelites, psammites, calc-silicates, and marbles (Brown and Psutka, 1980) overlies the Monashee Complex, a high-grade metamorphic core complex (Read and Brown, 1981; Scammell and Brown, 1990, and references therein; Armstrong *et al.*, 1991; Johnston *et al.*, 2000). The east-dipping Columbia River Fault Zone truncates the metasedimentary sequence on the east bank of the Columbia River valley across from the toe of Downie Slide, and a foliated biotite granodiorite pluton occupies the hanging wall (Brown and Psutka, 1980). Cataclastic fabrics across the Columbia River Fault Zone are evident in a zone up to 1 km thick, and more recent brittle faulting has resulted in localized zones of clay and graphitic gouge (Brown and Psutka, 1980).

Downie Slide occurs within a thick continuous sequence of pelites, semipelites, and minor psammites. Subtle variation in the rock type and complex interlayering of the schist, gneiss, and quartzite make it impossible to correlate lithological units between boreholes (Jory, 1974; BC Hydro, 1978). Quaternary sediments on the floor and slopes of the valley adjacent to the landslide include local alluvial and colluvial deposits, till, and glaciofluvial sediments (Fulton and Achard, 1985).

28.4.2 STRUCTURAL SETTING

The Downie Slide region has experienced three phases of deformation. Phase 1 deformation is defined by isoclinal folding with the same geometry as Phase 2 and, as such, exerts the same influence on the landslide. The location and attitude of the second and third phases of deformation and their associated fabrics control the slide geometry (Brown and Psutka, 1980). The axial surfaces of Phase 2 folds are parallel to the primary compositional layering (Brown and Psutka, 1980). Phase 2 deformation has produced a penetrative axial planar mica foliation which dips 20° eastward; the basal shear zone of the landslide has developed parallel to this foliation. Phase 3 deformation is defined by open, upright folds with moderate to steeply west-dipping axial surfaces (Brown and Psutka, 1980). Phase 3 deformation is prominent west of Downie Slide, but dies out to the east and has only minor significance at the site of the landslide. Its steep north–south-trending axial planes, however, may have produced release surfaces for the formation of the head scarp.

The rock mass is highly fractured, and three joint sets have been identified (Imrie *et al.*, 1992). Joint set 1 (J1) is parallel to foliation; two sub-vertical sets dip roughly east (J2) and north (J3). J1 joints are continuous, closely spaced, smooth, and wavy or planar. They may have micaceous gouge coatings and/or surface staining, and offset J2 and J3. J2 joints are closely to moderately spaced and are characterized as smooth and wavy to planar with surface staining, chloritization, and clay infilling. J3 joints are closely to moderately spaced and are characterized as planar and smooth to wavy and rough. All joint sets are well developed in the gneiss units. In the weaker, more ductile schist, J1 and J2 joints are poorly developed or widely spaced, and J3 joints are absent (BC Hydro, 1983).

28.4.3 GROUNDWATER

Multiple water tables are controlled by the low permeability of the shear zones at Downie Slide. A lower water table is confined below the basal slip surface; an upper water table is confined within the slide; and a perched water table may occur near the landslide toe (Kalenchuk *et al.*, 2009a). The well-defined, confined lower and upper water tables can be differentiated from each other by piezometric head, geochemistry of the groundwater (Bourne and Imrie, 1981), and the magnitude of the response to changing groundwater boundary conditions and seasonal fluctuations. The water table positions, shown in Figure 28.2, are drawn down within the lower half of the slope by the drainage works (see Section 28.6.2).

Groundwater levels peak seasonally between May and August, first at lower elevations and later in the year within the higher regions of the slope. Water levels decline through the fall, and minimum levels are reached during the winter months. The magnitude of seasonal fluctuations differs spatially (Fig. 28.3). Typical water table fluctuations on the upper part of the landslide are of the order of 10 m, but they locally exceed 25 m and, in rare instances, can be up to 85 m. These high-amplitude seasonal variations are controlled by infiltration in the highly fractured and dilated upper region of the landslide, which is a recharge zone. No streams cross this area. Through the central portion of the slope, 5-m fluctuations are normal, and near the toe, seasonal variations are nearer 1 m. Streams are observed through these regions, where infiltration is limited by thick ground cover. The small magnitude of seasonal fluctuations near the reservoir reflects relatively steady-state reservoir levels and the influence of drainage measures.

28.4.4 SHEAR ZONES

Shear zones at Downie Slide include a well-developed basal slip surface and secondary shears. The basal shear zone is associated with weak mica-rich horizons parallel to the foliation. Near the toe of the landslide, this shear zone curves upward and cross-cuts the foliation to outcrop near the

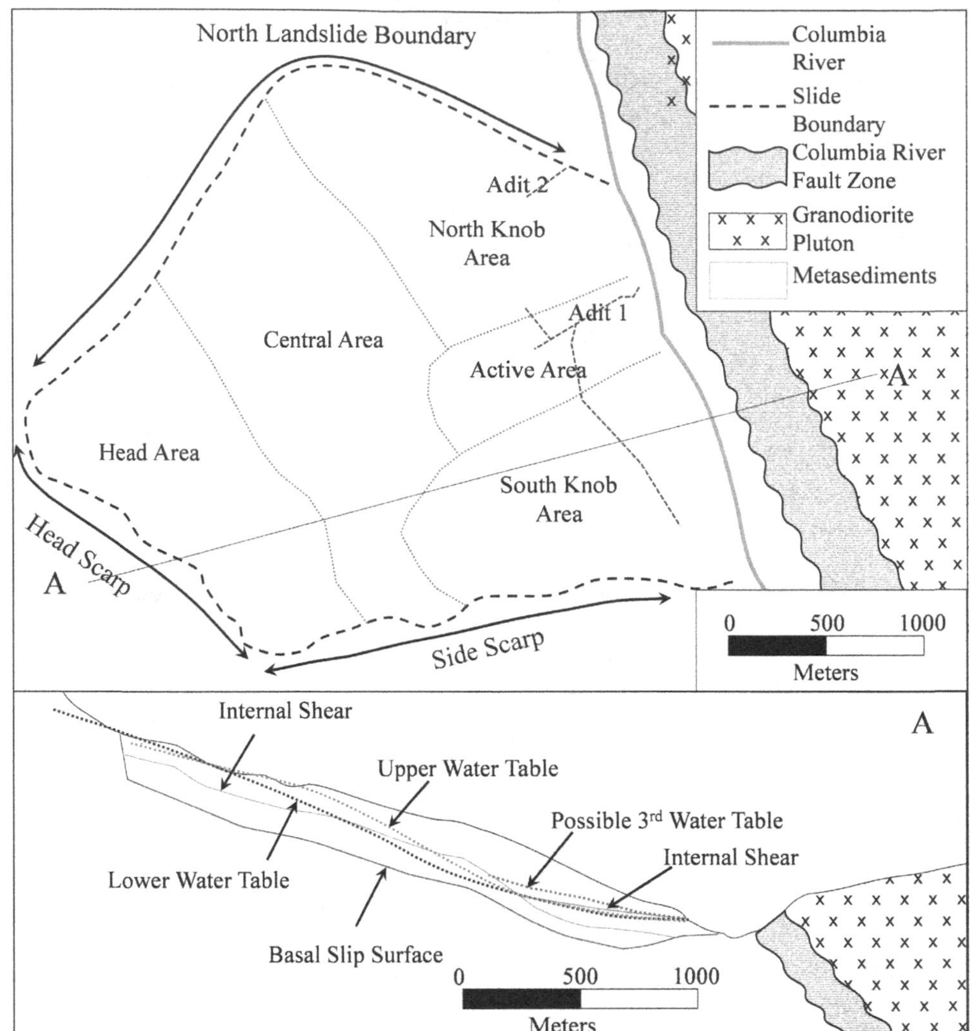

Fig. 28.2. Top: Downie Slide local geology (after Kalenchuk *et al.*, 2012a: after Patton and Hodge, 1975; Piteau *et al.*, 1978; Brown and Psutka, 1980). Bottom: Schematic illustrating the basal slip surface, internal shears, and multiple water tables (modified from Kalenchuk *et al.*, 2009a).

bottom of the river valley. This prominent shear is characterized by closely to very closely spaced, foliation-parallel joints, with areas of fractured and crushed material, clay and mica gouge, and chlorite-altered clay (BC Hydro, 1978, 1983; Bourne and Imrie, 1981).

The basal shear zone at Downie Slide ranges in thickness from less than 2 m to nearly 50 m (Fig. 28.4). The shear zone is typically thinner where slip surfaces follow foliation planes, and thicker where the shear cross-cuts foliation; these differences contribute to spatial variation in the magnitude of deformation observed across Downie Slide. Numerical modeling by Kalenchuk *et al.* (2012b) assumes that thin regions are effectively stiffer than thicker regions. Differences in shear zone stiffness induce localized stress concentrations, which contribute to complex three-dimensional slope deformation.

28.4.5 LANDSLIDE GEOMETRY

The irregular nature of displacements in large slow-moving landslides is an indication of complex landslide geometries. The three-dimensional extent of the landslide mass is defined by the geometry of the basal slip surface. The true geometries of slip surfaces within massive landslides are largely unknown and must be interpolated from the available data obtained from surface and subsurface site investigation programs. The three-dimensional geometry of the Downie basal slip surface has been interpreted from shear zone intercepts in borehole geology logs, inclinometer profiles, and the landslide boundary on the surface identified from aerial photographs and LIDAR data (Kalenchuk *et al.*, 2009b).

Figure 28.5 illustrates two different interpretations of the Downie Slide basal slip surface geometry. The first interpretation is an oversimplified bowl-shaped geometry. The second interpretation is a moderately undulating geometry, interpreted using a stepped minimum curvature algorithm. This undulating geometry is geologically reasonable for Downie Slide because the shears are associated with foliation, which has been gently warped by folding. The stepped feature along the south and west margins of the landslide matches the prominent side and head scarps, although it should be noted that the true depth of

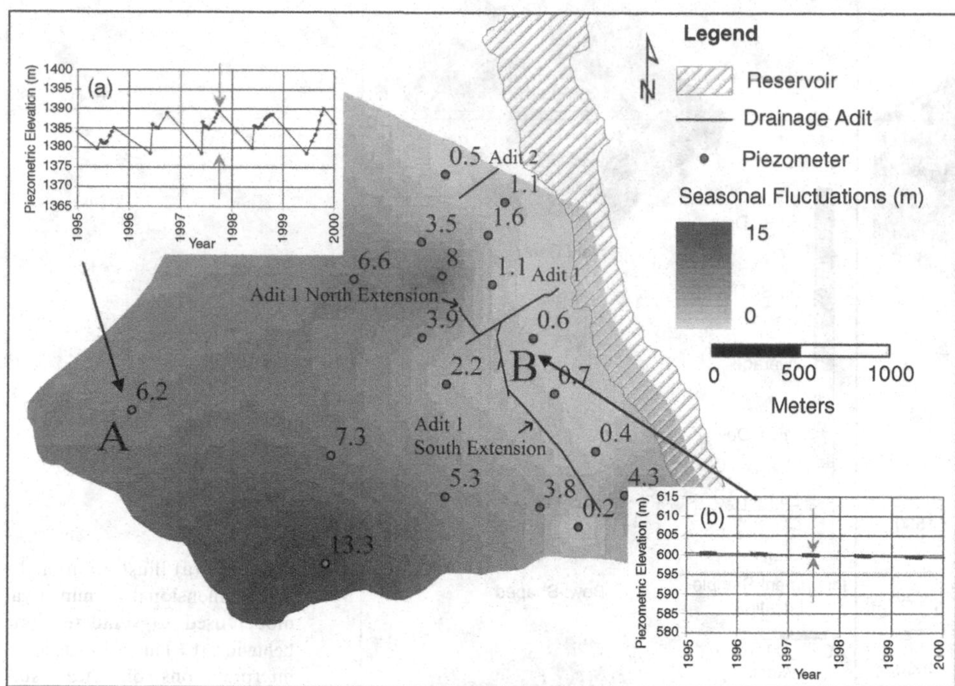

Fig. 28.3. Spatial variation in the magnitude of seasonal groundwater fluctuations in the upper water table at Downie Slide. Data from point A show high seasonal fluctuations near the head scarp recharge zone; and data from point B show very little annual variation (from Kalenchuk *et al.*, 2009a). The lower water table shows similar trends.

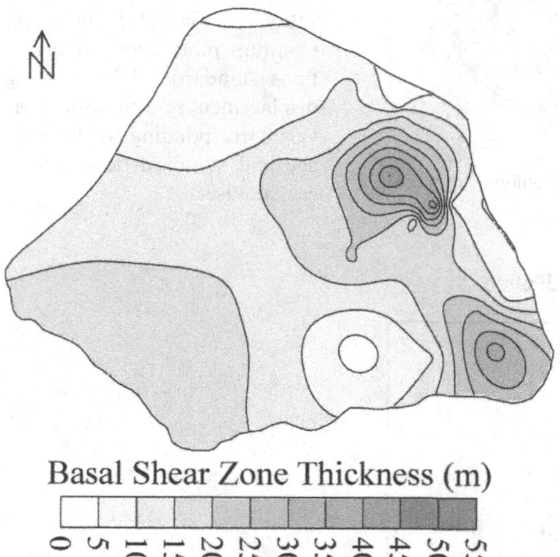

Fig. 28.4. Variability in the thickness of the Downie Slide basal shear zone.

the sub-vertical face extending below the topographic surface is unknown.

Three-dimensional interpretations of shear surface geometry are important for landslide studies. Knowledge of shear surface locations across the entire slope provides improved estimates of landslide volume, which could only otherwise be resolved from estimates of surface area and thickness. Furthermore, three-dimensional geometries are required for state-of-the-art numerical stability analyses. Figure 28.5 illustrates normalized contour plots of surficial deformation magnitudes. The undulating, geologically realistic, three-dimensional geometry is better able to reproduce observed slope deformation patterns than the oversimplified bowl-shaped geometry.

28.4.6 LANDSLIDE MORPHOLOGY

Kalenchuk *et al.* (2012a) have classified the modern Downie Slide as a composite rockslide, meaning that different styles of movement are observed across different areas of the displaced mass. They identified a number of zones based on morphological features identified from airborne LIDAR data, observations made during site visits, and analysis of spatially discriminated failure mechanisms and deformation patterns. The main slide body is divided into the Upper, Central, and Lower regions (Fig. 28.6). Secondary instabilities include the South Trough, Talus Slopes, Oversteepened Slopes, Toe Slump, Basin, Lobe, and Disturbed North regions. The North Knob is an inactive area. This landslide zonation is based on observations of the modern Downie Slide; the mechanisms of failure during initial instability may have differed significantly from those observed today.

- **Upper Slope**: This zone is characterized by hummocky terrain comprising large, partially disturbed rock blocks, separated by extensional features that are littered with jumbled talus.
- **South Trough**: The South Trough extends approximately parallel to the side and head scarps and contains numerous internal scarps and tension cracks. This depression has been interpreted to mark the southern boundary of the main landslide mass (Patton and Hodge, 1975; Kalenchuk *et al.*, 2009a). Patton and Hodge (1975) concluded that the total slide displacements are about 250–300 m based on the position of the South Trough.

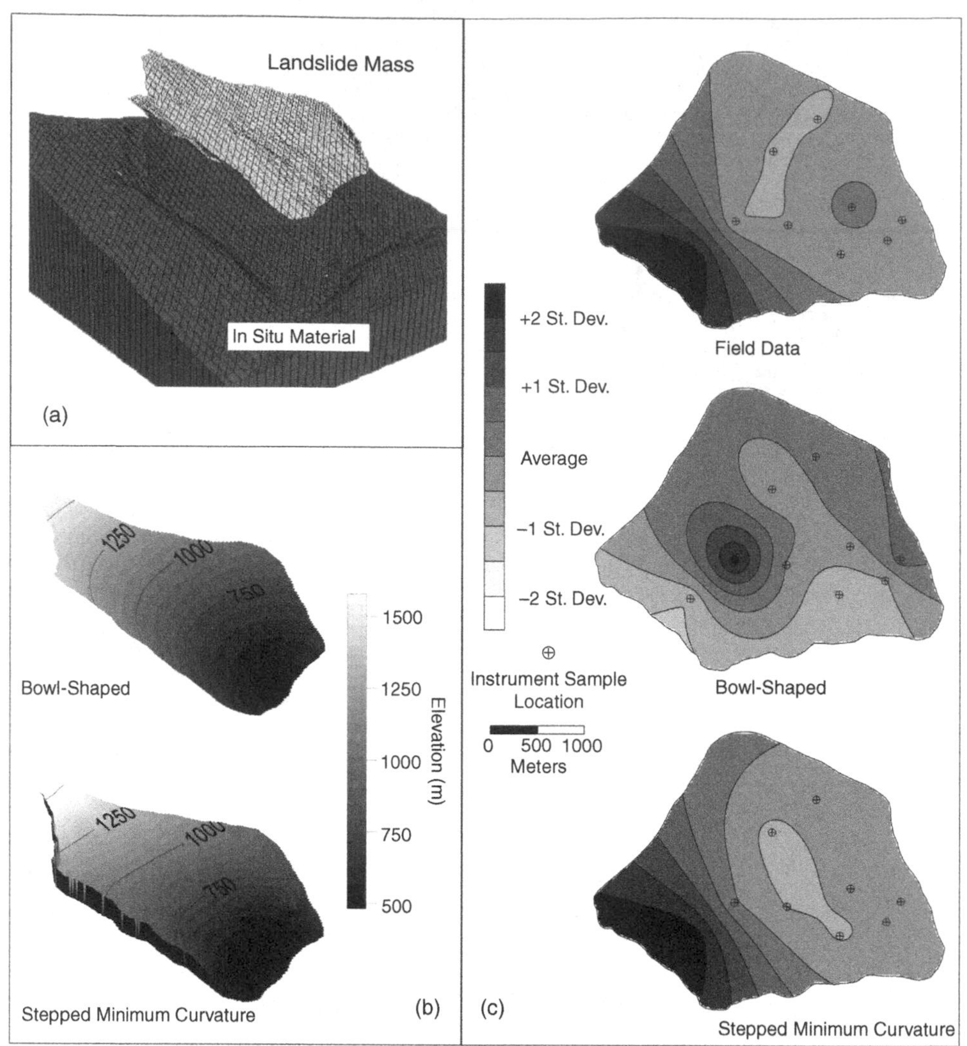

Fig. 28.5. (a) Illustration of the three-dimensional numerical model used to simulate slope behavior. (b) Three-dimensional interpretations of shear surface geometry. (c) Comparison of deformations measured in numerical models to field observations of landslide behavior. Contour plots show deformations standardized by plotting displacement rate contour intervals corresponding to the average and standard deviation of each dataset.

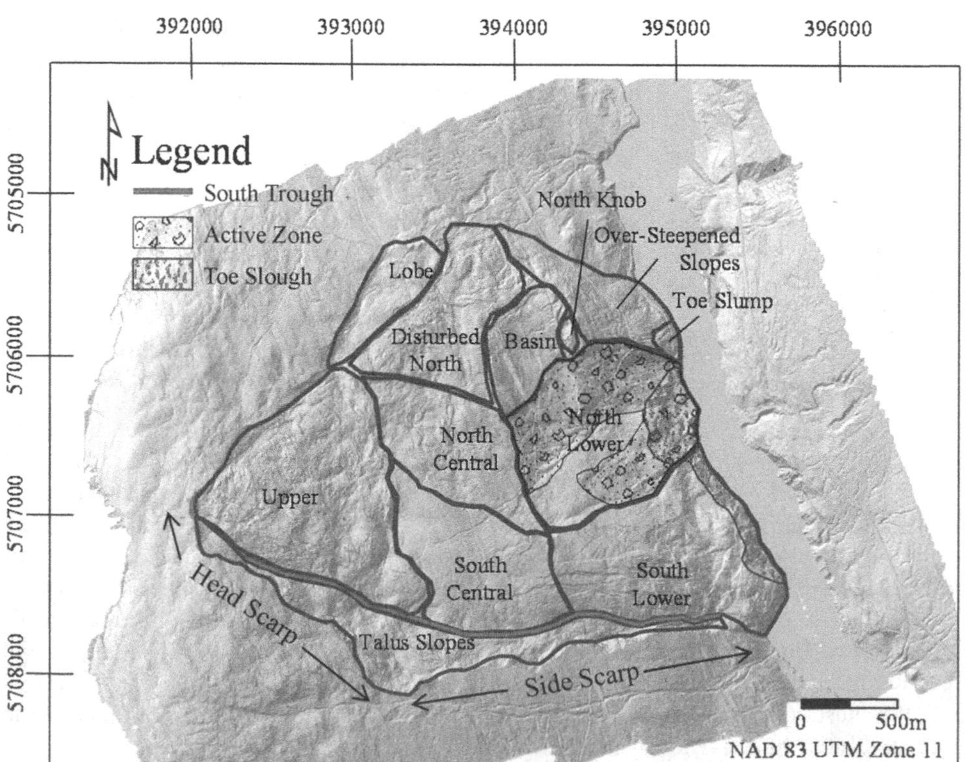

Fig. 28.6. Morphological zones at Downie Slide defined by distinct morphological features and observed slope behavior. The image is based on a LIDAR survey flown in 2009.

- **Talus Slopes**: This zone comprises talus derived from gradual raveling of the scarps. Raveling has been occurring for a long time, as is shown by the presence of both newly deposited debris and talus covered with mature coniferous forest.
- **Central Slope**: This zone features gentle slopes overgrown by mature forest. The South Central area is relatively featureless, suggesting that this portion of the slope has remained intact or that there has been little deformation since initial landslide activation. A subtle topographic break marks the boundary between the North and South Central regions, suggesting greater activity to the north, where the terrain is slightly hummocky.
- **Lower Slope**: This zone features irregular terrain marked by depressions, crevices, internal scarps, and fractures. The southern part of the Lower Slope is a broad, oversteepened ridge. East–west-trending lineaments indicate north–south extension through the southern part of the Lower Slope; however, without detailed mapping, it is difficult to conclude whether the movement is by pure extension or by translational extension, with some component of downslope shearing. Downslope deformation is, however, indicated by the slight clockwise rotation in lineament orientation at lower elevations and bulging of the toe. The northern portion of the Lower region is the Active Zone – a basin marked by hummocky disturbed terrain and by scarps at its edges.
- **North Knob**: This feature is a prominent topographic high, marked by talus-covered slopes and surrounded on all sides by extensional features. Data from a survey monument on the peak of this feature indicate that it has been stationary throughout the period of monitoring.
- **Oversteepened Slopes**: This zone is located east and north of the North Knob. Curved tree trunks and fresh surface colluvium suggest that this area is actively deforming, although there is no evidence of modern, deep-seated displacements.
- **Toe Slump**: An active area bounded by scarp features occurs near the landslide toe just north of the North Lower region.
- **Basin**: The Basin is an area bounded to the west, north, and east by extensional features including scarps and sinkholes. It slopes gently to the south, draining into the North Lower region. This area is not monitored, thus there are no data to define the magnitude and direction of modern, local deformation. However, material from the Basin may have gradually moved south-southeast toward the Active Zone.
- **Disturbed North Region**: This is a region of secondary failures, hypothesized to have initiated in response to the main instability to the south.
- **Lobe**: The Lobe is also interpreted to be a secondary failure. Its upper portion is not morphologically suggestive of landslide activity. The lower portion is terraced, giving the impression of shallow failure in glacial deposits or colluvium, rather than deep-seated failure in bedrock. Northeast-directed surficial movement in this zone is indicated by curved tree trunks.

Downie Slide is bounded to the south and southwest, respectively, by prominent side and head scarps (Fig. 28.7). These scarps are near-vertical and reach heights up to approximately 120 m. The west limit of the head scarp coincides with the hinge zone of a major monoclinal flexural fold associated with Phase 3 deformation (Brown and Psutka, 1980). The north landslide boundary, which can be clearly recognized in LIDAR imagery (Fig. 28.6), has a distinct northeast-trending linear alignment. Where this boundary curves to the southeast near the toe of the landslide, streams continue to flow northeast toward the river valley. This linearity suggests a structural control on the location of the north landslide boundary.

28.5 MONITORING

Groundwater and displacement monitoring was initiated at Downie Slide in 1965, prior to construction of the drainage system in order to provide baseline data to evaluate the effectiveness of the drainage. Intense monitoring was carried out during reservoir filling and for about a year after the reservoir reached full capacity. During the period of toe inundation, an automatic data acquisition system (ADAS) and an automatic early-warning system were deployed to provide immediate notification of any sudden changes in slide behavior. Currently there are nearly 200 instruments monitoring Downie Slide, 65 of which are automated, including piezometers, inclinometers, extensometers, GPS survey monuments, water volume flumes in the adits, and tunnel convergence pins.

Recently, more accurate and reliable sensors, dataloggers, and freewave radio data transmission have been added to the monitoring system. Enhanced real-time monitoring has been achieved with both in-house custom and commercial software. The automated instruments installed at Downie Slide are typically automatically read and stored every 6 minutes. When an instrument has neared or exceeded the threshold level, an e-mail and phone message are automatically sent to the surveillance engineer and on-call person in the BC Hydro Surveillance Department.

28.5.1 DISPLACEMENT MONITORING

Slope movements are the most important indicator of the stability of Downie Slide, and changes in deformation rates are indicative of the effectiveness of the drainage system and the effects of toe inundation by the reservoir. Once reservoir filling was complete, slope movement trends – and therefore warning bounds – became known, and monitoring frequency was reduced. The displacement monitoring network has been periodically expanded and upgraded, and currently consists of 17 inclinometers, 25 surface monuments, and 4 extensometer nests.

SURFACE MONUMENTS

Surface monuments have been surveyed since 1975. The initial surveys were triangulation–trilateration surveys that required up to 7 days of field measurements and several days of data processing. In 1998, the survey was converted to a GPS network; it now

Fig. 28.7. Left: View of the Downie Slide scarps, looking west. Right: The prominent head scarp reaches heights of up to 120 m.

takes 1 day to acquire the data and about a day to process the data and complete the survey. Of the 25 surface monuments currently monitored, 16 are located on the landslide, 6 are on the perimeter, and 3 are control points on the east bank of the Columbia River. The perimeter survey points are monitored to provide an alert of displacements that could indicate enlargement of the current slide mass; these show little displacement, as would be expected.

INCLINOMETERS

Full-length inclinometer profiles are measured once each year. If unusual readings are detected in the hole, a second set of readings is taken to verify the change in movement rate. Monitoring data are routinely reviewed to determine whether new zones of movement have developed, or to check previously defined movement zones for any changes in the thickness of the shear zone. Seven of the inclinometer holes contain multiple in-place probes that are linked to the ADAS system. These probes monitor movement rates at discrete zones throughout the slide, including the basal slip surface and internal shears, and also measure surface and near-surface displacements.

When monitoring slopes, it is important to recognize that there are inherent instrumentation errors that need to be corrected. Interpretation of inclinometer data must consider sensitivity to instrument drift, spiraled inclinometer casings, and sensor alignment. These corrections can be applied to known data errors, such as zero-shift errors and rotational errors. However, inclinometer data, particularly in deep holes, are best used to describe the nature of deformation through the hole and to provide a sense of where localized deformation is occurring rather than to develop an exact displacement profile (Mikkelsen, 1996).

28.5.2 WEIR MONITORING

Water flow from drain holes has been measured in the past, but these measurements are highly variable and are difficult to automate. Weir monitoring was initiated in 1975 in Adit 1, and in 1978 in Adit 2. In 2007, a second weir was installed in the south branch of Adit 1 so that the flows from the north and south extensions of the drainage adit could be monitored individually. Flow from the adits, as recorded in hourly readings, is highly affected by seasonal variations in the groundwater regime. The highest flow rates generally occur in late May, following a doubling of drainage water volume during the month. Through the summer and fall months, the water levels fluctuate and then drop back to their lowest levels in the winter. A significant decrease in the flow rate out of the adits could be a sign of diminishing drainage system effectiveness.

28.5.3 PIEZOMETER MONITORING

Currently there are about 100 operational piezometers monitoring groundwater conditions at Downie Slide. The ADAS system is used to collect data from 28 key piezometers within the slide area on a 6-hour interval throughout the year. Manual measurements of the non-automated piezometers onsite are taken several times during the spring, summer, and fall seasons.

28.5.4 LIDAR

Airborne LIDAR was flown over Downie Slide in 2009. The resulting "bare earth" image of the ground surface of Downie Slide proved to be an invaluable resource (Fig. 28.6). In the future, repeated LIDAR surveys will be used to evaluate differential movements, assuming that sufficient slide movement occurs to distinguish these changes from noise or inaccuracy of the data.

28.5.5 INSPECTIONS

The Downie Slide monitoring program also includes inspections of the slide surface and the drainage system. The site is inspected at least monthly between April and October by technical staff, and annually by a rockslide specialist. Winter conditions impede physical access to the site. Site activities include routine maintenance of the adit ventilation and lighting, site access roads, adit generator, fans and lighting, and the permanent site vehicle, and also repair and replacement of instrumentation sheds, vegetation clearing from helicopter landing pads and survey monuments, and maintenance and calibration of ADAS instrumentation.

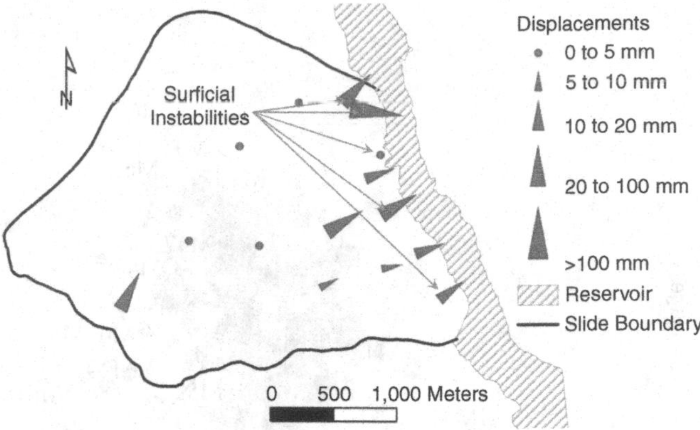

Fig. 28.8. Annual displacements defined by deformation measured since impoundment of the Revelstoke Reservoir (data from 1985–2003).

28.6 SLOPE BEHAVIOR

The pattern of current deformation at Downie Slide is complex, as the magnitude of deformation varies both spatially and temporally. Spatial variations in surficial deformation patterns of landslides can be influenced by the three-dimensional shape of the failure surface, different thicknesses of the moving mass, heterogeneous material properties, and differences in the state of stress associated with the geologic history, groundwater conditions, and previous movement. Temporal variations are influenced by changing boundary conditions, for example groundwater fluctuations or changing material properties due to deformation or weathering processes.

28.6.1 SPATIAL VARIATION IN MODERN SLOPE BEHAVIOR

Downie Slide does not behave as a monolithic instability; instead, different zones deform differently (Kalenchuk *et al.*, 2012a). In the following discussion, Cruden and Varnes' (1996) description of landslide deformation rates is used: rates less than 16 mm per year are extremely slow, and rates between 16 mm per year and 1.6 m per year are very slow.

The morphological zones described previously (Fig. 28.6) display the following deformation patterns.

- The Upper Slope shows evidence of retrogressive failure and has the highest movement rates (about 30 mm per year). Large blocks are interpreted to have broken off the scarp progressively, thereby increasing the spatial extent of the landslide over time.
- Translational sliding occurs over the Central Slope, but displacements are extremely slow (less than 5 mm per year). Inclinometer data from the northern part of the Central Slope indicates that the deformation is dominantly surficial, with minor slip on the basal slip surface and minor internal strain through the landslide profile. There is, however, evidence in the inclinometer records of deeper-seated creep below the basal slip surface. Close- to moderate-spaced jointing below the landslide basal slip surface is documented in borehole geology logs.
- The Lower Slope exhibits translational and rotational sliding. The active landslide zone is found within the North Lower Slope, and the Lower Slope features extensive toe sloughing along the reservoir shoreline. Deformation rates are extremely slow to very slow through the Lower Slope, in the order of 5–10 mm per year on the main slide mass and 20 mm per year in the Active Zone.
- Active deformation in the Disturbed North Region is probably surficial; however, there are no subsurface deformation data and there is no monitoring in this portion of the slope to quantify rates.
- The Lobe is a secondary failure and is active today. However, there is no monitoring on this portion of the slope; thus movement rates are unknown.
- The North Knob is inactive and makes no contribution to modern landslide behavior.
- Although material from the Basin has gradually moved toward the active area, there are no survey monuments to provide rates or directions.
- The Oversteepened Slopes show surficial deformation of about 20 mm per year since reservoir filling.
- The Toe Slump is a localized instability with anomalously high deformation rates – in excess of 200 mm per year since reservoir filling – that are not representative of the overall slide behavior.

Figures 28.5 and 28.8 illustrate spatial variation in the behavior of Downie Slide. With the exception of surficial instabilities near the reservoir, the highest deformation rates are on the upper part of the slope; the central region of landslide has the lowest rates; and moderate rates are observed near the toe.

28.6.2 TEMPORAL INFLUENCE OF GROUNDWATER LEVELS ON SLOPE DEFORMATIONS

Groundwater levels have a strong influence on the behavior of Downie Slide (Fig. 28.9), and have changed in response to

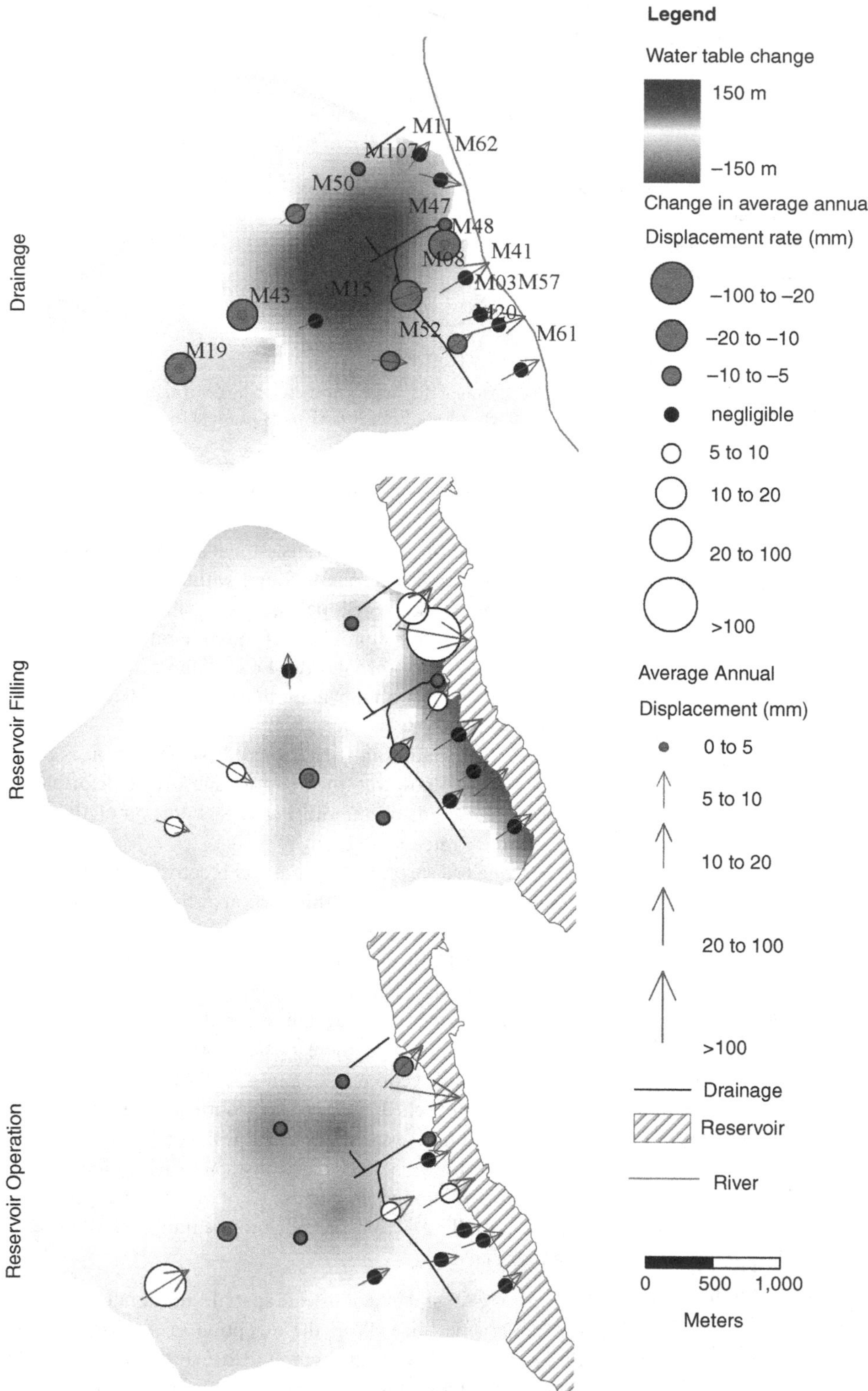

Fig. 28.9. Water table changes and temporal variation in slope behavior. (Note: Instruments M11, M62, M47, M41, M57, and M61 are tracking surficial instabilities.)

implementation of drainage works and reservoir operations. Dewatering resulted in a significant drawdown of the water table throughout the central portion of the slope. During reservoir filling, the landslide toe was inundated by about 70 m of water. Over the operating life of the reservoir, the effectiveness of the drainage system has decreased, perhaps due to a variety of factors: bacterial sludge accumulation, calcification, gas bubbles evolving from the groundwater, migration of fines into flow paths, closing of fractures as a result of increased effective stress, drain-hole wall collapses, or drain-hole blockage by slide movement (BC Hydro, 2010). The deterioration of the system has contributed to a slow rise in water table elevations, although the initial pre-drainage levels have not been exceeded.

The slope response to changing groundwater boundary conditions differs spatially (Kalenchuk, 2010). A marked reduction in slope movements through the central and lower portions of the landslide was achieved by construction of the drainage system. The most significant change is in the lower region of the slide. Instruments tracking surficial instabilities near the landslide toe show little change because the water table is now located below these shallow failures. Changes in rates observed on the upper slope are considered to be a function of localized behavior of large blocks attributed to scarp retrogression.

No changes in movement rates were detected during reservoir filling. The changes in the Lower Slope region were negligible, and the Central Slope region even experienced minor deceleration, although this change is not interpreted to be a response to reservoir filling, given its distance from the landslide toe. Surficial movements, particularly in the Oversteepened Slopes and the Toe Slump, accelerated following inundation. This change is a localized response to the rise in groundwater levels at the landslide toe. The Upper Slope of the landslide shows a minor acceleration following inundation, which we again interpret to be localized.

No significant change in landslide behavior has been observed at Downie Slide since reservoir filling, and the landslide, as a whole, is considered to be in steady state condition (Kalenchuk, 2010). Temporal variations in movement continue to be observed on the Upper Slope, which is most influenced by infiltration. These variations may be related to precipitation and snowmelt, but meteorological data are not available to test this possibility. Minor accelerations, likely a response to a small rise in groundwater levels, have been observed in the Active Zone (North Lower region). The overall rate of movement of the landslide is too small to raise concerns about overall slide stability.

28.6.3 IMPLICATIONS OF SLOPE STUDIES FOR LANDSLIDE HAZARD MANAGEMENT

Interpretation of the differing deformation patterns of a large landslide requires an understanding of how failure mechanisms differ between specific areas of the slope. At Downie Slide, deformation rates are anomalously high in the Toe Slump and Oversteepened Slopes, where surficial failure mechanisms are operating. Recognition of these secondary instabilities has significant implications for the interpretation of monitoring data for slope hazard management. If the secondary surficial deformation rates were misinterpreted as being representative of global behavior, the entire landslide would be perceived to be moving much more rapidly than it is. Furthermore, these secondary instabilities have been shown to be susceptible to changing boundary conditions that do not influence the entire landslide mass. Thus localized surficial accelerations are not necessarily a cause for alarm.

As information continues to become available at Downie Slide, consideration should be given to the spatial distribution of monitoring instruments. The retrogressive Upper Slope region exhibits some of the fastest movement on the landslide, yet characterization of this part of the slope is currently based on a single data point (M19). It is possible that movement rates at this single point may be anomalously low or high. Future additions to the instrumentation network at Downie Slide should take into consideration sparsely monitored slope regions, particularly those interpreted to be contributing to the overall landslide behavior. Areas that contribute less to the overall behavior of the landslide, for example the Lobe, might be given a lower priority for increased instrumentation.

28.7 STABILITY ASSESSMENTS

Detailed site investigations have provided the information necessary to analyze slope stability. A number of stability assessments have been completed for Downie Slide, including two-dimensional numerical modeling of the slope by Enegren (1995) and Kjelland (2004). More recently, Kalenchuk (2010) and Kalenchuk *et al.* (2012a, 2012b) have completed a rigorous analysis using three-dimensional numerical modeling that took into account complex three-dimensional shear surface geometry, interactions between discrete landslide zones, heterogeneity of material properties, and multiple water tables. This modeling has improved our understanding of massive landslide geomechanics.

Numerical simulation of changing groundwater conditions yielded reasonable correlations with field observations (Fig. 28.10). Specifically it simulated:

- the slowing of the landslide following drainage works
- accelerations of the lower portion of the slope after reservoir filling
- the negligible effect of the gradual loss in the performance of the drainage system over the operating life of the reservoir, with the exception of minor accelerations near the toe
- the different responses of the morphological regions to changing groundwater levels.

The development of calibrated numerical models capable of reproducing field observations enables experts to examine potential trigger scenarios and the effectiveness of engineered mitigation approaches. Forecasting has been completed for Downie Slide using three-dimensional, hybrid continuum–discontinuum numerical models (Fig. 28.11). The results suggest that overall simulated slope behavior would change little in response to rapid reservoir drawdown. It is likely, however, that surficial instabilities at the landslide toe would increase in such a scenario. Total loss in drainage capacity would likely result in accelerations in movement of the same magnitude, but in the

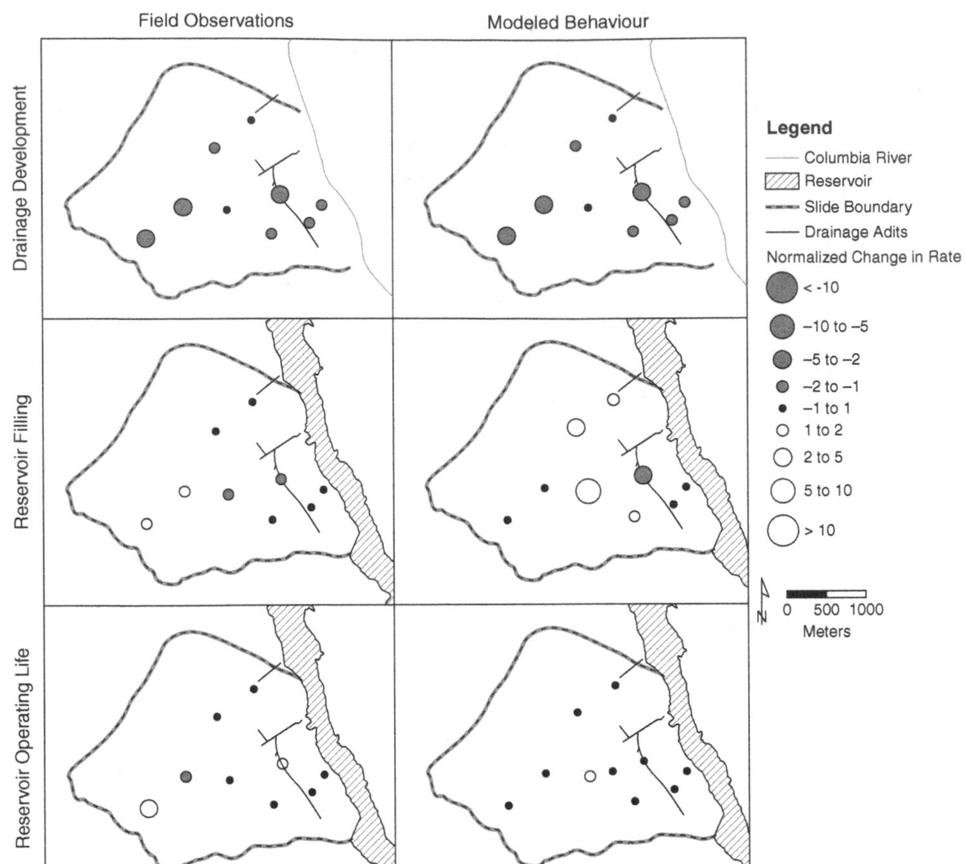

Fig. 28.10. Normalized change in deformation rates for (left) field observations and (right) numerical modeling results, through each phase of reservoir development and operation.

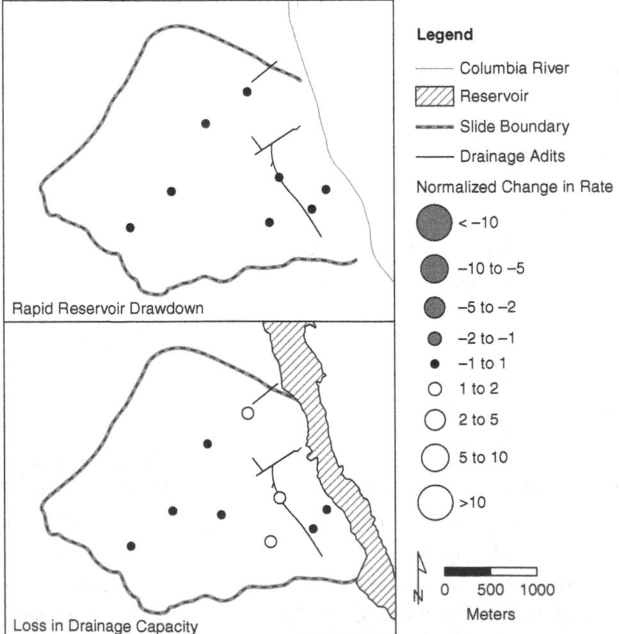

Fig. 28.11. Sophisticated numerical models have been used to forward test potential trigger scenarios at Downie Slide; hypothesized deformation rate change in response to (top) rapid reservoir drawdown, and (bottom) total loss of drainage capacity (after Kalenchuk, 2010).

opposite direction, as those observed during the original installation of the drainage system. Although total failure of the drainage system is extremely unlikely, particularly with continued site maintenance, simulations indicate that this worst-case scenario would only cause minor slope accelerations, at a maximum at the low end of the range of very slow slides.

28.8 CONTINUED WORK ON HAZARD ANALYSIS AND RISK MANAGEMENT

Regular and appropriate maintenance and inspection protocols have been developed over the 35 years that BC Hydro has monitored Downie Slide. With this experience, BC Hydro has added suitable instrumentation types to extend the spatial coverage of deformation monitoring.

The monitoring and assessment program was established before the construction of Revelstoke Dam, with the objective of determining whether the drainage measures would increase slope stability enough to more than offset any decrease in stability caused by submergence of the toe of the slide, as specified in the original conditional water license issued by the BC Comptroller of Water Rights.

Further investigation and analysis of data gathered from the landslide indicate that the potential for rapid catastrophic failure is even lower than originally thought. The overall rates of movement fall into the very slow to extremely slow categories, and the landslide can be divided into a number of different regions that display varying mechanisms of instability. Approximately one-fourth of the landslide mass lies on a horizontal slide surface near the toe, thereby providing considerable resistance to sliding (BC Hydro, 2010). Should any appreciable increase in rate occur in the future, a substantial lead time is anticipated to respond to and understand the change. Ongoing maintenance of the equipment and upgrades of the monitoring system will be required to retain this capability. In addition, monitoring of adit drainage flows and drainage system inspections must continue, in order to ensure that drainage efficiency is maintained.

To comply with the original requirements of the water permit, BC Hydro has been working to maintain the 10 percent increase in stability generated by implementation of the drainage works. The recent work to add more drainage capacity to the system in 2007 and 2008 was a response to the reduction in stability to an estimated 8 percent above the original conditions. However, given the very low probability of failure of a large volume of the slide mass, a partial lessening of these stringent conditions seems warranted before investment is made in more drainage works.

28.9 SUMMARY

Downie Slide has been intensely studied for 45 years. Work on the landslide has provided an evolving and impressive knowledge base of massive landslide geomechanics. Recent improvements in understanding the geomechanics of Downie Slide have significant implications for the continued interpretation of monitoring data and for hazard management. Defining a "significant" change in landslide behavior is only meaningful when the overall behavior of the slope can be modeled and understood.

Recognition of primary and secondary instabilities is important to the analysis of deformation data for slope hazard management. Surficial secondary instabilities near the toe of Downie Slide have higher deformation rates than those throughout the main body of the landslide. If these secondary deformation rates were misinterpreted as being representative of the overall behavior, the entire landslide would be perceived to be moving much more rapidly than it is. In addition, secondary instabilities are susceptible to changing groundwater conditions that do not influence the behavior of the entire landslide mass.

This case history provides an excellent example of the use of advanced numerical models in tandem with high-quality instrumental data. Once models simulating observed behavior have been calibrated effectively, they can be used to examine the possible influence of a variety of potential destabilizing or conditioning factors in order to evaluate the risk of lower-probability events that have not yet been experienced at the location of the landslide. Finally, this case history provides an excellent example of the use of drainage as a means of stabilizing large, slow-moving rockslides.

REFERENCES

Armstrong, R.L., Parrish, R.R., Heyden, P. *et al.* (1991). Early Proterozoic basement exposures in the southern Canadian Cordillera: Core gneiss of the Frenchman Cap, Unit I of the Grand Forks Gneiss, and the Vaseaux Formation. *Canadian Journal of Earth Sciences*, 28, 1169–1201.

BC Hydro (1974). *Summary of 1973 Exploration Program*. BC Hydro Engineering Hydroelectric Design Division, Report 725.

(1976). *Summary of 1974–1975 Exploration Program*. BC Hydro, Report 744.

(1978). *Downie Slide Investigations Report on 1976–1977 Field Work*. BC Hydro Hydroelectric Generation Projects Division, Report HE.C.925.

(1983). *Downie Slide Field Report on Contract CR-10A Geology and Construction*. Vancouver, BC: BC Hydro.

(1987). *Columbia River–Revelstoke Project, Downie Slide Final Report*. BC Hydro Report, H1900.

(2010). *Revelstoke Dam–Downie Slide Drainage Improvement Project, Stability Reassessment*. BC Hydro Engineering, Report E798.

Blown, I.G. (1966). *A geological investigation of the Downie Slide near Revelstoke, British Columbia*. B.A.Sc. thesis, Department of Geology, University of British Columbia, Vancouver, BC.

Bourne, D. and Imrie, A. (1981). *Revelstoke Project Downie Slide Investigations Report on 1981 Drill Program*. BC Hydro, Report 1469.

Brown, R.L. and Psutka, J.F. (1980). Structural and stratigraphic setting of the Downie Slide, Columbia River valley, British Columbia. *Canadian Journal of Earth Sciences*, 17, 698–709.

Cruden, A.M. and Varnes, D.J. (1996). Landslide types and processes. In *Landslides: Investigation and Mitigation*, ed. A.K. Turner and R.L. Schuster. Washington, DC: National Academy Press, Special Report 247, pp. 36–75.

Enegren, E.G. (1995). *Reassessment of the Static Stability Analysis of Downie Slide 1994*. BC Hydro, Report H2889.

Fulton, R.J. and Achard, R.A. (1985). *Quaternary Sediments, Columbia River Valley, Revelstoke to the Rocky Mountain Trench, British Columbia*. Geological Survey of Canada, Paper 84–13.

Imrie, A.S., Moore, D.P. and Enegren, E.G. (1992). Performance and maintenance of the drainage system at Downie Slide. *In Proceedings of the 6th International Landslides Symposium*, Christchurch, New Zealand, pp. 751–757.

Johnston, D.H., Williams, P.F., Brown, R.L., Crowley, J.L. and Carr, S.D. (2000). Northeastward extrusion and extensional exhumation of the crystalline rocks of the Monashee Complex, southeastern Canadian Cordillera. *Journal of Structural Geology*, 22, 603–625.

Jory, L.T. (1974). Appendix 2, Summary of geology. In *Revelstoke Project Downie Slide Investigations Summary of 1973 Exploration Program*. BC Hydro, Report 725.

Kalenchuk, K.S. (2010). *Multi-dimensional analysis of large, complex slope instability*. Ph.D. thesis, Queen's University, Kingston, ON.

Kalenchuk, K.S, Hutchinson, D.J. and Diederichs, M.S. (2009a). Downie Slide: Interpretations of complex slope mechanics in a massive, slow moving, translational landslide. In *GeoHalifax2009*. Canadian Geotechnical Conference, Halifax, NS, pp. 367–374.

(2009b). Application of spatial prediction techniques to defining three-dimensional landslide shear surface geometry. *Landslides*, 6, 321–333.

(2012a). Morphological and geomechanical analysis of the Downie Slide using 3-dimensional numerical models: Testing the influence of internal shears and interaction between landslide regions on simulated slope behavior.

Kalenchuk, K.S., Diederichs, M.S. and Hutchinson, D.J. (2012b). Three-dimensional numerical simulations of the Downie Slide to test the influence of shear surface geometry and heterogeneous shear zone stiffness. *Computational Geosciences*, 16, 21–38.

Kjelland, N. (2004). *Constraints on GIS-based decision support systems for slope stability analysis via geotechnical modelling*. M.Sc. thesis, Queen's University, Kingston, ON.

Mikkelsen, P.E. (1996). Field instrumentation. In *Landslides: Investigation and Mitigation*, ed. A.K. Turner and R.L. Schuster. Washington, DC: National Academy Press, pp. 278–316.

Patton, F.D. and Hodge, R.A.L. (1975). *Airphoto Study of the Downie Slide, British Columbia*. Report prepared for the Downie Slide Review Panel, British Columbia Hydro and Power Authority Revelstoke Dam Project.

Piteau, D.R, Mylrea, F.H. and Blown, I.G. (1978). Downie Slide, Columbia River, British Columbia, Canada. In *Rockslides and Avalanches*, ed. B. Voight, New York: Elsevier, pp. 365–392.

Read, P.B. and Brown, R.L. (1981). Columbia River Fault Zone: Southeastern margin of the Shuswap and Monashee complexes, southern British Columbia. *Canadian Journal of Earth Sciences* 18, 1127–1145.

Scammell, R.J. and Brown, R.L. (1990). Cover gneisses of the Monashee Terrane: A record of synsedimentary rifting in the North American Cordillera. *Canadian Journal of Earth Sciences* 27, 712–726.

Wheeler, J.O. (1965). *Big Bend Map-Area, British Columbia (82 M East Half)*. Geological Survey of Canada, Paper 64–32.

29 The 1963 Vaiont landslide, Italy

MONICA GHIROTTI

Anyone working in the field of slope stability may meet a similar problem and no pains should be spared to discover an explanation of the Mount Toc landslide.

V. Mencl (1966)

ABSTRACT

On October 9, 1963, about 270 million m^3 of rock slid into the newly created Vaiont Reservoir in northern Italy. The slide moved a 250-m-thick mass of rock 300–400 m horizontally with an estimated velocity of 20–30 m s^{-1}, before running up and stopping against the opposite side of the Vaiont Valley. Most of the slide mass moved as a block and reached the opposite side of the valley with little change in shape apart from a rotation that is evident from both the surface morphology and the stratigraphic sequence. The landslide generated a displacement wave that overtopped the arched Vaiont Dam and flooded settlements below, killing over 2000 people. The dam resisted the forces imposed by the landslide and suffered only minor damage. The Vaiont landslide helped lay the foundation for modern engineering geology by establishing a new vision for evaluating landslide hazard and applying rock mechanics to slope stability.

29.1 INTRODUCTION

The Vaiont landslide in northern Italy is one of the best-known and most tragic examples of a natural disaster induced by human activity. On October 9, 1963, a catastrophic landslide occurred on the north slope of Mount Toc; a rock mass of approximately 270 million m^3 collapsed into the Vaiont Reservoir at velocities up to 30 m s^{-1}, generating a wave that overtopped the Vaiont Dam and swept into the Piave Valley below, with the loss of about 2000 lives (Fig. 29.1). The arched concrete dam resisted the forces imposed by the landslide and suffered only minor damage.

Many questions arose, and remain, concerning the legal, economic, social, and scientific issues associated with the history of the dam and, in particular, management of the instability on the slope above the reservoir up to the time of catastrophic failure. The event stimulated a large body of research on the stability of natural rock slopes, especially on geotechnical risk protocols for hydroelectric projects in mountainous terrain. In the decades after the disaster, the scientific community has developed and disseminated knowledge which now suggests that a completely different approach should have been taken to manage the situation. The disaster is also a dramatic example of the consequences of having an insufficient amount of available data (parameter uncertainty) and a poor understanding of slope failure processes (model uncertainty). Geotechnical investigations carried out at Vaiont before 1963 were inadequate, mainly due to the limited development of the field of engineering geology at that time. As a consequence, engineers involved in the project failed to realize the gravity of the situation and did not engage in adequate countermeasures.

The story of Vaiont is more than a chronology of technical operations and events; it is a complex combination of factors that can be interpreted in different ways. Despite the many research investigations undertaken since the landslide, the Vaiont disaster remains of great scientific interest. Many contradictory statements and conclusions remain in the literature, and these are evident in the continued publication of scientific papers. Although some of the main issues have been addressed, such as the presence of clay-rich layers along the failure surface (Hendron and Patton, 1985) and whether the 1963 slide was a new slide or a reactivation of a prehistoric slide (Semenza and Ghirotti, 2000; Semenza, 2010), there are still more questions that require further investigation.

29.1.1 LITERATURE ON THE VAIONT SLIDE

During the early part of the twentieth century, reservoir slope stability was generally not considered during dam

Landslides: Types, Mechanisms and Modeling, ed. John J. Clague and Douglas Stead. Published by Cambridge University Press.

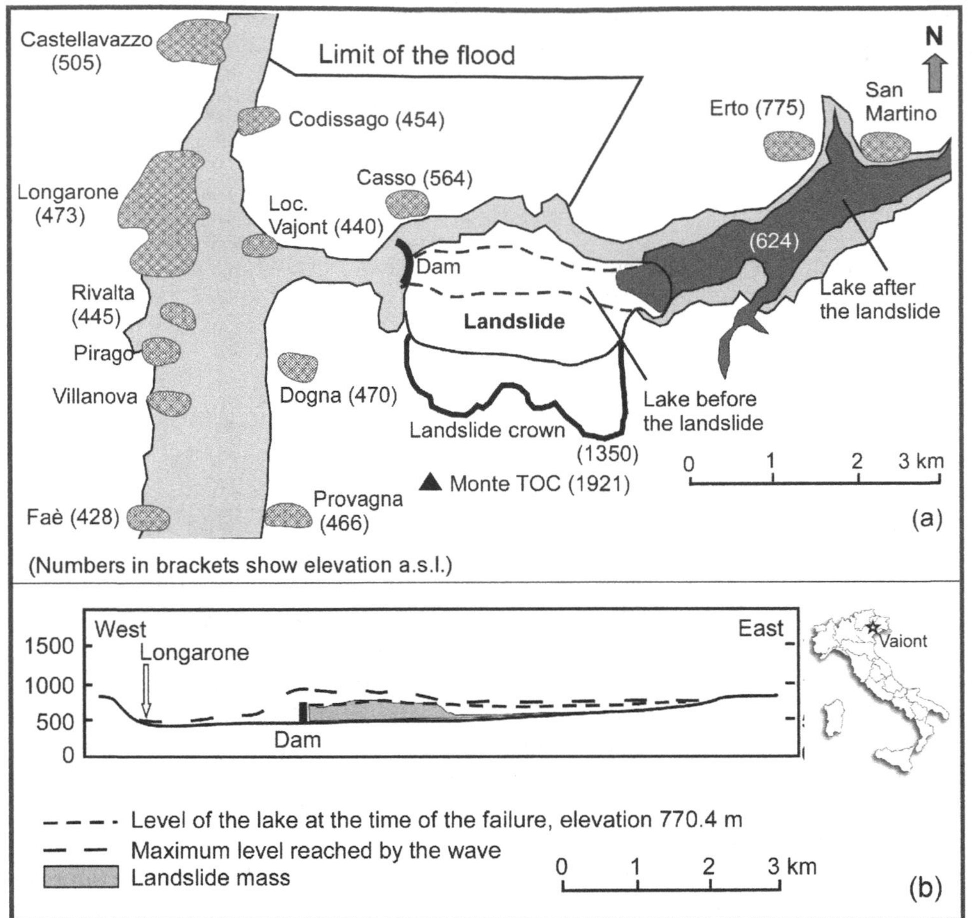

Fig. 29.1. (a) General plan showing the Vaiont landslide and the limit of the flood wave. (b) Schematic longitudinal section showing the original lake level and the elevation of the flood wave (after Selli and Trevisan, 1964).

construction. Studies in the Vaiont Valley at that time were limited to geological aspects (Boyer,1913; Dal Piaz,1928). In the years immediately preceding the landslide, the only available documents were reports prepared by engineers and technicians describing the progress of the work and the problems encountered (Caloi and Spadea, 1960; Giudici and Semenza, 1960; Müller, 1961). After 1963, many papers on Vaiont were published in the international literature. These can be grouped into four categories.

- Papers based on geological and geomorphological data collected at Vaiont (Carloni and Mazzanti, 1964a, 1964b; Frattini *et al.*, 1964; Kiersch, 1964, 1965; Müller, 1964, 1968, 1987a, 1987b; Selli and Trevisan, 1964; Selli *et al.*, 1964; Rossi and Semenza, 1965; Semenza, 1965, 2005, 2010; Broili, 1967; Martinis, 1979; Hendron and Patton, 1985; Riva *et al.*, 1990; Semenza and Melidoro, 1992; Semenza and Ghirotti, 1998, 2000; Mantovani and Vita-Finzi, 2003).
- Papers that provide detailed engineering geological descriptions of the landslide or that deal with specific aspects, including geotechnical properties of the material involved, physical and rheological behavior of the failed rock mass, and application of different types of stability analysis to better understand the role of the many factors involved in landslide triggering and development (Ciabatti, 1964; Kiersch, 1964; Jäeger, 1965a, 1965b, 1972; Caloi, 1966; Mencl, 1966; Skempton, 1966; Kenney, 1967; Nonveiller, 1967, 1987; Lo *et al.*, 1972; Habib, 1975; Chowdhury, 1978; Trollope, 1980; Corbyn, 1982; Voight and Faust, 1982, 1992; Hendron and Patton, 1985; Belloni and Stefani, 1987; Hutchinson, 1987; Leonards, 1987; Voight, 1988, 1989; Sitar and MacLaughlin, 1997; Tika and Hutchinson, 1999; Erismann and Abele, 2001; Vardoulakis, 2002; Crosta and Agliardi, 2003; Kilburn and Petley, 2003; Sornette *et al.*, 2003; Helmstetter *et al.*, 2004; Sitar *et al.*, 2005; Rose and Hungr, 2007; Veveakis *et al.*, 2007; Alonso and Pinyol, 2010; Alonso *et al.*, 2010; Ferri *et al.* 2010, 2011; Mufundirwa *et al.*, 2010; Pinyol and Alonso, 2010).
- Papers that report microseismic and other instrumental data (Belloni and Stefani, 1987; Kilburn and Petley, 2003).
- Papers on the landslide-generated impulsive wave (Panizzo *et al.*, 2005; Roubtsova and Kahawita, 2006; Pastor *et al.*, 2009; Bosa and Petti, 2011; Ward and Day, 2011).

Genevois and Ghirotti (2005) provide a comprehensive review of research on the Vaiont landslide, and Superchi *et al.* (2010) report on an up-to-date electronic bibliographic database for the event.

29.1.2 CHRONOLOGY OF EVENTS

The Vaiont Dam was constructed between 1957 and 1960 on the Vaiont River in northern Italy, about 100 km north of Venice. The doubly curved arch dam, which extends 261.1 m above the valley floor, was, in 1963, the highest thin-arch dam in the world. Its abutments are situated on the steep flanks of a deep canyon cut into limestones of Middle to Late Jurassic age (Cobianchi and Picotti, 2003). The plan was to operate the reservoir at a full capacity of 169 million m^3. The dam was completed in February 1960, at which time it began to impound water.

The catastrophic failure was preceded by a phase of accelerating creep that lasted almost three years, related both to the reservoir operation and to precipitation. During the initial filling of the reservoir, a 1-m-wide and 2.5-km-long M-shaped tension crack opened on Mount Toc. This instability supported the argument of Giudici and Semenza (1960) that a gigantic old landslide covered the north slope of Mount Toc. In the following three years, the technical commission in charge of the reservoir prepared and implemented different strategies to manage a possible emergency. The scenario envisioned by the engineers was based on the principle that a rock mass of such size would move gradually, with progressive but limited displacements and at slow rates. Within this conceptual framework, Müller (1961) proposed that the entire rock mass could be slowly mobilized by filling and drawing down the reservoir. This strategy was adopted and repeated a second time, while a bypass tunnel was excavated on the slope opposite Mount Toc to maintain the reservoir's functionality after the anticipated landslide. However, during a third filling phase, in the summer of 1963, the velocity of the moving rock mass exceeded the warning thresholds fixed by the technical commission. In late September the water level was slowly lowered to reduce the rate of creep. However, velocities of movement continued to increase, and rates of up to 20 cm per day were recorded. At 22:39 GMT on October 9, 1963 a sudden release of the entire rock mass occurred; the north of Mount Toc failed suddenly over a length of 2 km and a surface area of 2 km^2. A 250-m-thick mass of rock moved 300–400 m horizontally at an estimated velocity of 20–30 m s^{-1}, before running up and stopping against the opposite side of the Vaiont Valley. Most of the failed rock mass remained intact (Fig. 29.2), with little change in shape apart from a general rotation that is evident in both the surface morphology and the stratigraphic sequence, which was not changed by the movement (Rossi and Semenza, 1965). The landslide drove the water in the reservoir up to about 200 m up the north slope of the valley, pushing more than 30 million m^3 of water more than 100 m over the dam. The flood wave sped down the Vaiont Gorge to the Piave River, where it destroyed the villages of Pirago, Villanova, Rivalta, and Faé, and most of the town of Longarone (Fig. 29.1a). The landslide produced a seismic shock that was recorded throughout Europe. Remarkably, the dam remained intact, with only minor damage at its crest.

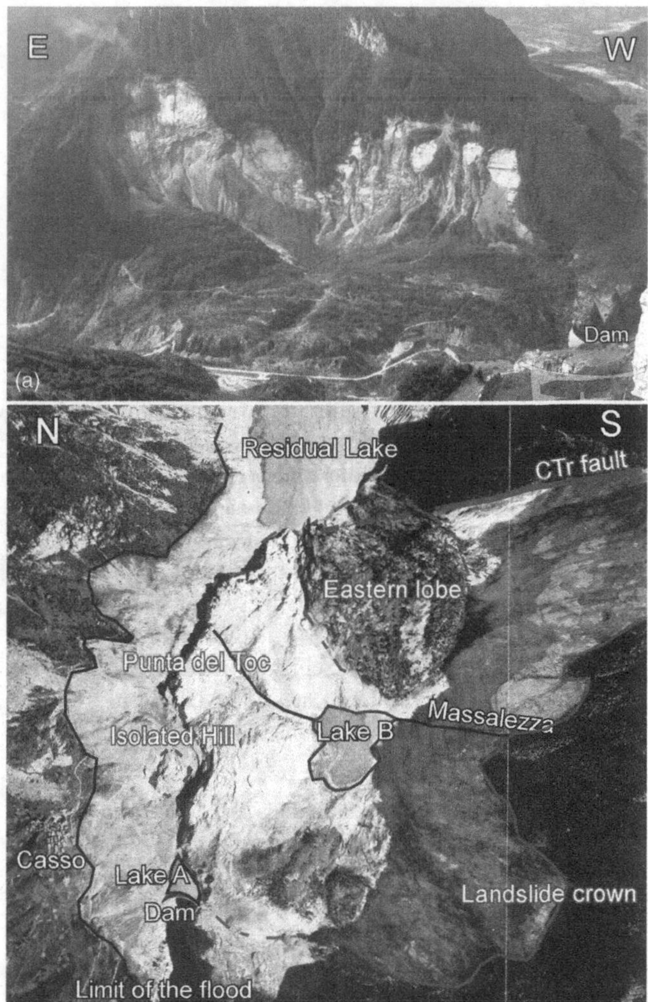

Fig. 29.2. (a) The failure scar and the deposit of the 1963 landslide; Mount Toc in the background. (b) Aerial photograph taken a few days after the landslide; place names are highlighted (modified after Selli and Trevisan, 1964).

29.1.3 GEOLOGICAL AND GEOMORPHOLOGICAL FEATURES OF THE LANDSLIDE

The landslide involved Jurassic and Cretaceous rocks, mainly fractured limestones and marls of the Socchér Formation. Movement occurred along a failure surface shaped like a chair, largely along 0.5–18-cm-thick clay-rich layers of the Fonzaso Formation (Hendron and Patton, 1985). The profile of the sliding surface becomes more circular to the east.

Geological and tectonic evidence suggests that parts of the perimeter of both the 1963 and prehistoric landslides correspond to one or more faults (Rossi and Semenza, 1965; Hendron and Patton, 1985; Mantovani and Vita-Finzi, 2003).

29.1.4 STRUCTURAL, GEOLOGICAL, AND GEOMORPHOLOGICAL FEATURES OF THE VAIONT VALLEY BEFORE 1963

Information on structural, geological, and geomorphological features of the Vaiont Valley prior to the 1963 landslide are required in order to understand the kinematics of the landslide. This information is contained within several unpublished engineering reports (Caloi and Spadea, 1960; Giudici and Semenza, 1960; Müller, 1961) and hundreds of photographs taken prior to the disaster (Masè *et al.*, 2004; Semenza, 2010).

The problem of slope stability in the Vaiont Valley arose in 1959, when SADE, the owner of the Vaiont hydroelectric power station, engaged Leopold Müller to carry out a rock engineering evaluation of the basin. A detailed geological survey of the area carried out by Giudici and Semenza (1960) led to the identification of an old, potentially dangerous, landslide on the south side of the valley, upstream of the dam (Fig. 29.3). Many geological and morphological features supported the existence of an older, large landslide, leading these geologists to raise the possibility that the mass could be reactivated during the filling of the reservoir (Giudici and Semenza, 1960; Semenza, 2010). This information was communicated to SADE during the summer of 1959. The main conclusions of Giudici and Semenza's study were:

- The Pian del Toc and the Pian della Pozza areas and an area east of the Massalezza Stream were part of an enormous ancient landslide that had slid down the north side of Mount Toc (Fig. 29.3a, d). The deposits of this landslide filled the Vaiont River valley at some time after terminal Pleistocene deglaciation.
- Incision of the Vaiont Gorge divided the prehistoric slide mass into two parts: a large preserved remnant south of the Vaiont River (Fig. 29.3a) and small preserved remnants on the north side of the river (Fig. 29.3a, c). One remnant on the south side of the valley could easily be distinguished from in-situ rock and was consequently called Colle Isolato ("Isolated Hill").
- Cataclasites, which were also described as mylonites, cropped out at an elevation of about 600 m asl and were assumed to be continuous over a distance of about 1.5 km along the south side of the Vaiont Valley (Fig. 29.3a). They were also found at the base of the Isolated Hill on top of stratified alluvial gravels (Fig. 29.3c). These cataclasites were subsequently exposed, together with many clay-rich horizons up to 15 cm thick, in adits constructed in 1961 along the west side of the Massalezza Ditch and in the adjacent stream bed (Fig. 29.3a). They were interpreted by Giudici and Semenza (1960) as the failure surface of the ancient landslide. Little attention was given to these clay-rich zones at that time and, in fact, for many decades.
- Some high-discharge springs were observed in the uncemented cataclasites at the base of the Pian del Toc and along the Vaiont Gorge. These springs confirmed the high degree of fracturing and permeability of the rock mass above the ancient basal failure surface that cropped out above the valley floor, but well below the maximum proposed reservoir level (Fig. 29.3a).
- The margin of the ancient landslide on the east is a north-northeast-trending fault (Col Tramontin Fault; see Fig. 29.3a).

Between late 1960 and 1963, Semenza ceased his collaboration with SADE and, except for the geological survey he carried out immediately after the landslide, he never again worked at Vaiont. He did, however, re-examine his documents in 1980 at the request of A. J. Hendron and F. D. Patton, who were working on the landslide at that time.

GEOLOGIC SETTING AND TECTONIC FEATURES OF THE VAIONT AREA

An important aspect to consider in trying to understand the landslide is the geologic structure of the northern slope of Mount Toc. The slide surface observable today is different from the surface immediately after the event, because erosion and weathering have removed material, especially clays, and colluvium now covers the lower part of the slope. Nevertheless, major and minor folds and faulted monoclinal structures can still be seen along much of the slope where the failure occurred.

The Vaiont Valley has been eroded along the axis of an east–west-trending syncline (Erto Syncline). This structure lies on the back side of a regional, south-verging asymmetric fold related to the Belluno Thrust. The hinge of the Erto Syncline is subparallel to the Vaiont River and not far from it, at the village of Erto. The hinge of the syncline plunges to the east from few degrees up to 20°, increasing eastward. Part of the Erto Syncline is a monoclinal flexure generated by a minor blind backthrust (Fig. 29.4; Doglioni and Carminati, 2008). It is responsible for the local north dip of the beds from 25° to 45°. The steepest beds form the back of the Vaiont Slide.

The Vaiont stability analyses cited above generally assumed a nearly horizontal seat of the slide. As noted, however, the Erto Syncline plunges to the east, and consequently the beds dip in that direction along the seat of the slide. The easterly dip of the beds had a significant effect on the behavior of the landslide, and thus the use of two-dimensional profiles by most previous researchers can provide only a limited explanation of the true failure kinematics. A realistic assessment of the landslide requires a consideration of the three-dimensional geologic structure. Many years after the preliminary 3D back analysis of Hendron and Patton (1985), it is time to reassess the problem using 3D numerical modeling. Work by Superchi *et al.* (2011) and Wolter *et al.* (2011) will provide the additional three-dimensional structural and geomechanical data required for such modeling.

After its formation, the Erto Syncline was affected by faulting and minor folding, which gave rise to transverse and oblique structures with significant strike–slip displacements (Riva *et al.*, 1990; Mantovani and Vita-Finzi, 2003). Among these structures are several that partly border, or are near to, the 1963 slide: on

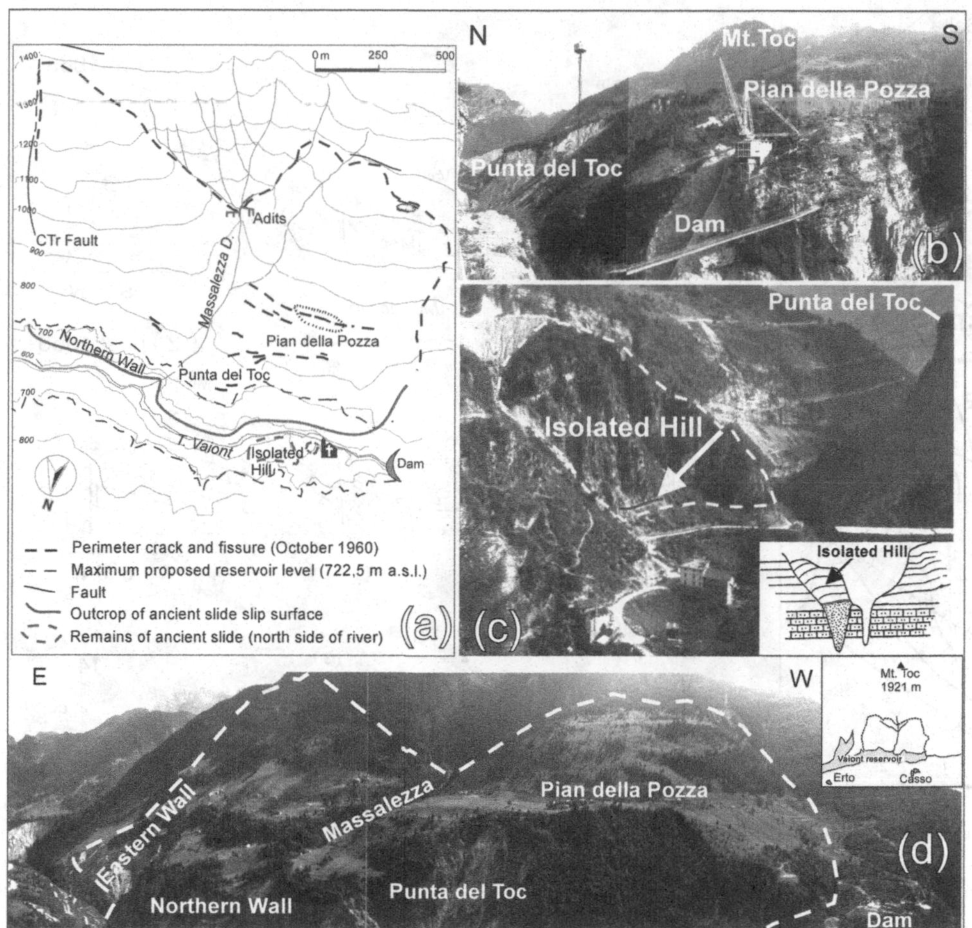

Fig. 29.3. Main geologic and geomorphological features of the Vaiont Valley before 1963. (a) Map of the pre-1963 Vaiont landslide area. The outcrop of an ancient failure surface along the Vaiont canyon walls is based on Giudici and Semenza (1960) (modified after Semenza and Ghirotti, 2000). (b) North slope of Mount Toc seen from the dam before 1963 (photograph by Semenza, August 25, 1959; modified after Semenza and Ghirotti, 2000). (c) North slope of the Vaiont Valley, seen from the dam. The dashed line delimits "Isolated Hill." Above the road, a thin horizontal layer of white cataclasites (arrow) separates the in-situ rock from the sub-horizontal layers of the overlying ancient landslide (photograph by Semenza, September 10, 1959; after Semenza and Ghirotti, 2000). The inset shows Semenza's sketch of the base of "Isolated Hill" alluvial deposits demarcating the old postglacial Vaiont River. (d) North slope of Mount Toc seen from the village of Casso; the dashed line delimits the ancient landslide deposit (Giudici and Semenza, 1960) (photograph by Semenza, September 1959; after Semenza and Ghirotti, 2000).

the east, the Col Tramontin (CTr) Fault, which strikes north-northwest and dips 55° west; on the south, the east-trending Col delle Erghene lineament; and west of the landslide, the north-trending Col delle Tosatte lineament. The complexity of the area can be clearly seen in the tectonic map and from the geological sections (Fig. 29.4; Riva *et al.*, 1990). In particular, pre- and post-1963 sections show the "chair-like" shape of the failure surface and the tectonic style of the oldest deformation phase (Miocene) of the Erto Syncline and the Mount Salta nappe; the longitudinal section (14) shows the true dips of the bedding planes toward the east, together with the younger north–south faults, which control the stability conditions of the diverse zones of the Vaiont Valley.

There are important differences in the morphology of the failure surface. The western part of the surface is concave and much smoother than the central folded area. It is marked by one large and several smaller steps, as well as conjugate discontinuities. The failure surface is at approximately the same stratigraphic position in this area and contains the clay-rich horizons. The central part of the landslide surface is incised by the Massalezza Stream. It is highly tectonized and characterized by a large fold with smaller parasitic folds along its limbs. The eastern part of the landslide surface has a pronounced convexity and an irregular shape with multiple steps. The failure surface in this area moves up the stratigraphic sequence progressively toward the east. Interpretation of the stair-stepped seat of the landslide in the east (Fig. 29.5a) is based on several local control points provided by drillholes and by part of one step observed in the field. The treads of the steps are the weakest clay units, whereas the risers are faults and major joints (Hendron and Patton, 1985).

Small folds and structures are present on much of the failure surface. These features were described as "cascata" (cascade

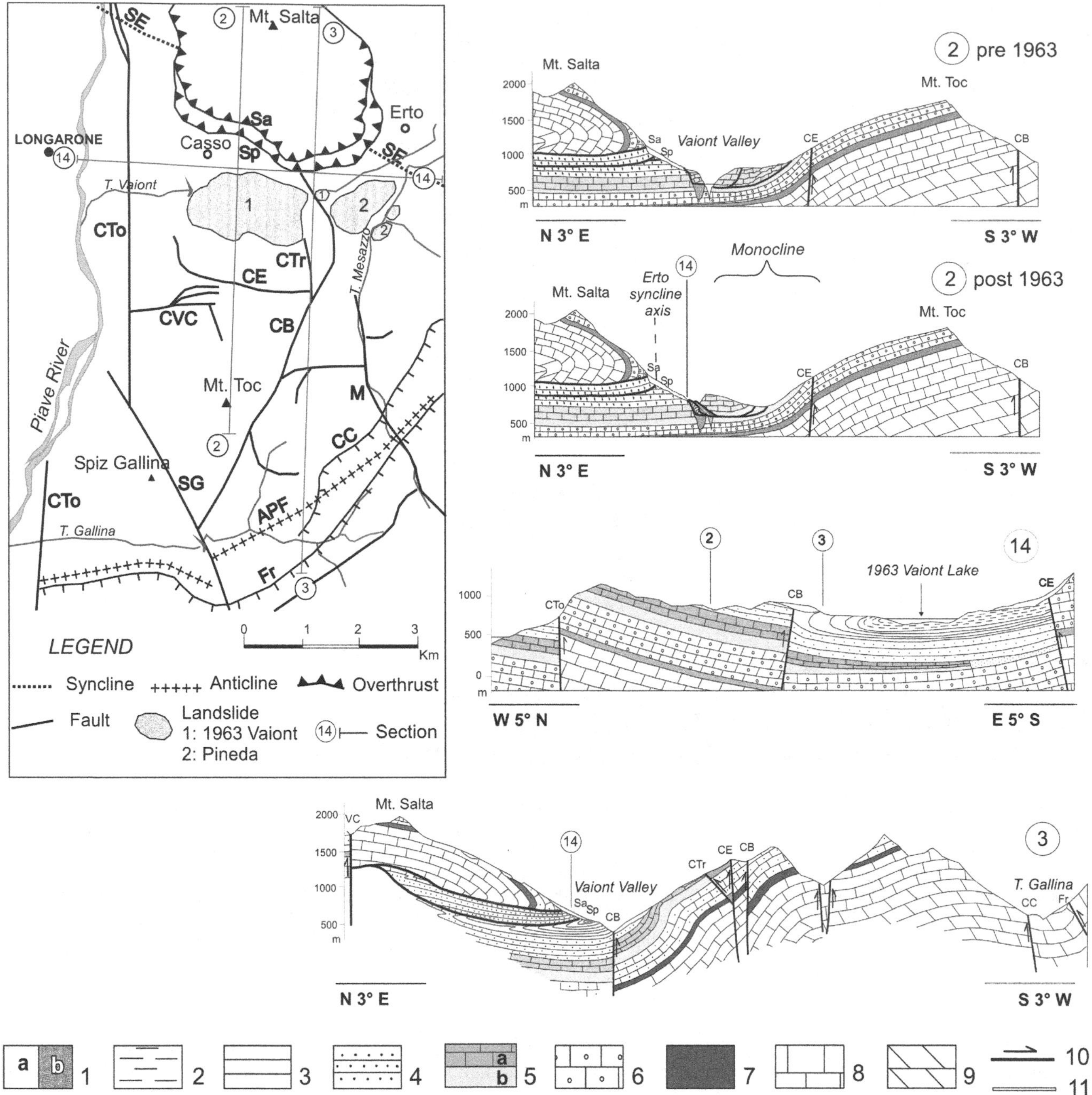

Fig. 29.4. Tectonic map of the Vaiont area (modified from Riva *et al.*, 1990) and geological sections of the Vaiont Valley (modified from Riva *et al.*, 1990, and Semenza and Ghirotti, 2000). CB: Croda Bianca Fault; CE: Col delle Erghene Fault; CTr: Col Tramontin Fault; Sa: Mt. Salta Overthrust; Sp: Le Spesse Overthrust; CVC: Costa Vasei-Calta faults; CTo: Col delle Tosatte Fault; SG: Spiz Gallina Fault; CC: Cima di Camp Fault; FR: Val Ferron Fault; M: Val Mesazzo Fault; APF: Pelf-Frugna Anticline; SE: Erto Syncline. Geological sections numbered as on the map. Both pre-1963 and post-1963 north–south sections are shown. Legend for the geological sections: 1a: Quaternary; 1b: stratified alluvial gravels; 2: Flysch Formation (Eocene); 3: Marls of Erto (Paleocene); 4: Scaglia Rossa Formation (Upper Cretaceous–Lower Paleocene); 5a: Socchér Limestone Formation; 5b: Ammonitico Rosso and Fonzaso formations (Cretaceous–Jurassic); 6: Vaiont Limestone (Middle Jurassic); 7: Igne Formation (Early Jurassic); 8: Soverzene Formation (Middle Jurassic); 9: Dolomia Principale (Upper Triassic); 10: faults and overthrusts; 11: failure surface of the 1963 landslide.

structures) by Semenza (1965). Hendron and Patton (1985) hypothesized that these cascading east-trending folds may have increased the normal stress slightly, but that their role in preventing motion was small.

The application of new remote sensing techniques has facilitated the characterization of previously inaccessible areas of the failure surface. Different scales of roughness and discontinuity sets that may have acted as release surfaces have been

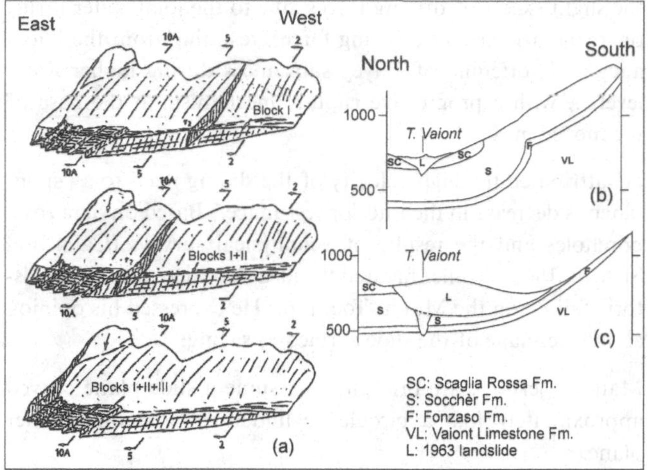

Fig. 29.5. (a) Schematic sketch showing the three-dimensional character of the landslide caused by the upstream dip of the strata (9–22° east) and the step-like shape of the east side of the basal failure surface that is assumed to step upward as the eastern limit of the slide is approached. The east boundary of the landslide probably coincides with one or more faults (CTr Fault) (Hendron and Patton, 1985). (b) North–south simplified geological profiles just east of the landslide. (c) North–south simplified geological profiles corresponding to section 2 of Figure 29.4b and c are separated by the north-northwest-trending CTr Fault, which dips 50° west-southwest (Semenza, 2010).

characterized by Wolter *et al.* (2011). This analysis suggests that the orientation of the failure surface is a product of several episodes of folding, and that the multiple steps are partially coincident with prominent discontinuity sets and partly result from fracturing of intact rock bridges. Moreover, analysis of east–west profiles derived from photogrammetric models confirms the variability of the failure surface over distances of even a few meters and thus the potential bias of two-dimensional profiles. All of these elements may have played a significant role in the failure and are the subject of ongoing 2D and 3D modeling.

Figure 29.5b, c shows two contiguous geological sections, separated by the Col Tramontin Fault, the eastern edge of the crown of the 1963 landslide (Semenza, 2010). Hendron and Patton (1985) assigned a friction angle of 36° to this fault and calculated that about 40 percent of the total resistance of the mass was provided by it. Semenza (2010) argued that the CTr Fault prevented the propagation of the landslide any farther eastward, in 1963 and in the past.

East of the CTr Fault, the Fonzaso Formation and the overlying Socchèr Formation dip steeply northward and therefore do not crop out in the Vaiont Gorge. They did, however, daylight in the valley west of the CTr Fault before the 1963 landslide (Figs. 29.5b, c). If the Vaiont Dam had been built at a previously chosen site, 1500 m upstream, the 1963 landslide probably would not have occurred (Semenza and Ghirotti, 2000). The CTr Fault is also a permeability boundary that could affect groundwater flow along the east boundary of the landslide mass. Hydrogeological investigations, which have never been carried out, might clarify the role played by groundwater in reducing the stability of the slope.

THE ORIGIN OF THE 1963 SLIDING SURFACE

The 1963 landslide surface had a complex origin related to previous periods of movement, including one or more prehistoric landslides and possibly a much older period of faulting. The loci of movement were the clay-rich layers in the Fonzaso Formation. Widely scattered outcrops of cemented breccia and the occurrence of fault-like grooves on the landslide surface are indications of past tectonic activity (Hendron and Patton, 1985; Kilburn and Petley, 2003; Mantovani and Vita-Finzi, 2003). Outcrops of these surfaces correspond to those mapped by Giudici and Semenza (1960) along the south side of the Vaiont Gorge (Fig. 29.3a). Significant slope movements, including rock avalanches, occurred periodically as alpine valleys were eroded and glacially loaded and unloaded (Pellegrini *et al.*, 2006). One of these landslides occurred in the Vaiont Valley and was responsible for the large remnant of slide material that was present in the valley prior to the 1963 landslide (Semenza and Ghirotti, 2000). Most researchers believe that the 1963 failure occurred at the site of a larger prehistoric landslide, but some (e.g., Mantovani and Vita-Finzi, 2003) hypothesized that the failure surface is a normal fault plane, marked by fault gouge, cataclasite, and mylonite.

Semenza (2005, 2010) provided a simplified palinspastic reconstruction of the landslide, starting from the postglacial mass movement and including the 1963 failure (Fig. 29.6). His reconstruction is based on a north–south section from the head of the 1963 landslide to Isolated Hill, passing through the Pian della Pozza area. Section 1 in Figure 29.6 shows the slope at the end of the last glacial period (Würm); a "fan" of faults matches the fold in the bedrock. This feature is important for subsequent movements of the entire mass (Mencl, 1966; Hutchinson, 1987). Semenza (2005, 2010) concluded that the ancient landslide started along clayey interbeds at the base of the northern part of Mount Toc and was accompanied by toppling and rotational slides in the calcareous rocks (Sections 2–5). Between deglaciation and 1960, the slope was eroded, especially at its base and top. Section 6 shows the situation in 1960. Section 7 shows the slope prior to the first filling of the reservoir in 1960, and Section 8 illustrates the situation just prior to the catastrophic failure in 1963.

The northern part of the slope appears unfolded and scarcely disturbed, whereas the rock mass in the west wall of Pian del Toc becomes increasingly fractured from north to south. These observations suggest that the 1963 landslide was the culmination of a progressive detachment that began in the south and extended, over time, toward the north. Semenza (2005, 2010) believed that major vertical fractures developed during the final phase of movement; it was his last contribution to Vaiont, made one year before his death in 2002.

29.2 IDEAS ABOUT THE LANDSLIDE MECHANISM

Since 1963, much has been learned about the causes of the Vaiont landslide. However, two crucial questions remain: how

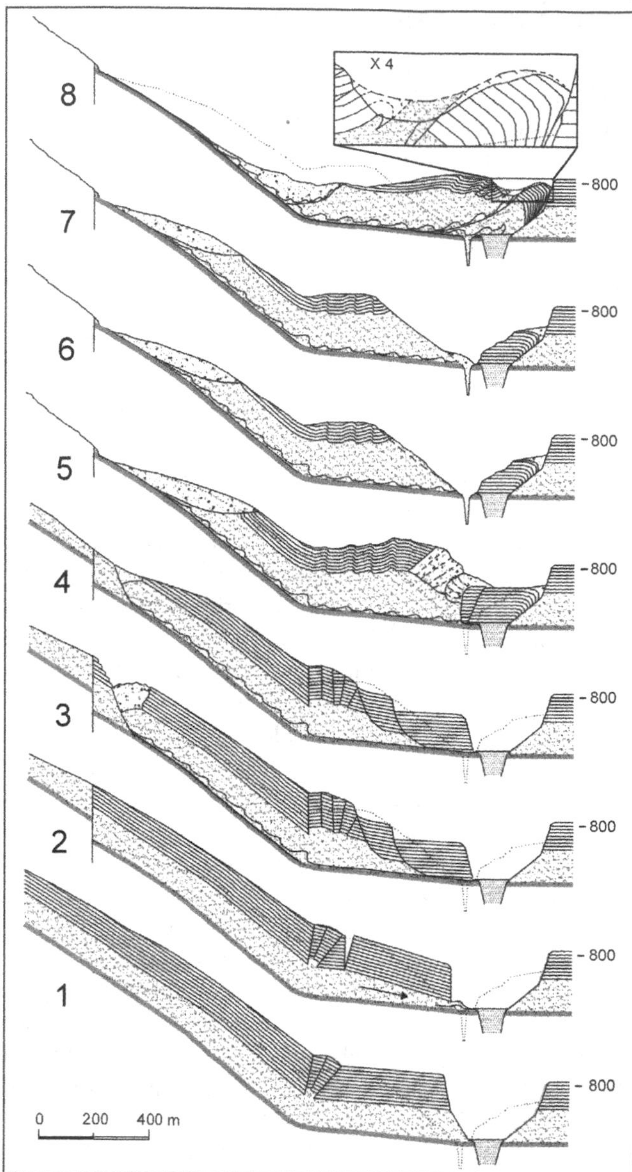

Fig. 29.6. Palinspastic reconstruction of the Vaiont landslide, beginning with (1) the early postglacial mass movement and ending with (8) the 1963 landslide. The simplified reconstruction is for a south–north section from the head of the 1963 landslide, through the Pian della Pozza area, to the "Isolated Hill" (Semenza, 2010).

was the landslide initially activated and why did it move so fast?

The first, and one of the most important, papers on the Vaiont landslide was by Müller (1964). After a detailed description of the studies carried out, and the phenomena observed, during the period of dam construction, he concluded that:

the interior kinematic nature of the mobile mass, after having reached a certain limit velocity at the start of the rockslide, must have been a kind of thixotropy.

The transition from creep to catastrophic failure, according to Müller, was caused by:

the slight excess of driving forces, due to the joint water thrust or to the decrease in resisting forces, resulting from the buoyancy and softening of clayey substances during higher water level … with a progressive rupture mechanism at the base of the moved mass.

He attributed the high velocity of the sliding mass to a "spontaneous decrease in the interior resistance." Based on data from boreholes and the results of geophysical surveys (Caloi and Spadea, 1960), Müller favored the idea that there was no prehistoric failure on the Mount Toc slope. He expressed his opinion about the shape of the slide surface by saying:

Many experts … are brought to assume a slide plane curved approximately like a circular cylinder or a spiral-cylinder plane.

This assumption, subsequently adopted by many scientists, provides a very simple and logical explanation of the behavior of the slide mass, but Müller underlined that:

exact kinematical observation and comparison of the slide mass before and after 1963, indicate that on the front of the mass a translation and shoving up have in fact taken place, and not simply a raising of the toe.

Kiersch (1964, 1965) took into account the existence of a prehistoric landslide and the presence of a weak zone of highly fractured rocks resulting from debuttressing of the slope following the last glaciation. He states:

Actual collapse was triggered by a rise in subsurface water level from bank infiltration with increased hydrostatic uplift and swelling pressures throughout an additional part of the subsurface.

Selli *et al.* (1964) published a comprehensive report on Vaiont, providing full details of the geology of the area and the hydraulic and seismic phenomena that accompanied the landslide. They concluded that the failed rock mass moved in a pseudo-plastic state, conditioned by the appearance of secondary shear surfaces at the base of the landslide. They further considered the main causes of the landslide to be the geologic structure, the morphology of the slope, and the variations in reservoir level. They performed a dynamic analysis of the movement, from which they concluded that a maximum velocity of 17 $m\,s^{-1}$ was achieved about 45 seconds after the initial failure.

Several researchers, including Müller (1964, 1968), Selli *et al.* (1964), and Mencl (1966) discussed whether a significant loss of strength is required to explain the high velocity of the landslide. Considerable uncertainty remains, even today, as to the mechanisms controlling both the rate of movement before catastrophic failure and the sudden acceleration of the failed rock mass.

Mencl (1966) applied the concept of a Prandtl wedge to the Vaiont landslide. Active and passive zones of the sliding mass likely developed due to the biplanar character of most of the failure surface. The passive lower seat impeded movement from the active upper back of the chair, and a transition zone was present between the two areas. Müller (1968) re-analyzed all available data and underlined the importance of the chair-like shape of the failure surface. In contrast to his 1964 publication, he concluded in 1968 that no clay beds existed on the slip surface and that the thin (1–3 mm) films of pelitic material along limestone bedding planes could not have played a significant role in the failure. Furthermore, he stated that the friction angle required to maintain a condition of limit equilibrium is too small compared to the strength properties attributed to the material involved in the movement. Because static calculations cannot explain the slope movements during the 3-year period leading up to the landslide, he focused on the influence of creep phenomena and related the reduction in frictional resistance to a progressive failure mechanism within the slope.

Today there is general agreement that failure occurred along clay-rich strata (5–18 cm thick) within the Fonzaso Formation. In 1961 these thin beds were exposed in adits constructed in the Massalezza area (Fig. 29.3a), but their importance with respect to landsliding was largely misunderstood or not appreciated.

Hendron and Patton (1985) re-examined Semenza's work and made significant progress in resolving several of the previously mentioned issues. The main results of their study are:

- the 1963 Vaiont event was a reactivation of an old landslide that probably occurred in the early Holocene
- the rock mass failed on one or more clay layers that were both impermeable and weak, with residual friction angles ranging from 5° to 16°
- two aquifers, separated by clay layers, were present in carbonate rocks on the north slope of Mount Toc.

Their hydrogeological model was supported only by measured piezometric levels and was adopted for the stability analysis. The water table in the highly fractured and permeable older landslide mass was influenced by the reservoir; the lower aquifer in the Vaiont Limestone was fed by both the reservoir and precipitation in the Mount Toc watershed. This hydrogeological model implies that high water pressures may have developed in the slope due to rainfall and snowmelt infiltration.

Many other researchers have contributed to our understanding of the mechanism of the Vaiont landslide. Habib (1967, 1975) proposed that the high velocity of the landslide was due to the conversion of mechanical energy into heat during frictional sliding. He argued that heating led to vaporization of pore water which lowered the friction to zero. A similar mechanism was later proposed for other large landslides (Anderson, 1985; Vardoulakis, 2002). Voight and Faust (1982) also proposed a thermal mechanism to explain the dynamic aspects of the Vaiont landslide. They considered a variable friction coefficient and a rise in pore water pressure due to frictional heating. They calculated acceleration, velocity (maximum: 26 m s^{-1}), and elapsed time as a function of the displacements.

Many two-dimensional, limit equilibrium analyses were performed after the failure. Lo *et al.* (1972) performed an analysis using Janbu's method for noncircular surfaces. They considered a sliding mass formed by two wedges that were separated by a vertical discontinuity located near the center of the slide mass. Using a groundwater level corresponding to the water level in the reservoir, they obtained friction angles at limit equilibrium as low as 13°.

Chowdhury (1978) also applied the limit equilibrium method, but modeled the failure as progressive. He observed behavior consistent with that proposed by Jäeger (1972), who noted the existence of a nonuniform zone of physical weakening separating the upper part of the slide mass from the lower part. Chowdhury's model was based on progressive creep of the unstable upper portion of the slope, which gradually increases the forces on the lower portion until they are high enough to cause sudden failure.

Hendron and Patton (1985) also back-calculated the friction angle required for stability and obtained values ranging from 17° to 28°. Direct shear-test data on clay along the failure surface gave residual friction angles ranging from 5° to 16°, with an average value of about 12°. These values are less than those required for stability, thus the slope should not have been stable even before reservoir filling. Because the slope was at least marginally stable for some time prior to failure, Hendron and Patton (1985) concluded that some factors affecting the stability of the slope had probably not been taken into account in the two-dimensional, limit equilibrium analyses. To resolve this discrepancy, they carried out three-dimensional stability analyses that accounted for the history of movement, the record of reservoir levels, the shape of the failure surface, the assumed distribution of water pressure, water levels, and appropriate shear strength values. They attributed the high velocity of the landslide to water pressures generated by vaporization along the failure surface, as a result of frictional heating during movement (Anderson, 1985). They concluded that the 1963 landslide occurred because of the combined effects of the rise in the reservoir level and the increase in piezometric levels as a result of rainfall and snowmelt.

Nonveiller (1987, 1992) regarded the frictional heat developed on the failure surface as necessary to explain the high velocity and the long trajectory of the Vaiont landslide. He estimated a maximum velocity of 15 m s^{-1} based on a total loss in the shearing resistance of the moving mass. He concluded, however, that this mechanism might induce such a decrease in the clay shear strength that the whole rock mass would reach very high velocities. The total loss of strength, however, would only happen some time after the initiation of movement.

Sitar and MacLaughlin (1997) introduced the technique of discontinuous deformation analysis (DDA) in their study of the Vaiont landslide. The main advantages of this method are: (1) the mode of failure does not have to be assumed *a priori*; and (2) displacements and velocities, which are not easily obtained using limit equilibrium methods, can be calculated. They used a simplified version of the cross-section of Hendron and Patton (1985), subdivided into a varying number of blocks. The results indicate that, under dry conditions, a single block would require a friction angle of only 8° for stability. If, however, the mass is divided by a single vertical discontinuity into two blocks, the required friction angle along the sliding plane rises to 8–14°, depending on the position of the vertical discontinuity and the assumed inter-block friction angle.

Tika and Hutchinson (1999) proposed a new hypothesis to explain the high velocity of the landslide based on the results of ring-shear tests carried out on two samples from the slip surface at slow and fast rates of shearing. Both samples showed a significant loss of strength as the shear rate increased; a minimum friction angle of 5° (up to 60 percent lower than the residual value) was obtained at rates greater than 100 mm min^{-1}. This mechanism of strength loss, alone or in combination with others, may explain the catastrophic failure and rapid movement.

Erismann and Abele (2001) examined selected "key events" of rock-slope failure processes. They considered the Vaiont landslide in terms of velocity, energy lines, and fahrböschung, and concluded that, based on the scientific knowledge of that time, the Vaiont catastrophe, especially the transition from slow to fast motion, could have been foreseen.

Vardoulakis (2002) showed that, during the pressurization phase, deformation was localized in a very thin shear band, only about 1 mm thick. The catastrophic pressurization phase of the Vaiont landslide should not have taken more than a few seconds to develop fully. He argued that the slide was moving at 20 m s^{-1} only 8 seconds after initial failure, corresponding to a total displacement of 74 m.

Veveakis *et al.* (2007) studied only the creep phase of the slide, using a rigid-block model moving over a thin zone of high shear strain rates. Introducing a thermal softening and velocity strengthening law for the basal material, they reformulated the governing equations of a water-saturated porous material, obtaining the critical time of failure up to 169 days before the actual failure. The total loss of strength in the failure surface during the last minutes prior to the slide is explained by the aforementioned thermal pressurization, triggered by the temperature rise within the clay-rich layers.

Ferri *et al.* (2010, 2011) describe an interesting and promising approach in the study of the behavior of clay-rich materials as a function of water content. They performed friction experiments at pressures of 1–5 MPa at a wide range of velocities, from 0.2 μm s^{-1} to 1.31 m s^{-1}, which included the transition from creep to final collapse (Veveakis *et al.*, 2007), with displacements up to *ca.* 36 m. They concluded that, in the presence of excess water, as was probably the case at Vaiont, thermal and thermochemical pressurization is not required to explain the high velocity achieved by the rock mass during the final collapse, at least for slip rates up to *ca.* 1.3 m s^{-1}, because friction is almost zero when velocity exceeds 0.01 m s^{-1}. Only at room humidity conditions (i.e., no water added to the sample) does the friction coefficient decay to *ca.* 0.09 at velocities exceeding 0.7 m s^{-1}, with the onset of thermochemical pressurization due to the release of water produced by the dehydration of smectite.

29.3 FORECASTING THE TIME OF THE VAIONT LANDSLIDE

Many researchers have attempted to predict the time of failure of the Vaiont landslide based on the 3 years of recorded surface movements prior to the catastrophic slope failure. Monitoring of slow, regularly increasing movement is probably the most widely used approach to forecast the time of slope failure. The method is based on the observation that, in some cases, the rate of surface movement increases as the slope approaches failure. In such cases, the remaining time to failure may be estimated from secondary or tertiary creep curves (Saito, 1965, 1969), even though these periods may be short (months or weeks). Extrapolation to estimate the time of failure is commonly based on empirical equations fitted to the observed curves. If the displacement velocity, v, at a point on the surface increases over time, its inverse number ($1/v$) must decrease. Failure occurs when $1/v$ approaches zero (Fukuzono, 1985). The curve of increasing velocity for the last phase of movement at Vaiont is illustrated in Figure 29.7a (Hutchinson, 2001). Following Fukuzono (1985) and re-plotting these data as the reciprocal of movement velocity against time, yield is reached (Fig. 29.7b). Even the simplest extrapolation technique of successively projecting to the time axis the best-fit tangents to the developing creep curve indicates that the Vaiont failure could have been forecast with a reasonable degree of accuracy about 1 month prior to failure (Hutchinson, 2001).

Kilburn and Petley (2003) and Petley and Petley (2004) report a similar inverse linear trend for the final pre-collapse acceleration at Vaiont (Fig. 29.7c). They interpreted the acceleration in movement in terms of a slow cracking mechanism, which assumes that mean rates of pre-collapse movement are proportional to rates of cracking. When a major failure plane develops, the shear resistance decreases abruptly, providing a sufficiently large force imbalance to trigger a catastrophic slope rupture. Because cracking triggers seismic events, it might be expected that rates of seismicity should increase prior to catastrophic failure. Although no relation can be seen between seismicity and movement in the 2 months preceding the Vaiont landslide (Belloni and Stefani, 1987), indirect evidence suggests that movement was accompanied by low-level seismicity below the recording threshold of the monitoring system (Kilburn and Petley, 2003).

Rose and Hungr (2007) applied the same method to the entire suite of Vaiont data and interpreted movements in relation to reservoir filling and precipitation (Fig. 29.7d). Their results support the conclusion of Hendron and Patton (1985) that the final trigger was a combination of high reservoir level and high precipitation. They also provide a useful discussion of the limitations of the inverse velocity method and recommend the use of linear fits, updated on an ongoing basis, to identify trend curvature or to detect the onset of trend change.

Voight (1988) found that accelerating displacements before some catastrophic landslides follow a time-to-failure power law, corresponding to a finite-time singularity of velocity. Finite-time singularity models have been used to characterize a wide variety of phenomena, including landslides, earthquakes, volcanic eruptions, fracture of structures, inflation, finance, economy, and ice avalanches from hanging glaciers. Finite-time singularities are caused by positive feedback processes that lead

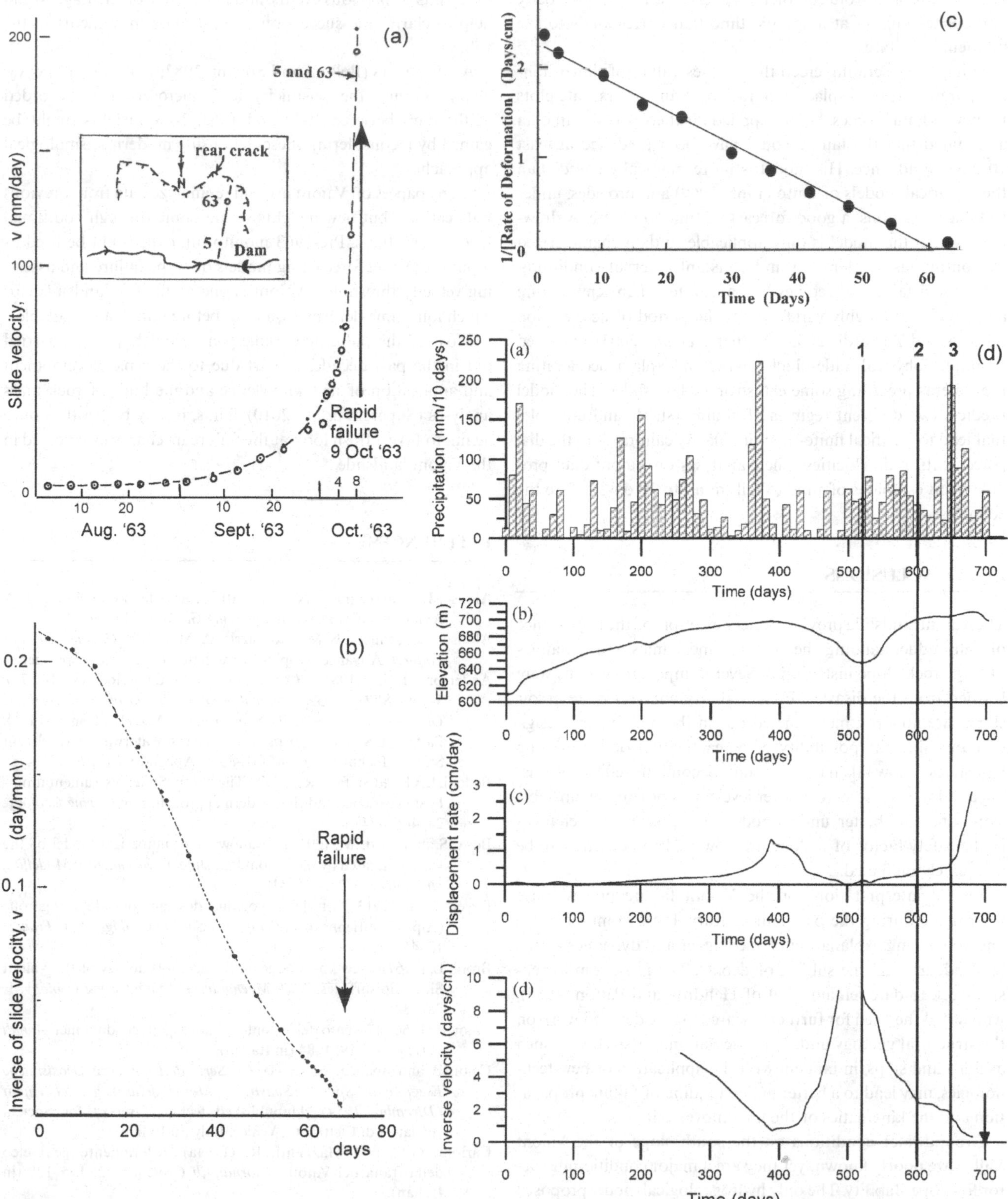

Fig. 29.7. Examples of the forecast time to failure of the Vaiont landslide based on three years of surface movements recorded before the collapse (Müller, 1964). (a) Displacement rate and (b) inverse rate against time (2 months before October 9, 1963) (Hutchinson, 2001). (c) Inverse rates of horizontal slope movements against time before October 9, 1963 (Kilburn and Petley, 2003). (d) Time series of observations over the two-year period before the Vaiont landslide: precipitation over 10-day periods; elevation of reservoir water surface; displacement velocity; and inverse velocity. Vertical lines 1–3 mark trend changes induced by reservoir filling and precipitation. The arrow shows the time of failure on October 9, 1963 (Rose and Hungr, 2007).

to a catastrophic evolution of the observed quantities. Velocity approaches infinity at a specific time that corresponds to the moment of failure.

Voight's accelerating creep theory uses values of dimensionless parameters on displacement rate-time and inverse rate plots to predict failure times. Voight applied his theory to Vaiont data and stated that the failure could have been predicted at least 10 days in advance. His model is more physically based than the empirical models of Saito (1965, 1969) and provides, under suitable conditions, a good forecast of time to failure. A drawback is that the model is only applicable to data characterized by continuous acceleration and constant external conditions. The model fails, or becomes less accurate, when temperature and rainfall are highly variable over the period of acceleration (Crosta and Agliardi, 2003). Sornette *et al.* (2003) proposed a simple, physical slide-block model to explain accelerating movements preceding some catastrophic landslides. The model predicts two different regimes of sliding (stable and unstable) that lead to a critical finite-time singularity calibrated to the displacements and velocities. Their analyses on Vaiont data provide good estimates of time to failure up to 20 days before the collapse.

29.4 CONCLUSIONS

The Vaiont landslide provides a clear example of the importance of fully understanding the complex mechanics and dynamics of large rock-slope instabilities. Several important lessons were learned from the disaster. First, stability analyses of reservoir slopes are now routinely carried out in the preliminary design of dams, and the possibility of a reactivation of pre-existing landslides is now taken into account. Second, the effects of raising and lowering reservoir water levels on a potentially unstable slope are now better understood and, consequently, changes in the safety factor of a slope are now understood and can be adequately predicted.

Several interpretations of the Vaiont failure process have been made during the past half century, but a comprehensive and convincing explanation of the trigger and dynamics of the landslide are still the subject of debate. The most comprehensive work to date remains that of Hendron and Patton (1985), who noted the need for further research. More detailed work on the structural geology and, in particular, minor structures such as folds and steps, in tandem with the application of new technologies, may lead to a better understanding of failure propagation and the kinematics of the final movement.

Groundwater conditions on the south slope of the Vaiont Valley are poorly known, yet they are a major conditioning factor for slope stability. The only hydrogeological model proposed is that of Hendron and Patton (1985).

Although breccias, identified as products of one or more episodes of reactivation, are exposed in a few areas along the headscarp and the east boundary of the landslide, no dating of these materials has ever been carried out. They are the fossil footprints of previous events, and knowledge of their age would help to clarify the succession of past mass movements in the valley.

A few papers (Belloni and Stefani, 1987; Kilburn and Petley, 2003) examine the seismicity and microseismicity recorded at the dam between 1960 and 1963. New insights might be gained by reconsidering these data using modern seismological approaches.

Many papers on Vaiont use and re-analyze data from previous publications, but new insights would come through additional field observations. Pre-1963 monitoring data could be used as input for testing forecasting models of slope failure and assessing velocity thresholds. Vaiont is one of the few landslides for which long-time deformation rates before failure are available.

Most of the published studies on Vaiont have been carried out in the past decade, in part due to the rapid development and application of new knowledge and methods of rock mass analyses (Superchi *et al.*, 2010). Thus, it may be fruitful, once again, to take a fresh look at the failure mechanisms involved in the Vaiont landslide.

REFERENCES

Alonso, E. E. and Pinyol, N. M. (2010). Criteria for rapid sliding. I. A review of Vaiont case. *Engineering Geology*, 114, 198–210.

Alonso, E. E., Pinyol, N. M. and Puzrin, A. M. (2010). *Geomechanics of Failures.* Advanced Topics. Dordrecht, Netherlands: Springer.

Anderson, D. L. (1985). Calculation of slide velocities. In *The Vaiont Slide: A Geotechnical Analysis Based on New Geologic Observations of the Failure Surface*, ed. A. J. Hendron and F. D. Patton. US Army Corps of Engineers Waterways Experiment Station Technical Report GL-85-5, Appendix E1–E5.

Belloni, L. G. and Stefani, R. (1987). The Vaiont Slide: Instrumentation – Past experience and the modern approach. *Engineering Geology*, 24, 445–474.

Bosa, S. and Petti, M. (2011). Shallow water numerical model of the wave generated by the Vajont landslide. *Environmental Modelling and Software*, 26, 406–418.

Boyer, R.A. (1913). Etude géologique des environs de Longarone (Alpes Vénitiennes). *Bulletin de la Societe Geologique de France*, 13, 451–485.

Broili, L. (1967). New knowledge on the geomorphology of the Vaiont Slide slip surfaces. *Rock Mechanics and Engineering Geology*, 1, 38–88.

Caloi, P. (1966). L'evento del Vajont nei suoi aspetti geodinamici. *Annali di Geofisica*, 19, 1–84 (in Italian).

Caloi, P. and Spadea, M.C. (1960). *Serie di Esperienze Geosismiche Eseguite in Sponda Sinistra a Monte della Diga del Vajont (Dicembre 1959).* Unpublished technical report for Società Adriatica di Elettricità, Venice, Italy (in Italian).

Carloni, G.C. and Mazzanti, R. (1964a). Rilevamento geologico della frana del Vaiont. *Giornale di Geologia*, 32, 105–138 (in Italian).

(1964b). Aspetti geomorfologici della frana del Vaiont. *Rivista Geografica Italiana*, 71, 201–231 (in Italian).

Chowdhury, R. (1978). Analysis of the Vaiont slide: New approach. *Rock Mechanics*, 11, 29–38.

Ciabatti, M. (1964). La dinamica della frana del Vaiont. *Giornale di Geologia*, 32, 139–154 (in Italian).

Cobianchi, M. and Picotti, V. (2003). The Vaiont Gorge section: The Toarcian to Bajocian Igne Formation and the unconformable base of the Vaiont Limestone. In *General Field Trip Guidebook: 6th International Symposium on the Jurassic System*, ed. M. Santantonio. Palermo, Italy, pp. 310–312.

Corbyn, J.A. (1982). Failure of a partially submerged rock slope with particular reference to the Vaiont rock slide. *International Journal of Rock Mechanics and Mining Sciences and Geomechanics Abstracts*, 19, 99–102.

Crosta, G.B. and Agliardi, F. (2003). Failure forecast for large rock slides by surface displacement measurements. *Canadian Geotechnical Journal*, 40, 176–191.

Dal Piaz, G. (1928). *Relazione di Massima su due Sezioni del Vajont Prese in Considerazione per Progetti di Sbarramento Idraulico*. Unpublished technical report for Società Adriatica di Elettricità, Venice, Italy (in Italian).

Doglioni, C. and Carminati, E. (2008). *Structural Styles and Dolomites Field Trip*. Memorie Descrittive Carta Geologica d'Italia 82. Turin, Italy: ENI.

Erismann, T.H. and Abele, G. (2001). *Dynamics of Rockslides and Rockfalls*. Berlin: Springer-Verlag.

Ferri, F., Di Toro, G., Hirose, T. and Shimamoto, T. (2010). Evidence of thermal pressurization in high-velocity friction experiments on smectite-rich gouges. *Terra Nova*, 22, 347–353.

Ferri, F., Di Toro, G., Han, R. *et al.* (2011). Low- to high-velocity frictional properties of the clay-rich gouges from the slipping zone of the 1963 Vaiont Slide (northern Italy). *Journal of Geophysical Research*, 116, B09208, doi:10.1029/2011JB008338.

Frattini, M., Arredi, F., Boni, A., Fasso, C. and Scarsella, F. (1964). *Relazione sulle Cause che Hanno Determinato la Frana nel Serbatoio del Vajont (9 Ottobre 1963)*. Frattini Commission Report prepared for Ente Nazionale per l'Energia Elettrica (ENEL), Rome (in Italian).

Fukuzono, T. (1985). A new method for predicting the failure time of a slope. In *Proceedings of the Fourth International Conference and Field Workshop on Landslides*. Tokyo: Japan Landslide Society, pp. 145–150.

Genevois, R. and Ghirotti, M. (2005). The 1963 Vaiont landslide. *Giornale di Geologia Applicata*, 1, 41–52.

Giudici, F. and Semenza, E. (1960). *Studio Geologico del Serbatoio del Vajont*. Unpublished technical report for Società Adriatica di Elettricità, Venice, Italy (in Italian).

Habib, P. (1967). Sur un mode de glissement des massifs rocheaux. *Comptes Rendus Hebdomadaires des Seances de l'Academie des Sciences*, 264, 151–153.

(1975). Production of gaseous pore pressure during rock slides. *Rock Mechanics*, 7, 193–197.

Helmstetter, A., Sornette, D., Grasso, J.-R. *et al.* (2004). Slider block friction model for landslides: Application to Vaiont and La Clapière landslides. *Journal of Geophysical Research*, 109, B02409, doi:10.1029/2002JB002160.

Hendron, A.J. and Patton, F.D. (1985). *The Vaiont Slide: A Geotechnical Analysis Based on New Geologic Observations of the Failure Surface*. US Army Corps of Engineers Waterways Experiment Station, Technical Report GL-85-5.

Hutchinson, J.N. (1987). Mechanisms producing large displacements in landslides on pre-existing shears. In *Proceedings of the First Sino-British Geological Conference, Taipei*. Memoir of the Geological Survey of China 9, pp. 175–200.

(2001). Landslide risk: To know, to foresee, to prevent. *Geologia Tecnica e Ambientale*, 3, 3–22.

Jäeger, C. (1965a). The Vaiont rockslide, Part 1. *Water Power*, 3, 110–111.

(1965b). The Vaiont rockslide, Part 2. *Water Power*, 4, 142–144.

(1972). The Vaiont slide. In *Rock Mechanics and Engineering*, ed. C. Jäeger. Cambridge, UK: Cambridge University Press, pp. 402–421.

Kenney, T.C. (1967). Stability of the Vajont valley slope [discussion of paper by L. Müller (1964) on the rock slide in the Vajont valley]. *Rock Mechanics*, 5(5), 10–16.

Kiersch, G.A. (1964). Vaiont reservoir disaster. *Civil Engineering*, 34(3), 32–39.

(1965). Vaiont reservoir disaster. *Geotimes*, May–June, 9–12.

Kilburn, C.R.J. and Petley, D.N. (2003). Forecasting giant, catastrophic slope collapse: Lessons from Vajont, northern Italy. *Geomorphology*, 54, 21–32.

Leonards, G.A. (1987). *Dam Failures: Proceedings*. Engineering Geology 24. Amsterdam: Elsevier.

Lo, K.Y., Lee, C.F. and Gelinas, P. (1972). Alternative interpretation of the Vaiont slide. In *Stability of Rock Slopes*, ed. E.J. Cording. New York: American Society of Civil Engineers, pp. 595–623.

Mantovani, F. and Vita- Finzi, C. (2003). Neotectonics of the Vajont dam site. *Geomorphology*, 54, 33–37.

Martinis, B. (1979). Contributo alla stratigrafia dei dintorni di Erto-Casso (Pordenone) ed alla conoscenza delle caratteristiche strutturali e meccaniche della frana del Vajont. *University of Padova Memorie di Scienze Geologiche*, 32, 1–33 (in Italian).

Masè, G., Semenza, M., Semenza, Pa., Semenza, P. and Turrini, M.C. (2004). *Le Foto della Frana del Vajont*. Ferrara: K-flash, CD-ROM (in Italian).

Mencl, V. (1966). Mechanics of landslides with non-circular slip surfaces with special reference to the Vaiont slide. *Géotechnique*, 16, 329–337.

Mufundirwa, A., Fujii, Y. and Kodama, J. (2010). A new practical method for prediction of geomechanical failure-time. *International Journal of Rock Mechanics and Mining Sciences*, 47, 1079–1090.

Müller, L. (1961). *Talsperre Vaiont: 15° Baugeologischer Bericht – Die Felsgleitung im Bereich Toc*. Unpublished technical report for Società Adriatica di Elettricità, Venice, Italy (in German).

(1964). The rock slide in the Vaiont valley. *Rock Mechanics and Engineering Geology*, 2, 148–212.

(1968). New considerations on the Vaiont slide. *Rock Mechanics and Engineering Geology*, 6, 1–91.

(1987a). The Vaiont catastrophe: A personal review. *Engineering Geology*, 24, 423–444.

(1987b). The Vaiont slide. *Engineering Geology*, 24, 513–523.

Nonveiller, E. (1967). Shear strength of bedded and jointed rock as determined from the Zalesina and Vaiont slides. In *On Shear Strength Properties of Natural Soils and Rocks*. Proceedings of the Geotechnical Conference, Oslo, Norway, pp. 289–294.

(1987). The Vaiont reservoir slope failure. *Engineering Geology*, 24, 493–512.

(1992). Vaiont slide: Influence of frictional heat on slip velocity. In *Proceedings of the Meeting on the 1963 Vaiont Landslide, Vol. 1*, ed. E. Semenza and G. Melidoro. International Association for Engineering Geology and the Environment, Italian Section, University of Ferrara, Ferrara, Italy, 1986, pp. 87–197.

Panizzo, A., De Girolamo, P., Di Risio, M., Maistri, A. and Petaccia, A. (2005). Great landslide events in Italian artificial reservoirs, *Natural Hazards and Earth System Sciences*, 5, 733–740.

Pastor, M., Herreros, I., Fernández Merodo, J.A. *et al.* (2009). Modelling of fast catastrophic landslides and impulse waves induced by them in fjords, lakes and reservoirs. *Engineering Geology*, 109, 124–134.

Pellegrini, G.B., Surian, N. and Albanese, D. (2006). Landslide activity in response to alpine deglaciation: The case of the Belluno Prealps (Italy). *Geografia Fisica e Dinamica Quaternaria* 29, 185–196.

Petley, D.N. and Petley, D.J. (2004). On the initiation of large rockslides; perspectives from a new analysis of the Vaiont movement record. In *Landslides from Massive Rock Slope Failures*, ed. S.G. Evans, G. Scarascia Mugnozza, A.L. Strom and R.L. Hermanns. Dordrecht, Netherlands: Kluwer Academic Publisher, pp. 77–84.

Pinyol, N.M. and Alonso, E.E, (2010). Criteria for rapid sliding. II. Thermo-hydro-mechanical and scale effects in Vaiont case. *Engineering Geology*, 114, 211–227.

Riva, M., Besio, M., Masetti, D. *et al.* (1990). La geologia delle valli Vaiont e Gallina (Dolomiti orientali). *Annali dell'Università di Ferrara, Sezione Scienze della Terra*, 2(4), 55–76 (in Italian).

Rose, N.D. and Hungr, O. (2007). Forecasting potential rock slope failure in open pit mines using the inverse-velocity method. *International Journal of Rock Mechanics and Mining Sciences*, 44, 308–320.

Rossi, D. and Semenza, E. (1965). *Carte Geologiche del Versante Settentrionale del M. Toc e Zone Limitrofe, Prima e Dopo il Fenomeno di Scivolamento del 9 Ottobre 1963*. Università di Ferrara, Istituto di Geologia, 2 maps, scale 1:5000.

Roubtsova, V. and Kahawita, R. (2006). The SPH technique applied to free surface flows. *Computers and Fluids*, 35, 1359–1371.

Saito, M. (1965). Forecasting the time of occurrence of slope failure. In *Proceedings of the Sixth International Conference on Soil Mechanics and Foundation Engineering, Vol. 2*. Toronto, ON: University of Toronto Press, pp. 537–542.

(1969). Forecasting time of slope failure by tertiary creep. In *Proceedings of the Seventh International Conference on Soil Mechanics and Foundation Engineering, Vol. 2*. Rotterdam: Balkema, pp. 677–683.

Selli, R. and Trevisan, L. (1964). Caratteri e interpretazione della Frana del Vajont. *Giornale di Geologia*, 32(I), 8–104 (in Italian).

Selli, R., Trevisan, L., Carloni, G.C., Mazzanti, R. and Ciabatti, M. (1964). La Frana del Vajont. *Giornale di Geologia*, 32(I), 1–154 (in Italian).

Semenza, E. (1965). Sintesi degli studi geologici sulla frana del Vaiont dal 1959 al 1964. *Memorie del Museo Tridentino Scienze Naturali*, 16, 1–52 (in Italian).

(2005). *La Storia del Vaiont Raccontata dal Geologo Che Ha Scoperto la Frana*. Published posthumously. Ferrara: K-flash [available at www.k-flash.it] (in Italian).

(2010). *The Story of Vaiont told by the Geologist who Discovered the Landslide*. Published posthumously. Ferrara: K-flash [available at www.k-flash.it].

Semenza, E. and Ghirotti, M. (1998). Vaiont-Longarone 34 anni dopo la catastrofe. *Annali dell'Università di Ferrara (Nuova Serie), Sezione Scienze della Terra*, 7(4), 63–94 (in Italian).

(2000). History of 1963 Vaiont Slide. The importance of the geological factors to recognise the ancient landslide. *Bulletin of Engineering Geology and the Environment*, 59, 87–97.

Semenza, E. and Melidoro, G. (ed.). (1992). *Proceedings of the Meeting on the 1963 Vaiont Landslide, Vol. 1*. International Association for Engineering Geology and the Environment, Italian Section, University of Ferrara, Ferrara, Italy, 1986.

Sitar, N. and MacLaughlin, M.M. (1997). Kinematics and discontinuous deformation analysis of landslide movement. In *Proceedings of the Second Pan-American Symposium on Landslides*, Rio de Janeiro, Brazil, pp. 65–73.

Sitar, N.M., MacLaughlin, M.M. and Doolin, D.M. (2005). Influence of kinematics on landslide mobility and failure mode. *Journal of Geotechnical and Geoenvironmental Engineering*, 131, 716–728.

Skempton, A.W. (1966). Bedding-plane slip, residual strength and the Vaiont landslide. *Géotechnique*, 16, 82–84.

Sornette, D., Helmstetter, A., Andersen, J.V. *et al.* (2003). Towards landslide predictions: Two case studies. *Physica A*, 338, 605–632.

Superchi, L., Floris, M., Ghirotti, M. *et al.* (2010). Implementation of a geodatabase of published and unpublished data on the catastrophic Vaiont landslide. *Natural Hazards and Earth System Sciences*, 10, 865–873.

Superchi, L., Wolter, A., Stead, D. *et al.* (2011). Comparison of photogrammetric and field survey data from the sliding surface of the 1963 Vajont Slide, Italy. *Geophysical Research Abstracts*, 13, 1632.

Tika, T.E. and Hutchinson, J.N. (1999). Ring shear tests on soil from the Vaiont landslide slip surface. *Geotechnique*, 49, 59–74.

Trollope, D.H. (1980). The Vaiont slope failure. *Rock Mechanics*, 13, 71–88.

Vardoulakis, I. (2002). Dynamic thermo-poro-mechanical analysis of catastrophic landslides. *Geotechnique*, 52, 157–171.

Veveakis, E., Vardoulakis, I. and Di Toro, G. (2007). Thermoporomechanics of creeping landslides: The 1963 Vaiont slide, northern Italy. *Journal of Geophysical Res*earch, 112, F03026, doi:10.1029/2006JF000702.

Voight, B. (1988). A method for prediction of volcanic eruptions. *Nature*, 332, 125–130.

(1989). Materials science law applies to time forecasts of slope failure. *Landslide News*, 3, 8–11.

Voight, B. and Faust, C. (1982). Frictional heat and strength loss in same rapid slides. *Geotechnique*, 32, 43–54.

(1992). Frictional heat and strength loss in some rapid landslides: Error correction and affirmation of mechanism for the Vaiont landslide. *Geotechnique*, 42, 641–643.

Ward, S.N. and Day, S. (2011). The 1963 Landslide and Flood at Vaiont Reservoir Italy: A tsunami ball simulation. *Bollettino della Società Geologica Italiana*, 130, 16–26.

Wolter, A., Superchi, L., Stead, D. *et al.* (2011). Preliminary results of a photogrammetric characterisation of the 1963 Vajont rockslide. In *Proceedings of the Fifth Canadian Conference on Geotechnique and Natural Hazards*, Kelowna, BC, Paper 173.

30 Hong Kong landslides

STEPHEN R. HENCHER AND ANDREW W. MALONE

ABSTRACT

Most landslides in Hong Kong are associated with rainstorms, and the greater the rainfall intensity the greater the number of landslides. Schematic, graphical hydrogeological models are presented to help explain the nature and cause of landslides that have been investigated in the past 30 years. The notional landslide types are associated with idealized geologic features and linked to times of landslide relative to transient rainfall.

30.1 INTRODUCTION

This chapter briefly summarizes the history of the study of landslides in Hong Kong and discusses the geologic features, weathering characteristics, and hydrogeologic circumstances that bear on landslide occurrence. A wide variety of landslides occur in the weathered terrain of Hong Kong because of the relatively complex geology and diverse geomorphological environments, which range from steep rock cliffs to deeply weathered foothills. Failures in man-made cuts, retaining walls, and embankments have had the greatest impact historically, but failures from natural slopes are also a significant hazard. The terrain is largely mountainous and is subject to severe rainstorms lasting several days, with total rainfall of the order of 500–1000 mm and short-term rainfall intensities that may exceed 100 mm h^{-1}. Most failures are less than 5 m^3 in volume, and 90 percent are less than 50 m^3. They typically detach and move rapidly upon failure; a few, mainly large, landslides stay relatively intact. Because of the dense population and steep terrain, damage and injuries have been common; even a small debris slide or rockfall can cause a fatality in a crowded area or on a busy highway. Records of severe landsliding impacting the population date back to 1889, and since the middle of the last century more than 480 people have been killed by landslides. A concerted effort has been made to reduce landslide risk to the population since the establishment of a Geotechnical Control Office (GCO) in 1977 (now the Geotechnical Engineering Office, GEO). Since that date, the designs of all new slopes in Hong Kong have been subject to geotechnical checking and approval. Furthermore all pre-existing man-made slopes are being investigated and upgraded as necessary following a risk-based priority system. There has been a marked reduction in fatalities since the establishment of the GCO, although much of this reduction can be attributed to a policy to rehouse people who had built their "squatter huts" in locations that were inherently hazardous (Fig. 30.1). The strategy for dealing with the risk of landslides in Hong Kong is discussed by Brand (1984) and has been updated by Malone (1997) and Wong (2005).

30.2 LANDSLIDE INVESTIGATIONS

Lumb (1975) identified the importance of rainfall intensity in triggering landslides in cuttings and embankments in Hong Kong. Three major landslides caused large numbers of deaths in 1972 and 1976 and were investigated, respectively, by a judge-led commission of inquiry and an independent panel of experts; the latter made recommendations that led to the establishment of the GCO in 1977. Near the end of the 1970s, there was considerable debate on the need for research specific to Hong Kong and its weathered rocks; the focus was on slope stability analysis rather than landslides as phenomena (Beattie and Chau, 1976; Sweeney and Robertson, 1979). It was recognized that many steep-cut slopes with low calculated factors of safety had survived severe rainstorms without collapsing. In 1981 the "Cut Slopes in Hong Kong – Assessment of Stability by Empiricism" (CHASE) study was undertaken by the GCO

Landslides: Types, Mechanisms and Modeling, ed. John J. Clague and Douglas Stead. Published by Cambridge University Press.

Fig. 30.1. Failure involving squatter huts at Yuen Mo Village, Lam Tin, 1982.

Fig. 30.2. Kowloon Peak (Fei Ngo Shan) is developed in volcanic rock, with weathered granite underlying the foothills. Note the large colluvial lobes on the side slopes. Lion Rock (resistant granite) is visible in distance.

Fig. 30.3. Mafic dike cutting granite.

(Brand and Hudson, 1982). About 200 failed and apparently stable slopes were investigated to try to differentiate them on the basis of mass-strength estimates derived from detailed logging, index testing, and other characteristics. The approach was similar to what might be attempted these days using a rock mass classification scheme such as the geological strength index (GSI; Hoek, 1999). The study, however, failed to achieve a clear differentiation; in the light of later landslide investigations, it is likely that the one-dimensional logging approach (where a narrow strip of surface cover was removed to allow inspection) failed to adequately characterize slope-specific geologic structures that probably controlled many of the failures. Other important studies carried out in the early 1980s included the Mid Levels Study (Geotechnical Control Office, 1982), which led to a much better appreciation of hydrogeologic processes, including soil suction and soil properties, and better methods of ground investigation, soil testing, and instrumentation. The North Point Study (Golder Associates UK Ltd., 1981) similarly led to improved methods for investigating and understanding rock-slope stability, and techniques for measuring shear strength along rock joints.

From 1978, records were kept of some reported landslide incidents (Malone and Shelton, 1982), and summary annual reports were compiled from 1982. In 1982, two major rainstorms caused hundreds of landslides, and a team of geotechnical engineers investigated some of these in detail. Six major landslides had occurred in spite of engineering design or assessment (Hencher, 1983). Based on these incidents, it was concluded that ground models used in design were often inadequate for identifying landslide mechanisms, and that piezometric pressures were underestimated in stability assessments, partly because piezometers were not installed at the depths where perched water tables developed. The major problem appeared to be in ground investigation and interpretation rather than in calculations of stability, and it was noted that adopting higher factors of safety could not compensate for incorrect geological models.

Landslides have been studied systematically in Hong Kong since the creation of a Landslide Investigation Division in the GEO in 1996. Between 1997 and 2009, about 3000 landslide records have been examined and 200 landslide studies carried out. Most reports are freely available to the profession and public for download from the Hong Kong Government website. Ho and Lau (2010) have reviewed the findings. Key factors contributing to relatively small failures include uncontrolled surface water flow, inadequate slope maintenance or drainage, and poor detailing of surface protection. The importance of adequate geological and hydrogeological models for understanding major failures (>50 m^3) has been confirmed. An additional factor that has become better appreciated is progressive slope deterioration in large slopes prior to failure (Malone, 1998, 2000; Hencher, 2006).

30.3 GROUND CONDITIONS

30.3.1 GEOLOGY

Current understanding of the geology of Hong Kong is summarized in two publications of the Hong Kong Geological Survey (Fyfe *et al.*, 2000; Sewell *et al.*, 2000). The major urban areas are underlain by Jurassic and Early Cretaceous volcanic rocks – mostly tuffs – and by granitic rocks of similar age but intrusive into the volcanic suite. Volcanic rocks make up the higher mountainous areas surrounding Hong Kong Harbor, as illustrated in Figure 30.2, although some of the rugged hills, such as Lion Rock, are granitic. The central harbor basin and many other low-lying areas are granitic and are commonly deeply weathered. Some of the contacts between the many igneous intrusions and volcanic rocks are faulted and may be associated with poor-quality rock, although these have rarely been cited as important factors for Hong Kong landslides, perhaps due to lack of recognition.

Mafic and intermediate dikes cut the major volcanic and granitic bodies (Fig. 30.3). Where fresh, the contacts tend to be sharp and fused, but where weathered, grain-size contrasts between the intrusion and host rock result in abrupt changes in permeability.

Much of the natural terrain is blanketed with colluvium, which is generally a few meters thick and comprises transported soil with rock fragments. The colluvium overlies in-situ weathered rock. Thicker deposits derived from landslides are commonly boulder-rich and can have high permeability, with many natural pipes and channels. Some colluvial deposits are ancient, as evidenced by the weathered state of boulders, induration, and cementation and, in some instances, the presence of geologic discontinuities (Fig. 30.4).

30.3.2 DISCONTINUITIES AND GEOLOGIC STRUCTURES

Studies over the past 30 years in Hong Kong have confirmed the importance of relict geologic discontinuities as contributing factors to many landslides in weathered rock profiles (Deere and Patton, 1971). Typical features include sheeting joints, tectonic and cooling joints, shear zones, primary volcanic fabrics, and dikes. Discontinuities are loci of weathering and may be partially infilled with transported clay and other sediments, commonly in association with dilation of the rock mass (Kirk *et al.*

Fig. 30.4. Ancient colluvium with very large weathered boulders along Clearwater Bay Road.

1997). Toward the center of the granitic plutons, where encountered at depth in tunnels, visible joints may be extremely widely spaced (tens of meters), but nearer the ground surface the rock tends to be more extensively jointed (Fig. 30.5). It is evident that many of these visible fractures have opened up through weathering and exhumation, as discussed by Hencher and Knipe (2007). The process is illustrated by the comprehensive disintegration into small fragments of a rock wedge defined by master joints in moderately weathered rock in Figure 30.6a, b. Within the granitic rocks, especially, joints tend to be steeply dipping and occur in orthogonal sets. Shallowly dipping joints also occur, and some of these are probably formed by shrinkage during cooling (Table 30.1). Sheeting joints are common in Hong Kong and occur as extensive features running roughly parallel to the steep terrain in rock that was unfractured at the time of slope formation (Hencher *et al.*, 2011). They are characteristic of many exposed, weakly weathered, granitic bodies, but are less common in volcanic terrain, probably because these rocks contain relatively close incipient jointing that allows stresses to be dissipated without the formation of new fractures. Sheeting joints are commonly associated with rockfalls and deeper-seated failures, especially where the joint walls are severely weathered or where the joints persist as relict discontinuities within a thick weathering profile. Such joints are commonly associated with channelized water flow, development of cleft and perched water pressures, and the influx and deposition of transported soil.

30.3.3 WEATHERING

DISTRIBUTION OF WEATHERED ROCK

Rocks in Hong Kong are locally deeply weathered, at some locations to depths of more than 100 m. Ruxton and Berry (1957) suggested that, over a long period, the weathering front descends to an equilibrium position controlled by the water table. They presented a suite of weathering profiles depicting

Fig. 30.5. Closely jointed granite, Anderson Road Quarry.

Fig. 30.6. (a) Well-defined wedge failure along master joints in volcanic rock, Kwai Shing Estate, Tsuen Wan. b) The wedge has broken up along closely spaced incipient joints.

the development of weathering profiles with time. The idealized "mature" stage includes remnant corestones of less-weathered rock surrounded by rock decomposed to a soil-like state that has weathered inward, away from major joints and other discontinuities that carry groundwater (Figs. 30.7 and 30.8). In the "old age" profile, nearly corestone-free saprolite overlies rock with little transition, and such profiles are common. Weathered rocks are described in this chapter according to the tried-and-tested material classification used in Hong Kong and published by the Geotechnical Control Office (1988). Residual soils (grade VI) in Hong Kong are typically red-brown, dense, and clay-rich, lack the parent rock fabric, and are only a few meters thick. Beneath the residual soil, completely weathered rock (grade V) retains the parent rock texture and fabric, but the dry density can be as low as 1.2 Mg m^{-3}; that is, less than half that of fresh rock. The material is open-textured, susceptible to disturbance, and prone to disaggregation (slaking) if placed in water. Highly decomposed material (grade IV) can be broken down by hand but, by definition, does not slake.

At the mass scale, the distribution of soil and rock fractions can be very complex, reflecting the parent rock structure and history of weathering. Thus, mass weathering classifications based on idealized corestone profiles are often difficult to apply. The depth and extent of weathering are difficult to estimate from landscape features alone. Some, but not all, valleys are associated with deep weathering. Depth of weathering differs considerably and abruptly laterally, especially across faults, which makes the extrapolation of ground investigation data difficult, not only for slopes but also for foundations. As a consequence, it is normal practice in Hong Kong to put down at least one borehole at every bored pile location to determine the location and nature of the weathering profile.

ENGINEERING PROPERTIES

Sampling and testing of soil-like weathered rock (saprolite) without significant disturbance is difficult (Vaughan, 1990); thus properties at a particular site may need to be estimated from experience elsewhere. This issue is one of the main reasons for employing a material weathering classification linked to index tests; in doing so, empirical rules and relationships can be applied (Irfan, 1996, 1999; Hencher, 2006). In granitic saprolite, peak angles of internal friction are typically greater than 30° (Martin, 1986). El-Ramly *et al.* (2005) report little difference between friction angle values for grade IV and V granite soils, with an average friction angle of 38°; the same value is typical for the basic dilation-corrected friction angle of natural rock joints in granite (Hencher and Richards, 1989). Some volcanic rock joints have lower basic friction angles, approaching 30° for fine-grained varieties. Coatings on joints can have an angle of friction as low as 17° at low stresses for chlorite (Brand *et al.* 1983), and even lower values have been measured for joints filled with clay. The patchy disposition of clay in depressions on joint surfaces, however, may mean that rock asperities still control shear strength of laterally persistent joints.

True cohesion of intact saprolite ranges widely, from zero for very weak grade V soils to a few MPa at the boundary between highly and moderately (grade III) weathered rock (Hencher, 2006). Reported values for true cohesion of saprolite are,

Table 30.1. *Typical discontinuities encountered in Hong Kong rocks (after Hencher, 2000).*

Discontinuity type	Occurrence	Geotechnical aspects
Petrological boundaries	Between different rock types especially minor intrusions such as dolerite, basalt or rhyolite dikes	May mark change in mass properties but, in fresh rock, contacts are commonly welded and not significant planes of weakness. Where weathered, and the parent rocks have different grain sizes, boundaries may be barriers to water flow because of permeability contrast.
Cooling joints	Often systematic and perpendicular to cooling surfaces especially in rhyolite and granitic rocks.	Steep orthogonal joint sets are common in granitic rocks; often act as release surfaces and loci for cleft water pressure in landslides.
	Doming joints in plutons.	Doming joints can be similar to sheeting joints (see below) but can occur at greater depths.
Tectonic joints	Fractures resulting from tectonic stresses including uplift and folding.	Often in roughly parallel or orthogonal sets – sometimes as conjugate sets or in spectra; often relatively planar. As with other discontinuities of geological origin, they commonly only develop as full mechanical fractures through weathering including unloading.
Sheeting joints	Develop parallel to natural slopes; more closely spaced nearer to ground surface (generally upper 10 m).	Rough and wavy tensile fractures; often persistent for many meters but terminate against cross-joints (being more recent).Short sections may increase in dip on an "up-wave" within the rock mass leading to local rock block failure.
	Most common in granitic rocks and tend to be locally developed where pre-existing mechanical fractures were absent at time of formation.	Focus for weathering and clay infill where dilated and they often act as channels for water flow.
Faults	Discrete fractures along which displacement has occurred.	Some faults in Hong Kong extend for hundreds of meters and are associated with zones of weak and weathered rock many meters wide. Other fault zones are much thinner. Depending on their nature, fault zones may act as barriers to water or as zones of high permeability. Shallowly dipping thrust faults have been associated with landslides.

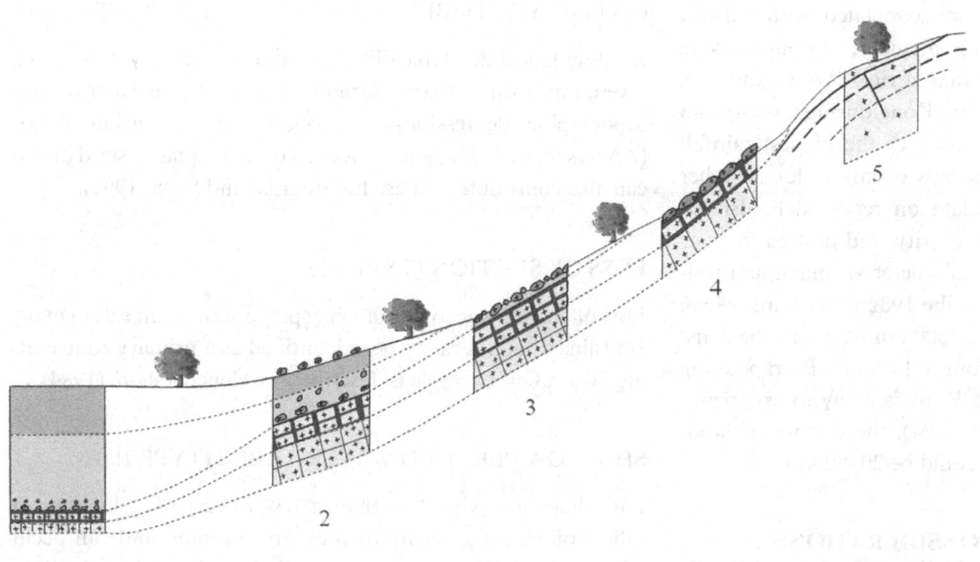

Fig. 30.7. Weathering profiles anticipated on different parts of a slope in Hong Kong. Sections 1–4 have been redrawn from the "mature" weathering profile of Ruxton and Berry (1957). At section 5, with sheeting joints, erosion exceeds weathering.

Fig. 30.8. Corestone-rich weathering profile in granite, Stubbs. Height of exposure is approximately 5 m.

however, remarkably low. El-Ramly *et al.* (2005) report a range of up to 25 kPa for the boundary between grades IV and V, while Ebuk (1991) measured higher values, up to 300 kPa for grade IV granite. The absence in the literature of even higher values probably reflects a lack of reported laboratory data on relatively strong saprolite.

Suction (water pressure lower than atmospheric pressure) in partially saturated soil is an ephemeral component of effective stress, and therefore strength (Fredlund, 1981; Shen, 1998). It cannot be used in design, however, because it is lost rapidly with wetting (Geotechnical Control Office, 1982; Rodin *et al.* 1982).

30.4 LINKS BETWEEN LANDSLIDES AND RAINFALL

Most landslides in Hong Kong are associated with periods of intense rainfall from April to September; although severe storms sometimes occur outside that period (Wong and Ho, 1995). Rainfall intensity across Hong Kong during a storm can be highly variable, and hilly regions with the highest rainfall intensity experience the greatest density of landslides. Hencher *et al.* (2006) analyzed available data on reported landslides, with no differentiation on size or severity, and plotted them as a function of number per square kilometer vs. maximum rolling 24-hour rainfall. In the case of the 1982 rainstorms, where the 24-hour rainfall exceeded about 500 mm in areas, the density of reported landslides was about 5–10 km^{-2}. Extrapolating to the heaviest rainfall that Hong Kong is likely to experience (Geotechnical Engineering Office, 2004), the density of landslides in areas of extreme rainfall could be 20–50 km^{-2}.

30.5 HYDROGEOLOGICAL CONSIDERATIONS

Theoretical calculations of pore pressure rise due to infiltration through uniform materials generally predict a slow response to rainfall (Lumb, 1962; Iverson, 2000). However, instruments show that rainfall-related rises in water pressure can be quick and dramatic. Richards and Cowland (1986) report rapid, sharp, but localized, rises in head of several meters of water pressure in sheeting joints in granite during rainstorms. Pope *et al.* (1982) present data from piezometers installed at different depths and report significant rises in water pressure at depths up to 12 m in bouldery colluvium within a few hours of the first rainfall. In comparison, piezometers installed in decomposed granite at depths between 20 and 40 m did not begin to respond until about 2 days after the first rainfall; water pressure then rose gradually, peaking about 5 days after the first rainfall and 2 days after the last rainfall.

The heterogeneity of weathered rock profiles mantled by colluvium is illustrated schematically in Figure 30.9. In general, hydraulic gradients run parallel to hillsides, and subsurface runoff may be dominated by pipe systems (Nash and Dale, 1984). Channel flow is also very important in less-weathered rock. These preferential flow paths can lead to relatively rapid infiltration and through-flow. Water is often observed issuing along these paths from the scarps of medium-sized and deep-seated landslides (Hencher, 2010). Water pressure in fractured bedrock at depth can cause upward flow into the overlying weathered mantle (Geotechnical Control Office, 1982; Leach and Herbert, 1982; Jiao *et al.*, 2005, 2006).

30.6 TIMING OF LANDSLIDES

Table 30.2 is a proposed classification of landslides in Hong Kong, based on time of occurrence relative to rainfall in a severe storm and failure mechanism (Fig. 30.10).

30.6.1 DURING AN INTENSE STORM

SHALLOW WASHOUTS, EROSION, AND ROCKFALL (TYPES IA AND IB)

Shallow landslides typically occur during, or very soon after, severe rainstorms. A concentrated flow of surface water toward topographic depressions is probably an important factor (Anderson *et al.*, 1983), and blocked or inadequately sized drains can also contribute to these failures (Au and Suen, 1991).

LOSS OF SUCTION (TYPE IC)

Loss of suction due to rainfall may play a significant role in many landslides, but has rarely been identified as a primary contributing factor. One example is described by Hencher *et al.* (1984).

SHALLOW PERCHED WATER TABLE (TYPE ID)

Landslides involving less than a few meters of colluvium or other soil above saprolite or rock are common and can occur relatively early during a storm. Failure is due to the development of perched water and cleft water pressure, as described by Devonald *et al.* (2009).

Table 30.2. *Classification of landslides based on timing and hydrologic processes.*

Timing	Size	Hydrogeologic process		Example	Comments
I During intense storm	Small to medium	Ia	Surface flow causing erosion and undermining	Gullying, shallow washouts, boulder fall	Convergence of flow toward topographic hollows
		Ib	Water pressure build-up in rock joint	Rockfall, rock-slab failure above adversely oriented joint	May be indicative of progressive, intermittent failure before a large scale failure
		Ic	Loss of suction or softening	Shallow failure in steep cut slope	Stability relies on apparent cohesion
		Id	Shallow perched water	Shallow slip in colluvium or saprolite overlying more competent rock or an aquitard such as a clay-rich seam	Very common in natural terrain, typically less than 3 m deep
II Late in storm or hours or days afterward	Medium to large	IIa	Deeper perched water table	Water perching above aquiclude within saprolite	Aquiclude may be a clay-infilled; discontinuity or geological unit of lower permeability, such as a weathered sill or gouge fault
		IIb	Rapid infiltration and through-flow by piping or along other high-permeability channels at depth	Natural pipes at base of thick colluvial deposits Permeable fault or shear zone	Water recharge from a higher catchment recharge can result in large failures, with pore pressures maintained until final detachment in spite of displacement and dilation of mass
III Delayed by days or weeks	Large	IIIa	Deep rise in water table, possibly by recharge from underlying bedrock	Fractured rock may channel water from remote catchment to location of landslide	Relatively rare. Delay can reflect a delayed rise in water table or a complex landslide mechanism that takes time to develop
		IIIb	Apparent trigger by minor rainfall may be illusory – only the 'final straw'	Progressive failure leads to increased vulnerability	

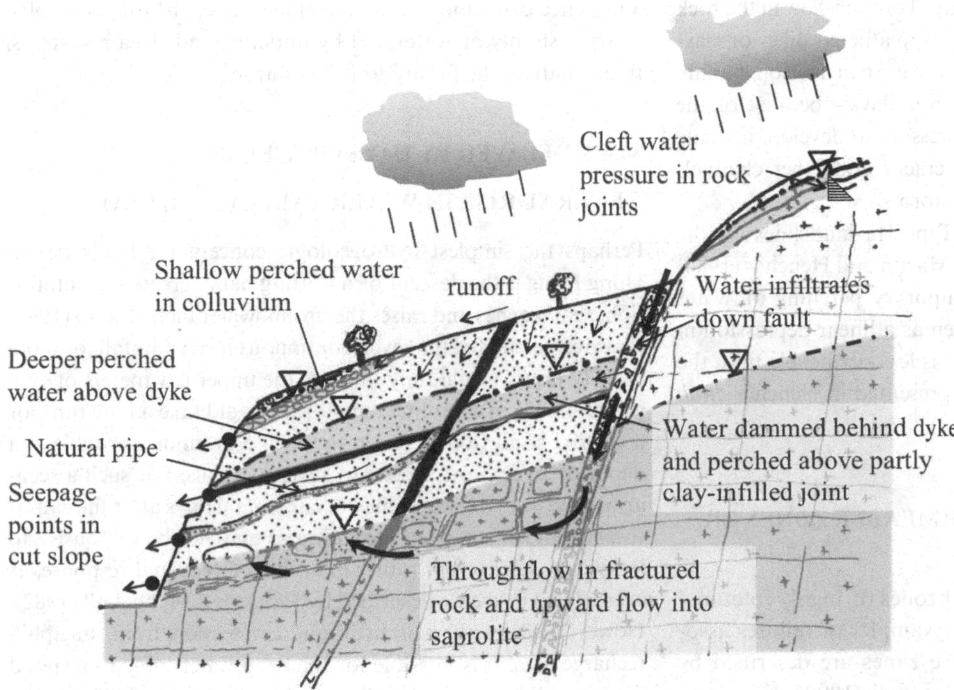

Fig. 30.9. Hydrogeologic processes that operate in saprolite and weathered rock in Hong Kong (modified from Hencher, 2010).

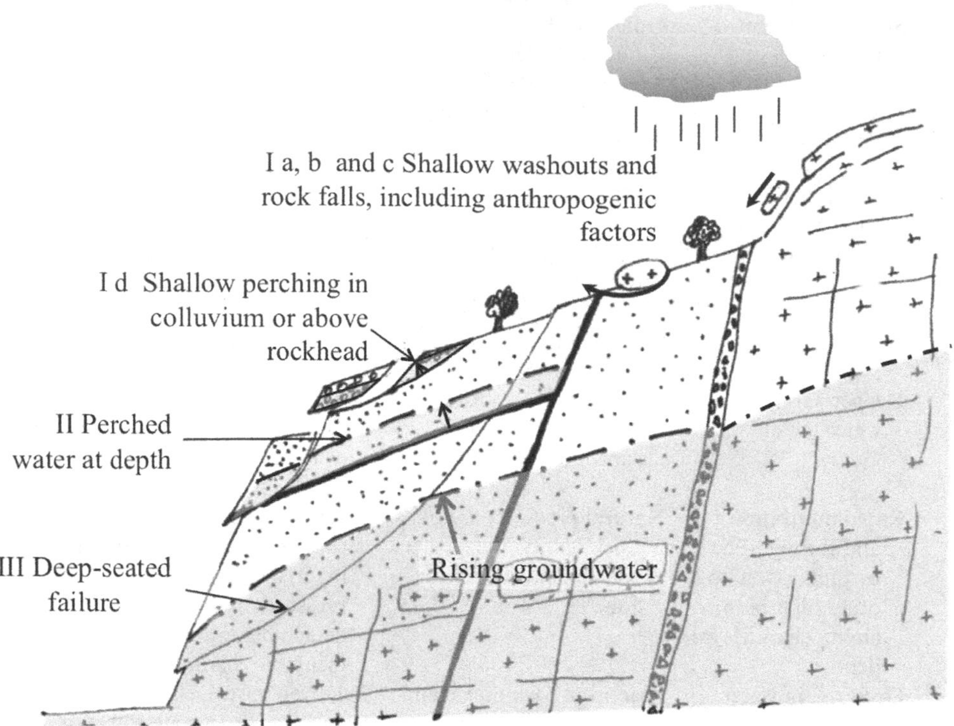

Fig. 30.10. Types of slope failure in Hong Kong (modified from Hencher and Lee, 2010).

30.6.2 LATE IN STORM OR HOURS OR DAYS AFTERWARD

PERCHING OF WATER AT GREATER DEPTH IN WEATHERED ROCK OR ABOVE AQUITARDS (TYPE IIA)

Type IIa landslides are distinguished from Type Id events mainly by their greater volume, by the involvement of relatively deep geologic structures, and by their timing. Through-flow in the rock mass is restricted by aquitards, such as weathered dikes or clay-filled joints, above which high water pressure can develop. Failure can be delayed by a few hours – or even days – because of the time it takes for the perched water pressure to develop; in some cases, however, pre-existing tension cracks and other channels facilitate a relatively fast response to storms.

Figure 30.11 shows a series of Type IIa landslides on the Tuen Mun Highway in Hong Kong. Martin and Hencher (1984) attributed the largest failure to temporary perching of water above a weathered dike, which is seen as a linear depression in the slope below. The weathered dike is less permeable than the granite. Several other examples are presented in Hencher *et al.* (1984) and in Hencher and Lee (2010).

HIGH THROUGH-FLOW IN PERMEABLE ZONES OR CHANNELS (TYPE IIB)

In other situations, shear zones and zones of highly fractured rock can allow rapid through-flow. Examples of failures associated with such highly transmissive zones are described by Hencher *et al.* (1984), Sun and Campbell (1999), Koor and Campbell (2005), and Hencher and Lee (2010). In many such cases, channels or zones of high permeability are observed in landslide scarps. These features continue to issue water after the landslide has occurred and the rainstorm has passed, which illustrates a distinctive feature of this landslide mechanism. Whereas in a shallow failure (e.g., Type I), initial dilation of the sliding mass might cause a reduction in water pressure, and hence cessation of the movement, deeper failures involve a larger supply of water, fed by underground stream systems, that can drive the failure to full collapse.

30.6.3 DELAYED BY DAYS OR WEEKS

GENERAL RISE IN WATER TABLE (TYPE IIIA)

Perhaps the simplest hydrogeologic concept for landslides in Hong Kong is the descent of a wetting band, created by infiltration, that reaches and raises the groundwater table. Lumb (1962) however, calculated that even continuous heavy rainfall for more than 12 hours would only saturate the upper few meters of typical Hong Kong saprolite; therefore it would take a long time for a wetting band from a storm to reach a groundwater table at a depth of 10 or more meters in such materials. For such a scenario, deep-seated failures may occur several days after the causative storm. In fact, deep groundwater may not be responsive to individual storms, but rather may show a seasonal response, as exemplified by a case described by Insley and McNicholl (1982). However, where there is hydraulic connection from an uphill recharge area, it is possible for the deep water table to respond more rapidly to storms, as discussed by Jiao *et al.* (2005, 2006).

Fig. 30.11. Slope failures on Tuen Mun Highway; a shallow-dipping dike runs up through main landslide.

PROGRESSIVE DETERIORATION (TYPE IIIB)

Large and disastrous failures can occur unexpectedly in Hong Kong, even during a relatively minor storm. One explanation is that the slope may have deteriorated progressively over a long period, making it susceptible to a final triggering storm. The Po Shan landslide of 1972 occurred after 3 days of heavy rainfall, but cracking in the road and local slips had been occurring for at least 11 months before the failure (Cooper, 1992). Careful mapping and investigation of the Lai Ping landslide of 1997 demonstrated that it had probably been moving intermittently for 20 years prior to failure (Sun and Campbell, 1999; Koor and Campbell, 2005).

30.7 CONCLUSIONS

There is a long history of study and engineering of slopes to reduce risk from landslides in Hong Kong. Investigations carried out 30 years ago showed that geologic structures commonly control hydrogeologic conditions and landslides. However, despite all the slope remediation work that has been completed, landslides still occur every year in Hong Kong, almost always during intense rainstorms. Shallow failures, including minor rockfalls, are the most common type of failure; they generally occur during storms. These failures are typically associated with surface erosion or saturation of surface layers of colluvium or residual soil, with perching of water above the underlying saprolite or bedrock. Deeper-seated landslides may be delayed, even by many days after the rainstorm. Studies in the 1990s in Hong Kong have identified the importance of precursory movements and ongoing deterioration prior to the final failure.

REFERENCES

Anderson, M.G., McNicholl, D.P. and Shen, J.M. (1983). On the effect of topography in controlling soil water conditions, with specific regard to cut slope piezometric levels. *Hong Kong Engineer*, 11(11), 35–41.

Au, S.W.C. and Suen, R.Y.C. (1991). The effect of road drainage and geometry in causing roadside slope failure. In *Proceedings of the 9th Asian Regional Conference on Soil Mechanics and Foundation Engineering, Vol. 1*, Bangkok, pp. 373–376.

Beattie, A.A. and Chau, E.P.Y. (1976). The assessment of landslide potential with recommendations for future research. *Journal of the Hong Kong Institution of Engineers*, 4(1), 27–44.

Brand, E.W. (1984). Landslides in Southeast Asia: A state-of-the-art report. In *Proceedings of the 4th International Symposium on Landslides, Vol. 1*, Toronto, pp. 17–59.

Brand, E.W. and Hudson, R.R. (1982). CHASE: An empirical approach to the design of cut slopes in Hong Kong soils. In *Proceedings of the Seventh Southeast Asian Geotechnical Conference, Vol. 1*, Hong Kong, pp. 1–16. (Discussion, *Vol. 2*, pp. 61–72, 77–79).

Brand, E.W., Hencher, S.R. and Youdan, D.G. (1983). Rock slope engineering in Hong Kong. In *Proceedings of the Fifth International Rock Mechanics Congress, Vol. 1*, Melbourne, pp. C17–C24.

Cooper, A.J. (1992). *Reassessment of the Po Shan Road landslide of 18th June 1972*. GEO Special Project Report, unpublished.

Deere, D.U. and Patton, F.D. (1971). Slope stability in residual soils. In *Proceedings of the 4th Pan-American Conference Soil Mechanics and Foundation Engineering, Vol. 1*, pp. 87–170.

Devonald, D.M., Thompson, J.A., Hencher, S.R. and Sun, H.W. (2009). Geomorphological landslide models for hazard assessment: A case study at Cloudy Hill. *Quarterly Journal of Engineering Geology*, 42, 473–486.

Ebuk, E.J. (1991). *The influence of fabric on the shear strength characteristics of weathered granites*. Ph.D. thesis, University of Leeds, UK.

El-Ramly, H., Morgenstern, N.R. and Cruden, D.M. (2005). Probabilistic assessment of a cut slope in residual soil. *Géotechnique*, 55, 77–84.

Fredlund, D.G. 1981. The shear strength of unsaturated soil and its relationship to slope problems in Hong Kong. *Hong Kong Engineer*, 9(4), 375–345.

Fyfe, J.A., Shaw, R., Campbell, S.D.G., Lai, K.W. and Kirk, P.A. (2000). *The Quaternary Geology of Hong Kong*. Hong Kong: Government of Hong Kong Special Administrative Region, Geotechnical Engineering Office, Civil Engineering Department.

Geotechnical Control Office (1982). *Mid-Levels Study: Report on Geology, Hydrology and Soil Properties*. Hong Kong: Geotechnical Control Office.

Geotechnical Control Office (1988). *Guide to Rock and Soil Descriptions (Geoguide 3)*. Hong Kong: Geotechnical Control Office.

Geotechnical Engineering Office (2004). *Rainstorm Severity and Landslide Potential*. Information Note 30/2004. Hong Kong: Geotechnical Engineering Office.

Golder Associates UK Ltd. (1981). *Final Report to Geotechnical Control Office Public Works Department Hong Kong on the North Point Rock Slope Study*. Four volumes, unpublished.

Hencher, S.R. (1983). *Summary Report on Ten Major Landslides in 1982*. Geotechnical Engineering Office, Special Project Report SPR 1/83.

(2000). Engineering geological aspects of landslides. In *Proceedings of the Conference on Engineering Geology HK 2000*. Hong Kong: Institution of Mining and Metallurgy, pp. 93–116.

(2006). Weathering and erosion processes in rock: Implications for geotechnical engineering. In *Proceedings of the Symposium on Hong Kong Soils and Rocks*. Hong Kong: Institution of Mining, Metallurgy and Materials and Geological Society of London, pp. 29–79.

Hencher, S. R. (2010). Preferential flow paths through soil and rock and their association with landslides. *Hydrological Processes*, 24, 1610–1630.

Hencher, S. R. and Knipe, R. J. (2007). Development of rock joints with time and consequences for engineering. In *Proceedings of the 11th Congress of the International Society for Rock Mechanics, Lisbon, Vol. 1*, pp. 223–226.

Hencher, S. R. and Lee, S. G. (2010). Landslide mechanisms in Hong Kong. In *Weathering as a Predisposing Factor to Slope Movements*, ed. D. Calcaterra and M. Parise. Geological Society of London, Engineering Geology Special Publication 23, pp. 77–103.

Hencher, S. R. and Richards, L. R. (1989). Laboratory direct shear testing of rock discontinuities. *Ground Engineering*, 22(2), 24–31.

Hencher, S. R., Massey, J. B. and Brand, E. W. (1984). Application of back analysis to some Hong Kong landslides. In *Proceedings of the 4th International Symposium on Landslides, Vol. 1*, Toronto, pp. 631–638.

Hencher, S. R., Anderson, M. G. and Martin, R. P. (2006). Hydrogeology of landslides. In *Proceedings of International Conference on Slopes*, Malaysia, pp. 463–474.

Hencher, S. R., Lee, S. G., Carter, T. G. and Richards, L. R. (2011). Sheeting joints: Characterisation, shear strength and engineering. *Rock Mechanics and Rock Engineering*, 44, 1–22.

Ho, K. K. S. and Lau, J. W. C. (2010). Learning from slope failures to enhance landslide risk management. *Quarterly Journal of Engineering Geology and Hydrogeology*, 43, 33–68.

Hoek, E. (1999). Putting numbers to geology: An engineer's viewpoint. *Quarterly Journal of Engineering Geology*, 32, 1–20.

Insley, H. T. M. and McNicholl, D. (1982). Groundwater monitoring of a soil slope in Hong Kong. In *Proceedings of the 7th Southeast Asian Geotechnical Conference, Vol. 1*, Hong Kong, pp. 63–75.

Irfan, T. Y. (1996). Mineralogy, fabric properties and classification of weathered granites in Hong Kong. *Quarterly Journal of Engineering Geology*, 29, 5–35.

(1999). Characterization of weathered volcanic rocks in Hong Kong. *Quarterly Journal of Engineering Geology*, 32, 317–348.

Iverson, R. M. (2000). Landslide triggering by rain infiltration. *Water Resources Research*, 36, 1897–1910.

Jiao, J. J., Wang, X.-S. and Nandy, S. (2005). Confined groundwater zone and slope instability in weathered igneous rocks in Hong Kong. *Engineering Geology*, 80, 71–92.

Jiao, J. J., Ding, G. P. and Leung, C. M. (2006). Confined groundwater near the rockhead in igneous rocks in the Mid-Levels area, Hong Kong, China. *Engineering Geology*, 84, 207–219.

Kirk, P. A., Campbell, S. D. G., Fletcher, C. J. N. and Merriman, R. J. (1997). The significance of primary volcanic fabrics and clay distribution in landslides in Hong Kong. *Journal of the Geological Society of London*, 154, 1009–1019.

Koor, N. P. and Campbell, S. D. G. (2005). *Geological Characterisation of the Lai Ping Road Landslide*. Hong Kong Geotechnical Engineering Office, GEO Report 166.

Leach, B. and Herbert, R. (1982). The genesis of a numerical model for the study of the hydrogeology of a steep hillside in Hong Kong. *Quarterly Journal of Engineering Geology*, 15, 243–259.

Lumb, P. (1962). Effect of rain storms on slope stability. In *Proceedings of the Symposium on Hong Kong Soils*, pp. 73–87.

(1975). Slope failures in Hong Kong. *Quarterly Journal of Engineering Geology*, 8, 31–65.

Malone, A. W. (1997). Risk management and slope safety in Hong Kong. *Transactions of the Hong Kong Institution of Engineers*, 4(2–3), 12–21.

(1998). Slope movement and failure: Evidence from field observations of landslides associated with hillside cuttings in saprolites in Hong Kong. In *Proceedings of the 13th Southeast Asian Geotechnical Conference, Vol, 2*, Taipei, pp. 81–90.

(2000). Panel Report: Slope movements associated with hillside cuttings in saprolites in Hong Kong. In *The Geotechnics of Hard Soils – Soft Rocks, Vol. 3*, ed. A. Evangelista and L. Picarelli. Rotterdam: Balkema, pp. 1593–1596.

Malone, A. W. and Shelton, J. C. (1982). Landslides in Hong Kong 1978–1980. In *Proceedings of the ASCE Specialty Conference on Engineering and Construction in Tropical and Residual Soils*, Honolulu, pp. 425–442.

Martin, R. P. (1986). Use of index tests for engineering assessment of weathered rocks. In *Proceedings of the Fifth International Congress of the International Association of Engineering Geology, Vol. 1*, Buenos Aires, pp. 433–450.

Martin, R. P. and Hencher, S. R. (1984). The failure of a cut slope on the Tuen Mun Road in Hong Kong. In *Proceedings of the International Conference on Case Histories in Geotechnical Engineering, Vol. 2*, St. Louis, pp. 683–688.

Nash, J. M. and Dale, M. J. (1984). Geology and hydrogeology of natural tunnel erosion in superficial deposits in Hong Kong. In *Geology of Superficial Deposits in Hong Kong*, ed. W. W. S. Yim. Geological Society of Hong Kong, Bulletin 1, pp. 61–72.

Pope, R. G., Weeks, R. C. and Chipp, P. N. (1982). Automatic recording of standpipe piezometers. In *Proceedings of the 7th Southeast Asian Geotechnical Conference, Vol. 1*, pp. 77–89.

Richards, L. R. and Cowland, J. W. (1986). Stability evaluation of some urban rock slopes in a transient groundwater regime. In *Proceedings of the Conference on Rock Engineering and Excavation in an Urban Environment*. Hong Kong: Institution of Mining & Metallurgy, pp. 357–363. (Discussion pp. 501–506.)

Rodin, S., Henkel, D. J. and Brown, R. L. (1982). Geotechnical study of a large hillside area in Hong Kong. *Hong Kong Engineer*, 10(5), 37–45.

Ruxton, B. P. and Berry, L. (1957). The weathering of granite and associated erosional features in Hong Kong. *Geological Society of America Bulletin*, 68, 1263–1292.

Sewell, R. J., Campbell, S. D. G., Fletcher, C. J. N., Lai, K. W. and Kirk, P. A. (2000). *The Pre-Quaternary Geology of Hong Kong*. Hong Kong: Government of Hong Kong Special Administrative Region, Geotechnical Engineering Office.

Shen, J. M. (1998). Soil suction in relation to slope stability: A summary of research carried out in Hong Kong in 1978–1997. In *Proceedings of the Geotechnical Division Seminar on Slope Engineering in Hong Kong*. Hong Kong: Hong Kong Institution of Engineers, pp. 93–99.

Sun, H. W. and Campbell, S. D. G. (1999). *The Lai Ping Road Landslide of 2nd July 1997*. Hong Kong Geotechnical Engineering Office, GEO Report 95.

Sweeney, D. J. and Robertson, P. K. (1979). A fundamental approach to slope stability problems in Hong Kong. *Hong Kong Engineer*, 7(10), 35–44.

Vaughan, P. R. (1990). Characterising the mechanical properties of in-situ residual soil. In *Proceedings of the 2nd International Conference on Geomechanics in Tropical Soils, Vol. 2*, pp. 469–487.

Wong, H. N. (2005). Landslide risk assessment for individual facilities. In *Landslide Risk Management* ed. O. Hungr, R. Fell, R. Couture and E. Eberhardt, London: Taylor and Francis Group, pp. 237–296.

Wong, H. N. and Ho, K. K. S. (1995). *General Report on Landslips on 5 November 1993 at Man-Made Features in Lantau*. Hong Kong: Geotechnical Engineering Office, GEO Report 144.

31 Landslides induced by the Wenchuan earthquake

MASAHIRO CHIGIRA, GONGHUI WANG, AND XIYONG WU

ABSTRACT

A large number of landslides were triggered by the 2008 M_w 7.9 Wenchuan earthquake, which was accompanied by 270 km of surface rupture along the Longmenshan fault at the eastern margin of the Tibetan Plateau. The landslides were strongly controlled by geology, geomorphology, and earthquake shaking. They were concentrated on the hanging walls of the fault ruptures, and their directions seem to have been affected by the directivity of seismic waves. Landslides were also concentrated within the inner gorge of Minjian River. Carbonate rocks failed preferentially, probably due to their low shear resistance, caused in part by dissolution. A power function defines the relationship between the cumulative number of landslides and landslide area. The largest landslide, with an estimated volume of 0.8 km^3, occurred on a cataclinal slope of carbonate rocks and was preceded by slow gravitational slope deformation. The second-largest landslide occurred in carbonate rocks on an undercut slope. Another large landslide had a long runout due to liquefaction and entrainment of sandy valley-bottom sediments. More than 800 of the landslides triggered by the earthquake blocked watercourses and impounded lakes.

31.1 INTRODUCTION

The Wenchuan earthquake occurred at 14:28 local time (18:28 UTC) on May 12, 2008, and caused devastating damage at the eastern margin of the Tibetan Plateau at its boundary with the Sichuan Basin of China (Fig. 31.1). The earthquake had a magnitude of M_w 7.9; its epicenter was located at 30.986° N, 103.364° E, and the focal depth was 19 km (US Geological Survey, 2008). The earthquake was accompanied by fault rupture over a distance of about 270 km (Burchfiel *et al.*, 2008; Li *et al.*, 2008a), and thus the affected area was far larger than that of recent earthquakes in mountainous areas. It induced numerous landslides and killed more than 69,000 people – the worst mountain disaster on Earth in the past two centuries (Huang and Li, 2008; Chigira *et al.*, 2010). The landslides induced by the earthquake created more than 800 lakes that threatened both downstream and upstream areas. Accelerometer records of the main shock indicate that the maximum acceleration was at the Wolong station, 18 km west-northwest of the epicenter: 957.7 Gal, 652.9 Gal, and 948.1 Gal for, respectively, the east–west, north–south, and vertical directions (Fig. 31.1; Li *et al.*, 2008b). Accelerations at Qingping were 824.1 Gal (east–west), 802.7 Gal (north–south), and 622.9 Gal (vertical) (Fig. 31.1; Li *et al.*, 2008b). In this chapter we summarize the characteristics of the landslides induced by this earthquake, based on Chigira *et al.* (2010) and supplemented with previously unpublished data.

We used satellite images taken after the earthquake by the Advanced Land Observing Satellite (ALOS) of the Japan Aerospace Exploration Agency (JAXA) to interpret the distribution of landslides in the affected area. The satellite images were acquired by an AVNIR-2 sensor with a resolution of 10 m and a PRISM sensor with a resolution of 2.5 m. We used AVNIR-2 images taken on June 4, 2008 to analyze the whole area, and PRISM images taken on May 18 and June 4 for areas of particular interest. We also used satellite images provided by Google Earth that were taken before and after the earthquake; most of the pre-earthquake images are French SPOT images and have a resolution of 3 m; in contrast, most of the post-earthquake images were taken by FORMOSAT II of Taiwan and have resolutions of 2 or 8 m. Digital elevation models (DEMs) with a 10-m mesh and 1:25,000-scale contour

Landslides: Types, Mechanisms and Modeling, ed. John J. Clague and Douglas Stead. Published by Cambridge University Press.

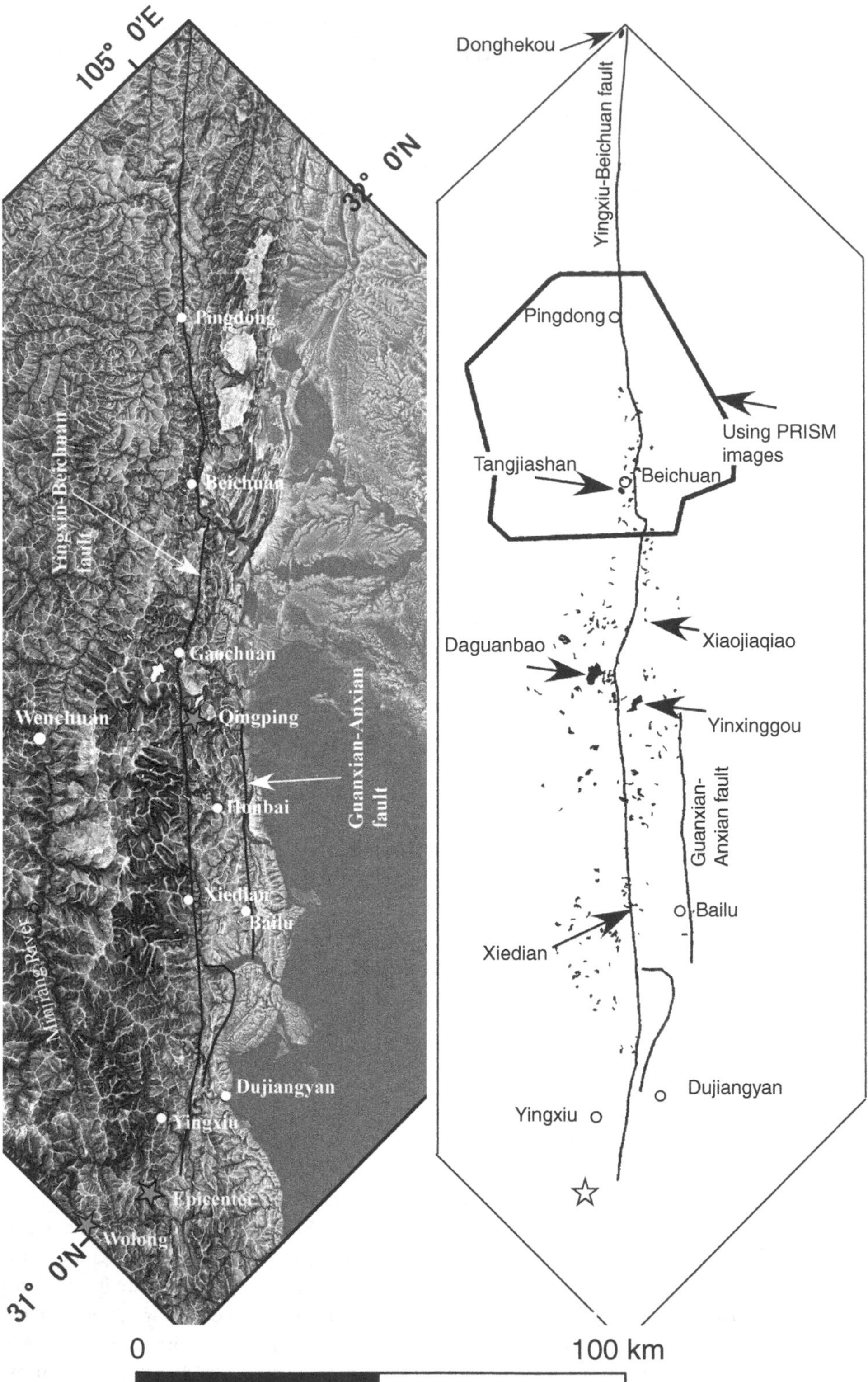

Fig. 31.1. Left: 3D image of the area around the Longmenshan fault zone. Right: distribution of coseismic landslides interpreted from ALOS/AVNIR-2 images (modified from Chigira *et al.*, 2010).

maps were made from the PRISM data for two landslides – Daguangbao and Xiejiedian; they were compared with DEMs of SRTM3 data that have a mesh size of 90 m. Interpretation of satellite images was supplemented with a 3-week field investigation.

31.2 GEOLOGIC SETTING AND EARTHQUAKE SURFACE RUPTURES

The area affected by the earthquake is mountainous, with elevations ranging from 1000 to 4500 m asl (Fig. 31.1). Ridges and valleys generally trend northeast, parallel to the geologic structure at the eastern margin of the Tibetan Plateau. Large rivers, including the Minjiang and Fujiang rivers, flow south or southeast across the topographic and structural grain. The northeast-trending Longmenshan fault zone extends along the boundary between the mountains and the Sichuan Basin (He and Tsukuda, 2003). One of its strands, the Yingxiu–Beichuan fault, is inferred to be the main structure that generated the 2008 earthquake (Fig. 31.1; Li *et al.*, 2008a). The rocks of the affected area range from Precambrian to Cretaceous in age and consist mainly of basalt, granite, phyllite, dolomite, limestone, sandstone, and shale.

The Longmenshan fault zone comprises three main faults: the Yingxiu–Beichuan, Guanxian–Anxian, and Wenchuan–Maowen faults (Figs. 31.1). Li *et al.* (2008a) documented major ruptures along the Yingxiu–Beichuan fault during the 2008 earthquake and minor ruptures along the Guanxian–Anxian fault; no ruptures were found along the Wenchuan–Maowen fault. Li *et al.* (2008a) reported that the northwest side of the Yingxiu–Beichuan fault, which is the hanging wall, rose up to 11 m, with right-lateral displacements of up to 12 m. At Qingping, where there was 3.5 m of vertical separation and 1.6 m of right-lateral separation, the rupture could be traced into a fault dipping 50° northwest in the bedrock.

The Yingxiu–Beichuan fault consists of southern and northern segments, which are connected by a step-over south of Beichuan (Li *et al.*, 2010). Hashimoto *et al.* (2010) analyzed InSAR data and proposed a displacement model with thrust slip along the Yingxiu–Beichuan fault in the southwest and dextral slip to the northeast, near Beichuan. Li *et al.* (2010) also reported that the northern segment of the Yingxiu–Beichuan fault is dominated by dextral slip, particularly at its northeast end. Surface ruptures also occurred along the Guanxian–Anxian fault from Changzhou to Hangwan, a distance of 30–40 km, with a maximum vertical separation of 4 m (Li *et al.*, 2008a).

31.3 TYPES OF LANDSLIDES INDUCED BY THE EARTHQUAKE

Landslides triggered by the Wenchuan earthquake include shallow and deep-seated rockslides, rockfalls, debris slides, and debris flows. Shallow landslides were the most common failures and occurred on convex slopes or ridge tops. Failures at isolated

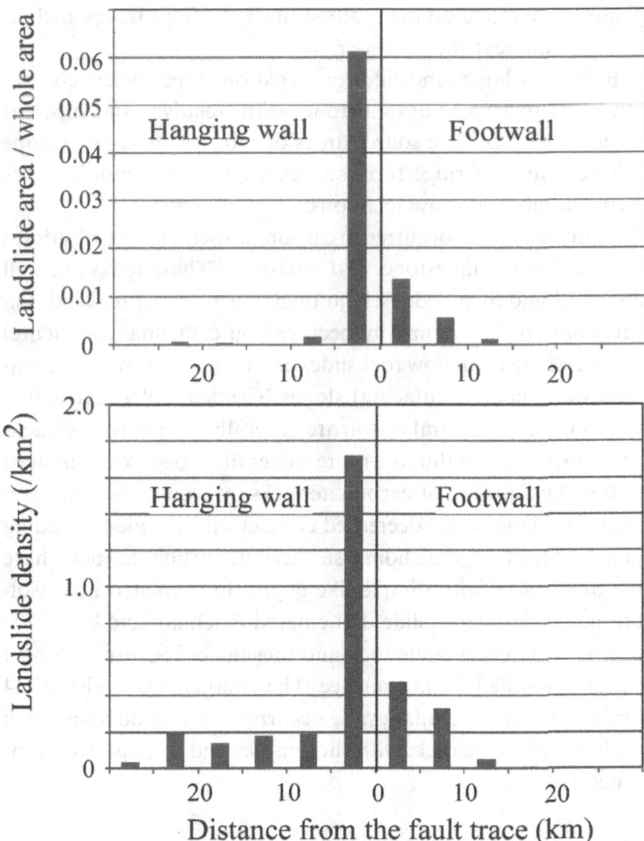

Fig. 31.2. Histograms showing the ratio of total landslide area to (top) the entire affected area and (bottom) landslide density as a function of distance from the Longmenshan fault between Bichuan and Pingdong (Chigira *et al.*, 2010).

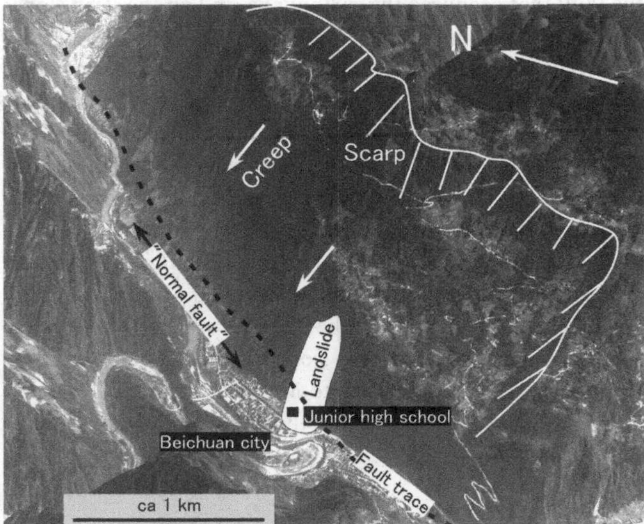

Fig. 31.3. Satellite image (Google Earth) of the Beichuan area. The fault at the foot of the slope may be a thrust that has been deformed into a normal structure by long-term creep.

ridge tops, where earthquake shaking was amplified, are notable (Pedersen *et al.*, 1994). Slopes composed of carbonate or granitic rocks had the largest number of landslides (Huang and Li, 2008). Alternating beds of sandstone and mudstone of

Triassic age failed on many slopes near the fault traces, including Yingxiu near the epicenter.

Relatively large landslides occurred on slopes where competent carbonate rocks or sandstone overlie relatively incompetent shale or phyllite. The source areas of most of these large slope failures are near ridge tops, suggesting that the seismic waves were amplified at those locations.

Many rockslides occurred in carbonate rocks, principally dolomite, dolomitic limestone, and limestone. These rocks are well stratified and locally densely jointed. Deep-seated rockslides in carbonate rocks commonly occurred on cataclinal (outfacing) slopes, whereas shallow rockslides and rockfalls were most common on anaclinal (infacing) slopes (Cruden, 1989). Anaclinal slopes in the epicentral region are generally steeper than cataclinal slopes, and are thus are more susceptible to shaking-induced failure. Dissolution of carbonate rocks by groundwater has created open cracks and decreased contact surfaces along bedding planes. Many of the sliding surfaces in carbonate rocks have rough surfaces with dimple-like depressions created by dissolution. The large rockslide that entered Beichuan and killed 700 people was in carbonate rock and presumably occurred along a bedding-parallel sliding surface. The Xiaojiaqiao landslide, 34 km northeast of Dujiangyan, occurred on a bedding-parallel fault in carbonate rock, with slickensides and a gouge a few centimeters thick.

31.4 DISTRIBUTION, DIRECTION, AND SIZE OF THE LANDSLIDES

Most of the landslides were concentrated in two areas: near the fault that produced the earthquake, and on the steep slopes adjacent to Minjiang River (Huang and Li, 2008). Landslides occurred over a distance of more than 270 km along the Longmenshan fault. The maximum width of the landslide zone is 30 km; the landslide zone is broader near the middle and southwest parts of the fault and narrower in the northeast (Fig. 31.1). Concentrations of landslides, except along Minjiang River, coincide approximately with the areas of Chinese seismic intensities IX or larger (Li *et al.*, 2008b), which correspond to IX or larger on the Modified Mercalli Intensity Scale and to VI or larger on the intensity scale of the Japanese Meteorological Agency. Steep slopes of the inner gorges along Minjiang River from Yingxiu to Maoxian experienced numerous landslides and were apparently within the area of Chinese seismic intensity IX.

31.4.1 LANDSLIDE DISTRIBUTION, FAULT RUPTURES, AND THE INNER GORGES

Landslides in the northeast part of the affected area were concentrated on the hanging wall of the Yingxiu–Beichuan fault, within a few kilometers of its rupture trace (Fig. 31.2). The area ratio of landslides to the affected ground is as high as 6 percent within 5 km of the rupture on the hanging wall, but decreases to 0.2 percent over the next 5 km, between Beichuan and Pingdong (Fig. 31.1). Scattered ridge-top failures occurred outside this zone. Far fewer landslides occurred on the footwall of the fault. The bedrock on both sides of the fault is the same – weathered phyllite with a northwest-dipping foliation. Between Pingdong and Beichuan, the fault trace has a compressional step-over, where a large number of landslides occurred over a wide area. The area of landsliding broadens from Beichuan to Honbai, perhaps reflecting differences in fault movement behavior: dextral slip northeast of Beichuan and thrusting to the southwest (Hashimoto *et al.*, 2010; Li *et al.*, 2010). Between Xiaojiaqiao and Honbai, landslides occurred on the hanging wall within 15 km of the trace of the Yingxiu–Beichuan fault and on the footwall within 13 km of the fault trace. In this area, however, the footwall of the Yingxiu–Beichuan fault is the hanging wall of the Guanxian–Anxian fault.

Fig. 31.4. Longmenshan fault trace, a gravitationally deformed slope, and the landslide that buried a junior high school in Beichuan.

As stated above, the Yingxiu–Beichuan fault is a thrust fault with a right-lateral component of displacement. However, it had exceptional normal displacements over a distance of 1 km northeast of Beichuan during the 2008 earthquake (Fig. 31.3; Li *et al.*, 2010), which can be attributed to gravitational slope deformation. In this area, the fault strikes northeast, dips 80° southeast, and lies at the foot of a 30° west-facing slope. A Google Earth image shows that the slope moved downward to the west before the earthquake. This movement must have loaded the fault surface, which is assumed to have originally dipped 70° northwest like the fault to the northeast (Li *et al.*, 2010), and probably overturned it to the southeast at the foot of the slope (Fig. 31.4). Surface rupture during the earthquake occurred along this section of the fault, which had previously been overturned. The landslide that struck the Beichuan junior high school occurred at the edge of this deformed slope. Similar bending of an active fault at the foot of a gravitationally deformed slope has been documented along the Median Tectonic Line in Japan (Kato and Chigira, 2009).

Relatively shallow landslides were concentrated on steep slopes in the valley of Minjiang River from Yingxiu to

Fig. 31.5. Landslides in the inner gorge of Minjiang River. Bedrock at this site is fractured granite. The granite is overlain by gravel, which directly underlies the gentle slope into which the gorge has been cut.

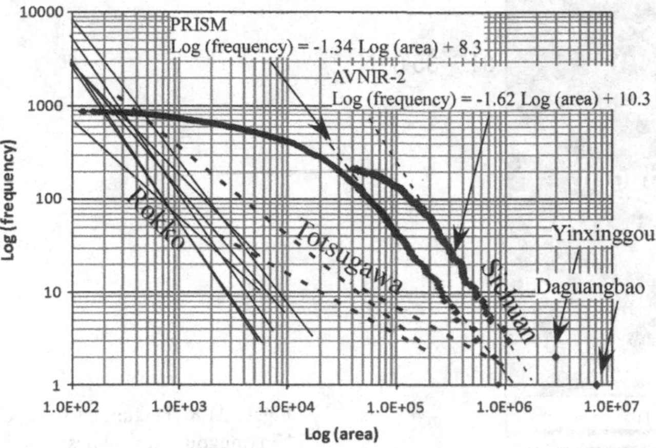

Fig. 31.6. Plots of cumulative coseismic landslide frequency and size for the Wenchuan earthquake (Chigira *et al.*, 2010). One plot is based on PRISM images and has a resolution of 2.5 m for the landslides northeast of Beichuan; the other plot is based on AVNIR-2 images with a resolution of 10 m for the entire study area. The Rokko and Totsugawa data are from Hirano and Ohmori (1989).

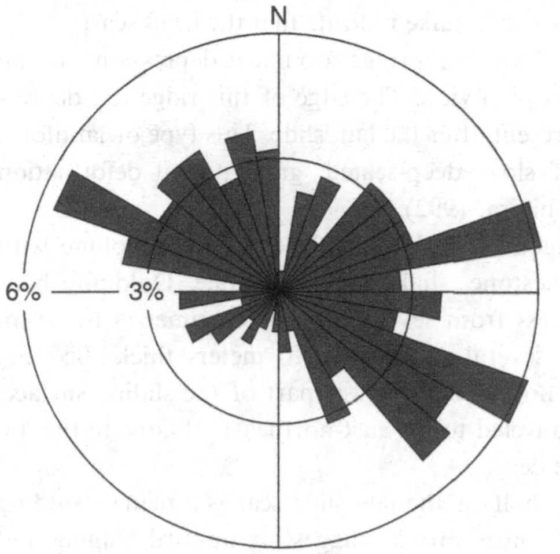

Fig. 31.7. Rose diagram showing movement direction for 212 landslides triggered by the Wenchuan earthquake (Chigira *et al.*, 2010).

Maoxian (Fig. 31.5). These slopes form the inner gorges or inner valleys described by Kelsey (1988). Riverbed elevations in the gorges are 1155–1250 m asl, and the elevations of convex slope breaks on the steep (36°), straight, valley side walls are 1750–1900 m asl north of Yingxiu. We observed well-rounded gravel in several places at the edges of gentle slopes above the convex slope breaks, which suggests that these slopes are terrace surfaces.

31.4.2 LANDSLIDE SIZES

The landslide size–frequency relationship for the Wenchuan earthquake is similar to the relationship between earthquake frequency and magnitude (Fig. 31.6). The relation between numbers of landslides and their areas, based on identification using AVNIR-2 images with a resolution of 10 m, can be described by the equation:

$$N(x) = 10^{a}10^{-bx} \text{ or } \log_{10}N(x) = a - bx \qquad (31.1)$$

where $N(x)$ is the cumulative number of landslides larger than or equal to x, which is the magnitude expressed by $\log_{10}A$, and A is the area of a landslide; a and b are constants. Landslides smaller than 2×10^5 m^2 deviate from the line defined by this equation (Fig. 31.6), suggesting incomplete sampling of small landslides on the AVNIR-2 images. Figure 31.6 indicates that the two largest landslides, the Yinxinggou and Daguangbao landslides, also deviate from the line. Similar relationships between landslide number and size have been reported before (Hatano, 1968; Hirano and Ohmori, 1989; Sugai *et al.*, 1994; Hovius *et al.*, 1997), but only for landslides induced by rainfall or those without specified triggers.

The relationship between landslide number and size for the area near Pingdong was analyzed more closely using PRISM images with a resolution of 2.5 m (Fig. 31.6). The linear part of the curve is parallel to the line based on analysis of the AVNIR-2 images, but it lies to the left in the graph. This result indicates that the landslides near Pingdong were smaller than those in the central and southwest parts of the affected area.

31.4.3 MOVEMENT DIRECTIONS OF LANDSLIDES

Movement directions of landslides identified on the AVNIR-2 images are shown in Figure 31.7. Landslides have preferred directions of movement, mainly to the northwest, west-northwest, southeast, or east-southeast. These directions are approximately normal to the rupture traces and the rupture propagation direction of the Yingxiu–Beichuan and Guanxian–Anxian faults (Li *et al.*, 2008b), suggesting a directivity effect of the seismic waves (Bolt and Abrahamson, 2003). The propagation direction of seismic waves during the Wenchuan earthquake was to the northeast (Li *et al.*, 2008a), but fewer landslides moved to the northeast than toward the northwest or southeast. It also is notable that two of the four large landslides described below occurred in directions parallel to the fault trace, which may be related to the behavior of the seismic waves. We were unable to analyze the

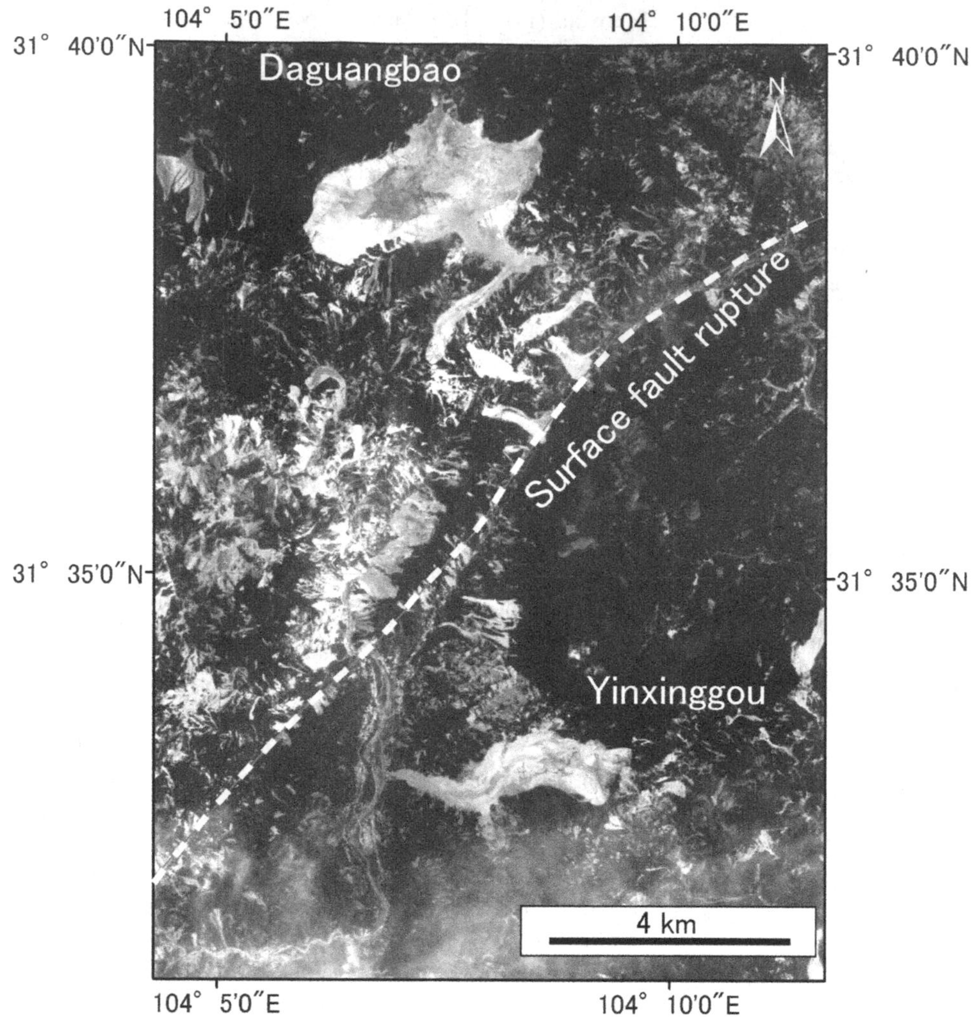

Fig. 31.8. Daguanbao and Yinxinggou landslides near Qingping (AVNIR-2 image by JAXA).

directions of all slopes, including some slopes with landslides, because of the lack of high-resolution DEM data. Many ridges, however, trend northwest and have northeast- or southwest-facing slopes, indicating that the orientations of landslides are not due to the slope directions. Movement directions of the landslides could also be affected by the bedding attitude, although we have insufficient data to test this hypothesis. However, the geologic structure in this area is complex, with folds and thrusts, and thus we infer that the direction of landslide movement is more affected by seismic waves than by the geologic structure.

31.5 LARGE SEISMICALLY TRIGGERED LANDSLIDES

31.5.1 DAGUANGBAO LANDSLIDE

The Daguangbao landslide, which was the largest mass movement triggered by the Wenchuan earthquake, occurred in a mountainous area 4 km northwest of the fault trace near Qingping (Figs. 31.1 and 31.8; Huang *et al.*, 2008). The area of the landslide, based on measurements made from the ALOS/PRISM images, is 7.35 km^2. We estimate its volume to be 0.8 km^3 based on comparison of the PRISM data and the SRTM DEM (Fig. 31.9). Examination of the ALOS and SPOT images taken before the earthquake indicate that the head scarp of the landslide was a 2-km-long, ridge-top linear depression with an open V-shape in plan view. The edge of this ridge-top depression was still present after the landslide. This type of landform is indicative of slow, deep-seated, gravitational deformation (Tabor, 1971; Chigira, 1992).

The Daguangbao landslide involved bedded dolomite with subordinate limestone, shale, and sandstone. Dolomite beds range in thickness from several tens of centimeters to 10 m; shale beds are several centimeters to meters thick. Bedding planes dip 35° north on the lower part of the sliding surface. The landslide traveled to the east-northeast, oblique to the dip direction of the beds.

The southern half of the landslide scar is a planar bedding surface with a convex profile, suggesting upward bulging and fracturing of the beds that may have resulted from a gravitational deformation coincident with the formation of a linear

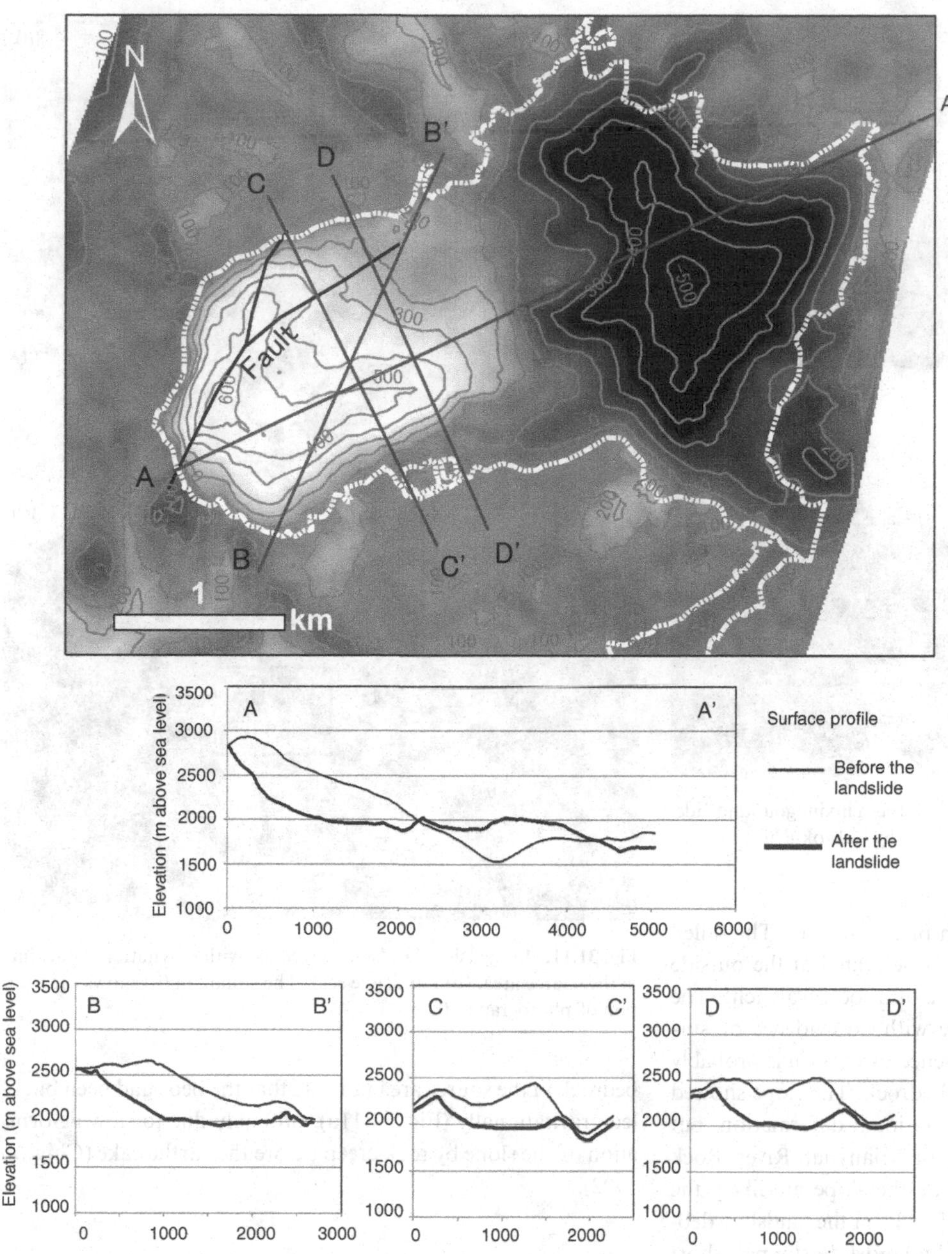

Fig. 31.9. Elevation change before and after the Daguanbao landslide. Pre-event DEM was made from SRTM3 data with an original mesh size of 90 m; post-event DEM was constructed from PRISM data with a resolution of 2.5 m. Contour interval is 100 m.

ridge-top depression. The bedding dips 35° on the lower part of the sliding surface, but less steeply higher on the slope. Upward bulging may be the mechanism by which the ridge-top depression mentioned above was formed. The head scarp of the Daguangbao landslide coincides with a fault that trends north-northeast and dips 75° east-southeast (Fig. 31.9). The north half of the ridge-top depression is located on this fault, suggesting that movement may have occurred along this fault at some time before the 2008 earthquake. As shown in the cross-sections in Figure 31.9, the beds that slid did not daylight on the slope before the landslide.

The Daguanbao landslide formed a huge deposit at the base of the affected slope (Fig. 31.9) and left a weakly laminated layer of airborne fine sediment up to about 7 cm thick on the slide deposit. This airborne sediment comprises particles of clay to very fine sand (0.5–200 μm), with two peaks at 20 and 100 μm.

31.5.2 YINXINGGOU LANDSLIDE

The second-largest landslide, the Yinxinggou or Wenjiagou landslide, occurred on a slope of carbonate rock 4 km southeast of the inferred trace of the fault, near Qingping (Figs. 31.8 and 31.10). It is up to 4 km long and 1 km wide, and has an area of 3 km². Its volume is estimated to be 50 million m³ (Tang Chuan, personal communication, 2010). The landslide scar is a

Fig. 31.10. The source area and the toe of the Yinxinggou landslide. The photograph of the source area was taken by T. Inokuchi.

west-facing dip slope with a smooth planar surface. The failed rock traveled down a valley and super-elevated at the outside bends in the path: behavior suggestive of a rock avalanche. The failure occurred on a smooth slope with no evidence of surface erosion, such as gullies; the absence of erosion is probably due to the high permeability of the bedrock. The slope showed no evidence of pre-earthquake gravitational deformation, but it had been undercut by a tributary of Mianyuan River. Rock dissolution and erosion of the toe of the slope are likely the main causes of the landslide. Many blocks in the landslide debris show evidence of dissolution. The landslide stopped short of Mianyuan River, but debris flows triggered by subsequent rainstorms reached the river, where they buried a village and formed a landslide dam that flooded the area upstream.

31.5.3 TANGJIASHAN LANDSLIDE

The Tangjiashan landslide near Beichuan formed the largest landslide dam in the affected area. The landslide is 560 m wide, 1100 m long, and has a volume of about 20 million m^3; it occurred on a dip slope of Cambrian shale and slate (Xu *et al.*, 2009). The bedrock is locally folded, but in most places strata dip into the valley. The failure occurred along bedding planes and joints parallel to the slope (Fig. 31.11 a, b). Formosat II and SPOT images show depressions with upslope-convex traces in plan view in the source area of the landslide. Exposures of bedrock in the source area indicate that the beds had been buckled gravitationally (Fig. 31.11a), probably due to slow deformation of the slope by rock creep before the earthquake (Chigira, 1992).

Fig. 31.11. Tangjiashan landslide. (a) Shale with gravitational buckling in the source area. (b) Landslide scar. The square in (b) shows the location of photograph (a).

31.5.4 XIEJIADIAN LANDSLIDE

The Xiejiadian landslide, which killed about 100 people, occurred in a small valley crossed by the Yinxiu–Beichuan fault, and has a long lobate shape, reflecting its high mobility (Fig. 31.12). It has a total volume of about 1.5 million m^3, a travel distance of about 1.7 km, and a drop in elevation of about 500 m. The bedrock is different on opposite sides of the fault. Precambrian granite is present on the northwest side (the hanging wall), whereas alternating strata of sandstone, shale, and black mudstone occur on the southeast side (the footwall). The main displaced materials, however, were valley-floor, colluvial sands that were entrained by the landslide. Tests of the landslide materials using an undrained ring-shear apparatus showed that sandy sediment is highly liquefiable under seismic loadings, whereas clayey material has a high resistance to cyclic loading,

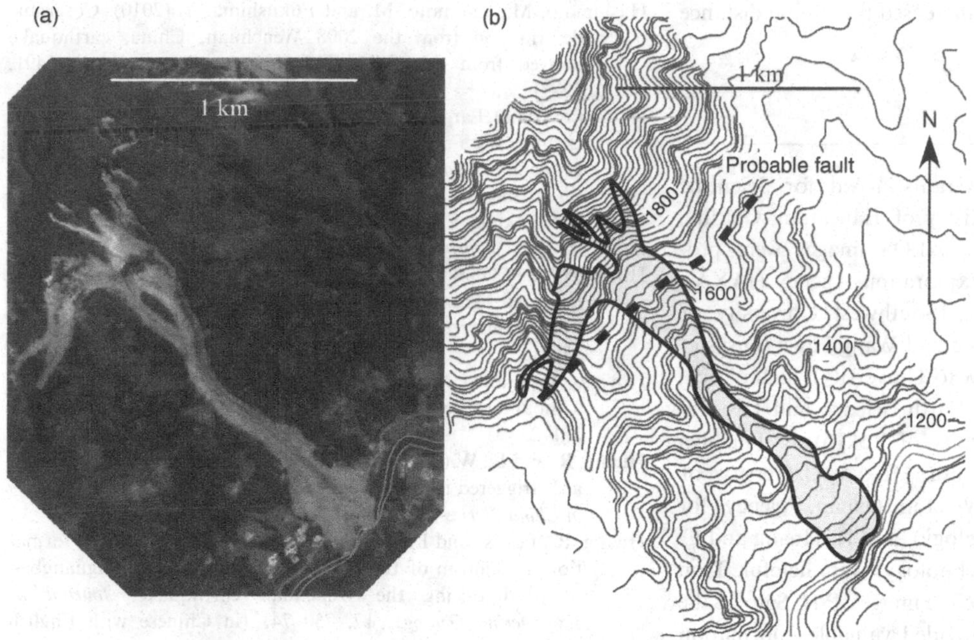

Fig. 31.12. Xiejiadian landslide (Chigira *et al.*, 2010). (a) PRISM image (© JAXA). (b) Contour map-made from the PRISM data.

although its shear resistance sharply decreased upon failure. The results suggest that liquefaction of sandy colluvium in the valley was the main cause of the long runout.

31.6 LANDSLIDE DAMS

More than 800 landslide dams formed during the Wenchuan earthquake, of which about 500 completely dammed rivers. About 45 percent of the dams failed within one week of formation, 60 percent failed within one month, and 90 percent failed within one year (Fan *et al.*, 2012).

The stability of the landslide dams was of great concern; thus, immediately after the earthquake, almost all of the large lakes were drained by excavating channels across low points on the dam surfaces. Through post-event analysis and geophysical surveys, we found that that the stability of the dams was controlled by material types and landslide runout distance. Dams consisting of large blocks of carbonate rocks and sandstone were stable, whereas those composed of finer materials, such as weathered mudstone, slate, or phyllite, were less stable. For example, the Donghekou landslide dam in Qingchuan Prefecture, which consisted mainly of weathered slate debris, was breached immediately after it was overtopped (Chang *et al.*, 2009).

Runout distance also had an effect on the stability of the dams. Long runout resulted in more intense fracturing and disintegration of clasts, and lithologic and textural segregation within the debris. We surveyed the shear-wave velocity structure of a landslide dam east of Qingping at different depths and analyzed the grain-size distribution of the debris by means of particle digital image and sieve analyses. This dam remained intact until an artificial outlet channel was excavated 2 years after the earthquake. The shear-wave velocity of the material forming the dam was generally higher than that of unstable landslide dams. A zone within the barrier has lower velocity and coincided with an area of finer materials. Sediment near the surface of the landslide dam and at its margins was coarser than sediment in the core of the dam. This structure retarded seepage and increased the resistance of the dam to breaching by overtopping.

The Tangjiashan landslide dam impounded a lake more than 6 km long. It was artificially trenched and the lake drained to prevent flooding downstream. During excavation, it was found that most of this dam consisted of fractured, but otherwise intact, shale – a consequence of the short runout distance of the landslide.

31.7 CONCLUSIONS

The 2008 Wenchuan earthquake triggered hundreds of landslides that caused much loss of life and damage in the mountains bordering the Sichuan Basin of southwest China. Landslides were concentrated on the hanging wall of the Longmenshan fault and in the inner gorges of Minjiang River. Carbonate rocks failed preferentially, probably due to their low shear resistance, resulting in part from dissolution of rock along bedding planes. Most landslides moved normal to fault traces, due to the directivity of the seismic waves. A power function describes the relationship between the cumulative number of landslides and landslide area. The largest landslide (7.35 km^2) occurred on a cataclinal slope of carbonate rock; we estimate its volume to be 840 million m^3. Its head scarp coincided with a ridge-top depression, indicating that slow, deep-seated, gravitational deformation preceded the landslide. The second-largest landslide (3 km^2) occurred in carbonate rocks on a smooth slope that had been undercut by a river. Liquefaction and entrainment of sandy valley-fill sediments greatly increased the runout distance of one large landslide.

ACKNOWLEDGEMENTS

This study was supported by Grants-in-Aid for Scientific Research from the Japanese Ministry of Education, Culture, Sports, Science, and Technology. ALOS images were provided by the Japan Aerospace Exploration Agency (JAXA). Discussions with Zhu Bao Long (Southwest University of Science and Technology), T. Kamai and Wang Fawu (Disaster Prevention Research Institute, Kyoto University), T. Inokuchi (National Research Institute for Earth Science and Disaster Prevention), K. Konagai (University of Tokyo), S. Tsuchiya (Shizuoka University), Y. Ishikawa (Tokyo University of Agriculture and Technology), W. Lin (Japan Agency for Marine-Earth Science and Technology), and R. Huang and C. Tang (Chengdu University of Technology) were helpful. Tsou Ching Ying helped process the satellite images. P. H. Sato of the Geographical Survey Institute provided valuable information about the satellite images. T. Chiba of the Asia Air Survey provided us with the red 3D images made from the SRTM DEM. O. Uchida of the Asia Air Survey calculated the volume of the Daguangbao landslide by comparing the PRISM and SRTM data. Part of our field investigation was performed as members of the reconnaissance team of the Japan Landslide Society.

REFERENCES

Bolt, B. A. and Abrahamson, N. A. (2003). Estimation of strong seismic ground motions. *International Handbook of Earthquake and Engineering Seismology*, 81B, 983–1001.

Burchfiel, B. C., Royden, L. H., Vander Gilst, R. D. *et al.* (2008). A geological and geophysical context for the Wenchuan earthquake of May 12, 2008, Sichuan, People's Republic of China. *GSA Today*, 18(7), 4–11.

Chang, D., Zhang, L., Xu, Y. and Huang, R. (2009). Analysis of overtopping failure of Hongshihe landslide dam after Wenchuan Earthquake. *Journal of Engineering Geology* 17, 50–55 (in Chinese with English abstract).

Chigira, M. (1992). Long-term gravitational deformation of rocks by mass rock creep. *Engineering Geology*, 32, 157–184.

Chigira, M., Wu, X., Inokuchi, T. and Wang, G. (2010). Landslides induced by the 2008 Wenchuan earthquake, Sichuan, China. *Geomorphology*, 118, 225–238.

Cruden, D. M. (1989). Limits to common toppling. *Canadian Geotechnical Journal*, 26, 737–742.

Fan, X., van Westen, C. J., Korup, O. *et al.* (2012). Transient water and sediment storage of the decaying landslide dams induced by the 2008 Wenchuan earthquake, China. Geomorphology doi:10.1016/j.geomorph.2012.05.003 (2012).

Hashimoto, M., Enomoto, M. and Fukushima, Y. (2010). Coseismic deformation from the 2008 Wenchuan, China, earthquake derived from ALOS/PALSAE images. *Tectonophysics*, 491, 59–71.

Hatano, S. (1968). Large slope failure and its geomorphic conditions: An example from the upstream of the Arida River. In *Proceedings of the 5th Symposium on Natural Disaster*, Tokyo, pp. 209–210 (in Japanese).

He, H. and Tsukuda, E. (2003). Recent progresses of active fault research in China. *Journal of Geography (Tokyo)*, 112, 489–520.

Hirano, M. and Ohmori, H. (1989). Magnitude-frequency distribution for rapid mass movements and its geomorphological implications. *Transactions of the Japanese Geomorphological Union*, 10, 95–111.

Hovius, N., Stark, C. P. and Allen, P. A. (1997). Sediment flux from a mountain belt derived by landslide mapping. *Geology*, 25, 231–234.

Huang, R. and Li, W. (2008). Development and distribution of geohazards triggered by 5.12 Wenchuan earthquake in China. *Science in China, Series-E Technical Science*, 52, 810–819.

Huang, R., Pei, X. and Li, T. (2008). Basic characteristics and formation mechanism of the largest scale landslide at Daguangbao occurred during the Wenchuan earthquake. *Journal of Engineering Geology*, 17, 730–741 (in Chinese with English abstract).

Kato, H. and Chigira, M. (2009). Gravitational slope deformation affecting the fault morphology of the Median Tectonic Line in the Ho-oh Range, central Japan. *Journal of the Japan Society of Engineering Geology*, 50, 140–150 (in Japanese with English abstract).

Kelsey, H. M. (1988). Formation of inner gorges. *Catena*, 15, 433–458.

Li, C., Wei, Z., Ye, J., Han, Y. and Zhen, W. (2010). Amounts and styles of coseismic deformation along the northern segment of surface rupture, of the 2008 Wenchuan Mw 7.9 earthquake, China. *Tectonophysics*, 491, 35–58.

Li, H., Fu, X., Van der Woerd, J. *et al.* (2008a). Co-seismic surface rupture and dextral-slip oblique thrusting of the Ms 8.9 Wenchuan Earthquake. *Acta Geologica Sinica* 82, 1623–1643.

Li, X., Zhou, Z., Yu, H. *et al.* (2008b). Strong motion observations and recordings from the great Wenchuan Earthquake. *Earthquake Engineering and Engineering Vibration*, 7, 235–246.

Pedersen, H., Brun, B. L., Hatzfeid, D., Campillo, M. and Bard, P. Y. (1994). Ground-motion amplitude across ridges. *Bulletin of the Seismological Society of America*, 84, 1786–1800.

Sugai, T., Ohmori, H. and Hirano, M. (1994). Rock control on magnitude-frequency distribution of landslide. *Transactions of the Japanese Geomorphological Union*, 15, 233–251.

Tabor, R. W. (1971). Origin of ridge-top depressions by large-scale creep in the Olympic Mountains, Washington. *Geological Society of America Bulletin*, 82, 1811–1822.

US Geological Survey (2008). Magnitude 7.9 – Eastern Sichuan, China 2008 May 12 06:28:01 UTC [available at http://earthquake.usgs.gov/eqcenter/eqinthenews/2008/us2008ryan/, accessed November 16, 2009].

Xu, Q., Fan, X., Huang, R. and van Westen, C. (2009). Landslide dams triggered by the Wenchuan Earthquake, Sichuan Province, southwest China. *Bulletin of Engineering Geology and the Environment*, 68, 373–386.

32 Landslides on other planets

MARKO H. K. BULMER

ABSTRACT

Using data from fly-by, orbiter, penetrometer, and lander missions, as well as from Earth-based telescopes, researchers have identified mass-movement features on the surfaces of Venus, Mars, and Mercury, the Martian moons of Phobos and Deimos, the Moon, the moons Io, Ganymede, Callisto, and Europa orbiting Jupiter, Iapetus orbiting Saturn, and the asteroids of 4 Vesta 433 Eros, 253 Mathilde, 951 Gaspra, 243 Ida, and 25143 Itokawa. Discussions of selected landslides on Mars, Venus, and Io highlight how knowledge of terrestrial landslide processes has been used to inform interpretation of geomorphic features on these bodies. New missions to bodies in the solar system continue to increase the availability of precise and accurate spatial information over large areas, enabling analysis of mass movements to extend beyond the Coulomb frictional model. With the availability of these data and the increasing sophistication of analyses that can be done, observations, interpretations, and explanations of mass-movement features on these bodies will increasingly inform our understanding of landslide triggering and emplacement on Earth.

32.1 INTRODUCTION

Our knowledge of the terrestrial planets, Mercury, Venus, and Mars, the gaseous giants Jupiter and Saturn, the ice giants Uranus and Neptune, and asteroids has been informed by fly-by, orbiter, penetrometer, lander, and rover missions plus Earth-based and near-Earth orbit telescopes. To date, there have been 2 missions to Mercury, 12 to Venus and Mars respectively, 5 to Jupiter and the Galilean satellites, 3 to Saturn and its satellites, 1 to Uranus and Neptune with their respective satellites, and 89 missions to the Moon. A recent focus has been the study of asteroids, and there have been four missions to date for this purpose.

These missions have carried sensors designed to obtain data over the range of the electromagnetic spectrum best suited to local conditions. Using data from these sensors as well as from Earth-based telescopes, researchers have identified features with the characteristics of mass-movement deposits on the planetary surfaces of Venus, Mars, and Mercury, and on the Martian moons of Phobos and Deimos, the Moon, the moons Io, Ganymede, Callisto, and Europa orbiting Jupiter, Iapetus orbiting Saturn, and the asteroids of 4 Vesta 433 Eros, 253 Mathilde, 951 Gaspra, 243 Ida, and 25143 Itokawa. Knowledge of mass-movement processes on Earth has been used to inform interpretation of geomorphic features identified on these surfaces, and comparisons are much debated within the planetary science community.

In this chapter, I define mass movement as the geomorphic process by which soil, regolith, and rock moves downslope under gravity. Mass-movement categories include falls, topples, slides, flows, and creep (see Dikau *et al.*, 1996, and references therein), each of which has its own characteristics. The process from failure to emplacement can occur over timescales of seconds to years. In keeping with the theme of this book, the term landslide, which is a type of mass movement, is used to describe a movement of a mass of rock, earth, or debris down a slope (Cruden, 1991). All slope-forming materials have a tendency to move downward under the influence of gravity, which is counteracted by a shearing resistance. Slope failures can be triggered either by internal changes in shearing resistance or by external factors that produce an increase in shear stress (Terzaghi, 1960). Mass movements, in general, including landslides on extraterrestrial bodies, are of interest as agents of surface modification and because their triggering and emplacement style reflect environmental conditions. Here specific focus

Landslides: Types, Mechanisms and Modeling, ed. John J. Clague and Douglas Stead. Published by Cambridge University Press.

is given to selected landslides on Mars, Venus, and on the moon of Io orbiting Jupiter, because they all have characteristics of long-runout landslides, but have been triggered and emplaced in very different environments.

32.2 BACKGROUND

All four terrestrial planets – Mercury, Mars, Earth, and Venus – show evidence of geologic activity. Mercury has a similar appearance to Earth's moon, with extensive mare-like plains and heavy cratering, but has been comparatively geologically inactive for the past *ca*. 3–4 billion years. On Venus, much of the surface is volcanic in origin and activity may be ongoing. Mars has had many epochs of geologic activity and there is evidence of currently active surface processes. Of the 19 known large moons, Io is the most volcanically active body in the solar system, with eruptions observed by spacecraft and the Hubble Space Telescope. Europa displays evidence consistent with tectonic activity and cryovolcanism, but it is not clear whether cryovolcanism is occurring at present. On Europa, geologic activity is related to tidal heating caused by its eccentric orbit around Jupiter. The Earth's moon, and Jupiter's moon Ganymede show features consistent with past geologic activity. On Europa, Ganymede, and Callisto, evidence exists that appears to support subsurface oceans of liquid water. The majority of known asteroids orbit in the main asteroid belt between the orbits of Mars and Jupiter, and these include 4 Vesta 253 Mathilde, 951 Gaspra, and 243 Ida. However there are other orbital paths with significant populations, including the Jupiter and Mars trojans, and near-Earth asteroids. Asteroid 25143 Itokawa is a Mars-and-Earth-crosser, while 433 Eros is a Mars-crosser. Asteroid 253 Mathilde, with its carbon-rich composition, is a C-type asteroid; while 951 Gaspra, 243 Ida, 25143 Itokawa, and 433 Eros are S-type asteroids with stony compositions. 4 Vesta is an evolved achondrite which has a stony composition but does not contain chondrules.

32.3 IDENTIFICATION AND INTERPRETATION

The ability to identify and interpret geomorphic features produced by mass-movement processes is essential for understanding the geologic and modification history of surfaces on terrestrial planets (Head *et al*., 1999), moons (Guest, 1971), and asteroids (Fujiwara *et al*., 2006). The mobility of mass movements on solar system bodies has been examined using plots of the coefficient of H/L (fall height/runout length) versus volume derived from available images and topography. Calculation of the average friction coefficient is given by the tangent of the slope connecting the pre- and post-event centers of gravity of a sliding mass (Cruden, 1980). Image geometry and poor-resolution topographic data commonly preclude determining the pre-and post-event centers of gravity for masses on most planets (Shaller, 1991), moons (Guest, 1971), and asteroids (Sullivan *et al*., 2002). Therefore, the approach that has been used is to examine the ratio between the height (H) and the length over which the material traveled (L) (Heim, 1932). This ratio equals the tangent of the slope angle of the line connecting the top of the scarp and the toe of the apron. The value derived using this method approximates the center-of-gravity gradient for those slope failures whose center of gravity lies near the toe (Shaller, 1991). Legros (2002) highlighted the limitations of this method, but acknowledged that, in the absence of other data, it enables comparisons between mass movements on the terrestrial planets, moons, and asteroids (McEwen, 1989).

In the absence of extensive ground-based observations on extraterrestrial surfaces, geomorphological and geologic interpretation must rely on remotely derived data. Contemporary imaging datasets of Mars are some of the most extensive and complete for bodies in the solar system, including Earth. Recently acquired Mars data has resolutions as high as 25 cm per pixel, and digital elevation models at *ca*. 1 m per pixel resolution are becoming the new norm. These data are essential for understanding the relationship between geologic processes and surface evolution. Reliable determinations of the processes that form surficial features are increasingly constrained by interpretation rather than data availability. High-resolution topographic data (Bulmer and Finnegan, 2010) provide the essential boundary conditions to quantify processes that can produce deposits on Earth with similar characteristics (Schumm, 1985), such as lava flows, landslide deposits, and fluvial landforms (Shepard *et al*., 2001).

32.4 MARS

Mars is the fourth planet from the Sun. It has a thin atmosphere and a surface with volcanoes, valleys, impact craters, deserts, and polar ice. The surface gravity is 3.716 m s^{-2}; surface temperatures range from 186 to 293 K; the mean surface pressure is 0.636 kPa; and the atmosphere is 95.32 percent CO_2. Between 1975 and 1978, Viking Orbiters 1 and 2 imaged Mars at a resolution of 150–300 m, with selected areas at 8 m resolution. Landslides evident in these images occur at the base of escarpments in crater interiors, at highland massifs, on the flanks of volcanoes, and within the 4000-km-long and 7-km-deep trough systems of the Valles Marineris. The largest landslide deposits on Mars may be the aureole deposits surrounding the Olympus Mons volcano. Their origin has been controversial; plausible emplacement mechanisms fall broadly into two categories: the first being catastrophic flank failure (Lopes *et al*., 1980, 1982); and the second, thrusting or gravity spreading of local sediments or flank material at low strain rates (Francis and Wadge, 1983; Tanaka, 1985).

Landslides in the Valles Marineris are also of interest because lobate aprons on the floor of the canyon have long runouts from their sources on the canyon walls. Failures in Coprates, Candor, and Ganges Chasmata show evidence of slumping at the base of the canyon wall, beyond which a lobate apron extends (Fig. 32.1). In plan view, the aprons are fan-shaped or lobate, with toes that extend >20 km from the slump blocks; aerial extents are commonly >100 km^2. The surfaces of these

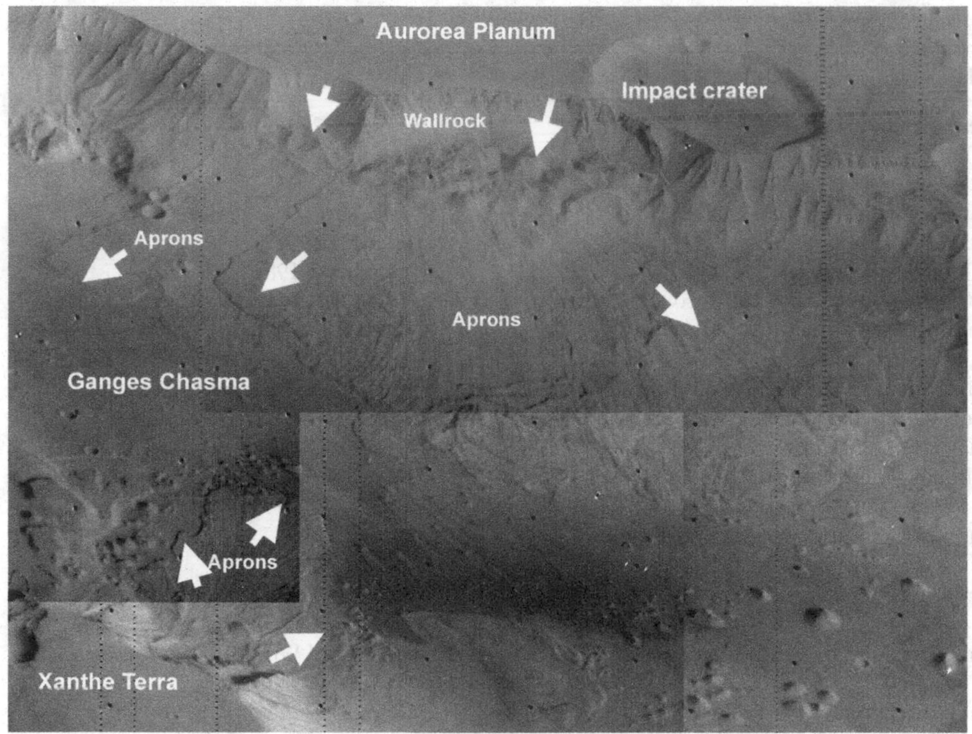

Fig. 32.1. Mosaic of Viking Orbiter 1 image 014A29–32 showing part of Ganges Chasma, looking southward. The aprons are interpreted to be giant landslide deposits that were emplaced at high velocity.

features commonly have regularly spaced ridges perpendicular to the direction of transport.

Based on analysis of Viking Orbiter images, the landslides in the Valles Marineris have been interpreted to be large (>10^6 m^3) catastrophic failures with long runouts (Lucchitta, 1978, 1979, 1987; McEwen, 1989; Shaller, 1991). A range of emplacement models based on the concepts of bulk fluidization, basal lubrication, and distributed mass loss (Table 32.1) have been proposed to explain their high mobility. However, the limited resolution of the Viking Orbiter data precludes conclusive validation of any of the models. Recent missions have obtained images with resolutions of *ca.* 10 m per pixel (High Resolution Stereo Camera) down to 1.4 m per pixel (Mars Orbiter Camera). Thermal Emission Mass Imaging Spectrometer images have resolutions of 100 m per pixel infrared and 19 m per pixel visible, and the Mars Orbiter Laser Altimeter topographic dataset has *ca.* 1 m horizontal and vertical resolution, with horizontal resolution limited by the along-track shot-to-shot spacing of 330 m and the vertical resolution by averaging over the 130-m diameter laser spot size. These higher-resolution images show new details of landslides in the Valles Marineris. Mars Orbiter Laser Altimeter topography and <20 m per pixel images have been used by Bulmer and Zimmerman (2005) to recalculate the dimensional values of published Martian landslides (Shaller, 1991) in the Ganges, Coprates, and Candor Chasmas. Viking Orbiter studies have concluded that the highest visible backscarp in the canyon wall was the source of the landslides and calculated fall height accordingly. Mars Orbiter Laser Altimeter and high-resolution image data, however, reveal lower secondary scarps on previously failed materials, as well as additional deposits. Volumes of failed masses have also been recalculated; values lie within, rather than above, the trend for terrestrial dry rock avalanches based on previous Viking Orbiter image analysis (Fig. 32.2; McEwen, 1989). Thus, Martian landslides display no unusual mobility trends relative to dry Earth rock avalanches. Using a comparison of *H*/*L* versus runout in an attempt to infer water content in Martian landslides during emplacement, Legros (2002) found nonunique distinctions between landslide types, which has motivated a search for an alternative approach and re-examination as to whether the deposits could be explained through a process other than catastrophic failure (Schumm, 1985).

32.4.1 FAILURE AND TRANSPORT MECHANISMS

Using high-resolution topography and image datasets from Mars Global Surveyor (Smith *et al.*, 2001), McGovern *et al.* (2004) invoked the mechanism described by Iverson (1995) to argue for flank spreading and failure at Olympus Mons volcano. On Hawaii, flank spreading occurs along a basal décollement with pore water trapped in an overpressurized basal sediment layer that reduces the effective friction at the base of the volcano. Applying such a model to Olympus Mons, a basal décollement is rooted in sediments deposited on top of an ancient basement in the lowlands. Assuming downhill transport, sediment thickness will tend to increase with decreasing elevation. A northwestward-thickening sediment wedge will lead to lateral weakening along the décollement, enhancing

Table 32.1. *Mechanical models proposed to explain the transport and emplacement of mass movements.*

Concept	Model	Author
Bulk fluidization	Kinematic flow	Heim, 1932
	Air fluidization	Kent, 1966
	Grain flow in fluid medium	Hsü, 1975
	Grain flow without a fluid medium	Davies, 1982
	Acoustic fluidization	Melosh, 1979
	Air fluidization and aerodynamic lift	Krumdieck, 1984
	Computer simulated flow, bi-viscous rheology	Dent, 1982; Trunk *et al.*, 1986
	Computer simulation of grain flow	Campell, 1989
	Bingham plastic rheology	McEwen, 1989; O'Brien *et al.*, 1993
	Computer simulated flow, power-law rheology	Fread, 1988; Potapov and Ivanov, 1991
Basal lubrication	Dissociated rock and melted rock lubrication	Erismann, 1979
	Air-layer lubrication	Shreve, 1968
	Earthquake fluidization	Solonenko, 1972; McSaveney, 1978; Hazlett *et al.*, 1991
	Basal lubrication by vaporized pore fluids	Habib, 1975; Goguel, 1978
	Distributed deposition and basal lubrication	Shaller, 1991
Distributed mass loss	Low-viscosity mass loss	Van Gassen and Cruden, 1989
Individual-case mechanism	Basal lubrication by weak stratum in landslide debris	Watson and Wright, 1969, Johnson, 1978
	Basal lubrication by overridden snow	McSaveney, 1978
	Basal lubrication by overridden mud	Hungr and Evans, 2004

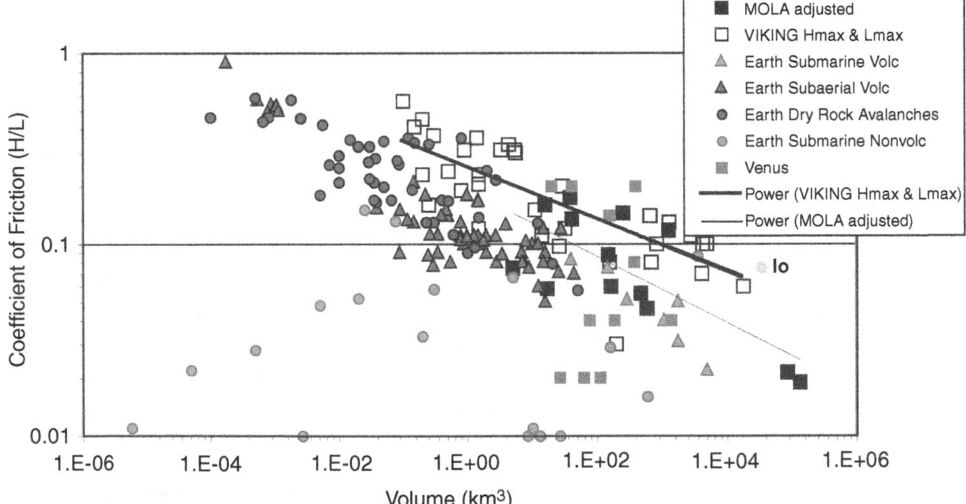

Fig. 32.2. Plot of H/L versus volume for landslides in Coprates, Candor, and Ganges Chasma on Mars, subaerial and submarine volcanic and nonvolcanic landslides on Earth, and landslides from modified domes on Venus. The debris apron at the foot of Euboea Montes is shown as a circle and labelled on the figure. Best-fit lines for sites using Viking and Mars Orbiter Laser Altimeter data are shown. Viking data shown on this plot were taken from previously published studies.

slip on the part of the flank that overlies the thicker sediments. The Iverson (1995) scenario requires sediments with very low hydraulic diffusivity to allow high pore pressures to develop; only clay sediments meet this requirement. The spectral signatures of phyllosilicate minerals have been detected on Mars (Mustard *et al.*, 2008). The geologic context of the observed outcrops implies stratigraphically low (early Noachian era) and areally extensive clays. These findings are consistent with the model of a detachment beneath Olympus Mons rooted in clay-rich Noachian sediments deposited on northern lowlands at least several hundred meters thick at the future location of Olympus Mons (Morgan and McGovern, 2005; McGovern and Morgan, 2009).

Large arcuate headscarps exist on the 5-km-high southern wall of Ganges Chasma in the Valles Marineris (Fig. 32.3). Layered rocks are visible in the headscarps (Fig. 32.4a, b). Scarps and rock chutes within each headscarp have been created by topples, falls, and slides. Large talus cones can be traced to individual rock chutes down which the failed rock moved (Fig. 32.4b). The strength characteristics of the rock exposed in the Valles Marineris are consistent with medium-strength terrestrial basaltic rock masses (Bieniawski, 1993; Hoek, 1998; Schultz, 2002).

Spectral observations from the Thermal Emission Spectrometer also suggest basaltic rocks (Banfield *et al.*, 2000). Analysis of available image and topography data supports evidence of fracturing and weathering of this basaltic wallrock. The extensive talus provides no evidence for draining of fluids (Schultz, 2002).

A topographic bench 5 km long and 15 km wide is present at the base of the headscarps at 2300 m below the datum point at which the atmosphere pressure is about that of the freezing point of water (Figs. 32.3 and 32.4a). This bench has many terraces, ridges, double ridges, ridge depressions, troughs, and anti-slope scarps (Fig. 32.4b), oriented approximately parallel to the canyon rim. The scarps of terraces are similar in appearance to those on the canyon rim. In places, rock chutes with arcuate headscarps appear to have grown laterally by incorporating smaller rock chutes at their margins.

The base of the topographic bench, at about −3300 m (Fig. 32.5a), slopes about 20°. Here talus deposits below rock chutes grade into lobes, sheets, and aprons (Fig. 32.5a, b). The latter originate from multiple source regions, display characteristics of flow including sinuous pathways, follow local topographic lows, and overlap, coalesce, and override one another. The different sizes of lobes reflect the size of the source zones from which they were fed. In many instances, multiple lobe fronts can be identified on the basis of the levees that extend along the margins of individual flows (Fig. 32.5b). The lobes traveled, on average, <5 km. Levees at the margins of flows appear to be the longitudinal structure on aprons noted by Lucchitta (1978) on Viking Orbiter images. Lobe interactions become more complex down the regional slope, which gradually lessens to 1°. In places, the flow margins cannot be delineated, but their tracks and transport directions can be identified by discrete sets of transverse en-echelon structures (Fig. 32.5b).

The age of the rocks exposed in the walls at Ganges Chasma is Upper Noachian (McEwen *et al.*, 1999); thus older than 3.5–4.3 Ga. Mapped superposition relations and counts of craters can be used to determine the relative ages of landslide deposits with respect to one another (Fig. 32.3b). For instance, lobes forming Ls2Aa,b,c are superposed on Lsla and therefore are younger in age. These deposits originated from Ls2A on the west edge of Ls2. The largest numbers of craters visible in the 100-m per pixel data are on Lsla, Ls2a, Ls3a, and Ls3b, which are the oldest. The use of impact crater densities on the aprons to compare times of emplacement to other features on Mars suggests that Ls3a and Ls3b are 1–2 Ga in age, placing them in the Hesperian epoch (Quantin *et al.*, 2003). The complexities of such studies are not addressed here, but center on the identification of deposit boundaries. In spite of these difficulties, the ages clearly indicate that the surfaces have existed for millions to billions of years. This antiquity, combined with geomorphology, led Bulmer and Zimmerman (2005) to propose that the terraces forming the topographic bench below the canyon rim are similar to terrestrial sackungen (deep-seated gravitational creep; Fig. 32.4c). The age, fractured nature, aspect, and thickness of the wallrock support a gravitational creep origin. Using

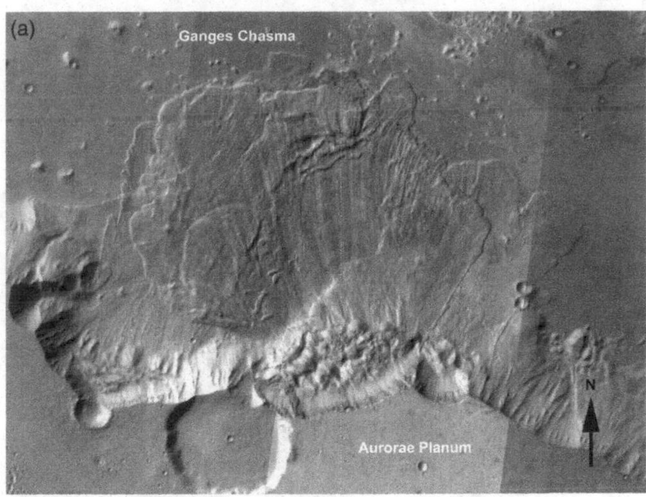

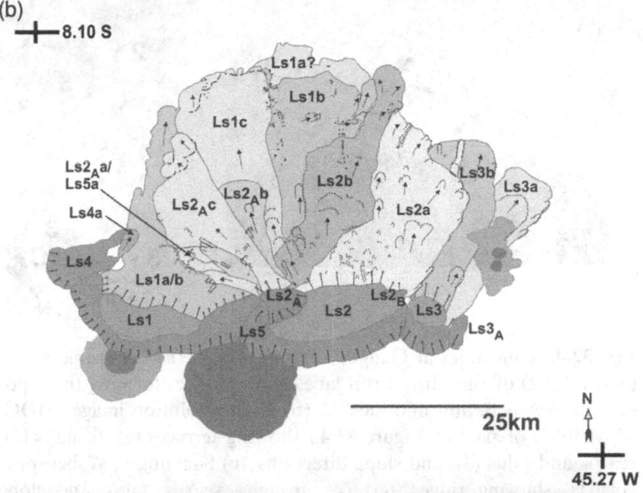

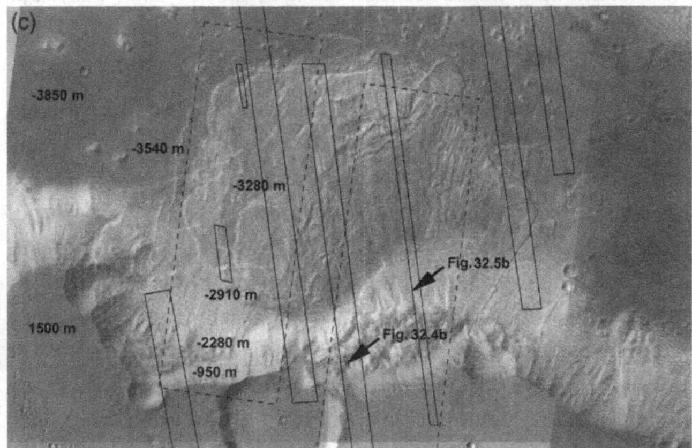

Fig. 32.3. (a) Mosaic of THEMIS daytime thermal infrared images of the landslide at Ganges Chasma. (b) Geomorphic sketch map; see text for discussion of units. Black arrows show the direction of movement. (c) THEMIS daytime thermal infrared images with Mars Orbiter Laser Altimeter topography. Black boxes and numbers refer to figures and THEMIS visible and MOC image footprints.

High Resolution Stereo Camera images, Mège and Bourgeois (2010) noted additional evidence for a gravitational origin, including horsts and ridge-top splitting resulting in narrow grabens. High-angle extensional shear planes in the upper part of

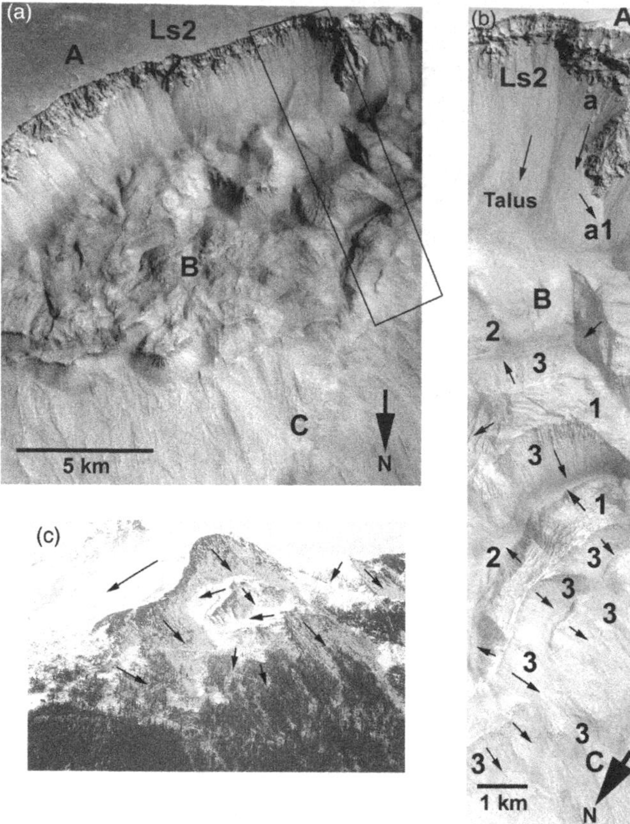

Fig. 32.4. Landslides at Ganges Chasma. (a) THEMIS visible image (V01001002) of the canyon rim labeled A, surfaces forming the topographic bench B, and midtrack C. (b) High-resolution image (MOC-M0904082) of part of Figure 32.4a showing terraces (1), troughs (2), scarps and talus (3), and slope directions. (c) Sackung at Wabenspitz, Austria, showing ridges, terraces, troughs, scarps, talus, and slope directions.

deep-seated, slowly deforming slopes produce the features seen in Figure 32.4a, b. In the basaltic wallrock on Mars, displacements have likely occurred along discontinuities in the fractured rock mass, and it is from these that terraces were created. Slopes creep down, with either a rotational or translational component, and over time have formed the complex morphology seen at the topographic bench. Gravitational creep therefore offers an alternative explanation to published ideas that the terraces are the product of large-scale catastrophic failures, although there are no features that definitively prove either explanation observable in images or Mars Orbiter Laser Altimeter topography data. We also know from Earth that slow slope deformation (Bisci *et al.*, 1996) can lead to catastrophic collapse, involving large rock avalanches (Petley *et al.*, 2002).

Along with large-scale slow gravitational deformation of canyon walls, Bulmer and Zimmerman (2005) suggested that physical and chemical weathering, and mechanical breakdown of the rock mass have enlarged scarps and developed rock chutes that shed talus to their base. Rockfalls, topples, and slides may also have been triggered by seismic shocks from impacts. Over millions of years, these talus slopes have deformed under gravity, moving colluvium downslope to the canyon floor in lobes or sheets that collectively have formed the large aprons seen today (Fig. 32.5a). The morphological features on the lobes and sheets are similar to those that characterize terrestrial rock avalanches, rock glaciers, and sand flows. The age of the wallrock, coupled with the morphologies and dimensions of talus lobes, the extended periods over which they accumulated, the environmental conditions, and the slope angles down which materials traveled, favor movements analogous to those characteristic of rock glaciers (Fig. 32.5c). On Earth, these features originate through deformation of the lower parts of talus slopes, move on low-angle slopes, and have velocities of 0.20 ± 0.11 m per year (Benn and Evans, 1997). At this rate it would take *ca.* 125,000 years for the front of the apron seen in Figure 32.3 to extend 25 km onto the floor of the canyon.

If lobes, sheets, and aprons at Ganges are talus-generated rock glaciers, flow occurred due to freeze–thaw and internal deformation of cores or lenses of ice. In the current environment, water in the solid state is unstable year-round on the surface in the equatorial regions of Mars (Mellon and Jakosky, 1993), but it may be stable at the surface in parts of the Valles Marineris (Haberle *et al.*, 2008). It is not known whether or not ice cores or lenses are present in the deposits at Ganges (Fig. 32.3), although Compact Reconnaissance Imaging Spectrometer data for Mars have revealed sulfates and hydrated silica on the floor of the Valles Marineris (Gendrin *et al.*, 2005; Roach *et al.*, 2009). These compounds have been interpreted to be the product of evaporates that formed in an ancient playa during a regional rise in the water table in the late Noachian (Grotzinger *et al.*, 2005; Andrews-Hanna *et al.*, 2010) or, alternatively, weathering of silicates in massive ice deposits (Niles and Michalski, 2009; Mège and Bourgeois, 2010). An explanation for the sulfate deposits is that, in the past, Mars experienced obliquity variations and that at >35° significant amounts of water could have been transported toward the equator and deposited at low latitudes (Jakosky and Phillips, 2001). Head *et al.* (2006) have presented geological evidence at mid-latitudes that supports an orbital pattern in which mean obliquity exceeded 45° during the Late Amazonian (*ca.* 400 million years to present) for a sufficiently long period to allow ice accumulation and glacier flow. Based on the superposition relations of aprons Ls2, Ls4, and Ls5 seen in Figure 32.3, wallrock materials were failing and materials were being transported downslope during this period. The environment during the Hesperian period (1.8–3.5 Ga) remains in doubt (Toon, 2010), but in the later part of the Noachian period, which ended *ca.* 3.7 Ga, geomorphological and geochemical features suggest that Mars was wet and relatively warm (Solomon *et al.*, 2005; Andrews-Hanna *et al.*, 2010); wallrock at Ganges dates to this time. Tian *et al.* (2009) have shown that early Mars could not have maintained a CO_2-dominated atmosphere and would have been cold and dry. Through most of the early Noachian, except perhaps during brief periods following large volcanic outgassing or impact events (Segura *et al.*, 2002), large amounts of H_2O and CO_2 could have been trapped as ice. A warm wet Mars could have

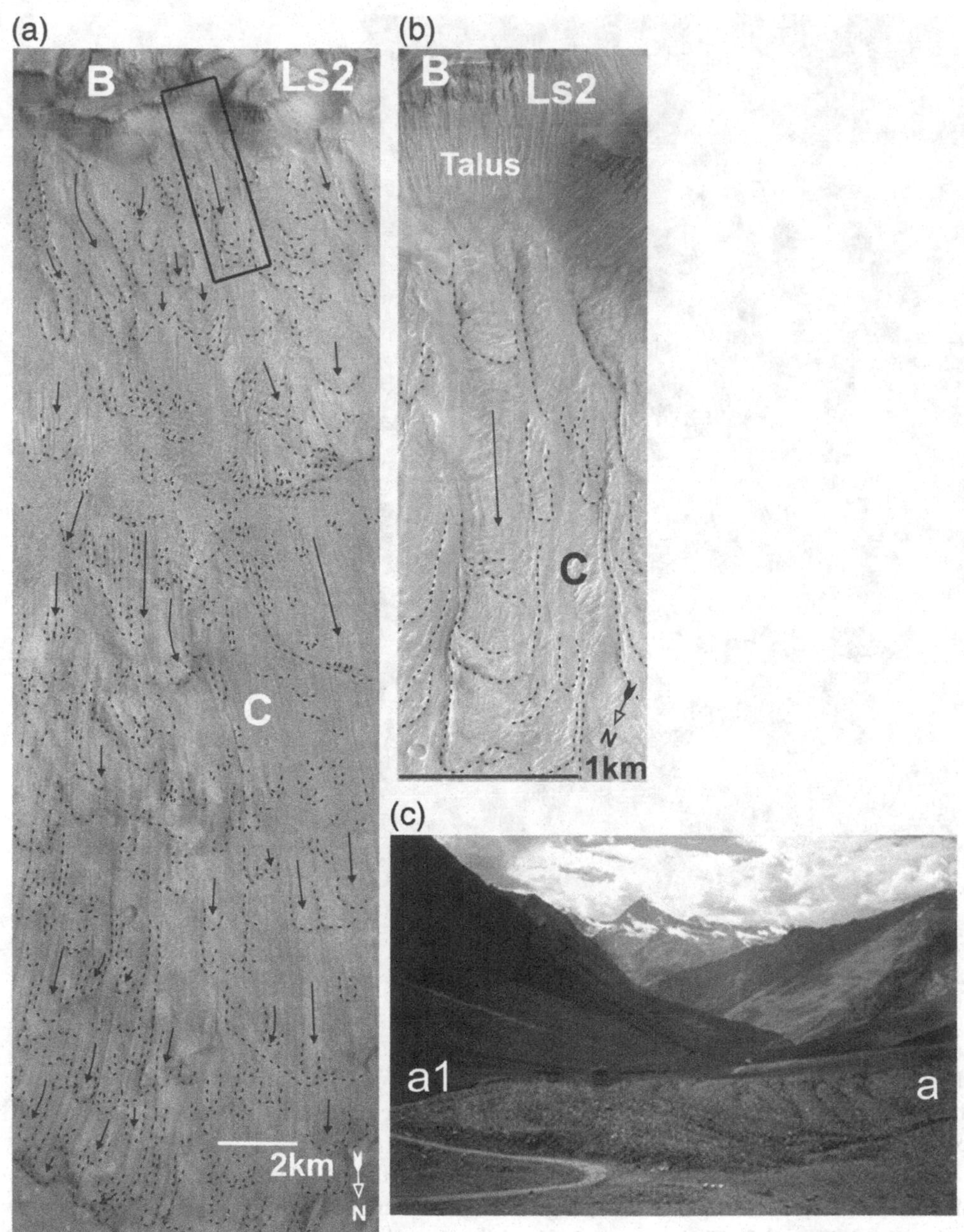

Fig. 32.5. Landslides at Ganges Chasma. (a) THEMIS visible image (V01001002) of the topographic bench B and the midtrack and lower slopes C. The surface of C shows sheets and lobes extending to the canyon floor. (b) High-resolution image (subscene M0703026) of part of Figure 32.4A showing the base of the topographic bench (B) from which lobes extend. The middle lobe is 3 km long, 30 m thick on average, and has a volume of 0.03 km³). (c) Talus-fed rock glacier moving from a to a1, showing lobes and transverse ridges; for scale the track is single lane.

been maintained for a few hundred million years in the late Noachian. Such conditions may have favored large catastrophic slope failures on steep canyon walls that produced long-runout landslides, but it appears that aprons continued to form long after the transition to colder conditions in the Amazonian, which supports the idea of gravitational creep and slow downslope movement of rock and ice-rich talus.

32.5 VENUS

Venus is the second planet from the Sun and is covered in an opaque layer of highly reflective clouds of sulfuric acid. It has a dense carbon dioxide atmosphere, and the surface shows evidence of extensive volcanism with relatively few impact craters, apparently due to volcanic resurfacing. The surface of the planet is between 300 and 600 million years old (Strom *et al.*, 1994; Nimmo and McKenzie, 1998), and appears to have been unaffected by plate movements (Nimmo and McKenzie, 1998). Almost 80 percent of the surface is lowland smooth basaltic plains. There are two mountainous highlands, one in the Northern Hemisphere and the other just south of the Equator. Mass movements have been identified in highland terrain (Malin, 1992) and in association with volcanic domes (Fig. 32.6; Guest *et al.*, 1992; Bulmer and Guest, 1996), based on analysis of Magellan radar images at 75 m per pixel.

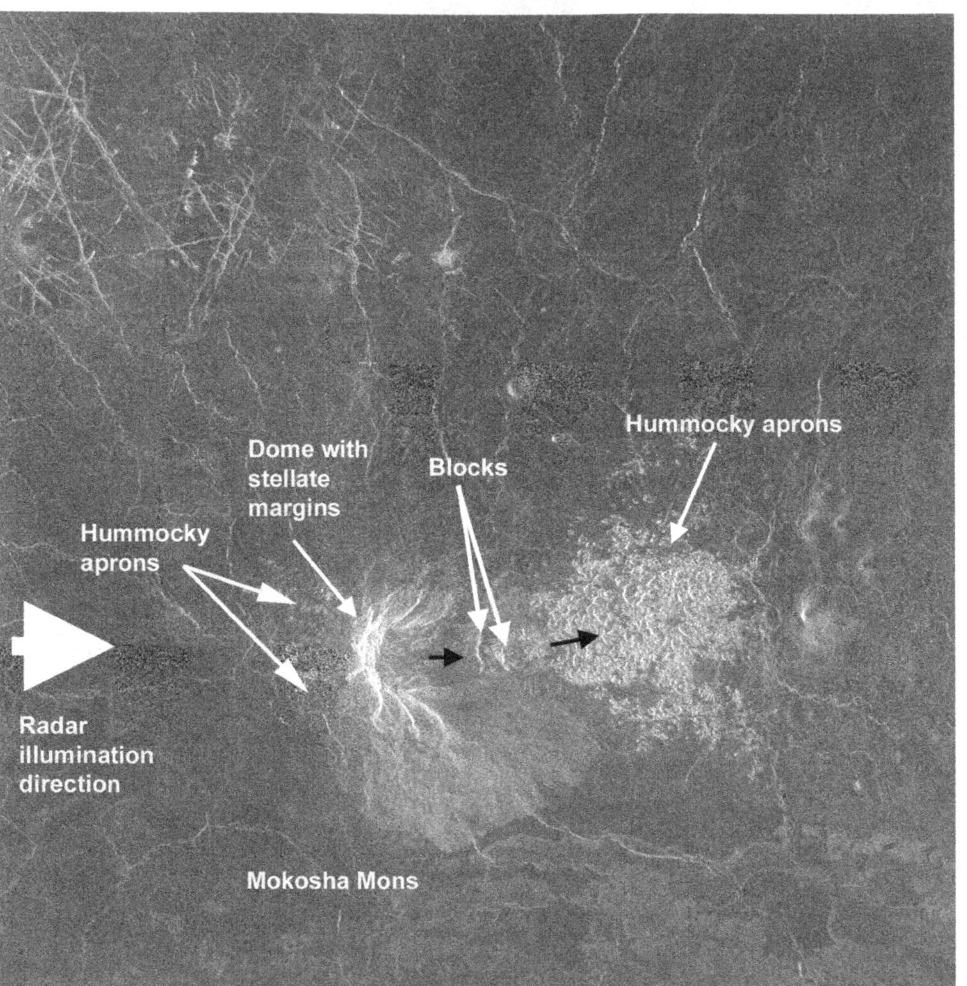

Fig. 32.6. A debris apron formed by the collapse of the eastern flank of a 15–20 km diameter volcanic dome visible in the center left (PIA00264). The flank of the dome illuminated by the radar from the left, at an incidence angle of 30°, is foreshortened. At the base of the east-facing slope is a large block of coherent rock, 8–10 km in length. The similarities in the shapes of the backscarp and block are typical of terrestrial slumped blocks. The rough hummocky surface of the debris apron is radar-bright compared with the smoother plains surrounding it. The bright lobe to the south of the volcano is either a lava flow or finer debris of other landslide deposits.

During the early stages of the Magellan mission, a stellate-shaped edifice was identified north of the mountainous region Alpha Regio (18.1° S, 5.5° E). Its origin was enigmatic, with no obvious terrestrial analog. Other similar stellate edifices were later found as more data were examined, and researchers began to note similarities to volcanic domes on Earth (Fig. 32.7). Their morphologies fell within a spectrum ranging from symmetric to highly degraded forms (Head *et al.*, 1992; Guest *et al.*, 1992). Over half of the total population of 375 domes have been modified; many of these have stellate planforms and deposits with a range of radar and morphologic characteristics (Bulmer, 1994; Bulmer and Guest, 1996). Domes range in diameter from <10 to 120 km; most are between 10 and 35 km in diameter and have slope angles between 15° and 28°. Domes >70 km in diameter are formed of superposed or coalesced domes whose margins are commonly indistinguishable. The diameters and heights (300–1700 m) of many domes on Venus are larger than subaerial domes on Earth (Pavri *et al.*, 1992; Bulmer and Guest, 1996), but surveys of Earth's seafloor near Hawaii using sonar imaging systems such as GLORIA (Geological Long-Range Inclined Asdic) and SeaBEAM have revealed flat-topped seamounts (Fig. 32.8) of similar diameter, volume, and slope angles, and many of these have stellate planforms (Bulmer and Wilson, 1999).

Aprons associated with the modified domes have lobate boundaries, large blocks, and hummocky surfaces (Fig. 32.9): features typical of landslide deposits (Siebert, 1984). These characteristics are distinct from those of lava flows, which exhibit a more uniform radar backscatter and irregular boundaries. Bulmer (1994) recognized four morphological groups of aprons with different radar characteristics and morphologies. Some aprons spread only near their termini, whereas others spread early along their track and have large widths. The maximum runout distances range from 12 to 80 km; maximum widths range from 5 to 62 km; and areas range from 52 to 2140 km^2. These morphologies and dimensions are comparable to those observed in GLORIA images off the Hawaiian Ridge (Moore *et al.*, 1989). Two groups of aprons are discussed here in more detail because they have characteristics consistent with catastrophic slope failures on the flanks of volcanic domes (Table 32.2).

Deposits in the first group (G1, Table 32.2) cover large areas (>100 km^2), have long travel distances (>20 km), and

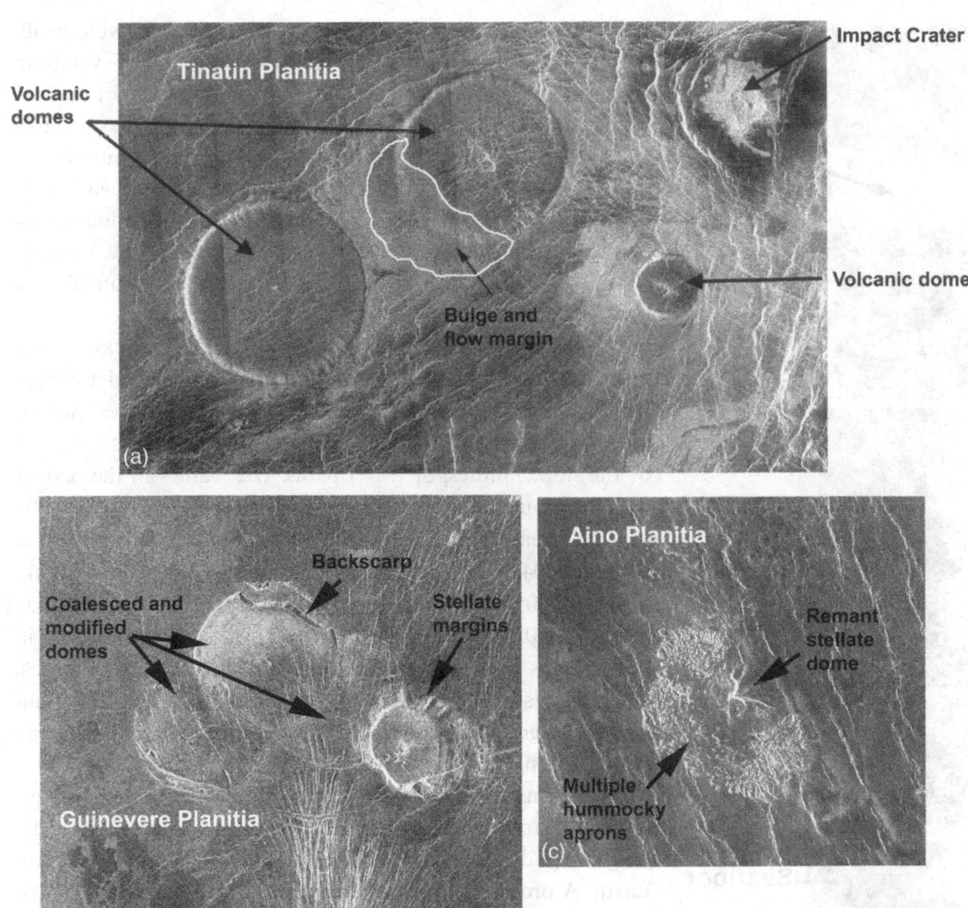

Fig. 32.7. Range of planimetric forms of domes on Venus. (a) Two near-symmetric domes (MD109 and MD107) located on Tinatin Planitia (120° N, 8° E). Both domes are 60 km in diameter and have steep convex slopes rising to extensive horizontal upper surfaces. The northeast dome shows some modification on its southwest flank. (b) Four coalesced domes in Guinevere Planitia (33.6° N, 312° E). The largest dome (MD50) is 50 km in diameter and has a well-defined failure on its northern margin. The smaller dome (MD51) is 25 km in diameter and has stellate margins. (c) Remnant dome west of Imdr Regio (45.2° S, 201.4° E). Only part of the dome remains and is 20 km across. Almost one-third of the edifice appears to have failed, producing multiple debris aprons and a deposit that is 70 km across. The remaining part of the domes has stellate margins. The identifying labels for domes (e.g., MD109) refer to a global database (Bulmer, 1994).

are characterized by hummocky topography. Surface textures are rough on a large scale, with large hummocks (>1 km) surrounded by finer material. In plan view, the deposits are fan-shaped or lobate. Some aprons extend from the base of a dome, whereas others are disconnected from the dome. An example of G1 is a dome in eastern Mokosha Mons (55° N, 266° E), from which a large portion of the eastern flank has detached (Fig. 32.6). Large hummocks occur at the base of this dome, and deposits extend outward across the basaltic plain. These large hummocks may record the initial movement of a coherent mass of rock before collapse and runout. The massive nature of the deposits and their limited disintegration imply that failure occurred after the eruptive activity had ceased and the dome had become solid.

Deposits of the second group (G2, Table 32.2) separated from the base of the source dome and have narrow proximal zones that broaden toward the distal margin. The deposits have a lower radar backscatter coefficient than the deposits in the first group, owing to smoother surfaces with smaller dispersed hummocks (<1 km). In plan view, the deposits have irregular lateral and distal margins. One dome in Navka Planitia (26° S, 296.7° E) has at least five debris aprons with the characteristics of this group G2. The long and narrow planforms of the deposits are characteristic of rock masses that traveled away from the dome at relatively high velocities. The presence of hummocks in the deposits indicates that the source was solid and the rock mass underwent only limited disintegration during runout. The lack of well-defined deposit boundaries and the relatively lower radar backscatter coefficient may reflect the influence of post-depositional reworking processes. For example, younger volcanic material may have partly buried some of the aprons, and chemical and physical erosion may have smoothed the surfaces.

32.5.1 FAILURE AND TRANSPORT MECHANISM

Analysis of image data suggests that two types of destructive process modify the domes: those accompanied by explosions, and those that are nonexplosive in origin. Given the constraints imposed on explosive activity by the atmosphere, which inhibits the expansion of exsolved bubbles in magmas, and the high magma volatile concentrations needed to cause an explosion (5 percent CO_2, is required in the lowlands, and 2 percent in the highlands; Head and Wilson, 1986), the most likely internal causes of slope failure on Venusian domes are localized volatile accumulations and oversteepening of slopes during dome growth. The formation of a chilled carapace composed of angular blocks during the growth of Venusian domes would

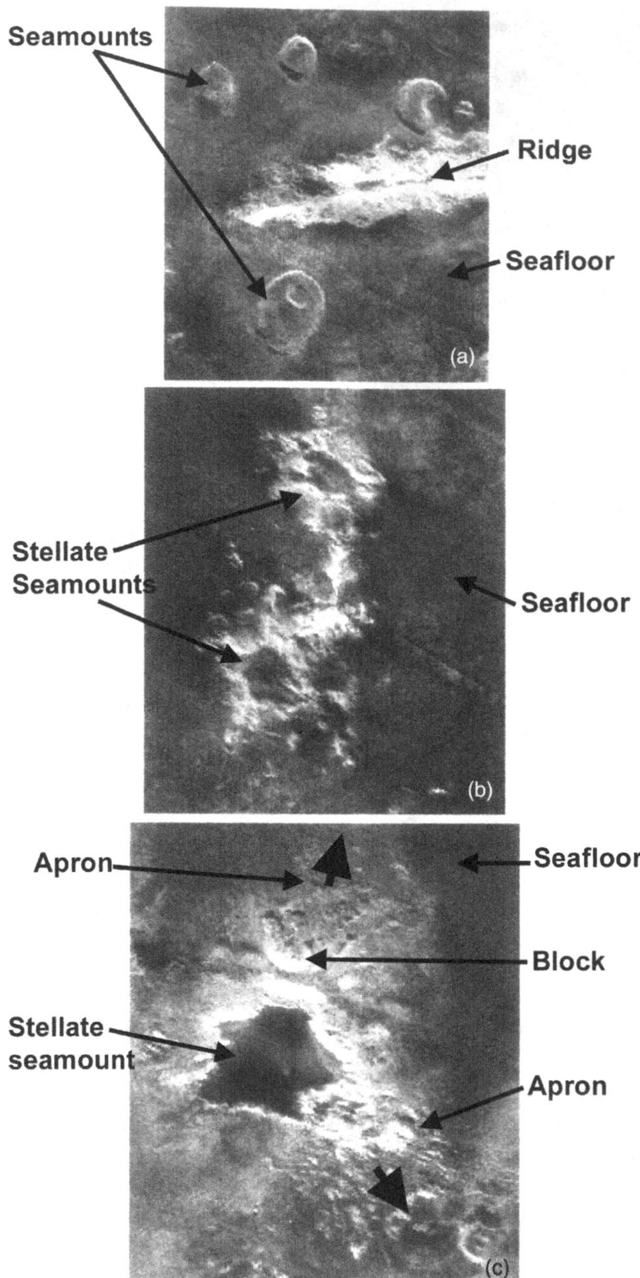

Fig. 32.8. Range of planimetric forms of submarine seamounts on the Hawaiian Ridge. (a) Four near-symmetric volcanoes with characteristics similar to those of unmodified domes on Venus. The seamounts range in diameter from 5 to 10 km (24° N, 160° W). (b) Two seamounts with stellate plan forms and diameters between 5 and 10 km (22° N, 162.5° W). (c) A highly modified stellate seamount (27° N, 168° W) surrounded by aprons of volcaniclastic material. The seamount is 15 km in diameter.

produce slopes predisposed to progressive failure along joints and other discontinuities. The internal stresses within a growing dome and external factors such as the local topography would control the magnitude and frequency of slope failures. Most of the landslides appear to have occurred after domes had cooled and were solid. Some large landslides were probably seismically triggered. Given the dimensions of Venusian domes, slope failure and subsequent conversion of potential energy into kinetic energy would cause the rock mass to brecciate and, with sufficient travel distance, become a hummocky deposit. Each hummock consists of one or a few mega-blocks (Siebert, 1984). These features have analogs in landslide deposits of the Pungarehu Formation at Egmont, New Zealand (Neall, 1979) and the deposits of some terrestrial nonvolcanic landslides (Fauqué and Strecker, 1988).

Based on data presented in Figure 32.2, hummocky aprons associated with modified volcanic domes appear to have been highly mobile (Bulmer, 1994). The high mobility cannot be adequately explained by the nature of the rock mass, the site of the slope failures, or travel paths. H/L values in the first of the two groups of deposits described above range from 0.001 to 0.01 (mean = 0.02); H/L values for the second group are between 0.04 and 0.1 (mean = 0.08) (Bulmer and Guest, 1996). Values of frictional resistance over the volcanic plains may range from 0.6 to 0.8 (Byerlee, 1978). The distance that debris in both groups traveled appears to require large volumes (Legros, 2002) but also a fluidizing agent or lubricant to enhance mobility and an emplacement process involving considerable energy conservation.

High atmospheric pressure (9.3 MPa) may have had an important role in the mobility of the debris aprons, given the long travel distances possible in submarine environments on Earth. A process allowing for ingestion into the debris of the hot (735 K) dense Venusian atmosphere (96.5 percent CO_2 with a surface gravity of 8.87 m s^{-2}), resulting in bulk fluidization, or a process of frictional melting that might produce a basal lubricant (Table 32.1), may explain the long runout distances of the landslides (Bulmer, 1994). Avalanches of disintegrating rock from the flanks of the domes could incorporate atmosphere into the debris, with resulting compression and heating causing an increase in the pore pressure and a reduction in effective stress and overburden pressure of the mass. However, a large volume of atmosphere would have to be ingested, and the mechanism by which this might occur and the gas be retained in a spreading mass is unknown, especially given the chemical and physical properties of CO_2. If a failed rock mass slid, at least until becoming significantly brecciated, the long travel distances can be explained by melting of rock through frictional heating at the base of the moving mass. There is evidence that rock can melt during the transport of very large rock masses on Earth (Heuberger *et al.*, 1984). Such a process is likely to be more efficient on Venus due to the high ambient temperature. An advantage of this model is that an abundant supply of lubricant is available during transit; the greater the thickness of a debris mass, the greater the heat derived through friction, which in turn should provide more lubricant and result in lower resistance to movement. If this model is correct, it would be expected that mobility is a function of landslide size; that is, the larger – and therefore thicker – the landslide, the greater the

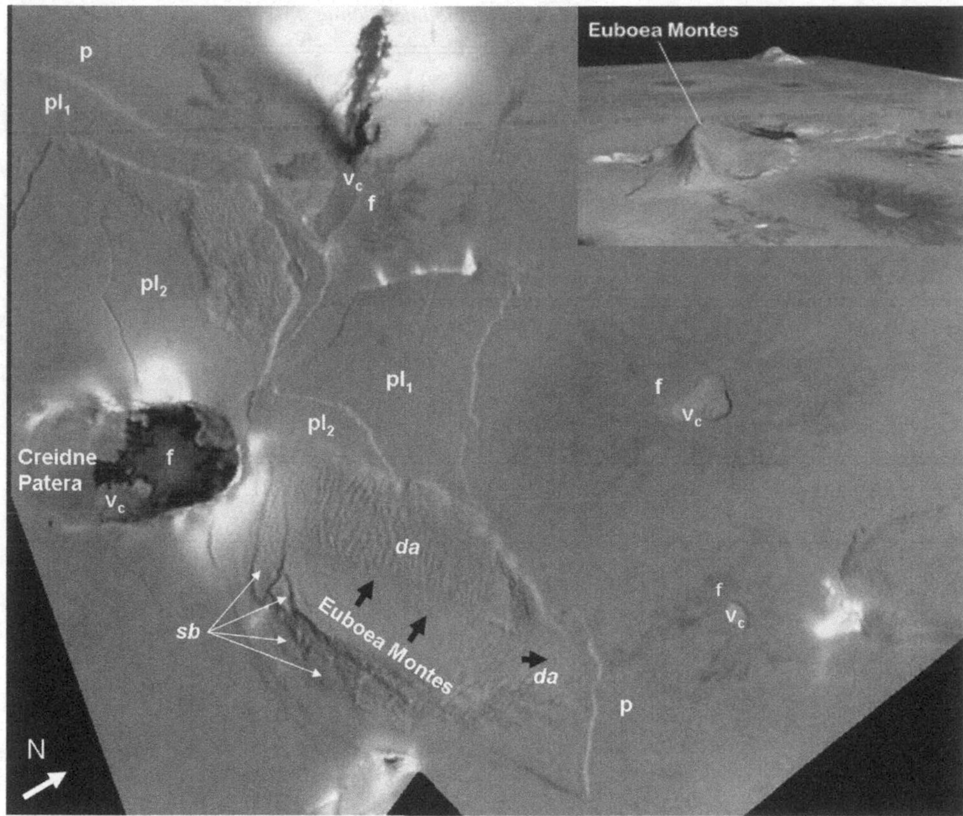

Fig. 32.9. Voyager stereoproduct (images 16390.38 and 16392.59) over Euboea Montes. The image is 650 km in width and has a horizontal resolution of 848 m per pixel and a vertical resolution of 252 m; the vertical exaggeration is 7.2. Geologic units are: lava flows (f), volcanic centers (vc), plains (p), layered plains (pl1 and pl2), basement massif (mb), debris aprons (*da*), and slump blocks (*sb*). Inset shows a perspective view of Euboea Montes from the east (modified from Schenk *et al.*, 1997).

runout distance. Testing this hypothesis will require improved image resolution and topographic data.

32.6 IO

Although vast featureless plains cover nearly half the surface of Io, mountains exist and are the sites of mass movements (Schenk and Bulmer, 1998; Turtle *et al.*, 2002). The surface gravity on Io is 1.796 m s^{-2}; surface temperatures range from 90 to 130 K; the surface pressure is at a trace level; and the atmosphere is 90 percent CO_2. The low number of impact craters on Io's surface indicates that it is geologically young, with continual reworking caused by ongoing mafic and ultramafic volcanic activity (McEwen, 2002). The presence of mountainous terrain on Io (Schaber, 1982) indicates that a material with greater strength than sulfur, and probably silicate-rich, forms much of Io's crust. McEwen *et al.* (1998) have suggested that sulfur and sulfur dioxide appear to play a role on Io similar to water and carbon dioxide on Earth.

Euboea Montes (47° S, 336° W) is a high massif on Io, with a base measuring about 175 km by 240 km. It can be divided into three geomorphic units (Fig. 32.9). A northeast–southwest-trending arcuate ridge crest forms the spine of the massif and reaches up to 10.5 ± 1 km above the surrounding plain. This ridge divides the massif into two sections. The southeast flank of the massif is relatively steep and characterized by a hummocky and striated surface. The northwest flank has a smooth planar surface that slopes uniformly at *ca.* 6°. At the base of the northwest flank is a thick ridged deposit with lobate margins. This deposit is about 70 km long and 200 km wide and has an aerial extent of *ca.* 14,800 km^2. Ridges are kilometers wide and up to 50 km long, oriented parallel to the northwest flank. Prominent ridges or levees form the margins of individual lobes within the deposit, similar to debris aprons on Mars. Along the proximal margin, the deposit stands *ca.* 6 km above the plain and slopes gently and uniformly toward the distal margin. Based on stereographic analysis, Schenk and Bulmer (1998) estimated the deposit to be 3.5 km thick near the midpoint, decreasing to 2 km toward the northeast and southwest margins.

32.6.1 FAILURE AND TRANSPORT MECHANISM

The thick ridged deposit on Io was originally interpreted to be either a viscous volcanic flow or a flow–creep feature produced by high heat flow (Schaber, 1982). Using stereo-images and topographic data, Schenk and Bulmer (1998) found the morphology to be more consistent with slope failure along the entire slope of the northwest flank of the Euboea Montes, forming a massive debris apron at the base of the mountain slope. Lobes and longitudinal ridges on the apron resemble those observed on landslide deposits on Earth, Mars, and the Moon (Shreve, 1966; Guest, 1971; Shaller, 1991). The width and thickness of the ridged deposits are consistent with the height and width of

Table 32.2. *Morphometric characteristics of G1 and G2 debris aprons associated with modified domes on Venus.*

Group	Area (km^2)	Length (km)	Width (km)	Height (km)	H/L	Population
Range						
G1	148–2128	22–81	4.0–62	0.2–4.0	001–0.01	32
G2	171–1460	14–69	7.0–26	2.2–2.8	0.04–0.1	11
Mean						
G1	805	41	31.0	0.6	0.02	32
G2	627	44	14.0	2.5	0.08	11

Table 32.3. *Largest mass movements in the Solar System identified to date.*

Name	Volume (km^3)	Area (km^2)	Length (km)	Drop (km)	H/L	Planet/ Moon
OMA NW	170,000	NA	750	6.5	0.009	Mars
OMA NE	150,000	NA	370	4	0.01	Mars
OMA N	68,000	NA	NA	6	NA	Mars
OMA W	65,000	NA	NA	5	NA	Mars
OMA SW	30,000	NA	510	5	0.01	Mars
Euboea	25,000	15,000	130	10.5	0.08	Io
Agulhas	20,331	79,488	700	NA	NA	Earth
Chamais	17,433	68,688	160	NA	NA	Earth
OMA SE	17,000	NA	425	3	0.007	Mars
OMA E	14,000	NA	300	4	0.01	Mars
Concepcion B	NA	51,000	75	NA	NA	Earth
Cape T N+S	9,920	47,952	300	NA	NA	Earth
SW Af un	4,549	33,696	180	NA	NA	Earth
Amazon C E	NA	32,500	NA	NA	NA	Earth

Notes: Data on Martian aprons from Lopes *et al.* (1980, 1982); on seafloor aprons from Jacobi (1976), Dingle (1977), Summerhayes *et al.* (1979), Damuth and Embley (1981), Lipman *et al.* (1988), Moore *et al.* (1989), and Hampton *et al.* (1996); and on Io from Schenk and Bulmer (1998).

the exposed northwest flank of the mountain, which would be unexpected for volcanic deposits and indicates that a layer of roughly uniform thickness failed and moved off the mountain, forming the deposit.

The presence of a debris apron at the base of a planar surface 200 km across suggests that the failure occurred along a distinct planar discontinuity within Io's crust after uplift and tilting of the mountain. Although the scales differ, the inferred style of failure is similar to that of a landslide that occurred along bedding surfaces at Saidmarreh, Iran (Harrison and Falcon, 1937). Material may have failed along a rheologic discontinuity between a weaker upper layer, such as a volcanic ash deposit, and a more competent lower crustal layer, thermally metamorphosed by Io's high heat flow. Alternatively, both of these layers may have been competent, and movement may have occurred along one or more mechanical or stratigraphic planar discontinuities.

Schenk and Bulmer (1998) estimated the volume of the landslides on Io using available image data and digital stereogrammetry. They estimated the pre-collapse surface by extrapolating the planar surface of the northwest flank of Euboea and the horizontal ground plane. The estimated volume of *ca.* 25,000 km^3 makes it the largest debris apron in the solar system, with the possible exception of the aureole deposits surrounding Olympus Mons on Mars (Table 32.3). Restoring this volume as a uniform layer over the entire surface of the tilted northwest flank of Euboea Montes, assuming it to be the pre-failure configuration, yields a layer of material *ca.* 2-km thick. Schenk and Bulmer (1998) proposed that layers pl_1 and pl_2, respectively 100–300 m thick and surrounding Euboea Montes (Fig. 32.9), were part of the original in-situ rock sequence before uplift. It is these sequences that failed on the slopes of the Euboea Montes and form the debris apron on the northwest slope.

A large portion of the toe of the debris apron, seen at 250-m resolution, has a homogeneous texture with an absence of blocks larger than *ca.* 500 m. This observation suggests that the material that failed was poorly consolidated, possibly volcanic ash from Io's active plumes or ash interbedded with lava flows. The maximum distance that materials in the apron traveled, if they did indeed move from the flank of Euboea Montes, is

130 km. This travel distance is greater than that of debris aprons observed on Mars, Venus, or the Moon (Shaller, 1991; Bulmer, 1994), but landslides on Earth's seafloor have traveled in excess of 150 km and on slopes of <5° (Table 32.3). The apron has characteristics that are consistent with the correlation between descent height and runout distance and the inverse relationship between volume and *H*/*L* observed on Earth. In addition to the large volume of the landslide and the huge derived kinetic energy, environmental conditions on Io probably contributed to the high mobility of materials that failed from Euboea Montes. Spectral properties of the moon's surface materials are consistent with low-temperature forms of sulfur, and with surface models that suggest that sulfur dioxide and sulfur exist in thermally stable stratified zones (Soderblom *et al.*, 1980). These zones provide distinct mechanical discontinuities along which failures on mountain slopes likely occurred. The failures may have been caused by earthquakes that were triggered by changes in stress in the upper lithosphere. These stress changes, in turn, may have been caused by vertical differences of as much as 100 m in the tidal bulge between times that Io, with its eccentric orbit, was at periapsis and apoapsis with Jupiter. The height of the bulge itself does not directly lead to mass wasting because it is a long-wavelength feature.

If the apron at the base of Euboea Montes is the product of one or more rock and debris avalanches from bedrock escarpments of basaltic materials and low-temperature forms of sulfur (layers pl_1 and pl_2), the landslide motion need not require an interstitial mixture for support, but can rely, if dry, on grain-to-grain interactions generating a matrix material through fracturing and grinding associated with the avalanche event (Yarnold, 1993). Sulfur deposited at the latitude of Euboea Montes will be weakened due to radiation damage, which causes breakage of the normally stable cyclic 8-chain sulfur. At the surface, sulfur dioxide is in vapor pressure equilibrium with frost, and failed materials in motion would generate a matrix of sulfur and sulfur dioxide frosts that may have behaved in a similar manner to terrestrial powder snow avalanches on Earth (Nishimura *et al.*, 1995). Very fine-grained particles could also generate electrostatic forces (Drake, 1990) that may have enhanced mobility. Once material was in motion, the thin atmosphere would offer little resistance to continued movement.

32.7 CONCLUSIONS

Discussions of selected landslides on Mars, Venus, and Io have highlighted how knowledge of terrestrial landslide processes has been used to inform interpretation of geomorphic features identified on the surface of planets, moons, and asteroids. Due to the resolution constraints of sensors that have imaged these bodies, studies of mass-movement processes have traditionally been limited to descriptions and comparisons of surface morphologies, and measurements of lengths, widths, fall heights and, where topographic data are available, first-order estimates of thickness. First-order estimates of landslide mobility have been made where it is possible to determine fall height, runout, and volume. Although rather simple, this approach has enabled examination of the diverse surface and environmental conditions in which mass movements have occurred on planets, moons, and asteroids. With the increasing availability of precise and accurate spatial information over large areas, it is becoming possible to extend the analysis of mass movements beyond the Coulomb frictional model. For example, improved estimates of landslide widths, lengths, and slopes, and of thickness variations along and across a deposit (Bulmer and Finnegan, 2010) provide input for a more in-depth examination of emplacement dynamics using Chezy-type modeling. Kinematic wave or hydraulic Chezy-type models have been used to examine the emplacement of debris flows (Takahashi, 1980; Bulmer *et al.*, 2002) and lava flows (Bruno *et al.*, 1996). With the availability of these data and the increasing sophistication of analysis that can be undertaken, observations, interpretations, and explanations of mass-movement features on these bodies will increasingly inform our understanding of landslide triggering and emplacement on Earth.

REFERENCES

Andrews-Hanna, J.C., Zuber, M.T., Arvidson, R.E. and Wiseman, S.M. (2010). Early Mars hydrology: Meridiani playa deposits and the sedimentary record of Arabia Terra. *Journal of Geophysical Research*, 115, E06002, doi:10.1029/2009JE003485.

Banfield, J., Hamilton, V.E. and Christensen, P.R. (2000). A global view of Martian surface compositions from MGS-TES. *Science*, 287, 1626–1630.

Benn, D.I. and Evans, D.J.A. (1997). *Glaciers and Glaciation*. New York: Oxford University Press.

Bieniawski, Z.T. (1993). Classification of rock masses for engineering: The RMR system and future trends. In *Comprehensive Rock Engineering, Vol. 3*, ed. J.A. Hudson and E. Hoek. Pergamon, New York, pp. 553–573.

Bisci, C., Dramis, F. and Sorriso-Valvo, M. (1996). Rock flow (Sackung). In *Landslide Recognition, Identification, Movement and Causes*, ed. R. Dikau *et al.*, Hoboken, NJ: John Wiley, pp. 150–160.

Bruno, B.C., Baloga, S.M. and Taylor, G.J. (1996). Modeling gravity driven flows on an inclined plane. *Journal of Geophysical Research*, 101, 11565–11577.

Bulmer, M.H. (1994). *Small volcanoes in the plains of Venus; with particular reference to the evolution of domes*. Ph.D. thesis, University of London, Senate House.

Bulmer, M.H. and Finnegan, D. (2010). Topography data on Mars: Optimizing its collection and application using laser scanning. *European Geophysical Union Conference, Vienna, Austria, Vol. 12*, EGU2010–7110.

Bulmer, M.H. and Guest, J.E. (1996). Modified volcanic domes and associated debris aprons on Venus. In *Volcano Instability*, ed. W.J. McGuire, A.P. Lones and J. Neuberg. Geological Society of London, Special Publication 110, pp. 349–371.

Bulmer, M.H. and Wilson, J.B. (1999). Comparison of stellate volcanoes on Earth's seafloor with stellate domes on Venus using side scan sonar and Magellan synthetic aperture radar. *Earth and Planetary Science Letters*, 171, 277–287.

Bulmer, M.H. and Zimmerman, B.A. (2005). Re-assessing landslide deformation in Ganges Chasma, Mars. *Geophysical Research Letters*, 32, L06201, doi:10.1029/2004GL022021.

Bulmer, M.H., Barnouin-Jha, O.S., Peitersen, M.N. and Bourke, M. (2002). An empirical approach to studying debris flows: Implications for planetary modeling studies. *Journal of Geophysical Research, Planets*, 107, E5, doi:10.1029/2001JE00153.

Byerlee, J. (1978). Friction of rocks. *Pure and Applied Geophysics*, 116, 615–626.

Campbell, C. (1989). The stress tensor for simple shear flows of a granular material. *Journal of Fluid Mechanics*, 203, 449–473.

Cruden, D. M. (1980). A large landslide on Mars: Discussion and reply. *Geological Society of America Bulletin*, 91, 63.

(1991). A simple definition of a landslide. *Bulletin of the International Association of Engineering Geology*, 43, 27–29.

Damuth, J.E. and Embley, R.W. (1981). Mass-transport processes on the Amazon Cone: western equatorial Atlantic. *American Association of Petroleum Geologists Bulletin*, 65, 629–643.

Davies, T.R.H. (1982). Spreading of rock avalanche debris by mechanical fluidization. *Rock Mechanics*, 15, 9–24.

Dent, J. (1982). *A bi-viscous modified Bingham model of snow avalanche motion*. Ph.D. thesis, Montana State University, Bozeman, MT.

Dikau, R., Brunsden, D., Schrott, L. and Ibsen, M.-L. (1996). Introduction. In *Landslide Recognition: Identification, Movement and Courses*, ed. R. Dikau, D. Brunsden, L. Schrott and M.-L. Ibsen. London: John Wiley and Sons, pp. 1–12.

Dingle, R. (1977). The anatomy of a large submarine slump on a sheared continental margin (SE Africa). *Journal of the Geological Society of London*, 134, 293–310.

Drake, T.G. (1990). Structural features in granular flows. *Journal of Geophysical Research*, 95, 8681–8696, doi:10.1029/JB095iB06p08681.

Erismann, T.H. (1979). Mechanisms of large landslides. *Rock Mechanics*, 12, 15–46.

Fauqué, L. and Strecker, M.R. (1988). Large rock avalanche deposits (Sturzströme, Sturzstroms) at Sierra Aconquija, northern Sierras Pampeanas, Argentina. *Eclogae Geologicae Helvetiae*, 81, 579–592.

Francis, P.W. and Wadge, G. (1983). The Olympus Mons aureole: Formation by gravitational spreading. *Journal of Geophysical Research*, 88, 8333–8344.

Fread, D.L. (1988). *Breach: An Erosion Model for Earthen Dam Failures*. NOAA, National Weather Service, Silver Spring, MD.

Fujiwara, A., Kawaguchi, J., Yeomans, D.K. *et al*. (2006). The rubble-pile asteroid Itokawa as observed by Hayabusa. *Science*, 312, 1330–1334.

Gendrin, A., Mangold, N., Bibring, J.-P. *et al*. (2005). Sulfates in Martian layered terrains: The OMEGA/Mars Express View. *Science*, 307, 1587–1591.

Goguel, J. (1978). Scale-dependent rock mechanisms. In *Rockslides and Avalanches, Vol. 1*, ed. B. Voight. Amsterdam: Elsevier, pp. 167–180.

Grotzinger, J.P., Arvidson, R.E., Bell, J.F., III, *et al*. (2005). Stratigraphy and sedimentology of a dry to wet eolian depositional system, Burns formation, Meridiani Planum, Mars. *Earth and Planetary Science Letters*, 240, 11–72, doi:10.1016/j.epsl.2005.09.039.

Guest, J.E. (1971). Geology of the farside crater Tsiolkovsky. In *Geology and Geophysics of the Moon*, ed. G. Fields. Amsterdam: Elsevier, pp. 93–103.

Guest, J.E., Bulmer, M.H., Aubele, J. *et al*. (1992). Small volcanic edifices and volcanism in the plains of Venus. *Journal of Geophysical Research*, 97, 15949–15966.

Haberle, R.M., Forget, F., Colaprete, A. *et al*. (2008). The effect of ground ice on the Martian seasonal CO_2 cycle. *Planetary and Space Science*, 56, 251–255.

Habib, P. (1975). Production of gaseous pore pressures during rocksliding. *Rock Mechanics*, 7, 193–197.

Hampton, M.A., Lee, H.J. and Locat, J. (1996). Submarine landslides. *Reviews of Geophysics*, 34, 33–59.

Harrison, J.V. and Falcon, N.L. (1937). The Saidmerrah landslip, southwest Iran. *Geographical Journal*, 89, 42–47.

Hazlett, R.W., Buesch, D., Anderson, J.L., Elan, R. and Scandone, R. (1991). Geology, failure, and implications of seismogenic avalanches of the 1944 eruption at Vesuvius, Italy. *Journal of Volcanology and Geothermal Research*, 47, 249–264.

Head, J.W. and Wilson, L. (1986). Volcanic processes and landforms on Venus: Theory, predictions and observations. *Journal of Geophysical Research*, 91, 9407–9446.

Head, J.W., Crumpler, L.S., Aubele, J.C. *et al*. (1992). Venus volcanism: Classification of volcanic features and structures, associations, and global distribution from Magellan data. *Journal of Geophysical Research*, 97, 13153–13197.

Head, J.W., III, Hiesinger, H., Ivanov, M.A. *et al*. (1999). Possible oceans in Mars: Evidence from Mars Orbiter Laser Altimeter data. *Science*, 286, 2134–2137.

Head, J.W., Marchant, D.R., Agnew, M.C. *et al*. (2006). Extensive valley glacier deposits in the northern mid-latitudes of Mars: Evidence for Late Amazonian obliquity-driven climate change. *Earth and Planetary Science Letters*, 241, 663–671.

Heim, A. (1932). *Bergsturz und Menschenleben*. Zurich: Fretz and Wasmuth AG.

Heuberger, H., Masch, L., Preuss, E. *et al*. (1984). Quaternary landslides and rock fusion in Central Nepal and in the Tyrolean Alps. *Mountain Research Development*, 4, 345–362.

Hoek, E. (1998). Reliability of Hoek–Brown estimates of rock mass properties and their impact on design. *International Journal of Rock Mechanics and Mining Science*, 35, 63–68.

Hsü, K.J. (1975). Catastrophic debris streams (sturzstroms) generated by rockfalls. *Geological Society of America Bulletin*, 86, 129–140.

Hungr, O. and Evans, S.G. (2004). Entrainment of debris in rock avalanches: An analysis of a long run-out mechanism. *Geological Society of America Bulletin*, 116, 1240–1252.

Iverson, R.M. (1995). Can magma-injection and groundwater forces cause massive landslides on Hawaiian volcanoes? *Journal of Volcanology and Geothermal Research*, 66, 295–308.

Jacobi, R.D. (1976). Sediment slides on the northwest continental margin of Africa. *Marine Geology*, 22, 157–173.

Jakosky, B.M. and Phillips, R.J. (2001). Mars' volatile and climate history. *Nature*, 412, 237–244.

Johnson, B. (1978). Blackhawk landslide, California, U.S. In *Rockslides and Avalanches 1: Natural Phenomena*, ed. B. Voight. Amsterdam: Elsevier, pp. 481–504.

Kent, P.E. (1966). The transport mechanisms in catastrophic rockfall. *Journal of Geology*, 74, 79–83.

Krumdieck, A. (1984). On the mechanics of large landslides. In *Proceedings of the 4th International Symposium on Landslides*. Toronto: University of Toronto Press, pp. 539–544.

Legros, F. (2002). The mobility of long-runout landslides. *Engineering Geology*, 63, 301–331.

Lipman, P.W., Normark, W.R., Moore, J.G. *et al*. (1988). The giant submarine Alika debris slide, Mauna Loa, Hawaii. *Journal of Geophysical Research*, 93, 4279–4299.

Lopes, R.M.C., Guest, J.E. and Wilson, C.J. (1980). Origin of the Olympus Mons aureole and perimeter scarp. *Moon and Planets*, 22, 221–234.

Lopes, R., Guest, J.E. and Wilson, C.J. (1982). Further evidence for a mass movement origin of the Olympus Mons aureole. *Journal of Geophysical Research*, 87, 9917–9928.

Lucchitta, B.K. (1978). A large landslide on Mars. *Geological Society of America Bulletin*, 89, 1601–1609.

Lucchitta, B.K. (1979). Landslides in the Valles Marineris, Mars. *Journal of Geophysical Research*, 84, 8097–8113.

Lucchitta, B.K. (1987). Valles Marineris, Mars: Wet debris flows and ground ice. *Icarus*, 72, 411–429.

Malin, M.C. (1992). Mass movements on Venus: Preliminary results from Magellan Cycle I observations. *Journal of Geophysical Research*, 97, 16337–16352.

McEwen, A.S. (1989). Mobility of large rock avalanches: Evidence from Valles Marineris, Mars. *Geology*, 17, 1111–1114.

(2002). Active volcanism on Io. *Science*, 297, 2220–2221.

McEwen, A.S., Keszthelyi, L., Spencer, J.R. *et al.* (1998). High-temperature silicate volcanism on Jupiter's moon Io. *Science*, 281, 87–90.

McEwen, A.S., Malin, M.C., Carr, M.H. and Hartmann, W.K. (1999). Voluminous volcanism on early Mars revealed in Valles Marineris. *Nature*, 397, 584–586.

McGovern, P.J. and Morgan, J.K. (2009). Volcanic spreading and lateral variations in the structure of Olympus Mons, Mars. *Geology*, 37, 139–142.

McGovern, P.J., Smith, J.R., Morgan, J.K. and Bulmer, M.H. (2004). Olympus Mons aureole deposits: New evidence for a flank failure origin. *Journal of Geophysical Research*, 109, E08008, doi:10.1029/2004JE002258.

McSaveney, M.J. (1978). Sherman Glacier rock avalanche, Alaska, U.S.A. In *Rockslides and Avalanches. 1. Natural Phenomena*, ed. B. Voight. Amsterdam: Elsevier, pp. 197–258.

Mège, D. and Bourgeois, O. (2010). Destabilization of the Valles Marineris wallslopes by retreat of ancient glaciers. In *41st Lunar and Planetary Science Conference*, Abstract 1713.

Mellon, M.T. and Jakosky, B.M. (1993). Geographic variations in the thermal and diffusive stability of ground ice on Mars. *Journal of Geophysical Research*, 98, 3345–3364.

Melosh, H.J. (1979). Acoustic fluidization: A new geologic process? *Journal of Geophysical Research*, 84, 7513–7520.

Moore, J.G., Clague, D.A., Holcomb, R.T. *et al.* (1989). Prodigious submarine landslide on the Hawaiian Ridge. *Journal of Geophysical Research*, 94, 17465–17484.

Morgan, J.K. and McGovern, P.J. (2005). Discrete element simulations of gravitational volcanic deformation. 1. Deformation structures and geometries. *Journal of Geophysical Research*, 110, B05402, doi:10.1029/2004JF003252.

Mustard, J.F., Murchie, S.L., Pelkey, S.M. *et al.* (2008). Hydrated silicate minerals on Mars observed by the Mars Reconnaissance Orbiter CRISM instrument. *Nature*, 454, 305–309.

Neall, V.E. (1979). *Sheets P19, P20 and P21, New Plymouth, Egmont and Mamaia (1st ed)*. Wellington: New Zealand Department of Science and Industrial Research, three maps and notes, 1:500,000 scale.

Niles, P.B. and Michalski, J. (2009). Meridiani Planum sediments on Mars formed through weathering in massive ice deposits. *Nature Geoscience*, 2, doi:10.1038/ngeo438.

Nimmo, F. and McKenzie, D. (1998). Volcanism and tectonics on Venus. *Annual Review of Earth and Planetary Sciences*, 26, 23–53.

Nishimura, K., Sandersen, F., Kristensen, K. and Lied, K. (1995). Measurement of powder snow avalanche. *Survey Geophysics*, 16, 649–660, doi:10.1007/BF00665745.

O'Brien, J.S., Julien, P.Y. and Fullerton, W.T. (1993). Two-dimensional water flood and mudflow simulation. *Journal of Hydrological Engineering, American Society of Civil Engineers*, 119, 244–261.

Pavri, B., Head, J.W., Klose, K.B. and Wilson, L. (1992). Steep-sided domes on Venus: Characteristics, geologic setting, and eruption conditions from Magellan data. *Journal of Geophysical Research*, 97, 13445–13478.

Petley, D., Bulmer, M.H. and Murphy, W. (2002). Patterns of movement in rotational and translational landslides. *Geology*, 30, 719–722.

Potapov, A.V. and Ivanov, B.A. (1991). Landslide motion: Numerical simulation for Earth and Mars. *Lunar and Planetary Science Conference*, 22, 1087–1088.

Quantin, C., Allemand, P. and Delacourt, C. (2003). Valles Marineris landslides: Morphologies, age and dynamics. In *6th International Conference on Mars*. Houston, TX: Lunar and Planetary Institute.

Roach, L.H., Mustard, J.F., Wyatt, B. *et al.* (2009). Hydrated mineral stratigraphy in Ius Chasma, Valles Marineris. In *40th Lunar and Planetary Science Conference*, Abstract 1834.

Schaber, P. (1982). The geology of Io. In *Satellites of Jupiter*, ed. D. Morrison. Tucson, AZ: University of Arizona Press, pp. 556–597.

Schenk, P. and Bulmer, M.H. (1998). Origin of mountains on Io by thrust faulting and large-scale mass movements. *Science*, 279, 1514–1518.

Schenk, P.M., McEwen, A.S., Davies, A.G. *et al.* (1997). Geology and topography of Ra Patera, Io, in the Voyager era: Prelude to eruption. *Geophysical Research Letters*, 24, 2467–2470.

Schultz, R.A. (2002). Stability of rock slopes in Valles Marineris, Mars. *Geophysical Research Letters*, 29, 1932–1953, doi:10.1029/2002GL015728.

Schumm, S.A. (1985). Explanation and extrapolation in geomorphology: Seven reasons for geologic uncertainty. *Transactions of the Japanese Geomorphological Union*, 6(1), 1–18.

Segura, T.L., Toon, O.B., Colaprete, A. and Zahnle, K. (2002). Environmental effects of large impacts on Mars. *Science*, 298, 1977–1980.

Shaller, P.J. (1991). *Analysis and implications of large Martian and terrestrial landslides*. Ph.D. thesis, California Institute of Technology, Pasadena, CA.

Shepard, M.K., Campbell, B.A., Bulmer, M.H. *et al.* (2001). The roughness of natural terrain: A planetary and remote sensing perspective. *Journal of Geophysical Research – Planets*, 106, 32777–32795.

Shreve, R.L. (1966). Sherman landslide, Alaska. *Science*, 154, 1639–1643.

(1968). Leakage and fluidization in air-layer lubricated avalanches. *Geological Society of America Bulletin*, 79, 653–658.

Siebert, L. (1984). Large volcanic debris avalanches: Characteristics of source areas, deposits and associated eruptions. *Journal of Volcanology and Geothermal Research*, 22, 163–197.

Smith, D.E., Zuber, M.T., Frey, H.V. *et al.* (2001). Mars Orbiter Laser Altimeter: Experiment summary after the first year of global mapping of Mars. *Journal of Geophysical Research*, 106, 23689–23722.

Soderblom, L.A., Johnson, T., Morrison, D. *et al.* (1980). Spectrophotometry of Io: Preliminary Voyager 1 results. *Geophysical Research Letters*, 7, 963–966, doi:10.1029/GL007i011p00963.

Solomon, S.C., Aharonson, O., Aurnou, J.M. *et al.* (2005). New perspectives on ancient Mars. *Science*, 307, 1214–1220.

Solonenko, V.P. (1972). Seismogenic destruction of mountain slopes. In *Proceedings of the International Geology Congress, Section 13*, Montreal, PQ, pp. 284–290.

Strom, R.G., Schaber, G.G. and Dawsow, D.D. (1994). The global resurfacing of Venus. *Journal of Geophysical Research*, 99, 10899–10926, doi:10.1029/94JE00388.

Sullivan, R.J., Thomas, P.C., Murchie, S.L. and Robinson, M.S. (2002). In *Asteroids III*, ed. W.F. Bottke. Tucson, AZ: University of Arizona Press, pp. 331–350.

Summerhayes, C.P., Bornhold, B.D. and Embley, R.W. (1979). Surficial slides and slumps on the continental slope and rise of South West Africa: A reconnaissance study. *Marine Geology*, 31, 265–277.

Takahashi, T. (1980). Debris flows on prismatic open channel. *Journal of Hydraulics Division, Proceedings of the American Society of Engineers*, 106, 381–396.

Tanaka, K.L. (1985). Ice-lubricated gravity spreading of the Olympus Mons aureole deposits. *Icarus*, 62, 191–206.

Terzaghi, K. (1960). Mechanics of landslides. In *Applications of Geology to Engineering Practice*. Geological Society of America, Berkeley Volume, pp. 83–123.

Tian, F., Kasting, J. F. and Solomon, S. C. (2009). Thermal escape of carbon from the early Martian atmosphere. *Geophysical Research Letters*, 36, L02205, doi:10.1029/2008GL036513.

Toon, O. B. (2010). The formation of Martian river valleys by impacts. In *Astrobiology Science Conference*, Houston, TX, Abstract 5161.

Trunk, F. J., Dent, J. D. and Lang, T. E. (1986). Computer modeling of large rockslides. *Journal of Geotechnical Division, American Society of Civil Engineers*, 112, 348–360.

Turtle, E. P., Jaeger, W. L., McEwen, A. S. *et al.* (2002). New Galileo observations of Ionian mountains. *Lunar and Planetary Science Conference*, 33, Abstract 1677.

Van Gassen, W. and Cruden, D. M. (1989). Momentum transfer and friction in the debris of rock avalanches. *Canadian Geotechnical Journal*, 26, 623–628.

Watson, R. A. and Wright, H. E. (1969). The Saidmarreh Landslide, Iran, In *United States Contributions to Quaternary Research*, ed. S. A. Schumm and W. C. Bradley. Geological Society of America, Special Paper 123, pp. 115–139.

Yarnold, J. C. 1993. Rock avalanche characteristics in dry climates and the effect of flow into lakes: Insights from mid-Tertiary breccias near Artillery Peak, Arizona. *Geological Society of America Bulletin*, 105, 345–360.

Index

Printed in the United States
By Bookmasters

Printed in the United States
By Bookmasters